Cardiac Glycosides 1785–1985

Boehringer Mannheim International Symposia

E. Erdmann, K. Greef, J. C. Skou (eds.)

Cardiac Glycosides 1785–1985

Biochemistry – Pharmacology – Clinical Relevance

With Contributions by C. Achenbach, K. van Ackern, T. Akera, M. F. Andersen, K. Andersson, J. K. Aronson, P. Bach, H. Bardenheuer, I. Blömer, H. H. Bodemann, D. Branco, E. Braunwald, D. Brent, L. Brown, D. L. Brutsaert, E. Bucha, C. Bufalino, V. P. Butler, M. B. Cannell, M. C. Chan, D. Charlemagne, K. Chatterjee, R. Dahlqvist, T. Daldrup, H. Daying, R. Dietz, W. Doelle, W. Doering, R. Doliwa, W. Drohn, D. A. Eisner, E. Erdmann, O. Faergeman, M. Fagoo, H. Fasold, B. Fichtl, K. Fischer, M. Fischer, J. L. Fleg, A. A. L. Fox, C. Frelin, L. Frey, L. Fridrich, A. Gaßner, M. H. Gault, R. A. Gillis, T. Godfraind, H. K. Gold, M. L. Gonçalves, B. Grabensee, K. Greeff, W. Grille, G. Grupp, I. L. Grupp, N. Gustafson, E. Haber, T. Hair, S. Halbach, O. Hansen, K.-O. Haustein, S. Herzig, T. Hickerson, P. M. Hooymans, E. Hug, P. G. Hugenholtz, G. Hüller, I. Hutton, A. S. Jaffe, W. Jenkel, T. Jogestrand, W. Johnsen, E. Johnson, T. Joreteg, P. H. Joubert, L. M. Jung, A. M. Katz, B. Kaufmann, T. L. Kirley, K. Kjeldsen, W. Knippenberg, K. Kochsiek, K.-D. Kolenda, M. L. P. Koten, M. Krekler, W. Kübler, R. Kuske, C. Lafleur, M. Lazdunski, W. J. Lederer, G. Lee, S. W. Lee, L. G. Lelièvre, J. Lindenbaum, J. J. H. M. Lohman, L. Longerich, B. Lorenz, H. Lüllmann, D. Magometschnigg, C. Maier, J. M. Maixent, D. T. Mason, W. Mäurer, E. Mayoux, W. Meister, F. W. H. M. Merkus, L. von Meyerinck, K. Mohr, K. Moreth, C. Mouas, R. G. Murray, J. E. Muller, M. S. Nieminen, G. Oltmanns, W. Osswald, H. J. Partanen, E. Passamani, T. J. Pellinen, E. Petzinger, M. Pichler, F. Piscione, J. Plum, W. K. Poole, R. Preisler, M. Preteseille, J. A. Quest, D. S. Raabe, D. S. Regal, C. Reithmann, D. Renner, K. R. H. Repke, T. Risler, R. E. Rude, U. Rüegg, J. Sander, H. Schadewaldt, S. G. Schäfer, W. Schaumann, M. Scheininger, K. Schenk-Gustafsson, P. Schlieper, W.-D. Schlott, A. Schmoldt, G. Scholtysik, H. Scholz, A. Schömig, W. Schoner, W. Schönfeld, G. Schönsteiner, H. Schühlen, K. P. Schüren, A. Schwartz, H. Schweitzer, O. M. Sejersted, P. W. Serruys, J. C. Skou, T. W. Smith, G. Sommer, G. Steinbeck, B. Sterndorff, P. H. Stone, H. Strobach, B. Swynghedauw, S. U. Sys, L. B. Tan, K. A. Taubert, H. Tawfik, K. Theisen, G. Tittel, Z. G. Turi, A. C. Tweddel, M. Valdeolmillos, S. Vasdev, E. Vigholt-Sørensen, P. Vigne, H. Waginger, H. Wagner, E. T. Wallick, E. Weber, K. T. Weber, J. Weiland, T. Wenger, K. Werdan, K. E. Wirth.

Springer-Verlag
Berlin Heidelberg GmbH

Prof. Dr. E. Erdmann
Klinikum Großhadern
der Universität München
Medizinische Klinik I
Marchioninistraße 15
8000 München 70
West Germany

Prof. Dr. K. Greeff
Institut für Pharmakologie
der Universität Düsseldorf
Moorenstraße 5
4000 Düsseldorf
West Germany

Prof. Dr. J. C. Skou
Biophysik Institut
Aarhus Universitet
Universitetparken
8000 Aarhus/Denmark

CIP-Kurztitelaufnahme der Deutschen Bibliothek

[Cardiac glycosides seventeen hundred and eighty-five to nineteen hundred and eighty-five]
Cardiac glycosides 1785–1985 : biochemistry – pharmacology – clin. relevance / E. Erdmann ... (eds.). With contributions by C. Achenbach ...
(Boehringer Mannheim international symposia)
ISBN 978-3-662-11294-6 ISBN 978-3-662-11292-2 (eBook)
DOI 10.1007/978-3-662-11292-2
NE: Erdmann, Erland [Hrsg.]; Achenbach, Christopher [Mitverf.]

Originally published by Dr. Dietrich Steinkopff Verlag, GmbH & Co. KG, Darmstadt in 1986
Softcover reprint of the hardcover 1st edition 1986

Medical Editorial: Juliane K. Weller – Copy Editing: Cynthia Feast – Production: Heinz J. Schäfer

Preface

The bicentenary of William Withering's now famous medical report entitled "An Account of the Foxglove and Some of its Medical Uses" has given us the occasion to analyze the state of the art. Cardiac glycosides in 1985 are considered to be the basis for medical treatment of myocardial failure, together with diuretics and vasodilators in the more severe cases. Nevertheless, the controversy as to their exact place in the treatment of heart failure with sinus rhythm has never ceased. Although cardiac glycosides are of unquestionable value in tachycardia caused by atrial fibrillation or atrial flutter, the chronic use of these drugs in sinus rhythm is not generally accepted. The development of tolerance has been reported. It is of interest that Withering himself wrote: "– let it (foxglove) be continued until it either acts on the kidneys, the stomach, the pulse, or the bowels; let it be stopped upon the first appearance of any of these effects, and I will maintain that the patient will not suffer from its exhibition, nor the practitioner be disappointed in any reasonable expectation."

In West Germany alone, more than three million patients (about 5% of the total population) are constantly taking cardiac glycosides. These drugs are the most prescribed medicaments in many countries. Thus, we considered it worthwhile and necessary to review in a critical way today's knowledge about the foxglove; to report new findings and to evaluate old statements.

Cardiac glycosides are used as tools in biochemistry and as model substances in pharmacokinetics. We know that they bind specifically and reversibly to the $(Na^+ + K^+)$-ATPase, the enzyme that actively transports Na^+ and K^+ through the cell membrane of all mammalian cells. Therefore, biochemists, physiologists and pharmacologists as well as clinicians have always been interested in the actions of cardiac glycosides at the molecular level. Their work has added greatly to the understanding of glycoside effects and side effects in our patients and recently has resulted in producing specific antibodies to cardiac glycosides (Fab-fragments). These antibodies have a higher affinity for these drugs than the myocardial receptor enzyme itself, the $(Na^+ + K^+)$-ATPase. Thus, the treatment of severe glycoside intoxication with these antibodies has now become possible.

Cardiac glycoside research is a good example of the value of mutual interaction and the interrelation of basic and clinical sciences. Problems of treating heart failure with glycosides can often be traced to changes in the quantity or affinity of cardiac glycoside receptors in the myocardial cell membrane. Therefore, Kurt Greeff, Jens C. Skou and I decided to assemble biochemists, physiologists, pharmacologists and clinicians for discussion and evaluation of the present knowledge 200 years after Withering, and to encourage theoreticians and clinicians to speak if not the same, at least a similar language. There are perhaps no other drugs about which we know so much – however, when going into detail, it seems to be the same as always in science – we know astonishingly little about the cardiac glycosides.

Therefore, this symposium was planned to demonstrate what we do know and what is still obscure about the most ingested drugs in medicine. Hopefully, the next anniversary symposium will succeed in clarifying these problems.

Last but not least, I should like to express my thanks to all who participated in this symposium. Especially, I thank all authors for submitting their manuscripts on time, although they are constantly pressed by important work.

The Steinkopff Verlag has made it possible for this book to appear so quickly by rapid, efficient, and expert work.

Most of all, I should like to express my sincere gratitude to Dr. Wenzel, Dr. Bergmeyer and Mr. Rühl from Boehringer Mannheim. They have generously supported this symposium and this book but moreover, they have themselves carried most of the heavy burden of organisational work. Without their skills, we would not have succeeded.

Even if William Withering may not have noticed that cardiac glycosides act by slowing the ventricular rate and by increasing the force and the speed of myocardial contraction, his last sentence of the "inferences" is certainly still valid: "That it (foxglove) has a power over the motion of the heart, to a degree yet unobserved in any other medicine, and that this power may be converted to salutary ends."

Munich 1986

Erland Erdmann

Contents

3. Digitalis-like substances

4. Mechanism of inotropic effects – cardiac glycosides and other positive inotropic agents

Contributors

Achenbach, Dr. C.
Department of Physiology II
Universität Bonn
Wilhelmstraße 31
5300 Bonn, West Germany

Ackern, Prof. Dr. K. van
Ludwig-Maximilians-Universität
Klinikum Großhadern
Institut für Amnesthesiologie
Postfach 701260
8000 München 70, West Germany

Akera, Prof. Dr. T.
Department of Pharmacology and Toxicology
B445 Life Science Building
Michigan State University
East Lansing
Michigan 48824-1317, U.S.A.

Aronson, Dr. J. K.
MRC Unit and University
Department of Clinical Pharmacology
Radcliffe Infirmary
Woodstock Road
Oxford, OX2 6HE, England

Bodemann, Prof. Dr. H.
Klinikum der Albert-Ludwigs-Universität
Medizinische Universitätsklinik
Hugstetter Straße 55
7800 Freiburg, West Germany

Brown, Dr. Lindsay
Medizinische Klinik I
der Universität München
Klinikum Großhadern
Marchioninistraße 15
8000 München 70, West Germany

Brutsaert, Prof. Dr. D. L.
University of Antwerp
Department of Physiology and Medicine
Groenenborgerlaan, 171
B-2020 Antwerp, Belgium

Charlemagne, Dr. D.
INSERM U 127
Hospital Lariboisiére
Université Paris 7
F-75010 Paris, France

Chatterjee, Prof. Dr. K.
University of California
Moffitt Hospital
San Francisco, California 94143, U.S.A.

Eisner, Prof. D.
Department of Physiology
University College London
Gower Street
London WC1E 6BT, England

Erdmann, Prof. Dr. E.
Klinikum Großhadern
der Universität München
Medizinische Klinik I
Marchioninistraße 15
8000 München 70, West Germany

Fleg, Dr. Jerome L.
Laboratory of Cardiovascular Science
Gerontology Research Center
4940 Eastern Avenue
Baltimore, Maryland 21224, U.S.A.

Frelin, Dr. C.
Centre de Biochimie
Centre National de la Recherche Scientifique – Université de Nice
Parc Valrose
F-06034 Nice, France

Fridrich, Dr. L.
Pensionsversichcrungsanstalt
der Angestellten
Rehabilitationszentrum Hochegg
Friedrich-Hillegeist-Straße 2
A-2840 Grimmenstein, Austria

Gault, Prof. M. M.
Memorial University of Newfoundland
St. John's, Newfoundland A1B 3V6, Canada

Gillis, Prof. Richard A.
Georgetown University
Schools of Medicine and Dentistry
3900 Reservoir Road N.W.
Washington D.C. 20007, U.S.A.

Godfraind, Prof. Dr. T.
Laboratoire de Pharmacodynamic
Generale et de Pharmacologie
Université Catholique de Louvain
Avenue Emmanuel Mounier 73
1200 Brüssel, Belgium

Gonçalves, Dr. Maria de Lurdes
Instituto Gulbenkian de Ciencia
Laboratorio de Farmacologia
Apartado 14-2781
Oeiras, Portugal

Grille, Dr. W.
II. Medizinische Universitäts-Klinik
Metzstraße 53
2300 Kiel, West Germany

Greeff, Prof. Dr. K.
Institut für Pharmakologie
der Universität Düsseldorf
Moorenstraße 5
4000 Düsseldorf, West Germany

Grupp, Prof. Dr. Ingrid
College of Medicine
University of Cincinnati
Cincinnati, Ohio 45267-0575, U.S.A.

Halbach, Dr. S.
Abteilung für Pharmakologie
Gesellschaft für Strahlen- und
Umweltforschung
Ingolstädter Landstraße 1
8042 Neuherberg, West Germany

Hansen, Prof. Dr. O.
Department of Animal Physiology
Royal Veterinary and Agricultural
University
1871 Copenhagen V, Denmark

Haustein, Prof. Dr. sc. K. O.
Abteilung für Klinische Pharmakologie
Medizinische Akademie Erfurt
Nordhäuser Straße 74
DDR-5010 Erfurt, East Germany

Jogestrand, Dr. Thomas
Department of Clinical Physiology
Karolinska Hospital
S-10401 Stockholm, Sweden

Joubert, Prof. P. H.
Department of Pharmacology and
Therapeutics
Medical University of Southern Africa
P O Medunsa 0204, South Africa

Katz, Prof. Dr. Arnold M.
Division of Cardiology
University of Connecticut
Farmington, Connecticut 06032, U.S.A.

Kjeldsen, Dr. Keld
Institute of Physiology
University of Aarhus
DK-8000 Aarhus C, Denmark

Kochsiek, Prof. Dr. K.
Medizinische Universitäts-Klinik
Luitpoldkrankenhaus
Josef-Schneider-Straße 2
8700 Würzburg, West Germany

Krekler, Dr. M.
Institut für Pharmakologie
der Universität Düsseldorf
Moorenstraße 5
4000 Düsseldorf 1, West Germany

Kübler, Prof. Dr. W.
Medizinische Klinik
Abteilung Innere Medizin III
Bergheimer Straße 58
6900 Heidelberg 1, West Germany

Lelièvre, Prof. Dr. L.
Unité de Recherches sur le
Métabolisme du Cœur et de Vaisseaux
Hôpital Lariboisiére
41, Boulevard de la Chapelle
F-75010 Paris, France

Lindenbaum, Prof. J.
Medical Service
Harlem Hospital Center
Lenox Avenue and 135th Street
New York, N.Y. 10037, U.S.A.

Lohmann, Dr. J. J. H. M.
Clinical Pharmacokinetics
and Toxicology Laboratory
Hospital of Sittard
Walramstraat 23
NL-6131 BK Sittard, The Netherlands

Lüllmann, Prof. Dr. H.
Klinikum der Christian-Albrechts-Universität Kiel
Abteilung für Pharmakologie
Hospitalstraße 4-6
2300 Kiel, West Germany

Maier, Dr. Ch.
Department of Anesthesia
Christian-Albrechts-Universität
Schwanenweg
2300 Kiel, West Germany

Mason, Prof. Dr. Dean T.
Physician-in-Chief
Western Heart Institute
St. Marys Hospital and Medical Center
450 Stanyan Street
San Francisco, California 94117, U.S.A.

Meister, PD Dr. W.
Medizinische Klinik Innenstadt
der Universität München
Ziemssenstraße 1
8000 München 2, West Germany

Osswald, Prof. Dr. W.
Laboratorio de Farmacologia
Faculdate de Medicina
4200 Porto, Portugal

Partanen, Dr. H. J.
Cardiovascular Laboratory and First
Department of Medicine
University Central Hospital
Helsinki, Finland

Petzinger, Dr. E.
Institut für Pharmakologie und
Toxikologie im Fachbereich
Veterinärmedizin
Frankfurter Straße 107
6300 Gießen, West Germany

Plum, Dr. J.
Med. Einrichtungen der Universität
Klinik F des Zentrum für Innere Medizin
und Neurologie
Moorenstraße 5
4000 Düsseldorf, West Germany

Repke, Prof. Dr. K.
Zentralinstitut für Molekularbiologie
der Akademie der Wissenschaften
der DDR
Robert-Rössle-Straße 10
DDR-1115 Berlin, East Germany

Risler, Prof. Dr. T.
Medizinische Klinik der Universität
Otfried-Müller-Straße 10
7400 Tübingen, West Germany

Schäfer, Dr. S. G.
Institut für Pharmakologie und Toxikologie
Medizinische Fakultät der Ludwig-Maximilians-Universität
Nußbaumstraße 26
8000 München 2, West Germany

Schaumann, Prof. Dr. W.
Boehringer Mannheim
Medizinische Forschung
Sandhoferstraße 116
6800 Mannheim 31, West Germany

Schenk-Gustafsson, Dr. Karin
Karolinska Institute
Division of Cardiology
Department of Internal Medicine
Huddinge University Hospital
S-141 86 Huddinge, Sweden

Schmoldt, Prof. Dr. A.
Institut für Rechtsmedizin
Universität Hamburg
Butenfeld 34
2000 Hamburg 54, West Germany

Scholtysik, Prof. Dr. G.
Sandoz AG
Präklinische Forschung
CH-4002 Basel, Switzerland

Scholz, Prof. Dr. Hasso
Abteilung für Allgemeine Pharmakologie
Universitäts-Krankenhaus Eppendorf
Martinistraße 52
2000 Hamburg 20, West Germany

Schoner, Prof. Dr. W.
Institut für Biochemie und Endokrinologie
Fachbereich Veterinärmedizin
Frankfurter Straße 100
6300 Gießen, West Germany

Schüren, Prof. Dr. K. P.
Innere Abteilung
Krankenhaus Moabit
Turmstraße 21
1000 Berlin 21, West Germany

Schwartz, Prof. Dr. A.
Department of Pharmacology and Cell Biophysics
University of Cincinnati
College of Medicine
231 Bethesda Avenue
Cincinnati, Ohio 45267-0575, U.S.A.

Sejersted, Prof. Dr. Ole
Institute of Muscle Physiology
PO Box 8149 Dep
0033 Oslo 1, Norway

Serruys, Dr. Patrick W.
Catheterization Laboratory
Thoraxcenter
PO Box 1738
NL-3000 DR Rotterdam, The Netherlands

Skou, Prof. Dr. J. C.
Institute of Biophysics
University of Aarhus
8000 Aarhus C, Denmark

Smith, Prof. Dr. Thomas W.
Cardiovascular Division
Brigham and Women's Hospital
75 Francis Street
Boston, Massachusetts 02115, U.S.A.

Steinbeck, PD Dr. G.
Medizinische Klinik I der Universität
Klinikum Großhadern
Marchioninistraße 15
8000 München 70, West Germany

Strobach, Prof. Dr. H.
Pharmakologisches Institut der Universität
Moorenstraße 5
4000 Düsseldorf, West Germany

Tan, Dr. L. B.
Department of Cardiovascular Medicine
University of Birmingham
Bordesley Green East
Birmingham B9 5ST, England

Taubert, Dr. Kathryn A.
American Heart Association
National Center
7320 Greenville Avenue
Dallas, TX 75231, U.S.A.

Tittel, Dr. G.
LAT GmbH Dr. Tittel
Labor für Analysen und Trenntechnik
Maria-Eich-Straße 7
80323 Gräfelfing, West Germany

Vigholt-Sørensen, Dr. Else
Medical Department I
Aarhus Amtssygehus and Specialist Clinics
8000 Aarhus C, Denmark

Waginger, Dr. H.
Institut für Botanik und Lebensmittelkunde
Veterinärmedizin Universität Wien
Linke Bahngasse 11
A-1030 Vienna, Austria

Weber, Prof. Dr. E.
Ruprecht-Karls-Universität Heidelberg – Klinikum
Bergheimer Straße 58
6900 Heidelberg, West Germany

Weber, Prof. Dr. K. T.
Division of Cardiology
Michael Reese Hospital
University of Chicago
Chicago, IL 60616, U.S.A.

Werdan, PD Dr. K.
Medizinische Klinik I der
Universität München
Klinikum Großhadern
Marchioninistraße 15
8000 München 70, West Germany

Wirth, Prof. Dr. K.
Pharmakologisches Institut der
Universität
Moorenstraße 5
4000 Düsseldorf, West Germany

William Withering – The man and his work

J. C. Skou

Institute of Biophysics, University of Aarhus (Denmark)

Introduction

In 1785 William Withering, a physician in Birmingham, published "An Account of the Foxglove and Some of its Medical Uses: with Practical Remarks on Dropsy and other Diseases" (6). This is a critical analysis of the effect of digitalis on dropsy based on ten years of observations with reports of 163 cases treated by Withering, and a number of cases treated by colleagues. Detailed instructions are given on how to use digitalis, on the dosage, on the clinical signs which can be used to follow the effect, and on toxicity.

The Account is a milestone in the history of medicine, not only because it gives digitalis a place in medicine as a therapeutic, but also because it shows the importance of careful observation and testing as a basis for evaluating the therapeutic effect of a drug. Unfortunately the lesson taught by Withering was not learned until 100 years later.

History of digitalis

Digitalis has a long history in medicine. It is mentioned as "Foxes glofa" in a Welsh pharmaceutical book "Meddygon myddmai" in 1250 as a component of ointments used as a treatment for headaches and spasms (4).

It was a botanist and physician from Tübingen, Leonard Fuchs, who in 1542 in his book on herbs, Historia Stirpium, gave the first detailed and illustrated description of the digitalis plant. He also gave the plant its name "Digitalis" derived from the German Fingerhut or in English thimble, which refers to the shape of the blossoms (4). Fuchs mentions its use as a laxative (4) and as an emetic (2).

Up to 1775 when Withering initiated his investigation, digitalis was used against a variety of diseases, in ointments for wound healing, against epileptic fits, phthisis, scrophulosis, and as an external application against goitre (1, 6). It was mentioned in the London pharmacopoeia of 1722, the Edinburgh pharmacopoeia of 1744, in the Paris pharmacopoeia of 1748 and 1758, and the Wittenberg pharmacopoeia of 1754 and 1771 (5).

Warnings were given against its toxic effects. As mentioned by Dr. Stokes in a letter to William Withering which is referred to in the Account, "it is a medicine which is proper only for strong constitutions, as it purges very violently, and excites excessive vomitings". H. Boerhaave in Leyden, who was famous for his course of lectures given each year and called "the Institutes of Medicine", the first instruction in what is now Physiology, regarded digitalis as a poison and warned against its use (3, 5). In 1748 a physician from Orleans, F. Salerne, reported on the toxic effect of digitalis to the French Academy; the report was published in the Annuals from the Academy and widely distributed. And even if the report was not an attempt to determine the therapeutic range but merely a report on the symptoms of an overdose of digitalis to a turkey, it apparently had an influence on the attitude towards digitalis as a therapeutic. The "experiment" and the toxic symptoms are described in the and which Boulton had brought home from a lead mine, and shoved that it was "terra ponderosa aerata". A paper "Experiments and Observations on the Terra Ponderosa" was read for the Royal Society and published in Philosophical Transactions. This gives the first chemical analysis of terra ponderosa aerata and distinguishes it from terra ponderosa vitriolata ($BaSO_4$). The paper also describes how to use terra ponderosa salita ($BaCl_2$) to remove impurities of vitriolic acid (H_2SO_4) from marine acid (HCl). From this Withering has been credited for the test for sulphates, the insolubility of $BaSO_4$ (7). $BaCO_3$ was named Witherite by a German geologist, Abraham Werner. As a result of this work Withering became recognized as a mineralogist (5, 7).

The Lunar Society (see[5])

In England it was the time of the industrial revolution. There was an interest in science, mechanics and applied science, which encouraged professionals and manufacturers to meet in informal groups for the exchange of information and to discuss new discoveries and the application of science to practical problems. The most famous of these groups was the Lunar Society in Birmingham. It was an informal gathering of men with common interests which dates back to around 1765 and lasted for about 40 years (5). The total number of members did not exceed 14 and at one time not more than 8–10. They met once a month for discussion at 2 o'clock in the afternoon at the house of one of the members. The meetings were on the Monday nearest the full moon in order to take advantage of its light for their return home; from this arose the name The Lunar Society (3). Members were men like Matthew Boulton, manufacturer; Joseph Priestly, physicist and chemist, discoverer of oxygen; Erasmus Darwin, physician and grandfather of Charles Darwin; James Watt, inventor of the steam engine; Josiah Wedgwood, manufacturer of ceramics; and William Small, mathematician and physician.

William Small had settled as a physician in Birmingham in 1765 and had a clinic together with a respected Birmingham physician John Ash. Small was Boulton's family physician. He was an invaluable member of The Lunar Society and his death in 1775 was a blow to the members. On that occasion Erasmus Darwin wrote a letter to William Withering (5).

"I am at this moment returning from a melancholy scene, the death of a friend who was most dear to me, Dr. Small of Birmingham, whose strength of reasoning, quickness of invention, learning in the discoveries of other men, and integrity of heart (which is worth them all), had no equal. Mr. Boulton suffers an inconceivable loss from the doctor's mechanical as well as medical abilities. . . . Now it occurred to me that if you should choose that situation your philosophical taste would give us a friendship of Mr. Boulton, which would operate for you what it did for Dr. Small . . . If you should think this prospect worth you going to see Mr. Boulton at Soho to inquire further into, I will take care to leave at home a proper letter for you to him if I should not see you."

Erasmus Darwin, who was a physician in Lichfield near Birmingham, apparently knew Withering from Withering's visits as a student to his uncle, Hector Brook, who also was a physician in Lichfield. The "philosophical taste" may refer to Withering's paper on marl in Philosophical Transactions.

The result of the letter was that Withering moved to Birmingham in 1775 and took over William Small's practice with John Ash. By the move to Birmingham Withering became physician for Boulton and entered The Lunar Society (5). He soon had a big practice and in 1779 Ash and Withering were appointed to the medical staff of the General Hospital in Birmingham.

Withering and digitalis

In the Account Withering writes:

"My worthy predecessor in this place, the very human and ingenious Dr. Small, had it made a practice to give his advice to the poor during one hour in a day. This practice, which I continued until we had a hospital opened for the reception of the sick poor, gave me an opportunity of putting my ideas into execution in a variety of cases; for the number of poor who thus applied for advice, amounted to between two and three thousand annually. I soon found the Foxglove to be a very powerful diuretic; but then and for considerable time afterwards I gave it in doses very much too large and urged its continuance too long, for mislead by reasoning from the effects of the squill, which generally acts best upon the kidneys when it excites nausea, I wished to produce the same effect by the Foxglove" (6).

The diuretic effect of squill, which was known in ancient times, was rediscovered and described in 1744 by Gerhard van Swieten, who was a physician to Maria Theresia of Austria (4).

From the preliminary experience with digitalis, Withering included in the first edition in 1776 of the Botanical Arrangement, that digitalis purpurea "merited more attention than modern practice bestowed on it".

A case reported to Withering by John Ash was an impetus to find a way to a regular mode for prescription of digitalis. A colleague, Dr. Cawley from Oxford, was cured from a hydrops pectoris "by an empirical exhibition of the root of Foxglove, after some of the first physicians of the age had declared they could do no more for him". Withering's reaction: "I was now determined to pursue my former ideas more vigorously than before . . ." (6).

A problem was to find an accurate way to prescribe digitalis. The problem is discussed in the Account in the chapter: "Of the Preparations and Doses of the Foxglove." The solution was to use the powder of the dried leaves collected "always in one condition of the plant, viz, when it was in its flowering state". A watery extract of the leaves was not as potent, and even if an alcoholic extract was as potent it had more undesired side effects. With the powder of the leaves "as it then became possible to ascertain its dosage, it was gradually adopted by the medical practitioners in the circle of my acquaintance".

In 1778 there was an epidemic of scarlet fever in Birmingham. In 1779 Withering described the epidemic in "An Account of Scarlet Fever and Sore Throat or Scarlatina Anginosa: Particularly as it Appeared in Birmingham in 1778". The result of the epidemic was that in 1779 "a number of dropsical cases afforded themselves to my attention . . . they were all without an exception cured by the Foxglove" (6).

Withering gradually reduced the dosage, "experience convinced me that the diuretic effect . . . do not all depend on its exciting nausea or vomiting". The result of this was that the incidence of side effects declined, and of the about 100 cases treated in the period from 1781 to 1784 the incidence of side effects was 16–20 per cent, which is comparable to results of treatment with digitalis today (2).

The Dr. Stokes mentioned above was a young physician Jonathan Stokes, who had obtained his degree in Edinburgh and had settled in Stourbridge. He was very able in botanics, became a close friend of Withering, became a son of the house, and took part in the revision of the Botanical Arrangements (5). In the Account Withering reports that in February 1779 Dr. Stokes communicated the results of Withering's experience so far of the treatment of dropsy with Foxglove to the Medical Society in Edinburgh, and that in a letter from November 1779 Dr. Stokes informed Withering that the information has been conveyed to Dr. Hope who had tried the Foxglove in the infirmary with success. He is also informed from a friend Dr. Duncan that Dr. Hamilton has "learned its use from Dr. Hope and has employed it very frequently in the hospital at Edinburgh", and Dr. Duncan furthermore informed Withering "that the late very ingenious and accomplished Mr. Charles Darwin, informed him of it being used by his father and myself (which is Withering), in cases of hydrothorax, and that he (Duncan) has ever since mentioned it in his lectures and sometimes employed it in his practice" (6). Withering's use of digitalis in the treatment of dropsy was thus known in 1779, 6 years before the publication of the Account.

The Charles Darwin referred to here was a son of Erasmus Darwin. He attended Medical School in Edinburgh, and after he had received his M. D., he died in 1778 from a dissection wound, at 21 years of age. In commemoration of his son, Erasmus Darwin edited an essay for which Charles Darwin had received a gold medal and Charles Darwin's thesis under the title "Experiments Establishing a Criterion between Mucaginous and Purulent Matter. And an Account of the Retrograde Motions of the Absorbent Vessels of Animal Bodies in Some Diseases". To this were added 9 case histories which described the therapeutic effect of digitalis on dropsy. This was published in 1780 (5). According to Withering's above mentioned report of the letter from Dr. Duncan, this must have been concerned with cases treated not by Charles Darwin but by Erasmus Darwin. Surprisingly, no credit was given to Withering, for Erasmus Darwin knew that Withering was testing the effect of digitalis on dropsy. In one of the cases Darwin had consulted Withering and Withering had suggested treatment with digitalis, and as Darwin had had no previous experience with digitalis, Withering prescribed the dosage. Darwin's essay was however the first published account on the treatment of dropsy with digitalis.

"An Account of the Foxglove" (6)

Withering writes in a letter to Stokes in September 1778 (6), "I wish it was as easy to write upon the Digitalis – I despair of pleasing myself or instructing others, in a subject so difficult. It is much easier to write upon a disease than upon a remedy. The former is in the hands of nature, and a faithful observer, with an eye of tolerable judgement, cannot fail to delineate a likeness. The latter will ever be subject to the whims, the inaccuracies, and the blunders of mankind". But finally after 10 years' experience with the use of digitalis Withering published his results in 1785 in the Account which begins:

"After being frequently urged to write upon this subject and as often declining to do it, from apprehension of my own inability, I am at length compelled to take up the pen, however unqualified I may still feel myself for the task.

The use of the Foxglove is getting abroad, and it is better the world should derive some instruction, however imperfect, from my experience, than the lives of men should be hazarded by its unguarded exhibition, or that a medicine of so much efficacy should be condemned and rejected as dangerous and unmanageable.

It is now 10 years since I first began to use this medicine. Experience and cautious attention gradually taught me how to use it. For the last 2 years I have not had occasion to alter the modes of management; but I am still far from thinking them perfect.

It would have been an easy task to give select cases, whose successful treatment would have spoken strongly in favour of the medicine, and perhaps been flattering to my own reputation. But Truth and Science would condemn the procedure. I have therefore mentioned every case in which I have prescribed the Foxglove, proper or improper, successful or otherwise. Such a conduct will lay me open to the censure of those who are disposed to censure, but it will meet the approbation of others, who are the best qualified to be the judges." And later

"There are men who will hardly admit of anything which an author advances in support of a favorite medicine, and I allow they may have some cause for their hesitation; nor do I expect they will wave their mutual modes of judging on the present occasion. I could wish therefore that such readers would pass over what I have said and attend only to the communications from correspondents, because they cannot be supposed to possess any unjust predilection in favour of the medicine; but I cannot advise them to this step, for I am certain they would then close the book, with much higher notions of the efficacy of the plant than they would have learnt from me. Not that I want faith in the discernment or in the veracity of my correspondents, for they are men of established reputation; but the cases they have sent me are, with some exceptions, too much selected. They are not upon this account less valuable in themselves, but they are not the proper premises from which to draw permanent conclusions . . ." The Preface ends

"After all, in spite of opinion, prejudice, or error, time will fix the real value upon this discovery, and determine whether I have imposed upon myself and others, or contributed to the benefit of science and mankind."

This preface shows an attitude to his work which Withering had ascribed to his teacher in medicine, William Cullen.

The mode of management Withering reached was "let it be given in doses of 1–3 grain 2 times a day of powder of the digitalis leaves collected at the time the blossoms are coming forth and let it be continued till it either acts on the kidneys, the stomach, the pulse, or the bowels; let it be stopped upon the first appearance of any one of these effects and I will maintain that the patient will not suffer from its exhibition, nor the practitioner be disappointed in any reasonable expectation".

The indication was dropsy. In Withering's view digitalis acted as a diuretic but it is emphasized "that the digitalis will not universally act as a diuretic, that it does so more generally than any other medicine, that it will often produce this effect after every other probable method has been fruitlessly tried . . . digitalis may be used with advantage in every species of dropsie, except the encysted".

It is also emphasized "that it has a power over the motion of the heart, to a degree yet unobserved in any other medicine, and that this power may be converted to salutary ends".

Withering was not aware of the connection between dropsy und congestive heart failure and of the connection between digitalis effect on the heart and the effect on dropsy. This was left to coming generations of physicians.

Dispute with Erasmus Darwin

In January 1785 Erasmus Darwin sent a letter to the Royal College of Physicians in London which was published in the 1785 volume of Medical Transactions with a notice "read March 16", in which Darwin refers to the cases of dropsy treated with Foxglove previously published in "Experiments Establishing a Criterion . . .", and reports some further cases of the use of Foxglove, but still without mentioning Withering (5). The Account was published in July 1785 after the letter was read but before the letter was published. There is no reference in the Account to Darwin's letter to the Royal College of Physicians, but the case (nr. IV in the Account) for which Darwin consulted Withering in 1776 is reported in detail (2, 6). After a description of the patient's symptoms and the previous treatment ". . . she had experienced no relief from any means that had been used . . . I proposed the digitalis to be tried; adding that I sometimes had found it to succeed when other, even the most judicious, methods had failed. Dr. Darwin very politely, acceded immediately to my proposition, and, as he had never seen it given, left the preparation in the dose to my direction . . ."

After a description of the result of the treatment, which was successful ". . . I have been more particular in the narrative of this case, partly because Dr. Darwin has related it rather imperfectly . . . trusting, I imagine, to memory, and partly because it was a case which gave rise to a very general use of the medicine in that part of Shrophshire".

Erasmus Darwin was the most prominent physician in England at that time. He had brought Withering to Birmingham and had close connection to him in the Lunar Society. It is difficult to understand why Darwin did not give credit to Withering in the essay in commemoration of his son or in the letter to the Royal College of Physicians. His view may have been that even if Withering was testing digitalis on dropsy, this was not Withering's discovery. Digitalis was used as a folk-medicine against dropsy, and doctors were using digitalis, some of them inspired by Withering, but others independent of Withering, as for example in the treatment of Dr. Cawley referred to above. On the other hand Darwin had consulted Withering for his first case. Futhermore on two later occasions, in 1789 and 1794, Darwin in writings on digitalis still did not refer to the Account (5). The episode spoiled their friendship, and even if they still met in the Lunar Society, the wound was never healed. On the contrary there were other episodes (see [5]) in which Darwin offended Withering and Withering offended Darwin, which added to the controversy. It is difficult to see whom to blame. Withering's reputation grew quickly and in the end overshadowed that of Darwin's, and this may have influenced Darwin's view on Withering. On the other hand there seems to be no doubt that Withering was not an easy person to get along with. He was shy and reserved, was very keen on his reputation, and reacted strongly to criticism. He was punctilious and his request for accuracy is seen from the title of his Botanical Arrangement which takes up 24 lines on the title page. With his close friend Jonathan Stokes, who had taken part in the revison of the second edition of the Botanical Arrangement, he had a dispute on priority and copyright and also on payment; the quarrel lasted for many years, the problems were never settled, and it spoiled that friendship too (see [5]).

In 1785 Withering was elected member of the Royal Society in London; this was the society with the greatest social prestige in England in the 18th century. Darwin had been elected as a member in 1761 and almost all the Lunar Society members became fellows of the Royal Society (5).

The Account was well received and many doctors in England and on the Continent took up the treatment of dropsy with digitalis. The Account was translated into German. Interest was also aroused in the U.S.A. and seeds of Foxglove, which is not indigenous to the U.S.A., were sent over.

Against this background it is odd to see that in less than 20 years digitalis was back to where it was before Withering, again being used as a panacea, and it was not until the second half of the 19th century that digitalis therapy started to find its proper place in medicine (1). Ackerknecht in "Aspects of History of Therapeutics" gives the following explanation: ". . . physicians . . . always have been and in spite of all progress still are, under terrific pressure to do something against disease, while the actual means to fight disease are limited. When then an effective drug appears, it is soon used, out of hope and despair, far beyond its original destination. In accordance with Corvisart's old saying that new drugs always work, favorable results are obtained, and because of the notorious "autistic" habits of humans, it takes decades till the actual situation is discovered. Here the true roots of the panacea tendency seem to be located" (1).

The Lunar Society members were supporters of the French revolution. In 1791, on 14th July, the Bastille day, there was a riot in Birmingham, the King and Church riot, which started as a riot against dissenters. Houses were destroyed, among them Priestley's house, and Priestley was driven out of Birmingham. Withering's house was ransacked and suffered damage.

Withering's last years

In 1792 Withering gave up his appointment at the Birmingham General Hospital and his practice. He suffered from pulmonary consumption. He spent his last years editing new editions of Botanical Arrangements, and he also published a new edition of the Account on Scarlet Fever (5). In 1792 he was painted in London by a Swedish artist, C. F. von Breda. He is sitting reading the Account with a Foxglove in his hand. The picture is in the National Museum in Stockholm.

He died in 1799, 58 years old, and a friend who saw him the day before he died is responsible for the celebrated pun, "The flower of English physicians is indeed Withering" (3).

Conclusion

Withering was a remarkable man, with a great reputation as a botanist, mineralogist, and physician. Considering that besides his work at the General Hospital in Birmingham he had a large practice, it is impressive how much he accomplished. The Account of the Foxglove was not an isolated episode but one of the expressions of his whole attitude, his curiosity, his interest in the practical application of knowledge, his quantitative approach, and with the words he used to describe his teacher in medicine, William Cullen, an expression of his modesty, his caution and thoroughness, and his slow consenting academic doubt. He is a man well worth commemorating, and a man from whom we can learn what he first learned from William Cullen. And even if we no longer test drugs on human beings as Withering did, and even if we now know more about the effect of digitalis, anyone who is responsible for the treatment of patients can learn from reading the Account of the Foxglove. This book ought to be part of the curriculum for all medical students.

References

1. Ackerknecht EH (1962) Aspects of the History of Therapeutics. Bull History Med 36: 389–419
2. Estes JW, White PD (1965) William Withering and the Purple Foxglove. Scientific American 212: 110–119
3. Fulton FF (1953) The Place of William Withering in Scientific Medicine. J Hist Med 8: 1–15
4. Greeff K, Schadewaldt H (1981) Introduction and Remarks on the History of Digitalis. In: Greeff K Handbook of Experimental Pharmacology. Springer-Verlag, Berlin, pp 1–13
5. Schofield RE (1963) The Lunar Society of Birmingham. A Social History of Provincial Science and Industry in Eighteenth-Century England. Clarendon Press, Oxford
6. Withering W (1785) An Account of the Foxglove and Some of its Medical Uses: with Practical Remarks on Dropsy and other Diseases. Birmingham
7. Zeman DF (1950) William Withering as a Mineralogist. Bull Hist Med 24: 530–538

Author's address:

Prof. Dr. J. C. Skou, Institute of Biophysics, University of Århus, 8000 Århus C (Denmark)

Fruitful routes and blind alleys taken by digitalis therapy since its introduction by Withering

H. Schadewaldt

Institute for the History of Medicine, University of Düsseldorf (West Germany)

Our very famous Düsseldorf colleague, Professor of internal medicine Franz Grosse-Brockhoff (1907–1981) would surely have been the best speaker to address the subject of digitalis therapy and all its problems. Not only was he a prominent expert in this field, but he had also studied the history of digitalis in the last years of his life. Unfortunately he died in 1981. Therefore, Professor Kurt Greeff (born 1920) and I published some time ago a short account of the development of the knowledge on digitalis effects, which was based on older similarly formed ideas like those in the German literature; the historical reports by Horst Wirth, Günther Baumgarten and Werner Förster, not to mention other very important papers in the English speaking literature.

A two hundred years digitalis jubilee was celebrated in 1975, since it was supposed that in the year 1775 William Withering (1741–1799), at that time physician at the Municipal Hospital in Birmingham, showed perhaps for the first time the diuretic effects of internal digitalis infusions and began to gather his own observations. However, according to other sources he had begun his first treatment of a patient with ascites and general edemas with a decoction of digitalis purpurea on the 14th of January 1776. Unfortunately, the only effect was one of the most typical adverse symptoms, vomiting, which was particularly taken note of, yet without any recovery of his condition.

In year 1985 we are nearly in the same situation, and on not too certain a base. Meanwhile it is clear that the edition of the classical digitalis monograph by Withering "An Account of Fox Glove and Some of its Medical Uses", published in Birmingham in 1785, was not at all the first publication by physicians concerning the internal effects of digitalis. Withering himself had already presented a paper on this subject and his first experiences before the Medical Society of Edinburgh on 20th February 1779, read by Dr. Stokes, which could be considered as the first publication of this kind. In the medico-historical literature there has been a long discussion on the question as to whether the other author who had published a paper on digitalis at that time, Erasmus Darwin (1731–1802) was informed on the special attributes of this drug by his younger colleague Withering or not, or if he was personally present when the paper in Edinburgh was read. Anyway, the first published article of this kind is the observation of Darwin in the "Medical Transactions of the Royal College of Physicians of London", 1780, which was translated into the German language one year later.

Our Düsseldorf colleague L. M. Jung describes this very complicated evolution in his article in this volume, p 17. He shows that digitalis was not at all a new drug, but was known under the scientific name since its mention in Leonhard Fuchs's (1501–1566) "Historia stirpium", published at Basle in 1542 as an external remedy, especially used to induce wound healing and as an internal drug, for instance as a diuretic and purgative. Unfortunately we only have the Lyons 1551 edition in our Düsseldorf medical library (Fig. 1). Shortly after, in the year 1546, the famous Hieronymus Bock (1498–1554) mentioned his herbal digitalis

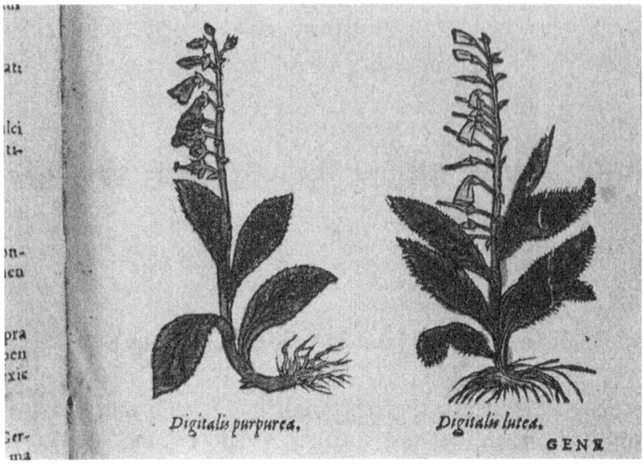

Fig. 1. First illustration of Digitalis plants from Leonhard Fuchs's book "Historia stirpium", Lyons 1551.

ointments and even digitalis pills. This drug became official in the London Pharmacopoeia of 1650 in both forms, ointments and pills, as can be seen in Figs. 2 and 3 of the pharmacopoeia of 1721.

The general diuretic effects were well described by the first authors, Darwin and Withering, but the effect on the heart was not known for a long time, and even in the first part of the 19th century was not the general knowledge of practitioners, although many other indications for digitalis therapy were discussed in those times. It is not clear if digitalis was known and used in classical antiquity; at least it is not mentioned in the common works on medicine. But in the early Middle Ages an ointment fabricated from Folia digitalis was known, used in tumefactions and abscesses. In a Welsh-Irish herbal "Meddygon myddwai", which was written between 1100 and 1250, the first mention of a digitalis ointment for headaches and convulsions can be found.

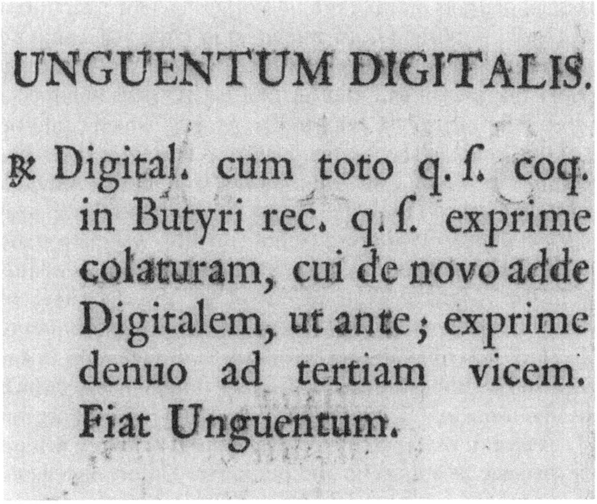
UNGUENTUM DIGITALIS.

℞ Digital. cum toto q. ſ. coq. in Butyri rec. q. ſ. exprime colaturam, cui de novo adde Digitalem, ut ante; exprime denuo ad tertiam vicem. Fiat Unguentum.

Fig. 2. Unguentum Digitalis, recipe from the Pharmacopoeia Londinensis of 1721.

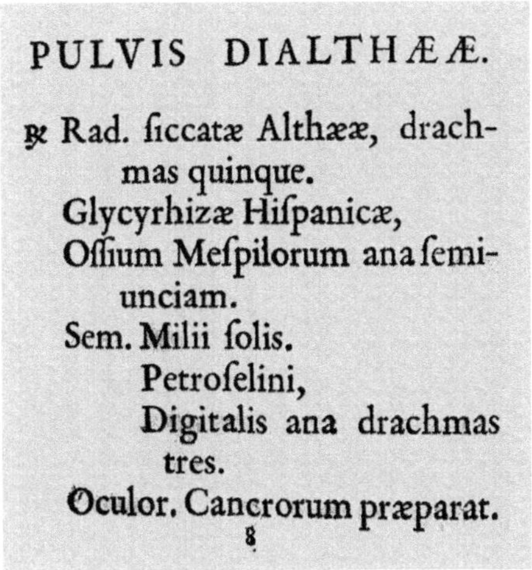

PULVIS DIALTHÆÆ.

℞ Rad. ſiccatæ Althææ, drachmas quinque.
Glycyrhizæ Hiſpanicæ,
Oſſium Meſpilorum ana ſemiunciam.
Sem. Milii ſolis.
Petroſelini,
Digitalis ana drachmas tres.
Oculor. Cancrorum præparat.
8

Fig. 3. Pulvis dialtaeae with digitalis powder for preparations of pills from the Pharmacopoeia Londinensis of 1721.

Introduced in the "Theatrum botanicum", written by John Parkison (1567–1650) and published in London in 1640, prescriptions of digitalis infusions for epileptic convulsions and ointments to treat goiters are introduced. Shortly afterwards, "phthysis" became a new indication, for instance in the "Botanologia" of William Salmon (1644–1713), London 1710, and again as internal indication of foxglove leaves. The small therapeutical spectrum soon became known, because no less than the famous herbalist Rembert Dodoens (1517–1585) had previously warned against frivolous use in his book which was published posthumously in 1616 in Antwerp. In 1748 experiments with turkeys, which were fed with foxglove leaves, resulted in the discovery of a very high toxicity, as was shown by François Salerne (died 1760) in the famous Parisian journal "Histoire de l'Académie Royale des Sciences" of that year. However, there is no doubt that it was Withering's monograph alone, with its clear casuistical descriptions, and which even today forms part of the general information of all foxglove experts, which led to the world-wide use of this drug and not the earlier report by Darwin. But even Darwin's first paper was soon translated into German. In 1786 Withering's original English book was published in German (Fig. 4), and a French version followed soon.

Even though Withering has without doubt described symptoms which were clearly consequences of a cardiac insufficiency, it took a long time until the supposed effects of digitalis on the heart were realized and could finally be confirmed, although even today the final details of their effects are not clear. In the 19th century all known indications of external and internal uses remained valuable. In addition to wound healing, phthisis and consumption, there were scrofulosis, acute fevers, which were considered until the era of bacteriology as special clinical units, such as nerve fever, intermittent fever, so-called inflammations such as pneumonia, and also asthma, rheumatism, goiter and even mental diseases. Therefore for the medical historian it is not a peculiar surprise that the mentally ill German poet Friedrich Hölderlin (1770–1843) was treated in the clinic of the University of Tübingen immediately after his admission on the 16th September 1806 with 6 grains of belladonna leaves and 2 grains of digitalis leaves dissolved in 2 ounces of aniseed-camomile water, as documented in the hospital's formulary. Shortly before his death, another famous

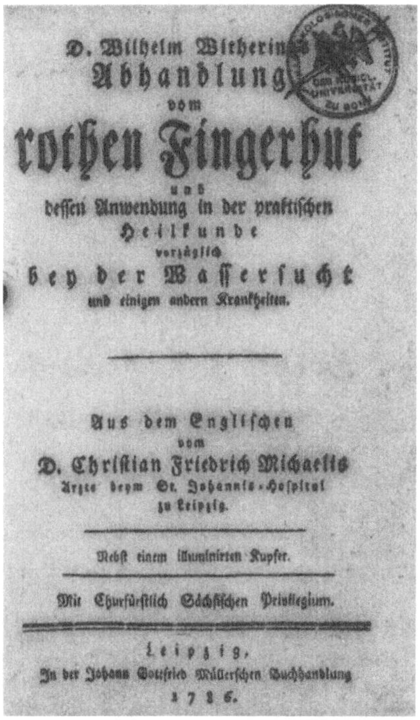
D. Wilhelm Witherings
Abhandlung
vom
rothen Fingerhut
und
dessen Anwendung in der praktischen
Heilkunde
vorzüglich
bey der Wassersucht
und einigen andern Krankheiten.

Aus dem Englischen
vom
D. Christian Friedrich Michaelis
Arzte beym St. Johannis-Hospital
zu Leipzig.

Nebst einem illuminirten Kupfer.

Mit Churfürstlich Sächsischen Privilegium.

Leipzig,
In der Johann Gottfried Müllerschen Buchhandlung
1786.

Fig. 4. First German translation of the classical work of William Withering by Christian Friedrich Michaelis, Leipzig 1786.

insane person, the painter Vincent van Gogh (1853–1890) had painted his treating physician Dr. Paul Gachet. He added to the portrait a special symbol of his therapy, a foxglove twig, but it is still discussed whether it is a specimen of digitalis purpurea or lutea.

The search for pure active agents in the phytotherapeutic drugs of that time led in 1845 to the production by Augustin-Eugène Homolle (1808–1883) of a concentrated mixture of "digitaline", which had certainly principally contained digitoxine and gitoxine. Since 1854 they have been termed digitale and digitaline. But before this it was clear that the actions of digitalis were especially cardiac ones, even though the striking bradycardiac effects were related to a special central vagus influence and the pretented antipyretic effects were particularly studied. For the substances in the foxglove, special new names were introduced under the main topic of "glycosides" in 1852, created by the Strasbourg pharmacist Charles Frederic Gerhardt (1816–1856), as some decades before Karl Friedrich Meissner (1792–1853) had done with a new group of "alkaloids" in 1818. These were substances with an alkaline character, which were sugar-like chemical compounds.

In 1842 Rudolf Boehm (1844–1926) discovered in fundamental experiments the direct effects of digitalis on the heart muscle, and he described the bigeminal rhythm as a special symptom of digitalis overdose and supersaturation. Some years later, in 1875, the famous Strasbourg pharmacologist Oswald Schmiedeberg (1838–1921), who had given the name "digitoxine" to a purified fraction he had won, could isolate further pharmalogically efficient constituents. With his work digitalis research in pharmacologic and chemical laboratories began on a larger scale.

I have attempted to answer the question of why only Withering had such a success with his book. He lived at the time when therapeutic scepticism was just starting, with the so-called first Vienna Medical School which had superseded the old polypharmacy, which had sometimes recommended drugs with more than 100 substances. Now in place of the traditional prescriptions of so many old-fashioned medicaments in the pharmacopoeias, began a very critical examination of the real effects of special individual drugs. In these cases the new generation of physicians preferred plants which showed a quick appearence of side-effects, because they hoped to find the same strong effect in illnesses, if used by prudent physicians. Among these new substances introduced especially by the personal physician of the Empress Maria Theresia (1717–1780) in Vienna, Anton Stoerck (1731–1803), was one of the most famous colchicum autumnale, meadow saffron, used for gout, and the foxglove, surely coming from the knowledge of popular medicine but very soon introduced into so-called scientific medicine, even if Withering himself had mentioned the famous but never identified "old dame" who was supposed to be the inaugurator of the diuretic and cardiac therapy. The year 1785 was in no manner the year of discovery of a hitherto unknown drug, but rather a renaissance which had very far-reaching consequences because the observations of Withering were based on very well and particularly finely described casuistics, and this was, I am convinced, the actual innovation of this period. Therefore William Withering must be considered as a great clinician, in the same rank as his British compatriot Thomas Sydenham (1624–1689) and the Professor of the famous Leiden University Herman Boerhaave (1668–1738).

References

1. Baumgarten G, Förster W (1963) Die herzwirksamen Glykoside. Thieme Leipzig
2. Baumgarten G (1969) 100 Jahre Digitalis. Pharmazie 24: 778–780
3. Bock H (1546) Kreuter-Buch, Strassburg fol, 335 (First edition New Kreutter Buch, Strassburg 1539)
4. Boehm R (1872) Untersuchungen über die physiologische Wirkung der Digitalis und des Digitalin. Arch ges Physiol 5: 153–191
5. Darwin E (1780) An account of the successfull use of foxglove in some dropsies. Med Transact Roy Coll Phys London 255–286 – Über die Wirkung des rothen Fingerhutes. Samml. auserl. Abb. z. Gebrauche praktischer Ärzte (1781) 6, 2 : 2
6. Dodoens R (1608) Cruydt-Boek. Leiden p 271
7. Drost R (1970) Die Digitalisglykoside, Ein historischer Rückblick. Fortschr Med 88: 1117–1120
8. Fuchs L (1542) Historia stirpium. Basel p 892

8a. Gerhardt, C. F. Aide-mémoire pour analyse chimique. Paris 1852

8b. Goerke, H. (1985) „Das Messer des Internisten" Bemerkungen zur Geschichte der Digitalistherapie. Münch. med. Wschr. 127. 938–941

9. Greeff K, Schadewaldt H (1981) Introduction and Remarks on the History of Cardiac Glycosides. In: Greeff K (ed) Handbook of Experimental Pharmacology. Springer-Verlag, Berlin Heidelberg New York, p 1
10. Grosse-Brockhoff F, Hausamen T-U (1975) 200 Jahre Herztherapie mit Digitalis. Dtsch med Wschr 100: 1980–1991
11. Grosse-Brockhoff F (1979) Digitalis. In: Blümchen G (ed) Beiträge zur Geschichte der Kardiologie. Rodenbirken p 193ff
12. Grosse-Brockhoff F (1980) Herzbehandlung mit dem 'Fingerhut' einst und jetzt. Vorträge der Natur-, Ingenieur- und Wirtschaftswiss. Klasse, Rheinisch-Westf Akad Wiss N 293: Opladen
13. Homolle A-E (1845) Mémoire sur la digitale pourprée. J pharm chim 3 sér 7: 57–83
14. Issekutz B (1971) Die Geschichte der Arzneimittelforschung. Akadémiai Kiadó Budapest p 251ff
15. Jung LM (1985) Therapeutic Use of Digitalis until the Time of William Withering. This volume
16. Lesky E (1965) Die Wiener Medizinische Schule im 19. Jahrhundert. Böhlaus & Nachf. Graz Köln p 101ff

17. Meissner KF (1818) Über das Morphium und die Mekonsäure. Trommsdorf N J Pharm 2: (1 St) 22
18. Overhamm H (1976) Zur Geschichte der Digitalis unter besonderer Berücksichtigung ihrer äußerlichen Anwendung. Quellen Stud Gesch Pharm Bd. 13 Jal-Verlag Würzburg
19. Parkinson J (1640) Theatrum Botanicum. London p 654
20. Salerne F (1748) Observation de botanique. Hist Acad Roy Scienc Paris 1748 (1752): 84
21. Schmiedeberg O (1875) Untersuchungen über die pharmakologisch wirksamen Bestandteile der Digitalis purpurea. Arch exp Path Pharmak 3: 16–43
22. Schmitz R, Overhamm H (1982) Einige Bemerkungen zur Geschichte des Roten Fingerhuts (Digitalis purpurea L.). In: Keil G (ed) Fachprosa-Studien Schmidt Berlin p 497ff
23. Schorderet M (1985) Deux cents ans de digitale: 1785–1985. Méd-Hyg 43: 290–295
24. Stenius R (1916) Die Geschichte der Digitalis purpurea und ihre Bedeutung in der Medizin bis etwa zum Jahre 1870. Vet med Diss Dresden 1916
25. Stenius R (1918) Die neue Geschichte der Digitalis purpurea. Arch wiss prakt Tierheilkd 44: H 5–6
26. Weese H (1936) Die Geschichte der Digitalis purpurea und der Strophantusglykoside. In: Digitalis. Thieme Leipzig p 1ff
27. Wirth H (1961) Der Rote Fingerhut und andere herzwirksame Heilpflanzen. Franckh'sche Verlagshandlung Stuttgart, Kosmos-Verlag p 106ff
28. Withering W (1785) An Account of the Foxglove and some of Its Medical Uses: Birmingham 1785 (Nachdruck Mannheim 1984); Bericht über den Fingerhut und seine medizinische Anwendung: German trans by F Johannessohn, Mannheim 1929, 1964 and 1984

Author's address:

Professor Dr. med. Hans Schadewaldt, Institute of History of Medicine, University of Düsseldorf, Moorenstraße 5, 4000 Düsseldorf (West Germany)

Therapeutic use of digitalis until the time of William Withering

L. M. Jung

Institute of Pharmacology, University of Düsseldorf (West Germany)

Summary

Digitalis has been known since the Middle Ages and used mainly by external application. Oral application, however, was mentioned before William Withering.

The priority of publication of the use of digitalis in dropsy is discussed between William Withering and Erasmus Darwin.

Our definite knowledge of the foxglove traces its genesis back to the renaissance of botany, but not – as in the case of so many of the more powerful of European vegetable remedies – to the mists of classic antiquity (2). The name digitalis was first – so far as it is known today – applied to the plant foxglove in the sixteenth century. The inventor of the appellation was Leonhart Fuchs, one of the pioneer fathers of modern scientific botany. It would at first sight appear – as popular opinion seems to have everywhere made it – to have been suggested by the crude resemblance of the shape of the flower to that of the finger of a glove. The familiar foxglove is one of the very few of our important vegetable agents of European growth of which no definite botanic history or quasi scientific lore has been transmitted to us from the classic fountains of the venerable antiquity of the semi-inspired pagan past.

The unrecorded employment of foxglove in the older centuries – which was, fairly surley, extensive and of which further cognition could possibly be obtained by investigating German manuscripts of these centuries following the key-word "Fingerhut" (4) – appears to have been mostly among the ordinaries, and practised after unskilled advice.

The herbalist and apothecary John Parkinson (5) depicts the "vertues" of digitalis, as known at that date, as follows: "The Italians have an usual proverbe with them concerning this herbe, called by them Aralda,... Aralda salveth all sores: for they use it familiarly to heale any fresh or greene wound or cut,... and sometimes they use the juyce in old sores to clean them, dry up their moysture,... which is performed by the bitter quality therin whereby it is found to be heating and drying...; so that when soever there is neede of a rarefying, or extenuating of thicke tough flegme and viscous humours troubling the chest or stomacke... and to open the obstructions of the liver and spleen;... it has been found... to be available for the King's Evil... and to be effectuall against the falling sickness;... after the taking of the decoction of two handfulls thereof, with four ounces of Polydody, of the oack made in Ale,..." – Thus we can see that the therapeutic use of digitalis and its reputation were even at that early date well founded and thus it is not surprising to find it named officially in the London Pharmacopoeia of 1650.

In 1710 William Salmon gave another account (6) of contemporary foxglove therapy; under the heading "The Qualities" he writes that "Fox Glove is hot and dry at least in the second degree, sulphureous and saline; aperitive, abstersive, adstringent, digestive and vulnerary; pectoral, hepatick and athritick; emetick, cathartick and analeptick." Under the heading "The Specification" he gives medical indications: "It cures consumption, King's – Evil, Green-sickness and Falling-sickness; also wounds, old sores and running ulcers."

Among other galaenical preparations are these: a liquid juice; an essence; a syrup; a decoction or infusion; an ointment or balsam; a cataplasm. Relating to the syrup "or rob of the juice of the herb and flowers made with honey" he wrote that "persons in deep consumptions..., have by the use of this syrup or rob been strangely recovered, and so perfectly restored as to grow fat again".

The awakening of interest in the foxglove as an individual herb is a symptom of tendencies towards enlightenment: clarifying the Materia Medica – and pharmacopoeias, so far existing – from inactive magic medicinal herbs and remedies; releasing therapeutics from the aristocratic tradition; changing over to a kind of medicine for all classes (see, for example, the ambulances for the poor in Birmingham at the beginning of the XVIIIth century! (1) and finally the increasing importance of the therapeutic value of the individual agent which had to be as exactly specified as possible and of a thoroughly proved efficacy (7). The school of Edinburgh figures here as the leading preceptor in Great Britain, while the School of Vienna was its equivalent for continental education: both originated in Boerhave's school in Leyden.

Thus it cannot be surprising that two medical men both generally acknowledged to have been successful doctors, both members of a scholarly society, and both of high social reputation had been disputing the merit of possessing the inventor's right of application of digitalis in dropsy (3).

References

1. Ackerknecht EH (1970) Therapie von den Primitiven bis zum 20. Jahrhundert. Stuttgart
2. Greeff K, Schadewaldt H (1981) Handbook of Pharmacology, Vol 56/1. Springer Verlag, Berlin Heidelberg New York, pp 1–12
3. Leclerc H (1926) Histoire de la digitale. Aesculap 16: 86–89
4. Overhamm A (1976) Zur Geschichte der Digitalis unter besonderer Berücksichtigung ihrer äußerlichen Anwendung. Würzburg
5. Parkinson J (1640) Theatrum botanicum. The Theater of Plants. Or an Herball of large Extent. London
6. Salmon W (1710) Botanologia; The English Herbal. London
7. Schadewaldt H, personal communication

Author's address:

Dr. L.M. Jung, Hansa Allee 176, D-4000 Düsseldorf 11 (West Germany)

The function of Na^+,K^+-ATPase and its importance for drug action

T. Akera

Department of Pharmacology and Toxicology, Michigan State University, East Lansing, Michigan (U.S.A.)

Summary

The primary function of Na^+,K^+-ATPase in cardiac and other excitable tissues is to maintain the resting, excitable state. In addition, the intracellular Ca^{2+} concentration is markedly affected by the balance between Na^+,K^+-ATPase activity (capacity of the sodium pump) and Na^+ influx rate. The sodium pump in cardiac muscle has a substantial reserve capacity under normal conditions; however, a relatively modest sodium pump inhibition or an increase in Na^+ influx rate elevates $[Na^+]_i$ and therefore intracellular Ca^{2+} via Na^+/Ca^{2+} exchange. The subsequent loading of the sarcoplasmic reticulum by Ca^{2+} and an enhanced Ca^{2+} release upon membrane excitation appear to be responsible for the positive inotropic effect of a Na^+,K^+-ATPase inhibitor.

Introduction

The action of cardiac glycosides to inhibit active Na^+ and K^+ transport was demonstrated in the early 1950s (12). This property of the glycosides was used to substantiate that Na^+,K^+-ATPase is the enzymatic representation of the sodium pump (13). Subsequently, it was established that the enzyme system undergoes a cycle of conformational transitions in the presence of Mg^{2+} associated with the transport of three Na^+ from the cytoplasm to the extracellular space and two K^+ in the opposite direction for each molecule of ATP hydrolyzed. This transport is the basis for the intracellular ionic composition; low Na^+ and high K^+ concentrations which are characteristics of most living cells and essential for cellular functions. Thus, the modulation of Na^+,K^+-ATPase alters many cellular functions, including contraction in myocardial cells.

Physiological functions of the Na^+,K^+-ATPase/sodium pump

Biochemical functions of Na^+,K^+-ATPase are to establish and maintain K^+ and Na^+ gradients across the cell membrane and to hydrolyze ATP. Additionally, the operation of the sodium pump causes one positive charge to be extruded during each cycle, and thereby generates outward currents. These functions have the following biochemical and physiological consequences.

In many cells, the balance between the ability of the membrane sodium pump to transport K^+ and the K^+ permeability of the cell membrane is such that the intracellular K^+ activity at equilibrium is 20 to 30 times higher than the extracellular K^+ concentration ($[K^+]_e$). This K^+ gradient is the basis for the transmembrane potential and therefore the excitability of nerve and muscle cells. A transmembrane potential of approximately −80 mV is required to keep the membrane ion channels closed and to maintain the resting state of nerve and muscle cells. The above mentioned electrogenicity of the sodium pump also contributes to the resting transmembrane potential. This contribution, however, is less than 5 mV in cardiac muscle, presumably because the Na^+ influx rate is small in this tissue and therefore the turnover rate of the sodium pump is low (see below). In addition to generating membrane potential, a high intracellular K^+ is necessary for the actions of many enzymes which require K^+ as cofactor.

A low intracellular Na^+ concentration ($[Na^+]_i$) is necessary for the generation of ATP by mitochondrial oxidative phosphorylation because the elevation of $[Na^+]_i$ uncouples this process via Na^+/Ca^{2+} exchange in mitochondrial membranes. Low $[Na^+]_i$ is also necessary for maintaining a low intracellular Ca^{2+} concentration. An elevation of $[Na^+]_i$ would increase intracellular Ca^{2+} via Na^+/Ca^{2+} exchange occurring in the cytoplasmic membrane. An elevation of intracellular Ca^{2+} is a signal for activation of cellular functions such as neurotransmitter release, secretion of endocrine and exocrine glands, and contraction of cardiac, skeletal or vascular smooth muscle.

A slight reduction in energy expenditure due to low sodium pump activity has been proposed to be the cause of obesity in several genetic models (8), and an increase in Na^+,K^+-ATPase units has been suggested to increase the rate of ATP hydrolysis and to elevate basal metabolism in hyperthyroidism (9). These proposals, however, have not been convincingly supported by experimental evidence. It appears that the rate of Na^+ influx rather than the number of Na^+,K^+-ATPase units determines the energy consumption by the sodium pump in cells maintaining ionic equilibrium. In cardiac muscle, hydrolysis of ATP per se does not appear to play an important role because energy utilization attributable to sodium pump activity probably accounts for less than 5 % of the total energy expenditure in this tissue.

Finally, regulation of cell volume has been suggested to be another function of Na^+,K^+-ATPase. Thus, it may be concluded that the physiological functions of Na^+,K^+-ATPase are to provide an adequate intracellular environment for many biochemical systems and to maintain the resting state of excitable cells.

Regulators of the sodium pump

Temperature and fluidity of membrane lipids. Na^+,K^+-ATPase has a relatively high temperature coefficient. Thus, an increase in the force of contraction observed in isolated cardiac muscle when the incubation temperature was lowered to 25–27°C has been cited as proof that sodium pump activity is inversely related to developed tension (10), although depression of the calcium pump could also be a contributing factor. Fluidity of membrane lipids surrounding Na^+,K^+-ATPase seems to affect activity of this enzyme. In crude enzyme preparations obtained from rat brain and heart muscle, break points on Arrhenius plots were 22 and 32°C, respectively (unpublished observations). In the same preparations, lipid phase transition temperatures estimated using an electron spin resonance probe, 5-doxylstearate, were 19 and 31°C. Changes in Na^+,K^+-ATPase activity due to the difference in lipid composition of the diet have been reported in rabbits and were attributed to changes in membrane lipids (7).

Extracellular K^+. A half-maximal activation of the sodium pump by external K^+ occurs at 1.5 mM (2) indicating that the sodium pump is sensitive to changes in $[K^+]_e$ at this range. Although physiological concentrations of K^+ are higher, changes in K^+ concentration have been shown to have a substantial effect on the rate of glycoside binding to Na^+,K^+-ATPase in beating heart muscle. These results indicate that changes in K^+ concentrations in the physiological range affect Na^+,K^+-ATPase in intact myocardial cells. In cardiac muscle isolated from most mammalian species, an elevation of $[K^+]_e$ decreases developed tension.

This has been explained as resulting from stimulation of the sodium pump and the ensuing decrease in $[Na^+]_i$ and thereby the intracellular Ca^{2+} via Na^+/Ca^{2+} exchange. In the rat heart, however, an elevation of $[K^+]_e$ from 3.5 to 9.5 mM increases developed tension, which is seemingly inconsistent with the above explanation. An increase in $[K^+]_e$ from 1 to 3.5 mM, however, does decrease developed tension in the rat heart muscle, consistent with activation of the sodium pump. Further elevation of K^+ increases developed tension in this species via stimulation of mechanisms unique to the rat myocardium. The latter actions, which include prolongation of action potential duration, mask the negative inotropic effect of K^+ (unpublished observations). Thus, changes in $[K^+]_e$ near the physiological range have substantial effects on sodium pump activity.

Intracellular Na^+. $[Na^+]_i$ is the most important factor in determining sodium pump activity. A half-maximal activation of isolated Na^+,K^+-ATPase occurs with approximately 15 mM Na^+. In intact cells, this concentration has not been precisely estimated; however it is likely to be even higher due to possible competition by high intracellular K^+. Because $[Na^+]_i$ is lower than the above concentration, and because the Na^+ activation curve is steep owing to positive cooperativity, the sodium pump is a sensitive feedback mechanism for maintaining $[Na^+]_i$. It also indicates that the turnover of the sodium pump in the quiescent myocardium or in muscles which are stimulated at a low frequency is substantially lower than the maximal rate that may be observed if $[Na^+]_i$ is elevated above 100 mM. In myocyte preparations isolated from ventricular muscle of guinea-pig heart, ouabain-sensitive $^{86}Rb^+$ uptake was increased 5-fold by 30 μM monensin, a sodium ionophore which increases the rate of Na^+ influx (Table 1). This would indicate that the level of sodium pump activity in quiescent myocytes is less than 20 percent of the maximal rate or capacity of the sodium pump, and therefore the sodium pump has a substantial reserve capacity. When the sodium pump is inhibited by cardiac glycosides or if the rate of Na^+ influx is increased by agents such as monensin or by high frequency stimulation, this reserve capacity would be decreased.

Table 1. $^{86}Rb^+$ uptake by isolated myocytes.

Monensin (μM)	Ouabain-sensitive $^{86}Rb^+$ uptake (nmol/mg protein/min)
0	15.5 ± 2.3
1	25.4 ± 2.9
3	38.0 ± 3.1
10	56.6 ± 3.2
30	71.8 ± 3.9

Myocytes were isolated from guinea-pig hearts by treatment with collagenase (0.52 mg/ml) plus hyaluronidase (0.2 mg/ml) and equilibrated in a Krebs-Henseleit solution containing 5 mM RbCl and no K^+ before the addition of ^{86}RbCl. Ouabain or monensin, if present, was added 35 or 5 min before the addition of ^{86}RbCl, respectively. After a 10 min incubation, myocytes were sampled into a cold stopping solution and centrifuged through a layer of 300 mM sucrose. Ouabain-sensitive $^{86}Rb^+$ uptake is the difference in values observed in the absence and presence of 1 mM ouabain. Mean ± S.E. of 7 experiments.

Other substrates. ATP and Mg^{2+} would appear to be relatively unimportant for the regulation of sodium pump activity. In the ischemic myocardium, sodium pump activity is well preserved as estimated from tissue Na^+ content, despite a reduction in tissue ATP concentration (3). This finding is consistent with the affinity of the enzyme for ATP; Km values for ATP are approximately 1 μM and 0.1 mM at the catalytic and regulatory sites, respectively

(11). Compared to these values, the ATP concentration in heart muscle is substantially higher even when it is reduced by ischemia. A complete removal of Mg^{2+} eliminates sodium pump activity; however, Na^+,K^+-ATPase activity appears relatively insensitive to small changes in Mg^{2+} concentration which are within the physiological range.

Tissue concentration of Na^+,K^+-ATPase. Hyperthyroidism has been shown to increase functional Na^+,K^+-ATPase units (9). Chemically induced diabetes or chemical sympathectomy with 6-hydroxydopamine or reserpine have been reported to reduce the number of cardiac Na^+,K^+-ATPase units, although the effect of chemical sympathectomy is still controversial. Right ventricular hypertrophy resulting from partial ligation of the pulmonary artery in cats also reduced the number of Na^+,K^+-ATPase units (unpublished observations). It should be noted, however, that the tissue concentration of Na^+,K^+-ATPase per se does not markedly influence the sensitivity of heart muscle to the cardiac glycosides. In guinea-pig heart, ventricular muscle has a higher Na^+,K^+-ATPase activity and a correspondingly larger number of ouabain binding sites per unit protein compared to atrial muscle (17). The positive inotropic effect of a given concentration of ouabain, however, is similar in atrial and ventricular muscle preparations. In hypertrophied right ventricular muscle of the cat heart, the magnitude of the positive inotropic effect of strophanthidin was reduced; however, there was no change in sensitivity to strophanthidin as estimated from the concentration needed to cause a half-maximal inotropic effect. A decrease in the maximal inotropic effect of strophanthidin is more likely to result from changes in contractile proteins.

Endogenous regulators. Endogenous ligands for the glycoside binding site on Na^+,K^+-ATPase have been sought by many investigators. Lipids or their derivatives appear to play some role in modifying Na^+,K^+-ATPase activity. These compounds, however, do not bind to the enzyme at the glycoside binding site. Convincing evidence has not been available to indicate the presence of a true digitalis-like principle. These aspects are discussed elsewhere in this book.

Drug-induced alterations of sodium pump activity

A current popular hypothesis for the mechanism of the positive inotropic action of digitalis glycosides is as follows. In therapeutic concentrations, the glycoside causes a modest (20 to 40%) inhibition of the sodium pump and thereby slightly elevates $[Na^+]_i$. During the diastolic phase of the cardiac cycle, Na^+/Ca^{2+} exchange mediates the influx of 3 Na^+ coupled with the efflux of 1 Ca^{2+}. Because both the Na^+ gradient and the transmembrane potential are driving forces for Ca^{2+} extrusion, an increase in $[Na^+]_i$ inhibits Ca^{2+} efflux, causing more Ca^{2+} to be taken up by the sarcoplasmic reticulum and more Ca^{2+} to be released during the subsequent membrane excitation.

As discussed above, the sodium pump has a reserve capacity unless the cardiac muscle is beating at an extremely high frequency. In spite of this reserve capacity, a modest inhibition of the pump of a magnitude less than the reserve capacity would elevate $[Na^+]_i$ as depicted in Fig. 1. Curve (1) represents a normal relationship between $[Na^+]_i$ and sodium pump activity. During the diastolic phase, the pump is operating not at its maximal rate (A), but at a rate (C) which is equivalent to the rate of Na^+ influx. An increase or a decrease in Na^+ influx rate increases or decreases $[Na^+]_i$, respectively, until the rate of Na^+ extrusion matches the rate of Na^+ influx at point B or D. A modest inhibition of the sodium pump shifts curve (1) to curve (2). If the rate of Na^+ influx is unchanged, $[Na^+]_i$ increases to point E, at which time the rate of Na^+ extrusion matches the rate of Na^+ influx. When the sodium pump is inhibited to a greater degree by a toxic concentration of the glycoside, a greater increase in $[Na^+]_i$ is required (point F) for the rate of Na^+ efflux to match the rate of Na^+ influx.

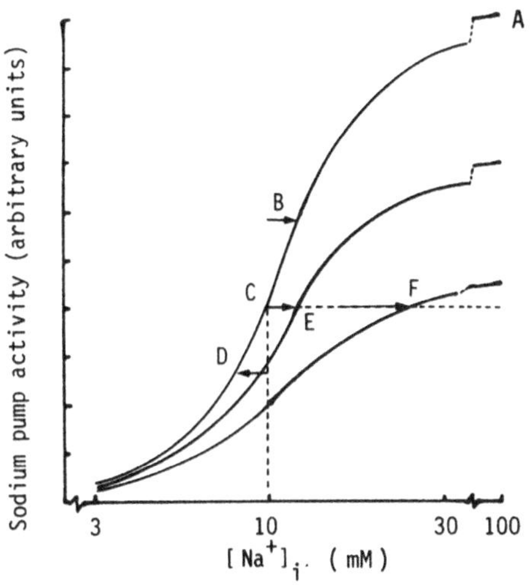

Fig. 1. Na^+ influx, $[Na^+]_i$ and sodium pump activity: a scheme.

The above hypothesis predicts that a given degree of Na^+,K^+-ATPase or sodium pump inhibition should cause a similar increase in $[Na^+]_i$ and developed tension regardless of the cause. An increase in the rate of Na^+ influx also increases $[Na^+]_i$ and developed tension. There is good experimental support for the positive inotropic effect of various Na^+,K^+-ATPase inhibitors or agents that increase the rate of Na^+ influx (1).

The above scheme also predicts a potentiating interaction between a drug which increases the rate of Na^+ influx and that which inhibits the sodium pump for their effects to elevate $[Na^+]_i$ and to cause toxicity. Consistent with this prediction, a combination of the sodium ionophore, monensin, and digoxin markedly shorten the time to onset of digoxin-induced arrhythmias (6). A part of this shortening can be explained from a faster binding of the glycoside to sarcolemmal Na^+,K^+-ATPase caused by an elevation of $[Na^+]_i$; however, the degree of Na^+,K^+-ATPase inhibition was smaller at the time of onset of arrhythmias indicating that monensin markedly reduced the tolerance of the heart to digoxin-induced arrhythmias. A reduced tolerance of senescent Fischer 344 rats to glycoside-induced toxicity also appears to result from a reduced reserve capacity of the sodium pump in heart muscle produced by the aging process (4, 5). Thus, the relationship between the capacity of the sodium pump and the rate of Na^+ influx, rather than the number of functional sodium pumping units per se, seems to determine the sensitivity of cardiac muscle to a Na^+,K^+-ATPase inhibitor.

Elevation of $[Na^+]_i$ and its effect on Na^+/Ca^{2+} exchange

The above discussion indicates that a modest sodium pump inhibition or an increase in Na^+ influx rate slightly elevates $[Na^+]_i$. This increase has been proposed to elevate intracellular Ca^{2+}during diastole, to increase the Ca^{2+} loading of the sarcoplasmic reticulum, and to increase Ca^{2+} release during membrane excitation. This indirect mechanism, rather than a direct augmentation of Ca^{2+} influx in exchange for elevated $[Na^+]_i$ during membrane excitation, appears to be supported by the following observations. The positive inotropic effect of vanadate observed in atrial muscle of rat heart is inhibited by Mn^{2+}, indicating an involvement of a superficial Ca^{2+} pool (14). In this tissue, vanadate increases the force of contraction observed under 0.5–2 Hz stimulations, but fails to alter the size of the potentiated postrest contraction observed after a brief quiescent period. In guinea-pig heart, agents which are shown to increase Ca^{2+} influx during membrane excitation such as BAY K 8644 also affect developed tension under 0.5 Hz, but have a relatively small effect on the potentiated postrest contractions observed under 0.003 Hz stimulation (16). In contrast to these agents, the cardiac glycosides increase potentiated post-rest contractions in addition to developed tension under 0.5 Hz stimulation (15). Compared to the effect of isoproterenol or to an elevation of extracellular Ca^{2+}, the effect of low concentrations of strophanthidin was greater on potentiated post-rest contractions compared to contractions under 1 Hz stimulations (unpublished observations). These results suggest that sarcolemmal Ca^{2+} loading is involved in the positive inotropic effect of a Na^+,K^+-ATPase inhibitor.

A combination of digoxin and BAY K 8644 produced a positive inotropic effect which was greater than the effect seen with either digoxin or BAY K 8644 alone (Fig. 2). In contrast, a combination of digoxin and a Ca^{2+} ionophore, A-23 187, failed to produce a positive inotropic effect which was greater than that caused by digoxin alone. These results also indicate that BAY K 8644, a dihydropyridine derivative which increases Ca^{2+} influx during membrane excitation, has a somewhat different mechanism from the cardiac glycoside whereas A-23 187, a Ca^{2+} ionophore which increases Ca^{2+} influx mostly during the diastolic phase, is simply additive with the glycoside. Thus, it may be concluded that an increase in $[Na^+]_i$ during the diastolic phase, and the ensuing inhibition of the Ca^{2+} efflux, play an important role in the mechanism of the positive inotropic action of a sodium pump inhibitor and probably of agents that enhance the rate of Na^+ influx.

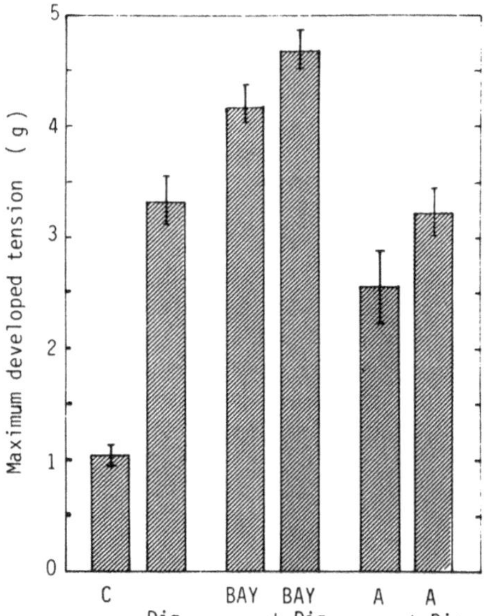

Fig. 2. Maximal positive inotropic effects observed in left atrial muscle isolated from guinea-pig heart. Preparations were incubated at 30°C under 1-Hz stimulation in Krebs-Henseleit bicarbonate buffer solution containing 0.1 μM nadolol. Developed tension observed in the presence of 3 μM BAY K 8644 (BAY) or 1 μM A-23 187 (A) or maximal developed tension produced by 3 μM digoxin (Dig) before onset of arrhythmias in the presence or absence of BAY K 8644 or A-23 187 was recorded. C: Control developed tension observed before drug addition. Mean of 8 or 12 experiments. Vertical lines indicate S.E.

Acknowledgements

Experiments described in this paper were performed by Dr. Richard H. Kennedy and Mr. Paul Stemmer. This work was supported by U.S. Public Health Service grants, AG-02398 and HL-16052. I also acknowledge the valuable suggestions and encouragement of Dr. Theodore M. Brody.

References

1. Akera T, Fox AL, Greeff K (1981) In: Greeff K (ed) Handbook of Experimental Pharmacology, Vol 56/I, Cardiac Glycosides. Springer-Verlag, Heidelberg, pp 459–486
2. Glitsch HG (1984) Basic Res Cardiol 79: 611–619
3. Jennings RB, Schaper J, Hill ML, Steenbergen C, Jr, Reimer K (1985) Circ Res 56: 262–278
4. Katano Y, Akera T, Temma K, Kennedy RH (1984) Eur J Pharmacol 105: 95–103
5. Katano Y, Kennedy RH, Stemmer PM, Temma K, Akera T (1985) Eur J Pharmacol 113: 167–178
6. Kennedy RH, Akera T, Brody TM (1983) Eur J Pharmacol 89: 199–209
7. Kutscherskij E, Gunther J, Mehley E (1984) Experientia 40: 812–815
8. Lin MH, Vander-Tuig JG, Romsos DR, Akera T, Leveille GA (1980) Amer J Physiol 238: E 193–E 199
9. Philipson KD, Edelman IS (1977) Amer J Physiol 232: C 196–C 201
10. Repke K (1964) Klin Wochensch 42: 157–165
11. Robinson JD, Flashner MS (1979) Biochim Biophys Acta 549: 145–176
12. Schatzmann HJ (1953) Helv Physiol Acta 11: 346–354
13. Skou JC (1960) Biochim Biophys Acta 42: 6–23
14. Takeda K, Temma K, Akera T (1982) J Pharmacol Exp Ther 222: 132–139
15. Temma K, Akera T, Brody TM (1981) Eur J Pharmacol 76: 361–370
16. Thomas G, Chung M, Cohen CJ (1985) Circ Res 56: 87–96
17. Yamamoto S, Akera T, Kim D-H, Brody TM (1981) J Pharmacol Exp Ther 217: 701–707

Author's adress:

Prof. Dr. T. Akera, Department of Pharmacology and Toxicology, B 445 Life Science Building, Michigan State University, East Lansing, Michigan 48824-1317 (U.S.A.)

Structural studies on the cardiac glycoside receptor[1]

E. T. Wallick, T. L. Kirley and A. Schwartz

Department of Pharmacology and Cell Biophysics, University of Cincinnati College of Medicine, Cincinnati (U.S.A.)

Introduction

The recent publication by Shull et al. (20) of the complete primary amino acid sequence of the alpha subunit of Na^+,K^+-ATPase of sheep kidney allows investigators in the field to address questions of structure/function relationships more directly. Based on the analysis of Shull et al., one can schematically represent the enzyme as shown in Fig. 1. Eight membrane spanning sequences (H1–H8) are predicted. Very few extramembranous sequences are predicted to be extracellular. The majority of the hydrophilic sequences are predicted to be intracellular, primarily between H4 and H5. There are five possibles sites for N-glycosylation but these are all predicted to be intracellular. There is very little internal homology suggesting that equivalent distinct separate sites for cations do not exist. Another striking observation is the high degree of homology between the Na^+,K^+-ATPase (20) of sheep kidney and the sarcoplasmic reticulum Ca^{2+}-ATPase from skeletal muscle (14). The complete primary

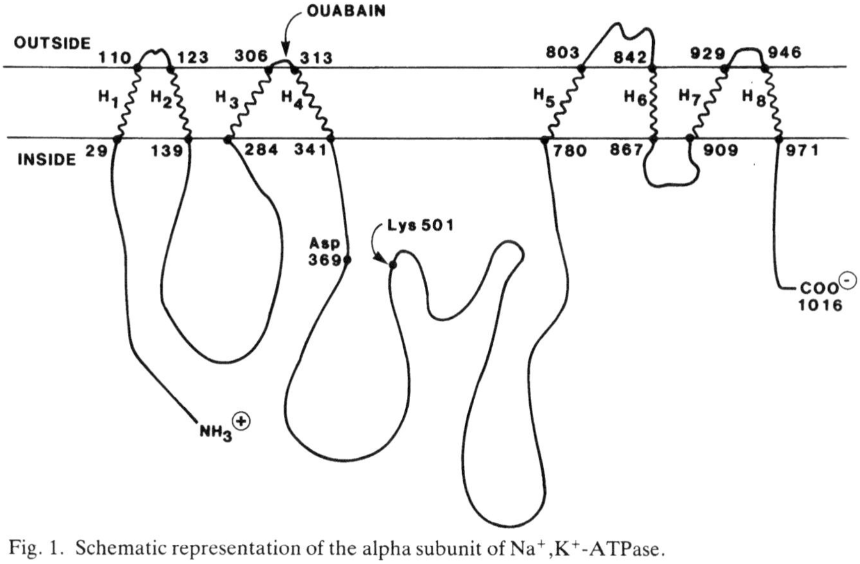

Fig. 1. Schematic representation of the alpha subunit of Na^+,K^+-ATPase.

[1]) Supported by PO1-HL-22, 619, RO1-HL-25, 545 and T32-HL-07382

amino acid sequence of the alpha subunit of Na^+,K^+-ATPase of the electric organ of torpedo ray has also been determined (7). There is a remarkable 87% sequence homology between the alpha subunit of the ray electric organ and sheep kidney Na^+,K^+-ATPases. One would expect even greater homology between Na^+,K^+-ATPases from more closely related sources, e.g., the alpha and alpha+ subunits detected in brain (23) and two putative forms of Na^+,K^+-ATPase indirectly detected (1) in rat ventricle. Consistent with this concept, a comparison (Fig. 2) of the small (<4%) sequence determined for Na^+,K^+-ATPase from rat brain (18) with the same region of sheep kidney and torpedo enzyme indicates that the mammalian enzymes are much more homologous to each other than they are to the torpedo enzyme. One puzzling feature is the amino acid substitution at position 489 of an alanine residue in the sheep kidney enzyme for a proline residue in rat brain enzyme. Since proline residues cause folds in secondary structure, one would have predicted a higher degree of conservation for enzymes with similar functions.

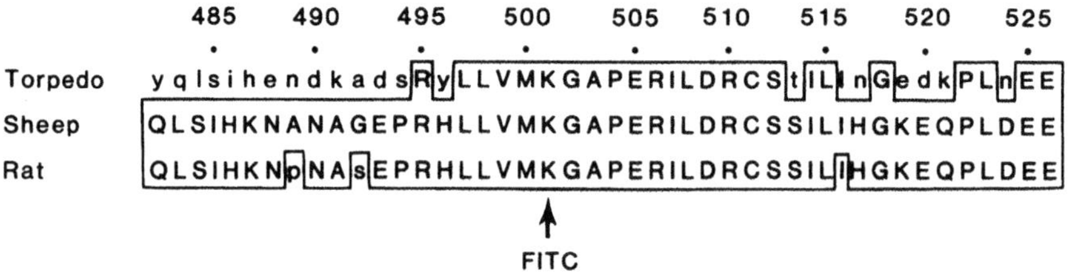

Fig. 2. Comparison of the primary amino acid sequence of sheep kidney, torpedo electric organ, and rat brain Na^+,K^+-ATPases.

Most detailed kinetic studies, e.g. those of Askari et al. (16) indicate that Na^+,K^+-ATPase isolated from different sources have similar affinities for most ligands such as Na^+,K^+, Mg^{2+} and ATP. The affinity for cardiac glycosides however differs widely in the same organ from different species. The I_{50} for inhibition of Na^+,K^+-ATPase by ouabain of sheep and rat kidney differs by a factor of 2400 (24). Even within a single species, the affinity of Na^+,K^+-ATPase isolated from different tissues for cardiac glycosides varies. For example, Na^+,K^+-ATPase isolated from rat brain, rat skeletal muscle, and rat kidney have K_D for ouabain binding of 16 nM, 175 nM and 10 μM, respectively (4). One would predict, therefore, that multiple forms of Na^+,K^+-ATPase exist, each of which could have differences in the primary sequence in the region which comprises the cardiac glycoside binding site.

Fortes (5) has suggested that the cardiac glycoside binding site is hydrophobic and that there is a tryptophan residue in the immediate vicinity of the site. Consistent with this suggestion is the observation by Goeldner et al. (6) that a photoaffinity analogue of ouabain can be activated by energy transfer from an excited tryptophan residue. Based on these and other considerations, Shull et al. (20) have suggested that residues 307–312 and probably some adjacent residues comprise the ouabain binding site on the alpha subunit. A comparison (Fig. 3) of this region of the primary amino acid sequence of the alpha subunit of Na^+,K^+-ATPase from lamb kidney and torpedo electric organ shows complete homology in the two transmembrane regions and a single substitution of a glycine for a glutamic acid at residue 307. Although this is a major substitution, the IC_{50} for inhibition of the enzyme by ouabagenin differs only by a factor of five to ten (unpublished observations). It is very possible that other extracellular regions also influence the cardiac glycoside binding site.

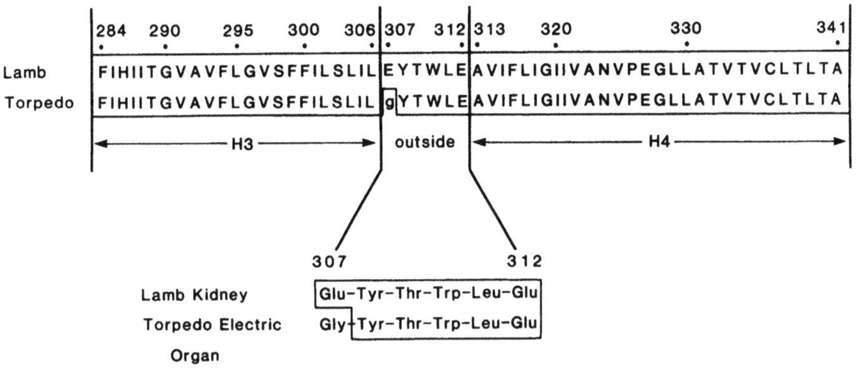

Fig. 3. Comparison of putative ouabain binding regions of sheep kidney and torpedo electric organ Na^+,K^+-ATPases.

Materials and methods

Tetranitromethane, N-acetylimidazole, 2-hydroxy-5-nitrobenzyl bromide, NADH, pyruvate kinase/ lactate dehydrogenase and ouabagenin were obtained from Sigma. Dithothreitol was obtained from Calbiochem. HPLC grade water, acetonitrile, methanol, ammonium acetate and reagent grade sodium dithionite were from Fisher. Dansyl chloride and trifluoroacetic acid were from Pierce. The 0.45 μm filters used in binding assays were obtained from Gelman, $[^3H]$ ouabain from New England Nuclear, and unlabeled ouabain from Boehringer Mannheim.

Lamb kidney Na^+,K^+-ATPase was isolated as described previously (11) and generously furnished by Dr. L. K. Lane. Na^+,K^+-ATPase activity was measured using the linked enzyme spectrophotometric assay (19), and para-nitrophenyl phosphatase activity (pNPPase) by the increase in absorbance at 410 nm (12). $[^3H]$ ouabain binding was measured after equilibration at 37° C in 50 mM Tris-C1 pH = 7.4 including 5 mM $MgCl_2$ and 5 mM Tris-$P0_4$ (ouabain binding medium). Maximal binding was measured using 100 nM $[^3H]$ouabain in the ouabain binding medium and incubating at 37°C. After 20 minutes, aliquots were filtered (0.45 μm) and washed 3 times with 5 ml of cold H_2O.

Modification of Na^+,K^+-ATPase by TNM (and other reagents) was normally done at protein concentration of 1.0 mg/ml at pH = 8.2 (100 mM Tris-C1, 5 mM $MgCl_2$, 5 mM Tris-$P0_4$), although some experiments utilizing TNM were done at pH = 5.9 (100 mM sodium acetate buffer containing 5 mM $MgCl_2$ and 5 mM Tris-$P0_4$).

Ouabagenin protection experiments were normally carried out by preincubating the enzyme with 5 μM ouabagenin for 10 minutes at room temperature in the modification buffer. At this concentration 98 % of the sites will be occupied by ouabagenin. The time course of the loss of activity and decrease in ouabain binding capability of the Na^+,K^+-ATPase caused by chemical modification was determined as follows. ATPase assays were carried out by diluting 10 μl of the 1 mg/ml protein solution into the 2.5 ml assay cuvette at varying time points (a 250-fold dilution) and allowing the activity to recover fully from ouabagenin inhibition (5 minutes). Control experiments, in absence of a modifying reagent, indicated more than 95 % recovery of both ATPase activity and $[^3H]$ouabain binding capability. Aliquots for ouabain binding were diluted 300-fold into cold binding medium and stored on ice until all aliquots were taken, at which time they were incubated in the $[^3H]$ouabain binding solution for 20 minutes at 37° C and filtered.

Reduction of the nitrotyrosine formed by the reaction of Na^+,K^+-ATPase with TNM to an aminotyrosine was accomplished by incubation of the TNM treated enzyme (at 1.0 mg/ml) with 100 μM sodium dithionite (in 50 mM Tris-Cl, pH = 8.0) for 15 minutes at room temperature (21). The enzyme was then centrifuged for 30 min at 150,000 x g, the supernatant discarded and the pellet used for dansylation. The aminotyrosine groups formed were subsequently dansylated by reaction of 1.0 mg/ml Na^+,K^+-ATPase with 100 μM dansyl-Cl in 100 mM sodium acetate pH = 5.0 for 1 hour at room temperature in the dark (8, 9). The reaction medium was then diluted with water and centrifuged (150,000 x g, 30 min). The dansylated material (2.0 mg/ml) was then digested with TPCK-trypsin for 90 minutes in the dark at 37°C in 25 mM imidazole-Cl and 1 mM EDTA, pH = 7.2, at a trypsin to Na^+,K^+-ATPase ratio of 1:50. The digested material was centrifuged at 150,000 x g for 30 min. The supernatant was lyophilyzed and the pellet was homogenized in 1 mM EDTA, pH = 7.3, and stored shielded from light.

All solutions of TNM (diluted into 95% ethanol), $Na_2S_2O_4$ (in H_2O), and dansyl chloride (in HPLC grade dry acetone) were made fresh and used within ten minutes of dilution/dissolution.

SDS-PAGE was carried out according to Laemmli (10), isoelectric focussing as described by O'Farrell (15), and protein concentrations were determined according to Lowry et al. (13).

Results

The inhibition by 50 μM TNM of ATPase and pNPPase activities, as well as maximal ouabain binding (MgPi binding conditions) of Na^+,K^+-ATPase does not follow first order reaction kinetics (Fig. 4). As has been found with a variety of chemical modifiers, the order of susceptibility of activities is ATPase > pNPPase > ouabain binding. As determined by Scatchard analysis, the tetranitromethane-induced decrease in ouabain binding is due to a 50% decrease in the number of sites, rather than to a change in the affinity of the receptor for ouabain (Fig. 5). The inhibition of ouabain binding by TNM can be blocked by preincubating the Na^+,K^+-ATPase (at 1.0 mg/ml) with 5 μM ouabagenin for ten minutes (Fig. 6). The inhibition of ATPase activity by TNM was not affected by ouabagenin preincubation, as long as care was taken to allow complete recovery (approximately 5 minutes) of ATPase and pNPPase activities after diluting an aliquot of the modified enzyme into the assay cuvette.

Subjection of TNM-labeled, reduced, and dansylated enzyme to SDS-PAGE revealed that the fluorescence of the dansyl group comigrated with the alpha subunit and also remained at the top of the gel (data not shown). Further experiments showed that the reaction of TNM, without reduction and dansylation, caused a small amount of crosslinking of the

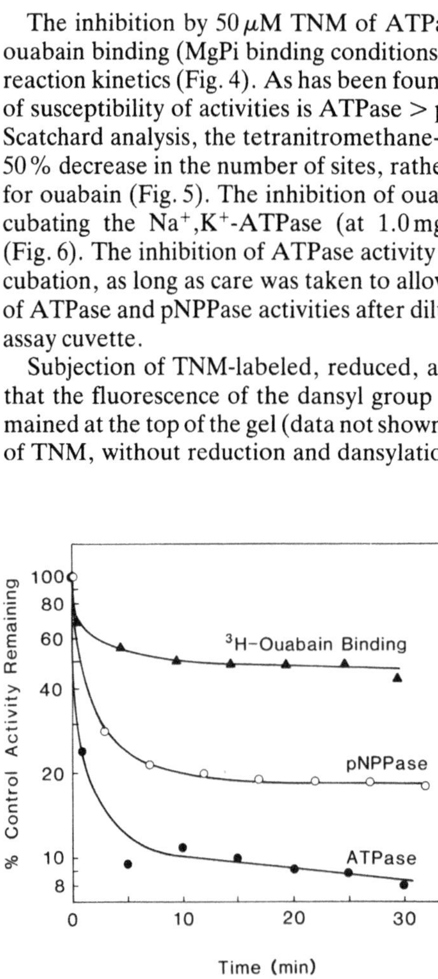

Fig. 4. The time course of inhibition of lamb kidney Na^+,K^+-ATPase by 50 μM tetranitromethane at pH 8.2. The remaining ATPase activity (●), pNPPase activity (o) and [^{3}H]ouabain binding (maximal) capacity (▲) are shown as a function of time of incubation of 1.0 mg/ml Na^+,K^+-ATPase with tetranitromethane. For experimental details refer to Methods.

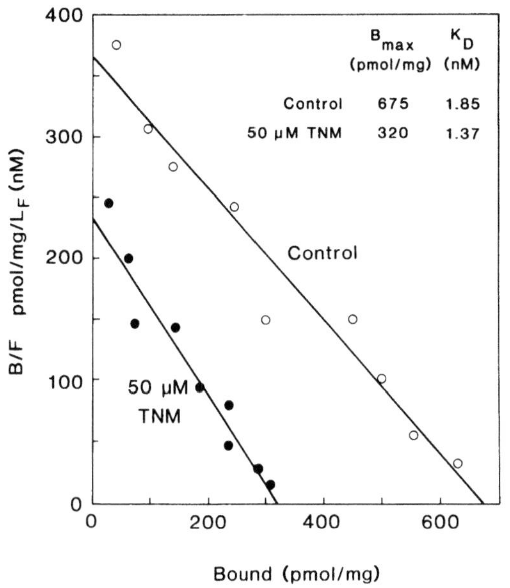

Fig. 5. The effect of TNM on equilibrium [^{3}H]ouabain binding to lamb kidney Na^+,K^+-ATPase. The enzyme was incubated with (•) or without (o) 50 μM TNM at pH 8.2 for 30 minutes, diluted 100-fold into ice cold 1 mM EDTA-imidazole, pH = 7.4, and diluted again 10-fold into binding media containing various concentrations of [^{3}H]ouabain. The binding solutions were incubated for 16 hours at 37°C, to allow equilibration at the very low concentrations of [^{3}H]ouabain, and filtered. The relatively low B_{max} values observed are due to the long incubation at 37°C.

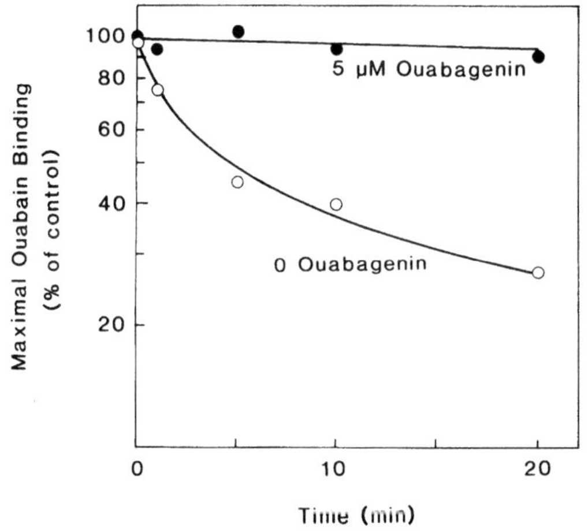

Fig. 6. The protection of TNM inhibition (100 μM) of [^{3}H]ouabain binding capacity of lamb kidney Na^+,K^+-ATPase by ouabagenin at pH 8.2. The enzyme in the absence (o) and presence (•) of 5 μM ouabagenin was assayed for maximal [^{3}H]ouabain binding at various times as described in Materials and methods.

enzyme protein, as evidenced by Coomassie blue staining material on top of the gel. This phenomenon has been shown to occur with other proteins (22). Reversed phase HPLC of the supernatants of the tryptic digests of Na^+,K^+-ATPase which had been labeled with TNM, plus or minus ouabagenin, and then reduced and dansylated yielded virtually identical patterns of fluorescent peptides. This indicates that the presence of ouabagenin before and during the TNM incubation does not protect any tyrosine residues from nitration and subsequent dansylation. Nearly all of the protein-associated dansyl fluorescence was released into the supernatant fraction by trypsin digestion.

The effect of TNM on ATPase activity and [^{3}H]ouabain binding was also determined at pH = 5.9, since it is believed that TNM reacts only with sulfhydryls at this pH (17). As shown in Fig. 7, treatment with 500 μM TNM inhibits both ATPase and [^{3}H]ouabain binding, but only the inhibition of ouabain binding exhibits pseudo first order kinetics. Control experiments carried out in the absence of TNM at pH = 5.9 clearly show that neither ATPase activity nor [^{3}H]ouabain binding is irreversibly inhibited by incubation under these solvent conditions. The inhibition of ouabain binding by TNM at pH = 5.9 also is due to a decrease in the number of sites. Ouabagenin still protects against inhibition, but due perhaps to the lower pH or to the higher concentration of TNM, the protection is no longer complete (data not shown).

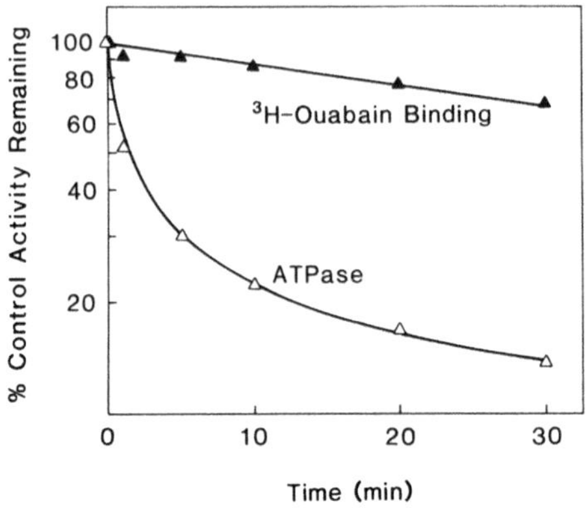

Fig. 7. The effect of 500 μM TNM on Na^+,K^+-ATPase at pH 5.9. The enzyme was assayed for ATPase activity (△) as well as for maximal [^{3}H]ouabain binding (▲) at various times of incubation with TNM as described in Materials and methods.

Discussion

To test the hypothesis that residues 307–312 comprise the ouabain binding site on the alpha subunit, the effect of several tyrosine and tryptophan directed reagents on [^{3}H]ouabain binding was determined. Tetranitromethane was chosen as a site directed reagent because of its hydrophobicity and its known reaction with tyrosine residues. However, under the conditions first used (pH = 8.2), TNM can also react with sulfhydryl groups (17) and this appears to account for at least part of the inhibition of ouabain binding by TNM at pH = 8.2. At pH = 5.9, where only sulfhydryls should react with TNM, ouabain binding is still inhibited, albeit at a higher concentration. At pH = 5.9, the inhibition is pseudo first order, suggesting that at least part of the inhibitory action of TNM at pH 8.2 is due to modification of cysteine residue(s). TNM also causes crosslinking (22) of the Na^+,K^+-ATPase, which may also account for part of the inhibition.

Attempts to label tyrosine or tryptophan residues within the ouabain binding site using other chemical modifying agents were unsuccessful (data not shown). The results obtained using NBD-Cl, a tyrosine directed ATPase inhibitor (3), were not very reproducible, due probably to instability of the reagent. In addition, several different sulfhydryls and/or amino groups are labeled by this reagent, as revealed by the presence of many strongly fluorescent tryptic peptides resolved by HPLC. Both thio and amino group adducts of NBD-Cl are known to be very fluorescent (2). Experiments utilizing N-acetylimidazole (tyrosine directed) and 2-hydroxy-5-nitrobenzyl bromide (tryptophan directed), even at very high concentrations (0.5–10 mM), failed to show significant inhibition of ouabain binding. We conclude from these results that there is probably not an essential, reactive tyrosine or tryptophan residue in the ouabain binding site of lamb kidney Na^+,K^+-ATPase.

References

1. Adams RJ, Schwartz A, Grupp G, Grupp I, Lee S-W, Wallick ET, Powell T, Twist VW, Gathiram P (1982) Nature 296: 167–169
2. Aeschbach R, Amado R, Neukom H (1976) Biochim Biophys Acta 439: 292–301
3. Cantley LC, Gelles J, Josephson L (1978) Biochemistry 17: 418–425
4. Erdmann E (1981) In: Greeff K (ed) Handbook of Experimental Pharmacology, Vol 56/I. Springer-Verlag, Berlin Heidelberg, pp 337–380
5. Fortes PAG (1977) Biochemistry 15: 531–540
6. Goeldner MP, Hirth CG, Rossi B, Ponzio G, Lazdunski M (1983) Biochemistry 22: 4685–4690
7. Kawakami K, Noguchi S, Noda M, Takahashi H, Ohta T, Kawamura M, Hiroshi N, Nagano K, Hirose T, Inayama S (1985) Nature 316: 733–736
8. Kenner RA (1971) Biochemistry 10: 545–550
9. Kenner RA, Neurath H (1971) Biochemistry 10: 551–557
10. Laemmali UK (1970) Nature 227: 680–685
11. Lane LK, Potter JD, Collins JH (1979) Prep Biochem 9: 151–170
12. Lane LK, Anner BM, Wallick ET, Ray MV, Schwartz A (1978) Biochem Pharm 27: 225–231
13. Lowry DH, Rosebrough NJ, Farr AL, Randall RG (1951) J Biol Chem 193: 265–275
14. MacLennan DH, Brandl CJ, Korczak B, Green NM (1985) Nature 316: 696–700
15. O'Farrell PH (1975) J Biol Chem 250: 4007–4021
16. Periyasamy SM, Huang W-H, Askari A (1983) Comp Biochem Physiol 76B: 449–454
17. Riordan JF, Vallee BL (1972) Meth Enzymol 25: 515–521
18. Schneider JW, Mercer RW, Caplan M, Emanuel JR, Sweadner KJ, Benz EJ jr, Levenson R (1985) Proc Natl Acad Sci USA 82: 6357–6361
19. Schwartz A, Allen JC, Harigaya S (1969) J Pharm Exp Ther 168: 31–41
20. Shull GE, Schwartz A, Lingrel JB (1985) Nature 316: 691–695
21. Sokolovsky M, Riordin JF, Vallee BL (1967) Biochem Biophys Res Comm 27: 20–25
22. Sokolovsky M, Harell D, Riordan JF (1969) Biochemistry 8: 4740–4745
23. Sweadner KJ, Gilkeson R, Chapman C (1985) J Biol Chem 260: 9016–9022
24. Wallick ET, Pitts BJR, Lane LK, Schwartz A (1980) Arch Biochem Biophys 202: 442–449

Author's address:

Prof. Dr. A. Schwartz, Department of Pharmacology and Cell Biophysics, University of Cincinnati, College of Medicine, 231 Bethesda Avenue, Cincinnati, Ohio 45267–0575 (U.S.A.)

Isoenzymes of Na,K-ATPase identified by pyrithiamin

Otto Hansen*

Institute of Physiology, Aarhus University, DK-8000 Århus C (Denmark)

Summary

It was assumed that Na,K-ATPase from grey matter of ox brain contains two independent populations of ouabain receptors and that the high affinity component is sensitive to pyrithiamin whereas the low affinity component resists treatment with the vitamin analogue. With these assumptions the two components of digitalis receptors and their affinities (the reciprocal of the slope) are calculated. The curves drawn from the calculated components and dissociation constants are seen to fit the experimental points well.

Ouabain binding data obtained before and after treatment of Na,K-ATPase from ox brain with pyrithiamin are thus compatible with a model assuming two independent populations of receptors (and thus enzymes) in this preparation.

Introduction

It has been known for years that some ligand binding isotherms, e. g. nucleotide binding in the presence of K^+ and ouabain binding, to Na,K-ATPase isolated from grey matter of the brain are incompatible with the existence of one homogeneous population of receptors (1, 2, 11). Hansen (2) proposed that two (or more) populations of ox brain Na,K-ATPase with different affinities for ouabain could explain curvilinear Scatchard plots, although apparently a conversion to one uniform population was seen in the presence of K^+. The principal interpretations of curvilinear Scatchard-type plots of ouabain binding data were recently reviewed (4).

In 1979 Sweadner emphasized that brain contains two distinct molecular forms of Na,K-ATPase differing in the gel electrophoresis mobility of their α-subunit and in their sensitivity to some cardiac glycosides. The bigger (α+) molecular form of the Na,K-ATPase has recently been claimed to be selectively inhibited by the vitamin analogue pyrithiamin (9, 10). If this were the case, a selective reduction of the high-affinity component of ouabain binding would be expected according to the two component model (2). Thus, an analysis of the two populations of ouabain binding receptors of grey matter Na,K-ATPase was performed before and after pyrithiamin treatment.

Materials and methods

Na,K-ATPase preparations from ox brain grey matter and from pig kidney outer medulla were prepared according to Klodos et al. (7) and according to Jørgensen (6) as modified by Ottolenghi and Jensen

---

* Present address: Department of Animal Physiology, Royal Veterinary and Agricultural University, DK-1871 Copenhagen V

(12), respectively. The ouabain-sensitive ($Na^+ + K^+$)-activated ATPase activity was 301 μmoles P_i · (mg protein)$^{-1}$ · h^{-1} for the ox brain enzyme and 395 μmoles P_i · (mg protein)$^{-1}$ · h^{-1} for the pig kidney enzyme.

(^{3}H)ouabain binding to Na,K-ATPase was carried out in the presence of Tris-HCl buffer (pH 7.25) at 37° C and (Mg^{2+} + P_i) or in the presence of (Mg^{2+} + orthovanadate), two well-known ligand combinations for ouabain binding. The determination of equilibrium binding of (^{3}H)ouabain was carried out as previously described (2, 3).

(^{3}H)ouabain was obtained from New England Nuclear Corp. and purified by chromatography with Na,K-ATPase as previously described (2). Unlabelled ouabain (g-strophanthin) was from Merck and pyrithiamin from Sigma Chemical Co.

Results

In preliminary experiments, it was established that preincubation of equal amounts (units) of enzyme activity with different concentrations of pyrithiamin was more deleterious to Na,K-ATPase from grey matter than to pig kidney enzyme. From Fig. 1 it is seen that the Na,K-ATPase activity of brain enzyme as well as its partial reaction, pNPPase, are halved by 15 min treatment with a concentration of pyrithiamin that hardly affects the kidney enzyme preparation.

A batch of ox brain enzyme was then treated with 1 mg pyrithiamin · ml^{-1} for 15 min at 37° C, cooled and washed by centrifugation in the cold. The Na,K-ATPase activity after this treatment was reduced to 60 %.

Equilibrium binding of (^{3}H)ouabain to the untreated brain enzyme and to the pyrithiamin treated derivative was determined for a range of ouabain concentrations both in the presence of (3 mM Mg^{2+} + 3 mM P_i) and in the presence of (3 mM Mg^{2+} + 10 μM vanadate). The symbols in Fig. 2 represent the binding isotherms in a Scatchard-type plot. Obviously, the binding data are incompatible with a model assuming a simple bimolecular reaction between a homogeneous population of receptors and purified (^{3}H)ouabain. Provided equilibrium exists between enzyme and the ligands (e. g. Mg^{2+}, P_i, etc.) which facili-

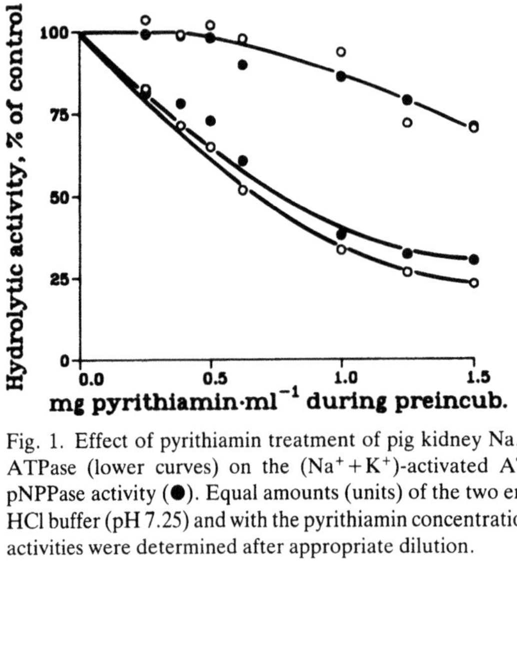

Fig. 1. Effect of pyrithiamin treatment of pig kidney Na,K-ATPase (upper curve) and ox brain Na,K-ATPase (lower curves) on the ($Na^+ + K^+$)-activated ATPase activity (○) and on the K^+-activated pNPPase activity (●). Equal amounts (units) of the two enzyme preparations were preincubated in Tris-HCl buffer (pH 7.25) and with the pyrithiamin concentrations indicated for 15 min at 37° C. The hydrolytic activities were determined after appropriate dilution.

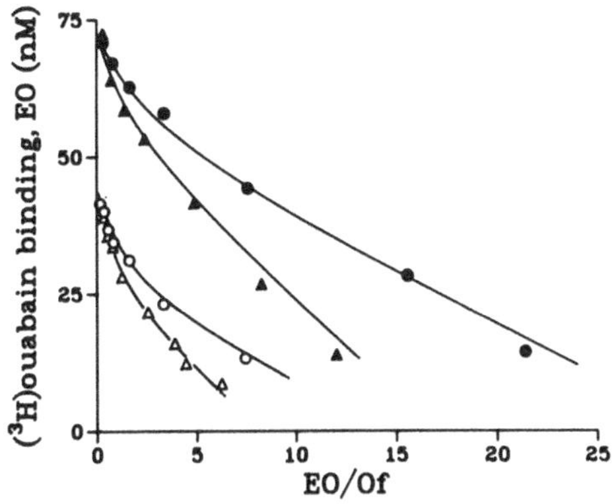

Fig. 2. Scatchard-type plot of (^{3}H)ouabain binding isotherms to ox brain Na,K-ATPase before (●, ▲) and after (○, △) pyrithiamin treatment and at equilibrium of binding. (^{3}H)ouabain binding was facilitated by 3 mM Mg^{2+} + 3 mM P_i (●, ○) or by 3 mM Mg^{2+} + 10 μM vanadate (▲, △) and binding was determined as described elsewhere (2, 3). The curves were drawn by using the calculated components and their affinities, see Figs. 3 and 4.

tate ouabain binding, as well as between different enzyme conformations, it can be shown that upward concave Scatchard plots can be due to 1) a dimeric structure of the enzyme and anticooperativity among primarily identical sites or 2) the presence of two or more distinct molecular forms of Na,K-ATPase, e. g. the presence of isoenzymes. In the former case, a resolution of binding data into two components must give two equal sized populations of sites, but not necessarily so in the latter case (4, 5).

All the ouabain binding data of Fig. 2 – irrespective of whether binding was obtained with one or another combination of ligands – was pooled and the computer was asked to resolve them into a high affinity component sensitive to destruction by pyrithiamin and a low affin-

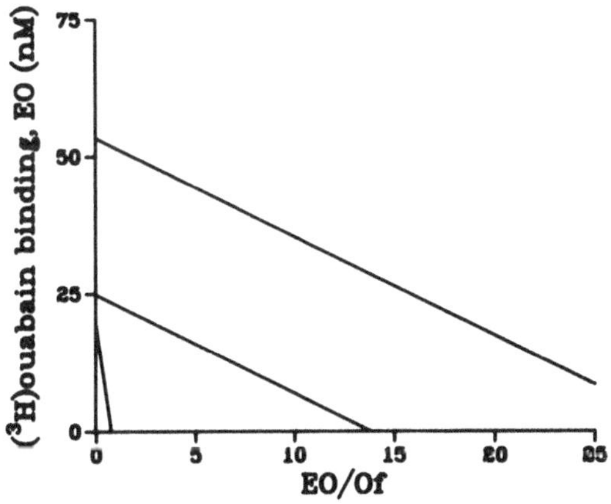

Fig. 3. Calculated high-affinity components (upper lines) and the low affinity component (lower line) of the (^{3}H)ouabain binding isotherms of Fig. 2 obtained with Mg^{2+} + P_i (●, ○) before (●) and after (○) pyrithiamin treatment (see text).

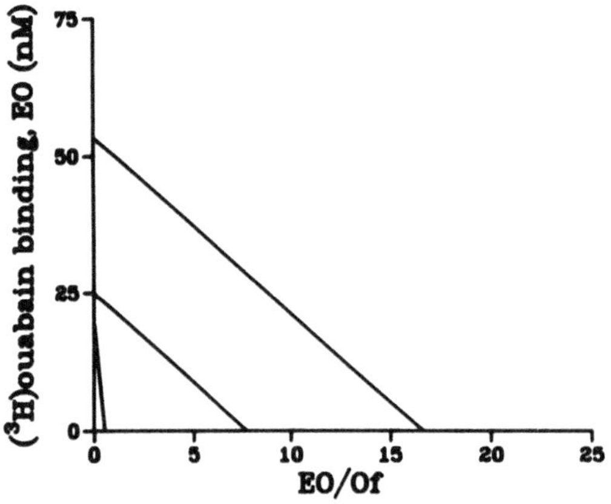

Fig. 4. Calculated high-affinity components (upper lines) and the low affinity component (lower line) of the (^{3}H)ouabain binding isotherms of Fig. 2 obtained with Mg^{2+} + vanadate (▲, △) before (▲) and after (△) pyrithiamin treatment (see text).

ity component insensitive to pyrithiamin. The apparent ouabain affinity of the receptors remaining after pyrithiamin treatment was assumed to have been unaffected by this treatment, but allowed to differ with the ligand conditions employed for facilitation of ouabain binding. The results of this analysis are given in Figs. 3 and 4 with either P_i- or vanadate-facilitation of ouabain binding: They show the changes in high-affinity component brought about by pyrithiamin treatment and the unaltered low-affinity component. The curves in Fig. 2 were drawn by using the calculated size of the two components and their affinities (Figs. 3 and 4).

It is seen that the experimental data (the symbols) fit the calculated curves well and that a division of the ouabain binding receptors into the same two high-affinity and low-affinity components seems possible irrespective of whether P_i- or vanadate is used to facilitate binding. Data are compatible with the simple reduction by pyrithiamin of the high affinity ouabain binding component.

Discussion

It should be emphasized that upward concave binding isotherms in Scatchard-type plots can be due to anticooperativity between two ligand binding subunits of a receptor and – excluding trivial artefacts – convincing evidence has recently been presented that nucleotide binding sites and, perhaps, also the ouabain binding sites of Na,K-ATPase exhibit anticooperativity (5, 12). Pure or purified Na,K-ATPase from pig kidney, however, does not show anticooperativity of ouabain receptors with the ligand conditions employed here for the promotion of glycoside binding (4). It thus seems likely that the non-linear Scatchard plots of ouabain binding isotherms obtained with grey matter Na,K-ATPase (2) were due to different affinities towards ouabain of two or more isoenzymes of the enzyme preparation.

It was the merit of Sweadner (13) to have demonstrated that brain Na,K-ATPase really contains two molecular forms of the enzyme with different gel electrophoretic mobilities and different affinities towards strophanthidin. The present analysis of the pooled ouabain binding data is compatible with a ratio between the high-affinity and the low-affinity component of ouabain receptors before pyrithiamin treatment of roughly 2:1. In support of the fitting procedure, it is worth mentioning that a similar ratio was found by Sweadner (13) from other criteria.

Why the $\alpha+$ peptide of brain Na,K-ATPase should be more sensitive to pyrithiamin than the α peptide found both in brain ATPase and in the enzyme elsewhere in the organism is unknown. Thiamin is said to have a specific function in nerve tissue independent of its coenzyme role (9) and pyrithiamin is an antimetabolite of the vitamin. Sweadner (13) pointed out that $\alpha+$ seems to have more exposed sulphydryl groups than α does, and the effect of pyrithiamin on brain Na,K-ATPase is antagonized by dithiothreitol (10).

The presence of the $\alpha+$ molecular form of Na,K-ATPase was recently demonstrated also in adipocytes and skeletal muscle plasma membranes (8), i. e. in tissues sensitive to insulin stimulation of K^+-uptake. The suggestion by these authors that $\alpha+$ could be responsible for the hormonal regulation seems hazardous, however, since the very low recovery of enzyme activity from their preparations may indicate a preferential isolation of neural tissue (for refs., see [4]). It would also be interesting to know whether $\alpha+$ is more sensitive to endogenous inhibitors or unsaturated free fatty acids than the α peptide.

Acknowledgements

Thanks are due to Jørgen Jensen for help with the design of the computer programme, to Toke Nørby for carrying out the curve-fitting, and to Kirsten Pedersen for technical assistance.

References

1. Erdmann E, Schoner W (1973) Ouabain-receptor interactions in (Na^+ + K^+)-ATPase preparations. II. Effect of cations and nucleotides on rate constants and dissociation constants. Biochim Biophys Acta 330: 302
2. Hansen O (1976) Non-uniform populations of g-strophanthin binding sites of (Na^+ + K^+)-activated ATPase. Apparent conversion to uniformity by K^+. Biochim Biophys Acta 433: 383–392
3. Hansen O (1982) Studies on ouabain-complexed (Na^++ K^+)-ATPase carried out with vanadate. Biochim Biophys Acta 692: 187–195
4. Hansen O (1984) Interaction of cardiac glycosides with (Na^+ + K^+)-activated ATPase. A biochemical link to digitalis induced inotropy. Pharmacol Rev 36: 143–163
5. Jensen J, Nørby JG, Ottolenghi P (1984) Sodium and potassium binding to the sodium pump: Stoichiometry and affinities evaluated from nucleotide-binding behaviour. J Physiol (London) 346: 219–241
6. Jørgensen PL (1974) Purification and characterization of (Na^+ + K^+)-ATPase. III. Purification from the outer medulla of mammalian kidney after selective removal of membrane components by sodium dodecylsulphate. Biochim Biophys Acta 356: 36–52
7. Klodos I, Ottolenghi P, Boldyrev A (1975) Large-scale preparation of Na,K-ATPase from ox brain. Anal Biochem 67: 397–403
8. Lytton J, Lin JC, Guidotti G (1985) Identification of two molecular forms of (Na^+ + K^+)-ATPase in rat adipocytes. Relation to insulin stimulation of the enzyme. J Biol Chem 260: 1177–1184
9. Matsuda T, Cooper JR (1983) Inhibition of neuronal sodium and potassium ion activated adenosinetriphosphatase by pyrithiamin. Biochemistry 22: 2209–2213
10. Matsuda T, Iwata H, Cooper JR (1984) Specific inactivation of $\alpha(+)$ molecular form of (Na^+ + K^+)-ATPase by pyrithiamin. J Biol Chem 259: 3858–3863
11. Nørby JG, Jensen J (1974) Binding of ATP to Na^+,K^+-ATPase. Ann NY Acad Sci 242: 158–167

12. Ottolenghi P, Jensen J (1983) The K^+-induced apparent heterogeneity of high-affinity nucleotide-binding sites in (Na^+ + K^+)-ATPase can only be due to the oligomeric structure of the enzyme. Biochim Biophys Acta 727: 89–100
13. Sweadner KJ (1979) Two molecular forms of (Na^+ + K^+)-stimulated ATPase in brain. Separation, and difference in affinity for strophanthidin. J Biol Chem 254: 6060–6067

Present address:

Dr. O. Hansen, Department of Animal Physiology, Royal Veterinary and Agricultural University, DK-1871 Copenhagen V (Denmark)

Interaction of cardiac glycosides with Na/K-ATPase in intact myocardium

S. Herzig, H. Lüllmann and K. Mohr

Department of Pharmacology, University of Kiel (West Germany)

Summary

An attempt was made to predict an appropriate equation for the binding of cardiac glycosides in intact myocardium. For this purpose, two features of the interaction of cardiac glycosides with Na/K-ATPase of intact myocardium were quantitatively considered and combined: 1. Cardiac glycoside binding depends on the turnover rate of Na/K-ATPase, and 2. occupation and inhibition of a fraction of Na/K-ATPases by cardiac glycoside leads to an increased turnover rate of the remaining pump molecules. The resulting binding equation predicts a concentration-proportional receptor occupation. Both the slope of the linear relationship and the maximum non-toxic cardiac glycoside binding depend on the turnover rate which is required to maintain the sodium and potassium balance in the absence of cardiac glycoside. 3H-ouabain binding experiments were performed with isolated guinea-pig left atria. The binding curves did not follow a saturation isotherm, but were in accordance with the delineated binding equation.

Introduction

The cardiac glycoside receptor is distinguished from common receptors by a peculiar feature. The binding site occurs with high affinity only at a particular conformation adopted during a transport cycle of Na/K-ATPase [for references see (6)]. Accordingly, the glycoside binding properties of the enzyme depend on its turnover activity.

This does not matter in binding studies with homogenized tissue, because Na/K-ATPase faces a constant environment. Under this condition, both the turnover rate and the glycoside binding properties are constant. Consequently, glycoside binding is often found to follow a Langmuir saturation isotherm [for references see (3)].

In intact myocardium, however, the situation is more complex. This conclusion is reached when two features are considered, which characterize the interaction between car-

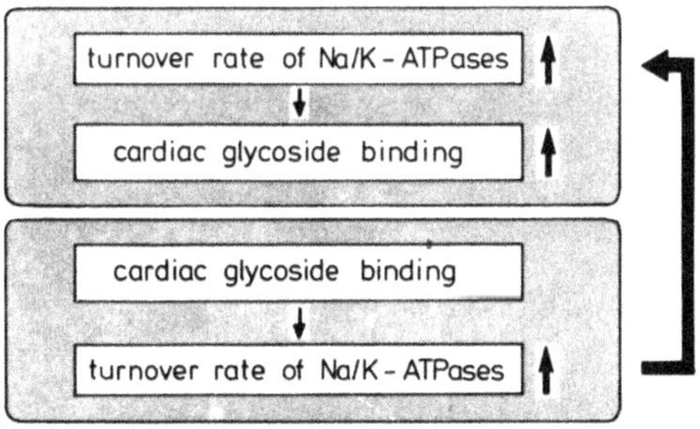

Fig. 1. Features which lead to a self-perpetuation of cardiac glycoside binding in intact myocardium.

diac glycosides and Na/K-ATPase in intact myocardium. These features, which have been recently reviewed by Akera and Brody [1], are stated in Fig. 1. The first feature is, that cardiac glycoside binding depends on the turnover rate of Na/K-ATPase. The turnover rate is defined as the transport activity of a single Na/K-ATPase-molecule. With increased turnover rate the amount of cardiac glycoside bound at a given concentration is elevated.

The second feature is that occupation and inhibition of a fraction of the cellular Na/K-ATPase leads to an increased turnover rate of the unoccupied enzyme molecules in order to maintain the appropriate transport of sodium and potassium.

When these two features are combined, it becomes evident that cardiac glycoside binding should promote further binding; it seems to be a self-perpetuating process [4]. Accordingly, it is unlikely that cardiac glycoside binding would follow a simple Langmuir saturation isotherm.

In the following it will be shown that cardiac glycoside binding in intact myocardium should be positively cooperative [4, 5]. The attempt is made to delineate in terms of receptor ligand interaction kinetics the appropriate relationship between cardiac glycoside concentration and binding. Finally, the predicted binding curve will be compared with results obtained in 3H-ouabain binding experiments in guinea-pig atria.

Theoretical considerations

The dependency of cardiac glycoside binding on the turnover rate of Na/K-ATPase is illustrated in Fig. 2a .

a

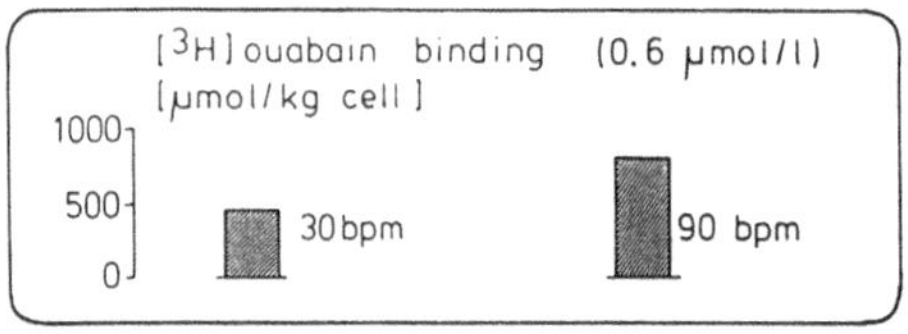

Fig. 2a. Dependency of 3H-ouabain binding on the stimulation frequency (data taken from Fig. 4): the increased equilibrium binding at 90 beats per minute (bpm) can be explained by an increased affinity, since the concentration of 3H-ouabain and the maximum number of receptors, i. e. Na/K-ATPases are identical under both conditions.

It has been frequently reported that cardiac glycoside binding in intact myocardium depends on the stimulation frequency of the cardiac preparation [e. g. 1, 2]. For instance, when 3H-ouabain was added in a concentration of 0.6 μM to isolated guinea-pig left atria, which were stimulated at rates of 30 and 90 beats per minute, binding of ouabain was found to be about twofold higher in the atria which were driven at the higher frequency. When this result is to be explained in terms of receptor ligand interaction kinetics, it must be noted that the increased receptor occupation occurred at a constant ligand concentration G and at a constant receptor concentration R_{max}. Consequently, the affinity of binding is increased. In Fig. 2b, two myocardial cells are schematically depicted. The sodium influx (thin arrows) is counterbalanced by the Na/K-ATPases (shaded circles) working at an appropriate turnover rate (thick arrows). The movements of potassium are not considered, being assumed to correspond to the sodium movements. At a doubled sodium load (right cell), the Na/K-ATPases have to turn at a twofold higher rate in order to counterbalance the sodium load. The stimulation of the Na/K-ATPase molecules is brought about by a slightly elevated cytosolic sodium-ion concentration.

b

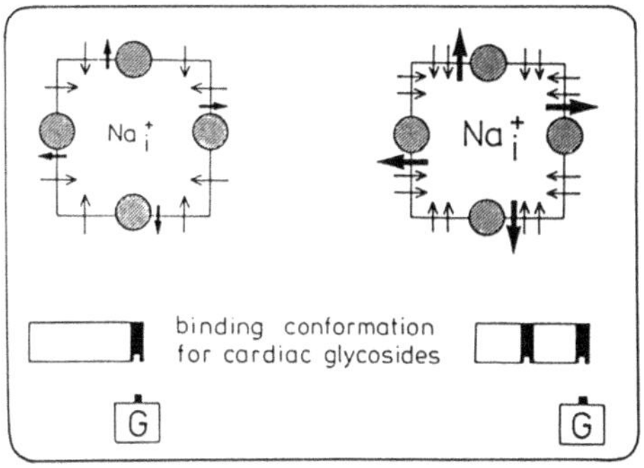

Fig. 2b. Schematic illustration of two myocardial cells under conditions of different sodium influx (thin arrows), e. g. induced by different stimulation frequencies. At a doubled sodium-load, the Na/K-ATPase (shaded circles) work at doubled turnover rate. Within a given time interval (total bar), the cardiac glycoside binding-conformation (black zone of the bar) occurs twice as often at the doubled turnover rate. Note that the lifetime of a binding conformation is taken to be constant. Thus, the probability of a cardiac glycoside molecule to associate to Na/K-ATPase is enhanced.

Since the binding conformation is only temporarily exposed during a turnover cycle, the increased turnover rate is accompanied by a more frequent appearance of the binding conformation.

Accordingly, the probability that a cardiac glycoside will associate to Na/K-ATPase is increased. In terms of receptor kinetics, the association rate constant k_{+1} is a function of the turnover rate TOR of the Na/K-ATPase molecules. For a quantitative analysis, assuming that the lifetime of a binding conformation is constant, k_{+1} is set proportional to the turnover rate:

$$k_{+1} \sim \text{TOR}$$

When the dissociation rate constant k_{-1} of the complex between cardiac glycoside and Na/K-ATPase is taken to be independent of the turnover rate and thus constant, the affinity constant, k_{+1}/k_{-1}, is proportional to the turnover rate.

It can be written:

$$\text{Affinity constant (Aff)} = \frac{k_{+1}}{k_{-1}} \sim \text{TOR} .$$

Introduction of a proportionality constant f results in

$$\text{Aff} = f \times \text{TOR} .$$

The maximum turnover rate TOR_{max} leads to the maximum affinity Aff_{max}:

$$\text{Aff}_{max} = f \times \text{TOR}_{max}$$

These equations combine to:

$$\text{Aff} = \text{Aff}_{max} \times \frac{\text{TOR}}{\text{TOR}_{max}} .$$

At binding equilibrium, the ratio between the concentration of the glycoside-receptor complexes (GR) and the product of the concentrations of the glycoside (G) and the unoccupied Na/K-ATPases (R) is determined by the affinity of binding. It can be written:

$$\frac{GR}{G \times R} = Aff = Aff_{max} \times \frac{TOR}{TOR_{max}} \quad (1)$$

Equation (1) describes the dependency of cardiac glycoside binding on the turnover rate of Na/K-ATPase.

As a further complexity, the actual turnover rate of unoccupied Na/K-ATPases is not constant, but is altered in the course of cardiac glycoside binding. This is illustrated in Fig. 2c.

c

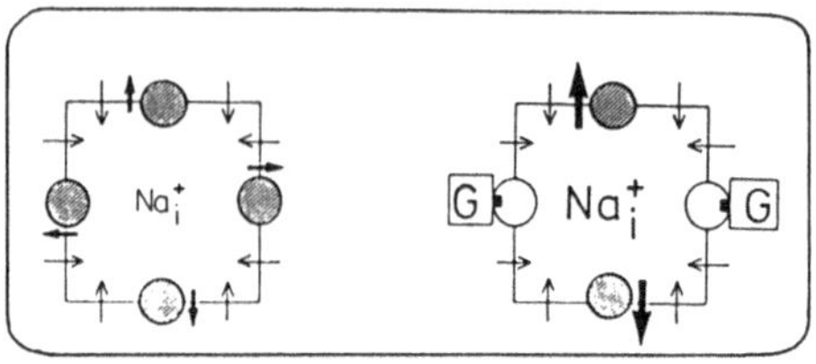

Fig. 2c. Schematic illustration of the effect of cardiac glycoside binding to a fraction of the cellular Na/K-ATPases on the unoccupied Na/K-ATPases.
The remaining pump molecules work at a doubled turnover rate in order to compensate for the inhibition of the occupied Na/K-ATPases. This stimulation is brought about by a slightly elevated intracellular sodium-ion concentration (Na_i).

When half of the cellular Na/K-ATPases are occupied and inhibited by cardiac glycoside (right cell), the remaining unoccupied Na/K-ATPases are working at a doubled turnover rate, thus maintaining the transport of the sodium load.

The stimulation of the remaining Na/K-ATPases is brought about by a slight elevation of the cytosolic sodium-ion concentration. As delineated above, an increased turnover rate means an increased affinity of Na/K-ATPase.

In general, the cellular Na/K-ATPase molecules are not independent of each other. Occupation and inhibition of a fraction of the cellular Na/K-ATPases by a cardiac glycoside results in an alteration of the binding affinity of the unoccupied enzyme molecules: the receptors are cooperative [4, 5]. In the following, the dependency of the turnover rate on the receptor occupancy will be quantified. For this purpose, one assumption is made: as long as the inhibition of cellular Na/K-ATPases does not evoke an intoxication, the same sodium transport as under control conditions is provided by the remaining unoccupied Na/K-ATPases.

The transport rate TR is defined as the amount of sodium transported out of the cell per unit of time. In the absence of cardiac glycoside, the transport rate TR_{contr} is given by the turnover rate of a single Na/K-ATPase molecule (TOR_{contr}) multiplied by the number of cellular Na/K-ATPases, which equals R_{max}:

$$TR_{contr} = TOR_{contr} \times R_{max} \;.$$

In the presence of bound cardiac glycoside, the transport rate is the product of the turnover rate of the unoccupied Na/K-ATPase molecules (TOR) and the number of unoccupied Na/K-ATPases, which is equal to the number of free receptors R:

$$TR = TOR \times R \;.$$

According to the above assumption both transport rates are set equal:

$$TR_{contr} = TR$$

$$TOR_{contr} \times R_{max} = TOR \times R$$

$$TOR = TOR_{contr} \times \frac{R_{max}}{R} \quad . \qquad (2)$$

The resulting equation (2) describes the dependency of the turnover rate on cardiac glycoside binding, or on the concentration of unoccupied receptors, respectively.

The two features stated in Fig. 1 are now quantitatively described by the Eqs. (1) and (2).

When the turnover rate TOR is substituted in Eq. (1), the final binding equation (3) is found:

$$GR = [R_{max} \times Aff_{max} \times \frac{TOR_{contr}}{TOR_{max}}] \times G \quad . \qquad (3)$$

Equation (3) describes the dependency of the cardiac glycoside binding GR on the cardiac glycoside concentration G, because all further unknown variables (TOR and R) have been eliminated. The maximum number of Na/K-ATPases R_{max}, their maximum affinity Aff_{max}, and the ratio of the initial turnover rate in the absence of cardiac glycoside TOR_{contr} over the maximum turnover rate TOR_{max} are constant for a given condition. Consequently, there is a concentration-proportional relationship between cardiac glycoside binding and the concentration, with the product of the constant parameters as the proportionality factor. It should be kept in mind that this equation is only valid for non-toxic conditions, i. e. the remaining Na/K-ATPases are still able to counterbalance the sodium load.

At the maximum non-toxic receptor occupancy GR_{max}, the unoccupied Na/K-ATPases R

$$R = R_{max} - GR_{max}$$

have to work at the maximum turnover rate TOR_{max} to keep up the required transport rate TR_{contr}:

$$TOR_{max} \times (R_{max} - GR_{max}) = TR_{contr} = TOR_{contr} \times R_{max} \quad .$$

The resulting equation indicates that the maximum tolerable receptor occupancy decreases with increasing initial turnover rate TOR_{contr}:

$$GR_{max} = R_{max} \times (1 - \frac{TOR_{contr}}{TOR_{max}}) \quad .$$

Three predictions can be made about cardiac glycoside binding curves in intact myocardium within the non-toxic concentration range. The computed curves are shown in Fig. 3.

1. The binding curves should be linear in a bilinear plot.
2. When the sodium load is increased, for instance by higher stimulation rates, the initial turnover rate will increase and the slope of the lines will steepen.
3. The higher the sodium load to be transported, the lower will be the Na/K-ATPase occupation which can be tolerated without intoxication. At the end-points of the curves, namely the final non-toxic receptor occupation, the free Na/K-ATPases work at their maximum turnover rate; the higher the sodium load, the more Na/K-ATPases have to remain unoccupied to maintain homeostasis.

Experimental results

The binding of 3H-ouabain was measured in electrically stimulated, isolated guinea-pig left atria as described previously [2]. In short, freshly dissected atria were mounted in organ

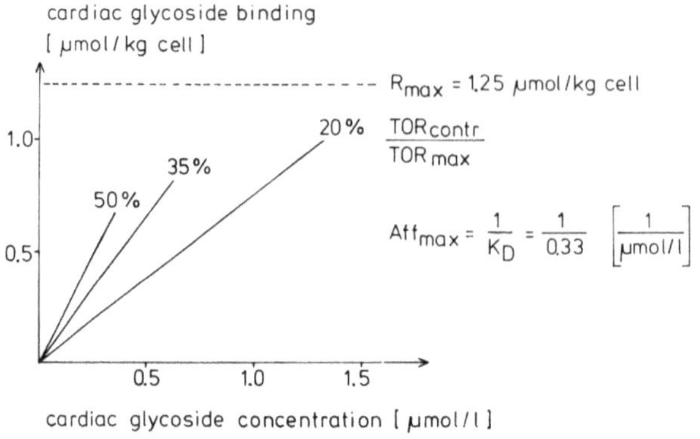

Fig. 3. Cardiac glycoside binding curves computed with equation (3) for conditions of different control turnover rates (TOR_{contr}), which are expressed in percent of the maximum turnover rate TOR_{max}. The indicated values for the number of Na/K-ATPases R_{max} and the maximum affinity Aff_{max} are chosen to correspond to the experimental data shown in Fig. 4.

baths containing a modified Tyrode solution (in mmol/l): NaCl 136.8, KCl 5.4, $NaHCO_3$ 11.9, NaH_2PO_4 0.21, $CaCl_2$ 1.8, $MgCl_2$ 1.1, glucose 5.5. The solution was gassed with O_2/CO_2 95 %/5 % and kept at a temperature of 32°C. After an equilibration period of 1 h, 3H-ouabain was added in a concentration of 1 to 3 nmol/l together with various concentrations of unlabelled ouabain. After different time intervals up to 3 h, atria were removed, weighed and prepared for liquid scintillation counting. Specific ouabain binding was calculated from the actual specific radioactivity and the difference between the total tissue bound radioactivity and the amount of 3H-ouabain present in the extracellular space. The extracellular space amounted to 30 % of tissue wet weight. Determination of unspecific 3H-ouabain binding in the presence of 10^{-4} mol/l unlabelled ouabain revealed similar results.

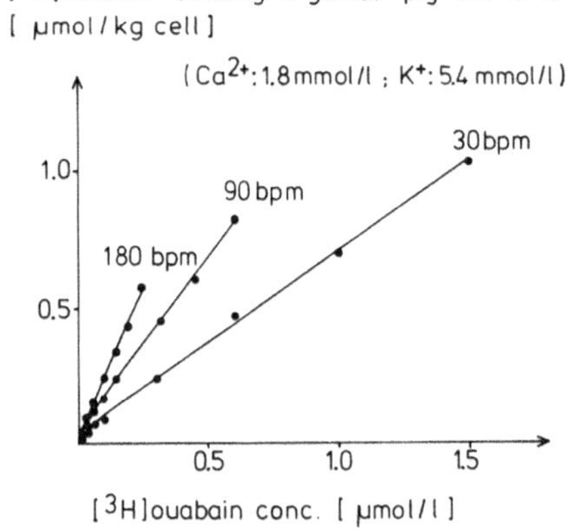

Fig. 4. 3H-ouabain binding curves covering the non-toxic concentration range were obtained with guinea-pig left atria stimulated at three different beat frequencies of 30, 90, and 180 beats per minute (bpm). For methodological details see text. The 30 bpm data were taken from [5], the 90 bpm data were taken from [4].

For every ouabain concentration, 8 atria were used to determine the time course of ouabain binding. The equilibrium value of ouabain binding was derived from an exponential function fitted to the uptake data by means of a non-linear regression analysis.

Equilibrium values of specific ouabain binding were measured at three beat frequencies of 30, 90, and 180 beats per minute (bpm) in the non-toxic range of ouabain concentrations.

The concentration-binding curves are depicted in Fig. 4.

1. All curves reveal a linear segment. This concentration-proportionality is particularly impressive with the curve obtained at 30 bpm. All curves deviate from linearity near the origin. This slightly curved segment is relatively pronounced in the 180 bpm curve. However, even in this curve the amount of ouabain bound proportional to the concentration represents the prevailing fraction of cardiac glycoside bound. At 30 bpm and 90 bpm, the positive inotropic effect of ouabain started to occur at those ouabain-concentrations at which the linear segment of the binding curves began. This indicates that the concentration-proportional binding is pharmacologically relevant.

2. Fig. 4 furthermore illustrates that the slopes of the binding curves markedly increase with increased beat frequency.

3. It can be taken from these curves that the maximum non-toxic binding declines with increasing beat frequency. Correspondingly, the maximum tolerated ouabain concentration falls when the beat frequency is raised.

Conclusion

■ Cardiac glycoside binding to intact myocardium is expected to deviate from a saturation isotherm.

■ A binding equation has been deduced, which predicts a concentration-proportional binding in the non-toxic concentration range.

■ 3H-ouabain binding experiments revealed that the predicted concentration-proportional binding does indeed occur in intact myocardium.

References

1. Akera T, Brody TM (1985) Sodium ion and the cardiac actions of digitalis. TIPS 7: 296–298
2. Bentfeld M, Lüllmann H, Peters T, Proppe D (1977) Interdependence of ion transport and the action of ouabain in heart muscle. Br J Pharmacol 61: 19–27
3. Erdmann E (1981) Influence of cardiac glycosides on their receptor. In: Greeff K (ed) Handbook of experimental Pharmacology, Vol 56/I: Cardiac Glycosides. Springer-Verlag, Berlin Heidelberg New York
4. Herzig S, Krey U, Lüllmann H, Mohr K (1985): Is digitalis binding to intact myocardium cooperative? TIPS 7: 432–433
5. Herzig S, Lüllmann H, Mohr K (1985) On the cooperativity of ouabain binding to intact myocardium. J Mol Cell Cardiol 17: 1095–1104
6. Lee KS, Klaus W (1971) The subcellular basis for the mechanism of inotropic action of cardiac glycosides. Pharmacol Rev 23: 193–261

Author's address:

Prof. Dr. H. Lüllmann, Klinikum der Christian-Albrechts-Universität Kiel, Abteilung Pharmakologie, Hospitalstraße 4–6, 2300 Kiel (West Germany)

The human red blood cell: A model for ouabain receptor regulation in the heart?

L. Brown, K. Werdan and E. Erdmann

Medizinische Klinik I, Ludwig-Maximilians-Universität München, München (West Germany)

Summary

The assumption that the human red blood cell can be used as a model for ouabain receptor regulation in the heart has now been tested. The following results were obtained:

1. The affinity of the ouabain receptor was similar in both erythrocytes and heart muscle of the same species (either human or guinea-pig), but the density of binding sites was much greater in the heart muscle. When measured in the same patient, the binding capacity of the heart and erythrocytes were not correlated.

2. Ouabain receptor occupation and inhibition of active Na^+/K^+-transport were closely correlated in human erythrocytes and chick heart non-muscle cells in culture. This correlation was not found in chick heart muscle cells where an occupation of 40% of the receptors decreased the Na^+/K^+-transport rate by only 10%.

3. Hypokalaemia increased the number of ouabain binding sites on human and guinea-pig erythrocytes but caused no change in guinea-pig heart muscle. A severe reduction in extracellular K^+ concentration was necessary to induce increases in ouabain binding capacity in chick heart muscle cells. Similar increases were measured after moderate K^+-deficiency in chick heart non-muscle cells in culture.

The differences in ouabain receptor regulation in erythrocytes and heart muscle cells may be attributable to the lack of a "sodium pump reserve" in erythrocytes. Such variations in receptor regulation indicate that the human erythrocyte is not a suitable model for the ouabain receptor in the human heart.

Introduction

Most of the evidence suggesting that the pharmacodynamics of cardiac glycosides may be altered in various physiological and pathological states and in chronic drug therapy has been obtained using studies of cardiac glycoside binding to human erythrocytes (3, 6). The use of the erythrocyte for such studies assumes that parallel changes occur in the cardiac glycoside receptor in both erythrocytes and heart muscle. This assumption has been tested as follows:

1. characterization of the ouabain receptors on human erythrocytes and right atrial appendage from the same patient;
2. correlation of receptor occupation and function in human erythrocytes as well as in cultured muscle and non-muscle cells from embryonic chick hearts; and
3. comparison of changes in receptor properties occurring in erythrocytes and heart muscle during hypokalaemia and digitalis administration.

Methods and materials

The methods used have been described in detail previously:
^3H-ouabain binding to erythrocytes (4) and to homogenates of cardiac tissue (2); preparation and culti-

vation of cardiac muscle and non-muscle cells from 10–14 day old chick embryos (8, 9, 12). A Scatchard analysis of bound ^{3}H-ouabain against the ratio of bound to free ouabain was used to calculate the K_D-value and the maximal binding capacity (7).

Measurements of ^{3}H-ouabain binding to human right atrial appendage homogenates were performed on samples obtained during cardiac surgery involving replacement of the aortic or mitral valve, or coronary artery bypass. The materials used for these experiments have been described in the previous publications cited above.

All values are given as the mean ± standard error of the mean.

Results

^{3}H-ouabain binds to a homogenous class of binding sites on both heart muscle and erythrocytes of humans and guinea-pigs. When measured in the same patient, the affinity of ouabain binding given as the dissociation constant, K_D, was similar in human erythrocytes ($2.5 \pm 0.2 \times 10^{-9}$ M) and homogenates of right atrial appendage ($4.3 \pm 0.2 \times 10^{-9}$ M, n = 29). The affinity of ^{3}H-ouabain binding to guinea-pig tissues is lower than for human but similar in both left ventricle homogenates (K_D, $1.0 \pm 0.05 \times 10^{-7}$ M, n = 9) and erythrocytes (K_D, $2.4 \pm 0.2 \times 10^{-7}$ M, n = 15) (1). The number of binding sites was much lower on the erythrocytes of both species than on the heart muscle. Human erythrocytes had an average of 280 ± 15 sites/cell (n = 29) while guinea-pig erythrocytes possessed about 420 ± 80 sites/cell (n = 15); representing about 1 site/μm^2.

The homogenates from the right atrial appendage of 29 humans contained about $4.8 \pm 0.3 \times 10^{14}$ sites/g protein (about 280 sites/μm^2); the left ventricle of the guinea-pig heart had $12.2 \pm 1.4 \times 10^{14}$ sites/g protein (about 1320 sites/μm^2). The number of binding sites in erythrocytes and heart tissue from the same patient varied markedly, with no correlation between both values (Fig. 1).

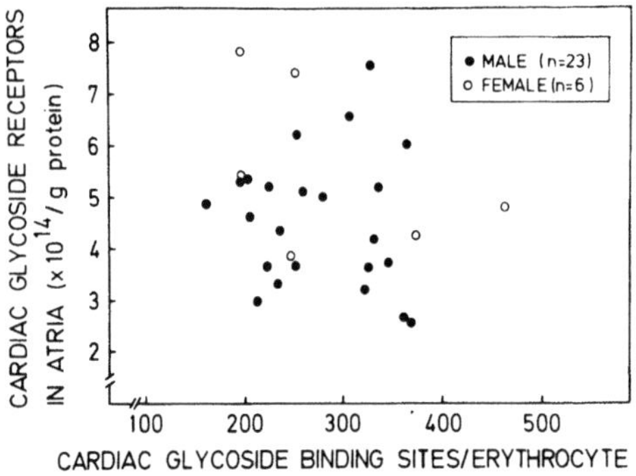

Fig. 1. Ouabain binding sites on human erythrocytes and right atrial appendage homogenates from 23 male and 6 female patients. The number of binding sites was calculated from Scatchard analysis of ^{3}H-ouabain binding (7).

Occupation of the receptor by ouabain leads to an inhibition of active Na^+/K^+-transport which can be measured using ($^{86}Rb^+ + K^+$)-uptake rates (12). This coupling between receptor occupation by ouabain and inhibition of active transport was determined in human erythrocytes as well as in cultured cardiac muscle and non-muscle cells from chick embryos (Fig. 2). Strict coupling of receptor occupation and inhibition of ($^{86}Rb^+ + K^+$)-influx is found in human erythrocytes as well as in chick heart non-muscle cells. In contrast, in heart

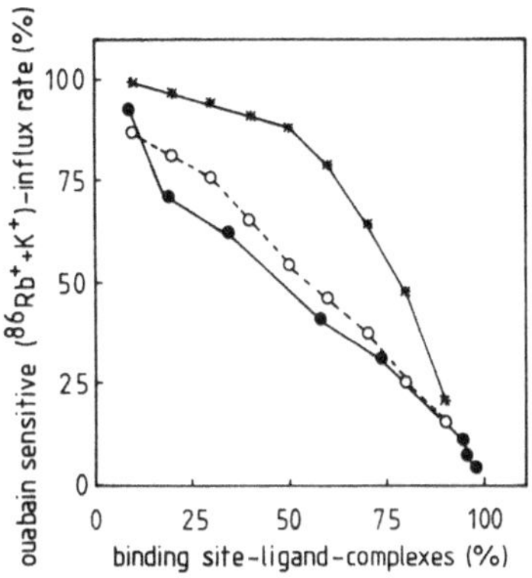

Fig. 2. Correlation of cardiac glycoside receptor occupancy and inhibition of active Na^+/K^+-transport in human erythrocytes (●—●), chick heart muscle cells (*—*) and non-muscle cells (o—o) in culture.
The dissociation constant of specific ^3H-ouabain binding was used to calculate the cardiac glycoside receptor occupancy at each concentration; for each concentration, inhibition of ($^{86}Rb^+$ + K^+)-uptake was determined ($[K^+]$ = 0.75 mM for chick heart cells and $[K^+]$ = 1.0 mM for human erythrocyte experiments).

Tissue	K_D	IC_{50} ($^{86}Rb^+$ + K^+)-uptake
Human erythrocytes	1.8×10^{-9} M	2.5×10^{-9} M
Chick heart muscle cells	1.5×10^{-7} M	5.8×10^{-7} M
Chick heart non-muscle cells	1.9×10^{-7} M	2.6×10^{-7} M

muscle cells, the relationship between occupation and inhibition was non-linear with an occupation of about 40% of the receptors being necessary to inhibit the ($^{86}Rb^+$ + K^+)-influx rate by 10%.

The regulation of the receptor affinity or capacity in chronic hypokalaemia has been investigated using human erythrocytes, guinea-pig erythrocytes and heart tissue homogenates, as well as chick heart muscle and non-mu scle cells in culture. Patients with chronic hypokalaemia, usually caused by diuretics or laxatives, showed an increased number of ouabain binding sites on their erythrocytes (5). In guinea-pigs fed a K^+-poor diet for 14–18 days, the blood potassium fell from 5.4 ± 0.3 mM to 3.3 ± 0.2 mM. No changes in the affinity of (Mg^{2+} + P_i)-supported ^3H-ouabain binding or in the ^3H-ouabain binding capacity of any chamber of the heart was found (1). However, this degree of hypokalaemia induced a different response in guinea-pig erythrocytes where a doubling of the binding capacity was observed without any change in affinity. Further, in the soleus muscle of these hypokalaemic guinea-pigs, the number of ouabain binding sites was reduced by about 33% (Fig. 3).

Cardiac muscle and non-muscle cells from chick embryos in culture respond to chronic exposure to low K^+ in a qualitatively identical but quantitatively different manner (11): in both cell types, incubation for three days at lowered extracellular K^+ increases the number of cardiac glycoside receptors (Fig. 4). However, in cardiac muscle cells, this process is initiated only at K^+-concentrations lower than 2.0 mM; in contrast, lowering the extracellular K^+ from 5.5 mM to 3.0 mM is sufficient to induce this adaptation mechanism in cardiac non-muscle cells. The clinical relevance of an increase in the receptor number is shown by the

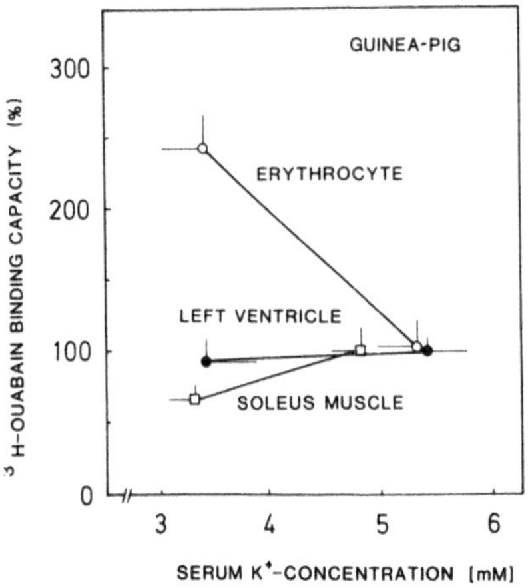

Fig. 3. Ouabain binding capacity after 14–18 days hypokalaemia in guinea-pig erythrocytes, left ventricular and soleus muscle homogenates. Scatchard analysis of ^{3}H-ouabain binding to erythrocytes and muscle homogenates of normo- and hypokalaemic guinea-pigs were performed. The affinity of the binding sites was unchanged in hypokalaemia.

	Binding capacity	
Tissue	Normokalaemia	Hypokalaemia
Erythrocytes	420 ± 80 sites/cell (n = 15)	1020 ± 100 sites/cell (n = 20)
Left ventricle	$1.32 \pm 1.4 \times 10^{14}$ sites/g wet weight (n = 9)	$1.21 \pm 0.20 \times 10^{14}$ sites/g wet weight (n = 7)
Soleus muscle	$1.5 \pm 0.2 \times 10^{13}$ sites/g wet weight (n = 8)	$1.0 \pm 0.1 \times 10^{13}$ sites/g wet weight (n = 11)

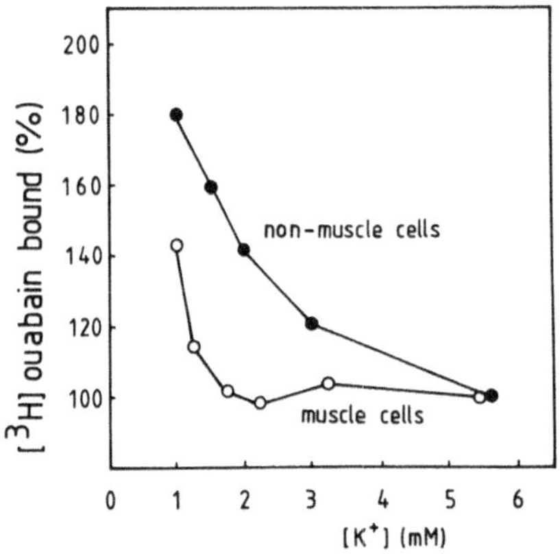

Fig. 4. Ouabain binding capacity in cultured chick heart cells after incubation in solutions containing various K^+-concentrations.

Cardiac muscle and non-muscle cells from 10–14 days old chick embryos were cultured for three days with the K^+-concentration shown.

The affinity of ouabain binding was not affected by cultivation at different K^+-concentrations.

rightward shift of the concentration-response curve for ouabain-induced positive inotropy in chick heart muscle cells cultured in 1 mM K^+ for three days (10).

Chronic digitalis treatment has been shown to increase the number of ouabain binding sites on human erythrocytes (5, 6). Studies with chick heart muscle cells in culture to determine whether a tolerance to cardiac glycosides can be developed in these cells are reported in the present volume by Werdan and colleagues.

Discussion

The affinity of ouabain for its receptor on human erythrocytes and heart tissue seems to be identical; guinea-pig heart muscle and erythrocytes also probably possess the same receptor. The lack of an intraindividual correlation between receptor capacity in humans in both cell types as shown in Fig. 1 indicates that different mechanisms may control the binding capacity in erythrocytes and heart tissue. Local influences such as myocardial ischaemia or cardiomyopathy may decrease the number of cardiac receptors but would not be expected to influence the erythrocyte binding site. Ouabain binding is closely related to inhibition of the active Na^+/K^+-transport in human erythrocytes and chick heart non-muscle cells in culture but not in chick heart muscle cells (Fig. 2). Human erythrocytes then resemble fibroblasts (non-muscle cells) rather than muscle cells in the relationship between receptor occupation and pharmacological effects.

Disease states such as hypokalaemia (3) as well as digitalis administration (5, 6) increase the number of cardiac glycoside receptors on human erythrocytes without changing their affinity. Our results show an increase in binding capacity of red blood cells but not of heart muscle in hypokalaemic guinea-pigs (Fig. 3). In chick heart non-muscle cells, a decrease in extracellular K^+ similar to that achieved in the guinea-pigs was sufficient to increase the number of ouabain binding sites in these cells. In contrast, much more severe hypokalaemia (to less than 2.0 mM) was necessary to stimulate chick heart muscle cells to synthesize more binding sites. Thus, the erythrocytes have reacted the same as the non-muscle cells, rather than the heart muscle cells, to the altered physiological conditions in hypokalaemia.

Is there a common basis for this discrepancy between erythrocytes and heart muscle concerning regulation of the cardiac glycoside receptor? Erythrocytes lack a "sodium pump reserve" which has been demonstrated to exist in heart muscle cells (12, 13). This "pump reserve" enables the heart muscle cells to tolerate a considerable reduction in extracellular K^+, one of the substrates of the sodium pump, without a similar reduction in the active Na^+/K^+-transport (Fig. 2). In heart muscle cells, only very severe impairment of active Na^+/K^+-transport by a high degree of K^+-depletion (Fig. 4) initiates the compensatory increase in the number of active sodium pump molecules. Due to the lack of a "pump reserve", erythrocytes and non-muscle cells increase their number of sodium pump molecules at much lower degrees of Na^+/K^+-transport reduction to compensate for the effects of moderate K^+-depletion (Fig. 4).

These comparisons between the cardiac glycoside receptor on erythrocytes and heart tissue indicate that caution is required when extrapolating the erythrocyte results to the heart. Differences in the receptor regulation would mean that reliable results for the heart can only be obtained from studies with cardiac tissue.

Acknowledgements

These investigations were supported by the Deutsche Forschungsgemeinschaft (Er 65/4–4) and the Wilhelm Sander-Stiftung (78.014.3).

We would like to thank B. Lorenz, A. Reichl, C. Reithmann, H. Schäffler, G. Schneider, A. Thomschke, B. Wagenknecht, G. Wagner and B. Zwißler for their valuable help with the experimental work.

Our special thanks are due to the cardiac surgeons (Prof. E. Kreuzer, Prof. B. Kemkes, Dr. C. Weinhold and their colleagues, Herzchirurgische Klinik der Universität and Dr. H. Holper, Dr. W. Klövekorn and their colleagues, Herzchirurgische Klinik des Deutschen Herzzentrums, München) who provided the human heart samples without which this study would not have been possible.

References

1. Brown L, Wagner G, Hug E, Erdmann E (1985) Ouabain binding and inotropy in acute potassium depletion in guinea-pigs. Cardiovasc Res (accepted for publication)
2. Brown L, Werdan K, Erdmann E (1983) Consequences of specific ^{3}H-ouabain binding to guinea-pig left atria and cardiac cell membranes. Biochem Pharmacol 32: 423–435
3. Erdmann E (1981) Influence of cardiac glycosides on their receptor. In: Greeff K (ed) Handbook of Experimental Pharmacology, 56/I. Springer-Verlag, Berlin Heidelberg, pp 337–380
4. Erdmann E (1982) Binding studies with ^{3}H-ouabain to red cell membranes. In: Ellory JC, Young JD (eds) Red cell membranes – a methodological approach. Academic Press London New York San Francisco, pp 251–262
5. Erdmann E, Werdan K, Krawietz W (1984) Influence of digitalis and diuretics on ouabain binding sites on human erythrocytes. Klin Wochenschr 62: 87–92
6. Ford AR, Aronson JK, Grahame-Smith DG, Carver JG (1979) The acute changes seen in cardiac glycoside receptor sites, ^{86}Rb-uptake and intracellular sodium concentrations in the erythrocytes of patients during the early phases of digoxin therapy are not found during chronic therapy: Pharmacological and therapeutic implications for chronic digoxin therapy. Br J Clin Pharmacol 8: 135–142
7. Scatchard G (1949) The attraction of proteins for small molecules. Ann N Y Acad Sci 51: 660–672
8. Werdan K, Bauriedel G, Bozsik M, Krawietz W, Erdmann E (1980) Effects of vanadate in cultured rat heart muscle cells. Vanadate transport, intracellular binding and vanadate-induced changes in beating and in active cation flux. Biochim Biophys Acta 597: 364–383
9. Werdan K, Reithmann C, Erdmann E (1985) Cardiac glycoside tolerance in cultured chicken muscle cells – a dose-dependent phenomenon. Klin Wochenschr 63: 1253–1264
10. Werdan K, Reithmann C, Schneider G, Erdmann E (1985) Modulation of cardiac glycoside sensitivity of beating heart muscle cells in culture. In: Ellory JC, Glynn IM (eds) The sodium pump, Proceedings of the fourth International Conference on Na,K-ATPase, Cambridge UK pp 679–683
11. Werdan K, Schneider G, Krawietz W, Erdmann E (1984) Chronic exposure to low K^{+} increases cardiac glycoside receptors in cultured cardiac cells: different responses of muscle and non muscle cells from chicken embryos. Biochem Pharmacol 33: 1161–1164
12. Werdan K, Wagenknecht B, Zwissler B, Brown L, Krawietz W, Erdmann E (1984) Cardiac glycoside receptors in cultured heart cells. I. Characterization of one single class of high affinity receptors in heart muscle cells from chick embryos. Biochem Pharmacol 33: 55–70
13. Werdan K, Zwissler B, Wagenknecht B, Krawietz W, Erdmann E (1983) Quantitative correlation of cardiac glycoside binding to its receptor and inhibition of the sodium pump in chicken heart cells in culture. Biochem Pharmacol 32: 757–760

Author's address:

Dr. Lindsay Brown, Medizinische Klinik I der Universität, Klinikum Großhadern, Marchioninistr. 15, 8000 München 70 (West Germany)

Binding and release of ouabain at low Mg^{2+} in reconstituted human red blood cell ghosts

H. H. Bodemann

Medizinische Universitätsklinik, Freiburg (West Germany)

Introduction

Various conformational states of Na,K-ATPase as pharmacological receptor for cardiac glycosides are recognized by the rate of ouabain binding and release in porous and reconstituted human red blood cell ghosts. At low Mg^{2+}, conformational changes become evident in the presence of various ligands which cannot be studied at physiological concentrations of Mg^{2+}. By using reconstituted human red cell ghosts containing ATP and little Mg^{2+}, it is possible to evaluate the side-specificity of those effects of Na^+ and K^+ which have been observed in porous ghosts. Some of the following data are detailed in previous papers (5, 6).

Methods

The preparation of reconstituted and porous hemoglobin-free human red cell ghosts, the experimental procedures, and analytical techniques used are the same as those previously described (3, 5, 6).

Results

(A) Binding of tritiated ouabain

Fig. 1 summarizes the effects of external K^+, $(K)_e$, and internal Na^+, $(Na)_i$, on the relative rates of binding of [^{3}H]-ouabain to reconstituted ghosts. The inhibition of ouabain binding is dependent, at high internal Mg^{2+}, on the combined presence of high $(Na)_i$, given a low concentration of internal K^+, and high $(K)_e$. On the other hand, the ouabain binding rates obtained at low internal Mg^{2+}, i. e. at a concentration of Mg^{2+} less than that of internal EDTA, indicate that the coupled action of $(Na)_i$ to $(K)_e$ is completely lost. While $(Na)_i$ no longer influences ouabain binding, there is still an inhibitory effect of $(K)_e$. Not included are data showing an increase of the ouabain binding rate by $(Na)_e$ in this situation.

(B) Release of tritiated ouabain

The release of [^{3}H]-ouabain in porous ghosts, in the presence of ATP, is slowed by Mg^{2+} (Fig. 2). The effect is half-maximal at about 1.5 mM-Mg^{2+}. In the absence of Mg^{2+}, the rate of ouabain release is increased by ATP. A half-maximal effect is obtained at about 0.1 mM-ATP (Fig. 3). ADP is about as effective as ATP while other nucleotides are ineffective (Fig. 3). Again, the rate of ouabain release is slowed by K^+. A half-maximal effect in the presence of 1 mM-ATP is exerted by about 1 mM-K^+ (results not shown). This stabilizing effect of K^+ is reduced by the addition of Na^+ (Fig. 4).

ATP and Mg^{2+} are known to act more or less exclusively at the inner aspect of the membrane. For evaluation of the side-dependency of the effect of K^+, further experiments have

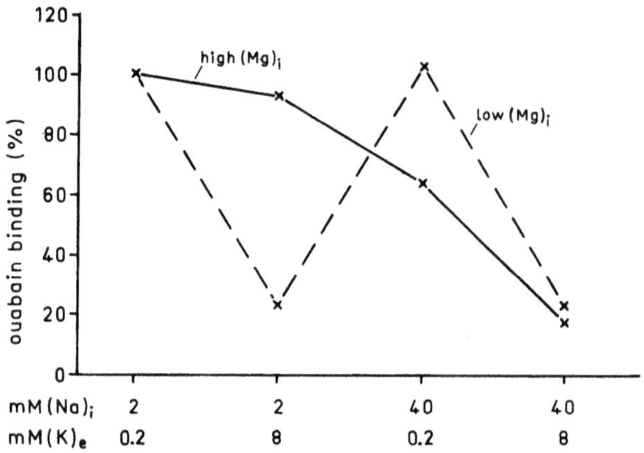

Fig. 1. Relative rates of ouabain binding to reconstituted red cell ghosts.

Reconstituted ghosts contained either 4 mM-$MgCl_2$ (high $(Mg)_i$) or 0.8 mM-$MgCl_2$ plus 1.5 mM-EDTA (low $(Mg)_i$) and in addition 2 mM-ATP, 4 mM-KCl, 7 mM-TrisCl, and 140 mM sum of (NaCl + choline Cl). Exposure to [^{3}H]-ouabain was carried out in a solution containing 20 mM-NaCl, 140 mM sum of (KCl + choline Cl) and 10 mM-TrisCl (pH 7.4). Incubation was carried out for 30 min at 37° C. The concentration of [^{3}H]-ouabain in the case of high $(Mg)_i$ was 5×10^{-8} M, for low $(Mg)_i$ it was 3×10^{-8} M. Further details are described elsewhere (5).

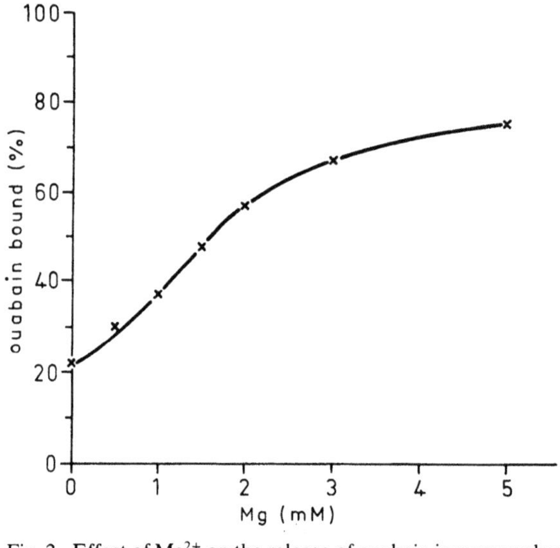

Fig. 2. Effect of Mg^{2+} on the release of ouabain in porous ghosts.

[^{3}H]-ouabain was bound to red blood cells as described previously (6). After the preparation of porous ghosts these were incubated for 60 min at 37°C in a solution containing (mM) 0.5 EDTA, 40 NaCl, 10 TrisCl (pH 7.4), 2 Na_2-ATP and the indicated concentrations of $MgCl_2$. [^{3}H]-ouabain bound was determined before (= 100%) and after incubation. The results were obtained by two experiments, each with duplicate analysis.

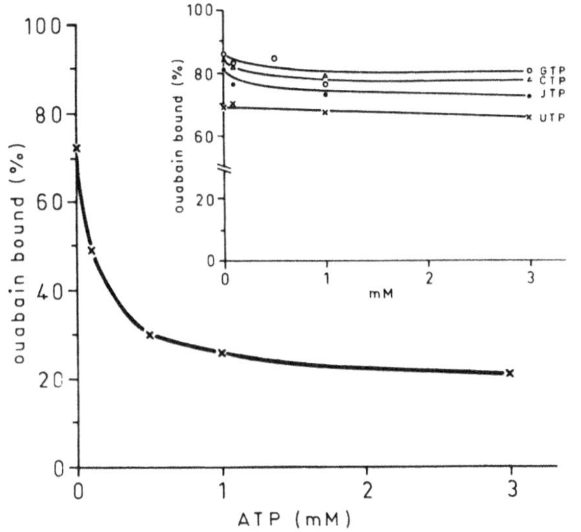

Fig. 3. Effects of nucleotides on the release of ouabain in porous ghosts in the absence of Mg^{2+}.
The experimental conditions were the same as described in the legend to Fig. 2. No Mg^{2+} was present. The nucleotide concentrations are indicated in the figure. The results of two experiments are shown.

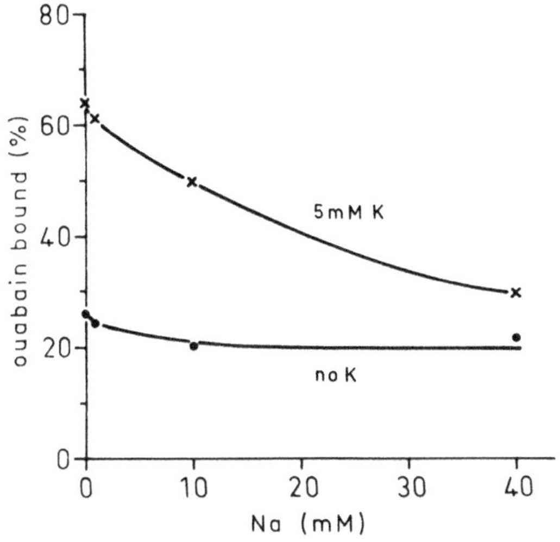

Fig. 4. Effect of K^+ and Na^+ on the release of ouabain in porous ghosts in the absence of Mg^{2+}.
The experimental design was the same as described in the legend to Fig. 2 with the following exceptions: incubation was performed for 90 min in a medium containing (mM) 0.4 EDTA, 50 (choline Cl + KCl + NaCl), 10 TrisCl (pH 7.4), 1 Tris-ATP. The results of two experiments are shown.

been performed using reconstituted ghosts. As shown in Table 1, the enzyme-ouabain complexes are stabilized by external K^+ but not by internal K^+. The addition of NaCl instead of choline chloride reduced this effect of external K^+ (results not shown).

Table 1. Effects of internal und external K^+ on the release of ouabain in reconstituted red blood cell ghosts. Binding of [^{3}H]-ouabain and preparation of reconstituted ghosts was performed as described previously (3, 6). These ghosts contained approx. (mM) 40 NaCl, 3 Na_2ATP, 1.5 $MgCl_2$, 5 TrisEDTA (pH 7.4), 7 TrisCl (pH 7.4) and 109 (choline Cl + KCl). In the case of variable extracellular K^+, the concentration of intracellular K^+ was 4 mM. In the case of variable internal K^+, the ghosts were incubated in (mM) 120 choline Cl, 40 NaCl, 0.5 KCl, 0.5 EDTA, 10 TrisCl (pH 7.4). In the case of variable external K^+, the medium contained (mM) 160 (cholineCl + KCl), 0.5 EDTA, 10 TrisCl (pH 7.4). [^{3}H] bound to the ghosts was determined in duplicates before (= 100 %) and after incubation for 90 min at 37° C. The data of two experiments are shown.

	(mM)	Release of [^{3}H]-ouabain (%)
Internal K^+	4	30
	14	35
	24	34
	44	37
	109	35
External K^+	0	38
	5	60
	20	67
	160	77

Discussion and conclusions

Ouabain binding occurs at the outer surface of the membrane with E_2P being the most relevant ouabain binding enzyme conformation (7, 10). Internal Mg^{2+} stimulates ouabain binding (5) and modifies the effects of the ligands Na^+ and K^+ on ouabain binding. Internal ATP, in this situation, is the only substrate to promote substantial ouabain binding and appears to counteract the inhibitory action of external K^+ on ouabain binding (4).

At high Mg^{2+}, the enzyme-ouabain complexes (type I complexes) are relatively stable and insensitive to actions of ligands like ATP, inorganic phosphate, Na^+ and K^+ (13). At low Mg^{2+}, the data of this contribution demonstrate an increased release of ouabain by adding ATP in agreement with previous reports (9, 12). Other nucleotides, with the exception of ADP, and inorganic phosphate are without effect (Fig. 3). In contrast, K^+ stabilizes the enzyme-ouabain complexes in accordance with results of Akera and Brody (1) and Repke et al. (11). Therefore, antagonistic effects of external K^+ (Table 1) and internal ATP may also regulate the rate of ouabain release at low Mg^{2+}.

In conclusion, the following reaction scheme is proposed: ouabain binds to E_2P (high affinity) at the outer surface with little release of ouabain after binding. E_2P is stabilized by internal Mg^{2+} and external Na^+ (2). At low Mg^{2+}, E_2-ouabain is stabilized by the occlusion of external K^+ as $E_2(K)$-ouabain. However, binding of ouabain to $E_2(K)$ is negligible. In this situation, ATP shifts $E_2(K)$-ouabain to E_1 with release of K^+ and ouabain (8) and stimulates ouabain binding via the formation of E_2P.

Acknowledgements

The technical assistance of Ms. Regine Baur is gratefully acknowledged. This study was supported by the Deutsche Forschungsgemeinschaft.

References

1. Akera T, Brody TM (1971) J Pharmacol Exp Ther 176: 545–547
2. Blostein R (1979) J Biol Chem 254: 6673–6677
3. Bodemann HH, Hoffman JF (1976) J Gen Physiol 67: 497–525
4. Bodemann HH, Hoffman JF (1976) J Gen Physiol 67: 527–545
5. Bodemann HH, Hoffman JF (1976) J Gen Physiol 67: 547–561
6. Bodemann HH (1985) In: Glynn I, Ellory C (eds) The Sodium Pump. The Company of Biologists Ltd Cambridge, pp 283–287
7. Cantley LC (1981) Curr Topics Bioenerg 11: 201–237
8. Forbush B (1983) Curr Top Membr Transp 19: 167–201
9. Huang W-H, Askari A (1975) Life Sci 16: 1253–1262
10. Post RL, Kume S, Tobin T, Orcutt B, Sen AK (1969) J Gen Physiol 54: 306s–326s
11. Repke KRH, Herrmann I, Portius HJ (1984) Biochem Pharmacol 33: 2089–2099
12. Tobin T, Akera T, Brody TM (1974) Ann NY Acad Sci 242: 120–131
13. Wallick ET, Pitts BJR, Lane LK, Schwartz A (1980) Arch Biochem Biophys 202: 442–449

Author's address:

Prof. Dr. H. Bodemann, Klinikum der Albert-Ludwigs-Universität, Medizinische Universitätsklinik, Hugstetter Straße 55, 7800 Freiburg i. Br. (West Germany)

The ouabain receptor in the myocardium and conduction system of the sheep heart

E. Hug, L. Brown and E. Erdmann

Medizinische Klinik I der Universität, Klinikum Großhadern, Munich (West Germany)

Summary

The ouabain receptor in the conducting system and the working myocardium of the digitalis-sensitive sheep heart has been characterized to determine the molecular basis for the increased digitalis sensitivity of the conducting system. ^{3}H-ouabain binding has the following characteristics:

(1) in sheep left ventricular cell membranes, specific binding was of high affinity (K_D 1.9 x 10^{-9} M at 37°); was coincident with an inhibition of (Na^+ + K^+)-ATPase activity; and was inhibited by unlabelled cardiotonic steroids; (2) in crude homogenates, the maximal binding capacity but not the affinity for ouabain varied in different parts of the sheep heart with Purkinje fibres containing markedly fewer binding sites (0.33 ± 0.1 x 10^{14}/g wet weight) than left ventricles (1.3 ± 0.1 x 10^{14} sites/g wet weight), and (3) in isolated, contracting Purkinje fibres and moderator band strips, concentration-response curves for ^{3}H-ouabain binding, increase in force of contraction and inhibition of $^{86}Rb^+$ uptake were coincident. In both contracting tissues, 3 x 10^{-7} M ouabain occupied about 50% of the specific binding sites, inhibited Na^+/K^+ transport by about 50% and gave approximately maximal inotropic effects. The maximal binding capacity was lower in contracting Purkinje fibres (2.0 x 10^{14} sites/g wet weight) than in moderator band strips (3.9 x 10^{14} sites/g wet weight). Maximal inotropy was reached faster in Purkinje fibres than in ventricular muscle as was toxicity. We conclude that the ouabain receptor mediating positive inotropy and inhibition of (Na^+ + K^+)-ATPase is identical in both the conducting system and working myocardium of the sheep heart. The reduced receptor number, rather than a change in affinity, may underlie the increased digitalis sensitivity of the Purkinje fibres.

Introduction

The conducting system of the heart has been reported to be more sensitive to the toxic effects of digitalis than the working myocardium (8, 12, 14, 17, 18). Differences between these tissues may be related to different properties or amounts of the digitalis receptor, the (Na^+ + K^+)-ATPase, in the tissues. This hypothesis has been tested in the present investigation by characterization of the ouabain receptor in Purkinje fibres and ventricular muscle of the digitalis-sensitive sheep heart using cell membrane preparations, crude homogenates and contracting heart tissues.

The semisynthetic digitalis derivative, 16-epi (α) gitoxin, has been reported to have an increased therapeutic index possibly due to an unusually flat curve for (Na^+ + K^+)-ATPase inhibition (8). This mechanism would explain the increased therapeutic index of 16-acetylgitoxin (12). This was tested by measuring the inhibition of ^{3}H-ouabain binding to homogenates of Purkinje fibres and ventricular muscle by 16 α-gitoxin.

Methods

Detailed descriptions of the measurement of ^{3}H-ouabain binding, ($Na^+ + K^+$)-ATPase activity and $^{86}Rb^+$-uptake have been given in earlier publications (1, 3, 5, 6).

Sheep hearts were obtained fresh from a local slaughter-house. Portions of about 60 g left ventricular tissue were frozen at −40°. Cardiac cell membranes enriched in ($Na^+ + K^+$)-ATPase were prepared using sodium deoxycholate and sodium iodide extraction as previously described (5). Homogenates were prepared from freshly dissected left or right ventricle, moderator band (right ventricular trabecula septomarginalis) or Purkinje fibres.

Free-running Purkinje fibres, usually from the left ventricle, and moderator band strips cut approximately parallel, were suspended in an organ bath (75 ml) in a modified Krebs-Henseleit solution (2) containing 2.5 mM $CaCl_2$ continuously gassed with 95 % O_2/5 % CO_2 at 35°. The tissues were field stimulated at 1 Hz for 5 msec at 10–20 % above threshold. Each contracting preparation was used for one ouabain concentration only which was added after an equilibration of at least 60 minutes. For the relevant experiments, ^{3}H-ouabain (about 1.8 x 10^7 cpm, about 1.2 x 10^{-8} M) was added together with unlabelled ouabain for 120 minutes while ^{86}RbCl (about 5 μCi, about 1 x 10^7 cpm) was added for 15 minutes after a stable inotropic effect had been achieved at each ouabain concentration. All values are given as the mean ± standard error of the mean. A p-value of less than 0.05 by Student's t-test was taken as the criterion of statistical significance.

Results

^{3}H-ouabain binding to sheep heart cell membranes and homogenates

In cell membranes prepared from sheep left ventricle, ^{3}H-ouabain bound to one site, as shown by straight Scatchard plots, with high affinity (K_D, 1.9 x 10^{-9} M at 37°). Human heart cell membranes have a lower affinity for ^{3}H-ouabain binding (K_D, 3.7 x 10^{-9} M (1)).

Simultaneous measurement of equilibrium ^{3}H-ouabain binding and ($Na^+ + K^+$)-ATPase activity (Fig. 1) showed a coincident increase in binding and enzyme inhibition. At a concentration of about 3 x 10^{-9} M ouabain, about 50 % of the specific binding sites were occupied while the enzyme activity was inhibited by about 50 %.

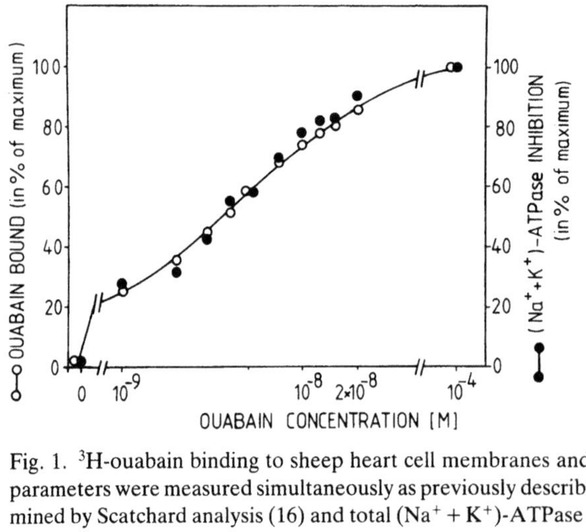

Fig. 1. ^{3}H-ouabain binding to sheep heart cell membranes and inhibition of ($Na^+ + K^+$)-ATPase. Both parameters were measured simultaneously as previously described (5). Maximal ^{3}H-ouabain bound determined by Scatchard analysis (16) and total ($Na^+ + K^+$)-ATPase activity were set at 100 %.

The number of receptors and their affinity was measured in homogenates of left and right ventricle, moderator bands and Purkinje fibres. The K_D-values were similar in all tissues but the number of binding sites in the Purkinje fibres (0.33 ± 0.1 x 10^{14} sites/g wet weight) was significantly lower than in the left ventricle (1.3 ± 0.1 x 10^{14} sites/g wet weight), right ventricle (0.86 ± 0.07 x 10^{14} sites/g wet weight) or moderator band (1.1 ± 0.2 x 10^{14} sites/g wet weight) (2).

The binding of ^{3}H-ouabain can be inhibited by other cardiotonic steroids. The concentration-response curves for unlabelled ouabain, dihydroouabain, gitaloxin, gitoxin and 16-epi (α)-gitoxin on homogenates of Purkinje fibres and left ventricle are given as Fig. 2. All curves are parallel suggesting an interaction of all substances with the same binding site. The concentrations which inhibited ^{3}H-ouabain binding by 50 % were similar in both tissues for all compounds. Using the method of Erdmann & Schoner (6), the K_D-value of the unlabelled derivatives can be calculated from the concentration inhibiting ^{3}H-ouabain binding by 50 %. A comparison of K_D-values for 14 cardiotonic steroids on sheep and human heart cell membranes is given in Fig. 3. There is a good correlation (r > 0.99), with sheep heart showing approximately 1.9 times higher affinity.

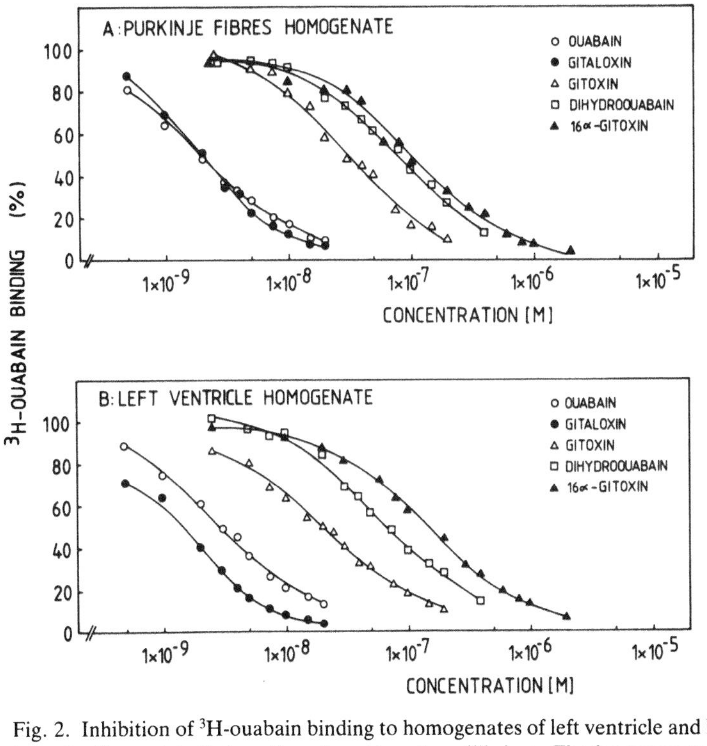

Fig. 2. Inhibition of ^{3}H-ouabain binding to homogenates of left ventricle and Purkinje fibres from sheep heart by five unlabelled cardiac glycosides at equilibrium. The homogenates were incubated in 3 mM $MgCl_2$, 3 mM PO_4/imidazole (pH 7.25), about 1.1 x 10^{-9} M ^{3}H-ouabain and increasing concentrations of the cardiac glycosides for 5 hours at 37°. Bound ^{3}H-ouabain was separated from free ^{3}H-ouabain by filtration through Millipore GF/C glass fibre filters.

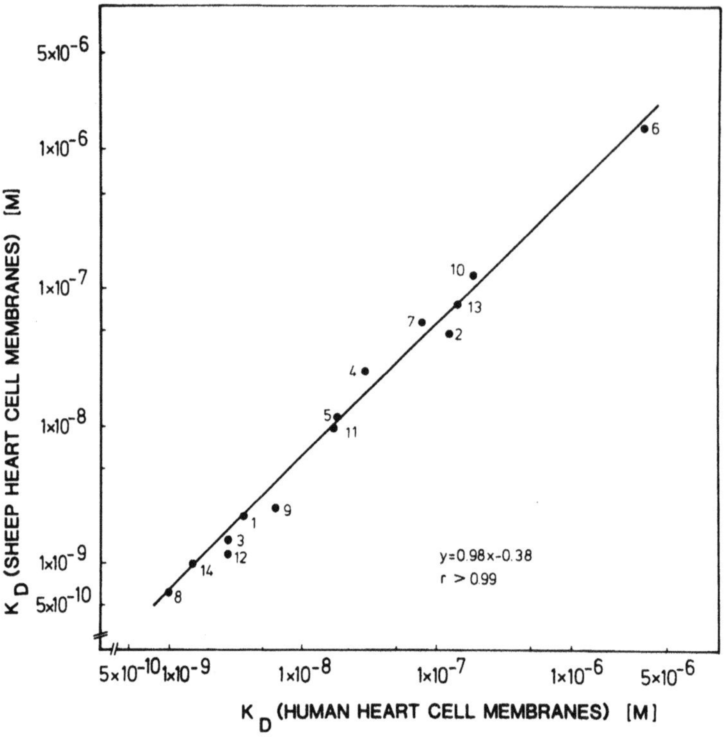

Fig. 3. K_D-values with sheep and human heart cell membranes. These values were calculated from the concentration of unlabelled drug which inhibited ^{3}H-ouabain binding by 50%. The curve was calculated from a least squares linear regression analysis. The compounds were:
1 – ouabain, 2 – dihydroouabain, 3 – digitoxin, 4 – dihydrodigitoxin, 5 – digitoxigenin, 6 – anhydrodigitoxigenin, 7 – 3 α-methyldigitoxigenin glucoside, 8 – digitoxigenin thevetoside, 9 – digoxin, 10 – dihydrodigoxin, 11 – gitoxin, 12 – 16-acetylgitoxin, 13 – pentaacetylgitoxin, 14 – gitaloxin.

^{3}H-ouabain binding to contracting sheep muscle

Isolated, contracting sheep heart Purkinje fibres and moderator band strips were used to simultaneously measure ^{3}H-ouabain binding and its effects on force of contraction and active Na^+/K^+-transport. The Scatchard plots of ^{3}H-ouabain binding to both contracting tissues are given in Fig. 4. Both tissues gave straight Scatchard plots for one site of similar affinity: Purkinje fibres, K_D 3.4 x 10^{-7} M; moderator band strips, K_D 2.5 x 10^{-7} M (2).

The maximal binding capacity was much lower on contracting Purkinje fibres (about 2.0 x 10^{14} sites/g wet weight) than in moderator band strips (about 3.9 x 10^{14} sites/g wet weight).

In both cardiac tissues, ouabain gave positive inotropic effects without toxicity at concentrations from 2 x 10^{-8} M to 2 x 10^{-7} M (Fig. 5). Half maximal inotropic effects were measured at about 8 x 10^{-8} M ouabain in both tissues. The maximal inotropic effect at 3 x 10^{-7} M was not stable, especially in Purkinje fibres, being followed by a gradual decline in developed force without an increase in the diastolic force. Higher ouabain concentrations gave similar effects in both tissues – decreases in developed force accompanied by a slow-developing contracture with variable periods of ectopic beats. At all concentrations, the

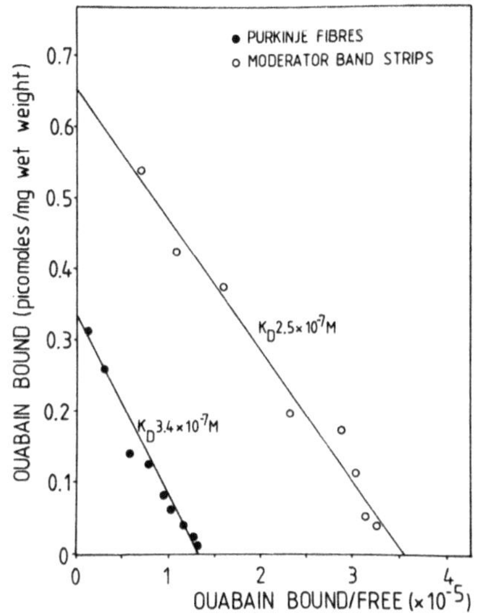

Fig. 4. ^{3}H-ouabain binding to contracting Purkinje fibres and moderator band strips from sheep heart. Experiments were performed as described in Methods and in (2). Data have been plotted according to Scatchard (16) with non-specific binding being defined as that in the presence of 1 x 10^{-4} M ouabain. The force of contraction measurements from these contracting tissues are included in Fig. 5.

time to maximal positive inotropic effects and onset of toxicity was lower in Purkinje fibres although this difference did not always reach statistical significance.

The Na^+/K^+-transport activity was measured using $^{86}Rb^+$-uptake for 15 minutes after a stable inotropic effect had been achieved at each ouabain concentration. Non-specific uptake unrelated to $(Na^+ + K^+)$-ATPase was defined as that in the presence of 1 x 10^{-4} M

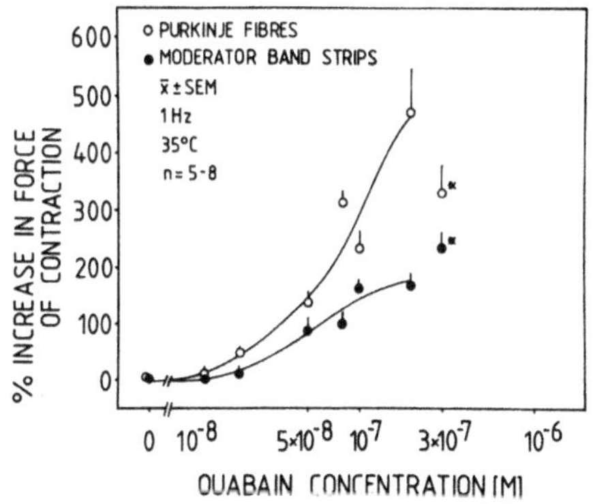

Fig. 5. Ouabain concentration-response curves on contracting sheep heart Purkinje fibres and moderator band strips. The inotropic effects of ouabain were measured as described in Methods. The Purkinje fibres had an average wet weight of 5.6 ± 0.9 mg; the moderator band strips weighed 8.2 ± 1.3 mg. At a concentration of 3 x 10^{-7} M ouabain, initial signs of toxicity were observed, especially in the Purkinje fibres (indicated by asterisk).

ouabain. The maximal specific uptake of $^{86}RbCl$ was 2.7 μmol/g wet weight/15 min in Purkinje fibres and 4.4 μmol/g wet weight/15 min in moderator band strips. Ouabain inhibited the $^{86}Rb^+$-uptake in both contracting Purkinje fibres and moderator band strips over the same concentration range (Fig. 6); at ouabain concentrations of about 3 x 10^{-7} M, specific $^{86}Rb^+$-uptake was inhibited by about 50 % in both tissues.

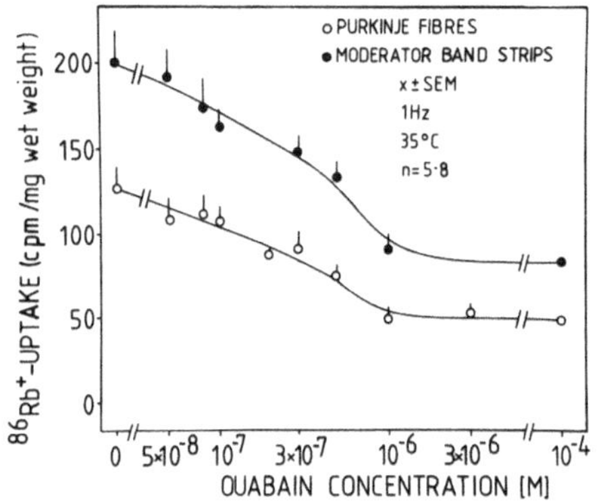

Fig. 6. Concentration-response curves for ouabain inhibition of $^{86}Rb^+$-uptake into contracting Purkinje fibres and moderator band strips from sheep heart. The uptake of $^{86}Rb^+$ was measured as described in Methods and in (2). The IC_{50} values for inhibition of active $^{86}Rb^+$-uptake were: Purkinje fibres, 2.5 x 10^{-7} M; moderator band strips, 3.3 x 10^{-7} M. The force of contraction values measured in these tissues have been given in Fig. 5.

Discussion

The increased ouabain sensitivity reported for Purkinje fibres relative to the working myocardium (8, 12, 14, 17, 18) may be explained by either qualitative or quantitative differences in the ouabain receptor present in these tissues. We have therefore characterized this receptor in contracting cardiac preparations and cardiac cell membranes from the digitalis-sensitive sheep.

In cardiac cell membranes, the specific ouabain binding site can be defined as the receptor mediating ($Na^+ + K^+$)-ATPase inhibition since concentration-response curves for binding and inhibition were coincident (Fig. 1). ^3H-Ouabain binding to contracting Purkinje fibres and moderator band strips (Fig. 4) was closely related to the ouabain-induced positive inotropy (Fig. 5) and inhibition of active Na^+/K^+-transport (Fig. 6). Ouabain concentrations occupying 50 % of the receptors gave almost maximal positive inotropic effects without marked toxicity and inhibited active Na^+/K^+-transport by about 50 %. The specific ouabain binding site can then be defined as the inotropic receptor which is part of the ($Na^+ + K^+$)-ATPase, the enzyme responsible for active Na^+/K^+-transport.

Further, the affinity of this receptor is the same in both the conducting system and ventricular muscle. However, the number of ouabain receptors varies considerably, with the Purkinje fibres having markedly fewer receptors than moderator band strips or ventricular muscle. These receptors are functional as shown by the reduced maximal $^{86}Rb^+$-uptake in Purkinje fibres. This decreased receptor capacity in Purkinje fibres may explain the increased digitalis toxicity in the conducting system. A decrease in the "receptor reserve" will lead to an earlier overload of the Na^+/K^+-transport capacity and thus an increase in intracellular Na^+ and Ca^{2+} above the toxicity threshold before such effects are noted in the working myocardium.

The semisynthetic derivatives, 16-acetyl and 16 α-gitoxin, have been proposed to have an increased therapeutic index because of a weaker interaction with the conducting system (8, 12). In the present study, 16 α-gitoxin was considerably less potent than its naturally occurring isomer, gitoxin, but the ^{3}H-ouabain binding inhibition curves were parallel and similar in both Purkinje fibre and left ventricle homogenates. If this compound binds specifically to the digitalis receptor, then any improvement in therapeutic index probably reflects differences in drug distribution within the heart.

Variable results have been presented from biochemical and pharmacological studies of ouabain effects on working myocardium and Purkinje fibres. Vassalle et al. (18) and Nowak & Haustein (12) showed that ouabain-induced toxicity occurred earlier in the Purkinje fibres. Isolated ventricular muscles were less prone to the development of toxicity, although arrhythmias could be induced (7).

Ventricular muscle is also less susceptible to cardiac glycoside-induced changes in action potential duration (18) and inhibition of ion transport (14). Several electrophysiological studies have shown qualitatively similar frequency-dependent changes in sodium activity in both tissues (4, 9). Resting sodium and calcium ion activities may be lower in ventricular muscle (10). Since increases in the sodium ion activity are closely related to the positive inotropic and toxic effects of ouabain (11, 19), then the increased resting values in Purkinje fibres may underlie the reported increase in toxicity. No significant difference could be found in the (Na^+ + K^+)-ATPase sensitivity to ouabain of beef heart Purkinje fibres and ventricular muscle but the enzyme yield was lower for the Purkinje fibres (13). The same ouabain concentration (6×10^{-7} M) inhibited $^{86}Rb^+$-uptake by 50% in both tissues from the dog heart (15). In contrast, Somberg et al. (17) showed a threefold higher ouabain sensitivity for inhibition of $^{86}Rb^+$-uptake as well as a threefold higher uptake in dog heart Purkinje fibres. However, these uptake studies were performed on tissue slices; any correlation between receptor occupancy and function can only be made using simultaneous measurement of these parameters under identical conditions, as in the present study.

Our studies indicate that the specific ouabain binding site is identical in all portions of the sheep heart. This binding site can be defined as the receptor mediating positive inotropic effects and inhibition of (Na^+ + K^+)-ATPase activity in contracting muscle and cell membranes from the sheep heart. The conducting tissue contains fewer receptors relative to its weight and surface area than the ventricular muscle. This reduction in receptor density, rather than any alteration in receptor affinity, may explain the increased toxicity of the Purkinje fibres for ouabain.

Acknowledgements

These studies were supported by the Deutsche Forschungsgemeinschaft (Er 65/4–4). We would like to gratefully acknowledge the excellent technical assistance of Miss Anita Reichl.

References:

1. Brown L, Erdmann E (1984) Binding of digitalis derivatives to beef, cat and human cardiac ($Na^+ + K^+$)-ATPase. Affinity and kinetic constants. Archs Int Pharmacodyn Ther 271: 229–240
2. Brown L, Hug E, Wagner G, Erdmann E (1985) Comparison of ouabain receptors in sheep myocardium and Purkinje fibres. Biochem Pharmacol 34: 3701–3710
3. Brown L, Werdan K, Erdmann E (1983) Consequences of specific ^{3}H-ouabain binding to guinea-pig left atria and cardiac cell membranes. Biochem Pharmacol 32: 423–435

4. Cohen CJ, Fozzard HA, Sheu S-S (1982) Increase in intracellular sodium ion activity during stimulation in mammalian cardiac muscle. Circ Res 50: 651–662
5. Erdmann E, Philipp G, Scholz H (1980) Cardiac glycoside receptor, $(Na^+ + K^+)$-ATPase activity, and force of contraction in rat heart. Biochem Pharmacol 29: 3219–3229
6. Erdmann E, Schoner W (1974) Ouabain-receptor interactions in $(Na^+ + K^+)$-ATPase preparations. IV. The molecular structure of different cardioactive steroids and other substances and their affinity to the glycoside receptor. Naunyn-Schmiedeberg's Arch Pharmacol 283: 335–356
7. Ferrier GR (1976) The effects of tension on acetylstrophanthidin-induced transient depolarizations and after-contractions in canine myocardial and Purkinje tissues. Circ Res 38: 156–162
8. Haustein K-O, Markwardt F, Repke KRH (1970) Different relationships between therapeutic and toxic actions of 16-epi-gitoxin, gitoxin, and ouabain on isolated cardiac preparations. Europ J Pharmacol 10: 1–10
9. January CT, Fozzard HA (1984) The effects of membrane potential, extracellular potassium, and tetrodotoxin on the intracellular sodium ion activity of sheep cardiac muscle. Circ Res 54: 652–665
10. Lado MG, Sheu S-S, Fozzard HA (1984) Effects of tonicity on tension, intracellular sodium and calcium activities in sheep heart. Circ Res 54: 576–585
11. Lee CO, Kang DH, Sokol JH, Lee KS (1980) Relation between intracellular Na-ion activity and tension of sheep cardiac Purkinje fibres exposed to dihydro-ouabain. Biophys J 29: 315–330
12. Nowak G, Haustein K-O (1976) Different toxic effects of ouabain and 16-epi-gitoxin on Purkinje fibres and ventricular muscle fibres. Pharmacology 14: 256–264
13. Palfi FJ, Besch HR jr, Watanabe AM (1978) Ouabain sensitivity of the $(Na^+ + K^+)$-ATPase activity from single bovine cardiac Purkinje fiber and adjacent papillary muscle. J Mol Cell Cardiol 10: 1149–1155
14. Polimeni PI, Vassalle M (1971) On the mechanism of ouabain toxicity in Purkinje and ventricular muscle fibres at rest and during activity. Am J Cardiol 27: 622–629
15. Rhee HM (1982) Ouabain sensitivity of Rb^+-uptake in canine Purkinje fibre and ventricular muscle. Naunyn-Schmiedeberg's Arch Pharmacol 318: 344–348
16. Scatchard G (1949) The attraction of proteins for small molecules and ions. Ann N Y Acad Sci 51: 660–672
17. Somberg JC, Barry WH, Smith TW (1981) Differing sensitivities of Purkinje fibres and myocardium to inhibition of monovalent cation transport by digitalis. J Clin Invest 67: 116–123
18. Vassalle M, Karis J, Hoffman BF (1962) Toxic effects of ouabain on Purkinje fibres and ventricular muscle fibres. Am J Physiol 203, 433–439
19. Wasserstrom JA, Schwartz DJ, Fozzard HA (1983) Relation between intracellular sodium and twitch tension in sheep cardiac Purkinje strands exposed to cardiac glycosides. Circ Res 52: 697–705

Author's address:

Prof. Dr. Erland Erdmann, Medizinische Klinik I der Universität München, Klinikum Großhadern, Marchioninistraße 15, 8000 München 70 (West Germany)

Electrophysiological effects of cardiac glycosides

D. A. Eisner*, M. Valdeolmillos*, W. J. Lederer+, and M. B. Cannell+

* Department of Physiology, University College London (England)
+ Department of Physiology, University of Maryland School of Medicine, Baltimore, Maryland (U.S.A.)

Summary

In this paper we review the effects of cardiac glycosides on the electrophysiological properties of mammalian cardiac muscle. Cardiac glycosides produce an initial prolongation of the action potential due to inhibition of the electrogenic Na-K pump current. This is then followed by a gradual decrease of action potential duration to less than control length. The origin of this decrease is uncertain. Cardiac glycosides also produce a transient depolarization which follows the action potential and which can produce a spontaneous action potential. This transient depolarization is produced by a transient inward current which results from a Ca-activated conductance. The triggering increase of $[Ca^{2+}]_i$ originates from the spontaneous release of Ca ions from the sarcoplasmic reticulum (s. r.). Both this spontaneous release and the resulting transient inward current can be abolished by inhibitors of s. r. function such as caffeine and ryanodine. As well as these oscillations of $[Ca^{2+}]_i$ which follow repolarization, there are also spontaneous oscillations in the absence of stimulation. These spontaneous oscillations, which also result from release of Ca from the sarcoplasmic reticulum, interfere with the systolic rise of Ca produced by stimulation. Thus the Ca oscillations are not only responsible for the cardiac arrhythmias produced by digitalis but also compromise the magnitude of the positive inotropic effect.

The effects of digitalis on the action potential

One of William Withering's principal contributions to the therapeutic use of digitalis was to lay down careful guidelines for its use and, in particular, to stress the importance of using the correct dosage. This is particularly important as digitalis has a very narrow therapeutic range. Of paramount importance is the fact that digitalis produces various cardiac arrhythmias. In this paper we will review the cellular mechanism of these toxic effects.

Fig. 1 shows the effects of the cardiac aglycone strophanthidin on the action potential recorded from an isolated calf Purkinje fibre. The control trace (0) shows a characteristic Purkinje fibre action potential. Trace 1 which was recorded 5 min after adding strophanthidin (1 μM) shows a marked prolongation of the action potential. This effect is probably due to the inhibition of an electrogenic Na-K pump current (18). This outward current will contribute to repolarization of the action potential. The prolongation of the action potential is only transient and trace 2 shows that, after 18 min exposure to strophanthidin, the action potential is considerably shorter than control. The mechanism of this shortening of the action potential is still controversial. One suggestion (17) is that it is produced by a Ca-activated potassium conductance which will be activated by the increase of systolic Ca produced by cardiac glycosides (1). However, although such Ca-activated K channels have been characterized in non-cardiac preparations, their existence in the heart is much more equivocal (11). Indeed a recent study has failed to find any evidence for them (27). Two

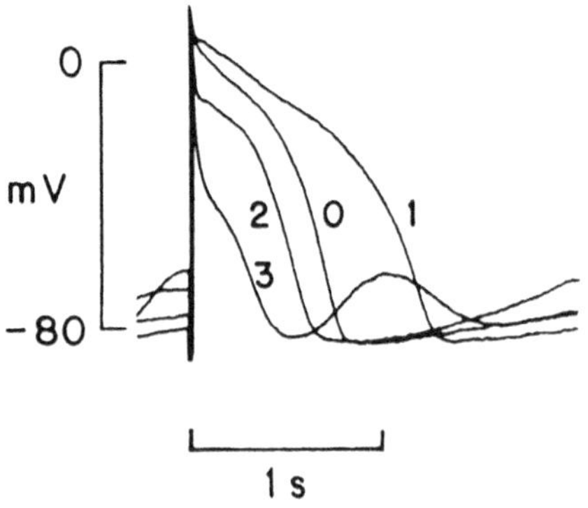

Fig. 1. The effects of strophanthidin on the action potential of a calf Purkinje fibre. The records show superimposed action potentials recorded during stimulation at 0.5 Hz. The trace labelled (0) is a control. The others were obtained at the following times after adding strophanthidin (1 μM): (1), 5 min; (2), 18 min; (3), 57 min. Taken from Lederer (25).

other explanations have been suggested by recent patch clamp work. (i) Kameyama et al. (20) have demonstrated a Na-activated K channel. This channel was not activated at normal $[Na^+]_i$ (10 mM) and was half maximally activated by about 70 mM $[Na^+]_i$. It may therefore be activated to some extent by the increase of $[Na^+]_i$ produced by toxic concentrations of cardiac glycosides. (ii) Another explanation is provided by the demonstration of an ATP-inhibited K channel. It has long been known that toxic concentrations of cardiac glycosides decrease $[ATP]_i$ (35) perhaps by elevating $[Ca^{2+}]_i$. In the patch clamp studies these K channels remained closed until ATP was reduced to less than 1 mM (19). It is not yet known whether ATP would have fallen to sufficiently low levels to open this channel.

Whatever its cause, this shortening of the action potential will have serious consequences for the heart. This is because a shortening of the action potential will decrease the refractory period and therefore make the heart more vulnerable to premature excitation. This can result initially in the production of extrasystoles and, eventually, in ventricular arrhythmias and fibrillation.

The final trace numbered 3 in Fig. 1, shows that, in addition to shortening the action potential, cardiac glycosides have other untoward effects. Following the action potential, a transient depolarization is produced. Although not shown, this transient depolarization can become large enough to produce a spontaneous action potential and thence an extrasystole. The appearance of these transient depolarizations has been correlated with the onset of toxicity as revealed by the electrocardiogram (31).

The cellular basis of the transient depolarization was investigated in voltage-clamp studies on isolated Purkinje fibres (22, 24, 26). A typical result is shown in Fig. 2. This shows the effects of a repolarizing voltage clamp pulse in a fibre which had been exposed to a toxic concentration of strophanthidin. Repolarization produces a transient inward current. If the fibre had not been voltage-clamped this would have produced a transient depolarization similar to that seen in Fig. 1. This transient inward current is accompanied by (and slightly precedes) an aftercontraction. The correlation between current and tension is emphasized by the bottom trace of Fig. 2 which shows that, if the tension trace is advanced by 80 msec and inverted, it has a similar time course to that of current. Since the aftercontraction presumably indicates an increase of $[Ca^{2+}]_i$, it was suggested that the transient inward current is accompanied by an increase of $[Ca^{2+}]_i$. Direct evidence of this correlation between mem-

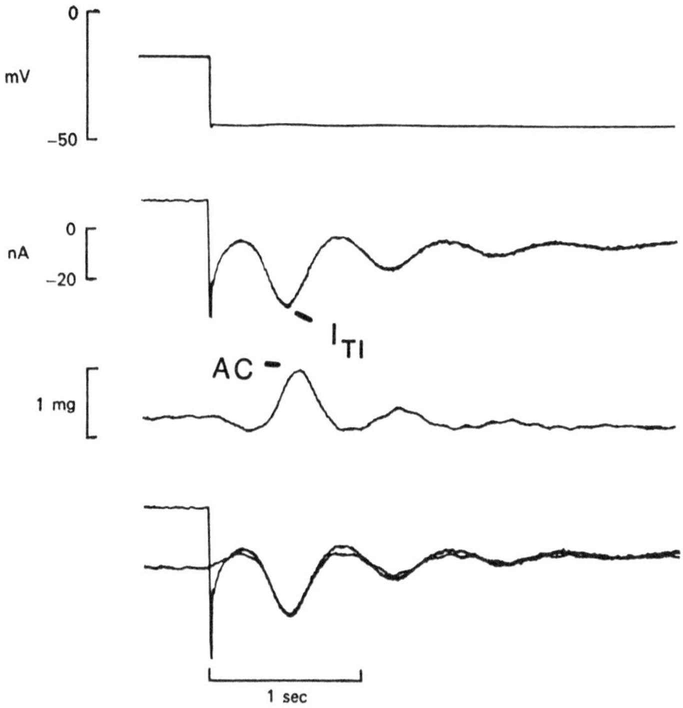

Fig. 2. Transient inward current (I_{TI}) recorded from a calf Purkinje fibre. The record shows the effects of a repolarizing voltage clamp pulse in the presence of strophanthidin (10 μM). The top three traces show: membrane potential; current; tension. The bottom trace shows the result of inverting the tension trace, advancing it by 80 ms, and then superimposing it on the current trace. Taken from Kass et al. (22).

brane current and $[Ca^{2+}]_i$ is provided by the experiment of Fig. 3. Here the photoprotein aequorin was used to measure $[Ca^{2+}]_i$. This figure shows that the transient inward current is, indeed, accompanied by an oscillatory increase of $[Ca^{2+}]_i$. The peak of the rise of $[Ca^{2+}]_i$ occurs at the same time as the maximum inward current. The fact that tension lags behind current (Fig. 2) is presumably due to the fact that tension development requires many steps following an increase of $[Ca^{2+}]_i$.

The correlation between current and $[Ca^{2+}]_i$ could be produced by two mechanisms. (i) The transient inward current carries Ca ions into the cell and thereby increases $[Ca^{2+}]_i$; (ii) The increase of $[Ca^{2+}]_i$ is produced by some other mechanism and this increase of $[Ca^{2+}]_i$ activates the transient inward current. The latter alternative was favoured by Kass et al. on the basis of various indirect experiments. In support of this interpretation, Kass & Tsien (23) have shown that intracellular injection of the Ca chelator EGTA abolishes the transient inward current confirming that the transient inward current is indeed Ca-activated.

Which ions carry the transient inward current?

Given that the transient inward current is Ca-activated it is of interest to investigate the ionic nature of the pathway which is activated by an increase of $[Ca^{2+}]_i$. Kass et al. (22, 23) measured the equilibrium potential of the transient inward current and found that it was at about 0 mV and was sensitive to changes of extracellular Na concentration. It was therefore

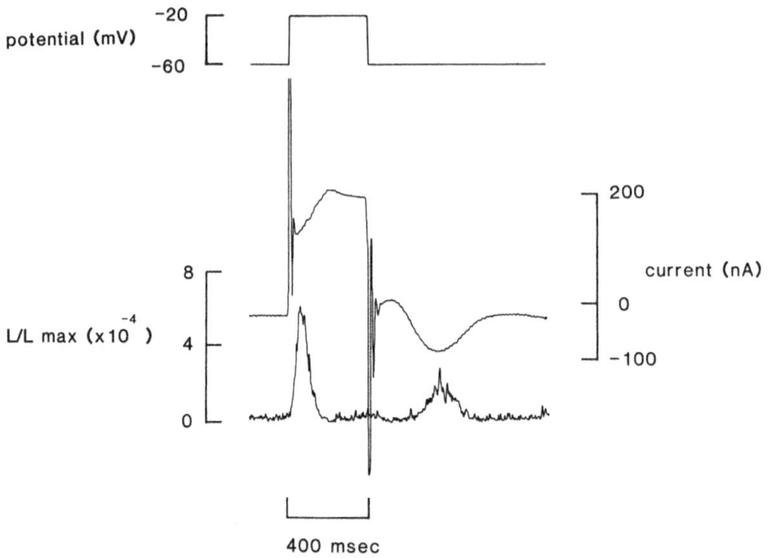

Fig. 3. Simultaneous measurement of transient inward current and $[Ca^{2+}]_i$ in a sheep cardiac Purkinje fibre. The record shows the effects of a depoarizing voltage-clamp pulse in the presence of stropanthidin (10 μM). Traces show (from top to bottom): membrane potential; membrane current; aequorin light emission (a measure of $[Ca^{2+}]_i$). The records are the average of 16 pulses. Unpublished record of Eisner & Valdeolmillos.

suggested that the current was carried partly by Na ions entering the cell. The fact that the reversal potential was 0 rather than the Na equilibrium potential of about + 60 mV was accounted for by suggesting that the transient inward current was produced by a Ca-activated increase of conductance to both Na & K ions. This hypothesis gained support from the demonstration of a Ca-activated nonspecific conductance in isolated patches of membrane (6). However, no reversal potential could be found in either ventricular muscle (3, 21) or sino atrial node (14). This work suggested that the transient inward current might be produced by an electrogenic Na-Ca exchange. On this model an increase of $[Ca^{2+}]_i$ would increase Ca efflux and Na entry on the exchange. If the exchange transports more than two Na^+ ions per Ca^{2+} then this would produce a current of appropriate polarity.

One must also question whether the demonstration of a reversal potential (22, 24) excludes the possibility that the transient inward current is entirely produced by an electrogenic Na-Ca exchange. It is generally assumed that an increase of $[Ca^{2+}]_i$ will increase the net inward current or decrease the net outward current. However, there is evidence that an increase of $[Ca^{2+}]_i$ can, for unknown reasons, increase the Ca influx in exchange for Na efflux (7). Such an effect would mean that, at some potentials, increasing $[Ca^{2+}]_i$ will increase outward current. This is illustrated in Fig. 4 which shows the current voltage relationships expected for a Na-Ca exchange. Fig. 4A shows a simple scheme. Here increasing $[Ca^{2+}]_i$ from 0.1 to 0.3 μM always shifts the current in the inward direction. In contrast, in the model of Fig. 4B increasing $[Ca^{2+}]_i$ is also assumed to have a "catalytic" role and to increase both Ca influx and efflux. This results in a clear reversal potential at which increasing $[Ca^{2+}]_i$ has no effect on current. Therefore the demonstration of a reversal potential for the transient inward current does not exclude the possibility of the current being produced by a Na-Ca exchange.

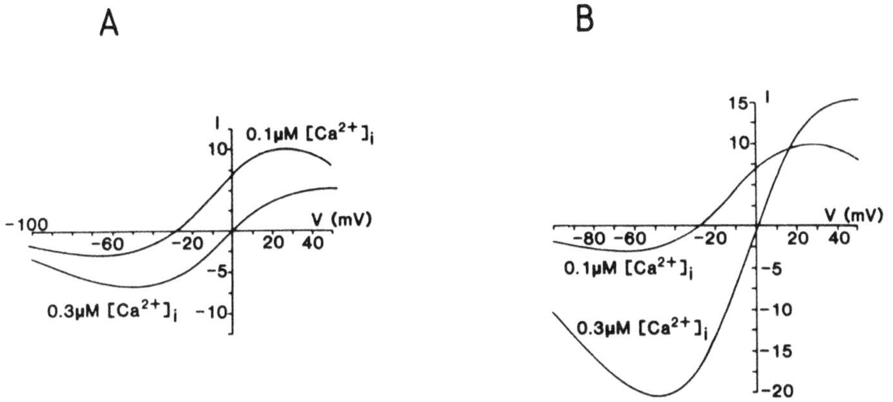

Fig. 4. A model of the effects of Na-Ca exchange on membrane current. *A*. Conventional model. In this scheme the Na-Ca exchange is represented by a carrier which can cross the membrane either with $3Na^+$ or $1Ca^{2+}$ bound but cannot cross in the unloaded form. The traces show the calculated effects of membrane potential on the current produced by a $3Na^+/1Ca^{2+}$ Na-Ca exchange. The two curves were calculated at 0.1 and 0.3 μM $[Ca^{2+}]_i$. Positive current denotes net Na efflux/Ca influx while negative current is net Ca efflux/Na influx. *B*. The predictions of a more complicated model. The model is similar to that in A except that elevating $[Ca^{2+}]_i$ increases the fluxes in both directions. For further details see Eisner & Lederer (8).

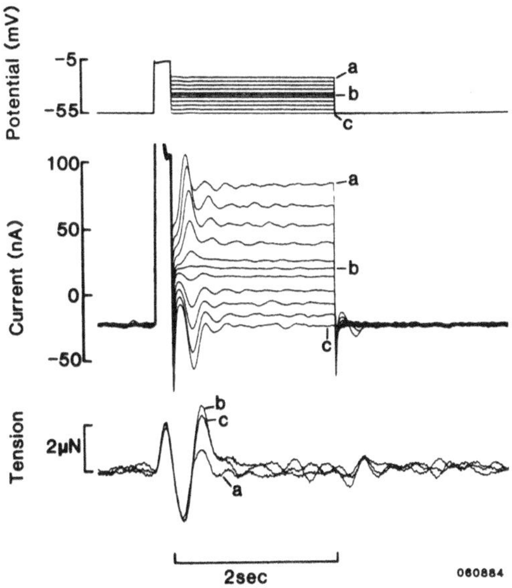

Fig. 5. Determination of the reversal potential of the transient inward current. Traces show (from top to bottom): membrane potential; current; tension. Each trace shows superimposed records. All records have a conditioning depolarization to −5 mV followed by a repolarization to various test potentials. This repolarization produces a transient outward current at the most positive potentials, no current at intermediate potentials and a transient inward current at more negative potentials. From Cannell & Lederer (5).

Cannell and Lederer (5) have recently used a different approach to examine the question of the ionic nature of the transient inward current. Fig. 5 shows an experiment to examine the reversal potential. In this experiment all external Na was replaced by Ca. Under these conditions the Na-Ca exchange will be eliminated. In this experiment a conditioning depolarization was applied to −5 mV and then the membrane potential was stepped to various levels. It is clear that, at sufficiently negative potentials, a transient inward current is produced whereas, at more positive potentials, a transient outward current is seen. A clear reversal potential is observed at intermediate voltages. This experiment therefore provides clear evidence that a Ca-activated conductance contributes to the transient inward current. It does not, however, exclude the possibility that a Na-Ca exchange may also contribute under normal conditions. In this experiment the inward current must have been carried by Ca^{2+} ions therefore showing that the channel is to some extent permeable to this ion. However, at physiological $[Ca^{2+}]_o$ the fraction of current carried by Ca^{2+} may be very small.

The source of the oscillations of $[Ca^{2+}]_i$

We have seen above that the transient inward current is produced by an oscillation of $[Ca^{2+}]_i$. There is now much evidence to support the idea that these oscillations result from oscillatory Ca release from the sarcoplasmic reticulum. Spontaneous force oscillations can

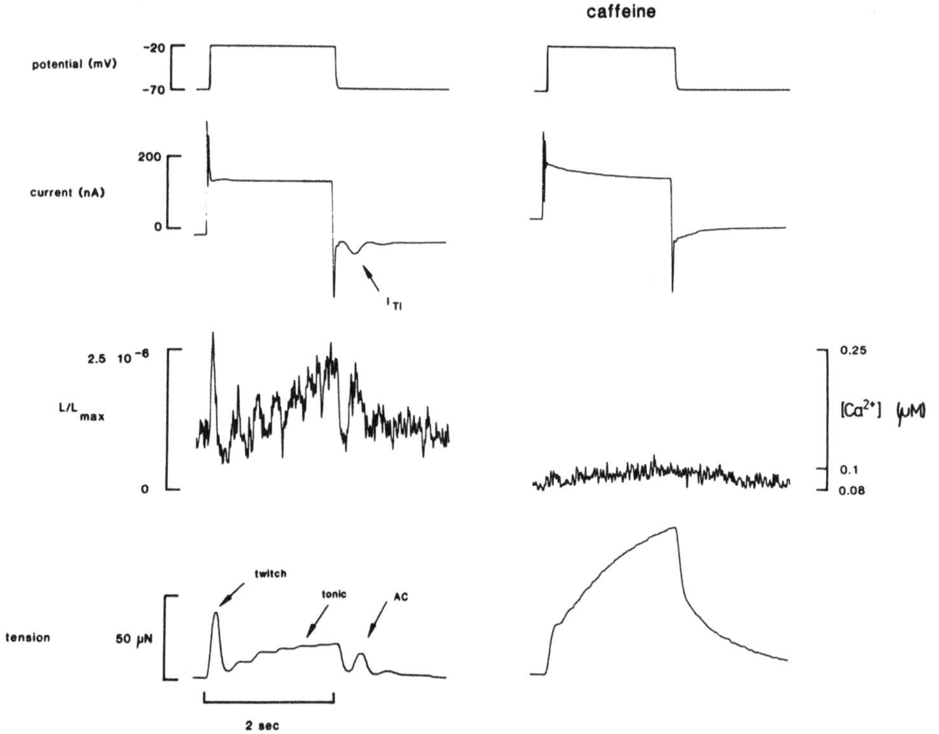

Fig. 6. The effects of caffeine on the transient inward current and oscillations of $[Ca^{2+}]_i$. In both panels traces show signal-averaged (n = 64) records of (from top to bottom): membrane potential; current; aequorin light; tension. The preparation had been exposed to strophanthidin (10 μM). Panels show: left, control; right, after 4 min exposure to 10 mM caffeine. Taken from Eisner & Valdeolmillos (10)

be seen in cardiac muscle cells which have been skinned to remove their surface membrane (12). Furthermore in intact preparations both the oscillations of force and current and the underlying Ca oscillations can be abolished by inhibitors of Ca release from the sarcoplasmic reticulum. An example of this is shown in Fig. 6 which shows records of current, aequorin light and force obtained after prolonged exposure to strophanthidin. A transient inward current and oscillations of force and $[Ca^{2+}]_i$ are apparent. The application of caffeine abolished all these oscillations.

Spontaneous oscillations of $[Ca^{2+}]_i$

The oscillations of $[Ca^{2+}]_i$ are particularly prominent following stimulation. There is, however, evidence that oscillations of $[Ca^{2+}]_i$ can be seen in unstimulated preparations. For example when $[Ca^{2+}]_i$ is elevated spontaneous oscillations of force can be measured (16, 23) and, more recently, spontaneous oscillations of $[Ca^{2+}]_i$ have also been demonstrated under these conditions (30, 34). One problem with studying these spontaneous oscillations is that they are unsynchronized between the different cells in a multicellular preparation. Furthermore they may also not be synchronized even within one cell (4). The former problem has been overcome by injecting aequorin into only one cell. A typical record is shown in Fig. 7. Fig. 7A shows a filtered record of aequorin light from this cell. The application of strophanthidin produced a slow increase of $[Ca^{2+}]_i$. The records in Fig. 7B show specimen traces at faster temporal resolution. It can be seen that the control aequorin light record is flat. However, oscillations of $[Ca^{2+}]_i$ develop during exposure to strophanthidin. During this period the frequency of the oscillations increases (from about 0.7 Hz in b to 3 Hz in d) and their magnitude also increases. These spontaneous oscillations of $[Ca^{2+}]_i$, like those following stimulation, can be abolished by inhibitors of the s. r. (9, 34)

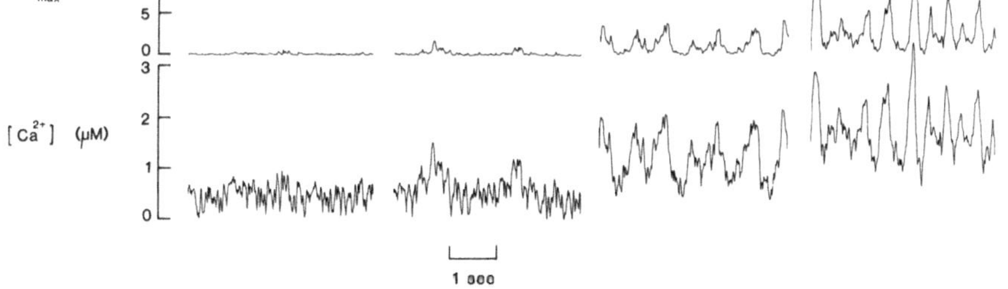

Fig. 7. Measurement of spontaneous oscillations of $[Ca^{2+}]_i$ in one cell in a sheep cardiac Purkinje fibre. *A*. Filtered (DC−0.4 Hz) record of aequorin light. Strophanthidin (10 μM) was applied for the period shown above. *B*. Specimen records obtained at higher temporal resolution at the points shown in A. In each panel the top trace shows aequorin light and the bottom the result of digitally converting the aequorin signal to $[Ca^{2+}]_i$. Taken from Eisner & Valdeolmillos (10a).

The effects of oscillations of $[Ca^{2+}]_i$ on force

The above results have shown that oscillations of $[Ca^{2+}]_i$ can produce some cardiac arrhythmias. More recent work has shown that they also compromise the ability of cardiac muscle to contract. Fig. 8 shows the effects of stimulating a ferret papillary muscle in the presence of Ca oscillations. Each time the preparation is stimulated there is a Ca transient. However, the magnitude of this Ca transient is very variable. Spontaneous oscillations of $[Ca^{2+}]_i$ can also be seen and it is obvious that the small Ca transients are produced when the stimulus occurs just after a Ca oscillation. In contrast, when there has not been an oscillation, the Ca transient is large (2). This effect of a previous oscillation on the magnitude of the Ca transient could arise because either: (i) Oscillatory release of Ca depletes the sarcoplasmic reticulum of Ca and means that less is available for release on a subsequent stimulation; (ii) Fabiato (13) has shown that Ca-induced release of Ca ions from the sarcoplasmic reticulum is inhibited by an excessive rise of $[Ca^{2+}]_i$.

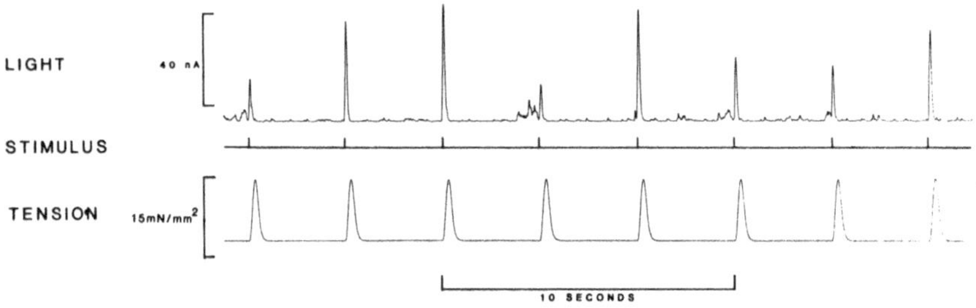

Fig. 8. The relationship between spontaneous oscillations and systolic $[Ca^{2+}]_i$ in a ferret papillary muscle. Traces show: top, aequorin light; bottom, tension. The stimuli are marked between the traces. Note that while the magnitude of the systolic Ca signal is very variable that of force is constant. This is because force is recorded from the tens of thousands of cells in the preparation. Therefore the variability is averaged out. Taken from Allen et al. (2).

The effects of Ca oscillations to decrease the systolic release of $[Ca^{2+}]_i$ may account for the fact that, although cardiac glycosides initially increase the force of contraction of cardiac muscle, they can eventually decrease it (33). This situation is called "Ca-overload" and, when it occurs, manoeuvres which under normal conditions increase $[Ca^{2+}]_i$ and thence force (e. g. elevating $[Ca^{2+}]_o$) now decrease systolic force. The hypothesis that Ca oscillations are involved in the reduction of force is supported by the observation that ryanodine, which inhibits these oscillations, can increase systolic force in Ca-overload (32).

Conclusions

The results reviewed in this paper have shown that the arrhythmogenic effects of digitalis result from the production of oscillations of intracellular calcium. The production of these oscillations, like the inotropic effects, is a consequence of an increase of $[Ca^{2+}]_i$. Furthermore these oscillations of $[Ca^{2+}]_i$ can also eventually interfere with the systolic rise of $[Ca^{2+}]_i$. Therefore the problem of balance between therapeutic and toxic concentrations of digitalis can be seen as that of increasing $[Ca^{2+}]_i$ sufficiently to increase the force of contraction while not reaching the level at which oscillations of $[Ca^{2+}]_i$ are produced. Digitalis would be a much more useful drug if it were possible to keep its positive inotropic effect while abolishing the spontaneous Ca oscillations. Unfortunately the drugs so far examined which abolish the oscillations (caffeine, ryanodine) appear to interfere with all Ca release from the sarcoplasmic reticulum and therefore to decrease developed force. However, as our knowledge of the processes controlling both stimulated and also spontaneous Ca release increases, so it may become possible to separate the therapeutic from the toxic effects.

Acknowledgements

Original work from the authors' laboratories was supported by the Medical Research Council and British Heart Foundation (D.A.E.) and the N.I.H. (W.J.L.). M.V. was in receipt of a M.E.C. Fleming award.

References

1. Allen DG, Blinks JR (1978) Calcium transients in aequorin-injected frog cardiac muscle. Nature, London 273: 509–513
2. Allen DG, Eisner DA, Pirolo JS, Smith GL (1985) The relationship between intracellular calcium and contraction in calcium-overloaded ferret papillary muscles. J Physiol 364: 169–182
3. Arlock P, Katzung BG (1985) Effects of sodium substitutes on transient inward current and tension in guinea-pig and ferret papillary muscle. J Physiol 360: 105–120
4. Berlin JR, Cannell MB, Goldman WF, Lederer WJ, Marban E, Wier WG (1986) Subcellular calcium inhomogeneity indicated by fura-2 develops with calcium-overload in single rat heart cells. J Physiol 371: 200p.
5. Cannell MB, Lederer WJ (1986) The arrhythmogenic inward current I_{TI} in the absence of electrogenic sodium-calcium exchange in sheep cardiac Purkinje fibres. J Physiol (in press)
6. Colquhoun D, Neher E, Reuter H, Stevens CF (1981) Inward current channels activated by intracellular Ca in cultured cardiac cells. Nature, London 294: 752–754
7. DiPolo RH, Rojas H, Beauge L (1982) Ca entry at rest and during prolonged depolarization in dialyzed squid axons. Cell Calcium 3: 19–41
8. Eisner DA, Lederer WJ (1985) Na-Ca exchange: stoichiometry and electrogenicity. Am J Physiol 248: C189–202
9. Eisner DA, Orchard CH, Allen DG (1984) Control of intracellular ionized calcium concentration by sarcolemmal and intracellular mechanisms. J mol cell Cardiol 16: 137–146
10. Eisner DA, Valdeolmillos M (1985) The mechanism of the increase of tonic tension produced by caffeine in sheep cardiac Purkinje fibres. J Physiol 364: 313–326

10a. Eisner DA, Valdeolmillos M (1986) A study of intracellular calcium oscillations in sheep cardiac Purkinje fibres measured at the single cell level. J Physiol 372: 539–556

11. Eisner DA, Vaughan-Jones RD (1983) Do calcium-activated potassium channels exist in the heart? Cell Calcium 4: 371–386
12. Fabiato A, Fabiato F (1975) Contractions induced by a calcium-triggered release of calcium from the sarcoplasmic reticulum of single skinned cardiac cells. J Physiol 249: 469–495
13. Fabiato A (1985) Time and calcium dependence of activation and inactivation of calcium-induced release of calcium from the sarcoplasmic reticulum of a skinned canine cardiac Purkinje cell. J gen Physiol 85: 247–289
14. Fedida D, Noble D, Spindler AJ (1985) Simultaneous recording of transient inward current and attraction in single guinea-pig ventricular cells. J Physiol 366: 119p.
15. Ferrier GR, Saunders JH, Mendez C (1973) A cellular mechanism for the generation of ventricular arrhythmias by acetylstrophanthidin. Circulation Res 33: 508–515
16. Glitsch HG, Pott L (1975) Spontaneous tension oscillations in guinea pig atrial trabecuale. Pflugers Arch 358: 11–25
17. Isenberg G (1975) Is potassium conductance of cardiac Purkinje fibres controlled by $[Ca^{2+}]_i$? Nature 253: 273–274
18. Isenberg G, Trautwein W (1974) The effect of dihydro-ouabain and lithium ions on the outward current in cardiac Purkinje fibres. Pflugers Arch 350: 41–54

19. Kakei M, Noma A, Shibasaki T (1985) Properties of adenosinetriphosphate-regulated potassium-channels in guinea-pig ventricular cells. J Physiol 363: 441–462
20. Kameyama M, Kakei M, Sato R, Shibasaki T, Matsuda H, Irisawa H (1984) Intracellular Na^+ activates a K^+ channel in mammalian cardiac cells. Nature 309: 354–356
21. Karagueuzian HS, Katzung BG (1982) Voltage-clamp studies of transient inward current and mechanical oscillations induced by ouabain in ferret papillary muscle. J Physiol 327: 255–271
22. Kass RS, Lederer WJ, Tsien RW, Weingart E (1978a) Role of calcium ions in transient inward currents and aftercontractions induced by strophanthidin in cardiac Purkinje fibres. J Physiol 281: 187–208
23. Kass RS, Tsien RW (1982) Fluctuations in membrane current driven by intracellular calcium in cardiac Purkinje fibres. Biophys Journal 38: 259–269
24. Kass RS, Tsien RW, Weingart R (1978b) Ionic basis of transient inward current induced by strophanthidin in cardiac Purkinje fibres. J Physiol 281: 209–226
25. Lederer WJ (1976) The ionic basis of arrhythmogenic effects of cardiotonic steroids in canine Purkinje fibres. *Ph D. Dissertation Physiology Department: Yale University, New Haven, CT*
26. Lederer WJ, Tsien RW (1976) Transient inward current underlying arrhythmogenic effects of cardiotonic steroids in Purkinje fibres. J Physiol 263: 73–100
27. Matsuda H (1983) Effects of intracellular calcium injection on steady-state membrane currents in isolated single ventricular cells. Pflugers Arch 397: 81–83
28. Matsuda H, Noma A, Kurachi Y, Irisawa H (1982) Transient depolarization and spontaneous voltage fluctuations in isolated single cells from guinea-pig ventricles. Calcium mediated membrane potential fluctuations. Circulation Res 51: 142–151
29. Noma A (1983) ATP-regulated K^+ channels in cardiac muscle. Nature 305: 147–148
30. Orchard CH, Eisner DA, Allen DG (1983) Oscillations of intracellular Ca^{2+} in mammalian cardiac muscle. Nature 304: 735–738
31. Rosen MR, Gelband H, Hoffman BF (1973) Correlation between effects of ouabain on the canine electrocardiogram and transmembrane potentials in isolated Purkinje fibres. Circulation 47: 65–72
32. Valdeolmillos M, Eisner DA (1985) The effects of ryanodine on calcium-overloaded sheep cardiac Purkinje fibres. Circ Res 56: 452–456
33. Vassalle M, Lin CI (1979) Effect of calcium on strophanthidin-induced electrical and mechanical toxicity in cardiac Purkinje fibres. Am J Physiol 236: H689–697
34. Wier WG, Kort AA, Stern MD, Lakatta EG, Marban E (1983) Cellular calcium fluctuations in mammalian heart: direct evidence from noise analysis of aequorin signals in Purkinje fibres. Proceedings of the National Academy of Sciences USA 80: 7367–7371
35. Wollenberger A (1949) The energy metabolism of the failing heart and the metabolic action of the cardiac glycosides. Pharmacol Rev 1: 311–352

Author's address:

Dr. D. Eisner, Department of Physiology, University College London, Gower Street, London WC 1E 6BT (England)

Electrophysiological assay of glycoside-induced sodium pump inhibition in isolated sheep heart Purkinje fibres at the onset of toxicity

C. Achenbach, Han Daying* and R. Preisler

Department of Physiology II, University of Bonn, Bonn (West Germany)

Summary

In cardiac tissue, the rate of sodium extrusion is linearly related to the intracellular sodium activity. By establishing this relationship using sodium-sensitive microelectrodes, sodium pump activities could be described quantitatively in sheep heart Purkinje fibres. In the presence of cardiac steroids, partial sodium pump inhibition caused intracellular sodium to rise, thus presumably stimulating non-inhibited pumping sites. At the transition to toxic drug action (onset of delayed after-potentials) the steady rate of net sodium uptake was used to determine the extent of pump inhibition. To do so, sodium turnover was calculated from experimental data on pump activity and changes of membrane potential, and the residual pump performance at the onset of toxicity was compared to that which the non-inhibited pump would have achieved had it been stimulated by corresponding amounts of intracellular sodium. Six different cardiac steroids were tested by this procedure and showed quite similar pump inhibitions of approx. 58 % (range: 52.8 %–65.5 %) at the onset of toxicity.

Introduction

Following the work of Schatzmann (16), a vast amount of experimental evidence has supported the view that cardioactive steroids bind to the Na/K-ATPase in the cell membrane of a variety of tissues and thereby inhibit monovalent cation transport. In cardiac tissue, there is an apparent link between sodium pump inhibition, enhanced intracellular sodium and myocardial inotropism. However, therapeutic glycoside doses often seem too low to effectively impede active sodium transport changes which may be difficult to quantify in vitro since the time course of glycoside action is subject to large variations at these concentrations. This problem is aggravated further by the fact that enhanced intracellular sodium stimulates the sodium pump so that the net effect may actually be negligible (see Discussion).

It is therefore necessary to measure intracellular sodium and to take into account the effects that a rise in intracellular sodium would have on the fraction of pump molecules that are not inhibited in the presence of low steroid concentrations. This requires that sodium pump activity needs to be defined with respect to its sensitivity to changes of intracellular sodium and that temporal resolution of the measurement can adequately disclose sodium turnover.

These requirements are jointly met by continuous measurements of intracellular sodium activities with sodium-selective micro-electrodes. Sodium pump activity and its sensitivity

* Second Medical College, Beijing (China)

to intracellular sodium can be determined by repetitive trains of beats (3) in individual fibres (sheep cardiac Purkinje fibres) in order to raise intracellular sodium under control conditions. The linear relationships between the rates of subsequent sodium extrusion and internal sodium levels are consistent with former views on active sodium transport in the heart (5, 6, 8, 9).

In the presence of cardiac steroids, this procedure is also applicable as will be shown in Fig. 1. In this communication, results are presented as to the extent of sodium pump inhibition at the onset of toxicity which is defined here as the first occurrence of delayed afterpotentials (7, 13). Repetitive trains of beats to determine sodium extrusion were generally not feasible at higher drug concentrations since the transition to toxic drug action rarely coincided with the experimental protocol. Instead, the rate of net sodium uptake occurring during steady beating of the preparations was employed to calculate the relative sodium pump inhibition, as will be described in the Results.

In this study, six glycosides were tested in order to detect possible differences in the relation between pump inhibition and drug toxicity arising from differences in chemical structure. Five of these were commercially available (digoxin, meproscillarin, k-strophanthin, ouabain and deslanoside), one was the new compound acrihellin (2).

Methods and materials

Sheep hearts were obtained at a slaughter-house (Kind, Grevenbroich) and transported to the laboratory in Tyrode solution oxygenated with 100% oxygen at approx 8°C. Purkinje fibres were excised from both ventricles; for work with ion-sensitive micro-electrodes, fibres of 5 mm length and of variable thickness were chosen. The Tyrode solution contained (in mmol/l): NaCl: 142; KCl: 4.0; $MgCl_2$: 0.5; $CaCl_2$: 1.8; HEPES-buffer: 2.1; glucose: 11. All salts and HEPES were standard analytical reagents of p. A. quality (Merck, Darmstadt; Sigma, St. Louis). Oxygenation with pure oxygen did not alter the pH-value of 7.4. Solutions containing cardiac steroids were prepared from the following drugs: digoxin (Lanicor, Boehringer, Mannheim), meproscillarin (Clift®, Knoll, Ludwigshafen), acrihellin (Degussa Pharma-Gruppe, Frankfurt), ouabain (Purostrophan®, Kalichemie, Hannover), deslanoside (Cedilanid®, Sandoz, Nürnberg), and k-strophanthin (Kombetin®, Boehringer, Mannheim).

The experiments were performed in a perspex chamber using standard electrophysiological appliances described in detail elswhere (1). Sodium-sensitive micro-electrodes were made using borosilicate glass capillaries (Hilgenberg, Malsfeld, # 11 032 27) cut to size and boiled 3–5 times in fresh double distilled water for 30 min. Micro-pipettes were pulled on a horizontal micro-electrode puller (PN 3, Narishige, Tokyo) with tip apertures below 1 μm and silanized with hexamethyldisilazane (HMDS, Sigma, St. Louis) at the comparatively low temperature of 120°C. They were first filled with the internal reference solution (150 mmol/l NaCl) and then held tip downwards into the sodium-sensitive resin (ETH 227 (17), Fluka, Neu-Ulm) which passed through the aperture to reach an optimal tip length of 200–300 μm within about 30 min.

Micro-electrodes were calibrated in the usual way and showed slopes of around 55 mV/decade. Amplification of the signal of the ion-sensitive electrode was carried out with an AD 311 J varactor bridge input amplifier (Analog Devices, Munich). The membrane potential was measured with a separate 3 mol/l KCl electrode and was subtracted from the signal of the selective electrode; with due regard to interfering ions, this voltage reflected intracellular sodium and was recorded continuously (cf. Figs. 1 and 2).

Results

After impaling both electrodes, fibres were stimulated at a steady rate of 0.4 Hz for at least 30 min to ensure stable impalements and equilibrated electrode potentials. Trains of stimuli were then applied at rates of between 0.5 and 3 Hz for periods of 3 to 5 min to obtain a variety of levels of intracellular sodium accumulation. After cessation of beating, the sodium signal returned to its resting value within several minutes depending on the extent of sodium uptake. The top panel of Fig. 1 shows trains of 4 min beats at 2 Hz in Tyrode; all trains were identical, indicating that the method yielded highly reproducible results.

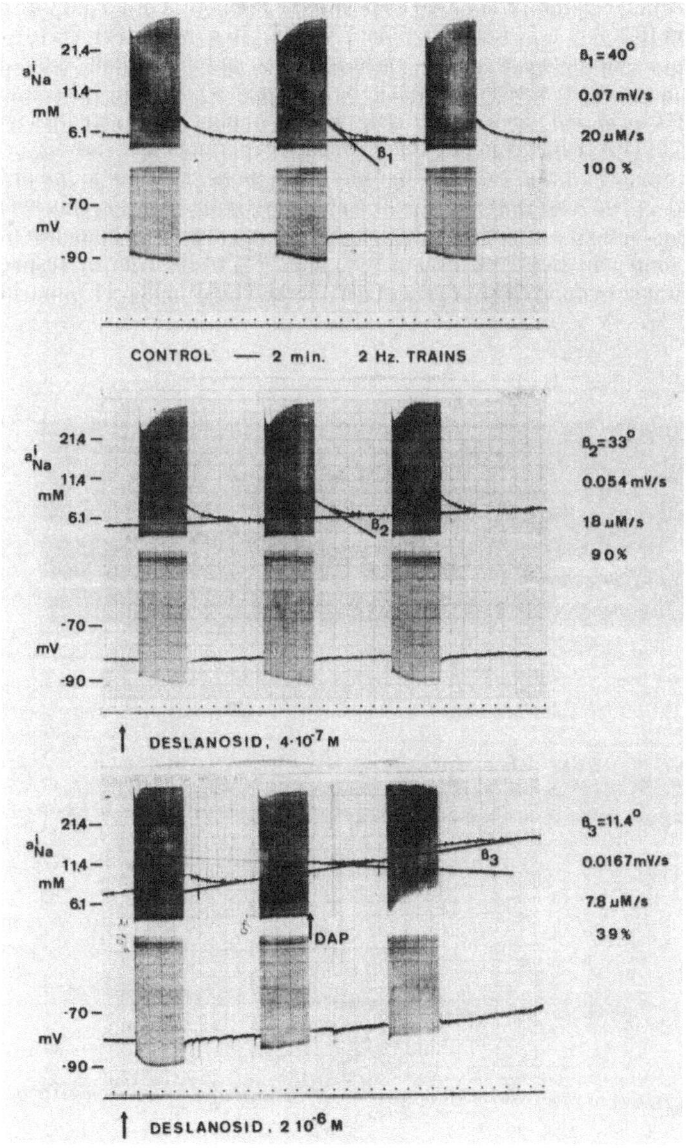

Fig. 1. Original recording of beat-dependent sodium uptake and extrusion and of diastolic membrane potentials before (top panel) and after exposure of a sheep heart Purkinje fibre to deslanoside (middle and lower panels). The arrow denotes the first occurrence of delayed after-potentials (DAP).

Referring to Fig. 1 the rates of sodium extrusion were evaluated graphically from the tangent of the decaying sodium signal drawn through the point obtained immediately after termination of cardiac activity. Plots of the rates of sodium extrusion versus intracellular sodium activity at different levels of sodium accumulation were strictly linear. The mean slope of the experiments discussed here was 7.9 μmol/l/s per mmol/l or $7.9 \times 10^{-3}\ s^{-1}$ (1). This value of the rate constant of sodium pump activity which also represented the sensitivity of the pump to intracellular sodium compared well with the findings of other authors in various cardiac tissues (in $10^{-3} s^{-1}$): 5.8–7.8 (3); 1.7 (4); 2.5–8 (5); 10.8 (6); 8.5 (8); 5.2 (9).

Due to the close relationship between sodium extrusion rates and intracellular sodium levels, resting sodium pump activity was considered to be determined largely by the resting sodium activity (cf. (6)). Calculated mean values of the resting sodium pump activities before application of digitalis (PA_0) are given in Table 1 for each experimental series.

The middle and lower panels of Fig. 1 display the rates of sodium extrusion in the presence of a non-toxic and a toxic dose of deslanoside. Rates were reduced noticeably even though stimulated sodium uptake proceeded to a higher level and resting sodium also increased. The residual pump activities were around 90% and 39% of the control, respectively. At the first appearance of delayed after-potentials (marked DAP in Fig. 1), pump inhibition amounted to 61%.

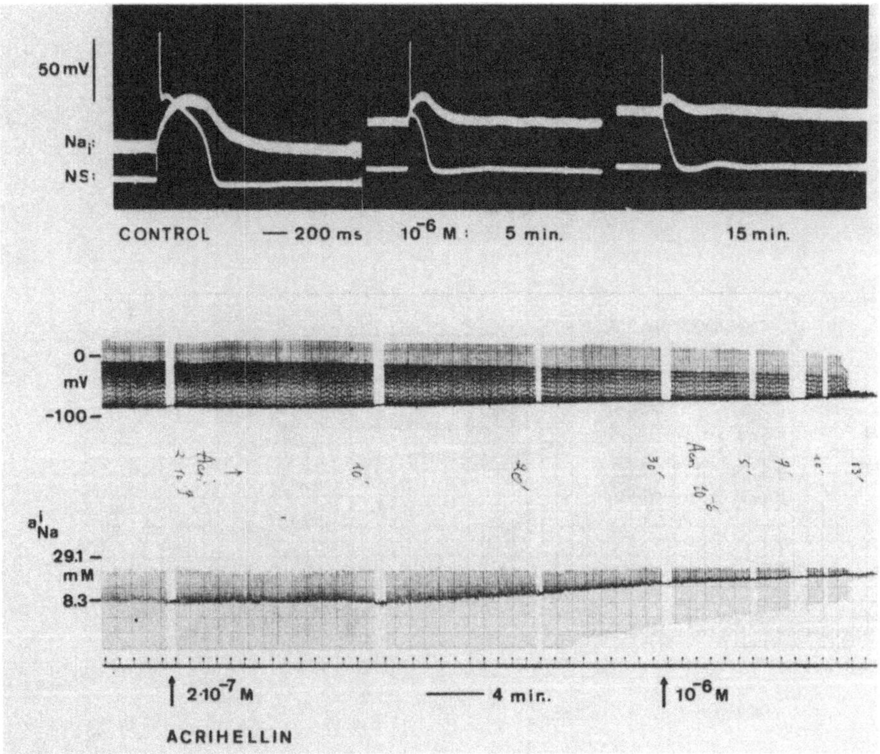

Fig. 2. Action potential recordings (top row) and continuous traces of membrane potential and intracellular sodium at a steady beating rate of 0.4 Hz in the presence of acrihellin.

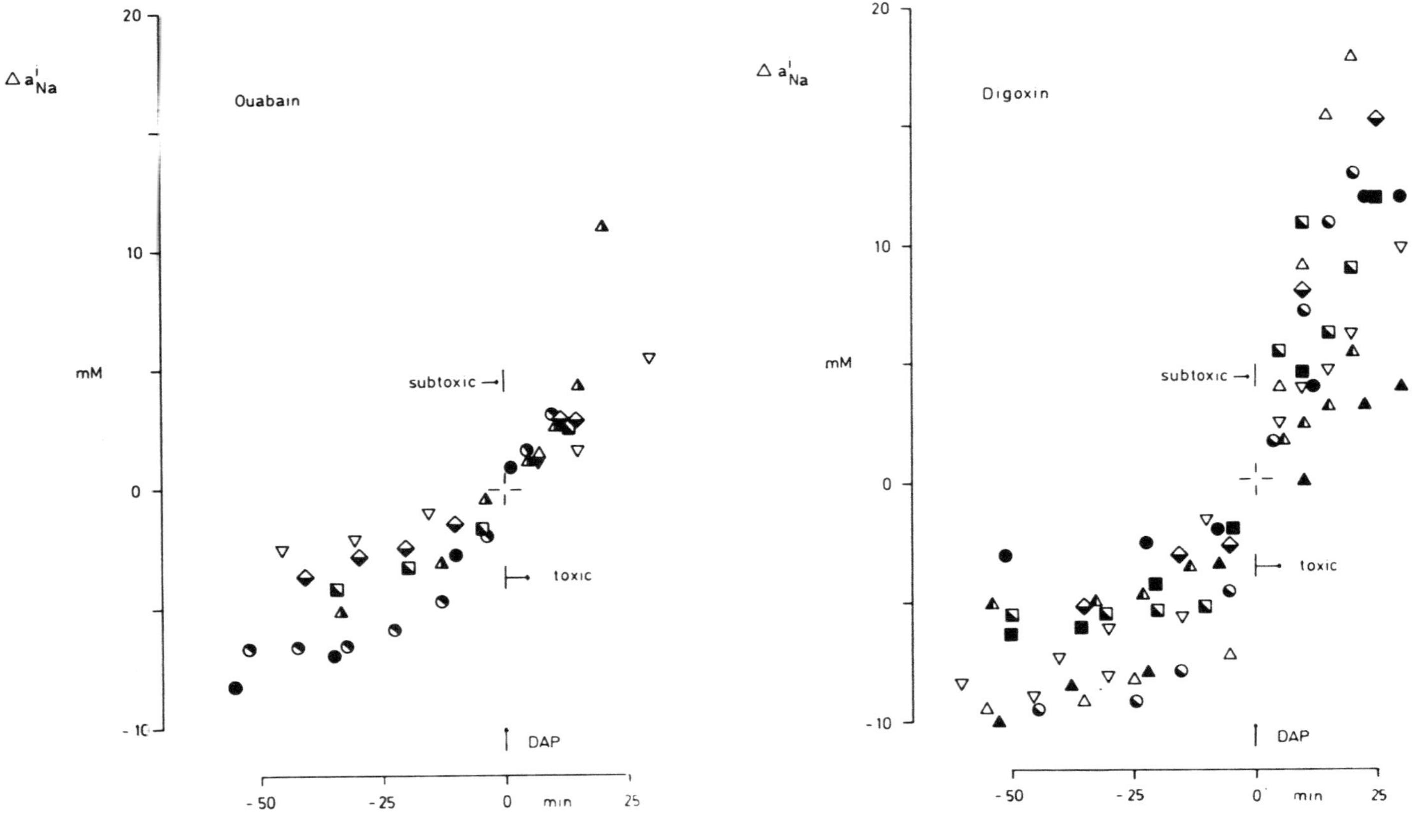

Fig. 3, a and b Changes of intracellular sodium with respect to the onset of delayed after-potentials (DAP) during exposure to ouabain (left) and digoxin (right).

Temporal resolution with respect to the first occurrence of after-potentials was not very good in these kind of experiments as a single run took about ten minutes. The experiments were therefore carried out at steady beating rates at which the onset of toxicity could be observed immediately. One of these experiments is shown in Fig. 2 using acrihellin. During glycoside action, intracellular sodium increased progressively after an initial delay; after-potentials occurred during early exposure to the high dose.

In Table 1 (upper now), all experimental results of intracellular sodium and resting potentials of the controls and of the increase of sodium and the membrane depolarisation at the onset of toxicity are listed. The third column of the upper row of Table 1 presents mean values of the rate of net sodium uptake at the onset of toxicity (NU_{DAP}). These values were determined from data shown in Fig. 3 for ouabain and digoxin. In Fig. 3, the changes of intracellular sodium before and after the onset of after-potentials were referred to corresponding time intervals preceding or following first toxic manifestations. This procedure of relating the data to the onset of toxicity as a new mutual point of reference removed much of the variability in the time courses of the changes of intracellular sodium. It also showed that intracellular sodium actually behaved quite similarly in different experiments with respect to the rates of sodium uptake (cf. Table 1).

To determine the average degree of sodium pump inhibition induced by the different cardiac steroids, it was necessary to compare sodium influx at the onset of delayed after-potentials (SI_{DAP}) to that before glycoside treatment (SI_o). First, SI_o was equated to PA_o since in the control sodium influx was balanced by sodium extrusion. Furthermore the ratio of SI_{DAP}/SI_o was considered to be equal to the ratio of sodium driving forces since sodium conductance would not be expected to change appreciably during membrane depolarisations of usually less than 6 mV evoked before transition to toxic drug action. Since the driving forces and SI_o were known, SI_{DAP} could be determined.

The residual pump activity at the onset of delayed after-potentials (PA_{DAP}) was thought to be equal to sodium influx SI_{DAP} minus net sodium uptake per unit of time NU_{DAP}. This is

Table 1. Experimental results

Steroid	a^i_{Na} (mmol/l) Controls	Increase	NU_{DAP} (μmol/l/s)	Resting potentials (mV) Controls	Increase	Number of experiments (fibras)
Digoxin	5.4 ± 1.83	5.8 ± 3.0	5.7 ± 2.5	−83.0 ± 5.2	5.4 ± 2.3	12
Meproscillarin	5.33 ± 1.16	4.74 ± 3.15	3.87 ± 0.42	−81.8 ± 7.0	4.8 ± 4.4	5
Acrihellin	5.7 ± 1.2	4.3 ± 2.5	4.6 ± 2.5	−82.6 ± 4.4	5.3 ± 3.5	8
Deslanoside	5.43 ± 0.64	3.75 ± 2.53	3.42 ± 1.2	−83.2 ± 5.7	6.0 ± 2.7	6
Ouabain	5.8 ± 1.2	5.0 ± 2.16	3.9 ± 1.54	−83.5 ± 6.2	5.3 ± 2.6	6
k-Strophanthin	5.7 ± 2.7	3.9 ± 2.15	3.6 ± 1.23	−82.5 ± 2.8	7.9 ± 4.2	8

means ± S.D.

	Pa_o (μmol/l/s)	SI_{DAP} (μmol/l/s)	PA_{DAP} (μmol/l/s)	$PA_{max,DAP}$ (μmol/l/s)	Inhibition %
Digoxin	42.7	36.2	30.5	88.5	65.5
Meproscillarin	42.1	36.3	32.6	79.55	59.1
Acrihellin	45.0	39.3	34.7	79.0	56.1
Deslanoside	42.9	37.6	34.2	72.5	52.8
Ouabain	45.8	39.6	35.7	85.3	58.2
k-Strophanthin	45.0	38.8	35.2	75.85	53.6

equivalent to saying that because PA_{DAP} did not balance SI_{DAP}, there was a steady net uptake of intracellular sodium equal to NU_{DAP}.

To conclude the analysis, one had to consider the magnitude of the pump activity one would have observed in the absence of glycosides if intracellular sodium had attained intracellular sodium levels corresponding to those at the onset of toxicity. This hypothetical value $PA_{max,DAP}$ represented maximal pump stimulation of the uninhibited pump at given intracellular sodium levels and followed from the sensitivity relationship of control pump activity. Mean values of $PA_{max,DAP}$ are tabulated in Table 1.

The degree of pump inhibition at the onset of toxicity could now be determined. The ratio $PA_{DAP}/PA_{max,DAP}$ was the fraction of pump molecules still active in the presence of the cardiac steroids whereas $(PA_{max,DAP} - PA_{DAP})/PA_{max,DAP}$ corresponded to the fraction of inhibited pumping sites. Values for the extent of pump inhibition (in %) are presented in the right hand lower column of Table 1.

Discussion

The results presented here indicate that different kinds of cardiotonic steroids do not differ significantly in the extent of sodium pump inhibition caused by them in cardiac tissue when toxic drug effects become apparent. A possible conclusion to this finding is that the link between toxicity and pump inhibition is very close and other toxic factors have been successfully suppressed in the drugs used.

We have compared our average value of 58 % pump inhibition at the onset of toxicity with values obtained by other authors who applied entirely different methods to this problem. The values agree very well: 62 % (10); 59 % (11); 56 % (12); 60 % (14); 60 % (15). Very recent studies on glycoside binding in conjunction with electrophysiological observations also revealed a receptor occupancy of slightly less than 60 % at the beginning of arrhythmia (18).

In all experimental series shown here, intracellular sodium increased by a factor of about 2 at the transition to toxic drug action whereas the residual pump activity decreased by only about 20 % compared to the control values. This suggests that, up to about 50 % pump inhibition, sodium uptake can effectively counteract pump inhibition by stimulating free pump molecules so that very little loss of performance ensues. Beyond this point, sodium uptake can no longer compensate pump inhibition and the ionic and electrical integrity of the tissue dissolves.

Acknowledgements

The authors gratefully acknowledge support granted to H. D. by the following institutions: Deutscher Akademischer Austausch-Dienst, Bonn; Schering A.G., Berlin; Hermann und Lilly Schilling-Stiftung im Stifterverband für die Deutsche Wissenschaft, Essen.

References

1. Achenbach C (1985) Elektrophysiologische Untersuchungen zur Hemmung der Natriumpumpe durch Kardiosteroide in isoliertem Herzgewebe. Habilitationsschrift, Bonn
2. Achenbach C, Engel J, Heusinger JH, Metzenauer P (1985) Acrihellin. In: Scriabine A (ed) New Drugs Annual: Cardiovascular Drugs, Vol 3. Raven Press, New York, pp 219–243
3. Cohen CJ, Fozzard HA, Sheu SS (1982) Increase in intracellular sodium ion activitiy during stimulation in mammalian cardiac muscle. Circ Res 50: 651–662
4. Daut J (1982) The role of intracellular sodium ions in the regulation of cardiac contractility. J mol cell cardiol 14: 189–192

5. Deitmer JW, Ellis D (1978b) The intracellular sodium activity of cardiac Purkinje fibres during inhibition and re-activation of the Na-K pump. J Physiol 284: 241–259
6. Eisner DA, Lederer WJ, Vaughan-Jones RD (1981) The dependence of sodium pumping and tension on intracellular sodium activity in voltage-clamped sheep Purkinje fibres. J Physiol 317: 163–187
7. Ferrier GR, Moe GK (1973) Effect of calcium on acetylstrophanthidin-induced transient depolarizations in canine Purkinje tissue. Circ Res 33: 508–515
8. Gadsby DC, Cranefield PF (1982) Effects of electrogenic sodium extrusion on the membrane potential of cardiac Purkinje fibres. In: Normal and Abnormal Conduction in the Heart. Eds: Paes de Carvalho A, Hoffman BF, Lieberman M; Futura Publishing Company, Mt Kisco, New York, pp 225–247
9. Glitsch HG, Pusch H (1980) Correlation between changes in membrane potential and intracellular sodium activity during K activated response in sheep Purkinje fibres. Pflügers Arch 384: 189–191
10. Hall RJ, Gelbart A, Silverman M, Goldman RH (1977) Studies on digitalis-induced arrhythmias in glucose- and insulin-induced hyperkalemia. J Pharmacol Exp Ther 201: 711–722
11. Hougan TJ, Lloyd BL, Smith TW (1979) Effects of inotropic and arrhythmogenic digoxin doses and of digoxin-specific antibody on myocardial monovalent cation transport in the dog. Circ Res 44: 23–31
12. Marsh AL, Lloyd BL, Taylor RR (1981) Age dependence of myocardial Na^+-K^+-ATPase activity and digitalis intoxication in the dog and guinea-pig. Circ Res 48: 329–333
13. Reiter M (1962) Die Entstehung von "Nachtkontraktionen" im Herzmuskel unter Einwirkung von Calcium und von Digitalisglykosiden von der Reizfrequenz. Naunyn-Schmiedeberg's Arch exp Path u Pharmak 242: 497–507
14. Repke K, Portius HJ (1963) Über die Identität der Ionenpumpen-ATPase in der Zellmembran des Herzmuskels mit einem Digitalis-Rezeptorenzym. Experimentia XIX, pp 452–458
15. Rhee HM (1982) Ouabain sensitivity of Rb^+ uptake in canine Purkinje fibre and ventricular muscle. Naunyn-Schmiedeberg's Arch Pharmacol 318: 344–348
16. Schatzmann H-J (1953) Herzglykoside als Hemmstoffe für den aktiven Kalium- und Natriumtransport durch die Erythrozytenmembran. Helv Physiol Acta 11: 346–354
17. Steiner RA, Oehme M, Ammann D, Simon W (1979) Neutral carrier sodium ion-selective microelectrode for intracellular studies. Anal Chem 51: 351–353
18. Werdan K (1985) Langzeittherapie mit Herzglykosiden – Gibt es einen Wirkungsverlust? Münch med Wschr 127: 946–949

Author's address:

Dr. med. C. Achenbach, Department of Physiology II, University of Bonn, Wilhelmstraße 31, 5300 Bonn (West Germany)

Rat cardiac hypertrophy: contribution of heteregeneous digitalis receptor forms

L. G. Lelièvre*, J. M. Maixent**, D. Charlemagne* and B. Swynghedauw**

* Inserm U 127 Hôpital Lariboisière, Université Paris.7. 75010 Paris, and **CERES Laboratoire Nativelle 1, Chemin Saulxier 91, Longjumeau (France)

Summary

It is generally accepted that the primary target of the cardiac glycosides is the sarcolemma-bound Na/K-ATPase activity. The inhibition of this function by ouabain occurs in the same range of concentrations as pharmacologically active doses in various species including rat. In this species, it has been shown that the sensitivity of Na/K-ATPase to ouabain was heterogeneous with high and low sensitivity forms (K_D 10^{-7}, 10^{-8} M and 10^{-6}, 10^{-5} M, respectively). According to experimental and clinical reviews the cardiac glycosides were inotropic in both normal and failing hearts. However, none of these studies were comparative.

Thus the question arises as to whether modifications of the responsiveness to ouabain of hypertrophied rat heart may be related to alterations of the Na/K-ATPase properties.

The inotropic effects of ouabain (10^{-9} to 10^{-4} M) were studied on both normal and pressure-overload induced hypertrophied rat hearts working under isovolumic conditions in the presence of 0.25 mM $[Ca]_o$.

Under these conditions, as compared to normal hearts, the inotropic response of hypertrophied hearts to ouabain appeared to be depressed and mainly prolonged. This delayed recovery of normal contractile function after the inotropic response paralleled the delayed release of ouabain from both high and low sensitivity Na/K-ATPase forms present in normal and hypertrophied cardiac sarcolemma preparations.

In order to determine the relative contributions of high and low sensitivity Na/K-ATPase forms in the inotropic responses, either in normal or hypertrophied rat hearts, the respective rates of dissociation in vitro were compared to the biphasic release observed with isolated hearts. According to the dose-response curves in vivo, the low sensitivity enzyme form would be mainly involved in the inotropic response in both heart groups.

The present results clearly show that, in hypertrophied rat heart, the digitalis receptor binds and releases ouabain at slower rates than in normal hearts. Furthermore, it clearly appeared that the determination of the contributions of the high and low sensitivity forms in hypertrophied or normal hearts was highly dependent upon the criteria chosen: onsets and amplitudes of the inotropic responses, dose-effect curves and recoveries of normal contractile performances.

Many reported data have evidenced the cause-effect relationship between the inotropic action of cardiac glycosides and the concomitant inhibition of the Na/K-ATPase, in vitro manifestation of the pharmacological receptor (2, 14). This assertion accepted in most species is still a subject of controversy in rat (1, 3, 4, 5, 10, 11, 13, 15).

Indeed, the Na/K-ATPase activity exhibited a low sensitivity to ouabain (10 μM) whereas high and low affinity ouabain-binding sites did exist in the same cardiac preparation (1, 4, 5). Their apparent affinities were 0.1–0.4 μM and 10–60 μM, respectively. Both site types detected in vitro corresponded to the heterogeneous cardiotonic effects of ouabain as revealed with isolated right ventricular strips (1, 4, 5). The observed inotropic effect would be due to a high sensitivity component (apparent affinity of 0.3 μM) and a low sensitivity component (apparent affinity ca 10 μM). According to the data presented by Adams et al. (2) and Finet et al. (5), the high sensitivity inotropic component would account for 30% of the maximal cardiotonic effect. These groups reported either no enzyme inhibition (1) or a very slight inhibition (approximately 10%) (5, 15).

Taken together, all these results supposed that, in rat heart, the Na/K-ATPase activity was linked to low-affinity sites. Chiefly, the inotropic effects observed at low doses of ouabain would not be related to an inhibition of the enzyme. Note that these results have been obtained with right ventricular strips stimulated at 1 Hz at a physiological calcium level (millimolar concentration) (1, 4, 5, 15).

It was of interest to know whether the characteristics presented above were still valid in hypertrophied rat heart. The inotropic effect of ouabain on this muscle was evaluated on an isolated Langendorff rat heart paced at 6 Hz at 0.25 mM external calcium. A systematic and comparative study of ouabain binding sites and Na/K-ATPase activity in normal and hypertrophied rat hearts has been reported (3). The results showed the existence of two families of ouabain binding sites with apparent dissociation constants of 25 ± 7 nM and 4.5 ± 3.5 μM, respectively, which were similar to those found in normal hearts. The presence of high affinity receptor in hypertrophied rat heart was correlated to a detectable inhibition of the Na/K-ATPase (IC_{50} = 20 ± 10 μM). However, the high and low affinity sites in hypertrophied hearts bound and released ouabain at 4–5-fold slower rates than the corresponding sites in normal hearts (3).

The physiological consequences of these observations would be:

- a biphasic inotropic action of ouabain in hypertrophied hearts and
- a delayed recovery of normal contractile performance when the inotropic effect was washed out.

Materials and Methods

Pressure overload hypertrophy. Aortic stenosis has been produced in male Wistar rats (body weights about 200 g) according to the procedure previously described (9). The diameter of the clips was calibrated so as to produce cardiac hypertrophy (+40 to 60%) after two weeks in most of the animals and much more pronounced hypertrophy (around 100%), in one or two out of every ten. Animals with a high degree of hypertrophy have been selected on the basis of a ventricular weight/body weight (mg/g) above 3.

Protocols for contractile experiments (8). Male Wistar rats weighing approximately 200 g were killed and the hearts rapidly removed. The retrograde coronary perfusion at 37° C was maintained at a pressure of 90 mm Hg. The perfusate consisted of modified Krebs-Henseleit buffer containing 118 mM NaCl, 4.96 mM KCl, , 25 mM $NaHCO_3$, 1.2 mM $MgSO_4$, 1.17 mM NaH_2PO_4, 11 mM glucose and 0.25 mM $CaCl_2$. After prolonged oxygenation in 5% CO_2, 95% O_2, the pH was 7.4 and the pO_2 around 600 mm Hg. Hearts were paced at 6 Hz by two atrial electrodes (Philips TP 300) with rectangular unipolar pulses of 1 msec duration (4 mA). A drain was placed in the apex of the left ventricle. The left atrium was opened and a collapsed latex ballon was inserted into the left ventricle.

The temperature was kept constant at 37° C. The coronary flow averaged 14 ml/min and was not maintained constant by a pump. The perfusate efflux was not recirculated. Pressures were recorded by a P 23 ID Statham gauge transducer. Measurements were made every five minutes. The collapsed balloon was filled with water in order to obtain a diastolic pressure of 10–15 mm Hg which usually corresponds to 110 or 120 μl of water.

In this model, records were made in isovolumetry with balloon volumes and diastolic pressures similar from heart to heart. Nine sham-operated rats (controls) and 8 with aortic stenosis were compared. After 20 minutes of equilibration, ouabain (from 1 nM to 100 μM) was continuously infused for sequential 5 minute periods. This protocol was chosen to rapidly obtain a dose-response curve using a rather rare material. When the inotropic effect was washed out, the time course of the recovery of normal contractile performance was studied for 30 or 40 minutes.

Analysis of the results. Inotropy was expressed either in terms of developed pressure or according to the (dP/dt)/P sys. value which was a good index in this model (6, 12). The experimental data were analyzed assuming that the relation between the ouabain concentrations and the inotropic action could be considered as the sum of two independent saturable components:

$$I\ obs = I\ obs_H + I\ obs_L$$

$$I\ obs = \frac{I\ max_H}{\frac{EC\,50_H}{Ouab} + 1} + \frac{I\ max_L}{\frac{EC\,50_L}{Ouab} + 1}$$

where H or L: contribution of high or low sensitivity inotropic components,
- I obs: observed inotropy
- I max: maximal inotropy
- EC 50: concentration of ouabain producing a half maximal effect
- ouab: concentration of ouabain in the perfusate.

Data were analysed using the computerized non-linear regression program which adjusted the parameters in order to minimize the sum of the squarred errors.

Results and discussion

If one refers to the developed pressure in isolated hypertrophied rat heart working under isovolumic conditions, the log. concentration-effect curve was biphasic. The experimental values fitted the model implying two independent, non interconvertible, saturable components (Fig. 1). A component, highly sensitive to ouabain ($EC_{50} = 0.3\ \mu M$) would contribute 27% of the maximal response observed with 100 μM drug.

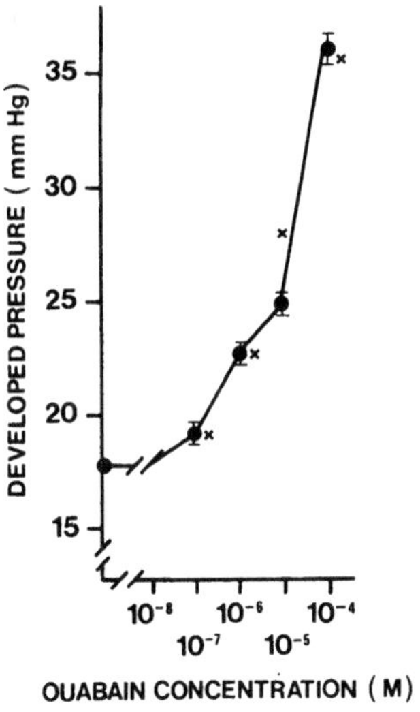

Fig. 1. Biphasic effect of ouabain on the developed pressure in the hypertrophied rat heart. N = 8

This inotropic component sensitive to low doses of ouabain shared the same characteristics (EC_{50} value and contribution) as those reported in the literature using ventricular strips from normal rat heart as a model (1, 4, 5, 13).

Note the differences between the two models: right ventricular strips stimulated at 1 Hz in the presence of millimolar calcium concentrations and isolated working rat heart paced at 6 Hz in the presence of 0.25 mM calcium (8). As illustrated in Fig. 1, the second component would be of low sensitivity ($EC_{50} = 20\,\mu M$) and would account for 70 % of the maximal response. Again, these results fit the published values for this type of site.

If the inotropic response was expressed according to the (dP/dt)/P sys. index, the computed parameters of each site type were different. Indeed, as depicted in Fig. 2, in hypertrophied rat heart, 50 % of the maximal inotropism would be due to the high sensitivity component. The computed apparent affinity for ouabain would be 20 nM. This highly sensitive inotropic component was also detected in control experiments with normal rat hearts (L.G.L., data not shown).

In contrast, the EC_{50} value for the low sensitivity component remained unchanged ($EC_{50} = 20\,\mu M$).

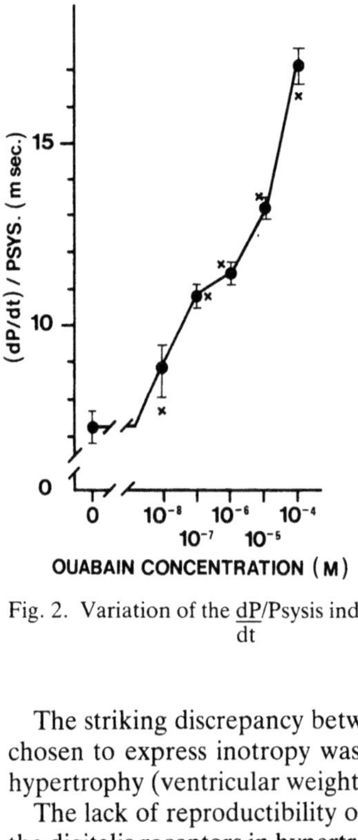

Fig. 2. Variation of the $\frac{dP}{dt}$/Psysis index as a function of ouabain concentration (logarithmic scale).

The striking discrepancy between these two types of results as a function of the criteria chosen to express inotropy was only observed with rat hearts exhibiting at least a 80 % hypertrophy (ventricular weight/body weight > 3).

The lack of reproductibility of these analyses could be due to the intrinsic properties of the digitalis receptors in hypertrophied hearts.

According to in vitro experiments carried out with cardiac sarcolemma vesicles (3, 8), it has been demonstrated that the digitalis receptors bound and released ouabain at 5-fold slower rates in hypertrophic than in normal hearts (3, 8 and Table 1).

This was confirmed with experiments performed in vivo (Fig. 3).

When inotropic effect was washed out, the recovery of the original contractile performance was delayed in hypertrophied hearts. From these data, it can be calculated that the time required for a half maximal recovery was 14.9 ± 1.1 min and 57 ± 3.8 min in normal

Table 1. Dissociation process in vitro and reversal of inotropy in vivo expressed in terms of t 1/2.

	Normal	Hypertrophied
3H-ouabain binding sites	0.61 ± 0.07 min (a)	3.2 ± 0.2 min
Na/K-ATPase activity	1.05 ± 0.11 min (b)	3.2 ± 0.2 min
Inotropic effect (d)	14.9 ± 1.1 min (c)	57 ± 3.8 min

(a) = time required to release 50 % of the amount of 3H-ouabain previously bound to sarcolemma vesicles (3); (b) = time required to recover 50 % of the low affinity Na/K-ATPase form previously inhibited by ouabain (8); (c) = time required to recover 50 % of the original contractile performances (this study); (d) = since the high sensitivity inotropic component could not be determined accurately in vivo, only the low affinity component was depicted in this Table.

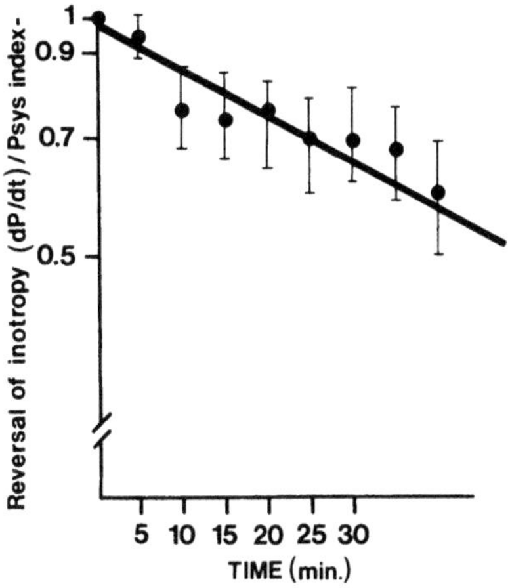

Fig. 3. Washout of inotropic effect.

and hypertrophied rat hearts, respectively. The reversal of inotrop was fourfold slower in hypertrophied hearts.

Note that the phenomenon described in Fig. 3 only took the low sensitivity inotropic component into account. Indeed, by analogy with low affinity ouabain binding sites (3, 16) or low sensitivity Na/K-ATPase forms (3, 8) in vitro, the low sensitivity inotropic component could be characterized by a slow association and a fast dissociation process.

Thus during the recovery period, the initial decrease in inotropism will be due, mainly if not exclusively, to the dissociation of ouabain from the low sensitivity component.

In normal hearts, 50 % of the inotropic response was reversed in 15 min whereas no more than 25 % could be reversed in 30 min with hypertrophic cardiac preparations. Due to the limited stability of the isolated heart preparation, a period of recovery lasting more than 40 min has to be avoided.

In Table 1 are summarized the half-time (t 1/2) measured in vitro (dissociation from ouabain binding sites, see (3) and in vivo (reversal of inotropy, Fig. 3). In each heart group, as well in vitro as in vivo the release of ouabain from the digitalis receptor was slower in hypertrophied hearts. The recovery (in vivo data) was much slower than the release of ouabain from its specific sites in vitro. A 15 to 20-fold difference was found. Such a difference could be explained by the complex physiological events implied in the action of ouabain, the kinetic parameters of these events being much slower than the simple dissociation of ouabain from its sites.

Conclusions

In terms of inotropy,

1. the involvement of both high and low sensitivity inotropic components in the effect of ouabain, as demonstrated by Adams et al. (1) and Finet et al. (5) in normal rat heart also seems valid for the rat heart hypertrophied by pressure overload;
2. the affinity of the low sensitivity component appeared to be constant: 20 μM;
3. neither the exact contribution of the high-sensitivity component nor its apparent affinity for ouabain could be determined accurately in hypertrophied hearts;
4. the prolonged inotropic effect of ouabain in hypertrophied hearts as compared to normal was mainly due to a slower release of ouabain from the digitalis receptor.

The physiological role of such a change of the digitalis receptor in the adaptation of the cardiac muscle to a pressure overload remains an open question.

Acknowledgements

We thank Mr. C. Mouas for his excellent technical assistance. This work was supported by grants from the M.I.R. P.R.C. Medicaments. Convention n° 131 030 (L. G. L.), the Université Paris 7. n°C. 32R01 (L.G.L.), the Fondation des Maladies Cardiovasculaires (B.S.) and the M.G.E.N. (B.S.).

References

1. Adams RJ, Schwartz A, Grupp G, Grupp I, Lee SW, Wallick ET, Powell T, Twist VM, Gathiram P (1982) Nature 296: 167–169
2. Akera T, Brody T (1978) Pharmacol Rev 29: 187–220
3. Charlemagne D, Maixent JM, Preteseille M, Lelièvre L (1986) J Biol Chem (in press)
4. Erdmann E, Philipp G, Scholz H (1980) Biochem Pharmacol 29: 3219–3229
5. Finet M, Godfraind T, Noel F (1983) Br J Pharmacol 80: 751–759
6. Grossmann W, Haynes F, Paraskos JA, Saltz S, Dalen JE, Dexter L (1972) Circ Res 31: 83–94
7. Herzig S, Mohr K (1984) Br J Pharmac 82: 135–142
8. Lelièvre LG, Maixent JM, Lorente P, Mouas C, Charlemagne D, Swynghedauw B (1986) Am J Physiol (in press)
9. Lompré AM, Schwartz K, D'Albis A, Lacombe G, Nguyen VT, Swynghedauw B (1979) Nature 282: 105–107
10. Mansier P, Cassidy P, Preteseille M, Charlemagne D, Lelievre L (1983) FEBS Lett 153: 357–360
11. Mansier P, Lelièvre L (1983) Nature 300: 535–537
12. Mason DT (1969) Am J Cardiol 23: 516–527
13. Noël F, Godfraind T (1984) Biochem Pharmacol 33: 47–53
14. Schwartz A, Adams RJ (1980) Circ Res 46: 1154–1160
15. Schwartz A, Whitmer K, Grupp G, Grupp I, Adams RJ, Lee SW (1982), Ann NY Acad Sci 402: 253–271
16. Wellsmith N, Lindenmayer GE (1980) Circ Res 47: 710–720

Author's address:

Prof. Dr. L. Lelièvre, Unité de Recherches sur le Métabolisme du Cœur et de Vaisseaux, Hôpital Lariboisière 41, Boulevard de la Chapelle, 75010 Paris (France)

Digitalis receptors in normal and hypertrophied rat hearts. Differential effects of a Ca^{2+} free perfusion

D. Charlemagne, E. Mayoux, M. Preteseille, J. M. Maixent*, Ch. Mouas, B. Swynghedauw and L. G. Lelièvre

Inserm U 127 Hôpital Lariboisière, Université Paris 7, 75010 Paris (France), and *CERES Nativelle, Longjumeau (France)

Summary

Recently, it has been demonstrated that more than one receptor type would be responsible for the inotropic effect of ouabain. This effect is linked to an increased availability of Ca_{in}. In rat heart two classes of ouabain binding sites have been described: a high affinity (H. A.) class with a low capacitiy, and a low affinity (L. A.) class with a high capacity. Although both sites are involved in the inotropic response, these H. A. sites may not be related to inhibition of Na/K-ATPase activity. However, we have reported that a Ca-free perfusion of rat heart revealed a Na/K-ATPase activity highly sensitive to ouabain by modifying the sarcolemmal calcium pool.

The purpose of this study is to further characterize the mechanism(s) by which Ca affects the digitalis receptors. This pharmacodynamical study has been performed on sarcolemmal vesicles isolated from normal and hypertrophied rat hearts. The pressure-overload induced hypertrophy represented a suitable model in that the movements of Ca_{in} appeared to be slowed.

When the heart was maintained at a physiological calcium level (2.5 mM), hypertrophied and normal cardiac preparations had heterogeneous ouabain binding sites: H. A. (K_D 3 x 10^{-8} M) associated with 2–4 pmol/mg protein and L. A. sites (K_D 4–18 x 10^{-6} M) associated with 115 ± 25 pmol/mg protein. Although both cardiac preparations had the same total number of sites, in the hypertrophied group the number of H. A. sites is two fold higher ($p < 0.02$) than in normal group and the dissociation and association rate constants were five times slower. This strongly suggested the apparition of new forms of Na/K-ATPase in hypertrophied rat heart.

As revealed by Ca-free perfusion, the normal heart is greatly affected by the calcium present at the sarcolemmal level: the number of H. A. sites (from 2 to 4 pmol/mg) is doubled, the number of L. A. sites is decreased from 140 to 90 pmol/mg ($p < 0.01$). As a consequence the relative proportion of H. A. sites reached 22 % of the total number of sites. Moreover the pharmacodynamic parameters of the L. A. affinity sites are modified: the apparent affinity was increased by a 4 fold factor. These latter observations were consistent with the fact that the inotropic effect of ouabain is increased at low external Ca (0.25–0.5 mM, compared to 2 mM).

Strikingly, in hypertrophied hearts, a Ca-free perfusion had no apparent effect on the H. A. and L. A. ouabain binding site characteristics (dissociation constant, number of sites, rate constants). This independancy toward Ca is a strong argument in favor of adaptational changes of Na/K-ATPase in hypertrophied rat heart, in relation to the depressed and prolonged effect of the drug in vivo.

Introduction

It is well known that *adult rat heart* is relatively insensitive to the positive inotropic effect of ouabain (1, 4, 12). This could be related to a small systolic transsarcolemmal Ca^{2+} influx and to a highly developed and very active sarcoplasmic reticulum (5).

In fact, a positive inotropic response was observed in isolated perfused rat heart at low Ca^{2+} level (8). This effect was partly associated with a form of (Na^+,K^+)-ATPase highly sensitive to ouabain in sarcolemma vesicles (10). This form was detected only after a perfusion of the intact heart with a low Ca^{2+} containing buffer (9).

In *hypertrophied rat heart* perfused with a physiological Ca^{2+} concentration, we showed that the (Na^+,K^+)-ATPase sensitivity to ouabain and the properties of the ouabain binding sites were different from those of the normal rat heart and similar to that found in neonatal rat heart: an increased sensitivity of (Na^+,K^+)-ATPase and a five fold increase in the dissociation rate constants were observed in sarcolemma vesicles (2).

In isolated perfused hypertrophied heart, the positive inotropic effect of ouabain observed at low Ca^{2+} level (0.25 mM) (8) was more prolonged than in normal rat heart. This correlated well with an increased time of release of the drug in in vitro experiments (2, 8, 9, 10).

In hypertrophied and in neonatal heart, a longer action potential has been observed. It might be related to a higher transsarcolemmal Ca^{2+} influx and Ca^{2+} binding capacity (11) as already described in neonatal rat heart.

If a different Ca^{2+} regulation mechanism exists in the sarcolemma of hypertrophied rat heart, it would be interesting to test the following:

What are the effects of the perfusion at low Ca^{2+} concentration on the hypertrophied rat heart?

Is the high sensitivity (Na^+,K^+)-ATPase form of hypertrophied rat heart dependent on the presence of Ca^{2+}, as observed in normal rat heart?

Materials and methods

Biological model. Abdominal aortic constriction was performed in adult male Wistar rats (180–200 g) to produce a 60–130 % increase in the left ventricular weight.

Isolation of sarcolemma fraction. Sarcolemmal vesicles were isolated according to our previously published procedure from normal and hypertrophied rat hearts after a perfusion of the heart with a Ca^{2+}-free medium (5 mM EDTA) or a medium containing 2 mM Ca^{2+} (9, 10).

The studies reported here have been carried out with sarcolemmal vesicles subfractionated by centrifugation on sucrose. They were kept at −40° C and used within two weeks. Protein concentrations were estimated according to Lowry.

The vesicles were then subjected to repeated freezings and thawings or to SDS treatment (0.2 mg/mg of protein, 30 min at 20° C) to make them permeable to substrates and ligands before (^{3}H)-ouabain binding studies and (Na^+,K^+)-ATPase assays.

(Na^+,K^+)-ATPase inhibition. The enzymatic activity in the absence or presence of various concentrations of ouabain was determined using the coupled assay method previously described. The activity was measured by continuously recording NADH oxidation in an ATP regenerating assay medium. Each cell contained (final volume 0.6 ml): 100 mM NaCl, 5.86 mM KCl, 4 mM $MgCl_2$, 4 mM ATP, 40 mM Imidazole/HCl (pH 7.4), 0.4 mM NADH, 2 mM phosphoenolpyruvic acid, 3.5 units pyruvate kinase, 5 units lactate dehydrogenase and 2 to 5 ug microsomal protein. After 5 min, various amounts of ouabain were added and the corresponding values of NADH oxidation were continuously monitored (up to 60 min) using a Gilford spectrophotometer. The final concentrations of ouabain in the assay medium varied from 10^{-9} M to 10^{-4} M. Inhibition percentage was calculated by comparing the activity in the presence of ouabain wiih that in the control after correcting for the ouabain-insensitive ATPase activity measured in the presence of 2 x 10^{-3} M ouabain. This basal Mg^{2+}-ATPase activity was similiar to the (Na^+,K^+)-ATPase activity measured in a K^+-free assay medium (K^+ ions being compensated by choline chloride).

The (Na^+,K^+)-ATPase activities accounted for about 70 % of the total ATPase activities of the preparations from either normal or hypertrophied hearts. There was no significant difference in either the yield of sarcolemmal proteins (0.3 mg/g of heart) or in the average specific activity of (Na^+,K^+)-ATPase (105 ±

16 μmol of phosphate liberated per mg of protein per hour) for sarcolemma preparations from normal and hypertrophied hearts.

(^3H)-Ouabain binding assays. Equilibrium binding of (^{3}H)-ouabain was measured after 60 min at 37°C in one ml of incubation medium containing 0.04 mg protein, 4 mM $MgCl_2$, 4 mM ATP, 100 mM NaCl and 40 mM imidazole-HCl pH 7.4 (buffer I) with increasing concentrations of ouabain (6 x 10^{-9} M up to 3 x 10^{-6} M and specific radioactivity varying from 19 Ci/mmol to 0.6 Ci/mmol). Duplicate aliquots of 450 μl were filtered under vacuum on HA WP Millipore® filters (0.45 μm). The filters were rinsed three times with 4 ml of ice-cold buffer I pH 7.4. The radioactivity that remained bound to the filters was measured by liquid scintillation counting (Bio Fluor, NEN) after addition of 1 ml of ethylene glycol monomethyl ether. Specific binding was calculated by subtracting the binding observed in the presence of 10^{-3} M unlabelled ouabain. Binding experiments were analysed by Scatchard plots which indicate the existence of high and low affinity sites. Curvilinear plots were fitted with an Apple IIe computer according to a two sites model assuming the existence of two independent classes of ouabain sites and allowing the calculation of dissociation constants and numbers of binding sites. The free concentration of ouabain was assumed not to be significantly reduced by binding sites, since the concentration of binding sites was less than 10% of the K_D.

Kinetics of (^3H)-ouabain dissociation. Cardiac cell membranes (0.04 mg protein/ml) and 10^{-8} M (^{3}H)-ouabain (specific activity of 19 Ci/mmole) were incubated for 60 min at 37°C as for (^{3}H)-ouabain binding assays. Dissociation of (^{3}H)-ouabain was started by a ten fold dilution of the incubation medium with unlabelled ouabain (10^{-3} M final). At the appropriate points of time, aliquots of 4.5 ml were filtered and the unspecific binding amounted to less than 10% of total radioactivity bound to the membranes.

Immunodetection of the α chain of the (Na^+,K^+)-ATPase. Sarcolemma vesicles (40 μg) were submitted to an S. D. S. electrophoresis according to Sweadner (2) with a 5–10% polyacrylamide gradient. After electrophoresis, the bands were transferred on nitrocellulose sheets and immunodetection of the α and α^+ bands were performed with an anti α antibody (gift from K. Geering, Lausanne). Detection was realised by colloidal gold labelled antibodies (Janssen).

Inotropic response to ouabain was performed on isolated perfused heart under isovolumic conditions (cf. Ref (8) and L. G. Lelièvre, this volume) at 10^{-5} M ouabain.

Results

The effects of a low level of Ca^{2+} were studied on sarcolemma preparation as well as on isolated perfused hearts.

(Na^+,K^+)-ATPase sensitivity to ouabain (Fig. 1 and Table 1)

The sarcolemma vesicles obtained from normal rat heart perfused under physiological Ca^{2+} conditions exhibited only a low sensitivity to ouabain (IC_{50} = 1–3 x 10^{-5} M) whereas the vesicles obtained after a perfusion of the heart with a low level of Ca^{2+} showed a detectable inhibition (40%) of the (Na^+,K^+)-ATPase at 10^{-7} M ouabain (8, 9).

The hypertrophied heart was insensitive to the perfusion medium (low Ca^{2+} or physiological Ca^{2+} concentration), as there was no difference in the dose response curve of the (Na^+,K^+)-ATPase to ouabain whatever the origin of the vesicles.

Characteristics of the high-affinity ouabain binding sites (Table 1)

On normal rat heart, the low Ca^{2+} level perfusion induced an increase in the number of high affinity sites (x 2) and in the relative proportion of H. A./L. A. (x 10) without any change of the K_D or of the dissociation rate constants.

On hypertrophied rat heart there was only a slight increase in the number of H. A. sites (x 1.2) with a similar proportion of H. A./L. A. sites without any change of the K_D or of the dissociation rate constants. -α form of the (Na^+,K^+)-ATPase (Fig. 2).

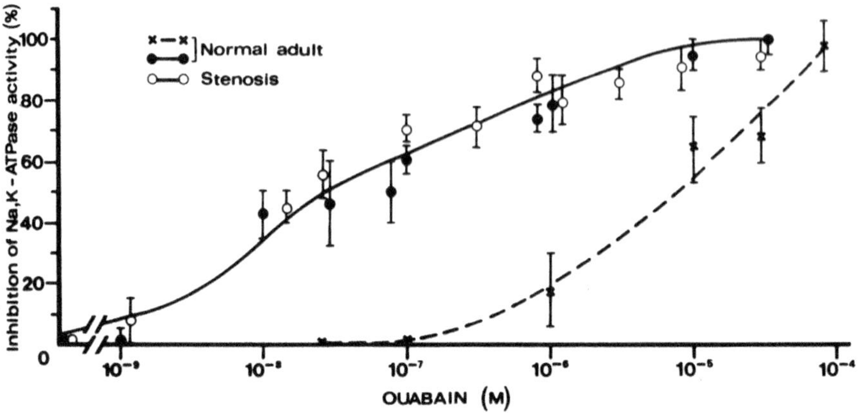

Fig. 1. Ouabain inhibition of the (Na^+,K^+)-ATPase from sarcolemma vesicles of rat hearts. Normal adult heart perfused with 2.5 mM (x–x) or 0.25 mM (●–●) Ca^{2+} concentrations.
Hypertrophied rat heart perfused either with 2.5 or 0.25 mM Ca^{2+} concentration (O–O).

Table 1. Comparison of the IC_{50} (concentration of ouabain which induces a half-maximum inhibition of the (Na^+,K^+)-ATPase) and of the results obtained with ouabain binding assays between sarcolemma vesicles from normal or hypertrophied rat hearts perfused with a physiological Ca^{2+} concentration ($+Ca^{2+}$) or with a low Ca^{2+} concentration ($-Ca^{2+}$).

Perfusion medium		Normal heart $+Ca^{2+}$	Normal heart $-Ca^{2+}$	Hypertrophied heart $+Ca^{2+}$	Hypertrophied heart $-Ca^{2+}$
IC_{50}	(10^{-8} M)	–	2.0	1–3	1–3
	(10^{-5} M)	0.7	0.7	0.1	0.1
Ouabain	K_D (10^{-8} M)	2.0–4.0	2.2–3.8	1.8–3.2	3.0–5.0
Binding	Bmax (pmol/mg)	1.6–2.6	3.2–4.4	3.3–4.3	4.3–5.0
Assays	k^{-1} (10^{-4} s^{-1})	10	10	2.8	2.6
	% H. A./L. A.	2	22	3.2	4.1

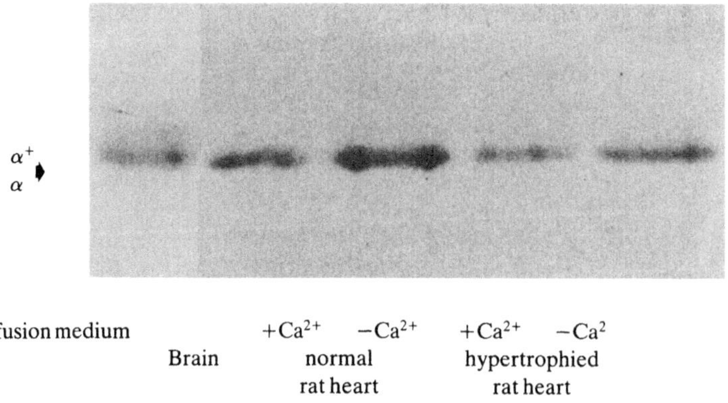

Fig. 2. Immunodetection of the α chain of (Na^+,K^+)-ATPase.

In sarcolemma vesicles from normal or hypertrophied heart only the α form of the enzyme was detected by the specific antibody whereas the α^+ form of the brain appeared under the same experimental conditions.

– Inotropic effect of ouabain at 10^{-5}M ouabain (Fig. 3).

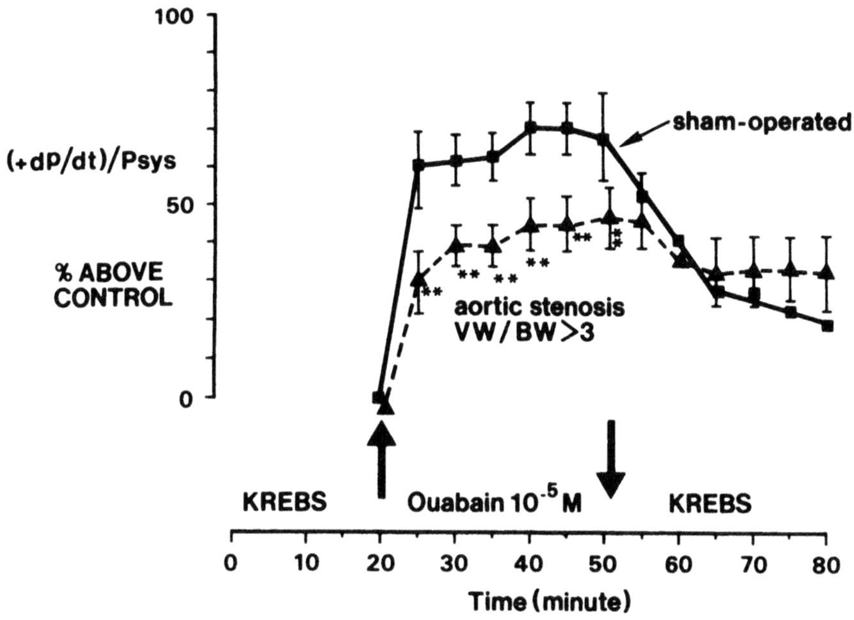

Fig. 3. Positive inotropic effect of ouabain (10^{-5} M) on isolated perfused rat hearts at 0.25 mM Ca^{2+} (cf. L. G. Lelièvre, this volume).

On normal rat heart the low Ca^{2+} concentration (0.25 mM) of the perfusion medium allowed the detection of a positive inotropic effect which was not observed at 2.5 mM Ca^{2+} at which only the toxic effect appeared.

On hypertrophied rat heart the detection of a prolonged positive inotropic effect was observed. This increase in the contractile force would still exist at physiological Ca^{2+} level (C. Mouas, preliminary results).

Conclusion

From these results, we could conclude that a low Ca^{2+} concentration in the perfusion medium of the heart revealed a high sensitivity form in normal heart but did not modify the properties of the (Na^+,K^+)-ATPase forms in hypertrophied heart.

1. It seemed unlikely that the high sensitivity forms (in normal or hypertrophied heart) came from a different membrane domain selected during the purification procedure as:

– Identical protein recovery and enzyme activities enrichment have been found in all sarcolemma preparations.

– Identical numbers of nitrendipine binding sites/mg of protein have been quantified (E. Mayoux, personal communication).
– Only the α form of the enzyme can be detected in normal or hypertrophied heart, whatever the perfusion medium.
2. There is no direct relationship between the presence of a high sensitivity form of (Na^+,K^+)-ATPase and a low Ca^{2+} concentration: the high sensitivity form of normal heart was revealed in sarcolemma preparation which contained 7 nM Ca^{2+}/mg of protein whereas the high sensitivity form of hypertrophied heart was present in sarcolemma preparations containing 65 nM Ca^{2+}/mg of protein.
3. We propose that:

In normal heart, a complex regulatory mechanism would exist, implying Ca^{2+} and a putative regulatory factor (3, 6, 7) absent at low Ca^{2+} concentration.

In hypertrophied heart, this factor would be absent or inefficient on the new form of (Na^+,K^+)-ATPase whatever the concentration of Ca^{2+}.

Acknowledgements

This work was supported by grants from the M. I. R. P. R. C. Medicaments. Convention n° 131 030 (L. G. L.), the Université Paris 7. n° C. 32RO 1 (L. G. L.), the Fondation des Maladies Cardiovasculaires (B. S.) and the M. G. E. N. (B. S.).

References

1. Allen JC, Schwartz A (1969) J Pharmac exp Ther 168: 42–46
2. Charlemagne D, Maixent JM, Preteseille M, Lelievre L (1986) J Biol Chem (in press)
3. English LH, Epstein J, Cantley L, Housman D, Levenson R (1985) J Biol Chem 260: 1114–1119
4. Erdmann E, Philipp G, Scholz H (1980) Biochem Pharmacol 29: 3219–3229
5. Fabiato A (1982) Federation Proc 41: 2238–2244
6. Geny B, Paraf A, Fedon Y, Charlemagne D (1982) Biochim Biophys Acta 692: 345–354
7. Lelievre LG et al (1985) Eur J Biochem 148: 13–19
8. Lelievre LG et al (1986) Am J Physiol (in press)
9. Mansier P, Lelievre L (1982) Nature 300: 535
10. Mansier P, Charlemagne D, Rossi B, Preteseille M, Swinghedauw B, Lelievre LG (1983) J Biol Chem 258: 6628–6635
11. Narayanan N (1981) Biochim Biophys Acta 678: 442–459
12. Repke K, Est M, Portius HJ (1965) Biochem Pharmac 14: 1785–1802
13 Sweadner KJ (1979) J Biol Chem 254: 6060–6067

Author's address:

D. Charlemagne, M. D., INSERM U 127, Hôpital Lariboisière, Université Paris 7, 75010 Paris (France)

Biphasic contractile response to ouabain: Species specific? Calcium dependent? Altered sensitivity?

G. Grupp, I. L. Grupp, T. Hickerson, S. W. Lee* and A. Schwartz

Departments of Pharmacology and Cell Biophysics, Physiology and Biophysics, and Medicine, University of Cincinnati, College of Medicine, Cincinnati (U.S.A.)

Summary

Up to now, only the rat, an "ouabain insensitive" rodent, has shown a consistent biphasic positive inotropic response to ouabain, "low-dose" effect $ED_{50} \sim 0.3–0.5\,\mu M$, "high dose" effect $ED_{50} > 10\,\mu M$. The "low-dose" (high affinity?) effect was masked by high Ca^{++} (2 mM) and emphasized in low Ca^{++} (0.5 mM). The "low-dose" inotropic effect in rat ventricular strips was significantly diminished by repeated exposure to high ouabain concentrations as was ^{3}H-ouabain binding (K_D and B_{max}). The decrease of the "low-dose" effect could be desensitization, tachyphylaxis (quick tolerance) or continued receptor occupancy by residual ouabain. The contractile activity of the preexposed strips did not indicate the presence of residual ouabain since their basal contractile force was decreased 12% compared to initial control instead of an expected 20% increase if ouabain occupancy had continued. The maximal response to high ouabain concentrations was identical in both exposures. In Langendorff perfused rat heart preparations the "low-dose" response was not diminished by repeated ouabain exposure and ^{3}H-ouabain binding was not altered. The "low-dose" effect was not present in rat atrial strip preparations.

Although the I_{50} of rat Na^+,K^+-ATPase (NKA) inhibition by ouabain is very high ($\sim 50\,\mu M$) and the low dose effect occurred with an ED_{50} of $0.3–0.5\,\mu M$, the low dose effect was nevertheless related to inhibition of NKA since it was accompanied by an increase in Na_i activity (12, 16). The minimal effect of Na_i was observed at $< 0.1\,\mu M$ ouabain.

Up to now, no other species has consistently shown a biphasic positive inotropic response to ouabain, although episodic observations have been reported (15). In rabbit, cat and guinea-pig Langendorff hearts, and atrial and ventricular strip preparations only monophasic, not biphasic dose-response curves (10^{-11} to 10^{-6} M) could be obtained. Low calcium (0.8 mM) did not bring about an appearance of the lacking "low-dose" response in these "glycoside-sensitive" species. In guinea-pig, rabbit and cat hearts (9) no biphasic or other "low-dose" effects were produced by alteration of temperature (30–37° C); stimulation rate (1–4 Hz); time (immediate exposure or 2 to 4 hour equilibration with possible loss of external Ca pools); alterations of on concentrations of Mg (0.1 to 1.2 mM), K^+ (2.7 to 9 mM) or Ca^{++} (0.5–2.5 mM); and repetitive exposure to ouabain.

Introduction

The interest in and the intensity of the quest for two or more site glycoside receptor(s) is underscored by the 1983 Munich Symposium on "Cardiac Glycoside Receptors and Positive Inotropy: Evidence for more than one receptor?" (4). These questions were based on re-

* Present address: Dr. S. W. Lee, College of Pharmacy, Yeungnam University, Gynangsan 632, Korea.

ports about high and low affinity ^3H-glycoside binding sites; dissociation of glycoside inotropy and directly or indirectly determined Na^+,K^+-ATPase (NKA) inhibition; positive or negative inotropy at nM concentrations of glycosides clearly outside of the range of monophasic, steep dose-response curve; and others (see Erdmann editorial in (4)).

A major upsurge for the quest was provided when Erdmann and his coworkers (3) reported in 1980 about a positive inotropic effect in rat heart strips (ED_{50} of 0.3 μM), far below the previously established rat heart ED_{50} of > 10 μM. By now it has become clear that the rat ventricle (5, 8, 13) and the whole heart of the rat (14, 15), but not rat atria (6, 8), respond with a biphasic positive inotropic effect to cardiac glycosides. Observation of biphasic inotropic responses to glycosides in other species have been episodically reported ((15), see also (9)) – but have not yet been systematically reproduced in other laboratories. Many of the other discrepancies in the inotropy receptor relationship could not be repeated by independent laboratories or found other explanations not requiring a correction of the generally accepted glycoside-receptor hypothesis (9). Recently, Mansier and LeLievre (18, 19) demonstrated that rat NKA displayed two or more sensitivities to glycosides when Ca^{++} was removed from the perfusion fluid. We therefore felt that it was important to find out (a) whether there is generally an appearance of two or more inotropic effects in heart tissue strips or in perfused hearts of other species which appear more sensitive to ouabain than the rat, e. g. guinea-pig, rabbit and cat; (b) whether low calcium would bring about a biphasic inotropic response curve; and (c) whether the desensitization of the "low-dose" inotropic effect produced by repeated exposure to ouabain reported in rat ventricular strips (8, 10) can be modified by low calcium.

Methods

Pharmacology: Isolated heart and ventricular strip preparations: An extensive description of the methods was previously published (11). Briefly, the animals were anesthetized with pentobarbital, the hearts quickly removed and perfused retrogradely through the aorta with heparinized and oxygenated (95% O_2:5% CO_2) Krebs-Henseleit solution at 37°C (2 mM or 0.8 mM or 0.5 mM Ca^{2+}, 5.9 mM HK^+), pH 7.4. The left ventricular pressure was recorded through a thin catheter, and the transducer signal differentiated to record dP/dt. Based on our experience (11), we decided to avoid the unnecessary trauma to the ventricle produced by the introduction of an intraventricular balloon. The hearts were paced at a frequency about 20% above spontaneous heart rate. The perfusion pressure was adjusted to about 60 mm Hg. Coronary flow was continuously measured by weight with an electronic balance. After 30 min equilibration, ouabain 10^{-10} to 10^{-4} M in rat hearts, 10^{-11} to 10^{-6} M in cat, rabbit and guinea-pig hearts was infused using a Harvard infusion pump, each concentration for 10 min for the rat, guinea-pig, and rabbit heart and for 30 min for the cat heart at a constant rate through catheters inserted into the aortic flow line 10 cm above the heart.

Ventricular strips from rat hearts (11) were used for the strip experiments at 35°C in Tyrode solution (mM): 137 NaCl, 5.0 KCl, 1.8 $CaCl_2$, 1.05 $MgSO_4$, 0.42 KH_2PO_4, 11.9 $NaHCO_3$ and 10 glucose, saturated with 95% O_2 and 5% CO_2 at a pH of 7.0. The strips were studied under isometric conditions with an initial load of 1 g and a stimulation rate of 1 Hz (stimulus duration of 2 msec, voltage 20% above threshold, usually 3 to 5 V). Resting and active tension were measured. The preparations were equilibrated for 60 to 120 minutes. All dose-response curves were obtained on a cumulative basis and the ED_{50} was determined both graphically and estimated using a non-linear least square regression program (14). Values given are means ± SEM.

Ouabain binding: Ventricular *homogenate* was prepared by homogenization of 30 to 40 mg of perfused ventricle or electrically stimulated right ventricular strip with glass homogenizer in the cold, in 1–2 mls of a medium containing 50 mM Tris-Cl (pH 7.4, 37°C), 5 mM $MgCl_2$, and 5 mM Tris Pi. *Microsomal* fractions were prepared from perfused rat hearts after atria, connective tissue, and major vessels were removed.

The hearts were homogenized with a polytron PT-20 at one-half maximal speed in 5 volumes of a solution containing 10 mM $NaHCO_3$, 5 mM NaN_3, and 0.25 M sucrose at a pH of 7.0 and centrifuged at 87,000 x g for 20 min. The microsomal fraction was prepared by the method of Van Alstyne et al. (21) and protein was estimated by the method of Lowry et al. (17). The microsomal protein yields in mg per g of ventricle for the control and ouabain perfused heart were 2.96 ± 0.13 and 2.95 ± 1.10 mg protein per gram of heart (n = 3), respectively.

Binding. Aliquots of 0.1 and 0.2 mg protein per ml for microsome and homogenate, respectively, were added to a solution containing 50 mM Tris-Cl, 5 mM $MgCl_2$, 5 mM Tris Pi, and 0.1 μM [^{3}H]-ouabain (mCi/mol) and various concentrations of unlabeled ouabain. The reaction mixture was rapidly filtered on 0.45 μM methylcellulose filters (Gelman) by a filtration method (22). Preliminary experiments indicated that this incubation time was enough to reach equilibrium. The filters were then rinsed three times with 5 ml of cold 100 mM KCl. The filters were placed in scintillation fluid and radioactivity was measured in a liquid scintillation counter. Nonspecific binding measured in the presence of 4 mM unlabeled ouabain was subtracted from the total to obtain specific binding. The ouabain binding site concentration and the K_D for binding were estimated using a displacement assay as described by Akera and Cheng (2).

Results and discussion

It has by now been generally accepted that in the rat heart the dose-response curve of the positive inotropic effect produced by ouabain is biphasic. Table 1 summarizes the reports of 6 independent laboratories. Note: some of the values in Table 1 have been estimated by us from published data. The positive inotropic effect contains a "low-dose" component, ED_{50} < 1 μM and maximum about 110–140 % of control, and a "high-dose" component, ED_{50} > 10 μM, maximum about 130–210 % of control. The "low-dose" effect may be related to a high affinity binding site (12) of the receptor NKA. The "high-dose" effect is generally considered to be well correlated with NKA inhibition (20). Biphasic curves (examples see Fig. 1) were obtained in rat ventricular strips (5, 8, 13) and in isolated rat hearts (14, 15) but not in rat atrial strips (5, 8). It is possible that there is also a very small "low-dose" effect in

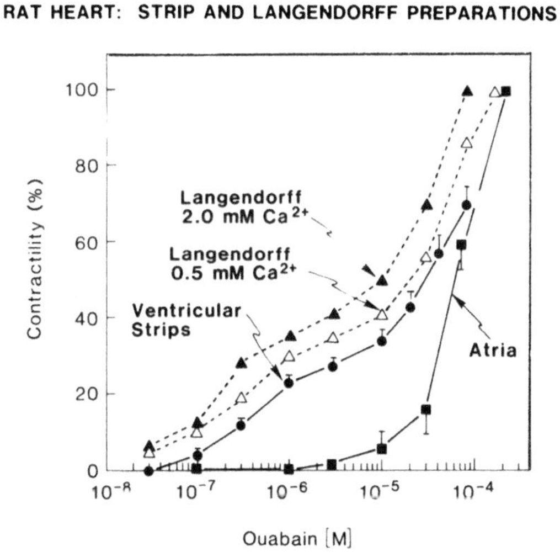

Fig. 1. Dose-response curve (10^{-8} to 10^{-4} M) of ouabain effects on contractility of the rat heart. Ordinate: percent of control. ● = ventricular strips (n = 83); ■ = atrial strips (n = 20); ▲ = rat hearts perfused at 2.0 mM Ca^{++} (n = 22); △ = rat hearts perfused at 0.5 mM Ca^{++} (n = 18).

Table 1. Positive inotropic effects of ouabain on isolated rat Langendorff or ventricular strip preparations (a = values estimated from published data)

Authors and conditions	Preparations	Ouabain effects on cardiac contraction			
	VS = ventr. strips LP = Langendorff prep. IMT = isometr. tens. ITT = isotonic tens. dP/dt = LVP dev. tens.	"Low dose" ED_{50} μM	"Low dose" Maximum % of Cont.	"High dose" ED_{50} μM	"High dose" Maximum % of Cont.
Fricke et al., 1975, Fig. 1 0.9 Ca, 5.4 K, pH 7.3, 37°C	LP-IMT n = 62; 4 Hz	–	(120%[a] at 8μM)	24	196% at 60μM
Erdmann et al., 1980, Fig. 7b, 8 1.8 Ca, 5.9 K, pH 7.4, 35°C	VS-IMT n = 52; 1 Hz	0.3	(130%[a] at 3μM)	–	140% at 30μM
Grupp et al., 1981, Tabl. 1A, 1 1.8 Ca, 5.42 K, pH 7.4, 35°C	VS-IMT n = 43; 1 Hz	0.55	137% at 3μM	20	198% at 80μM
same Table 1A, 2 2.0 Ca, 5.9 K, pH 7.4, 37°C	VS-IMT n = 28; 3 Hz	0.46	131% at 3μM	22	197% at 80μM
Koomen et al., 1982, Fig. 1 1.4 Ca, 4.7 K, pH 7.35, 37°C	LP-ITT n>40; 5 Hz	0.1[a]	115%[a] at 3μM	30[a]	155 at 100μM
Finet et al., 1982, Fig. 1 30°C	VS-IMT n>40; 1 Hz	<1[a]	140%[a] at 10μM	90[a]	210%[a] at 1000μM
same, 1983, Fig. 2a 1.82 Ca, 6 K, 30°C	VS-IMT n = 20; 1 Hz	0.61	140% at 10μM	270	193% at 1000μM
Herzig et al., 1984, Fig. 1 1.8 Ca, 5.4 K, 35°C	VS-IMT n<17; 1 Hz	<1[a]	~140% at 3μM	>10[a]	160% at 30μM
Hickerson, 1985, Table 5 2.0 Ca, 5.9 K, pH 7.4, 37°C	LP-dP/dt n = 21; ~4 Hz	0.3	110% at 3μM	50	130% at 140μM
same, Table 15 0.5 Ca, 5.9 K, pH 7.4, 37°C	LP-dP/dt n = 18; ~3 Hz	0.3	168% at 3μM	80	330% at 140μM

the rat atria (6). But this effect, if present, does not visibly influence the dose-response curve. Godfraind's group (6) reported the capacity of a high affinity binding site they observed in rat atria to be 4 times lower than the one they observed in the ventricle occupying only 3% of the total number of ouabain binding sites.

We analyzed the ouabain dose-response curves of 22 rat heart Langendorff preparations obtained in the presence of 2 mM external Ca^{++} to determine whether the obtained curves would be fitted better with a one-site or a two-site model (14). Table 2, upper part, shows that 8 of the hearts had dose-response curves fitted best with the one-site model where the "low-dose" effect was absent or negligible, whereas 14 hearts could better be fitted with the two-site model. Since Mansier and colleagues (18, 19) observed a better expression of a high affinity site of the rat heart NKA in the absence of Ca^{++}, we repeated the experiments in 18 hearts at 0.5 mM Ca^{++}, the lowest Ca^{++} concentration possible under which the rat hearts would still function. We found that there were no monophasic curves and all dose-response curves could be fitted best with the two site model. Furthermore, the variation (Table 3) of the "low-dose" effect around its mean value was reduced from 11.2% at 2 mM Ca^{++} to 2.8% at 0.5 mM whereas the variation of the "high-dose" effect remained the same (10.6%). From this we concluded (14) that high external calcium suppresses the expression

Table 2. Distribution in rat Langendorff hearts of predominantly monophasic or biphasic ouabain/inotropy dose-response curves at low and high external calcium

2 mM Ca^{++}		"Low dose" ED_{50} mM	"Low dose" % of Control	"High dose" ED_{50} mM	"High dose" % of Control
Monophasic	computed	–	–	120	152%
8 of 22 exp.	graphic.	–	–	60	134%
Biphasic	computed	0.2	112%	150	151%
14 of 22 exp.	graphic.	0.2	112%	20	125%
0.5 mM Ca^{++}					
Monophasic	computed	–	–	–	–
0 of 18 exp.	graphic.	–	–	–	–
Biphasic	computed	0.27	170%	87	395%
18 of 18 exp.	graphic.	0.3	169%	80	331%

Table 3. Isolated rat Langendorff hearts. Effects of ouabain on contractility (+dP/dt)

	At low Ca^{++} (0.5 mM) (basal contractility 200–600 mm Hg/sec) Low dose (at 3μM)		High dose (10–100μM)	At high Ca^{++} (2.0 mM) (basal contractility 1400–2350 mm Hg/sec) Low dose (at 3μM)		High dose (10–100μM)
	I*	II*	I	I	II	I
	44%	35%	125%	0%	0%	33%
	45%	34%	133%	0%	0%	20%
	50%	33%	152%	0%	0%	57%
	50%	36%	138%	3%	10%	29%
	53%	39%	135%	3%	18%	17%
	57%	44%	130%	3%	15%	20%
	58%	44%	132%	6%	32%	19%
	65%	45%	145%	6%	19%	31%
	68%	41%	165%	7%	18%	38%
	72%	28%	260%	7%	23%	31%
	79%	42%	186%	7%	30%	23%
	90%	33%	270%	7%	54%	13%
	100%	47%	213%	7%	50%	14%
	100%	47%	380%	10%	50%	20%
	100%	41%	242%	10%	37%	27%
	138%	46%	300%	11%	25%	44%
	175%	48%	362%	13%	48%	27%
	200%	47%	428%	14%	33%	43%
				21%	64%	33%
				21%	40%	53%
				23%	74%	31%
				27%	68%	40%
Mean	85.8%	40.6%	216.4%	12.1%	34.8%	30.1%
±SEM	±10.6	±1.4	±22.9	±1.6	±3.9	±2.6
	n=18	V=2.8%		n=22	V=11.2%	

* I = Percent increase over basal contractility
II = Percent of maximal ouabain effect
V = Variation

of the "low-dose" contractile force effect as it does the high affinity ouabain sensitive site of the NKA (18, 19).

We reported earlier (8, 10) that in rat ventricular strips the "low-dose" inotropic effect disappears or is diminished significantly when the ouabain dose-response curve (10^{-7}–10^{-4} M) is repeated after the washout of the effects of the first curve, whereas the maximal "high-dose" effect remained the same in both exposures to ouabain. We repeated these experiments in rat ventricular strips (Fig. 2, left panel) and in high and low calcium Langendorff perfused rat hearts (Fig. 2, middle and right panel). We again found the previously re-

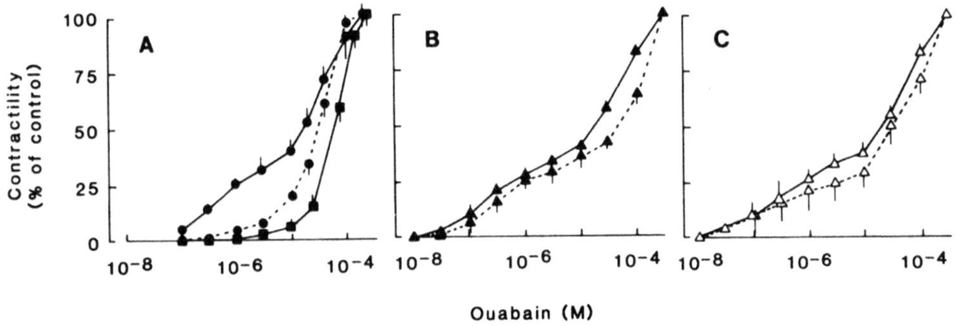

Fig. 2. Repetition of ouabain dose-response curves in rat heart. Solid line = first dose response curve; broken line = second dose-response curve obtained 120 min (strips) or 60 min (perfused hearts) after washout of the first ouabain exposure (10^{-8} to 10^{-4} M). Left panel: ● = rat ventricular strips (n = 26); ■ = atrial strips (n = 20); Middle panel: ▲ = rat hearts perfused at 2.0 mM Ca^{++} (n = 22); Right panel: △ = rat hearts perfused at 0.5 mM Ca^{++} (n = 18).

ported desensitization in the ventricular strip preparations but there was *no* change in the "low-dose" effects in both sets of the Langendorff perfused hearts. One possible explanation for this difference could be that the hearts perfused through their own coronary systems could more effectively remove the ouabain from myocardial cells compared with the strip preparations, where the glycoside had to reach and leave its site of action in a much more complicated manner. We compared, therefore, the ^{3}H-ouabain binding in control Langendorff and ventricular strip preparations which were not previously exposed to ouabain, with Langendorff and ventricular strip preparations that had been first exposed to a complete ouabain dose-response curve (10^{-7} to 10^{-4} M). In these hearts ^{3}H-ouabain binding was determined 120 min after ouabain had been removed from the preparations. We found that ^{3}H-ouabain binding was significantly reduced in the rat ventricular strips previously exposed to ouabain. In homogenates of the control strips (n = 5) the K_D of ^{3}H-ouabain binding was 500 ± 53.4 nM, B_{max} 1.66 ± 1.9 pmol/mg protein. In the strips pre-exposed to ouabain, K_D and B_{max} were significantly ($p > .01$) decreased to 240 ± 67.2 nM and B_{max} 0.78 ± 0.16 pmol/mg protein. On the other hand, ^{3}H-ouabain binding was not changed significantly comparing microsomes of control Langendorff hearts at 2 mM external Ca^{++} with hearts pre-exposed to ouabain. The control K_D was 990 ± 66 nM, B_{max} 4.94 ± 0.69 pmol/mg protein; the K_D of the pre-exposed heart was 963 ± 67.2 nM, B_{max} 5.40 ± 0.22 pmol/mg protein. It is not likely that the decrease in ^{3}H-ouabain binding after ouabain exposure in ventricular strips signifies that the ouabain receptor was still occupied by residual ouabain left behind from the first dose-response curve, because the functional evidence speaks against this assumption: If ouabain would still occupy the receptor, active tension should still be about 20–40% above the basal tension of 0.45 g, in actuality it was 0.40 g or 12% lower (10). The conclusion, therefore, has to be entertained that in rat ventricular strips a perfusion-dependent desensitization or tachyphylaxis of the "low-dose" (high affinity) effect is a distinct possibility, which is absent in Langendorff perfused hearts.

There have been occasional reports (for summary see (9)) about the presence of biphasic positive inotropic effects of ouabain in other species than rat (15). We have therefore exposed cat, rabbit and guinea-pig hearts in Langendorff preparations at 2 mM Ca^{++} to a wide range of increasing concentrations (10^{-11}–10^{-6} M) of ouabain. Since we found in the rat heart that reduction of Ca^{++} to 0.5 mM accentuated the "low-dose" effect of ouabain, we also obtained ouabain dose-response curves in cat, rabbit and guinea-pig hearts at low Ca^{++}. We used 0.8 mM Ca^{++} since these hearts did not function well at 0.5 mM Ca^{++}. Fig. 3 summarizes all experiments with these 3 species at high and low calcium. We could not observe a single biphasic dose-response curve either in cat, rabbit or guinea-pig hearts, whether at high or low calcium (Table 4).

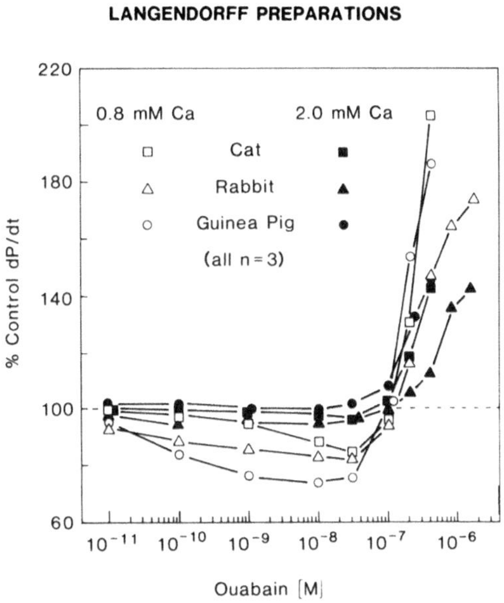

Fig. 3. Dose response curves (10^{-11}–10^{-6}) of ouabain inotropic effect on perfused cat, rabbit and guinea-pig hearts (all n = 3) at 2 mM and at 0.8 mM Ca^{++}.

The continued fall in contractility in some of the dose-response curves between 10^{-11} and 10^{-8} M in Fig. 3 are not negative inotropic effects produced by ouabain but rather continuations of the falling trend of basal contractility (see (9), Fig. 1 and 2). These observations in Langendorff hearts (Fig. 3 and Table 4) are in keeping with our earlier studies applying low ouabain concentrations to guinea-pig atrial strips and papillary muscles under multiple temperature, rate of stimulation, ionic and pH conditions (9). If there are "low-dose" effects in the so-called glycoside sensitive species they occur either only occasionally under specific conditions, or they are too small to detect, or they are so close to the major ("high-dose") effect that they are indistinguishable.

Table 4. Ouabain dose-response curves +dP/dt (mm Hg/sec)

	Ouabain concentrations										
0.8 mM Ca^{++}	Control	10^{-11}	10^{-10}	10^{-9}	10^{-8}	$3x10^{-8}$	10^{-7}	$2x10^{-7}$	$4x10^{-7}$	$8x10^{-7}$	$1.6x10^{-6}$
Rabbit	445	400	375	360	325	320	450	650	850	980	1050
	700	650	625	600	600	600	620	700	850	930	960
	550	530	500	480	480	480	550	680	800	900	940
mean±SEM	565±74	526±72	500±72	480±70	468±80	460±81	540±49	677±61	833±29	936±40	983±59
Cat	565	565	557	537	515	530	560	900	1450		
	750	750	730	715	700	690	745	1000	1350		
	475	470	450	425	370	290	400	450	850		
mean±SEM	597±81	595±82	579±82	559±84	528±96	503±116	568±99	783±105	1217±186		
Guinea-pig	450	430	340	300	290	300	390	575	700		
	305	300	255	235	215	220	320	488	590		
	375	350	340	330	330	340	458	675	821		
mean±SEM	376±42	360±38	312±28	288±28	278±34	286±35	390±40	579±54	704±67		
2 mM Ca^{++}											
Rabbit	1400	1400	1400	1400	1365	1350	1380	1400	1600	1750	1800
	1000	950	900	900	900	900	950	1100	1200	1500	1650
	1200	1200	1150	1150	1150	1200	1280	1350	1300	1650	1700
mean±SEM	1200±115	1183±130	1150±144	1150±144	1138±134	1150±117	1203±82	1283±92	1366±120	1633±73	1716±44
Cat	1180	1180	1180	1200	1200	1150	1250	1500	1850		
	825	860	855	825	800	825	850	950	1100		
	1025	1020	1015	1015	1000	990	1050	1225	1475		
mean±SEM	1025±88	1020±99.5	1015±95	1015±108	1000±115	1050±93	1050±115	1225±158	1475±216		
Guinea-pig	1150	1160	1150	1120	1150	1180	1300	1500	1700		
	1000	1050	1050	1030	1020	1020	1060	1350	1420		
	1075	1105	1100	1075	1050	1100	1130	1425	1560		
mean±SEM	1075±43	1105±32.0	1100±28.9	1075±26.0	1073±39.3	1100±46.2	1163±71	1425±43.30	1560±80.8		

Conclusions

So far the rat heart, considered to be insensitive to glycosides, is the only mammalian species with a distinct biphasic inotropic response to ouabain. The other, more glycoside-sensitive species show only monophasic inotropic dose-response curves. Whether the biphasic contractile response is absent in the other species, or it is insignificant, or it is so close to the main effect that it is difficult to distinguish, cannot be decided. When the biphasic inotropic effect is there as in the rat heart ventricle, it is significant, separable, and both phases reach over 2–3 orders of magnitude. The rat heart atria do not show a biphasic inotropic response to ouabain.

The "low-dose" effect in the rat heart (10^{-7} to 5×10^{-6} M) can be enhanced by lowering extracellular Ca^{++} to 0.5 mM, and it can be desensitized (tachyphylaxis?) in ventricular strips by repeated exposure to ouabain. The desensitization can not be observed in rat Langendorff preparations.

The "low-dose" inotropic effect is accompanied by an increase in Na_i (12, 16) indicating that the sodium pump may be involved in the "low-dose" effect as well as in the "high-dose" effect.

A model for the mechanism of action of ouabain in the rat heart ventricle (Fig. 4) is proposed for considering two ouabain binding sites, both located on the NKA enzyme (14). Binding to either of these sites results in inhibition of NKA and positive inotropy. One site has a high affinity for ouabain binding at low concentrations. This high affinity site is fully expressed only in low but not in high extracellular calcium. The second, high concentration, low affinity ouabain binding site is the classical site, not modulated by extracellular calcium, but by potassium (20).

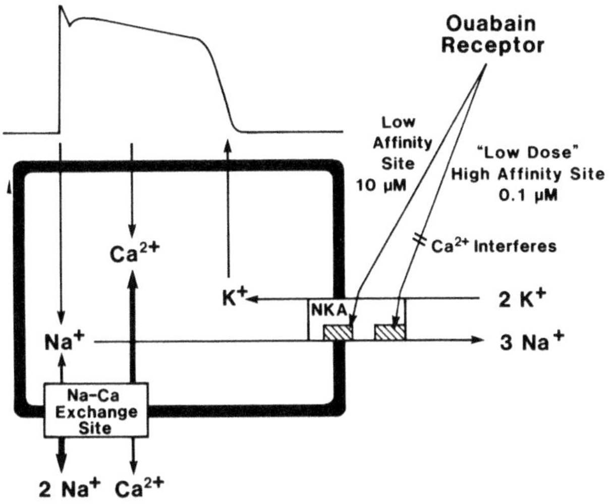

Fig. 4. Model of the proposed ouabain-NKA interaction in the cell membrane of the myocardial cell of the rat heart.

Acknowledgements

We appreciate the advice received from Dr. Earl T. Wallick, the editorial assistance of Ms. Julie Camele and Gwen Kraft and the technical assistance of Mr. Gilbert Newman and Robert Benza. This research was supported by a grant of the American Heart Association, Southwestern Ohio Chapter and by NIH Grant P01-HL-22 619 (IC and Core I).

References

1. Adams RJ, Schwartz A, Grupp G, Grupp IL, Lee SW, Wallick ET (1982) High-Affinity Ouabain Binding Site and Low-Dose Positive Inotropic Effect in Rat Myocardium. Nature 296: 167–169
2. Akera R, Cheng VJK (1977) A Simple Method for the Determination of Affinity and Binding Site Concentration in Receptor Binding Studies. Biochim Biophys Acta 470: 412–423
3. Erdmann E, Philipp G, Scholz H (1980) Cardiac Glycoside Receptor, $(Na^+ + K^+)$-ATPase Activity and Force of Contraction in Rat Heart. Biochem Pharmac 29: 3219–3229
4. Erdmann E (ed) (1984) Cardiac Glycoside Receptors and Positive Inotropy: Evidence for more than one receptor? Supplement Bas Res Cardiol 79: 7–160
5. Finet M, Noel F, Godfraind T (1982) Inotropic Effect and Binding Sites of Ouabain to Rat Heart. Arch Int Pharmacodyn 256: 168–170
6. Finet M, Godfraind T, Noel F (1983) The Inotropic Effect of Ouabain and its Antagonism by Dihydroouabain in Rat Isolated Atria and Ventricles in Relation to Specific Binding Sites. Br J Pharmac 80: 751–759
7. Fricke U, Hollborn U, Klaus W (1975) Inotropic Action, Myocardial Uptake and Subcellular Distribution of Ouabain, Digoxin and Digitoxin in Isolated Rat Hearts. Naunyn Schmiedeberg's Arch Pharmacol 288. 195–214
8. Grupp IL, Grupp G, Schwartz A (1981) Digitalis Receptor Desensitization: Ouabain produces two inotropic effects. Life Sci 29: 2789–2794
9. Grupp G, Grupp IL, Ghysel-Burton J, Godfraind T, Schwartz A (1982) Effects of Low Concentrations of Ouabain on Contractile Force of Isolated Guinea-Pig, Rabbit and Cat Atria and Right Ventricular Papillary Muscles. J Pharmacol Exp Therap 220: 145–151

10. Grupp G, DePover A, Grupp IL, Schwartz A (1984) Analysis of the Inotropic Action of Ouabain in Rat Ventricles: Two apparent ouabain inotropic responses. Proc Soc Exper Biol Med 175: 39–43
11. Grupp IL, Grupp G (1984) Isolated Heart Preparations Perfused or Superfused with Balanced Salt Solutions. In: Schwartz A (ed) Methods in Pharmacology, Vol 5. Plenum Publ, 111–128
12. Grupp IL, Im WB, Lee CO, Lee SW, Pecker MS, Schwartz A (1985) Relation of Sodium Pump Inhibition to Positive Inotropy at Low Concentrations of Ouabain in Rat Heart Muscle. J Physiol 360: 149–160
13. Herzig S, Mohr K (1984) Action of Ouabain on Rat Heart: Comparison with its effect on guinea-pig heart. Br J Pharmac 82: 135–142
14. Hickerson TW (1985) The Biphasic Inotropic Response of Ouabain in the Isolated, Retrogradely Perfused Rat Heart at High and Low Extracellular Calcium Concentrations. Inaug Diss Univ of Cincinnati Ohio, Univ Library
15. Koomen JM, Van Gilst WH, Zimmerman ANE, Van Noordwijk J (1982) A Concentration-Dependent Biphasic Positive Inotropic Action of Ouabain on Isolated Hearts of Rat and Guinea-Pig. Arch Int Pharmacodyn 255: 212–219
16. Lee CO, Im WB, Pecker MS, Grupp IL, Schwartz A (1983) Effect of Ouabain on Intracellular Sodium Activity and Twitch Tension of Rat Ventricular Muscle. Fed Proc 45: 581
17. Lowry OH, Rosebrough MJ, Farr AL, Randall RL (1951) Protein measurement with the Folin phenol reagent. J Biol Chem 193: 265–275
18. Mansier P, LeLievre LG (1982) Ca^{++} Free Perfusion of Rat Heart Reveals a Na + K^+-ATPase Form Highly Sensitive to Ouabain. Nature 300: 535–537
19. Mansier P, Cassidy PS, Charlemagne D, Preteseille M, LeLievre LG (1983) Three Na^+,K^+-ATPase Forms in Rat Heart as Revealed by K^+/Ouabain Antagonism. FEBS Lett 153: 357–360
20. Schwartz A, Matsui H, Laughter AH (1968) ^3H Digoxin Binding to NKA: Possible allosteric site. Science 195: 323–325
21. Van Alstyne E, Burch RM, Knickelbein RG, Hungerford RT, Gower EJ, Webb JG, Poe SL, Lindenmayer GE (1980) Isolation of Sealed Vesicles Highly Enriched with Sarcolemma Markers from Canine Ventricle. Biochim Biophys Acta 602: 131–143
22. Wallick ET, Anner BM, Ray MV, Schwartz A (1978) Effect of Temperature on Phosphorylation and Ouabain Binding to N-ethylmaleimide-treated (Na^+,K^+)-ATPase. J Biol Chem 253: 8778–8786

Author's address:

Prof. Dr. Ingrid Grupp, College of Medicine, University of Cincinnati, Cincinnati, Ohio 45267–0575 (U.S.A.)

The influence of α-tocopherol-nicotinate (Renascin®) on cardiac glycoside effects

H. Tawfik* and P. Schlieper**

Institute of Pharmacology, University of Düsseldorf (West Germany)

Summary

The influence of α-tocopherol-nicotinate and related compounds on cardiac glycoside effects on isolated guinea-pig atria and on the glycoside binding to guinea-pig heart homogenate was investigated.

α-tocopherol-nicotinate and dodecanoic acid had an inhibitory effect on the positive inotropic action of digoxin but not on that of ouabain. This effect was parallel to their inhibitory effect of digoxin and not ouabain binding to heart homogenate. The tested compounds showed a protective effect against cardiac glycoside induced arrhythmias which is not in correlation with their effects on the positive inotropic action of the glycosides or on their binding to heart homogenate. The protective effect may be due to a direct action of the drugs on the cell membrane.

1. Introduction

α-tocopherol-nicotinate is an ester of dl-α-tocopherol and nicotinic acid. It combines the beneficial effects of both vitamin E and nicotinic acid. α-tocopherol as antioxidant inhibits oxidation of unsaturated fatty acids and prevents the production of toxic lipoperoxides (2, 27–30), consequently cell membrane structure, its stability and function are maintained (11). Vitamin E is given in case of arterial vascular diseases of different genesis (22, 30–32). It has been reported that nicotinic acid was beneficial in the treatment of vascular dysfunction (26). Nowadays α-tocopherol-nicotinate is frequently used by aged persons to improve physical capacity (3) and in the therapy of diseases such as intermittent claudication (13), arteriosclerosis, peripheral blood circulation disturbances (17), cerebro-vascular insufficiency (21) and in labyrinthine deafness (23) to improve blood circulation. Hochrein and Helwig reported a protective effect of α-tocopherol against digitalis induced cardiac arrhythmias (10). The treatment of digitalis intoxications still appears to be a problem. In this work detailed investigations were carried out to study the effect of α-tocopherol-nicotinate on cardiac glycoside induced arrhythmias in isolated guinea-pig atria. Biochemical studies were performed to study the effect of α-tocopherol-nicotinate on glycoside binding to cardiac tissue.

* Present address: Department of Pharmacology, Faculty of Pharmacy, Cairo University, el Kasr el Aini street, Cairo (Egypt)

** Present address: Woelm Pharma GmbH & Co., P. O. Box 840, D-3440 Eschwege (West Germany)

2. Materials and methods

2.1 Materials and drugs
α-tocopherol-nicotinate (Renascin^R) was a generous gift from Mack, Illertissen, F.R.G.
Ouabain and digoxin were obtained from Boehringer GmbH, Mannheim, F.R.G.
^3H-labelled ouabain (specific activity = 14 Ci/mmol) and digoxin (specific activity = 13 Ci/mmol) were obtained from Amersham Buchler, Düsseldorf, F.R.G.
Scintillation liquid, Unisolve I, was obtained from Werner Zinsser, Frankfurt, F.R.G.
All other chemicals were obtained from Merck, Darmstadt, F.R.G.

2.2 Isolated guinea-pig atria
Guinea-pigs (male, 300–450 g) were killed by a blow on the back of the neck, the hearts were quickly excised and the atria dissected and fixed on two platinum electrodes. The preparations were suspended in a 50 ml organ bath containing Krebs-Henseleit solution at 31°C and aerated with carbogen (95 % O_2 and 5 % CO_2). The composition of Krebs-Henseleit solution in mmol was: NaCl 118.04, KCl 4.70, $CaCl_2$ 2.52, $MgSO_4$ 1.19, NaH_2PO_4 1.18, $NaHCO_3$ 24.90, glucose 10.09. Left atria were electrically stimulated using a square wave stimulator by impulses of 1 Hz, 1 ms pulse duration and intensity of 2.5 fold the threshold value. The preload applied to the atria was 1 g. The force of contraction was recorded under isometric conditions by an inductive force transducer (Statham-UC-2), and the signals appeared after amplification on a Watanabe recorder.

The atria were allowed to equilibrate for one hour before adding the cardiac glycosides to the bath cumulatively. As the response to each concentration reached a steady state of force of contraction the concentration in the organ bath was elevated by adding the next dose to the bathing medium. To study the effect of α-tocopherol-nicotinate, α-tocopherol, nicotinic acid and dodecanoic acid on the action of cardiac glycosides on isolated guinea-pig atria, the atria were preincubated with 100 μmol of the drug to be tested for one hour. Afterwards concentration response curves of the glycosides were performed. The compounds were dissolved in ethanol apart from nicotinic acid, which is soluble in water. Control experiments were carried out in the presence of the same amount of ethanol which was added to the organ bath with the dose of α-tocopherol-nicotinate, α-tocopherol or dodecanoic acid (100 μl).

2.3 Binding studies to guinea-pig heart homogenate
In accordance to Yamamoto et al. (33) guinea-pig hearts were cut with scissors, added to a 1 mmol EDTA/10 mmol Tris HCl solution (pH 7.5) (30 mg tissue/1 ml buffer) and homogenized with Ultra Turrax homogeniser for 30 s at 4°C. Samples were prepared containing radioactive ligand, ATP, different concentrations of unlabelled glycoside and buffer solution of the following constituents in mmol: NaCl 200, $MgCl_2$ 5 and Tris-HCl 50 (pH 7.5). The non-specific binding was estimated in samples devoid of ATP. To each sample 100 μl homogenate was added and incubated at 37°C for 20 min in experiments with ouabain and for 60 min in experiments with digoxin. The incubation time, previously determined in separate time-dependent binding studies, was the time at which equilibrium between radioactively labelled and unlabelled glycoside was reached.

Bound and free glycosides were separated by filtering the aliquots through nitrocellulose filters (type AA, pore size 0.8 μm, Millipore Corp.). The filters were dissolved in ethyleneglycol, 7.5 ml scintillation liquid was added to the samples, vigorously shaken and the radioactivity was measured in a β-liquid scintillation counter (Packard Tri-Carb.).

To study the effect of tested drugs on the binding behaviour of cardiac glycosides, heart homogenate was incubated with 100 μmol α-tocopherol-nicotinate or dodecanoic acid for 2 hours and 5 hours, then the experiments were continued as mentioned before. To exclude the effect of the solvent, control experiments were carried out after preincubation of the homogenate with the same amount of ethanol used to dissolve 100 μmol of the drug (100 μl).

2.4 Water solubility of long chain hydrocarbons
α-tocopherol-nicotinate, α-tocopherol and dodecanoic acid are sparingly soluble in water. When added to watery media the compounds form micelles. Aeration of this suspension with carbogen leads to opti-

cally clear dispersion. To ensure a homogenous dispersion of the drug molecules in Krebs-Henseleit solution, the particle size of the drug was measured by laser light scattering (24). These measurements showed that the multimolecular α-tocopherol-nicotinate structures are converted by time to much smaller structures, probably to tri-, di- or monomers, which are probably more easily taken up by biological tissues.

In all studies the results were the mean of at least 6 experiments and expressed as mean ± s.e.m.

3. Results

3.1 Isolated guinea-pig atria

The effect of α-tocopherol-nicotinate, α-tocopherol, nicotinic acid and dodecanoic acid on the rate of contraction of right atria and force of contraction of left atria were studied. A concentration up to 100 μmol of the mentioned drugs had neither chronotropic nor inotropic effects on guinea-pig atria. These drugs did have, however, different effects on the positive inotropic action of cardiac glycosides.

Fig. 1 shows concentration response curves of ouabain alone and in the presence of 100 μmol α-tocopherol, nicotinic acid, α-tocopherol-nicotinate and dodecanoic acid. None of the tested compounds except dodecanoic acid had any significant effect on ouabain action on left guinea-pig atria. The curve in the presence of 100 μmol dodecanoic acid shows a plateau falling about 30% below the control value. Similarly the effect of the same compounds on the action of digoxin on left atria is demonstrated in Fig. 2. The concentration response curves of digoxin in presence of α-tocopherol and nicotinic acid are not significantly different from the control curve. In the presence of α-tocopherol-nicotinate or dodecanoic acid the action of digoxin started at a concentration level about 10 times higher than in the case of digoxin alone. The concentration causing maximal increase in force of contraction, however, was the same as in control conditions. As in the case of ouabain, the curve in the presence of 100 μmol dodecanoic acid shows a plateau falling about 30% below the control value. The EC_{50} (the glycoside concentration causing 50% increase in force of contraction

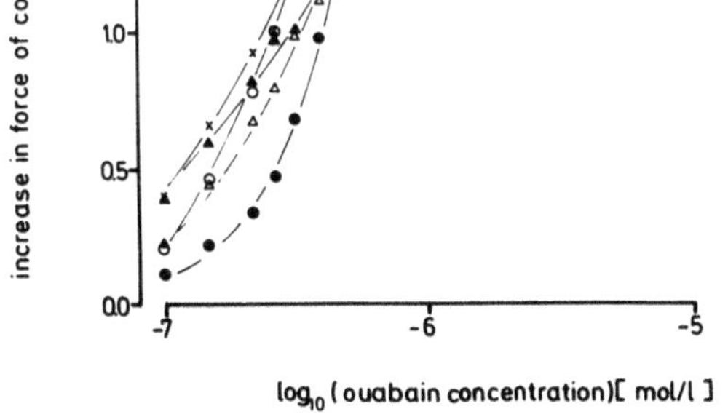

Fig. 1. Increase in force of contraction (Δ g) of guinea-pig left atria as a function of ouabain concentration under control conditions (X) and after pretreatment with different compounds (100 μmol/l each): ● α-tocopherol, ▲ nicotinic acid, Oα-tocopherol-nicotinate, Δ dodecanoic acid.

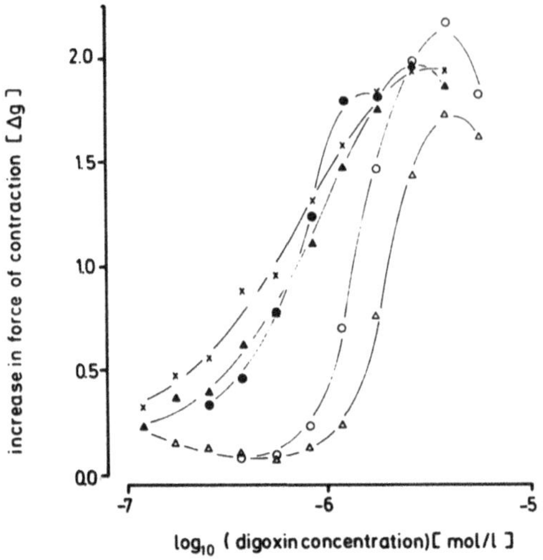

Fig. 2. Increase in force of contraction (Δ g) of guinea-pig left atria as a function of digoxin concentration under control conditions (X) and after pretreatment with different compounds (100 μmol/l each) ● α-tocopherol, ▲ nicotinic acid, ○ α-tocopherol-nicotinate, Δ dodecanoic acid.

of left atria) of ouabain was only slightly influenced by the presence of α-tocopherol, nicotinic acid and dodecanoic acid (Table 1). The EC_{50} of digoxin was highly influenced by the presence of α-tocopherol-nicotinate and dodecanoic acid. The two compounds respectively caused a shift in digoxin EC_{50} to values about 3 times and about 4 times the control value. Under normal conditions a subsequent elevation of the glycoside concentration in the bathing medium and exceeding the maximal inotropic concentration caused the appearance

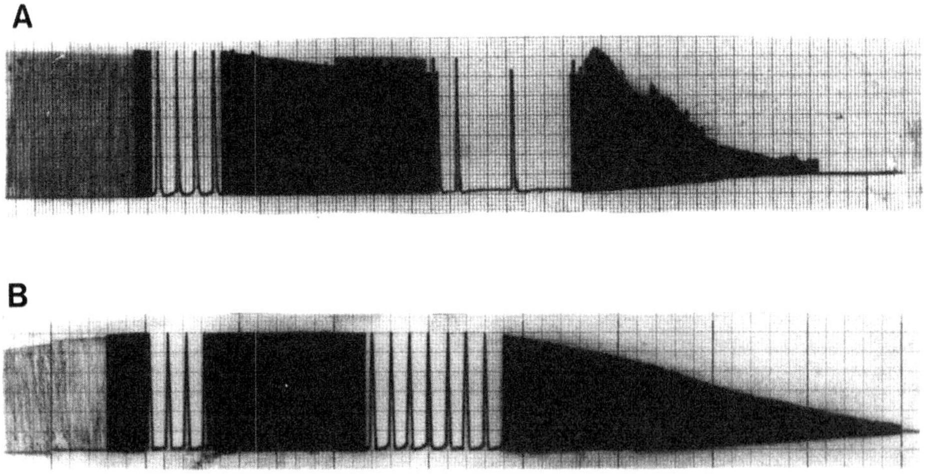

Fig. 3. Influence of digoxin on force of contraction of guinea-pig left atria in the absence (A) and presence (B) of α-tocopherol-nicotinate (100 μmol/l). (Original registrations).

of arrhythmias and an increase in diastolic tension as shown in Fig. 3 A. Toxic concentrations of ouabain and digoxin caused arrhythmias in 63 % and 69 % of the control experiments (Table 1). In the presence of α-tocopherol-nicotinate, α-tocopherol, nicotinic acid and dodecanoic acid the action of cardiac glycosides on the diastolic tension was not influenced. The development of arrhythmias, however, was markedly suppressed in the presence of α-tocopherol-nicotinate (Fig. 3B), which abolished the ouabain-induced arrhythmias and reduced the digoxin-induced arrhythmias to 6 %. Nicotinic acid had at least a protective effect against glycoside-induced arrhythmias in left atria, while α-tocopherol and dodecanoic acid had an intermediate protective effect (Table 1).

Table 1. The effective concentration which causes a 50 % increase in force of contraction (EC_{50}) in isolated guinea-pig left atria for ouabain and digoxin after pretreatment with different compounds and the percentage of arrhythmias observed (n = number of atria used).

Compound for pretreatment (100 μmol/1)	n	Ouabain EC_{50} (10^{-7}) mol	% arrhythmias	n	Digoxin EC_{50} (10^{-7}) mol	% arrhythmias
Control	8	2.50 ± 0.07	63	13	4.48 ± 0.5	69
α-tocopherol-nicotinate	6	2.62 ± 0.09	0	11	14.2 ± 0.11	6
α-tocopherol	10	3.62 ± 0.09	10	9	5.64 ± 0.12	25
Nicotinic acid	8	3.07 ± 0.06	50	8	6.02 ± 0.05	60
Dodecanoic acid	10	2.68 ± 0.07	20	8	18.5 ± 0.11	10

3.2 Binding studies to guinea-pig heart homogenate

The percentage of bound radioactive glycoside was plotted against the concentration of unlabelled glycoside (Fig. 4 and Fig. 5). The binding behaviour of ouabain was not greatly affected by the presence of α-tocopherol nicotinate and dodecanoic acid as shown in Fig. 5. In the case of digoxin the presence of α-tocopherol-nicotinate and dodecanoic acid caused a 7 % and 20 % shift respectively in the displacement curves. The maximal displacement, however, was not affected (Fig. 4) by the presence of any of the two drugs. Table 2 summarizes the K_D values of ouabain and digoxin in the presence and absence of α-tocopherol-nicotinate and dodecanoic acid (K_D = glycoside concentration which causes a displacement of 50 % of the radioactive ligand).

Prolonged preincubation of the homogenate with the tested compounds for 5 hours before starting the binding period did not give other results than those obtained after a 2 hour preincubation period.

Discussion

Our results demonstrate that α-tocopherol-nicotinate and dodecanoic acid partly inhibit the positive inotropic action of digoxin and not of ouabain while α-tocopherol and nicotinic acid had no effect on the positive inotropic action of the glycosides. All mentioned compounds suppressed the cardiac glycoside-induced arrhythmias in isolated guinea-pig atria to different degrees.

The antagonism to the positive inotropic action was not typically competitive as shown by the shape of the concentration response curves. Although shifted to the right, the curves in the presence of 100 μmol α-tocopherol-nicotinate and dodecanoic acid had their plateau at the same concentration as the controll curve. In presence of 100 µmol/l dodecanoic acid the curve showed a plateau falling about 30 % below the control value. The inhibitory effect of

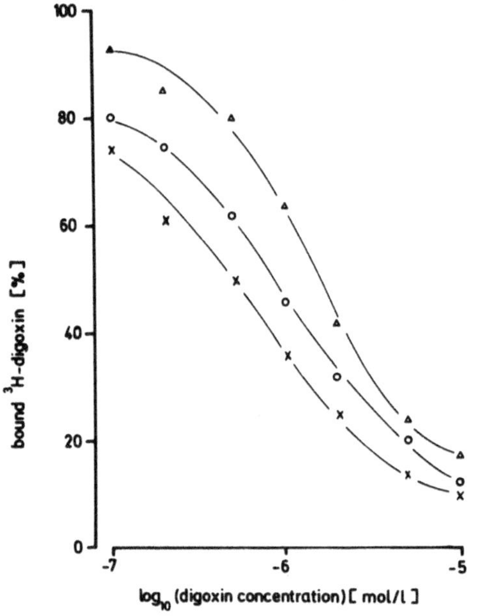

Fig. 4. Percent displacement of ^{3}H-digoxin by "cold" digoxin under control conditions (X) and after pretreatment with 100 μmol/l α-tocopherol-nicotinate (O) or dodecanoic acid (Δ).

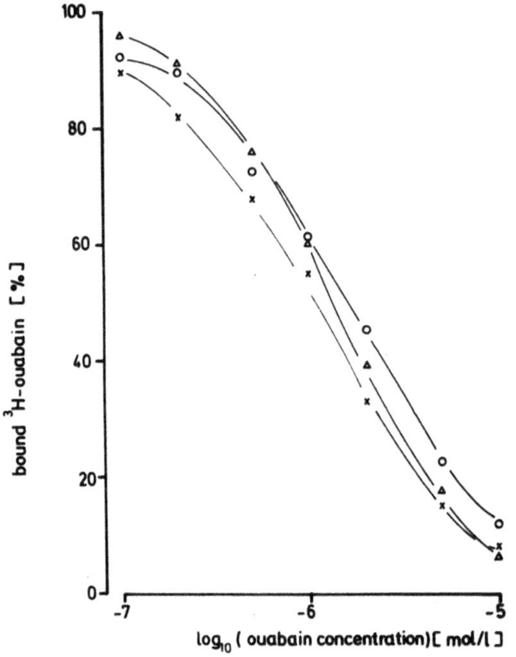

Fig. 5. Percent displacement of ^{3}H-ouabain by "cold" ouabain under control conditions (X) and after pretreatment with 100 μmol/l α-tocopherol-nicotinate (O) or dodecanoic acid (Δ).

Table 2. Glycoside concentrations which cause a 50% displacement of radioligand from guinea-pig heart homogenate (K_D) for ouabain and digoxin after pretreatment with α-tocopherol-nicotinate and dodecanoic acid (n = number of experiments).

Compound for pretreatment	Ouabain K_D (10^{-7} mol/l)	n	Digoxin K_D (10^{-7} mol/l)	n
Control	16.2 ± 0.70	8	4.45 ± 0.26	8
α-tocopherol-nicotinate	13.8 ± 0.64	10	8.03 ± 0.42	12
Dodecanoic acid	14.9 ± 1.42	12	16.8 ± 0.77	13

α-tocopherol nicotinate and dodecanoic acid on the positive inotropic action of digoxin and the protective effect against cardiac glycoside-induced arrhythmias may be due to interference with the glycoside action in different ways, e.g. interference with the glycoside binding to heart tissue, an effect on the Na^+,K^+-ATPase activity or via an effect on fluxes.

It has been reported by several authors that fatty acids (to which dodecanoic acid and α-tocopherol-nicotinate are related) inhibited Na^+,K^+-ATPase activity (1, 4, 14). This effect could not be the mechanism of the inhibitory effect of these compounds on the action of cardiac glycosides, since the positive inotropic and arrhythmogenic effects of cardiac glycosides on isolated guinea-pig atria are related to their inhibition of the Na^+,K^+-ATPase activity (5, 8, 28, 34). The effects of α-tocopherol-nicotinate and dodecanoic acid on the binding of cardiac glycosides to heart homogenate correlate with their effects on the inotropic action of the glycosides. The binding of digoxin was significantly inhibited while that of ouabain was not affected in the presence of either dodecanoic acid or α-tocopherol-nicotinate. Kim et al. (12) reported that long chain fatty acids antagonise the binding of ouabain competitively. This antagonism was only relevant at very high concentrations of fatty acids, which were not used in our experiments. The inhibitory effect of α-tocopherol-nicotinate and dodecanoic acid on the inotropic action of digoxin at low concentration levels could therefore be explained by the interference of the two compounds with the binding of lower concentrations of digoxin to heart tissue. By increasing digoxin concentration this interference was overcome and digoxin showed its maximum action in the presence of α-tocopherol-nicotinate or dodecanoic acid with the same concentration as under control conditions. This interference was not observed in case of ouabain. This phenomenon may be explained by the fact that digoxin is more lipophilic than ouabain. Fatty compounds in the reaction medium (dodecanoic acid and α-tocopherol-nicotinate) may bind to digoxin and retard its diffusion to its receptor sites. Assuming that there are two receptor sites for cardiac glycosides one intracellularly (6, 7, 20) and the other extracellularly (9, 25, 29), one can postulate that the less lipophilic ouabain acts mainly through the extracellular receptors, therefore its positive inotropic action was not affected by α-tocopherol-nicotinate or dodecanoic acid, while the more lipophilic digoxin acts mainly through intracellular receptors. Dodecanoic acid inhibits the maximal inotropic action of both digoxin and ouabain. This observation could be explained by a direct action of dodecanoic acid on the glycoside receptor sites. It has been reported that C_{12} and C_{14} hydrocarbon chains had a rigidifying effect on the acyl chains of the surrounding phospholipids (15, 16). A similar effect may be postulated for the hydrocarbon chains of dodecanoic acid (C_{12}), which may be assumed as a possible mechanism for its direct influence on the receptors.

The possibility that α-tocopherol-nicotinate and dodecanoic acid exert their effect on digoxin action via an interference with Ca^{2+} fluxes can be excluded, since in such a case these compounds should have similar effects on the action of both digoxin and ouabain, which is not shown in our results. Although the positive inotropic effect of digoxin and not of ouabain was inhibited by α-tocopherol-nicotinate and dodecanoic acid, the arrhythmogenic activity of both glycosides was similarly suppressed by the mentioned drugs. This observation may be an evidence for the previously reported dissociation between the positive inotropic action and the arrhythmogenic activity of cardiac glycosides (18, 19, 34). α-tocopherol-nicotinate had a high protective effect against cardiac glycoside-induced arrhythmias, α-tocopherol and dodecanoic acid had an intermediate effect and nicotinic acid had no significant protective effect. The mechanism through which these drugs exert their anti-arrhythmic action is not via an effect on Na^+,K^+-ATPase activity, nor via inhibition of glycoside binding, nor via an effect on Ca^{2+} fluxes as explained before.

The structure of these three compounds as long hydrocarbon chains may be the source of their protective effect against cardiac glycoside-induced arrhythmias in isolated atria by changing the physico-chemical properties of the membrane in a quinidine-like manner. Nicotinic acid itself, which is not a long chain hydrocarbon molecule, had no protective effect against cardiac glycoside-induced arrhythmias. The ester linkage in α-tocopherol-nicotinate may be responsible for the increased antiarrhythmic property compared with α-tocopherol and dodecanoic acid. Further investigation should be done to clarify the mechanism of the antiarrhythmic activity of these compounds.

Acknowledgements

The authors would like to thank Miss S. Schimmelpfennig for technical assistance. This work was supported by the Deutsche Forschungsgemeinschaft, SFB 30, Kardiologie.

References

1. Ahmed K, Thomas BS (1971) J Biol Chem 246: 103
2. Alfin-Slater RB (1963) In: Padetti R, Kritchevsky D (eds) Vitamin E and Lipid Metabolism. Advanced Press, New York, pp 183–210
3. Böhlau V (1971) Arzneim-Forsch/Drug Res 21: 674
4. Dahl DR (1968) J Neurochem 15: 815
5. Erdmann E, Bolte HD, Lüderitz B (1971) Arch Biochem Biophys 145: 121
6. Fricke U, Klaus W (1977) Br J Pharmacol 61: 423
7. Fricke U, Klaus W (1978) Br J Pharmacol 62: 255
8. Gelbert A, Goldman RH (1977) Biochem Biophys Acta 481: 689
9. Hilden S, Rhee HM, Hokin LE (1974) J Biol Chem 249: 7432
10. Hochrein H, Helwing HP (1970) Int Sympos Vit E, Hakone (Jap.)
11. Horwitt MK (1962) Vitamin Horm 20: 541
12. Kim RS, Labella FS (1980) J Molec Cell Cardiol 12: 847
13. Krause D, Dittmar K (1973) Münch Med Wschr 115: 1742
14. Lawers JM, Hülsmann WC (1977) J Molec Cell Cardiol 9: 343
15. Lee AG (1976) Biochemistry 15: 2448
16. Lee AG (1977) Molec Pharmac 13: 474
17. Michael RJ (1971) Dtsch Med J 20: 654
18. Okita GT (1977) Federation Proc 36: 2225
19. Okita GT, Richardson F, Roth-Schlechter BF (1973) J Pharmac Exp Ther 185: 1
20. Park MK, Vincenzi FF (1975) J Pharmac Exp Ther 195: 140
21. Poggi E (1971) Sist Nerv 23: 126
22. Reifferscheid M, Matis P (1951) Med Welt 20: 1168
23. Reinecke M (1974) Z präklin Geriatr 11: 257
24. Schlieper P, Medda PK, Kaufmann R (1981) Biochim Biophys Acta 644: 273

25. Schwart A, Lindenmayer GE, Allen JC (1975) Pharmac Rev 27: 3
26. Svedmyr, N (1969) Clin Pharmac Ther 10: 559
27. Tappel AL (1970) Am J Clin Nutr 23: 1137
28. Tawfik H, Fox AAL, Greeff K (1985) J Biochem Pharmac 34: 254
29. Temma K, Akera T, Brody TM, Kondo H (1982) Eur J Pharmac 77: 289
30. Vogelsang A (1951) Rec Med 164: 353
31. Vogelsang A (1970) Angiology 21: 275
32. Williams HTG (1962) Can Med Ass J 87: 538
33. Yamamoto S, Fox AAL, Greeff K (1981) Eur J Pharmac 71: 437
34. Zavecz JH, Dutta S (1977) Naunyn-Schmiedeberg's Arch Pharmac 297: 91

Author's address:

Priv. Doz. Dr. P. Schlieper, Woelm Pharma GmbH & Co., P. O. Box 840, D-3440 Eschwege (West Germany)

On the connection between cardiac glycosides and calcium action

H. Waginger and D. S. Regal

Department of Botany and Department of Pathology, University of Veterinary Medicine, Vienna (Austria)

Summary

Using precipitation with lead salts, the activities of Mg^{++}, Ca^{++}, and Na^+,K^+-stimulated ATPases in the cardiac muscles of rats and pigs were determined ultrahistochemically. The effect of cardiac glycosides on the Ca^{++}-stimulated ATPase was to increase its activity, especially in the sarcolemma, the mitochondrial membrane and between the myofibrils of the cardiac muscle. The inhibitory effect of cardiac glycosides on the Na^+,K^+-stimulated ATPase in the sarcolemma of the cardiac muscle could be shown. This enzyme has proved to be the receptor of cardiac glycosides. These findings complimented the biochemical findings, in as much as they made visible the active transport of the cations in the heart muscle of pigs, cats, dogs and horses. On the one hand the important role of calcium in digitalis therapy was demonstrated, in that under the effect of the cardiac glycoside the ionisable intracellular calcium fraction increased and so enhanced the positive inotropic effect. On the other hand, cardiac glycosides block the Na^+,K^+-ATPase and so create a higher intracellular sodium concentration level, with a resultant increase in the available calcium.

Introduction

At the beginning of the century, just as today, an essential feature of digitalis research was an attempt to explain the influence of electrolytes on the mechanism of positive inotropy. For this purpose the isolated heart of a cold-blooded animal was used. Inspired by Ringer's research, Loewi (2) described in several publications the connection between the recovery of a frog's heart isolated in a calcium-free solution and positive inotropy, and thereby established that calcium is the essential cell constituent for the action of heart contractions. Moreover he was able to render it comprehensible, that in a calcium-free Ringer solution the potassium content initially decreases, and that the recovery is essentially dependent upon the state of nutrition, and hence upon the pH-value of the cells of the frog's heart used for the experiment. Hearts with an acid cell-pH tended to be hypersensitive. But in the light of more recent research, a shift of the cell-pH towards more alkaline values can also occur through intervention in the Cl^--HCO^-_3-ion-exchange system, when chloride replacement causes an increased direct influx of calcium into the cells with a concomitant positive inotropic action.

At about the same time as Loewi (2), von Weizsäcker (6) supplemented these findings with the observation that the start of the action of strophanthin takes place all the earlier, the faster the heart beats, and that the number of contractions is more decisive for intoxication than the duration of the strophanthin action. In this sense the bradycardiac action of the cardiac glycosides is to be understood as a protection against intoxication. Both findings can be explained in the light of modern knowledge as a consequence of changed electrolyte transport conditions.

In this manner, the cardiac glycosides do not directly intervene in the mechanism of positive inotropy. In fact, a binding site with a high affinity for divalent cations (Ca^{++}, Mg^{++}) is responsible for this mechanism. Such sites occur both in the heart and in the skeletal musculature. Typical in this process are the light chains of the actomyosin complexes, which control the binding of the meromyosin heads with the actin filaments. The cause of this phenomenon may be found in the variously stimulated membrane-ATPases (Na^+,K^+-ATPase, Mg^{++}-ATPase, Ca^{++}-ATPase), of which the Na,K-activated ATPase is the most important receptor for cardiac glycosides, as Schatzmann (4) recognized. It is therefore appropriate that the following factors are decisive for a continuously positive inotropic action.

1. a constant receptor population;
2. a constant glycoside concentration on the receptor side;
3. balanced intracellular electrolyte concentrations and pH-values.

Material and Methods

In the first set of experiments using precipitation with lead salts, the activities of Mg^{++}-, Ca^{++}- and Na^+,K^+-stimulated ATPase in the cardiac muscle of rats were studied ultrahistochemically and the effect of strophanthin on the various ATPases' activities was examined.

For a second line of experiments, the porcine stress syndrome (PSS) of cross-bred pietrain pigs and wild boar was determined by a halothane exposure test, and animals were grouped into stress-resistant and stress-susceptible populations.

Tissue samples were taken from cardiac muscle and various ATPases' activities were traced ultrahistochemically by means of precipitation with lead salts (3).

For a third line of experiments by means of the Rb-transport through the erythrocyte membrane, the blocking effect of the cardiac glycosides on the effect of ATPase concerning membrane transport was demonstrated. Furthermore, the different electrolyte compositions comparing the dose-response curves of different cardiac glycosides were evaluated and showed the expected dependence on the concentrations in pigs, cats, dogs and horses.

The modified chemico-clinical methods as well as the pharmacokinetic measurements were described by Waginger et al. (5).

Results and Discussion

Under the influence of strophanthin no activity of Mg^{++}-stimulated ATPase was detected.

The effect of cardiac glycosides on the Ca^{++}-stimulated ATPase was to cause a marked increase in enzyme activity, especially in the sarcolemma, the mitochondrial membrane and the myofibrils (Fig. 1). After administration of glycosides an increased concentration of intracellular calcium, as well as an enhanced influx of extracellular calcium into the cell can be observed. Simultaneously inhibition of Na^+,K^+-stimulated membrane ATPase in the sarcolemma raises intracellular sodium concentration (Fig. 2).

These mechanisms directly or indirectly lead to an increase in the intracellular concentration of the free calcium ions. Thereby, high energy phosphates are utilized more efficiently.

In veterinary medicine digitalis therapy, such as is possible in the dog, and also on occasions in cattle and swine, creates essential problems during practical application. For example, for a continuously positive inotropic action one must ensure that the number of receptors at the heart musculature remains unchanged, and that no great differences in the

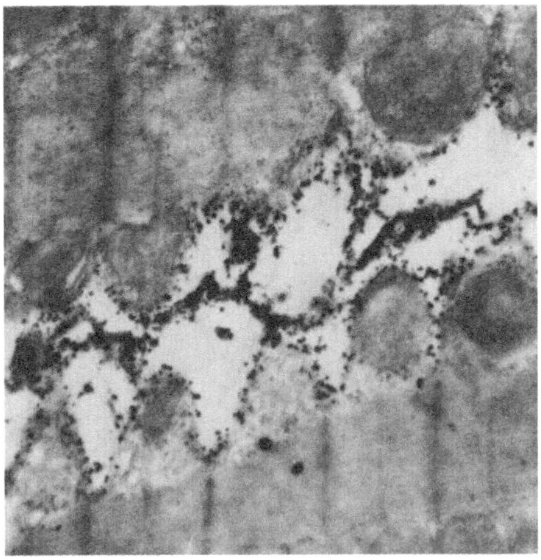

Fig. 1. Rat heart muscle: increase of Ca^{++}-stimulated ATPase activity in the sarcolemma, the myofibrils and the mitochondrial membrane after strophanthin treatment. (20000×)

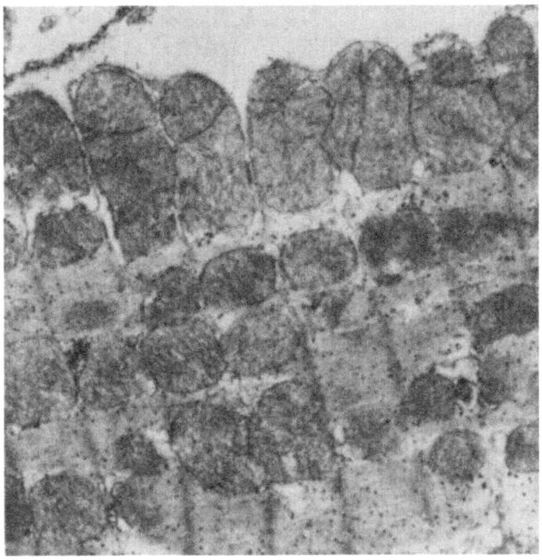

Fig. 2. Rat heart muscle: no activity of Na^{+},K^{+}-stimulated ATPase in the sarcolemma after strophanthin treatment; low enzyme activity remaining in the myofibrils. (15000×)

content of receptors arise between the various breeds of a species to be treated. While the first criterion is extensively fulfilled in the horse, nevertheless great differences in the contents of membrane ATPases determined by breed can be established. At this point cold-blooded horses showed a substantially lower ATPase activity than warm-blooded ones. Similar relationships are valid for swine. Stress-susceptible breeds with porcine stress syndrome (PSS) display higher ATPase activities within the erythrocytes than those not susceptible to stress. It proved possible to observe such differences directly at the cardiac musculature between wild stress-resistant boars and stress-susceptible pigs (Figs. 3, 4). From

Fig. 3. Wild boar heart muscle: no activity of Na^+,K^+-stimulated ATPase in the sarcolemma of the stress-resistant animal during halothane exposure test. (20000 ×)

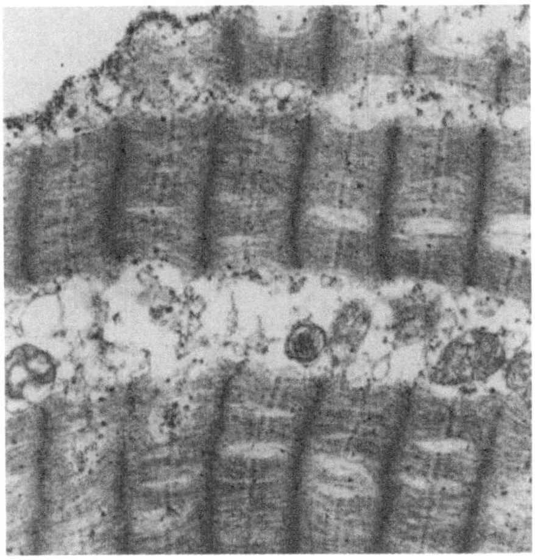

Fig. 4. Pietrain pig heart muscle: increased activity of Na^+,K^+-stimulated ATPase in the sarcolemma and the myofibrils of the stress-susceptible animal during halothane exposure test. (15000 ×)

repeated determinations of the ATPase activities in sattled pigs, it can be assumed that receptors in these animals are subject to great variability during their growth. In this way cardiac glycosides may quickly lose their action in pigs.

In connection with the initially outlined experiments of Loewi we have determined membrane ATPase activity levels dependent upon different electrolyte concentrations. It was found that after an hour's incubation the erythrocytes of horses acquire characteristic forms, dependent upon the state of metabolism (e.g. acanthocyte structure). It seems to us to be credible that this morphological process is caused by an intracellular change in the cytoskeleton (Fig. 5).

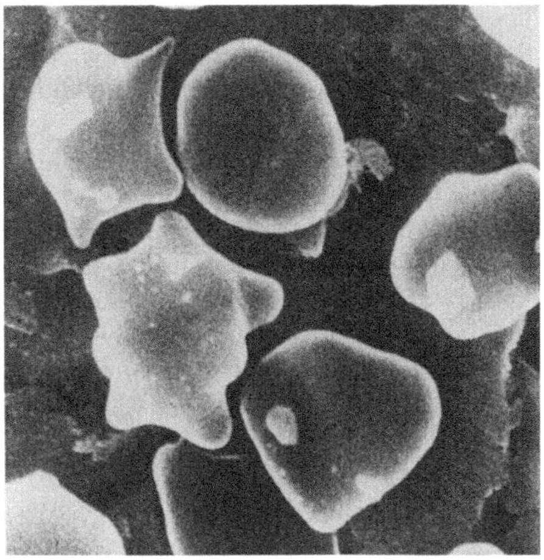

Fig. 5. Scanning electron microscopy: acanthocytic structure of erythrocytes in horses after treatment with 135mM Na^+, 5mM K^+ and 5mM Ca^{++}. (1000×)

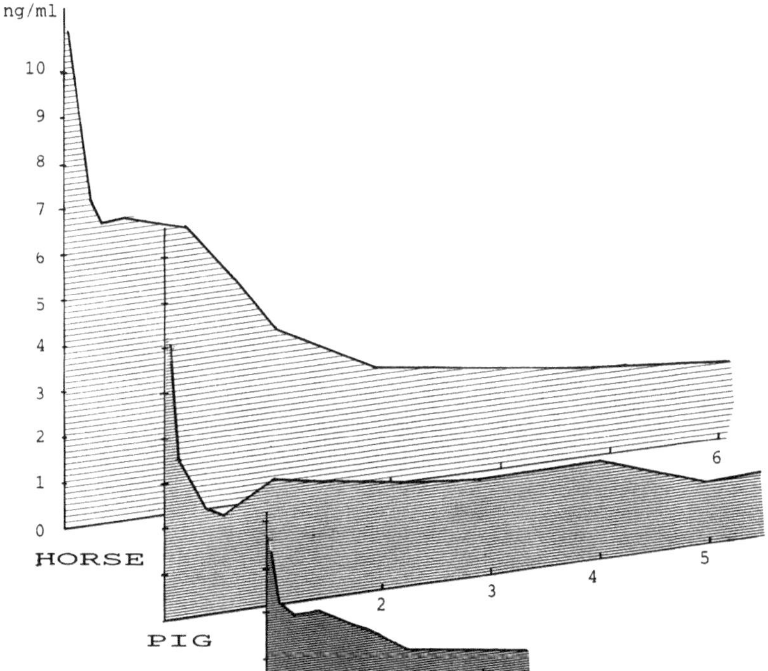

Fig. 6. Plasmadigoxin concentration in different species after doses of 0.020 mg/kg body mass administered intravenously to groups of six horses, pigs and dogs (mean values).

While the problems hitherto outlined cannot generally be explained by clinico-chemical methods as routine procedure at tolerable cost, there still exists the possibility of investigating the pharmacokinetic properties of the application of cardiac glycosides in individual animals (Fig. 6).

During such investigations it is worth noting that above all in pigs, the dosage should be calculated less on the basis of body weight than on that of muscle weight. In the above dog, all the short half-life elimination time of digitoxin is noteworthy. In the horse the apparently uniform pharmacokinetic behaviour should not lead us to the false conclusion that the glycoside level in the blood is essentially more decisive for the receptor, but that a non-uniform susceptibility of the receptors in various animals can nevertheless be responsible for side effects. On comparing digitoxin administration with that of digoxin, it is striking that the pharmacokinetic properties of methyldigoxin are as a rule clearly superior to those of digitoxin (Fig. 7).

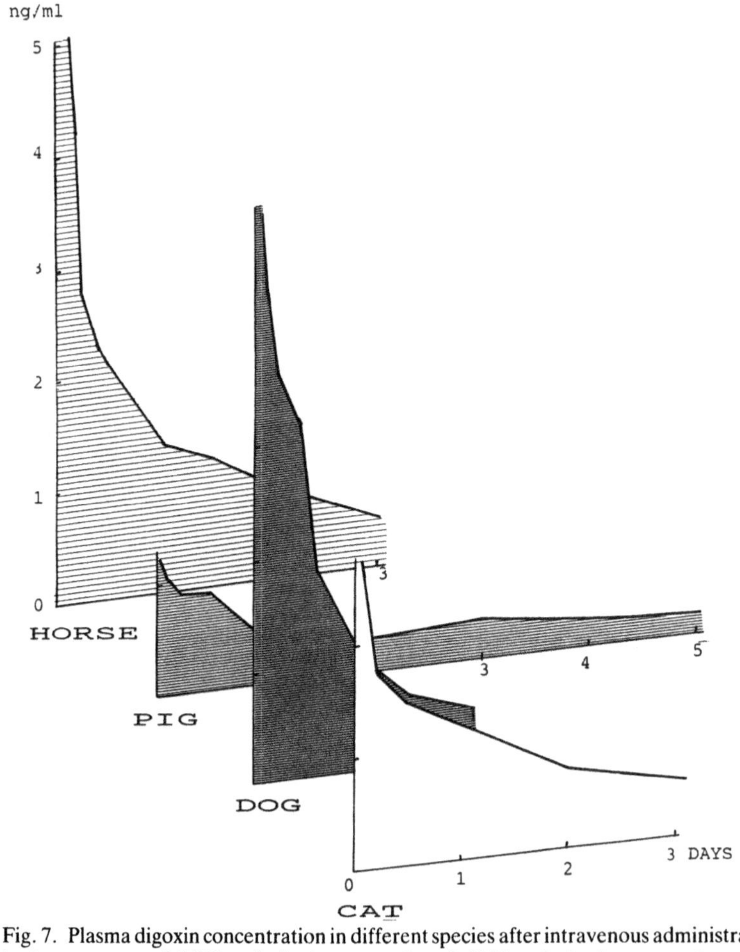

Fig. 7. Plasma digoxin concentration in different species after intravenous administration of 0.0285 mg/kg body mass to cats, 0.0175 mg/kg body mass to dogs, 0.0124 mg/kg body mass to pigs and 0.0100 mg/kg body mass to horses (mean values of six animals).

Above all this could be attributable to differences in the rates of solubility. At all events a high rate is not advantageous in all animals. In ruminants solubilized cardiac glycosides in the rumen can lead to noticeable losses in action as a result of hydrolytic splitting of the digitoxin. On the other hand digoxin reduction product formation did not occur in the model we studied; Heinz and Flasch (1), observed such a reaction in animals, especially in the cat. Additional methods of controlling the glycoside spectrum must be sought, such as above all electrolyte spectrum determinations of saliva (Mg^{++}, K^+, Ca^{++}), or determination of the serum electrolyte: erythrocyte electrolyte ratio. A fall in the K^+-serum level can thereby be recognized and treated easily. Such a fall would lead to a sensitization to cardiac glycosides. Nevertheless there remains a certain residual risk that veterinarians usually recommend treatment with digitalis glycosides less rapidly than doctors in human medicine.

References

1. Heinz N, Flasch H (1978) Comparison of pharmacokinetics of digoxin and dihydrodigoxin in cats in single dose studies. Naunyn-Schmiedeberg's Arch Pharmacol 303: 181–187
2. Loewi O (1918) Über den Zusammenhang zwischen Digitalis- und Kalziumwirkung. Naunyn-Schmiedeberg's Arch exp Path Pharmak 82: 131–139
3. Regal DS (1977) Elektronenoptischer Nachweis spezifischer Adenosintriphosphatasen im Herzmuskelgewebe von Ratten nach Einwirkung von Strophanthin unter besonderer Berücksichtigung des Kalziumstoffwechsels. Zbl Vet Med A 24: 833–847
4. Schatzmann HJ (1965) The role of Na^+ and K^+ in the ouabain inhibition of the $Na^+ + K^+$-activated membrane adenosinetriphosphatase. Biochim Biophys Acta 94: 89–96
5. Waginger H, Regal DS and Bonelli J (1977) Contribution to the biochemical characterization of cardiac glycosides concerning their influence on ATPases. Int J Clin Pharmacol 15: 526–529
6. von Weizsäcker V (1913) Über die Abhängigkeit der Strophanthinwirkung von der Intensität der Herztätigkeit. Naunyn-Schmiedberg's Arch exp Path Pharmak 72: 282–294

Author's address:

Dr. H. Waginger, Institut für Botanik und Lebensmittelkunde, Veterinärmedizinische Universität Wien, Linke Bahngasse 11, 1030 Vienna (Austria)

Biochemical basis for the targeted synthesis of cardiac glycosides with selective action

W. Schönfeld, J. Weiland and K. R. H. Repke

Energy Conversion Unit in the Central Institute of Molecular Biology, Academy of Sciences of G.D.R., Berlin (East Germany)

Summary

The paper briefly surveys the recent state of the authors' research into the biochemical basis for the targeted synthesis of cardiac glycosides with selective action.

Introduction

In the multi-stage development of the digitalis research in Berlin-Buch since 1957, which has included biological evaluations and metabolic studies as briefly reviewed in (14) and (15), we have attained the targeted strategy in drug research (2) which became possible to apply due to recognition of Na^+/K^+-transporting ATPase as the digitalis target enzyme in 1961 (9–11).

The targeted strategy is characterized by a cyclical refinement process involving the design, directed synthesis, and determination of inhibitory potency of selected derivatives on Na/K-ATPase. Of course, the quantitative nature and rapid generation of enzyme inhibition data are ideal for guiding chemical synthesis and for providing types of structure-activity relationships which are interpretable in terms of the kind of binding forces and the chemotopology of the receptor binding site. – Methods and materials used have been detailed in a previous paper (19).

Results and discussion

In the framework of the targeted strategy, the first critical decision to be made is the choice of the source of Na/K-ATPase. Since the initiatory work of Portius and Repke on structure-activity relationships in cardiotonic drugs through using the enzyme of guinea-pig cardiac muscle (8), numerous studies by various groups have been also done on the guinea-pig enzyme. However, this choice no longer appears to be fully suitable.

This judgement is based on the results of the comparative determination of the inhibitory potency for 68 steroidal compounds on the Na/K-ATPases of guinea-pig and human cardiac muscle (Fig. 1). The range of dissociation (inhibition) constants of their complexes with either enzyme covers 6 orders of ten. In general, the paired dissociation constants measured for the individual compounds with guinea-pig and human enzyme prove to be correlated, but the correlation coefficient is 0.91 only. This rather poor correlation is caused by a few derivatives.

Thus, the human enzyme shows for prednisolone-3,20-bisguanyl-hydrazone *(53)* an affinity lower by about 50-fold, and for 3β-O-glucosyl-digitoxigenin *(22)* an affinity about 100-fold higher than for the guinea-pig enzyme (throughout the text the italic numbers refer

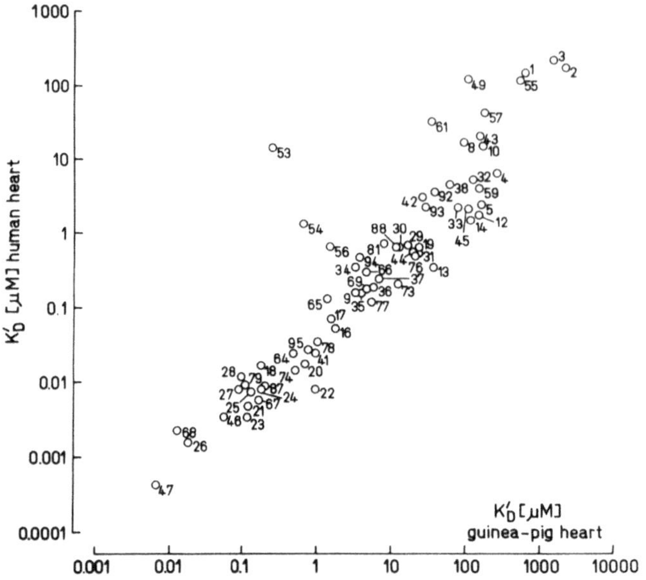

Fig. 1. Analysis of the correlation between the apparent dissociation constants (K_D^1) of the inhibitory complexes between 68 steroidal compounds and the Na/K-ATPases of guinea-pig heart (g.-p. h.) or human heart (h. h.), respectively. The regression equation has the form log K_D^1 h. h. = 0.97 log K_D^1 g.-p. h – 1.38 with the correlation coefficient r = 0.91. The numerals in the figure refer to the diverse compounds applied. So far as not given in the text, their systematic names will be listed in a forthcoming full-length paper.

to the numbering in Figs. 1 and 5). The latter finding suggests that the sugar-binding receptor subsite is different in the two enzyme variants.

A closer correlation is found for the derivatives containing the steroid nucleus of digitoxin, i. e., 5β, 14β-androstane-3β, 14-diol (Nos. *1–47* in Fig. 1). Here, the human enzyme shows on average an affinity about 30-fold higher than the guinea-pig enzyme. This finding suggests that the geometry of the steroid nucleus of digitoxin-like compounds in the human enzyme meets a fuller complementary fit to the steroid-binding receptor subsite.

In fact, the Na/K-ATPases of guinea-pig and human cardiac muscle differ in the geometry of their steroid binding receptor subsites, as predicted in 1965 (12). This conclusion emerges from the divergence of the interaction energies of steroid nuclei with a bent or flat A/B ring junction when reacting with the guinea-pig or human enzyme (Table 1). Thus, the steroid nucleus with bent A/B ring junction *(2)* as present in digitoxin-like glycosides, develops with guinea-pig enzyme an interaction energy which is about – 7 kJ/mole smaller than that developed with human enzyme. However, the steroid nucleus with a flat A/B ring junction *(49)* as present in uzara glycosides shows the same interaction energy with the guinea-pig enzyme as with the human enzyme.

Owing to differences in complementary fit and thus in spatial orientation of the steroid nuclei in their binding receptor subsites, the interaction energy increment through attachment of β-butenolide is much higher for the steroid nucleus with a bent A/B ring junction on both enzymes than for the steroid nucleus with a flat A/B ring junction.

Table 1. The Na/K-ATPases of guinea-pig and human cardiac muscle differ in the geometry of their steroid nucleus-binding receptor subsites as revealed by the difference in the Gibbs energy (Δ G°′) of interaction of steroid nuclei and the difference in the interaction energy increment (δ Δ G°′) through β-butenolide attachment.

Interaction energy source	Interaction energy (-kJ/mole) in	
	guinea-pig enzyme	human enzyme
Δ G°′ of steroid nuclei with bent[1] (flat[2]) A/B ring junction	15.5 (23.3)	22.2 (23.1)
δ Δ G°′ through β-butenolide attachment at C17β-H of steroid nuclei with bent[1] (flat[2]) A/B ring junction	18.5 (6.6)	21.0 (15.0)
Δ G°′ of composed cardenolide = sum of structural components	34.0 (29.9)	43.2 (38.1)

[1] 5β, 14β-androstane-3β, 14, 17β-triol *(2)*
[2] 5α, 14β-androstane-3β, 14, 17β-triol *(49)*

Taken together, the data prove, firstly, that differences in the geometry of the steroid-binding receptor subsites underlie the well-known differences in the digitalis sensitivity of guinea-pig and man, and, secondly, that the inhibitory potency of steroidal compounds estimated on the guinea-pig cardiac Na/K-ATPase cannot be either quantitatively or qual-

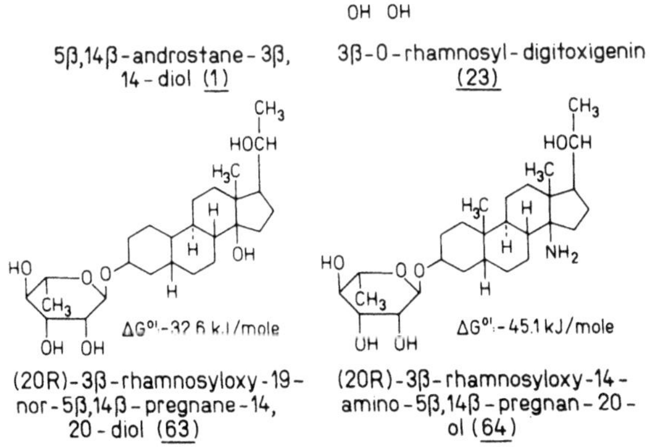

Fig. 2. Apparent Gibbs energies of interaction (Δ G°′) of the steroid nucleus *(1)* in digitoxin or some derivatives of *1* modified at C3β-OH, C14β-OH and C17β-H, with Na/K-ATPase from human brain cortex.

itatively extrapolated easily to the human cardiac enzyme, so that the latter enzyme variant deserves preference in the analysis of structure-activity relationships.

Almost any structural modification of the steroid nucleus in digitoxin, except substituents at C3β and C17β, results in weakening of the interaction energy. Among the two exceptions we know is the replacement of C14β-OH by C14β-NH_2 carried out in the Nativelle laboratories (4), which increases the interaction energy of a corresponding pregnanediol derivative by as much as −13 kJ/mole (Fig. 2). Thus, even in the absence of the butenolide ring, the integral interaction energy of the C14β-aminopregnane derivative is not much smaller than that of one of the most potent digitalis glycosides. The astonishing increase of inhibitory potency could be related to the existence of the amino group as ammonium cation at physiological pH.

In the electrostatic molecular potential map of digitoxin-like compounds as that of digitoxigenin shown in Fig. 3, the hydroxy group at C14β forms a negative well intruding into the positive shell surrounding the steroid nucleus (17). Thus, the replacement of C14β-OH by C14β-NH_3^+ seems not only to eliminate an interfering negative potential well, but also to complement the positive potential shell around the steroid nucleus and thus to strengthen the electrostatic binding forces between steroid nucleus and the complementary binding receptor subsite of Na/K-ATPase.

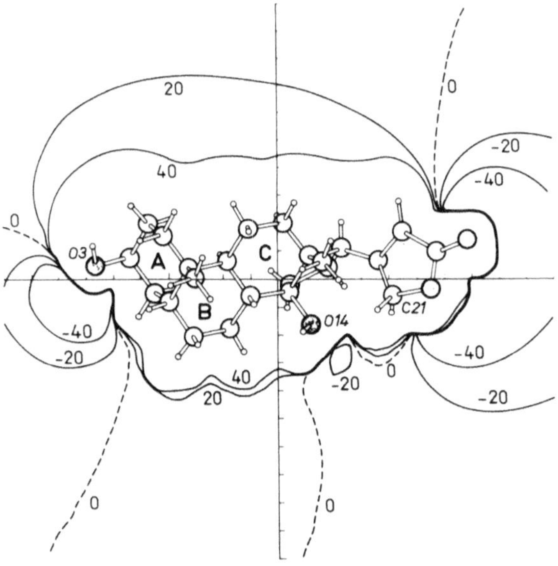

Fig. 3. Electrostatic molecular potential map of digitoxigenin. Modified and redrawn from (17). Data from A. Hache.

The circumscribed negative potential shell surrounding the butenolide ring suggests that one or the other or both oxygen atoms serve as hydrogen acceptors in hydrogen bonding to the side chain-binding receptor subsite.

The proposal, however, made in 1964 by Portius and Repke (8), that the carbonyl oxygen is the sole hydrogen acceptor can no longer be maintained (Fig. 4). The butenolide side chain may in fact be energetically replaced by furane or pyridine (19). Apparently, the

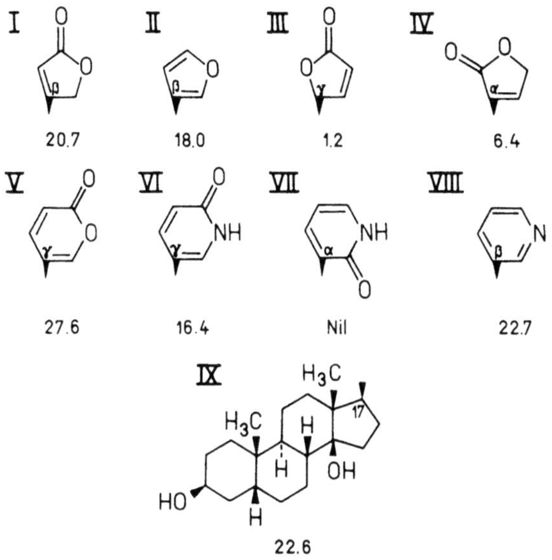

Fig. 4. Increment of apparent Gibbs interaction energy (–δ Δ G°′) of steroid nucleus in digitoxin, 5β, 14β-androstane-3β, 14-diol (IX), effected through attachment of the heterocycles I-VIII at C17β.

bridge oxygen atom in β-butenolide (I), β-furane (II) and γ-pentadienolide (V) and the nitrogen atom in β-pyridine (VIII) are the major hydrogen acceptors. This being so, the drastic changes of the interaction energy contributions by varying side chains (I-VIII) at C17β of the steroid nucleus (IX) are then related to differences in the spatial exposure of the acceptor atoms, since the potential energy of a hydrogen bond (maximally up to −45 kJ/mole) is critically dependent on the distance between hydrogen acceptor and donor atoms. Studies to examine this supposed relation through computerized receptor model building are in progress in our institute.

The hydrogen bonding hypothesis can also account for the very low increment of interaction energy contributed by the α-butenolide side chain (IV) of actodigin which is about −15 kJ/mole less than that contributed by the β-butenolide side chain (I) of cardenolides.

Actodigin *(12)*, 3β-O-glucosyl-3α-methyl-digitoxigenin *(30)* and (21R)-3β-O-glucosyl-21-methyl-digitoxigenin *(44)* have been reported to excel digitalis glycosides by a more favourable relation between therapeutic and toxic actions in animal tests (2, 6, 22). However, these synthetic derivatives show the same inhibitory power on Na/K-ATPases of human cardiac muscle and human brain cortex (Fig. 5).

Altogether, 76 compounds widely differing in the structure of steroid nucleus and side chains have been compared with regard to their inhibitory potency on the cardiac and cerebral enzyme. The apparent dissociation constants of their complexes with either enzyme, covering 7 orders of ten, are nevertheless strongly correlated (correlation coefficient r = 0.99). The absolute term in the regression equation calculated for the total of K'_D values reveals on an average about a 1.5 fold higher inhibitor-sensitivity of the cardiac enzyme.

Conclusions

As for the results presented, one may be tempted to draw the following deductions. Firstly, the similarity of the digitalis affinity of human cardiac and cerebral Na/K-ATPases

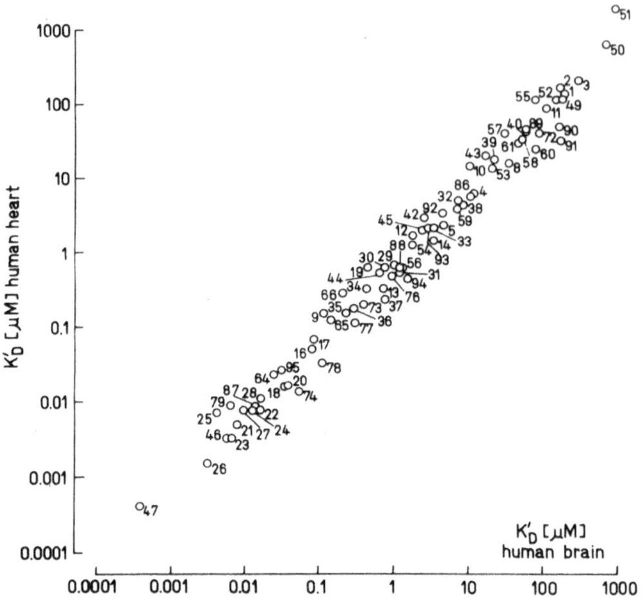

Fig. 5. Analysis of the correlation between the apparent dissociation constants (K_D^1) of the inhibitory complexes between 76 steroidal compounds and the Na/K-ATPases of human heart (h. h.) or human brain (h. b.), respectively. The regression equation has the form log K_D^1 h. h. = 1.00 log K_D^1 h. b. – 0.17 with the correlation coefficient r = 0.99. Further explanations in Fig. 1.

is a major reason for the high incidence of neuroexcitatory effects under digitalis-treatment of cardiac failure (1). Secondly, modification of the structure of digitalis compounds never produces selectivity of action on cardiac enzyme.

For the present, however, such deductions do not appear to be warranted for four major reasons.

(I) The Na/K-ATPase preparation of human brain cortex studied could have been a mixture of digitalis-insensitive and -sensitive enzyme variants which are known to exist in brain of various animal species (7, 20, 21). Studies with isolated variants from axolemma, midbrain, other cerebral structures and autonomous ganglia could lead to results divergent from those reported here.

(II) The similarity of the apparent dissociation constants does not exclude the possibility that the rate constants for the formation and dissociation of the inhibitor-complexes with the cardiac and cerebral enzyme variant significantly diverge, thus exerting a differentiating impact on the in vivo parameters (13).

(III) The structural modifications of digitalis compounds to be compared with regard to their inhibitory power on enzyme variants of cardiac muscle and central, peripheral and autonomous nervous system are far from exhausted. Surprises would not surprise us.

(IV) The structure-activity relationships described here are valid for the receptive intermediary state of Na/K-ATPase when cycling under optimized turnover conditions. In the cell, however, Na/K-ATPase exists in a two-conformation equilibrium between catalytically competent E_1- and incompetent E_2-enzyme conformers (Fig. 6). The equilibrium is poised by the ratio of the intracellular Na^+ and K^+ concentrations when operating as heterotropic allosteric activator or inhibitor, respectively (18). In the two limit enzyme conformations, the stereoelectronic properties of the digitalis binding domain are different

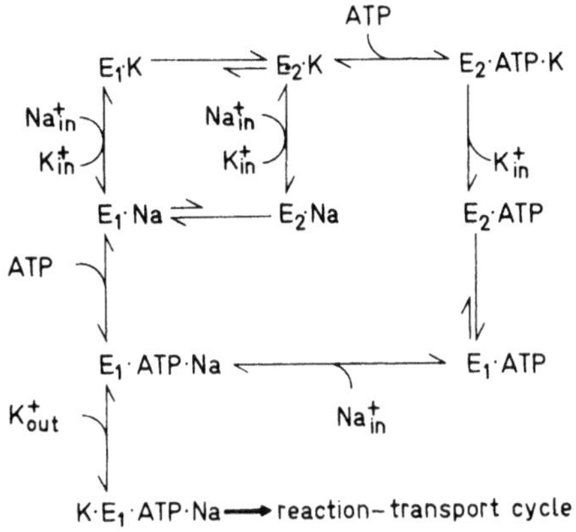

Fig. 6. Allosteric regulation of Na/K-ATPase activity and Na^+/K^+ pump power by the ratio of the intracellular concentrations of sodium ions [Na^+_{in}] and potassium ions [K^+_{in}] when interacting as heterotropic allosteric activator with the catalytically incompetent $E_2 \cdot K$ conformer of the enzyme, or as heterotropic allosteric inactivator with the catalytically competent $E_1 \cdot Na$ conformer, respectively. From (18).

(16). The catalytically incompetent conformer, the so-called $E_2 \cdot K$, comprising as reserve enzyme capacity the major part of enzyme molecules in the cell (18), can thus be expected to show structure-affinity relationships different from those found in the present study. In line with this reasoning, the two conformational states of dihydrofolate reductase have been proved to differ in ligand structure-affinity relationships, thus providing antifolates conformation selectivity (3).

Studies to elucidate the four unknowns enumerated above are in progress in our group.

References

1. Gillis RA, Quest JA (1980) The role of the nervous system in the cardiovascular effects of digitalis. Pharmacol Rev 31: 19–97
2. Glicklich JI, Gaffney R, Rosen MR, Hoffman BF (1975) Effects of AY 22 241 (Actodigin) on electrical and mechanical activity of cardiac tissues. Eur J Pharmacol 32: 1–9
3. Gronenborn A, Birdsall B, Hyde E, Roberts G, Feeney J, Burgen A (1981) ^{1}H and ^{31}P NMR characterization of two conformations of the trimethoprim-NADPH$^+$-dihydrofolate reductase complex. Mol Pharmacol 20: 145–153
4. Jarreau FX, Koenig JJ, Fénard S (1983) A new inotropic aminosteroid: LND 623. J Mol Cell Cardiol 15 (Suppl 2): Abstr 44
5. Krenitsky TA, Elion GB (1982) Enzymes as tools and targets in drug research. In: Keverling Buisman JA (ed) Strategy in Drug Research. Elsevier Publishing Company, Amsterdam, pp 65–87
6. Lüllmann H, Mohr K (1982) On the binding of a 3-α-methylated digitoxigenin-glucoside to ouabain receptors in heart muscle homogenate. Biochem Pharmacol 31: 2489–2494
7. Marks MJ, Seeds NW (1978) A heterogeneous ouabain-ATPase interaction in mouse brain. Life Sci 23: 2735–2744
8. Portius HJ, Repke KRH (1964) Versuch einer Analyse der Beziehungen zwischen chemischer Struktur und digitalis-ähnlicher Wirksamkeit auf der Rezeptorebene. Arzneim-Forsch 14: 1073–1077
9. Repke KRH (1963) Metabolism of cardiac glycosides. In: Wilbrandt W (ed) Proceedings of the First International Pharmacological Meeting, Stockholm 1961, Vol III. New aspects of cardiac glycosides. Pergamon Press, Oxford, pp 47–74
10. Repke KRH, Portius HJ (1963) Über die Identität der Ionenpumpen-ATPase in der Zellmembran des Herzmuskels mit einem Digitalis-Rezeptorenzym. Experientia 19: 452–458

11. Repke KRH (1964) Über den biochemischen Wirkungsmodus von Digitalis. Klin Wschr 42: 157–165
12. Repke KRH, Est M, Portius HJ (1965) Über die Ursache der Speziesunterschiede in der Digitalisempfindlichkeit. Biochem Pharmacol 14: 1785–1802
13. Repke KRH, Herrmann I, Kunze R, Portius HJ, Schön R, Schönfeld W (1973) Mechanism of digitalis action and the importance of the kinetics of the formation and decomposition of glycoside-receptor complex for understanding of overall pharmacokinetics of digitalis compounds. In: Storstein O (ed) Symposium on Digitalis. Gyldendal Norsk Forlag, Oslo, pp 143–157
14. Repke KRH, Dittrich F (1980) Thermodynamics of information transfer from cardiotonic steroids to receptor transport ATPase. Trends Pharmacol Sci 1: 398–402
15. Repke KRH, Schönfeld W (1984) Na^+/K^+-ATPase as the digitalis receptor. Trends Pharmacol Sci 5: 393–397
16. Repke KRH, Herrmann I, Portius HJ (1984) Interaction of cardiac glycosides and Na/K-ATPase is regulated by effector-controlled equilibrium between two limit enzyme conformers. Biochem Pharmacol 33: 2089–2099
17. Repke KRH (1985) Novel developments in cardiac glycoside structure-activity relationships. Trends Pharmacol Sci 6: 275–278
18. Repke KRH (1985) Allosteric regulation of Na^+/K^+-transporting ATPase activity. Submitted for publication
19. Schönfeld W, Weiland J, Lindig C, Masnyk M, Kabat MM, Kurek A, Wicha J, Repke KRH (1985) The lead structure in cardiac glycosides is 5β, 14β-androstane-3β,14-diol. Naunyn-Schmiedeberg's Arch Pharmacol 329: 414–426
20. Specht SC (1984) Development and regional distribution of two molecular forms of the catalytic subunit of the Na/K-ATPase in rat brain. Biochem Biophys Res Comm 121: 208–212
21. Sweadner KJ (1979) Two molecular forms of $(Na^+ + K^+)$-stimulated ATPase in brain. J Biol Chem 254: 6060–6067
22. Wiesner K, Tsai TYR, Kumar R, Sivaramakrishnan H (1984) A stereoselective synthesis of (21R)-21-methyldigitoxigenin, a fully active cardenolide with a wide margin of safety: a contribution to the topology of the digitalis receptors. Helv Chim Acta 67: 1128–1135

Author's address:

Professor K. R. H. Repke, M. D., Zentralinstitut für Molekularbiologie der Akademie der Wissenschaften der DDR, Robert-Rössle-Str. 10, DDR 1115 Berlin

The "endogenous cardiac glycoside"

W. Schoner, K. Moreth, R. Kuske and D. Renner[1]

Institut für Biochemie und Endokrinologie, Justus-Liebig Universität, Giessen, and [1]Medizinische Klinik, Krankenhauszweckverband Augsburg, West Germany

Evidence is accumulating that, in addition to the second messengers of hormone action cAMP, Ca^{2+} and inositoltriphosphate, alterations of the intracellular Na^{+}-concentration may also mediate hormonal effects on cellular processes (66). Provided this concept is right, the sodium pump must also be considered as a receptor of hormones and as a mediator of hormonal action (3). Alterations of the intracellular sodium concentration are considered to be involved in the mechanism of inotropy of the cardiac muscle cells (2) and in differentiation processes (66). With this in mind, it is not astonishing that a hormone of insects, which stimulates the differentiation of tissues (50), uses the sodium pump as a target in the follicle cells of the bug *Rhodnius prolixus* (28) (Fig. 1): isolated (Na^{+}+K^{+})-ATPase from follicle cells is activated below 10^{-8} M juvenile hormone but is inhibited above this concentration. The effect is specific for the follicle enzyme and is not seen with the brain enzyme of *Rhodnius prolixus*. Swelling of vitellogenic follicle cells is activated by juvenile hormone, but inhibited by ouabain (1). It is therefore likely that the juvenile hormone binds to the cardiac glycoside receptor site and that the activating effect of the juvenile hormone reflects the sometimes observable effect of a stimulation of (Na^{+}+K^{+})-ATPase by cardiac glycosides (18, 19, 42) at very low concentrations.

The cardiac glycoside binding site of the sodium pump is by no means specific for cardenolides or bufadienolides. Amongst a considerable number of inhibitors of the sodium pump which seem to bind to the cardiac glycoside binding site (58), the recent work of Habermann and his coworkers in Giessen (6), and that of Ishida et al. (29) and Ozaki et al. (48) are of special interest. They found that palytoxin, the most poisonous toxin we know, of the coral *Palythoa toxica* binds with much higher affinity to the cardiac glycoside receptor site of (Na^{+}+K^{+})-ATPase than bufadienolides or cardenolides. A comparison of the chemical structures of all these substances shows no structural relationship (Fig. 1). The only common characteristic is the existence of double bonds. Repke and coworkers (51) showed that the lead structure of all cardiac glycosides is 5β,14β-androstane-3β,14-diol. The structure of the cardiac glycoside binding site of (Na^{+}+K^{+})-ATPase has been conserved via long distances of the evolution (30, 49, 56). Apparently, all these substances acting at the cardiac glycoside binding site – and even unsaturated fatty acids, which have been isolated in the search for endogenous cardiac glycosides (4, 61) – are able to imitate this lead structure.

The blood plasma of amphibia and mammals contains circulating factors which partly cross react with digitalis antibodies and which inhibit by binding to the cardiac glycoside receptor site of the sodium pump (9, 16, 31). These endogenous cardiac glycosides, which may be identical to the natriuretic hormone (7, 12, 13, 34), have been found in increased concentrations in essential hypertension, in DOCA salt rats, volume expanded animals and after salt treatment (7, 12, 13, 34, 36, 61). The endogenous cardiac glycosides can be extracted from the hypothalamus (8, 25), the whole brain (15, 41), heart muscle (14, 17) and

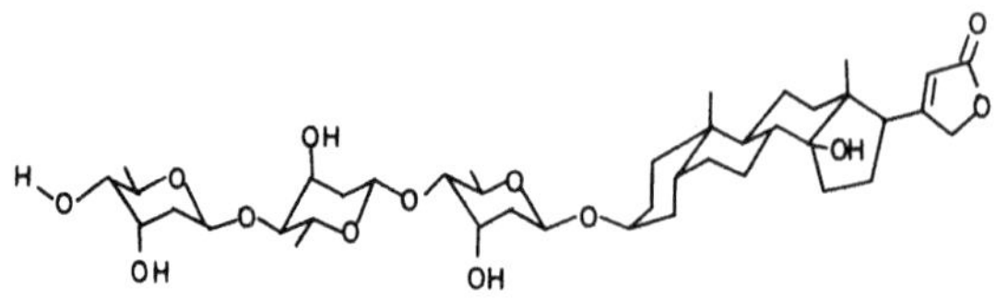

DIGITOXIN

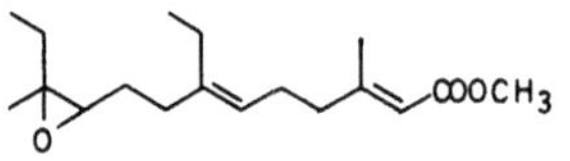

JH-I

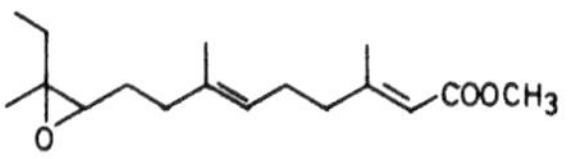

JH-II

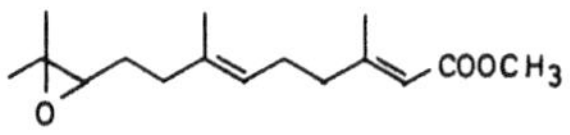

JH-III

JUVENILE HORMONES

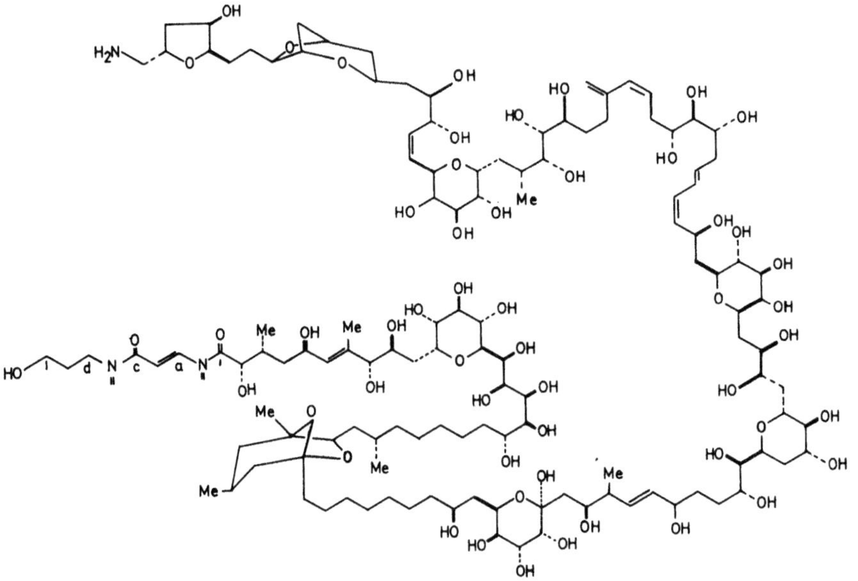

PALITOXIN

Fig. 1 Substances binding to the cardiac glycoside receptor site of $(Na^{+}+K^{+})$-ATPase.

the electric organ of *Electrophorus electricus* (4). Reports on the chemical and physical properties of the isolated factors give a confusing picture: all factors of low molecular weight (500–2000) have been reported to be heat resistant (7, 12, 13) and to show polar (12) or nonpolar characteristics (4, 61). The partially purified factors have been described as being resistant to proteases (12, 13, 23, 41) or not (7), to be destroyed by acid (20, 22) or not (12, 13, 15, 32, 36) and to lose activity upon treatment with alkali (32). Although cardiac glycoside do not inhibit Ca^{2+}-ATPase from sarcoplasmic reticulum, which has no cardiac glycoside receptor sites (43, 56), some of the fractions containing an endogenous cardiac glycoside have been reported to inhibit this enzyme (8, 53). It is therefore understandable that some authors claim that the "endogenous cardiac glycosides" are preparatory artefacts (52, 60). It has been proposed that the endogenous cardiac glycoside might be a peptide (7), a sugar attached to a ring structure (23), a steroid (55), an unsaturated fatty acid (4, 61) or a derivative thereof (54). Despite the fact that elevated concentrations of the circulating endogenous digitalis are found at elevated blood pressure (7, 9, 11, 12, 13, 16, 21, 24, 31), linoleic acid and oleic acid, which are known to decrease the blood pressure (27, 47, 62), have been identified as circulating endogenous inhibitors of the sodium pump in volume expanded hogs in increased concentrations (61). In contrast to the expectation of a hormone, the incredibly high concentrations of $2\text{–}5 \times 10^{-4}$ M fatty acids are needed to displace ^{3}H-ouabain from $(Na^+ + K^+)$-ATPase or radioactive digoxin from digoxin antibodies. This shows that linoleic acid or oleic acid are probably not the substances whose rise is responsible for the increased blood pressure, provided Baustein's hypothesis is valid (5). This hypothesis postulates that the rise in blood pressure is the consequence of the inhibition of the sodium pump of smooth muscle cells of the arteriolae, whose contraction is exerted by an increase of the intracellular Ca^{2+} concentration, which in turn is the consequence of the enhanced $Na^+ - Ca^{2+}$ exchange at elevated intracellular Na^+ concentrations. Consistent with this hypothesis, ouabain increases the peripheral vascular resistance in the forearm after intravenous infusion at concentrations which inhibit the sodium pump (26, 35).

One of the major obstacles to obtaining clearer view of the physiology and the regulation of the circulating "endogenous cardiac glycosides" and the method to purify the substance(s) is the unavailability of an assay which allows the quantitation of these factors in blood plasma. All data reported so far stem from heat-treated plasma (Table 1). Because some inhibitory activity survives after boiling, the substance is considered to be heat stable. Although most authors find increased concentrations of endogenous cardiac glycoside(s) in essential hypertension by measuring the displacement of radioactive ouabain from its receptor site of the sodium pump of red blood cells, the group of Meyer in Paris is nevertheless unable to correlate its concentration to the blood pressure (10). Moreover, as is evident from Table 1, the concentrations of the inhibitor determined so far differ considerably. Quantitation of the endogenous cardiac glycoside by a radioimmunoassay with a digoxin antibody is apparently difficult to reproduce (21, 24, 31) (Table 1). Furthermore, the meaning of this digoxin-like circulating substance is unclear, since upon fractionation of plasma, immunoreactive fractions have been obtained which do not inhibit the sodium pump (9, 31). In order to obtain better information on the validity of the reported findings, we repeated some of the previously published experiments and tried to develop a quantitative test to measure the concentration of "endogenous cardiac glycosides" in untreated serum. Such a test may be helpful at least for diagnostic purposes. It could also give a better understanding of the reasons for the considerable scatter of the concentrations reported so far (Table 1).

In our own experiments, we succeeded in seeing an inhibition of the $^{22}Na^+$ efflux from lymphocytes of normotensive donors upon their incubation in the serum of patients with

Table 1 Concentrations of endogenous digitalis in blood plasma

Authors	Species	Ouabain equivalents[a)] (nmoles/l) normo-tensive	hyper-tensive	H/N	Digoxin equivalents[b)] (nmoles/l) normo-tensive	hyper-tensive	H/N
Flier et al. (1978) [16]	Bufo marinus	20–50000			100		
untreated serum	Rana pipiens	not detectable			not detectable		
Gruber et al. (1982) [21]	Rhesus monkey				0.34	1.03	
boiled plasma at					$p < 0.01$		
pH 5.5 for 10 min,							
filtration through							
UM 10 membranes							3.0
Kojima (1984) [32]	DOCA.salt rat	1.32 ± 0.16	4.96 ± 0.88			0.58 ± 0.12	
boiled plasma for		(10)	(18)				
5 min, desalted		$p < 0.01$		3.7			
Hamlyn et al. (1982) [24]	Man				0.047	0.051	
boiled plasma at					$p > 0.1$		1.1
pH 5.5 for 2 min							
Devynck et al. (1984) [11]	Man	10.5 ± 2	36.0 ± 14				
boiled plasma for		(21)	(23)				
15 min		$p < 0.01$		3.5			
Kelly et al. (1984) [31]	Man	0.39 ± 0.25	1.25 ± 0.18		no significant difference		
boiled for 15 min		(6)	(10)				
		$p < 0.02$		3.2			
Moreth et al. (1985) [45]	Man	76.32 ± 9.29	243.82 ± 48.69		no significant difference		
untreated serum							
		(22)	(18)				
		$p < 0.001$		3.2			
boiled serum for		16.53 ± 2.15	41.65 ± 8.41				
15 min		(8)	(11)				
		$p < 0.05$		2.5			
boiled serum at		27.34 ± 5.07	22.8 ± 4,76				
pH 5.5 for 10 min		(11)	(11)				
		$p > 0.1$					
filtration of serum		9.41 ± 1	8.73 ± 0.5				
through PM 10		(5)	(5)				
membranes		$p > 0.1$					

a) Receptor assay with $(Na^{+}+K^{+})$-ATPase
b) Radioimmunoassay

essential hypertension (Fig. 2). Furthermore, we saw that even untreated serum from patients with essential hypertension, inhibits upon incubation isolated $(Na^{+}+K^{+})$-ATPase from pig kidney stronger than serum from normotensives (39, 44); boiling is apparently not necessary to see this difference (24). In addition, we learned that serum from hypertensive donors displaced ^{3}H-ouabain stronger from the cardiac glycoside receptor site of isolated $(Na^{+}+K^{+})$-ATPase than serum from normotensives (45). In contrast to the findings of the group of Meyer (10), all these effects correlated with the blood pressure of the donors (Fig. 2). The analysis of the type of inhibition of ^{3}H-ouabain displacement under equilibrium

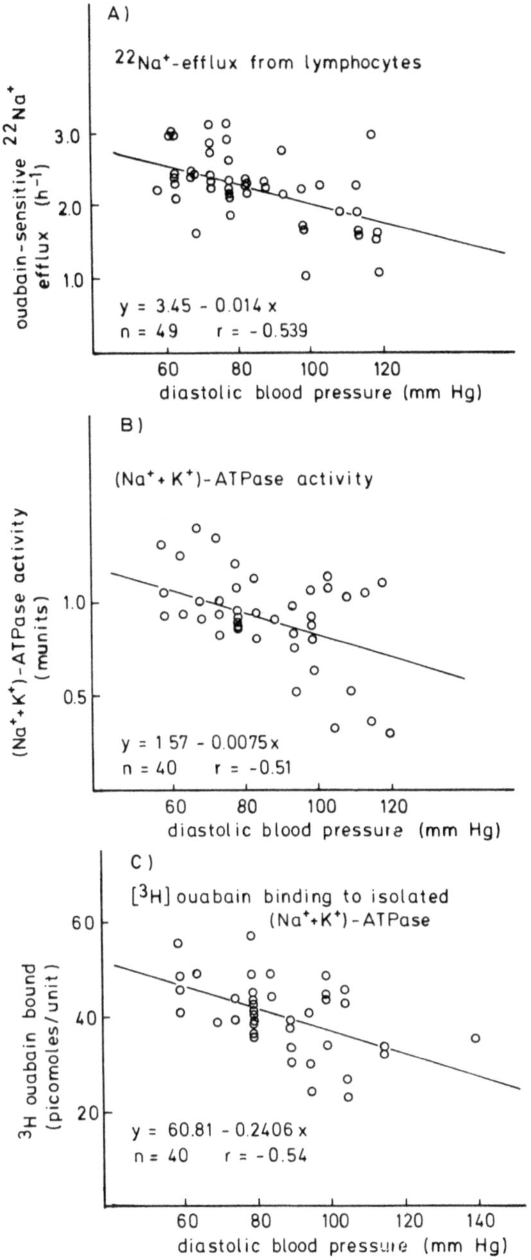

Fig. 2. Correlation of several parameters of the sodium pump with the diastolic blood pressure of normotensive donors of serum and of donors with essential hypertension. A) Ouabain-sensitive $^{22}Na^+$-efflux from lymphocytes of normotensive donors upon incubation in the sera of normotensives and hypertensives. B) Activity of isolated $(Na^+ + K^+)$-ATPase from pig kidney in the sera of normotensives and hypertensives. C) ^{3}H-ouabain binding to isolated $(Na^+ + K^+)$-ATPase from pig kidney in sera of normotensives and hypertensives (from Kuske et al. (38, 39), Moreth et al. (44, 45)).

conditions by the serum factor revealed that it competed with ouabain for its binding site on $(Na^+ + K^+)$-ATPase. The factor circulating in increased concentration in the serum of patients with essential hypertension therefore behaves like an endogenous cardiac glycoside. Although these data confirm the postulate of the existence of an endogenous cardiac glycoside, using test kits supplied by Amersham or New England Nuclear we were unable to see any increased concentrations of a digoxin-like substance in sera of patients with essential hypertension.

Since it was possible to measure the displacement of binding of ^3H-ouabain to the isolated $(Na^+ + K^+)$-ATPase by a factor of the serum, we were also able to quantitate its concentration by a receptor assay: assuming that the affinity of the unknown endogenous cardiac glycoside is identical to that of ouabain, we measured concentrations of 76.32 ± 9.29 nM ouabain equivalents in the serum of normotensives and 243.82 ± 48.69 nM ouabain equivalents in the serum of hypertensives (45) (Table 1). These data corroborate the earlier findings of about 3 times higher concentrations of circulating endogenous cardiac glycosides in essential hypertension (11, 31). However, the concentrations of endogenous cardiac glycosides determined by our procedure are 7 (11) or 195 (31) times higher than those determined earlier (Table 1). Provided that circulating endogenous cardiac glycosides bind with the same affinity to $(Na^+ + K^+)$-ATPase as ouabain, their serum concentration is in the range of steroid hormones (40). It is furthermore evident, that after boiling of serum only 21–35% of endogenous cardiac glycoside remains in the supernatant (Table 1). It is unclear so far, whether boiling or acid treatment plus boiling destroys the factor. The factor did not pass membranes which excludes substances of a molecular weight above 10,000. This might be due to the fact that PM 10 membranes adsorb steroids.

Our former trials to isolate endogenous cardiac glycosides from hemofiltrate of uremic patients with malignant hypertension (39) have so far been unsuccessful. We had to learn that we had isolated a hydrophobic quarternary ammonium compound which behaved like a cardiac glycoside. This was a constituent of an antiseptic which was used by the personnel of the dialysis ward to clean the bottles and to prevent bacterial growth in the hemofiltrate – a fact which was unknown to us. Since endogenous cardiac glycoside has been described to be hydrophilic (12) and since its quantitation by digoxin antibodies is unclear, we isolated a hydrophilic inhibitor of $(Na^+ + K^+)$-ATPase and a competitive inhibitor of ouabain binding to this enzyme from beef brain cortex. This anionic substance, which was purified to homogeneity, shows a maximum absorbance at 265 nm, is destroyed by acid hydrolysis and loses its inhibitory potency upon methylation. The substance of a molecular weight around 200 turned out to be ascorbic acid according to mass spectroscopy and enzymatic analysis. Inhibition by ascorbic acid was seen with the enzymes from beef brain and sheep kidney, and also Ca^{2+}-ATPase from sarcoplasmic reticulum was inhibited. This is of interest in so far as some authors claim that impure fractions of endogenous digitalis also inhibit Ca^{2+}-ATPase (8, 53). It has never been examined whether these fractions contain ascorbate. Since ascorbate survives better in organic solutions, it could easily be that ascorbate is a contaminant. Inhibition of ^3H-ouabain binding and of $(Na^+ + K^+)$-ATPase activity depends on the presence of a heavy metal in commercial ATP. Since AMP-PCP protected the enzyme competitively against the inactivation (37), the ascorbate-induced irreversible inhibition of the transport ATPases is not caused by lipid peroxidation (59) but by reduction of a component in the ATP binding site. Half-maximal inhibition of ATPase activity is seen at 40 μM; this is in the range of the ascorbate concentration of blood plasma (34–68 μM) but considerably lower than the intracellular concentration found in leucocytes (1.7 mM) (65). It remains to be clarified whether ascorbate can regulate its own Na^+-dependent uptake (57, 63) by the inhibition of the sodium pump.

Our own failure parallels that of Ng, Akera and others (46), who isolated ascorbate from bovine adrenal glands in their attempts to isolate endogenous digitalis. Unsaturated fatty acids have also to be excluded as circulating inhibitors of the sodium pump in essential hypertension, because unsaturated fatty acids lower the blood pressure (27, 47, 62). It is therefore evident that our efforts have to be continued to isolate the endogenous cardiac glycoside(s).

Acknowledgements

Our work has been supported by the Deutsche Forschungsgemeinschaft, Bonn-Bad Godesberg (Scho 139/16-2) and by the Fond der Chemischen Industrie, Frankfurt/M.

References

1. Abu-Hakima R, Davey KG (1979) Insect Biochem 9: 195–198
2. Akera T, Brody TM (1982) Ann Rev Physiol 44: 375–388
3. Anner BM (1985) Biochem J 227: 1–11
4. Bidard J-N, Rossi B, Renaud J-F, Lazdunski M (1984) Biochem Biophys Acta 769: 245–252
5. Blaustein MP (1977) Am J Physiol 232: C165–173
6. Böttinger H, Habermann E (1984) Naunyn-Schmiedeberg's Arch Pharmacol 325: 85–87
7. Buckalew jr, VM, Gruber KA (1984) Ann Rev Physiol 46: 343–358
8. Carilli CT, Berne M, Cantley LC, Haupert GT (1985) J Biol Chem 260: 1027–1031
9. Crabos M, Wainer JM, Cloix J-F (1984) FEBS-Letters 176: 223–228
10. De Thé H, Devynck M-A, Rosenfeld J, Pernollet M-G, Elghozi J-L, Meyer P (1984) J Cardiovasc Pharmacol 6: S49–S54
11. Devynck MA, Pernollet MG, Cloix JF, De Thé H, Kamal L, Elghozi JL, Rosenfeld J, Meyer P (1984) Clin and Exper Hyper – Theory and Practice A6 (1 & 2): 441–453
12. De Wardener HE (1982) Ann Clin Biochem 19: 137–140
13. De Wardener HE, Mac Gregor GA (1982) Lancet (June 26): 1450–1454
14. Fagoo M, Godfraind T (1985) FEBS-Letters 184: 150–154
15. Fishman MC (1979) Proc Natl Acad Sci 76: 4661–4663
16. Flier JS, Maratos-Flier E, Palotta J, McIsaac D (1979) Nature 279: 341–343
17. Fox AAL, Strobach H, Greef K (1985) Nauny-Schmiedeberg's Arch 329, R 57, Abstr 227
18. Giunta C, De Bortoli M, Sanchini M, Stacchini A (1985) Gen Pharmacol 16: 183–188
19. Godfraind T, Castaneda-Hermandez G, Ghysel-Burton J, De Pover A (1983) CVurr Top Membr Transp 19: 913–915
20. Gruber KA, Whitaker JM, Buckalew VM (1980) Nature 287: 743–745
21. Gruber KA, Rudel LL, Bullock BC (1982) Hypertension 4: 348–354
22. Gruber KA, Whitaker JM, Buckalew jr VM (1983) Curr Top Membr Transp 19: 917–921
23. Haddy FJ (1982) Biochem Pharmacol 31: 3159–3161
24. Hamlyn JM, Ringel R, Schaeffer J, Levinson PD, Hamilton BP, Kowarski AA, Blaustein MP (1982) Nature 300: 650–652
25. Haupert GT, Sancho JM (1979) Proc Natl Acad Sci 76: 4658–4660
26. Hultén UL, Bolli P, Kiowski W, Bühler FR (1984) J Cardiovasc Pharmacol 6: S75–81
27. Iacono JM, Dougherty RM, Puska P (1982) Hypertension 4 (Suppl III): 34–42
28. Ilenchuk TT, Davey KG (1982) Insect Biochem 12: 675–679
29. Ishida Y, Takagi K, Takahashi M, Satake N, Shibita S (1983) J Biol Chem 258: 7900–7902
30. Kawakami K, Noguchi S, Noda M, Takahashi H, Ohta T, Kawamura M, Nojima H, Nagano K, Hirose T, Inayama S, Hayashida H, Miyata T, Numa S (1985) Nature 316: 733–736
31. Kelly RA, O'Hara DS, Goldszer RC, Mitch WE, Smith TW (1984) in Proc IVth Int Conf on Na,K-ATPase (Ellory JC, Glynn IM, eds) p 31: Cambridge
32. Kojima J (1984) Biochem Biophys Res Commun 122: 129–136

33. Kracke GR (1983) Curr Top Membr Transp 19: 927–930
34. Kramer HJ (1981) Klin Wochenschr 59: 1225–1230
35. Kramer HJ, Glänzer K, Freitag T, Schönfeld J, Sorger M, Schlebusch H, Düsing R, Krück F (1985) Klin Wochenschr 63: 32–36
36. Kramer HJ, Heppe M, Weiler E, Bäcker A, Liddiard C, Klingmüller D (1985) Renal Physiol 8: 80–89
37. Kuske R, Schoner W (1985) unpublished results
38. Kuske R, Renner D, Schoner W (1984) Biomed Biochim Acta 43: 21–22
39. Kuske R, Moreth K, Wizemann V, Renner D, Schoner W (1985) In: Ellory J, Glynn I (eds), The Sodium Pump. Cambridge, The Company of Biologists, pp 651–656
40. Labhart A (1974) Clinical Endocrinology, Theory and Practice. Springer Verlag, Berlin
41. Lichstein D, Samuelov S (1980) Biochem Biophys Res Commun 96: 1518–1523
42. Lichstein D, Samuelov S, Bourrit A, Katchalski S (1982) Neurosci Lett 10 (Suppl): 292
43. MacLennan DH, Brandl CJ, Bozena K, Green NM (1985) 316: 696–700
44. Moreth K, Kuske R, Renner D, Schoner W (1986) Klin Wochenschr (in press)
45. Moreth K, Renner D, Schoner W (1985) unpublished results
46. Ng Y-C, Akera T, Han C-S, Braselton WE, Kennedy RH, Temma K, Brody TM, Sato PH (1985) Biochem Pharmacol 34: 2525–2530
47. Oster P, Arab L, Schellenberg B, Heuck CC, Mordasini R, Schlierf G (1979) Res Exp Med 175: 287–291
48. Ozaki H, Nagase H, Urakawa M (1984) FEBS-Letters 173: 196–198
49. Repke K, Est M, Portius HJ (1965) Biochem Pharmacol 14: 1785–1802
50. Riddiford LM, Truman JW (1978) In: Rockstein M (ed) Biochemistry of Insects. Academic Press, New York, pp 324–326
51. Schönfeld W, Weiland J, Lindig C, Masnyk M, Kabat MM, Kurek A, Wicha J, Repke KRH (1985) Naunyn-Schmiedeberg's Arch 329: 414–426
52. Schwartz A (1983) Curr Top Membr Transp 19: 825–841
53. Schwartz A, Whitmer K, Grupp G, Grupp I, Adams RJ, Lee S-W (1982) Ann N. Y. Acad Sci 402: 2253–2271
54. Schwartzman M, Ferreri NR, Caroll MA, Songu-Mize E, McGiff JC (1985) Nature 314: 620–622
55. Shimoni Y, Gotsman M, Deutsch J, Kachalsky S, Lichstein D (1984) Nature 307: 369–371
56. Shull GE, Schwartz A, Lingrel JB (1985) Nature 316: 691–695
57. Siliprandi L, Vanni P, Kesseler M, Semenza G (1979) Biochim Biophys Acta 552: 129–142
58. Song Kim R-S, Bella FS (1981) Pharmac Ther 14: 391–409
59. Svoboda P, Mosinger P (1981) Biochem Pharmacol 30: 427–432
60. Sweadner K (1983) Curr Top Membr Transp 19: 765–780
61. Tamura M, Kuwano H, Kinoshita T, Inagami T (1985) J Biol Chem 260: 9672–9677
62. Tobian L, Ganguli M, Johnson MA, Iwai J (1982) Hypertension 4 (Suppl II): 149–153
63. Toggenburger G, Häusermann M, Mütsch B, Genoni G, Kessler M, Weber F, Horning D, O'Neill B, Semenza G (1981) Biochim Biophys Acta 646: 433–443
64. Whitmer KR, Epps D, Schwartz A (1983) Curr Top Membr Transp 19: 945–950
65. Wissenschaftliche Tabellen Geigy (1979) Teilband Hämatologie und Humangenetik, 8. Auflage, Basel
66. Varon S, Skaper SD (1983) Trends Biochem Sci 8: 22–25

Author's address:

Prof. Dr. W. Schoner, Institut für Biochemie und Endokrinologie des Fachbereichs Veterinärmedizin, Justus-Liebig-Universität Giessen, Frankfurter Straße 100, 6300 Giessen (West Germany)

Endogenous digitalis, cardiodigin and other regulators of the sodium pump

T. Godfraind and M. Fagoo

Laboratoire de Pharmacodynamie Générale et de Pharmacologie, Université Catholique de Louvain, Bruxelles (Belgium)

Introduction

The identification of specific receptors for cardiac glycosides has been performed in cardiac muscle using intact tissue and microsomal preparations by measuring the binding characteristics of radioactive ouabain and of other cardiac glycosides. There is a large body of evidence suggesting that these specific binding sites have characteristics of functional receptors responsible for the transduction of an extracellular signal into a positive inotropic response. It is now becoming evident that these sites are related to Na^+,K^+-ATPase, the enzyme constitutive of the sodium pump. Therefore, digitalis receptors control not only cardiac inotropy but also the gradient of monovalent cations between the intracellular milieu and extracellular spaces. They are responsible for a wide variety of cellular functions including excitability, secretion, transmission etc.... The rate at which the sodium pump is operating is influenced largely by the relative distribution of ions across the plasma membrane. It is well established that Na activates from intracellular sites and K from extracellular ones. The purpose of this paper is to review the possible role of other factors which could be responsible for the physiological and pathophysiological control of the sodium pump. In view of the specificity of the binding sites for cardiac glycosides, we will also discuss the possible existence of endogenous specific ligands of the sodium pump.

Receptors for cardiac glycosides in myocardial cells

The analysis of the binding of ouabain and other cardiac glycosides to cardiac preparations has shown both in intact preparations and in microsomes the existence of high and low affinity receptor sites.

In guinea-pig atria, a high affinity, low capacity binding of ouabain has been identified in the nanomolar range and a low affinity, high capacity binding has been identified in a range of concentrations 100 or 1000 times higher (19). The high affinity sites represent 5 to 10 per cent of the total number of sites and correspond to a maximum binding capacity of about 10,000 molecules per cardiac cell.

The binding of cardiac glycosides to microsomal fractions containing Na^+,K^+-ATPase (37) is influenced by various ligands: ATP, Na, K, Mg. Enzyme phosphorylation increases its affinity for ouabain. According to the ligands supporting the binding, two types of complexes have been proposed to exist. Binding to type I (formed in the presence of Mg and ATP) is enhanced by Na. Binding to type II (formed in the presence of Mg and Pi) is antagonized by Na. Binding to both complexes is antagonized by K (3).

There are several experimental arguments suggesting that high and low affinity binding sites are related to two Na,K-ATPases. In both guinea-pig and rat microsomes, the ligands supporting the binding of ouabain are the same (41). In rat heart microsomes, the existence of two different Na^+,K^+-ATPase inhibitory processes can be identified when a wide range of ouabain concentrations is studied. Noël and Godfraind (41) have concluded that two Na^+,K^+-ATPase isozymes coexist in rat heart microsomes. The high and low affinity sites are in a ratio of 1 to 10. KCl has little influence on equilibrium binding of ouabain to the high affinity site, but strongly reduces equilibrium binding to the low affinity site. For a given species, the relative importance of the high and low affinity binding sites may vary according to the anatomical part of the heart. Rat heart ventricles contain more high affinity binding sites than rat atria (18). Furthermore, the affinity ratio between the high and low affinity sites may be different in different species, for example, taking ouabain as a ligand, it is equal to 60–70 in rat heart microsomes and only to 7 in human heart microsomes (12, 18).

Tissular regulators of the sodium pump of known structure

In the preceding section, we have emphasized evidence for the existence of subtypes of digitalis receptors, probably associated with two Na^+,K^+-ATPase isozymes. In rat brain, two enzyme forms have been identified by SDS electrophoresis according to the apparent molecular weight of their catalytic subunit and have been named alpha + for the high affinity form and alpha for the other (54). However, only few data are available on a possible differentiation of these subtypes by the various regulators (36, 51). Thyroid hormones and corticosteroids exert a long-term regulation of the sodium pump by stimulating synthesis of Na^+,K^+-ATPase molecules. This induction might be a secondary consequence of a change in membrane permeability resulting in greater cellular sodium load (25). Tissular organic and inorganic agents have an effect on the sodium pump in short-term studies. They may act either directly or indirectly as inhibitors or stimulants. They will be briefly reviewed below.

A) Inorganic agents

The role of Na, K, and Mg has already been considered in the previous section. Besides these, we may consider other agents, like Fe^{2+}, Pb^{2+}, Cu^{2+}, Zn^{2+}, Ca^{2+}, VO_3^- (28, 45, 46).

Calcium

Godfraind et al. (20) have reported that calcium ions inhibit Na^+,K^+-ATPase, at concentrations which are likely to be achieved in heart cell during the contractile cycle. This inhibition could be enhanced by acidification of the intracellular medium which occurs during the excitation-contraction cycle. As most of the experimental data suggest, it is likely that, in vivo, the Ca^{2+} inhibitory sites are facing intracellular fluid. The mechanism of Ca^{2+} inhibition is still unclear. Ca^{2+} might compete with Mg^{2+}, either directly or via ATP complexes. Alternatively, Ca^{2+} may act as an antagonist of Na^+ (5). It seems established that Ca^{2+} inhibits the Na^+,K^+-ATPase by acting at more than one step in the hydrolytic step. Recently, a Ca^{2+}-stimulated phosphatase activity sensitive to ouabain has been described (29), indicating the existence of a site of Ca^{2+} activation different from the site at which Ca^{2+} inhibits Na^+,K^+-ATPase.

Calmodulin enhances the inhibitory effect of calcium on Na^+,K^+-ATPase. This effect of calmodulin is dose-dependent and detectable within a range of concentrations between 10^{-8} and 10^{-6} M, as found for other cellular effects. The decrease in ouabain-sensitive ATPase activity induced by the Ca^{2+}-calmodulin complex might reflect either a lowered turnover of

Na^+,K^+-ATPase or a decrease in its sensitivity to ouabain. Since Ca^{2+}-calmodulin activates a particular type of protein kinase, it could also reduce the ouabain-sensitive ATPase by affecting its phosphorylation. The activity of the putative ouabain-like factor might also be modulated by intracellular level of calmodulin (10).

Vanadate

Vanadate has been reported to have actions on many physiological and biochemical processes, including a positive inotropic action on ventricular muscle as well as species-dependent positive or negative inotropic action on atrial muscle. These effects may be related to the ability of vanadate to inhibit various enzymes, including Na^+,K^+-ATPase and to stimulate adenylate cyclase. Vanadium in the +5 oxydation state is the most active species, the more reduced forms (+4, +3) being less effective (46). The vanadate site (a low affinity binding site for ATP) is located on the inner surface of the plasma membrane. Vanadate, at concentrations which have been reported to induce a positive inotropic action on mammalian ventricular muscle does not increase adenylate cyclase or inhibit the sodium pump activity in intact myocytes (2).

B) Organic agents

Several organic agents listed below are interfering with the activity of the sodium pump. Only few of them have a direct interaction with the specific binding sites for cardiac glycosides.

Lipids

The Na^+,K^+-ATPase is integrated in the structure of the plasma membrane, and there is increasing evidence that the state of the cell membrane in which it is embedded, may affect ATPase activity and its sensitivity to cardiac glycosides.

Fatty acids:

Free fatty acids have a potent in vitro inhibitory activity, which depends on the chain length and on the degree of insaturation. Free fatty acids inhibit Na^+,K^+-ATPase specifically and reversibly and they compete with K^+ (33). Swann (52) has suggested that free fatty acids could affect enzyme conformation, preventing the transition to E_2, the K^+ sensitive form of the enzyme.

In recent studies, particular interest has been devoted to unsaturated fatty acids as potential digitalis-like factors found in tissue extracts or plasma (6, 55). Metabolism disturbances as resulting from ischemia produce an accumulation of lipid intermediates, including fatty acylCoA thioesters, fatty acylcarnitine esters and lysophospholipids. These compounds inhibit the Na^+,K^+-ATPase activity in vitro and they might influence the enzyme activity in ischemic tissues (1).

Phospholipids and cholesterol:

Na^+,K^+-ATPase requires the presence of phospholipids for full activity; their removal causes partial to complete loss of activity. There seems to be no absolute requirement for a specific phospholipids class. It has been suggested that phospholipids could alter the membrane fluidity, causing functional alterations in the allosteric properties of integral enzymes.

Cholesterol which appears interdigitated between the fatty acids of the phospholipids has an action on the structural characteristics of the fatty acid regions (26).

Prostaglandins:

Some prostaglandins like PGE_2, PGF_{2alpha}, PGI_2 reduce basal Na^+,K^+-ATPase activity of rat heart sarcolemma. The nature of this inhibition appears complex, involving a reduction in Vmax and an elevation of the respective K_m values for either substrate or activator. It seems unlikely that the Pg effects are due to their fatty acid moiety. It might be possible that Na^+,K^+-ATPase has a specific binding site for Pg's. Pg's may be of physiological importance as endogenous modulators of Na^+,K^+-ATPase activity (30).

In cells derived primarily from the medullary portion of the thick ascending limb of the loop of Henle, Schwartzman et al. (48) have identified a cytochrome P_{450}-dependent monooxygenase system which metabolizes arachidonic acid to two biologically active oxygenated products; one of them inhibits Na^+,K^+-ATPase and the other relaxes blood vessels. This metabolization is influenced by arginine vasopressin and salmon calcitonin. So, the cytochrome P_{450}-dependent pathway of arachidonate metabolism could act as a common mechanism mediating natriuresis in response to natriuretic substances, including endogenous circulating inhibitors of Na^+,K^+-ATPase activity.

Bile acids:

Bile acids which are synthesized from cholesterol and conjugated with taurine or glycine in the liver, inhibit Na^+,K^+-ATPase (49). In sera of patients with fulminant hepatic failure (FHF), there is an increase of a wide variety of toxic substances like bilirubin (31), fatty acids and mercaptans; they inhibit rat brain Na^+,K^+-ATPase (49). Other potentially toxic substances include bile acids, phenols and endotoxin. Among them, chenodeoxycholic acid, p-hydroxyphenyllactic acid, p-hydroxyphenylpyruvic acid endotoxin cause 30–45% inhibition of partially purified rat brain Na^+,K^+-ATPase, at concentrations known to occur in serum of patients with (FHF). Although, the interaction of Na^+,K^+-ATPase with chenodeoxycholic acid and endotoxin showed competitive inhibition kinetics with respect to ATP, it is questionable whether compounds so chemically dissimilar would compete for the same binding site. The inhibition could be produced by binding at a site close to the ATP binding site and so preventing ATP binding. Alternatively, such compounds could affect membrane structure through interaction with the lipid components, thus rendering the ATP binding site less available.

Neurotransmitters

Catecholamines:

It has been proposed that Na^+,K^+-ATPase could be involved in the regulation of neurotransmitters release and reuptake. The inhibitory effect of noradrenaline on neurotransmitters release has been correlated with its stimulatory action on Na^+,K^+-ATPase. This stimulatory effect could be indirect, due to an activation of the Ca pump or to the reversal of divalent metal ions inhibition, for instance by complex formation. However, catecholamine stimulation could occur directly, or be mediated by adrenoreceptors (22, 42, 57).

Recent studies carried out by Swann (51) suggest that the stimulation previously reported in vitro may correspond to that observed after increased sympathetic nervous system activity in vivo. His data indicate that part of the stimulation occurred on the postsynaptic membrane. Norepinephrine stimulation appears to affect predominantly the enzymatic form with high affinity for ouabain. An interaction between norepinephrine and endogenous digitalis-like factor is suggested by the observation that Na^+,K^+-ATPase inhibitor from brain extracts prevents stimulation of enzyme by norepinephrine in vitro (53). In an attempt to further analyze such a relation, Swann has investigated the effect of a high NaCl diet on the sensitivity to norepinephrine stimulation. High sodium intake, which increases apparent levels of circulating inhibitor of Na^+,K^+-ATPase (8), prevents Na^+,K^+-ATPase stimulation by amphetamine in cerebral cortex.

Acetylcholine:

Secretion by the avian salt gland which depends on the sodium pump activity is activated by cholinergic nerves. According to Stewart and Sen (50), the cholinergic activation of the sodium pump occurs by a mechanism where cGMP is likely involved. Acetylcholine promotes an increase in Na^+ influx into the salt gland cells, perhaps, by a Na^+ ion carrier dependent on the presence of calcium. In addition, cholinergic stimulation could also promote increase in cytosolic Ca^{2+} concentration mediated by phosphatidylinositol turnover. Ca^{2+} activation of guanylate cyclase increases cGMP which activates the sodium pump.

Pancreatic hormones

Insulin:

Cell integrity is required for insulin stimulation of Na^+,K^+-ATPase activity. Several cellular mechanisms, including increased enzyme synthesis, enzyme recruitment from an intracellular pool, increased enzyme affinity for Na^+, have been proposed.

Recent findings have shown that insulin stimulation is mediated by the amiloride-sensitive Na^+/H^+ exchange (43). In rat adipocytes (36), regulation by insulin of ouabain-inhibitable ^{86}Rb uptake is mediated exclusively via the (alpha (+)) form of Na^+,K^+-ATPase.

Glucagon:

The mechanism underlying glucagon's stimulation of Na^+,K^+-ATPase has been studied in isolated rat hepatocytes. Like insulin, glucagon is able to increase the uptake of ^{86}Rb but this stimulation is only partially inhibited by amiloride. Glucagon could enhance K^+ efflux possibly through an increase in K^+ permeability (17).

Peptides

Histidyl-proline diketopiperazine:

Formed by metabolism of TRH, histidyl-proteine diketopiperazine inhibits the Na^+,K^+-ATPase activity. The site of that action is the vanadate site, located on the inner surface of the cell (4).

Angiotensin:

Angiotensin stimulates Na^+,K^+-ATPase activity in rat hypothalamus, mucosa of colon and bovine adrenal cortex. The action of angiotensin is dependent on the simultaneous presence of Na^+ and K^+ (24).

Vasopressin:

Vasopressin stimulates Na^+,K^+-ATPase; this effect seems mediated by an increase of the intracellular Na^+ concentration (38).

Sexual hormones

Mammalian tissues achieve the 14-hydroxylation of steroids. 14-OH Androstanes and estranes (namely 14-OH estradiol) show digoxin-like immunoreactivity (47). A screening has been carried out by Chow et al. (9), over 150 steroids derivatives of estrane, androstane and pregnane series. Only semisynthetic progesterone derivatives, having a 17alpha-acetoxy substituent and unsaturation and/or substitution in the B ring of the steroid nucleus, share properties with the digitalis family. Among them, chlormadinone acetate is the most potent. It is active in radioreceptor assay for cardiac glycosides in heart homogenates. It binds to the glycosides sites on purified Na^+,K^+-ATPase showing a competitive inhibition with respect to K^+ and a non competitive inhibition with respect to ATP. It also shows the species specificity of ouabain (56). Endogenous progestins, metabolites of progesterone could serve as physiological ligands of the digitalis receptor.

Miscellaneous

Ascorbic acid:

Ascorbic acid is an essential substance in the biochemistry of living cells. It serves as a reductant or a cofactor for certain enzymatic reactions and is a component of a general redox buffer system. According to Lohmann (35), at physiological pH, ascorbic/isoascorbic acids are present as radicals with a cyclic side-chain structure able to engulf Na^+ or K^+ in the case of the ascorbyl or the isoascorbyl radical respectively. These mobile carriers are electroneutral and can transport Na^+ and K^+ across artificial membranes. Since such a radical configuration could allow lipid peroxidation, it will be converted to a non-radical structure by GSH in natural membranes. The lipid peroxidation process requires the presence of membrane-bound metal ions, thus accounting for the antagonistic effect of EDTA on the activity of ascorbic acid, potent inhibitor of isolated Na^+,K^+-ATPase, and of specific ouabain ^{3}H binding (40).

Delta-aminolevulinic acid:

Acute phases of hereditary hepatic porphyrias, lead poisoning and hereditary tyrosinemia are characterized by similar neurological manifestations associated with an overproduction of delta-aminolevulinic acid. This acid inhibits Na^+,K^+-ATPase activity in brain membrane preparations (44).

Histamine:

Histamine stimulates Mg^{2+}-ATPase and inhibits Na^+,K^+-ATPase activity from membrane fractions of human gastric mucosa (39).

Adenosine and derivatives:

Adenosine contributes to the genesis of cardiac arrythmias during acute myocardial ischemia which can reduce transmembrane potentials of the myocardial cells and may increase myocardial adenosine level. Under hypoxic or ischemic conditions, myocardial adenosine reaches 10^{-4} M and may inhibit sarcolemmal Na^+,K^+-ATPase activity. This effect might be responsible for the alterations of transmembrane potentials under pathophysiological conditions (34).

S-adenosylmethionine:

S-adenosylmethionine is an inhibitor of Na^+,K^+-ATPase and a methyl donor for transmethylation of various compounds including porphyrins, fatty acids, phospholipids, polysaccharides, proteins, nucleic acids, and biogenic amines. The N-methylation of phospholipids regulates membrane fluidity. Methylation provides a general mechanism for signal transmission across biological membranes; it is not responsible for S-adenosylmethionine inhibition of Na^+,K^+-ATPase activity (27).

Cardiodigin, endogenous digitalis-like factor

We have proposed that an endogenous digitalis-like factor should fulfil most of the pharmacological criteria of cardiac glycosides themselves. Therefore:

1. it should interact with Na^+,K^+-ATPase showing
   a) reversible inhibition
   b) competition with KCl and dependence of this effect on Na/K ratio
   c) competition with (^3H) ouabain binding
   d) species sensitivity (namely a weak activity in rat heart as compared to guinea-pig)
2. it should inhibit the Na-K pump and evoke a positive inotropic effect in isolated heart preparations
3. it could interact with some of the specific cardiac glycosides antibodies, competing with the radioligand in a way similar to the glycoside itself (13).

The presence of a factor controlling the sodium pump activity in several states including hypertension has been proposed by de Wardener (14). Godfraind et al. (21) have proposed that such a factor termed cardiodigin could be a physiological regulator of the sodium pump. Evidence for the existence of cardiodigin comes from the following observations.

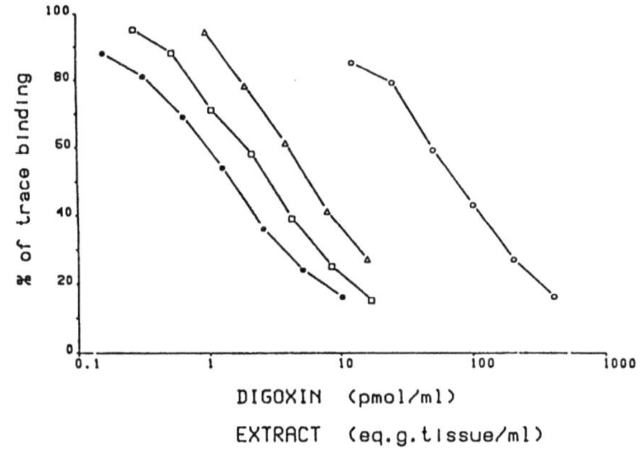

Fig. 1. Typical experiment showing the effect of water extracts from guinea-pig heart (□), rat heart (△), beef heart (○) and of unlabelled digoxin (●) on ^{125}I-antigen binding to antidigoxin specific antibodies (21).

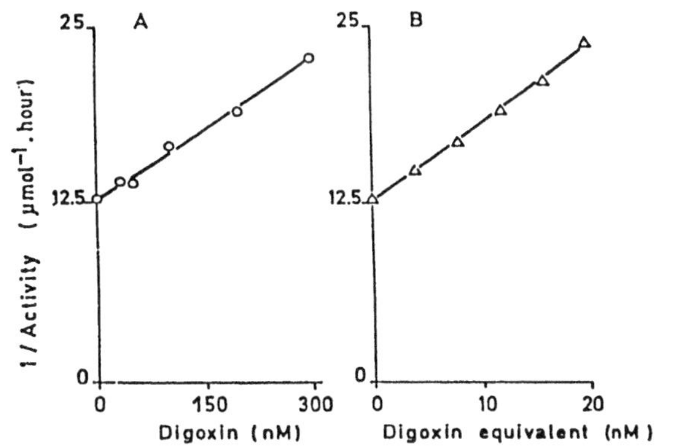

Fig. 2. Effect of digoxin (A) and of guinea-pig heart extract (B) on guinea-pig heart Na^+,K^+-ATPase (Dixon plot). Data are means from triplicate determinations in a typical experiment (13).

Water extracts of mammalian heart contain a factor that cross-reacts with antidigoxin antibodies, displacing ^{125}I digoxin analogue tracer in a way similar to digoxin (Fig. 1). This cross-reactivity allows not only to detect the presence of cardiodigin but also to express its amount in digoxin RIA equivalent (13). In guinea-pig heart, in 14 successive preparations, the yield of the extraction procedure was 0.22 ± 0.05 pmol/g wwt tissue (mean ± SEM). A dose-dependent inhibition of the Na^+,K^+-ATPase is evoked by guinea-pig heart extracts. As illustrated in Fig. 2, Dixon plots are straight lines, suggesting that inhibition obeys mass-action law. This inhibition appears reversible and specific: Mg-ATPase and Ca,Mg-ATPase from cardiac tissue are not affected. This inhibition is antagonized by potassium and shows species sensitivity.

In addition, this factor displaces ouabain specifically bound to its receptors in cardiac and in brain microsomes (Fig. 3). Recently, in an attempt to further characterize cardiodigin, a two-step reverse-phase HPLC analysis was carried out on a guinea-pig heart extract. As shown in Fig. 4, digoxin-like immunoreactivity peaked in the fraction which eluted between 10 and 12 min. There was a correlation between the immunological and the Na^+,K^+-ATPase inhibitory responses (15). Such a correlation, although not always observed, has already been reported by Gruber et al. (23), after HPLC analysis of plasma of volume expanded dogs.

In human plasma, Kelly et al. (32) have shown the presence of at least three fractions which inhibit Na^+,K^+-ATPase and cross-react to a variable degree with different digoxin-specific antibody populations.

We have recently reported that cardiodigin cross-reacts specifically not only with antidigoxin, but also with antidigitoxin antibodies (16). However, in the last case, the apparent affinity was about 10 times lower than for antidigoxin antibodies. So, cardiodigin appears to share molecular determinants critical for specific biological and immunological properties with digitalis glycosides, but whether they are structurally related is still an open question.

In all the chromatography systems so far tested (HPLC/TLC), digoxin and cardiodigin show similar retention properties. Only the cross-reactivity analysis with antidigoxin and

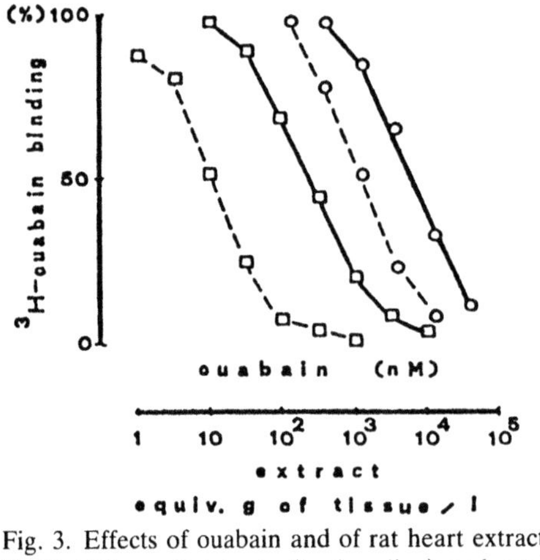

Fig. 3. Effects of ouabain and of rat heart extract (fraction M-60) on specific [^3H]ouabain binding to Na,K-ATPase. Rat brain (broken line) and rat heart (solid line) Na,K-ATPase preparations were incubated in the presence of various concentrations of ouabain (square) or of extract (circle) (11).

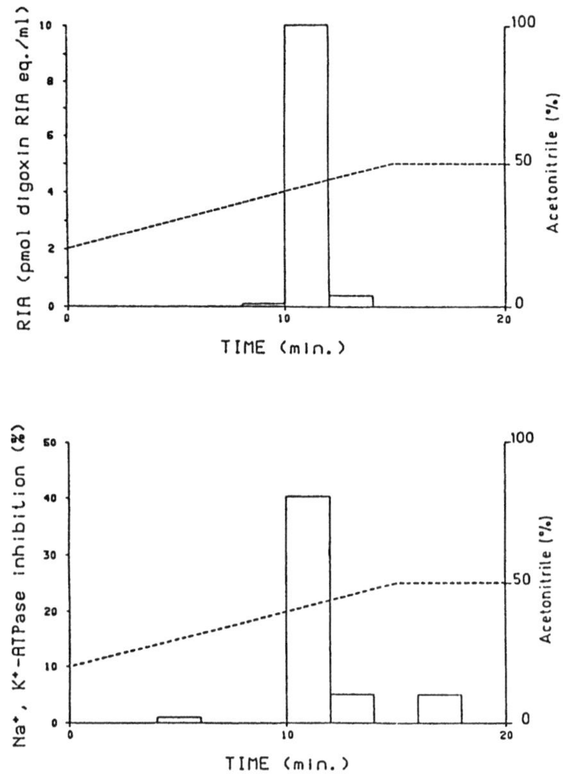

Fig. 4. Characterization of a guinea-pig extract purified by a two-step reverse-phase HPLC.
a) The RIA response, corresponding to 27.3 gt/ml is expressed in pmol digoxin RIA eq./ml.
b) Human heart Na^+,K^+-ATPase inhibition of each 2 min fraction, obtained as in (a). The inhibitory response, expressed in percent, was obtained with 20μl of each HPLC fraction. This volume of fraction corresponds to 41 g tissue (15).

antidigitoxin antibodies allows a clear distinction between them. Perhaps, the analysis of the nature of cardiodigin interaction with Na^+,K^+-ATPase will increase this distinction. As proposed by Carilli et al. (7) and Kelly et al. (32), the ligand requirements for optimal activity may be distinct from those of cardiac glycosides.

To date, in the light of immunological and chromatographic criteria, different compounds of known structure may be eliminated i.e., steroids like aldosterone, corticosterone, 11-dehydrocorticosterone, 11-deoxycorticosterone, progesterone, lipidic compounds like fatty acids, bile acids, cardiolipin, prostaglandins, phospholipids and other substances like ascorbic acid, serotonine, histamine, bilirubin.

The site of production also remains to be determined. Liver, adrenals and brain have been regarded as potential sources (14). In a distribution study, digoxin-like activity, detected by digoxin-RIA, has been found in all the rat tissues tested. A higher response is found with adrenal extracts (Fig. 5). The digoxin-like immunoreactivity present in rat plasma increases during high Na intake (8).

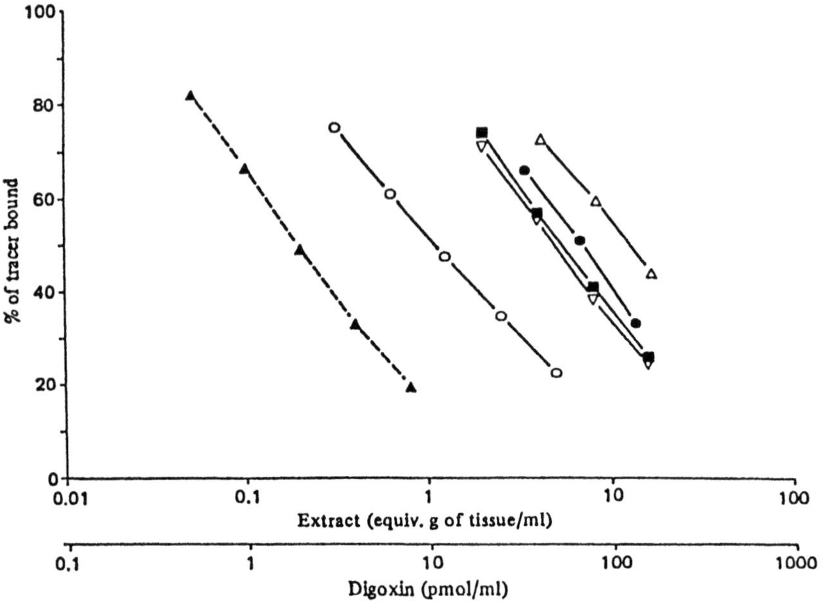

Fig 5. Typical digoxin radioimmunoassay. Effect of several dilutions of extracts from adrenals (0), brain (▽), heart (●), kidneys (△) and liver (■) and of digoxin (▲) on ^{125}I-labelled antigen binding to antidigoxin specific antibodies. Each point is the mean of two determinations; SD did not exceed the diameter of the symbol (8).

Conclusions

Despite intense investigation, the existence of an endogenous digitalis-like factor is still not proved. Since several endogenous agents may affect the sodium pump, their role in the physiological regulation should be clearly established. It might be that specific inhibitors of the sodium pump could reach active concentrations in blood only in pathological states.

Acknowledgments

M. Fagoo is Aspirant du Fonds National de la Recherche Scientifique.

References

1. Adams RJ, Cohen DW, Gupte S, Johnson JD, Wallick ET, Wang T, Schwartz A (1979) In vitro effects of palmitylcarnitine on cardiac plasma membrane Na,K-ATPase, and sarcoplasmic reticulum Ca^{2+}-ATPase and Ca^{2+}-transport. J Biol Chem 254 (24): 12404–12410
2. Aiton JF, Cramb G (1985) The effects of vanadate on rabbit ventricular muscle adenylate cyclase and sodium pump activities. Biochem Pharmacol 34 (9): 1543–1548
3. Akera T (1981) Effects of cardiac glycosides on (Na + K)-ATPase. In: Greeff K (ed). Cardiac Glycosides, Handb Exp Pharm Vol 56/1. Springer-Verlag, Berlin Heidelberg New York, pp 287–383
4. Battaini F, Peterkofsky A (1980) Histidyl-proline diketopiperazine, an endogenous brain peptide that inhibits (Na^+ + K^+)-ATPase. Biochem Biophys Res Commun 94 (1): 240–247
5. Beaugé L, Campos MA (1983) Calcium inhibition of the ATPase and phosphatase activities of (Na^+ + K^+)-ATPase. Biochim Biophys Acta 729: 137–149

6. Bidard JN, Rossi B, Renaud JF, Lazdunski M (1984) A search for an ouabain-like substance from the electric organ of Electrophorus electricus which led to arachidonic acid and related fatty acids. Biochim Biophys Acta 769: 245–252
7. Carilli CT, Berne M, Cantley LC, Haupert GT Jr (1985) Hypothalamic factor inhibits the (Na,K)-ATPase from the extracellular surface. J Biol Chem 260 (2): 1027–1031
8. Castaneda Hernandez G, Godfraind T (1984) Effect of high sodium intake on tissue distribution of endogenous digitalis-like material in the rat. Clin Sc 66: 225–228
9. Chow E, Kim RSS, Labella FS, Queen G (1979) Ouabain receptor binding of hydroxyprogesterone derivatives. Br J Pharmacol 67: 345–352
10. Cirillo M, David-Dufhilo M, Devynck MA (1984) Calmodulin reduces ouabain-sensitive ATPase of cardiac sarcolemmal membranes: high reduction in spontaneously hypertensive rats. Clin Sci 67: 535–540
11. De Pover A (1984) Endogenous digitalis-like factor and inotropic receptor sites in rat heart. Eur J Pharmacol 99: 365–366
12. De Pover A, Godfraind T (1979) Interaction of ouabain with (Na^+ + K^+)-ATPase from human heart and from guinea-pig heart. Biochem Pharmacol 28: 3051–3056
13. De Pover A, Castaneda Hernandez G, Godfraind T (1982) Water versus acetone-HCl extraction of digitalis-like factor from guinea-pig heart. Biochem Pharmacol 31 (2): 267–271
14. de Wardener HE, Clarkson EM (1985) Concept of natriuretic hormone. Physiol Rev 65 (3): 658–759
15. Fagoo M, Godfraind T (1985) Further characterization of cardiodigin, Na^+,K^+-ATPase inhibitor extracted from mammalian tissues. Febs Lett 184 (1): 150–154
16. Fagoo M, Godfraind T (1985) Interaction of cardiodigin, endogenous inhibitor of Na^+, K^+-ATPase, with antidigoxin and antidigitoxin antibodies. Biochem Biophys Res Commun 129: 553–559
17. Fehlman M, Freychet P (1981) Insulin and glucagon stimulation of (Na^+,K^+)-ATPase transport activity in isolated rat hepatocytes. J Biol Chem 256 (14): 7449–7453
18. Finet M, Godfraind T, Noël F (1983) The inotropic effect of ouabain and its antagonism by dihydroouabain in rat isolated atria and ventricles in relation to specific binding sites. Br J Pharmac 80: 751–759
19. Godfraind T, Ghysel-Burton J (1977) Binding sites related to ouabain-induced stimulation or inhibition of the sodium pump. Nature 265: 165–166
20. Godfraind T, De Pover A, Verbeke N (1977) Influence of pH and sodium on the inhibition of guinea-pig heart (Na^+ + K^+)-ATPase by calcium. Biochim Biophys Acta 481: 202–211
21. Godfraind T, De Pover A, Castaneda Hernandez G, Fagoo M (1982) Cardiodigin: endogenous digitalis-like material from mammalian heart. Arch Int Pharmacodyn Ther 258 (1): 165–167
22. Godfraind T, Koch MC, Verbeke N (1974) The action of EGTA on the catecholamines stimulation of rat brain Na-K-ATPase, Biochem Pharmac 23: 3505–3511
23. Gruber KA, Whitaker JM, Buckalew VM Jr (1980) Endogenous digitalis-like substance in the plasma of volume expanded dogs. Nature 287: 743–745
24. Gutman Y, Shamir Y, Glushevitzky, Hochman S (1972) Angiotensin increases microsomal (Na^+-K^+)-ATPase activity in several tissues. Biochim Biophys Acta 273: 401–405
25. Haber RS, Loeb JN (1984) Early enhancement of passive potassium efflux from rat liver by thyroid hormone: relation to induction of Na,K-ATPase. Endocrinol 115 (1): 291–297
26. Harris WE (1985) Modulation of (Na^+,K^+)-ATPase activity by the lipid bilayer examined with dansylated phosphatidylserine. Biochem 24: 2873–2883
27. Hattori H, Kanfer JN (1984) Inhibition of rat brain microsomal Na^+,K^+-ATPase by S-Adenosylmethionine. J Neurochem 42: 204–208
28. Hexum TD (1974) Studies on the reaction catalyzed by transport (Na,K)adenosine triphosphatase – I. Effects of divalent metals. Biochem Pharmacol 23: 3441–3447
29. Hųang, WH, Askari A (1984) Interaction of Ca^{2+} with (Na^+ + K^+)-ATPase: properties of the Ca^{2+}-stimulated phosphatase activity. Arch Biochem Biophys 231 (2): 287–292
30. Karmazyn M, Tuana BS, Dhalla NS (1981) Effect of prostaglandins on rat heart sarcolemmal ATPase. Can J Physiol Pharmacol 59: 1122–1127
31. Kashiwamata S, Goto S, Semba RK, Suzuki FN (1979) Inhibition by bilirubin of (Na^+ + K^+)-activated adenosine triphosphatase and K^+-activated p-Nitrophenylphosphatase activities of NaI-treated microsomes from young rat cerebrum. J Biol Chem 254 (11): 4577–4584

32. Kelly RA, O'Hara DS, Canessa ML, Mitch WE, Smith TW (1985) Characterization of digitalis-like factors in human plasma. J Biol Chem 260 (21): 11396–11405
33. Kim RSS, Labella FS (1980) Specific binding of (^{3}H)ouabain to heart is inhibited by long-chain fatty acids and alcohols. J Mol Cell Cardiol 12: 847–856
34. Lee YS, Chang CH (1985) Effects of adenosine on transmembrane potential and sarcolemmal Na^+-K^+-ATPase activity of human atrial myocardium. Jpn Heart J 26 (1): 69–80
35. Lohmann W (1984) Structure of ascorbic acid and its biological function. Biophys Struct Mech 10: 205–210
36. Lytton J, Lin JC, Guidotti G (1985) Identification of two molecular forms of (Na^+,K^+)-ATPase in rat adipocytes. J Biol Chem 260 (2): 1177–1184
37. Matsui H, Schwartz A (1968) Mechanism of cardiac glycoside inhibition of the (Na^+-K^+)dependent ATPase from cardiac tissue. Biochim Biophys Acta 151: 655–663
38. Mendoza SA, Wigglesworth NM, Rozengurt E (1980) Vasopressin rapidly stimulates Na entry and Na-K pump activity in quiescent cultures of mouse 3T3 cells. J Cell Physiol 105: 153–162
39. Mozsik G, Nagy L, Tarnok F, Javor Z, Kutas J (1974) Effects of histamine on transport (Na^+-K^+-dependent) ATPase system prepared from human gastric mucosa. Pharmacol 12: 193–200
40. Ng YC, Akera T, Han CS, Braselton WE, Kennedy RH, Temma K, Brody TM, Sato PH (1985) Ascorbic acid: an endogenous inhibitor of isolated Na^+,K^+-ATPase. Biochem Pharmacol 34 (14): 2525–2530
41. Noël F, Godfraind T (1984) Heterogeneity of ouabain specific binding sites and (Na^+ + K^+)-ATPase inhibition in microsomes from rat heart. Biochem Pharmacol 33: 47–53
42. Powis DA (1981) Does Na,K-ATPase play a role in the regulation of neurotransmitter release by prejunctional alpha-adrenoceptors? Biochem Pharmacol 30 (17): 2389–2397
43. Rosic NK, Standaert ML, Pollet RJ (1985) The mechanism of insulin stimulation of (Na^+,K^+)-ATPase transport activity in muscle. J Biol Chem 260 (10): 6206–6212
44. Russell VA, Lamm MCL, Taljaard JJF (1983) Inhibition of Na^+,K^+-ATPase activity by delta-aminolevulinic acid. Neur Res 8 (11): 1407–1415
45. Schaefer A, Seregi A, Kombos M (1974) Ascorbic acid-like effect of the soluble fraction of rat brain on adenosine triphosphatases and its relation to catecholamines and chelating agents. Biochem Pharmacol 23: 2257–2272
46. Schmitz W, Scholz M, Erdmann E, Krawietz W, Werdan K (1982) Effect of vanadium in the +5, +4 and +3 oxidation states on cardiac force of contraction, adenylate cyclase and (Na^+ + K^+)-ATPase activity. Biochem Pharmacol 31: 3853–3860
47. Schreiber V, Stepan J, Pribyl T, Starka L (1981) Digitalis glycoside-like biological activity: inhibition of $^{86}Rb^+$ uptake by red blood cells in vitro of certain steroids and other hormones. Biochem Pharmacol 30 (21): 3001–3002
48. Schwartzman M, Ferreri NR, Carroll MA, Songu-Mize E, McGiff JC (1985) Renal cytochrome P450-related arachidonate metabolite inhibits (Na^+ + K^+)-ATPase. Nature 314: 620–622
49. Seda HWH, Gove CD, Hughes RD, Williams R (1984) Inhibition of partially purified rat brain Na^+,K^+-dependent ATPase by bile acids, phenolic acids and endotoxin. Clin Sci 66: 415–420
50. Stewart DJ, Sen AK (1981) Role of cyclic GMP in cholinergic activation of Na-K pump in duck salt gland. Am J Physiol 240: C207–C214
51. Swann AC (1984) (Na^+,K^+)-Adenosine triphosphatase regulation by the sympathetic nervous system: effects of noradrenergic stimulation and lesions in vivo. J Pharmacol Exp Ther 228 (2): 304–311
52. Swann AC (1984) Free fatty acids and (Na^+,K^+)-ATPase: effects on cation regulation, enzyme conformation, and interactions with ethanol. Arch Biochem Biophys 233 (2): 354–361
53. Swann AC (1985) (Na^+,K^+)-ATPase regulation and NaCl intake: effects on circulating inhibitor and sensitivity to noradrenaline. Clin Sc 69: 441–447
54. Sweadner KJ, Gilkeson RC (1985) Two isozymes of the Na,K-ATPase have distinct antigenic determinants. J Biol Chem 26 (15): 9016–9022
55. Tamura M, Kuwano H, Kinoshita T, Inagami T (1985) Identification of linoleic and oleic acids as endogenous Na^+,K^+-ATPase inhibitors from acute volume-expanded hog plasma. J Biol Chem 260 (17): 9672–9677

56. Wehling MA, Schwartz A, Whitman K, Grupp G, Grupp IL, Wallick ET (1981) Interaction of chlormadinone acetate with the ouabain binding site of Na^+,K^+-ATPase. Mol Pharmacol 20: 551–557
57. Wiglusz Z, Nasal A, Damasiewicz B, Radwanska A (1984) Possible role of ascorbic acid in catecholamines-stimulated Na^+,K^+-adenosine triphosphatase activity of the central nervous tissues. Pol J Pharmacol Pharm 36: 313–321

Author's address:

Prof. Dr. T. Godfraind, Laboratoire de Pharmacodynamie Générale et de Pharmacologie, Université Catholique de Louvain, UCL 7350, Avenue Emmanuel Mounier 73, 1200 Brussels (Belgium)

Digitalis-like factors in plasma of healthy human adults

S. Vasdev, L. Longerich, E. Johnson, D. Brent* and M. H. Gault

Department of Lab Medicine, The General Hospital and Faculty of Medicine, Memorial University of Newfoundland, St. John's, Newfoundland (Canada) and *Burroughs Wellcome Co., Research Triangle Park, NC 27709

Summary

Digitalis-like factors (DLF) are defined as endogenous compounds which react positively in digoxin radioimmunoassays (RIA) in the absence of digoxin. Plasma from 11 healthy subjects was assayed for DLF and dehydroepiandrosterone sulfate (DHEA-S). DLF were extracted on C-18 Bond Elut columns with methanol and measured by the New England Nuclear (NEN) digoxin RIA and DHEA-S by RIA. Plasma values (means ± SD) were for DLF 44 ± 6 pmol/L digoxin equivalents, and for DHEA-S 11 ± 3 μmol/L.

We studied DLF (NEN) and DHEA-S for the presence of 3 digitalis-like properties. Both plasma DLF, and DHEA-S (25–1600 μmol/L) displaced ^{125}I-digoxin from antibody to digoxin, inhibited hog brain Na,K-ATPase, displaced (^{3}H)-ouabain from ATPase in a concentration dependent manner and gave curves parallel to those found with digoxin.

Increasing concentrations of DHEA-S were added to plasma and assayed for DLF by NEN and Kallestad digoxin RIAs and standard curves of DHEA-S vs. DLF concentration plotted. Using these curves to calculate the DLF equivalents for each plasma DHEA-S concentration, we found that 62–100% of the plasma DLF by NEN could be accounted for by the DHEA-S in the 11 subjects and similar proportions using Kallestad reagents.

A plasma extract from one subject was fractionated by HPLC and showed several DLF peaks when the 20 fractions were assayed for digoxin (NEN). The largest peak had the same retention time as the DHEA-S standard and when the fractions under this peak were assayed for DHEA-S, it was found that DHEA-S could account for 80% of the DLF. FAB mass spectrometry and accurate mass measurement confirmed the presence of DHEA-S in the fraction forming the major DLF peak.

DHEA-S has an apparent weak cross-reactivity (0.00024%) with NEN antidigoxin antibody and has other digitalis like properties only in high concentration. Yet its relatively high concentration in plasma appears to make it an important component of the total plasma digitalis-like factors in healthy subjects.

Endogenous digitalis-like factors (DLF), measured by radioimmunoassay (RIA) of digoxin, have been reported in plasma of neonates (21, 22), pregnant women (6), patients with hepatic (19) and/or renal failure (11, 14) and in healthy and hypertensive subjects (1, 12). These digitalis-like factors are reported to have one or more of the following characteristics: interaction with antidigoxin antibody, inhibition of Na,K-ATPase, displacement of labelled ouabain from Na,K-ATPase and natriuretic activity (1, 3, 8, 11, 12, 17, 23, 30). DLF have been suggested to be steroid, lipid, or peptide (3, 8, 11, 17, 23, 30), but the exact chemical nature of these substances in healthy adults has not been established. We report that dehydroepiandrosterone sulfate (DHEAS), an important circulating androgen, forms a major portion of the DLF measured by digoxin radioimmunoassay (RIA) in plasma of healthy adults, inhibits Na,K-ATPase and displaces [^{3}H]-ouabain from Na,K-ATPase.

Methods

The materials used have been previously described (27).

Blood (20 ml) from 8 healthy male and 3 female volunteers, with a mean age of 35 years (range 24–60), was collected for DLF in chilled heparinized tubes and the plasma was separated immediately at 4° C. Two ml of plasma was extracted in duplicate on C-18 Bond Elut columns with methanol. The methanol was evaporated under nitrogen, the dried material dissolved in digoxin free plasma and assayed for DLF by radioimmunoassay (RIA) for digoxin using New England Nuclear (NEN) reagents. Digoxin free plasma was spiked with increasing concentrations of DHEAS (2.5–80 μmol/L) and 2 ml plasma extracted and assayed for DLF. The DLF value in each plasma sample of 11 healthy subjects due to its DHEAS concentration assayed by RIA (Pantex, Santa Monica, CA) was calculated from a standard curve of DHEAS versus DLF concentrations.

DHEAS, DHEA, androsterone sulfate (AS), androsterone (A), epiandrosterone sulfate (ES), and cortisol, were also added to digoxin free plasma in 7 concentrations (6.25 to 400 μmol/L). DLF were assayed directly on unprocessed plasma using NEN digoxin reagents.

The above 6 steroids, except for androsterone, were also added to digoxin free serum in 9 concentrations from 0.312 to 80 μmol/L (Table 1). DLF were assayed after extraction as described above.

Table 1. DLF in plasma spiked with various steroids

Steroid concentration μmol/L	DLF value pmol/L due to: DHEAS	DHEA	Androsterone sulfate	Epiandrosterone sulfate	Cortisol
0.312	*	*	*	*	9
0.625	*	*	*	*	15
1.25	*	6	*	*	37
2.5	12	14	8	*	56
5.0	23	31	12	7	95
10	33	44	24	10	146
20	54	65	45	18	256
40	108	128	61	35	315
80	186	240	128	64	333

* Undetectable

For comparison with the steroid concentrations note that reference ranges in μmol/L are for DHEAS 7–16 (27), DHEA < 0.3, androsterone sulfate 0.3–3 (9) epiandrosterone sulfate 0.1–1.0 (9), and cortisol 0.1–0.6.

Two women under investigation for hirsutism were given 0.5 u ACTH (Cortrosyn) IV over 30 min, and one man and one woman suppected of having a pituitary neoplasm were given insulin 9 u, TRH 0.2 mg and LH-RH 50 μg. Blood was drawn over the next 30 to 120 min for assay of DLF, cortisol and DHEAS.

Extracts of two ml plasma from the 11 subjects were also assayed for DHEAS inhibitory activity on hog brain Na,K-ATPase (4) and 4 ml plasma extracts for displacement of [^{3}H]-ouabain from Na,K-ATPase (28). Digoxin was used as a standard for all three assays and results expressed as pmol or nmol digoxin equivalents/L plasma. The contribution of DHEAS to Na,K-ATPase inhibition and [^{3}H]-ouabain displacement each plasma sample was calculated from standard curves of DHEAS concentration versus Na,K-ATPase inhibition and [^{3}H]-ouabain displacement. Plasma from these subjects was also assayed directly for DHEAS by RIA.

All samples were assayed in duplicate and the mean value used.

For HPLC analysis, 40 ml plasma was extracted on Bond Elut columns, the solvents evaporated and the extract split into equal halves. One half was used as control and the other half was incubated at 37° C with a mixture of β-glucuronidase and sulfatase (Sigma) in acetate buffer, pH 5.0, for 18 hours. After enzyme hydrolysis, DLF was again extracted on Bond Elut columns. Both the control and enzyme hydrolyzed extracts were reconstituted in 100 μl of mobile phase (45 % methanol : 55 % 0.02 M ammonium formate, pH 6.5) and injected onto a C-8 reversed phase column (5 micron, 2.5 cm, Brownlee Labs, Santa Clara, CA) with flow rate 1 ml/min and monitored at 215 nm. After 50 minutes, any material still remaining on the column was flushed off with 15 ml of 100 % methanol. Forty, 2 ml fractions were collected, an aliquot taken from each for DHEAS RIA and the remainder assayed for DLF by RIA for digoxin. A DHEAS standard was also run on the same high performance liquid chromatography (HPLC) system and monitored by U.V. at 215 nm and by RIA for DHEAS. The HPLC eluate in the fractions collected under each DLF peak was pooled, dried, dissolved in a thioglycerol liquid matrix and analyzed by Fast Atom Bombardment mass spectrometry (10) using a VG 7070H double focussing mass spectrometer equipped with a FAB 11 N source from Ion Tech. Ltd.

Results

Relationship of concentrations of DHEAS and other hormones to DLF. DHEAS (0.3 to 80 μmol/L) when added to digoxin free plasma and extracts of 2 ml assayed for DLF, gave values of 0 to 186 pmol digoxin equivalents/L (Table 1). The digoxin equivalents found for the same concentrations of the DHEAS, androsterone sulfate, epiandrosterone sulfate and cortisol are also shown in Table 1. DHEAS is the major DLF, whereas androsterone sulfate and cortisol gave appreciably lower DLF equivalents with this method in concentrations found in the normal ranges.

Plasma DHEAS values in 11 healthy adults were (mean ± SD) 11 ± 3 (range 7.7–15.9 μmol/L) and DLF values were 44 ± 6 (range 39–53) pmol digoxin equivalents/L (Table 2). DHEAS accounted for 62–100 % of the total DLF value in plasma of these subjects. The percent contribution of DHEAS to plasma DLF using Kallestad antibody for digoxin RIA was similar. Plasma DLF (extracts of 1–32 ml plasma) and DHEAS (10–320 μmol/L) displaced [^{125}I]digoxin from digoxin antibody and gave curves parallel to those found with digoxin.

Table 2. DLF, DHEAS, Na,K-ATPase inhibition, and [^{3}H]-ouabain displacement from Na,K-ATPase in plasma of 11 healthy adults expressed as digoxin equivalents*.

					Calculated contribution of DHEAS to:		
	DLF pmol/L	ATPase inhib. nmol/L	Ouabain displac. nmol/L	DHEAS μmol/L	*DLF pmol/L	ATPase inhibition nmol/L	Ouabain displacement nmol/L
1	50	1.4	1.4	9.2	34	0.22	0.53
2	51	1.9	1.9	14.8	36	0.28	0.79
3	41	1.6	1.6	15.9	41	0.29	0.84
4	36	1.5	1.7	11.0	35	0.23	0.63
5	40	1.9	1.8	7.6	30	0.19	0.49
6	44	1.9	1.9	12.7	36	0.24	0.73
7	47	1.9	2.0	7.8	29	0.19	0.50
8	47	1.7	1.9	12.7	36	0.26	0.73
9	41	1.9	1.8	8.3	31	0.20	0.52
10	36	1.4	1.4	7.7	30	0.19	0.49
11	53	1.9	2.2	14.2	37	0.27	0.79
mean	44	1.7	1.8	11.0	34	0.23	0.64
±SD	±6	±.22	±.25	±3.0	±4	±.04	±.14

* Values are expressed as digoxin equivalents except for DHEAS which is μmol DHEAS/L plasma.

Measured directly on unprocessed serum, DHEAS in concentrations of 6.25, 12.5, 25, 50, 100, 200 and 400 μmol/L gave digoxin RIA values of 0.04, 0.13, 0.18, 0.23, 0.40, 0.73 and 1.31 nmol digoxin equivalents/L respectively. DHEA in similar concentrations gave digoxin values of 0.3, 0.49, 0.72, 1.02, 1.42 and 1.98 nmol digoxin equivalents/L respectively. DHEA gave about twice the apparent crossreactivity with antibody to digoxin in the RIA compared with DHEAS at the same concentrations. Serum spiked with 6.25 μmol/L of androsterone sulfate and of cortisol, gave digoxin equivalent values of 0.09 and 0.14 nmol/L respectively. Androsterone and epiandrosterone sulfate gave no measurable DLF value below 25 μmol/L.

Plasma DLF increased at least five fold within 30 min of receiving both ACTH and insulin with TRH and LH-RH (Table 3).

Table 3. Response of plasma DLF to ACTH and to insulin with TRH and LH-RH.

Time, minutes after administration	Digoxin equivalents pmol/L 0	10	15	20	30	60	90
Patient 1, ACTH	39	48	*	*	226	*	*
Patient 2, ACTH	60	*	*	135	298	*	*
Patient 1, Insulin TRH, LH-RH	33	*	129	*	221	*	25
Patient 2, Insulin TRH, LH-RH	17	*	49	*	117	148	240

* Not studied

Inhibition of Na, K-ATPase and [^{3}H]-ouabain displacement. Plasma extracts from the 11 subjects inhibited hog brain Na,K-ATPase and displaced [^{3}H]-ouabain from Na,K-ATPase with mean value ±SD of 1.8 ± 0.25 and 1.7 ± 0.22 nmol digoxin equivalents/L plasma respectively. DHEAS accounted for 10–19% of Na,K-ATPase inhibition and for 25–52% of the [^{3}H]-ouabain displacement (Table 2). Inhibition of hog brain Na,K-ATPase by DHEAS was 4, 7, 12 and 18% of control at 25, 50, 100 and 200 μmol/L concentrations respectively. With DHEA it was 8, 17, 22 and 38% respectively for the same concentrations.

Identification of DHEAS as a major plasma DLF. When a plasma extract was fractionated by HPLC and the eluted fractions assayed for DLF using digoxin RIA, there was one major DLF peak and several small peaks (Fig. 1B). The major peak had the same retention time (12 min) as the DHEAS standard and was the only peak positive for DHEAS by RIA (Fig. 1A). This peak accounted for 50% of the total plasma DLF and when analyzed by mass spectrometry showed a spectrum similar to the DHEAS standard. The principal feature was m/z 367, corresponding to the dehydroepiandrosterone sulfate anion (Fig. 2). The other DLF peaks failed to show this feature. Identification was supported by accurate mass measurement. The DLF peak which followed the methanol wash accounted for 20% of the total DLF.

After enzyme hydrolysis of an extract of this plasma using β-glucuronidase and sulfatase and HPLC fractionation, the DLF associated with the DHEAS peak was almost completely gone and there was a great increase in the DLF in the methanol wash (Fig. 1D), which accounted for 70% of the total DLF. Assay of the HPLC fractions by RIA for DHEAS also

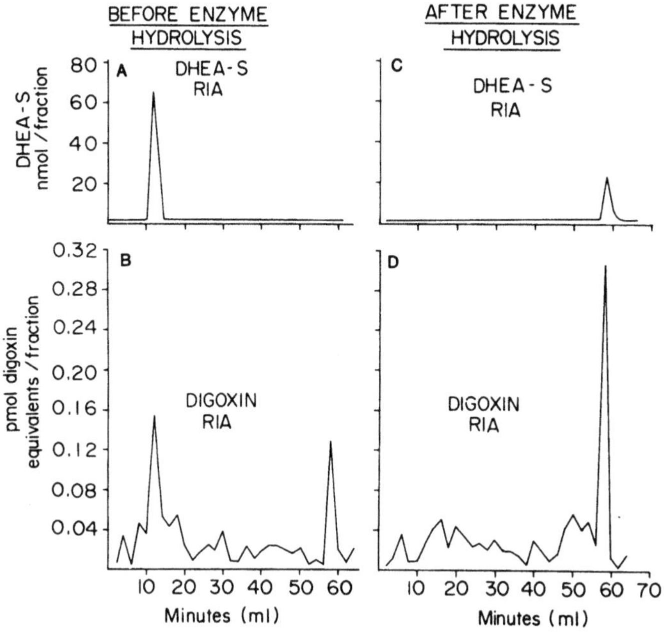

Fig. 1. HPLC separation of an extract of plasma containing DLF from a healthy adult before [A, B], and after hydrolysis with β-glucuronidase and sulfatase [C, D]. Two ml fractions were collected and aliquots assayed by RIA for DHEAS [A, C] and by RIA for digoxin to give DLF [B, D]. Results are plotted as nmol DHEAS and pmol digoxin equivalents per fraction. The peak at 12 min was identified as DHEAS by retention time based on a standard monitored at 215 nm, by RIA for DHEAS and by mass spectrometry. The increase in the peak at 58 min in D compared with B, is believed to be due to an increase in DHEA after hydrolysis, as a DHEA standard had a retention time of 58 min in this system.

showed disappearance of the DHEAS peak while material reactive to DHEAS antibody appeared in the methanol wash, which, based on its rentention time, was due to free DHEA (Fig. 1C). DHEA crossreactivity with the Pantex DHEAS antibody was 52% of DHEAS.

Discussion

This study has identified dehydroepiandrosterone sulfate as a major digitalis-like substance in the plasma of healthy adults. It is a major secretory product of human adrenals (2) and its concentration in adult plasma is greater than any other steroid (5, 18). No obvious biological function has been found for DHEAS in healthy subjects. DHEA, although more reactive with digoxin antibody, is present at only 1/1000 the concentration of DHEAS (20) and will not contribute appreciably to plasma DLF.

For the digoxin assay performed directly on serum in the usual way, none of the hormones when present in physiological concentrations have any clinical significance. However, infusion of massive doses of certain steroids such as 6-methyl prednisolone (13), and cortisol can give clinically important values. Plasma spiked with 25 μmol/L cortisol gave a concentration of 0.4 nmol/L digoxin equivalents.

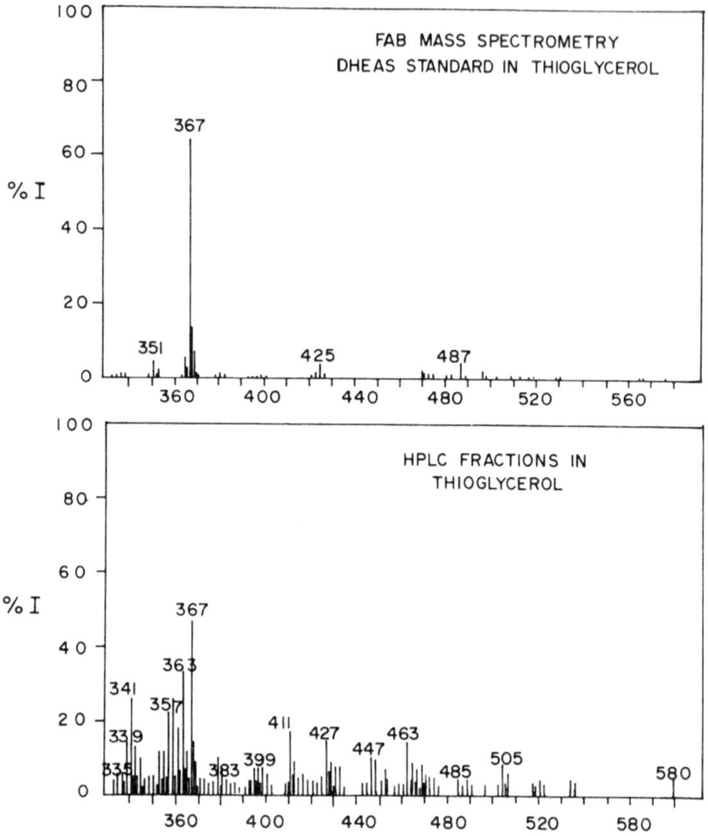

Fig. 2. Fast atom bombardment mass spectrometry of a DHEAS standard in the upper panel and of the HPLC eluate in fractions under the peak at 12 min (Fig. 1) which gave maximum reactivity with antibody to digoxin (DLF) in the lower panel.

Considering our assay for DLF, which concentrates plasma by a factor of 20, of the hormones studied only DHEAS, androsterone sulfate and cortisol in physiological concentrations, will have an effect on values with the NEN digoxin RIA. Roughly, if the DLF value were 50 pmol/L, DHEAS could account for 30, androsterone sulfate for 7 and cortisol for 12 pmol/L. Thus, in healthy subjects, DLF may be accounted for largely by known steroids crossreacting with the antibody to digoxin in our assay. However, this may not be the case when values for DLF are elevated as in renal failure where we have found DHEAS concentrations to be reduced in spite of more than an average two fold increase in values for DLF compared with those in healthy subjects (26). Also, in contrast to the 5 fold rise in DLF after ACTH, DHEAS did not rise in one patient and only doubled in the other. Neither did the increase in cortisol account for the rise in DLF.

DHEAS inhibited hog brain Na,K-ATPase, displaced [^{3}H]-ouabain from Na,K-ATPase on our studies and gave concentration dependent curves parallel to digoxin (27). However, to be of any significance in the regulation of membrane Na,K-ATPase, the effective tissue concentration of DHEAS would have to be about 5 times higher than plasma concentration. The DHEAS tissue : plasma ratio has been found up to 20 in rat brain (7), and for digoxin the ratio is 5 or more in many tissues including up to 70 for heart (15). In essential hypertensives, DHEA and DHEAS secretion rates have been reported to be increased by five and six fold respectively when compared to normal subjects (23). Increased plasma levels of DHEAS have also been reported in hypertensives (24). We found that DLF increased significantly with salt loading in essential hypertensives (12). DHEA has been shown to be natriuretic in rats (29).

DHEA and DHEAS levels in human plasma increase with administration of ACTH (16, 20) and we have shown DLF increased at least five fold after ACTH and also after insulin with TRH and LH-RH. The response to ACTH supports the hypothesis that DLF is primarily of adrenal origin.

In conclusion, dehydroepiandrosterone sulfate displaced [^{125}I]digoxin from digoxin antibody, inhibited Na,K-ATPase, displaced [^{3}H]-ouabain from Na,K-ATPase and is a major digitalis-like factor in plasma of healthy adult subjects.

Acknowledgements

Grant support: Medical Research Council of Canada (84–693–7960, 84–638–6760), Kallestad Labs, Austin, TX. and Burroughs Co., NC.

References

1. Balzan S, Clerico A, Del Chicca MG, Montali U, Ghione S (1984) Digoxin-like immunoreactivity in normal human plasma and urine, as detected by a solid-phase radioimmunoassay. Clin Chem 30: 450–451
2. Baulieu E, Corpechot C, Dray F, Emiliozzi R, Lefeau M, Mauvais-Jarvis P, Robel P (1964) An adrenal secreted androgen: dehydroisoandrosterone sulfate. Its metabolism and a tentative generalization on the metabolism of other steroid conjugates in man. Recent Prog Horm Res 21: 411–495
3. Buckalew VM, Gruber KA (1984) Natriuretic hormone. Ann Rev Physiol 46: 343–358
4. Burnett GH, Conklin RL (1971) The enzymatic assay of plasma digoxin. J Lab Clin Med 78: 779–784
5. Buster JE, Abraham GE (1972) Radioimmunoassay of plasma dehydroepiandrosterone sulfate. Analyt Letters 5: 543
6. Byers AD, Spruyt LL, Seifart HI, Kriegler A, Parkin DP, Van Jaarsveld PP (1983) Endogenous digoxin-like material. S Afr Med J 64: 42
7. Corpechot C, Robel P, Axelson M, Sjovall J, Baulieu E (1981) Characterization and measurement of dehydroepiandrosterone sulfate in rat brain. Proc Natl Acad Sci USA 78: 4704–4707
8. Diamandis EP, Papanastasiou-Diamandi A, Solidin SJ (1985) Digoxin immunoreactivity in cord and maternal serum and placental extracts. Partial characterization of immunoreactive substance by high-performance liquid chromatography and inhibition of Na^+,K^+-ATPase. Clin Biochem 18: 48–55
9. Feher T, Bodrogi L, Feher KG, Poteozin E, Kolosey IS (1977) Free and solvolysable dehydroepiandrosterone and adrosterone in blood of mammals under physiological conditions and following administration of dehydroepiandrosterone. ACTA Endocrinologica 85: 126–133
10. Gaskell SJ, Brownsey BG, Brooks PW, Green BN (1983) Fast atom bombardment mass spectrometry of steroid sulphates: Qualitative and quantitative analyses. Biomedical Mass Spectrometry 10: 215–219
11. Gault HM, Vasdev SC, Longerich LL (1984) Endogenous digoxin-like substance(s) in combined hepatic and renal failure. Ann Intern Med 101: 567–568
12. Gault MH, Vasdev SC, Longerich L, Fernandez P, Prabhakaran V, Dawe M, Maillet C (1984) Plasma digitalis-like factor(s) increase with salt loading. N Engl J Med 309: 1459
13. Gault MH, Longerich L, Dawe M, Vasdev SC (1985) Combined liquid chromatography/radioimmunoassay with improved specificity for serum digoxin. Clin Chem 31: 1272–1277

14. Graves SW, Brown B, Valdes R Jr (1983) An endogenous digoxin-like substance in patients with renal impairment. Ann Intern Med 99: 604–608
15. Hartel G, Kyllonen K, Merikallio E, Ojala K, Manninen V, Reissell P (1976) Human serum and myocardium digoxin. Clin Pharmacol 19: 153–1157
16. James VHT (1984) Physiology and pathology of the pituitary-adrenal axis. In: Makin HLJ (ed) Biochemistry of Steroid Hormones. Blackwell Scientific Publications, Oxford, pp 563–564
17. Kim RS, La Bella FS (1981) Endogenous ligands and modulators of the digitalis receptor: Some candidates. Pharmac Ther 14: 391–409
18. Metcalf MG, Espiner EA, Donald RA (1979) Lack of effect of prolactin suppression on plasma dehydroepiandrosterone sulphate. Clin Endo 10: 539–544
19. Nanji AA, Greenway D (1985) Falsely raised plasma digoxin concentrations in liver disease. Brit Med J 290: 432–433
20. Nieschlag E, Loriaux DL, Ruder HJ, Zucker IR, Kirschmer MH, Lipsett MB (1973) The secretion of dehydroepiandrosterone and dehydroepiandrosterone sulphate in man. J Endocr 57: 123–134
21. Pudek MR, Seccombe DW, Jacobson BE, Whitfield MF (1983) Seven different digoxin immunoassay kits compared with respect to interference by a digoxin-like immunoreactive substance in serum from premature and full-term infants. Clin Chem 29: 1972–1974
22. Pudek MR, Seccombe DW, Whitfield MF, Ling E (1983) Digoxin-like immunoreactivity in premature and full-term infants not receiving digoxin therapy. N Engl J Med 308: 904–905
23. Schreiber V, Stephan J, Atarka I (1983) Digoxin-like immunoreactivity of certain steroid and other hormones. Physl Bohem 30: 569–571
24. Sekihara H, Ohsawa N, Kosaka K (1975) Serum dehydroepiandrosterone sulfate and dehydroepiandrosterone levels in essential hypertension. J Clin Endocrinol Metab 40: 156–157
25. Shao A, Nowaczynski W, Kuchel O, Genest J (1970) Secretion rate of dehydroepiandrosterone and dehydroepiandrosterone sulfate in benign essential hypertension as compared to normal subjects. Can J Biochem 48: 1308–1313
26. Vasdev SV, Longerich LL, Johnson E, Madhavankutty K, Gault MH (1985) Endogenous digitalis-like factors (DLF) in plasma of normal and dialysis dependent subjects. Clin Chem 31: 965
27. Vasdev S, Longerich L, Johnson E, Brent D, Gault MH (1985) Dehydroepiandrosterone sulfate as a digitalis-like factor in plasma of healthy human adults. Res Comm in Chemical Pathology and Pharmacology 49:387–399
28. Wallick ET, Pitts BJR, Lane LK, Schwartz A (1980) A kinetic comparison of cardiac glycoside interactions with Na^+,K^+-ATPase from skeletal and cardiac muscle and from kidney. Arch Biochem Biophys 202: 442–449
29. Williamson HE (1965) Natriuretic action of certain adrenocortical androgens. Steroids 6: 365–369
30. Wood JM, Bush B, Pitts BJR, Schwartz A (1977) Inhibition of bovine heart Na^+,K^+-ATPase by palmitylcarnitine and palmityl-CoA. B. B. R. C. 74: 677–684

Author's address:

Professor M. M. Gault, Memorial University of Newfoundland, St. John's, Newfoundland, Canada A1B 3V6

A Na^{+},K^{+}-ATPase inhibitor isolated from pig kidney

M. Krekler, A. A. L. Fox, H. Strobach and K. Greeff

Institut für Pharmakologie, Universität Düsseldorf, Düsseldorf (FRG)

Summary

Pig kidney alcoholic extracts contain various substances which cross-react with antibodies for cardiac glycosides. Further fractions were found which inhibited Na^{+},K^{+}-ATPase activity, displaced ^{3}H-ouabain binding and had a positive inotropic effect on guinea-pig left atria.

The Na^{+},K^{+}-ATPase inhibitory factors did not correlate with cardiac glycoside antibody cross-reactivity.

One fraction A_4, which was radioimmunologically inert and extremely lipophilic, was not significantly contaminated with ions, showed a potent inhibition of the Na^{+},K^{+}-ATPase, displaced ^{3}H-ouabain binding and increased cardiac contractile force.

A further fraction, B_1, was radioimmunologically active, but was found to inhibit the Na^{+},K^{+}-ATPase, displace ^{3}H-ouabain binding and increase contractile force due to a contamination with a significant amount of calcium.

Introduction

Since the successful elucidation of naturally occurring ligands for the opiate receptor, the search has intensified for other endogenous substances whose actions mimic potent pharmacological agents, in this case cardiac glycosides (for review see (4)). Several investigators have detected substances which cause false positive results in the cardiac glycoside radioimmunoassay, but the biological activity of these factors has not been satisfactorily clarified. Our aim was to analyse organ extracts for cardiac glycoside immunoreactive substances and to determine whether these substances behaved in other test systems like cardiac glycosides.

Methods

A radioimmunoassay was performed using a polyvalent antibody developed in our laboratories, modified from Butler and Chen (2). Bound from free radioactivity was separated using dextran coated charcoal. The antibody had a high affinity for ouabain and ^{3}H-ouabain was therefore used as radioactive ligand in our assay. The antibody was more sensitive to radioimmunologically active substances in kidney extracts than digoxin specific antibody.

Na^{+},K^{+}-ATPase was prepared from pig kidney and guinea-pig heart by extraction with 0.1 % deoxycholate and 2 mmol/l sodium iodide as previously described (3). Enzyme activity was measured using a coupled optical assay, measuring the disappearance of NADH. ^{3}H-ouabain was bound to pig kidney Na^{+},K^{+}-ATPase in the presence of 3 mmol/l $MgCl_2$, 3 mmol/l Tris-PO_4 and 10 nmol/l ^{3}H-ouabain. Bound radioactivity was separated on millipore nitrocellulose filters.

Guinea-pigs were treated with reserpine, 10 mg/kg 24 h prior to the experiment. Left atria were suspended in Krebs-Henseleit solution, 1.2 mmol/l $CaCl_2$, at 31°C, with 1 g preload and stimulated at 1 Hz, 1 ms duration, 2.5 times threshold value.

Pig kidney (approx. 8 kg) was extracted with alcohol, centrifuged at 6,000 x g and the supernatant reduced at 35°C under reduced pressure. A preliminary separation was performed on silica gel (60) from Merck, Darmstadt, using a mixture of ethyl acetate/methanol/water (70:25:5). Further separation was achieved using reverse phase C18 chromatography (bonded phase-octadecyl, Baker, U.S.A.) eluted with acetonitrile/water with increasing concentrations of acetonitrile.

Results

Preliminary separation of alcohol extracts on silica gel yielded two major radioimmunologically active peaks, which both inhibited isolated Na^+,K^+-ATPase; these fractions will be called A and B.

The further separation of fraction A, using reverse phase C18 chromatography, is depicted in Fig. 1. Three radioimmunologically active fractions, A_{1-3}, were found, which had

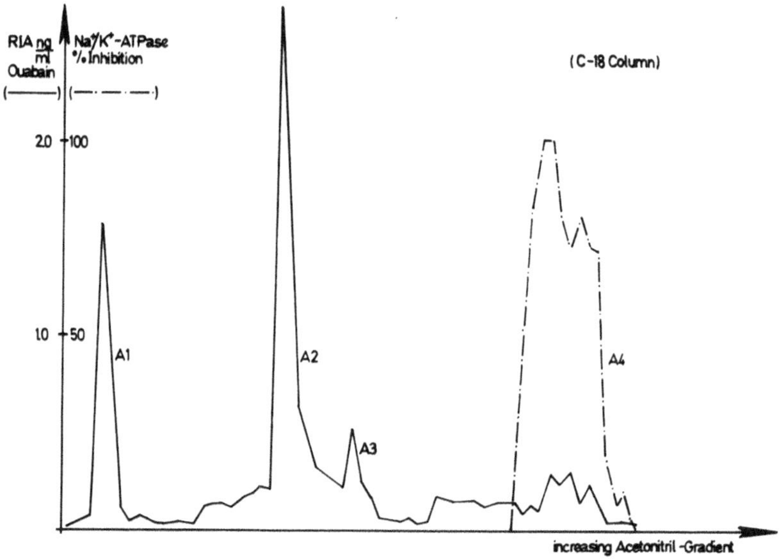

Fig. 1. Separation of fraction A on C18 column showing three immunologically active peaks (A_{1-3}) and one Na^+,K^+-ATPase inhibitory fraction (A_4).

Table 1. Inhibition of Na^+,K^+-ATPase: A_4, in alcohol, the effects of alcohol alone were tested in control experiments.

Extract	2.5 μl	5 μl	10 μl
Inhibition of pig kidney Na^+,K^+-ATPase	33.0 ± 7.1	81.7 ± 1.8	99.0 ± 1%
Guinea-pig heart Na^+,K^+-ATPase	30	49	66%
Inhibition of ^3H-ouabain binding to pig kidney Na^+,K^+-ATPase			
	14.0 ± 6.0	42.8 ± 6.3	65.7 ± 12.5%

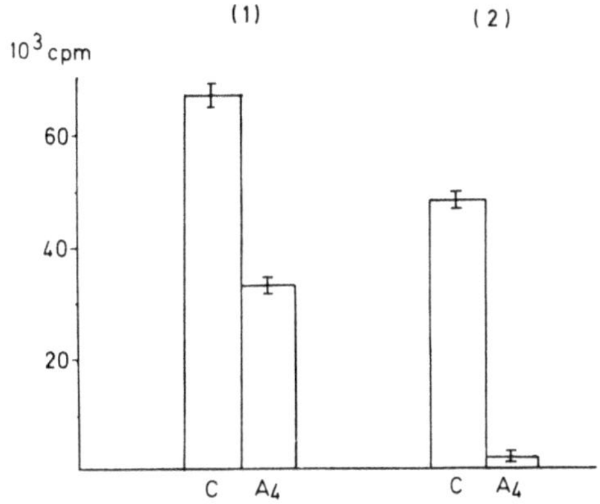

Fig. 2. Effect of preincubation with ^{3}H-ouabain and A_4 respectively on ouabain binding to pig kidney Na^+,K^+-ATPase.
(1) 10 nmol/l ^{3}H-ouabain was bound for 90 min, then treated with 5 μl A_4 for a further 90 min. In the control experiment (C) ^{3}H-ouabain was bound for 180 min, 5 μl alcohol was added after 90 min.
(2) 5 μl A_4 was bound for 90 min, then 10 nmol/l ^{3}H-ouabain was added and incubated for 90 min. In the control experiment (C) ^{3}H-ouabain was added for 90 min, subsequent to a 90 min treatment with alcohol.

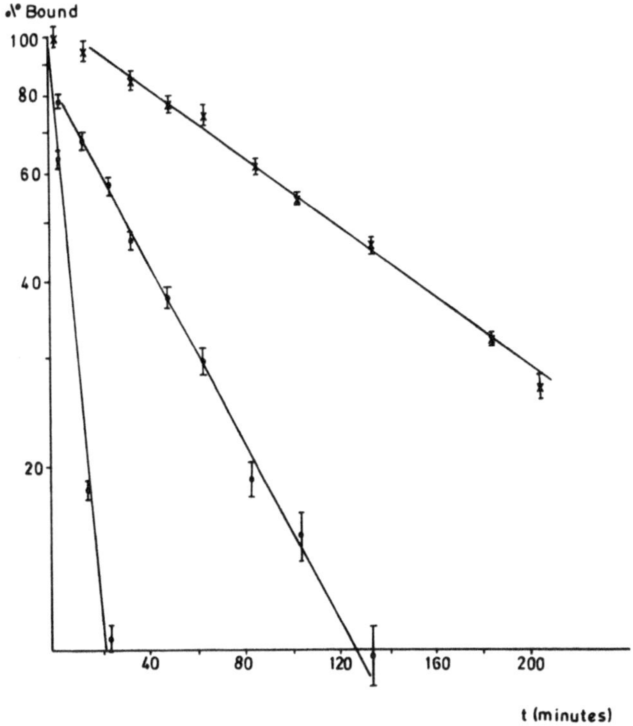

Fig. 3. Time-dependent dissociation of ^{3}H-ouabain receptor complex (n = 3).
x – 50 nmol/l ouabain
● – 5 μl A_4
○ – 10 μl A_4

no effect on the Na^+,K^+-ATPase. As can be seen in Fig. 1, all fractions were not only tested for their reactivity in the RIA, but also for their enzyme inhibitory potency. The last fraction A_4 was found to potently inhibit Na^+,K^+-ATPase, as can be seen in Table 1, but showed no cross-reactivity with the cardiac-glycoside antibody. Fraction A_4 also inhibited ^3H-ouabain binding to pig kidney Na^+,K^+-ATPase, 90 min incubation with 5 μl inhibits the subsequent binding of ^3H-ouabain by 96% (Fig. 2) and 5 μl A_4 causes a dissociation of the ouabain-receptor complex of approximately 50% (Table 1 and Fig. 2). The dissociation of the ouabain-receptor complex by A_4 is both time and concentration dependent, as can be seen from Fig. 3.

Fraction A_4, which was dissolved in alcohol, had a transitory positive inotropic action on guinea-pig left atria, which was also present when guinea-pigs were treated with reserpine (Fig. 4) to deplete catecholamine stores and in the presence of 0.3 μmol/l pindolol.

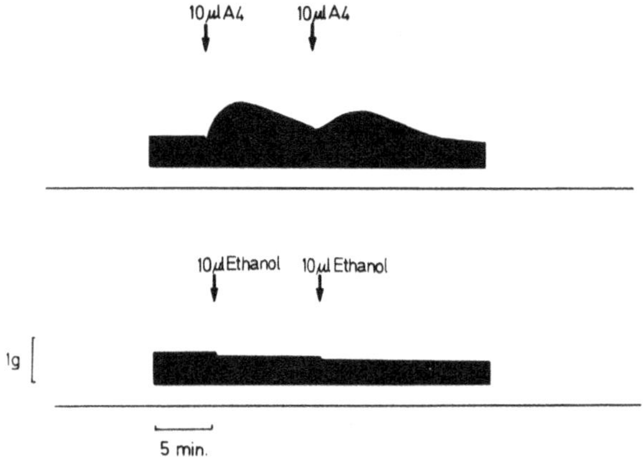

Fig. 4. The effect of A_4 on the contractile force of guinea-pig left atria (reserpine pretreated). Bath volume 3 ml.

The separation of fraction B using reverse phase C18 chromatography yielded two fractions which reacted positively in the RIA, B_1 and B_2. B_1 alone was found to inhibit the Na^+,K^+-ATPase, displace bound ^3H-ouabain and have a positive inotropic action on isolated guinea-pig atria. All active fractions were tested for the presence of sodium and potassium using flame photometry and calcium using atomic absorption spectroscopy. Fraction B_1 was found to contain 102 mmol/l sodium, 5 mmol/l potassium and 35 mmol/l calcium. In separate experiments we found exactly this amount of calcium could account for the inhibition of the Na^+,K^+-ATPase, for the decrease in the affinity of the Na^+,K^+-ATPase for ouabain and also for positive inotropic action of the extract, which reduces the chances of a further Na^+,K^+-ATPase active factor being present in this radioimmunologically active fraction.

Fraction A_4 on the other hand was not found to be significantly contaminated with ions. Furthermore this fraction was highly lipohilic, was dissolved in alcohol and had a limited solubility in water.

Discussion

Pig kidney alcoholic extracts contain several substances which cross-react with a RIA for ouabain. None of these fractions however possessed significant biological activity in our test systems, apart from one fraction, whose inhibitory action on Na^+,K^+-ATPase activity, cardiac glycoside binding, and positive inotropic effect was due to the presence of calcium. This indicates that factors detected by our RIA were not endogenously active cardiac glycoside-like substances as they did not fulfil any further criteria for a cardiac glycoside action. On the other hand the routine analysis of all fractions, radioimmunoactive or not, for Na^+,K^+-ATPase inhibitory action led to the discovery of the A_4 fraction, which fulfilled criteria for a cardiac glycoside-like action, without being active in the radioimmunoassay. In our opinion it may sometimes be erroneous to talk about endogenous cardiac glycoside-like factors on the strength of false positive reactions in RIA alone. On the other hand there are many endogenous ions and molecules which inhibit isolated Na^+,K^+-ATPase activity. In our experience calcium may often mimic the effects of a so-called endogenous factor in several test systems. Further factors have been summarized by Kim and LaBella (4). We are of the opinion that it is not necessary for an endogenous substance, which has several actions similar to those of cardiac glycosides, to be structurally related to cardiac glycosides to such an extent that it binds to cardiac glycoside specific antibodies. For structurally unrelated substances which possess a cardiac glycoside effect without necessarily binding to the same receptor, see the review by Akera, Fox and Greeff (1).

References

1. Akera T, Fox AAL, Greeff K (1981) In: Greeff K (ed) Cardiac Glycosides Part 1. Experimental Pharmacology Handbook. Exp Pharm 56: 459, Springer-Verlag, Berlin
2. Butler VP, Chen JP (1969) Proc Natl Acad Sci 57: 71
3. Fox AAL, Greeff K (1981) Biochem Pharmacol 30: 611
4. Kim RS, LaBella FS (1981) Pharmac Ther 14: 391

Author's address:

M. Krekler, Institut für Pharmakologie der Universität, Moorenstraße 5, D-4000 Düsseldorf 1 (West Germany)

Cardiac glycoside-like effects of a bufadienolide extracted from Kalanchoe Daigremontiana

G. Scholtysik[1], H. Wagner[2], M. Fischer[2] and U. Rüegg[1]

1) Preclinical Research, Sandoz Ltd., Basle (Switzerland) and
2) Institut für Pharmazeutische Biologie der Universität München (West Germany)

Summary

A bufadienolide (compound A_2) was isolated from Kalanchoe Daigremontiana Hamet et Perr. by methanol extraction and purification by column chromatography. Its pharmacological effects were investigated.

The force of contraction of rabbit papillary muscles and left guinea-pig atria was increased in a concentration dependent manner between 3×10^{-7} and 10^{-5} M. The effects were reversible upon washing. Rested state contractions were not increased and carbachol was ineffective in the presence of compound A_2 in rabbit papillary muscle. Thus, cyclic AMP seems not to be involved in the mechanism of the positive inotropic effect. Concentrations of up to 10^{-5} M did not alter the spontaneous beating rate, functional refractory period or the contraction and relaxation time of right guinea-pig atria. Arrhythmias followed by tonic contracture occurred with concentrations above 10^{-5} M. Electrophysiological measurements in guinea-pig papillary muscle revealed a decrease in $\dot{V}_{max}$ (18%) and overshoot (30%), associated with a shortening of action potential duration (14%) and increase in force of contraction (117%). Dog kidney Na^+/K^+-ATPase activity was inhibited with an IC_{50} of 1.4×10^{-7} M. Intravenous infusion of compound A_2 at a rate of 20 μg/kg/min in anaesthetized guinea-pigs caused ventricular arrhythmias and death after accumulated doses of about 760 μg/kg and 860 μg/kg respectively.

The results show that compound A_2 has a pharmacological profile similar to that of digitalis glycosides.

Introduction

Wagner and Fischer (4) were recently the first to desribe a new chemically defined bufadienolide designated compound A_2, obtained from the whole-plant extract of Kalanchoe Daigremontiana. The compound A_2 (Bersaldegenin-1, 3, 5-orthoacetate) is characterized by a OH-substitution in position 14 only, an aldehyde group at C-10 and an orthoester group at 1, 3, 5. The structure fulfils the conditions declared by Haustein (5) of glycosides to have cardioactivity. For this reason compound A_2 has been pharmacologically investigated in comparison to ouatbain for cardiac glycoside-like effects. Regardless of its central sedative component (4) compound A_2 possesses activities similar to those of ouabain.

Chemistry

The extraction and purification were described in full detail in the thesis of Fischer (4) and can be summarized as follows (scheme from (4)):

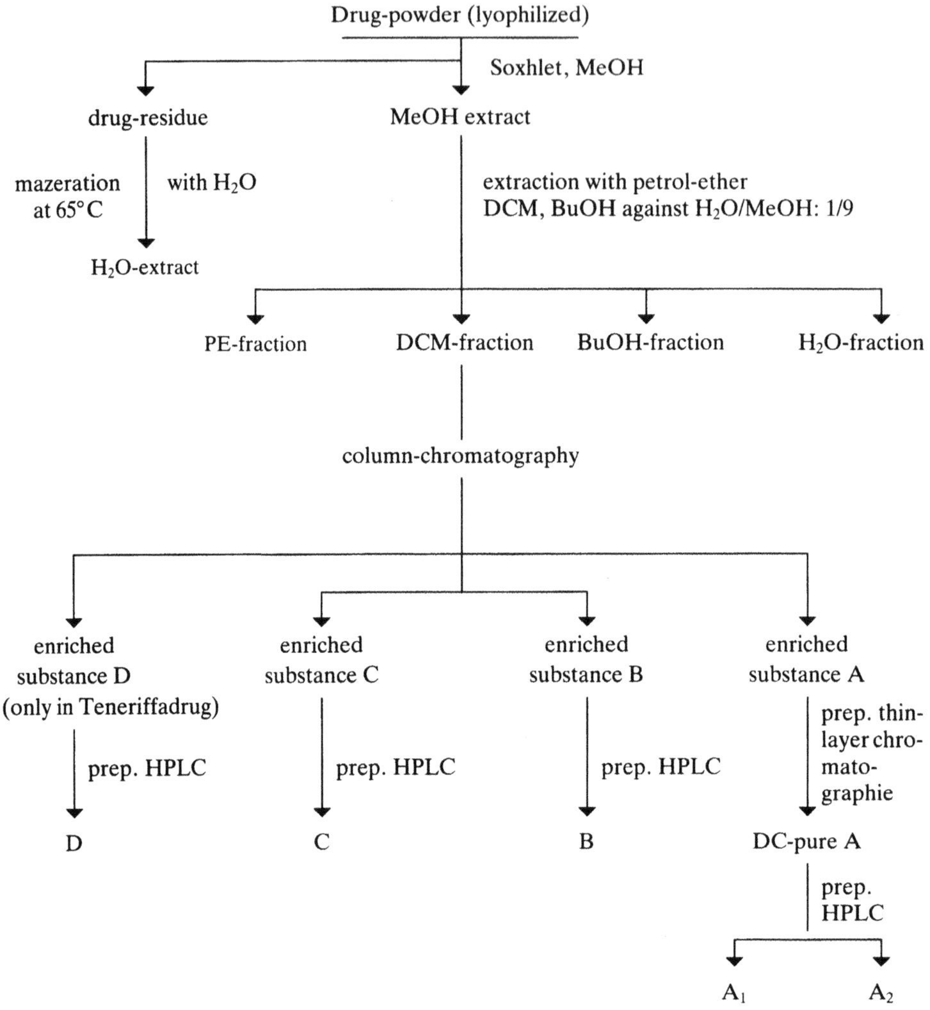

By means of IR- and NMR-spectra the chemical structure of compound A_2 presented in Fig. 1 (from (4)) could be identified. Compound A_2 has the composition $C_{26}H_{34}O_8$ and a molecular weight of 474.550. Stock solutions of 10^{-2} M were made using EtOH and N-methyl pyrrolidone and further diluted with saline or buffer before the experiments.

Pharmacological methods

Rabbit papillary muscles, force of contraction (FC) measurements

Papillary muscles from the right ventricles of adult male white New Zealand rabbits were mounted in Krebs-Henseleit solution of the following composition (mM): NaCl 118, KCl 4.7, $MgSO_4$ 1.2, $CaCl_2$ 2.5, KH_2PO_4 1.2, $NaHCO_3$ 25, glucose 10. The temperature was 31°C. Oxygenation was maintained by continuously bubbling with 95% O_2 and 5% CO_2. The muscles were connected to a Statham strain gauge UC 3 transducer for isometric force measurements. They were stimulated via bipolar platinum-iridium

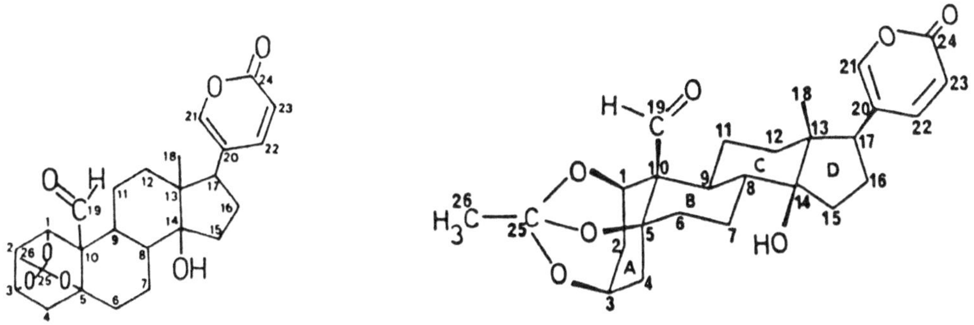

Fig. 1. Compound A_2 (Bersaldegenin-1, 3, 5-orthoester-acetate), with steric conformation (from (4)).

punctate electrodes placed at the basal part. Impulses of 3 ms duration, twice threshold and various frequencies were generated. The functional refractory period (FRP) was determined by delivering a test stimulus identical to the drive stimulus, with increasing delays. FRP was reached as soon as a positive inotropic effect occured in response to the paired stimulation.

Potentiating paired stimulation (PPS) as a standard positive inotropic intervention was elicited by applying successive paired stimuli with a delay of FRP plus 10 ms. The maximum positive inotropic response was reached after 5 beats. Electrical stimulation was then terminated. After a resting period of 20 minutes a rested-state contraction was elicited and this was followed by recording the frequency-force relationship. Compound A_2 was applied for 15 minutes during the resting period. Rested-state and steady-state contractions at different frequencies were measured.

In further experiments in the presence of TTX 3×10^{-8} M, increase in FC in response to PPS was used as standard reference and the positive inotropic effect of cumulative concentrations of compound A_2 and ouabain was expressed in % of the reference.

Refractory period, inotropic and chronotropic measurements in atria

In spontaneously beating right and stimulated left guinea-pig atria the functional refractory period (FRP), force of contraction (FC) and heart rate were measured as described elsewhere (8).

Electrophysiology

Papillary muscles from the right ventricle of guinea-pigs were used to record transmembrane electrical activity by a conventional electrophysiological method (for details see (8)). From the action potentials initiated by electrical stimulation with 0.5 Hz we measured the action potential amplitude (APA), resting membrane potential (RMP), overshoot (OS), maximal upstroke velocity ($\dot{V}_{max}$) and the duration of the AP at 30% and 70% repolarization (APD_{30} and APD_{70}).

Toxicity in guinea-pigs

Guinea-pigs of 400–500 g were anaesthetized with urethane (1.5 g/kg i. p.) and compound A_2 or ouabain infused intravenously at a rate of 20 μg/kg/min. Surface ECG and arterial blood pressure was recorded.

Assay for Na^+/K^+-ATPase

Dog kidney membranes enriched in Na^+/K^+-ATPase (2.5 m units per tube) (EC 3.6.1.3; Sigma) were incubated as described (2) with test compounds in a final volume of 0.12 ml in the following buffer (in mM): Tris 80, NaCl 100, KCl 5, $MgCl_2$ 3, EGTA 1, ATP 2.4 [γ-^{32}P]ATP (Amersham; 20 nCi), pH 7.4 (with HCl). After 45 min at 37° C, the suspension was cooled on ice, and 0.1 ml of 22% (w/v) trichloroacetic acid was added. After a further addition of 0.5 ml of a 20% (w/v) aqueous suspension of charcoal, the tubes were centrifuged at 20 000 x g for 2 min and the phosphate hydrolysed was measured in a 350 μl aliquot of the supernant by liquid scintillation counting. Ouabain or compound A_2 at 0.1 mM inhibited 85% of ATP hydrolysis.

Pharmacological results

Positive inotropic effect

In rabbit papillary muscles, compound A_2 did not increase rested-state contractions (Fig. 2). Steady-state contractions in response to 0.5 and 1 Hz were increased in the presence of compound A_2 10^{-6} M. The positive inotropic action of compound A_2 was concentration dependent up to 10^{-5} M. Arrhythmia occured with 3 x 10^{-5} M.

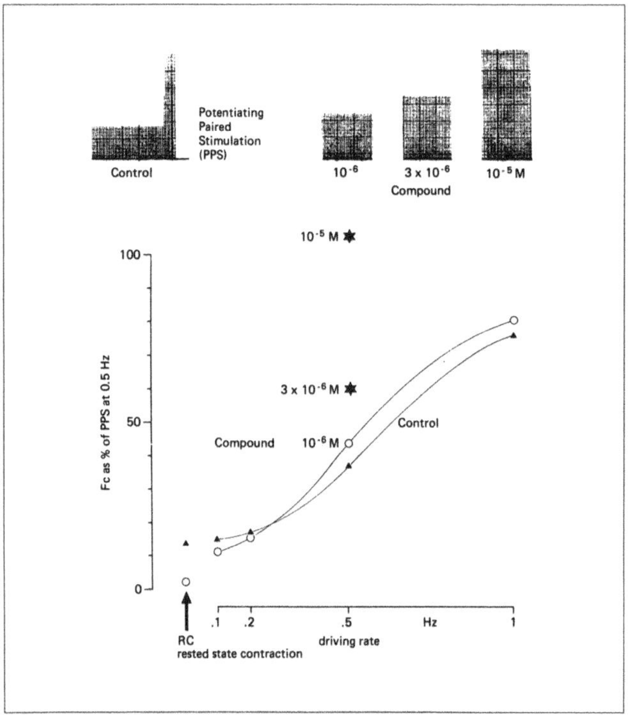

Fig. 2. Rabbit papillary muscle. Interval-force relationship as affected by compound A_2 (compound), n = 2.

When compound A_2 was applied to rabbit papillary muscles (driven with 0.5 Hz) in increasing concentrations, it increased force of contraction in a concentration dependent manner (Fig. 3). TTX 3 x 10^{-8} M was present to block adrenergic neurons (7). Ouabain was tested under similar conditions (Fig. 3). If the positive inotropic drug effect is expressed in percent of the maximum increase in FC induced by the potentiating paired stimulation (PPS), the efficacy of compound A_2 was 87 ± 4 % and of ouabain 65 ± 5 % of the PPS effect. The EC_{50} (50 % of the PPS effect) of compound A_2 is 4.3 x 10^{-6} M and of ouabain 8 x 10^{-7} M. Arrhythmia and contracture occured with 3 x 10^{-5} M of compound A_2 and with 3 x 10^{-7} M of ouabain.

In 5 muscles treated with compound A_2 10^{-5} M, FC amounted to 15.4 ± 3 mN. When carbachol 3 x 10^{-6} M was added to the medium, FC was 15.8 ± 3.6 mN. The difference was

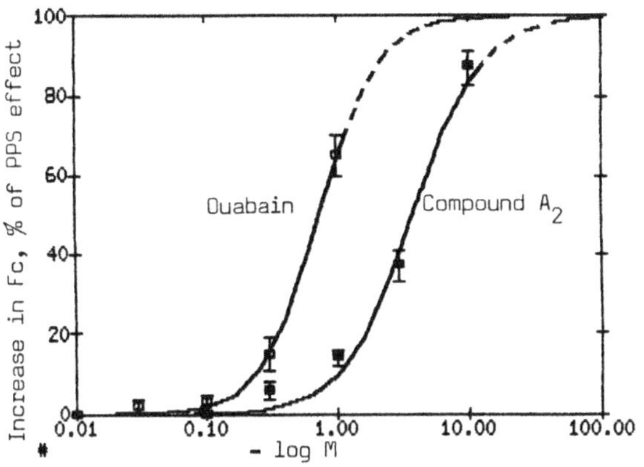

Fig. 3. Rabbit papillary muscles. Increases in force of contraction (Fc) expressed as % of the effect of potentiating paired stimulation (PPS). Cumulative concentration-response curves in the presence of TTX 3×10^{-8}M. Mean ± SEM of n = 5 (ouabain) and n = 13 (compound A_2).

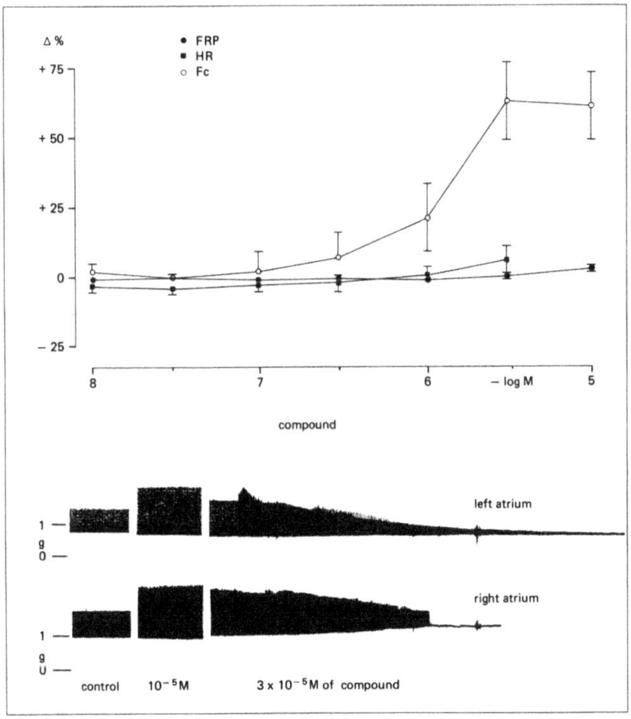

Fig. 4. Effects of compound A_2 (compound) on guinea-pig atria in vitro. Pretreatment values, mean ± SEM, n = 4; FRP = 93 ± 2 ms; FC = 1000 ± 215 mg; HR = 110 ± 6 bpm; Below: arrhythmia and contracture.

statistically not significant. Thus, carbachol did not influence the positive inotropic effect of compound A_2.

In guinea-pig atria compound A_2 increased FC in a concentration-dependent manner between 3 x 10^{-7} and 3 x 10^{-6} M (Fig. 4). Time to peak tension and relaxation time were not significantly changed in driven left atria up to 10^{-5} M (data not shown). Concentrations of compound A_2 higher than 10^{-5} M caused arrhythmia and tonic contractions. The functional refractory period (left atrium) was not influenced. Spontaneous beating rate (right atrium) decreased slightly after low concentrations (3 x 10^{-8} M) and increased slightly after higher concentrations (3 x 10^{-6} M) of compound A_2.

Electrophysiological effects

In right ventricular papillary muscles from guinea-pigs compound A_2 (10^{-6} M) caused in parallel to the increase in FC an abbreviation of the APD and decreases in APA, OS and $\dot{V}_{max}$ (Fig. 5). RMP remained unchanged. An explanation of the abbreviations is given in the method section. All effects are reversible after washout for 30 min with drug-free medium.

Toxicity in guinea-pigs

Anaesthetized guinea-pigs respond with typical ECG-signs of digitalis-intoxication when compound A_2 was infused i. v. until death (Fig. 6). In comparison to ouabain the following symptoms were recorded (mean ± SEM, doses in µg/kg i. v.):

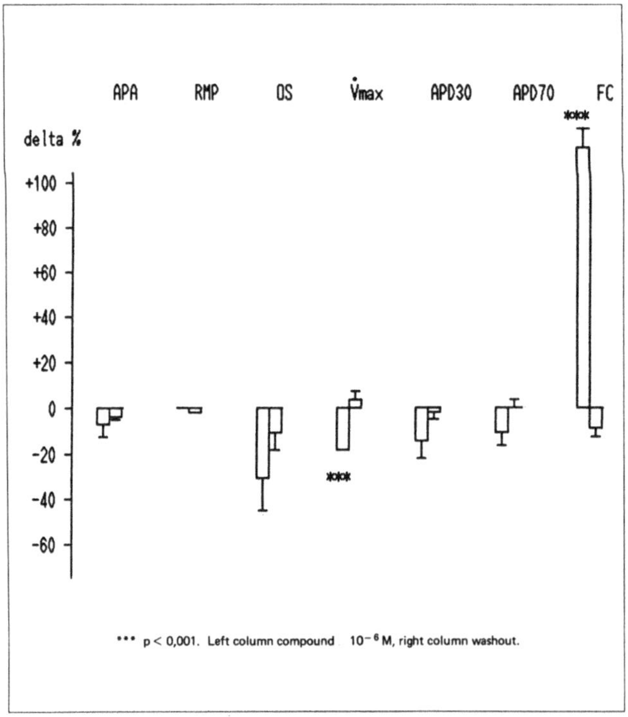

Fig. 5. Guinea-pig papillary muscles action potential characteristics as effected by compound A_2 (compound). Mean values ± SEM, n = 5.

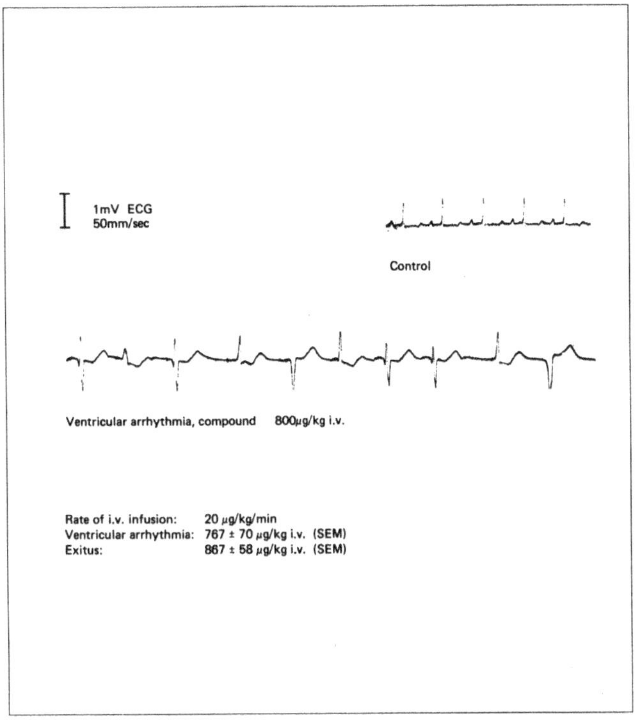

Fig. 6. Toxicity of compound A_2 (compound) in guinea-pigs (n = 3).

	Ventricular tachycardia	Ventricular fibrillation	Death	N
Ouabain	158 ± 34	177 ± 45	241 ± 41	8
Compound A_2	767 ± 70	not observed	867 ± 58	3

Compound A_2 is 5 (VT) until 3.5 (death) times less toxic than ouabain. Increase in blood pressure, increase in heart rate, PQ-prolongation, salivation and convulsions were seen with compound A_2 in sublethal doses.

Inhibition of Na/K-ATPase

The ATP hydrolysis by dog kidney membrane ATPase was inhibited in the presence of compound A_2 and ouabain in a concentration dependent manner (Fig. 7).

Discussion

Haustein (5) stated: "The following conditions must be fulfilled if the glycoside is to have cardioactivity: (i) A cis-cis-junction of the rings C and D of the genin moiety (ii) An unsaturated lactone moiety (Δ^{α}, $^{\beta}$-structure) (iii) The lactone ring must be attached to C17 in β-position (iv) A β-standing OH group at both C3 and C14."

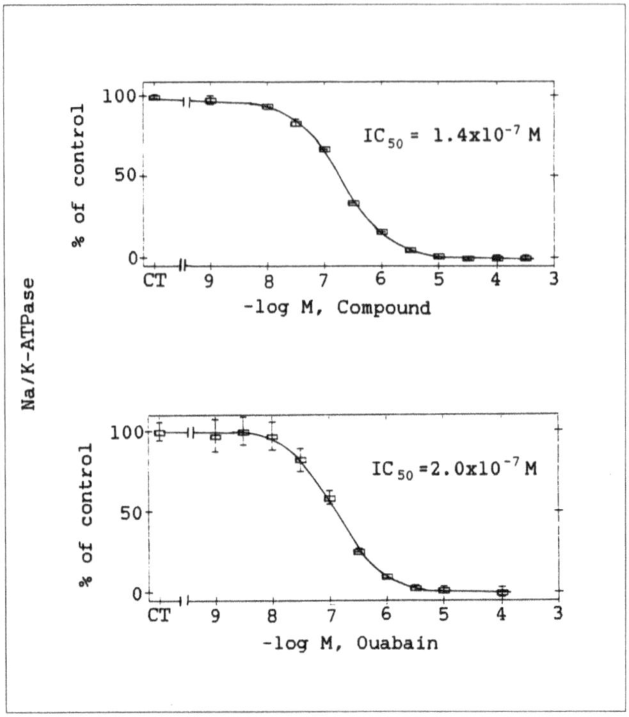

Fig. 7. Effect of compound A_2 (compound) and ouabain on ATPase (dog kidney). Stimulation 45 min at 37°C.

The bufadienolide compound A_2 from Kalanchoe Daigremontiana Hamet et Perr. which was recently described by Wagner and Fischer (4) fulfils the four conditions as can be seen from the structure given in Fig. 1. The pharmacological results confirm that compound A_2 has a profile of activity similar to that of the cardioactive glycoside ouabain. The force of contraction of rabbit papillary muscles and guinea-pig atria and papillary muscles was increased concentration dependently between 3 x 10^{-7} and 10^{-5} M. The effects were reversible upon washing. Positive inotropic effects are well known for digitalis glycosides (for review see (5)).

We have investigated whether cAMP is involved in the mechanism of the positive inotropic action of compound A_2. If so, rested state contractions would be increased (7) and the muscarinic agonist carbachol, which lowers elevated cAMP (6), would reverse the inotropic action. Our results show that rested state contractions were not increased and carbachol was ineffective in the presence of compound A_2 in rabbit papillary muscles. Thus, cAMP appears not to be involved in the mechanism of the positive inotropic effect. Further similarities with digitalis glycosides (5) were found with the results that concentrations of up to 10^{-5} M did not alter the contraction and relaxation times of right guinea-pig atria, arrhythmias followed by tonic contracture occurred with toxic concentrations. Intravenous infusion of compound A_2 in anaesthetized guinea-pigs caused ventricular arrhythmias and death. Na/K-ATPase activity was inhibited as known for glycosides (1). The action potential duration tended to decrease as with glycosides (3).

In conclusion, compound A_2 has cardiac pharmacological properties similar to those of digitalis glycosides.

References

1. Akera T, Brody TM (1977) The role of Na^+,K^+-ATPase in the inotropic action of digitalis. Pharmac Rev 29: 187–220
2. Crabos M, Wainer IW, Cloix JF (1984) Measurement of endogenous Na^+,K^+-ATPase inhibitors in human plasma and urine using high-performance liquid chromatography. FEBS Lettr 176: 223–228
3. Dudel J, Trautwein W (1958) Elektrophysiologische Messungen zur Strophanthinwirkung am Herzmuskel. Naunyn-Schmiedeberg's Arch Pharmacol 232: 393–407
4. Fischer M (1985) Über die toxischen und sedierend wirkenden Inhaltsstoffe (Bufadienolide) aus Kalanchoe daigremontiana Hamet et Perr. Thesis, Ludwig-Maximilians-University, München (1985); Wagner H, Fischer M: Helv Chim Acta (in press)
5. Haustein KO (1982) Digitalis. Pharmacol Ther 18 (No 1): 1–89
6. Holck M, Thorens R, Muggli R, Eigenmann R (1984) Studies on the Mechanism of Positive Inotropic Activity of Ro 13–6438, a Structurally Novel Cardiotonic Agent with Vasodilating Properties. J Cardiovasc Pharmacol 6: 520–530
7. Honerjäger P, Schäfer-Korting M, Reiter M (1981) Involvement of cyclic AMP in the inotropic action of amrinone; biochemical and functional evidence. Naunyn-Schmiedberg's Arch Pharmacol 318: 112–120
8. Scholtysik G, Salzmann R, Berthold R, Herzig JW, Quast U, Markstein R (1985) DPI 201–106, a novel cardioactive agent. Combination of cAMP–independent positive inotropic, negative chronotropic, action potential prolonging and coronary dilatory properties. Naunyn-Schmiedeberg's Arch Pharmacol 329: 316–325

Author's address:

Prof. Dr. G. Scholtysik, Sandoz AG, Präklinische Forschung, CH-4002 Basel (Switzerland)

Positive inotropic agents: Different mechanisms of action

H. Scholz

Abteilung Allgemeine Pharmakologie, Universitäts-Krankenhaus Eppendorf, Hamburg (West Germany)

Summary

An increase in the force of myocardial contraction is ultimately due to an increase in myoplasmic free Ca^{2+} to interact with the contractile proteins or to an increased sensitivity of the myofilaments for Ca^{2+}, or both. An increased myoplasmic Ca^{2+} concentration is mainly brought about through an increase in intracellular cAMP level, a cAMP-independent activation of Ca^{2+} channels or an increase in myoplasmic Na^+ concentration. An increased sensitivity of the myofibrils for Ca^{2+} is discussed for some new agents, e. g. for sulmazole and pimobendan. The importance of the latter mechanism is as yet not fully understood but it is conceivably of advantage because an increase in myoplasmic Ca^{2+} concentration may not only lead to positive inotropic but also to positive chronotropic or arrhythmogenic effects.

Agents which increase the force of contraction of the heart are referred to as positive inotropic or cardiotonic drugs. The increase in the force of myocardial contraction is ultimately due to one of two mechanisms: an increase in intracellular free Ca^{2+} $[Ca^{2+}]_i$ to interact with the contractile proteins or to an increased sensitivity of the myofilaments for Ca^{2+}, or both. This report discusses how these variables might be affected by various inotropic drugs.

1. Increased myoplasmic Ca^{2+} concentration

Mechanisms leading to an increase in myoplasmic Ca^{2+} concentration are summarized in Table 1.

Table 1. Mechanisms leading to an increased myoplasmic Ca^{2+} concentration

1. Increase in intracellular cAMP level
– Stimulation of adenylate cyclase
– Inhibition of phosphodiesterase
2. Cyclic AMP-independent activation of Ca^{2+} channels
– Alpha-adrenergic agents
– Ca-agonists
3. Increase in myoplasmic Na^+ concentration
– Inhibition of Na^+/K^+-ATPase
– Prolongation of the open state of Na^+ channels
4. Direct inhibition of Na^+/Ca^{2+} exchange
5. Inhibition of K^+ channels

1.1 Increase in intracellular cAMP level

An increase in intracellular cAMP level may be brought about by a stimulation of adenylate cyclase (increase in cAMP formation) or by an inhibition of phosphodiesterase (decrease in cAMP breakdown). Positive inotropic agents which affect these steps are summarized in Table 2.

Table 2. Positive inotropic agents which lead to an increase in intracellular cAMP level

1. Stimulation of adenylate cyclase
– Beta-adrenergic agonists: noradrenaline, adrenaline, isoprenaline, dopamine, dobutamine, prenalterol
– Glucagon
– Histamine-H_2-receptor agonists: histamine, impromidine
– Others: forskolin
2. Inhibition of phosphodiesterase
– Theophylline, caffeine, 3-isobutyl-1-methylxanthine (IBMX), amrinone, milrinone, sulmazole, pimobendan, UD-CG 212 Cl, enoximone (MDL 17043), piroximone (MDL 19205), OPC-8212

The main effect of cAMP on myocardial Ca^{2+} movements is to increase the slow Ca^{2+} inward current, I_{si}, during the action potential (Fig. 1, step 1). This effect leads to an increased Ca^{2+} release from the sarcoplasmic reticulum (Fig. 1, step 2) either because it serves as a

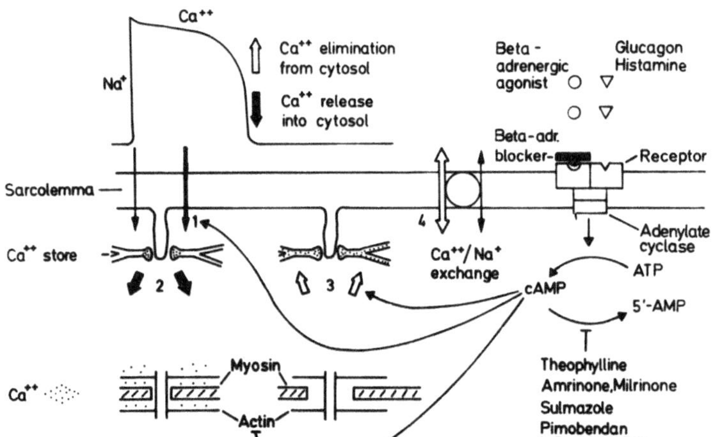

Fig. 1. Schematic summary of the possible mechanism of the cAMP-dependent positive inotropic effect of $beta_1$-adrenergic agents, glucagon, histamine, theophylline and several other agents. All drugs increase the cyclic adenosine monophosphate (cAMP) level, either through stimulation of adenylate cyclase (increase in cAMP formation) or through inhibition of phosphodiesterase (PDE; impairment of cAMP degradation). The increase in cAMP leads to an increase in slow Ca^{2+} inward current (1) and to an increased filling of intracellular Ca^{2+} stores (presumably the sarcoplasmic reticulum, SR) with releasable Ca^{2+}. This results in an increase in Ca^{2+} release (2) and, hence, in Ca^{2+} concentration at the contractile proteins and in force of contraction. In addition, cAMP increases the uptake of Ca^{2+} into the SR (3) and decreases the Ca^{2+} sensitivity of the contractile proteins. This explains that some of these drugs decrease the duration of the contraction (so-called relaxant effect). ⟶ and ⊣ denote stimulation and inhibition respectively.

greater trigger for Ca^{2+}-dependent Ca^{2+} release or because it increases the filling of these stores with Ca^{2+} that can be released during subsequent beats. In addition, cAMP leads to an increase in Ca^{2+} uptake into the sarcoplasmic reticulum (Fig. 1, step 3), which also produces an increase in the amount of releasable Ca^{2+} within, and an increased Ca^{2+} release from, these stores.

The increase in Ca^{2+} uptake by the sarcoplasmic reticulum not only contributes to the positive inotropic properties, but also explains the relaxant effects of some of the cAMP-increasing agents, e. g. of the beta-adrenergic agonists. However, there is some evidence that the latter might, in part, also be due to a decrease in the Ca^{2+} sensitivity of the contractile proteins.

It is widely accepted that cAMP leads to an activation of protein kinases and, as a result, to phosphorylation of several proteins. This changes the functional properties of the proteins, for example, their ability to bind Ca^{2+} ions, and may thus explain the effects on the calcium movements in the sarcoplasmic reticulum and possibly also in the sarcolemma.

Another characteristic effect found with cAMP-increasing agents is that their positive inotropic action in *ventricular* cardiac muscle is antagonized by carbachol or adenosine (cf. (6)). This means that carbachol or adenosine are negatively inotropic also in the ventricle in the presence of cAMP-increasing drugs. Such an effect is not found with drugs acting independently of the cAMP-system, e. g. Ca^{2+}, cardiac glycosides, alpha-adrenergic agents or "Ca-agonists" (see below). The reason for this difference is not known.

In principle, it does not matter whether the increase in cAMP is produced by a stimulation of adenylate cyclase or by an inhibition of phosphodiesterase. This explanation is not necessarily contradicted by the fact that the classic PDE inhibitors, the methylxanthines caffeine and theophylline, especially at high concentrations, have some effects which are opposite to those of the beta-adrenergic agonists (37). Most importantly, millimolar concentrations usually prolong rather than reduce the duration of the contraction. However, at concentrations of 1 mM and more, theophylline and other methylxanthines directly inhibit the uptake of Ca^{2+} by the sarcoplasmic reticulum (Fig. 1, step 3). This effect, which is independent of the cAMP system, counteracts the cAMP-dependent increase in Ca^{2+} uptake and leads to a deceleration of relaxation and, hence, to a prolongation of contraction.

It is important to note that PDE inhibitors act independently of beta-adrenoceptor blocking drugs and that they produce not only positive inotropic but also vasodilator effects.

Finally it is worth mentioning that cyclic AMP-increasing positive inotropic agents produce a variety of undesired effects. Most importantly, these include positive chronotropic and arrhythmogenic effects (38). Moreover, beta-adrenoceptor agonists decrease the number of beta-adrenoceptors and thus tolerance develops rapidly (8, 14). Whether or not other cAMP-increasing agents also produce tolerance is not known. Probably of even greater importance is that a deterioration of the underlying disease has been discussed in context with new phosphodiesterase inhibitors such as amrinone (34) or enoximone (39).

1.2 Cyclic AMP-independent activation of Ca^{2+} channels

An increase in slow Ca^{2+} inward current without elevation of the cAMP concentration can be brought about by alpha-adrenergic agents such as phenylephrine (9) or by the so-called "Ca-agonists" Bay K 8644 and CGP 28 392 (19, 26). The positive inotropic effects of these agents are not antagonized by carbachol or adenosine (7, 12).

The *phenylephrine*-induced increase in I_{si} is smaller than that of an equally effective inotropic concentration of isoprenaline. Thus, it is likely that other mechanisms are also involved in the alpha-adrenoceptor-mediated increase in myocardial force of contraction. One further mechanism might be an altered calcium sensitivity of the myofibrils (11). Both mechanisms, an increase in the intracellular Ca^{2+} concentration via I_{si} and an increase in the calcium sensitivity of the myofibrils, could thus be responsible for the positive inotropic response mediated by alpha-adrenoceptors.

Another aspect of the stimulation of alpha-adrenoceptors is an increased phosphatidylinositol turnover. In a variety of tissues, e. g. smooth muscle, salivary gland, blood platelets, liver and pancreas, it has been found that activation of calcium-mobilizing receptors is accompanied by the breakdown of membrane phosphoinositides (2, 3, 9, 13, 30, 32), but the role of the phosphatidylinositol turnover in heart tissue is as yet undefined.

The mechanism of action of the *Ca-agonists* Bay K 8644 and CGP 28 392 has been studied with patch clamp recordings of single Ca^{2+} channel activity (19, 26). These channels have three modes of gating behaviour, yielding current records (a) with brief openings, (b) with no openings because of channel unavailability and (c) with long-lasting openings and very brief closings. The Ca-agonists enhance Ca^{2+} channel current by stabilizing mode (c).

1.3 Increase in myoplasmic Na^+ concentration

The transsarcolemmal movement of Ca^{2+} and hence the inotropic state of cardiac muscle is controlled by the intracellular Na^+ concentration (4, 10, 20). The link in the interrelationship between Na^+ and Ca^{2+} is the Ca^{2+}/Na^+ exchange system in which transsarcolemmal Ca^{2+} movements are coupled to opposite movements of Na^+. Thus, Na^+ can exit from the cell in exchange for Ca^{2+}, or Ca^{2+} can exit in exchange for Na^+, depending upon the prevailing Na^+ and Ca^{2+} electrochemical gradients. The Ca^{2+}/Na^+ exchange system is the main mechanism to maintain the free intracellular Ca^{2+} concentration at the physiologically low level. Under equilibrium conditions, it follows the equation

$$[Ca^{2+}]_i = [Ca^{2+}]_o \frac{[Na^+]_i}{[Na^+]_o} \exp \frac{(n\text{-}2)\,FV}{RT}$$

where n is the number of Na^+ that exchanges for one Ca^{2+} ion, V is the membrane potential and F, R and T are the Faraday constant, the universal gas constant and the absolute temperature, respectively (10). Its driving force is the concentration gradient for Na^+ across the sarcolemma which, in turn, is accomplished largely by the sarcolemmal sodium pump, Na^+/K^+-ATPase, and the excitable properties of the membrane. The exact stoichiometry of the Ca^{2+}/Na^+ exchange system has not yet been resolved but in cardiac muscle n has a value close to 3. This means that only a very small increase in $[Na^+]_i$ is required to achieve a large increase in $[Ca^{2+}]_i$ and, hence, in force of contraction. The steep dependence of tension on $[Na^+]_i$ has been shown with Na^+-sensitive intracellular microelectrodes (27, 42). Quantitatively, a 1 mM increase in intracellular Na^+ activity was accompanied by about a 100 % increase in force of contraction. The relationship between $[Na^+]_i$ and $[Ca^{2+}]_i$ has also been verified with ion-selective microelectrodes (40).

The foregoing discussion shows that an increase in $[Na^+]_i$ may constitute an important mechanism of a positive inotropic intervention. As yet two mechanisms have been detected which are likely to produce an increase in myocardial force of contraction via an increase in $[Na^+]_i$:

a) inhibition of Na^+/K^+-ATPase and
b) prolongation of the open state of Na^+ channels.

1.3.1 Inhibition of Na^+/K^+-ATPase

The cardiac glycosides are the most important agents to inhibit Na^+/K^+-ATPase activity which is widely believed to increase $[Na^+]_i$ even at low concentrations and, hence, to enhance force of contraction. Accordingly, Blaustein (4) has recently summarized the mechanism of action of cardiac glycosides as follows (slightly modified): "The cardiac glycosides bind to the cardiac 'receptor' Na^+/K^+-ATPase. Consequent partial inhibition of the Na pump increases $[Na^+]_i$ and reduces the Na^+ electrochemical gradient. Then, as a result of Na^+/Ca^{2+} exchange, "background" $[Ca^{2+}]_i$ rises during diastole, and an increased amount of Ca^{2+} is stored in the SR; more than the normal amount of Ca^{2+} is then released from the SR when the cells are activated. In addition, during the long cardiac action potential, more Ca^{2+} than normal enters the cells via Na^+/Ca^{2+} exchange because of the reduced Na^+ gradient. The larger-than-normal transient increase that then occurs during cell activation enhances the force of cardiac musle contraction."

It should be noted, however, that the Na^+ pump inhibition theory is as yet not generally accepted. For a detailed discussion the reader is referred to a variety of other recent reviews (1, 15, 29, 33).

1.3.2 Prolongation of the open state of Na^+ channels

Table 3 summarizes a number of drugs which probably increase $[Na^+]_i$ and hence $[Ca^{2+}]_i$ and force of contraction via a prolongation of the open state of Na^+ channels (20–22). These drugs keep the Na^+ channels open, leading to intracellular Na^+ accumulation and increased Ca^{2+} influx via the Ca^{2+}/Na^+ exchange system. The mechanism of action of these agents is thus similar to that of the cardiac glycosides in that both increase the intracellular Na^+ concentration. They differ, however, in the way this increase is brought about. The agents summarized in Table 3 increase passive Na^+ influx whereas the cardiac glycosides impair active Na^+ extrusion.

Table 3. Agents which prolong the open state of Na^+ channels

1. *Alkaloids*
Ceveratrum alkaloids (e. g. veratridine, germitrine)
Batrachotoxin
Aconitine
2. *Diterpenoids*
Grayanotoxins
3. *Polypeptides*
Sea anemone toxins
Scorpion toxins
Coral toxin
4. Pyrethroids

Modified from P. Honerjäger (22), with kind permission from the author.

The main characteristics of the effect of, for example, the ceveratrum alkaloids are as follows. They decelerate the repolarisation of the cardiac action potential by prolonging the open state of Na^+ channels. This and the positive inotropic effect are abolished by tetrodotoxin (Fig. 2). The drugs increase the $^{22}Na^+$ influx in cultured cardiac cells and the positive inotropic effect lags behind the electrical events (21).

A prolongation of the open state of Na^+ channels has also been discussed for the new drug DPI 201–106 (36).

1.4 Direct inhibition of Na^+/Ca^{2+} exchange

A direct inhibition of Na^+/Ca^{2+} exchange, leading to a decrease in Ca^{2+} efflux and thus an increase in $[Ca^{2+}]_i$ and force of contraction has been discussed for amiloride and amiloride analogs (28). The importance of the mechanism, however, requires further experimentation.

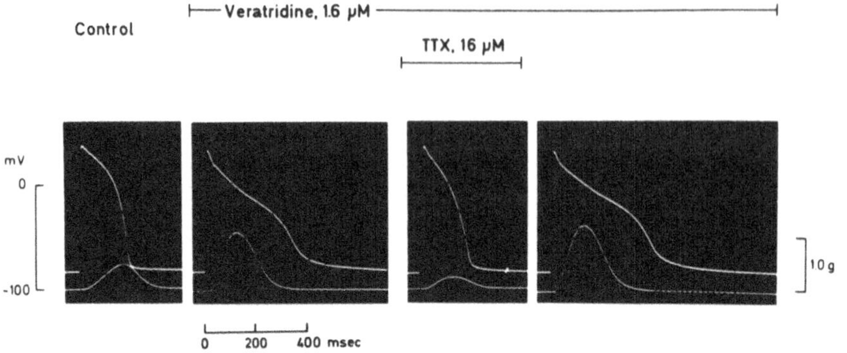

Fig. 2. Effect of veratridine on action potential (upper trace) and force of contraction (lower trace) in a guinea-pig papillary muscle in the absence and presence of tetrodotoxin (TTX). From Honerjäger and Reiter (24), with kind permission from the authors.

1.5 Inhibition of K^+ channels

K^+ channel-inhibiting agents such as 4-aminopyridine lead to a prolongation of the action potential and thus, at least theoretically, to a prolongation of the slow Ca^{2+} inward current. However, the positive inotropic effect of this agent is but moderate so that this mechanism does not seem to be very important (cf. (22)). A similar mechanism has also been discussed to contribute to the positive inotropic effect of sotalol (22), phenylephrine (16) or pimobendan (23).

2. Increased Ca^{2+} sensitivity of the contractile proteins

An increased Ca^{2+} sensitivity of the contractile proteins (Ca^{2+} sensitization) can be revealed by studying the contractile behaviour of isolated myocardial contractile structures or by measuring intracellular Ca^{2+} signals with the Ca^{2+}-sensitive bioluminescent protein aequorin in intact cardiac muscle. A Ca^{2+} sensitization has been discussed for sulmazole (5, 17, 18, 41), pimobendan (35), methylxanthines at millimolar concentrations (31, 43), alpha-adrenoceptor agonists (11) and DPI 201–106 (36). In theory, this effect can contribute to the overall inotropic action of a drug, provided the drug is able to cross the intact cell membrane or an intracellular messenger is involved in the Ca^{2+} sensitizing effect. To my knowledge, this has not yet been shown clearly for any of the above mentioned agents. Thus, the importance of Ca^{2+} sensitization as a positive inotropic mechanism in intact cardiac muscle or even more so in vivo remains to be elucidated. It is likely, however, that Ca^{2+} sensitization is a desirable property of a positive inotropic drug because it conceivably lacks the arrythmogenic effect of agents acting through an increase in myoplasmic Ca^{2+} concentration.

References

1. Akera T, Brody TM (1978) The role of Na^+,K^+-ATPase in the inotropic action of digitalis. Pharmacol Rev 29: 187–220
2. Berridge MJ (1984) Inositol trisphosphate and diacylglycerol as second messengers. Biochem J 220: 345–360
3. Berridge MJ, Irvine RF (1984) Inositol trisphosphate, a novel second messenger in cellular signal transduction. Nature 312: 315–321

4. Blaustein MP (1985) The cellular basis of cardiotonic steroid action. Trends Pharmacol Sci 6: 289–292
5. Blinks JR, Endoh M (1984) Sulmazol (AR-L 115 BS) alters the relation between [Ca^{2+}] and tension in living canine ventricular muscle. J Physiol 353: 63P
6. Böhm M, Brückner R, Hackbarth I, Haubitz B, Linhart R, Meyer W, Schmidt B, Schmitz W, Scholz H (1984) Adenosine inhibition of catecholamine induced increase in force of contraction in guinea-pig atrial and ventricular heart preparations. Evidence against a cyclic AMP- and cyclic GMP-dependent effect. J Pharmacol Exp Ther 230: 483–492
7. Böhm M, Burmann H, Meyer W, Nose M, Schmitz W, Scholz H (1985) Positive inotropic effect of Bay K 8644: cAMP-independence and lack of inhibitory effect of adenosine. Naunyn Schmiedeberg's Arch Pharmacol 329: 447–450
8. Bristow MR, Ginsburg R, Minobe W, Cubicciotti RS, Sagemann WC, Lurie K, Billingham ME, Harrison DC, Stinson EB (1982) Decreased catecholamine sensitivity and β-adrenergic-receptor density in failing human hearts. New Engl J Med 307: 205–211
9. Brückner R, Mügge A, Scholz H (1985) Existence and functional role of alpha$_1$-adrenoceptors in the mammalian heart. J Mol Cell Cardiol 17: 639–645
10. Chapman RA, Coray A, McGuigan JAS (1983) Sodium-calcium exchange in mammalian heart: the maintenance of low intracellular calcium concentration. In: Drake-Holland AJ, Noble MIM (eds) Cardiac metabolism. Wiley, Chichester, pp 117–149
11. Endoh M, Blinks JR (1984) Regulation of the intracellular Ca^{++} transient and Ca^{++}-sensitivity of myofibrils via α- and β-adrenoceptors in the rabbit papillary muscle. Cell Calcium 5: 301
12. Endoh M, Motomura S (1979) Differentiation by cholinergic stimulation of positive inotropic actions mediated via alpha-adrenoceptors and beta-adrenoceptors in the rabbit heart. Life Sci 25: 759
13. Farese RV (1984) Phospholipids as intermediates in hormone action. Mol Cell Endocrinol 35: 1–14
14. Ginsburg R, Bristow MR, Billingham ME, Stinson EB, Schroeder JS, Harrison DC (1983) Study of the normal and failing isolated human heart: decreased response of failing heart to isoproterenol. Am Heart J 106: 535–540
15. Greeff K (ed) (1981) Cardiac glycosides. Handbook of experimental pharmacology 56/1. Springer-Verlag, Berlin Heidelberg New York, pp 1–682
16. Handa Y, Wagner J, Inui J, Averesch H, Schümann HJ (1982) Effect of α- and β-sympathomimetic agonists on calcium-dependent slow action potential and force contraction in the rabbit papillary muscle. Naunyn Schmiedeberg's Arch Pharmacol 318: 330–335
17. Herzig JW (1984) Contractile proteins: possible targets for drug action. Trends Pharmacol Sci 5: 296–300
18. Herzig JW, Feile K, Rüegg JC (1981) Activating effects of AR-L 115 BS on the Ca^{2+} sensitive force, stiffness and unloaded shortening velocity (V_{max}) in isolated contractile structures from mammalian cardiac muscle. Arzneimittelforsch 31: 188–191
19. Hess P, Lansman JB, Tsien RW (1984) Different modes of Ca channel gating behaviour favoured by dihydropyridine Ca agonists and antagonists. Nature 311: 538–544
20. Honerjäger P (1982) Cardioactive substances that prolong the open state of sodium channels. Rev Physiol Biochem Pharmacol 92: 1–74
21. Honerjäger P (1983) Ceveratrum alkaloids: progress in understanding their membrane and inotropic actions. Trends Pharmacol Sci 4: 258–262
22. Honerjäger P (1986) Regulation of myocardial force of contraction by sarcolemmal ion channels, the sodium pump, and sodium calcium exchange. In: Rupp H (ed) Regulation of heart function – basic concepts and clinical applications. Thieme-Stratton, New York (in preparation)
23. Honerjäger P, Heiss A, Schäfer-Korting M, Schönsteiner G, Reiter M (1984) UD-CG 115 – a cardiotonic pyridazinone which elevates cyclic AMP and prolongs the action potential in guinea-pig papillary muscle. Naunyn Schmiedeberg's Arch Pharmacol 325: 259–269
24. Honerjäger P, Reiter M (1975) The relation between the effects of veratridine on action potential and contraction in mammalian ventricular myocardium. Naunyn Schmiedeberg's Arch Pharmacol 289: 1–28
25. Isenberg G, Ravens U (1984) The effects of the anemonia sulcata toxin (ATX II) on membrane currents of isolated mammalian myocytes. J Physiol (Lond) 357: 127–149

26. Kokubun S, Reuter H (1984) Dihydropyridine derivatives prolong the open state of Ca channels in cultured cardiac cells. Proc Natl Acad Sci 81: 4824–4827
27. Lee CO, Dagostino M (1982) Effect of strophanthidin on intracellular Na ion activity and twitch tension of constantly driven canine cardiac Purkinje fibers. Biophys J 40: 185–198
28. Luciani S, Floreani M (1985) Na^{+}-Ca^{2+} exchange as a target for inotropic drugs. Trends Pharmacol Sci 6: 316
29. Lüllmann H, Peters T (1979) Action of cardiac glycosides on the excitation-contraction coupling in heart muscle. A new concept. Prog Pharmacol 2/2: 1–58
30. Michell RH (1983) Phosphatidylinositol turnover and cellular function. V. Polyphosphoinositide breakdown as the initiating reaction in receptor-stimulated inositol phospholipid metabolism. Life Sci 32: 2083–2085
31. Morgan JP, Blinks JR (1982) Intracellular Ca^{2+} transients in the cat papillary muscle. Can J Physiol Pharmacol 60: 524–528
32. Nishizuka Y (1983) Phospholipid degradation and signal translation for protein phosphorylation. Trends Biochem Sci 8: 13–16
33. Noble D (1980) Mechanism of action of therapeutic levels of cardiac glycosides. Cardiovasc Res 14: 495–514
34. Packer M, Medina N, Yushak M (1984) Hemodynamic and clinical limitations of long-term inotropic therapy with amrinone in patients with severe chronic heart failure. Circulation 70: 1038–1047
35. Rüegg JC, Pfitzer G, Eubler D, Zeugner C (1984) Effect on contractility of skinned fibres from mammalian heart and smooth muscle by a new benzimidazole derivative, 4,5-dihydro-6-[2-(4-methoxyphenyl)-1H-benzimidazol-5-yl]-5-methyl-3 (2H)-pyridazinone. Arzneimittelforsch 34: 1736–1738
36. Scholtysik G, Salzmann R, Berthold R, Herzig JW, Quast U, Markstein R (1985) DPI 201–106, a novel cardioactive agent. Combination of cAMP-independent positive inotropic, negative chronotropic, action potential prolonging and coronary dilatory properties. Naunyn Schmiedeberg's Arch Pharmacol 329: 316–325
37. Scholz H (1984) Inotropic drugs and their mechanisms of action. J Am Coll Cardiol 4: 389–397
38. Scholz H, Meyer W (1985) Phosphodiesterase-inhibiting properties of newer inotropic agents. Circulation (in press)
39. Shah PK, Amin DK, Hulse S, Shellock F, Swan HJC (1985) Inotropic therapy for refractory congestive heart failure with oral fenoximone (MDL-17,043): poor long-term results despite early hemodynamic and clinical improvement. Circulation 71: 326–331
40. Sheu S-S, Fozzard HA (1982) Transmembrane Na^{+} and Ca^{2+} electrochemical gradients in cardiac muscle and their relationship to force development. J Gen Physiol 80: 325–351
41. Solaro RJ, Rüegg JC (1982) Stimulation of Ca^{++} binding and ATPase activity of dog cardiac myofibrils by AR-L 115 BS, a novel cardiotonic agent. Circ Res 51: 290–294
42. Wasserstrom JA, Schwartz DJ, Fozzard HA (1983) Relation between intracellular sodium and twitch tension in sheep cardiac Purkinje strands exposed to cardiac glycosides. Circ Res 52: 697–705
43. Wendt IR, Stephenson DG (1983) Effects of caffeine on Ca-activated force production in skinned cardiac and skeletal muscle fibres of the rat. Pfluegers Arch 398: 210–216

Author's address:

Prof. Dr. Hasso Scholz, Abteilung Allgemeine Pharmakologie, Universitäts-Krankenhaus Eppendorf, Martinistraße 52, D-2000 Hamburg 20 (West Germany)

Significance of calcium ions in the regulation of myocardial contractility*

A. M. Katz

Cardiology Division, Department of Medicine, University of Connecticut, Farmington (U.S.A.)

Introduction

Calcium ions (Ca^{2+}) represent the most important of the intracellular messengers in excitable cells. This messenger function occurs when a rise in cytosolic Ca^{2+} transmits a signal generated at the cell surface to an effector system within the cell. While Ca^{2+} can be viewed essentially as "universal" intracellular messenger, the signals that lead to the appearance of messenger Ca^{2+} inside the cell, the sources of this Ca^{2+}, and the receptors that recognize messenger Ca^{2+} can all differ among various cell types in the body. Specialized adaptations in these three aspects of the processes of excitation-contraction and stimulus-secretion coupling give rise to a complex and fascinating series of regulatory patterns in different tissues and organs of the body.

Signals which lead to a rise in cytosolic Ca^{2+}

Two general types of signal can initiate the sequence of reactions which leads ultimately to a rise in cytosolic Ca^{2+} concentration. The first of these are *electrically generated signals,* which are initiated by the changing transmembrane potential that occurs when an action potential is propagated along the plasma membrane at the cell surface. A more diverse group consists of a variety of *chemically mediated signals,* in which an interaction of the plasma membrane with one of a variety of different types of chemical serves to initiate a process which causes Ca^{2+} to appear in the cytosol. These chemical signals generally begin when a ligand, such as a neurotransmitter or a hormone, binds to a plasma membrane receptor. This ligand-receptor interaction then causes Ca^{2+} to enter the cytosol, either by crossing the plasma membrane from the extracellular space or by triggering the release of Ca^{2+} from an internal store (see below).

Sources of messenger Ca^{2+}

The Ca^{2+} that serves as an intracellular messenger can be derived from either or both of two sources. The first of these is the extracellular fluid, which is separated from the cytosol by the plasma membrane and where Ca^{2+} is present at millimolar concentrations. An alternate, or supplementary source of messenger Ca^{2+} is often found in internal stores, usually

* Supported by Research Grants HL-21812, HL-22135 and HL-33026 from the National Institutes of Health

present in membrane, delimited structures such as the sarcoplasmic reticulum of muscle, where Ca^{2+} concentration is also high.

Regardless of the source of the messenger Ca^{2+}, the movements of Ca^{2+} during activation represent "downhill" ion fluxes in which Ca^{2+} moves passively down its concentration gradient into the cytosol, where resting Ca^{2+} concentration is extremely low, usually in the micromolar range. For example the Ca^{2+} concentration in the cytosol of the resting heart is approximately 0.1–0.2 μM, which is many thousand-fold less than the Ca^{2+} concentration in the extracellular space and probably also in the sarcoplasmic reticulum.

The large Ca^{2+} concentration gradient between the sources for messenger Ca^{2+} and the cytosol of the resting cell accounts for two fundamentally important differences between the processes of activation and deactivation in excitable tissues such as the heart (6). The first arises because the processes which allow messenger Ca^{2+} to enter the cytosol do not require the expenditure of energy; as a result the Ca^{2+} fluxes which initiate cardiac systole are downhill, passive, ion movements. In contrast, the removal of Ca^{2+} from the cytosol, which leads to diastole by terminating the Ca^{2+}-initiated signal to contract, requires the direct utilization of energy for the active transport of Ca^{2+} against its concentration gradient. It is for this reason that relaxation is especially vulnerable to such abnormalities as ischemia and hemodynamic overloading, which deplete myocardial energy stores.

A second major difference between the Ca^{2+} fluxes involved in contraction and relaxation lies in their intrinsic rates. Activation, which delivers messenger Ca^{2+} to the interior of the cell, utilizes very rapid processes of diffusion, whereas the removal of this messenger from the cytosol is brought about by the much slower active transport of Ca^{2+} against a concentration gradient.

Receptors which recognize messenger Ca^{2+} within the cell

The complexity of the signalling processes which use Ca^{2+} as an intracellular messenger also reflects the diverse nature of the cytosolic receptors to which activator Ca^{2+} binds to serve its messenger function. Within eukaryotic cells, these Ca^{2+} receptors are proteins which appear to represent members of a single class of intracellular Ca^{2+}-binding proteins (7). These Ca^{2+}-binding proteins include both soluble proteins (e. g. calmodulin) and proteins that are bound into the structures whose function they regulate (e. g. troponin C, which is an integral part of the thin filament of muscle). All of the known intracellular Ca^{2+}-binding proteins are oligomers derived from two, or more commonly four subunits, each of which is believed to have evolved from a primitive ancestral Ca^{2+}-binding protein. The intracellular Ca^{2+} receptors, therefore, are probably members of a "family" of proteins which share a common ancestor that appeared some 600 million years ago, during the pre-Cambrian Era when multicellular animals developed in the primitive seas.

The common ancestor to these Ca^{2+}-binding proteins may have been a polypeptide chain of approximately 30 amino acids which contains two a-helical regions (designated E and F) separated by a non-helical sequence, or loop (7). The latter includes the Ca^{2+}-binding site, a non-helical region of the amino acid chain which contains five oxygen-containing amino acids that form an important spatial arrangement with one oxygen atom from the main peptide chain of the Ca^{2+}-binding protein. These electronegative oxygen atoms are arranged so as to form the vertices of an octahedron which surround a "pocket" capable of coordinating Ca^{2+} with very high affinity. As the two helical regions of the Ca^{2+}-binding proteins (called E and F) can be viewed as the extended index finger and thumb of a right hand, while the other folded fingers and palm correspond to the non-helical loop, this structure is often referred to as an "E-F hand", and the family of proteins which contain this structure as "E-F hand proteins".

A number of important proteins which have been derived from the family of Ca^{2+}-binding ("E-F hand") proteins have lost their ability to bind Ca^{2+}; however, many have retained their regulatory functions. For example, the *myosin light chains* in smooth muscle, which participate in the Ca^{2+}-mediated process of excitation-contraction coupling, are E-F hand proteins which serve as substrates for a Ca^{2+}-dependent protein kinase. These proteins have amino acid sequences which identify them as members of the family of Ca^{2+}-binding proteins, although amino acid substitutions that have taken place during evolution have caused these proteins to lose their ability to bind Ca^{2+}. In spite of their having lost the ability to bind Ca^{2+}, the myosin light chains of smooth muscle still participate in excitation-contraction coupling, although this function does not involve the binding of Ca^{2+} directly to this protein. Instead, these myosin light chains are substrates for a phosphorylation reaction that is initiated when Ca^{2+} binds to *calmodulin,* another of the Ca^{2+}-binding proteins. In other words, smooth muscle myosin light chains which have lost their ability to bind to Ca^{2+} still participate in a Ca^{2+}-initiated process, but they do so not by binding Ca^{2+}, but instead by serving as substrates for Ca^{2+}-dependent phosphorylation.

Role of Ca^{2+} in the regulation of myocardial contractility

It is now clear that a rise in cytosolic Ca^{2+} initiates the final steps in the process of excitation-contraction coupling in the working cells of the myocardium. This messenger function of Ca^{2+} arises from its ability to bind to troponin C, one of the Ca^{2+}-binding proteins referred to above (Fig. 1). This binding of Ca^{2+} to troponin C initiates a series of conformational changes among the proteins of the thin (actin) filament of muscle which allow active sites on actin to interact with myosin so as cause the muscle to contract.

The proteins of the thin filament include actin, tropomyosin and the troponin complex. The latter is made up of three discrete proteins: troponin C, troponin I, and troponin T (3). When the messenger Ca^{2+} which appears in the cytosol binds to troponin C, it activates the thin filament by allowing actin molecules to interact with myosin of the thick filament. This, in turn causes the contractile proteins to convert the chemical energy of ATP into the mechanical work expressed as shortening and tension development by the walls of the heart.

Each of the three components of troponin, along with tropomyosin, participates in the process by which Ca^{2+} regulates the interactions between actin and myosin. *Tropomyosin* inhibits the interaction between actin and myosin in resting muscle, so that this protein serves mainly as an inhibitor of the contractile process. *Troponin I,* along with tropomyosin, inhibits the interactions between actin and myosin in resting muscle, while *Troponin T* serves primarily to bind the troponin complex to tropomyosin. *Troponin C,* as already mentioned, is one of the Ca^{2+}-binding proteins which contains the important binding sites that recognize the appearance of Ca^{2+} in the cytosol as a signal which initiates contraction.

The ability of the interaction between Ca^{2+} and troponin C to initiate the contractile process is believed to result from a conformational change among the proteins of the thin filament which begins when troponin C binds to this intracellular messenger. This initiates a series of changes in the interactions between these proteins which reverses the inhibitory effects of tropomyosin and the troponin complex that block the actin-myosin interactions which generate tension and shortening in the walls of the heart.

Interactions between Ca^{2+} and troponin C

The polypeptide chain of troponin C contains four amino acid sequences, each of which has an amino acid sequence similar to the E-F hand structure of the putative ancestor of the

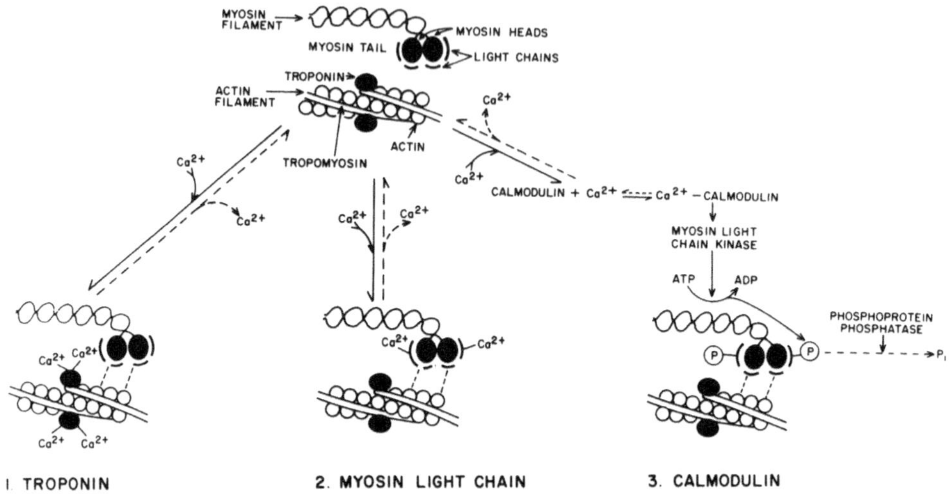

Fig. 1. Three mechanisms by which Ca^{2+} can activate muscle contraction. 1. *Troponin-linked regulation:* Ca^{2+} binding to troponin on the thin filament reverses an inhibitory effect of troponin and tropomyosin, thereby permitting actin in the thin filament to interact with the myosin crossbridges projecting from the thick filament. This mechanism is found in cardiac and most skeletal muscle. 2. *Myosin-linked regulation:* Ca^{2+} binding to one of the myosin light chains reverses an inhibitory effect of the latter which permits the crossbridge to interact with actin. This is an uncommon mechanism, found for example in scallop adductor muscle. 3. *Calmodulin-linked regulation:* Ca^{2+}-binding to calmodulin leads to the activation of myosin light chain kinase, a Ca^{2+}-dependent protein kinase which phosphorylates one of the myosin light chains Ⓟ. This phosphorylation activates the myosin-actin interaction and thereby leads to contraction. This mechanism is commonly found in smooth muscle and non-muscle motile systems. Relaxation (dashed lines) occurs in 1 and 2 when Ca^{2+} is removed from its binding sites on either troponin or the myosin light chain. Relaxation in 3 occurs when the rate of myosin light chain dephosphorylation, which is catalyzed by a phosphoprotein phosphatase, exceeds that of myosin phosphorylation as may occur when Ca^{2+} is dissociated from calmodulin.
Reprinted from (5).

family of Ca^{2+}-binding proteins described above. In the troponin C of fast skeletal muscle, all four of these amino acid sequences bind Ca^{2+}; however, two of these also bind Mg^{2+} and so are designated as *Ca^{2+}-Mg^{2+} sites*. The other two sequences bind only Ca^{2+}, and thus are called *Ca^{2+}-specific sites*. Although the Ca^{2+}-affinity of the Ca^{2+}-Mg^{2+} sites is much higher than that of the Ca^{2+}-specific sites, the Mg^{2+} concentration in the cells of the myocardium is much higher than that of Ca^{2+}. As a result, the Ca^{2+}-Mg^{2+} sites of troponin C in vivo are normally occupied mainly by Mg^{2+}, and are not influenced by the small rise in cytosolic Ca^{2+} that occurs during excitation-contraction coupling. As the Ca^{2+}-Mg^{2+} sites do not appear able to recognize this increase in cytosolic Ca^{2+} concentration as a signal which initiates the interactions between actin and myosin, it is likely that they do not serve this critical regulatory function. Instead, the Ca^{2+}-specific sites, which are able to bind to Ca^{2+} even in the presence of physiological concentration of Mg^{2+}, probably represent the functional Ca^{2+} receptors of troponin C.

Mammalian cardiac troponin C, like skeletal troponin C, contains four amino acid sequences which are similar to those of the E-F hand regions of the Ca^{2+} binding proteins. However, amino acid replacements in one of these sites, in which two aspartic acid residues have been replaced by a leucine and an alanine, have caused this site to have lost its ability to bind Ca^{2+}. Of the remaining three sites in cardiac troponin C, two are Ca^{2+}-Mg^{2+} sites and one is a Ca^{2+}-specific site (4). The latter probably represents the physiological Ca^{2+} receptor of the cardiac contractile proteins.

Calmodulin

One mechanism by which Ca^{2+} serves its role as an intracellular messenger is by initiating the phosphorylation of a number of intracellular proteins. These phosphorylation reactions are catalysed by a group of enzymes, called *Ca^{2+}-dependent protein kinases,* which play an important role in cell regulation. A single group of proteins serves as the Ca^{2+}-receptor which allows Ca^{2+} to activate many of these protein kinases. These proteins, called *calmodulin* (2), serve as Ca^{2+} receptors participating in a number of reactions which phosphorylate a diverse group of enzymes and functional proteins involved in such processes as cyclic nucleotide metabolism, intermediary metabolism, muscle contraction, and neurotransmitter release.

Calmodulin, like troponin C, contains four E-F hand regions. The ability of calmodulin to mediate signals initiated by a rise in cytosolic Ca^{2+} concentration begins when the formation of a Ca^{2+}-calmodulin complex allows this complex to interact with the protein whose function is to be modified. This interaction appears to be made possible when the binding of Ca^{2+} to calmodulin exposes a hydrophobic surface on the calmodulin molecule; this hydrophobic surface of the Ca^{2+}-calmodulin complex then interacts with the protein or membrane whose function is altered. A number of drugs, for example the tricyclic antidepressants and phenothiazines, interfere with the functions of the Ca^{2+}-calmodulin complex, and thus inhibit many cellular responses to Ca^{2+}. This action, sometimes called "calcium antagonism", is quite different from that of calcium channel blocking drugs which inhibit Ca^{2+} fluxes through plasma membrane calcium channels.

Activation of muscle contraction by non-troponin linked mechanisms

In some muscle types, the ability of Ca^{2+} to initiate contraction utilizes mechanisms which do not involve troponin as the intracellular Ca^{2+}-receptor. In smooth muscle, for example, the activating effect of Ca^{2+} is mediated by its ability to initiate a Ca^{2+}-dependent phosphorylation reaction (1). This activating effect of Ca^{2+} begins with its binding to calmodulin, which activates a Ca^{2+}-dependent protein kinase called *myosin light chain kinase.* This enzyme catalyses the phosphorylation of one of the light chains of smooth muscle myosin (Fig. 1) and thereby initiates interactions of smooth muscle myosin with actin which lead to contraction. Yet another mechanism (myosin-linked regulation) involves a Ca^{2+}-binding myosin light chain which serves as the Ca^{2+} receptor in a few invertebrate muscles, for example the adductor muscle of the scallop (Fig. 1).

Inositol triphosphate and the mobilization of Ca^{2+}

A messenger system which utilizes phosphatidylinositol, one of the membrane phospholipids, to generate at least two intracellular messengers has recently been characterized (8). These signalling mechanisms begin with the phosphorylation of the inositol moiety of phosphatidylinositol. The subsequent breakdown of this phospholipid forms a diacylglycerol and inosine trisphosphate, both of which are now recognized to play important messenger roles in cellular regulation. Diacylglyerols can promote the activity of Ca^{2+}-activated phospholipid-depended protein kinases, while inosine trisphosphate has been shown to relase Ca^{2+} from intracellular stores in a number of tissues. The role of these messengers in the regulation of myocardial function is not yet clear, however.

References

1. Adelstein R, Sellers JR, Conti MA, Pato MD, deLancrolle P (1982) Regulation of smooth muscle contractile proteins by calmodulin and cyclic AMP. Fed Proc 41: 2873–2878
2. Cheung WY (1982) Calmodulin: An overview. Fed Proc 41: 2253–2257
3. Ebashi S (1979) The Croonian Lecture, 1979. Regulation of muscle contraction. Proc R Soc Lond B 207: 259–286
4. Holroyde MJ, Robertson SP, Johnson JD, Solaro RJ, Potter JD (1980) The calcium and magnesium binding sites on cardiac troponin and their role in the regulation of myofibrillar adenosine triphosphatase. J Biol Chem 255: 11688–11693
5. Katz AM, Hager WD, Messineo FC, Pappano AJ (1984) Cellular actions and pharmacology of the calcium channel blocking drugs. Am J Med 77 (2B): 2–10
6. Katz AM (in press) Potential deleterious effects of inotropic agents in the therapy of chronic heart failure. Circulation
7. Kretsinger R (1977) Evolution of the informational role of calcium in eukaryocytes. In: Wassermann RH, Corradino RA, Carafoli E et al (eds) Calcium-Binding Proteins and Calcium Function. North Holland, New York, pp 63–72
8. Rasmussen H, Barrer PQ Calcium messenger system: An integrated view. Physiol Rev 64: 938–984

Author's address:

Arnold M. Katz, M. D., Division of Cardiology, University of Connecticut, Farmington, Connecticut 06032 (U.S.A.)

Additive and non-additive positive inotropic effects in human and guinea-pig myocardium

L. Brown, B. Lorenz and E. Erdmann

Medizinische Klinik I, Ludwig-Maximilians-Universität München, München (West Germany)

Summary

The positive inotropic effects of β-agonists (isoprenaline, dobutamine, dopamine), histamine, phosphodiesterase inhibitors (isobutylmethylxanthine, milrinone, theophylline), ouabain and calcium were measured using contracting papillary muscles from healthy guinea-pig hearts as well as on contracting papillary muscle strips from patients undergoing mitral valve replacement. In the guinea-pig myocardium, the potencies varied markedly but all compounds gave the same maximal increase in force of contraction. In contrast, compounds acting through cAMP gave maximal positive inotropic effects which were significantly reduced compared with ouabain and calcium in human papillary muscle strips. Further, the potency of β-agonists was decreased in the diseased human ventricular myocardium.

Either β-agonists, histamine or phosphodiesterase inhibitors were added after a stable, almost maximally effective but non-toxic effect of ouabain had been achieved. The maximal inotropic response to combined treatment was not greater than to either compound alone in the guinea-pig heart; the only increase was in the incidence of toxicity. In human papillary muscle strips, the maximal effect of ouabain could not be increased by any other compound but the reduced maximal effect of dobutamine could be increased by the addition of ouabain.

These results indicate that additional inotropic effects are not solely responsible for the haemodynamic improvements after combined inotropic therapy in optimally digitalis-treated patients. Other cardiovascular effects such as cAMP-induced vasodilatation must also be considered.

Introduction

The final intracellular messenger for all inotropic compounds is calcium. An increase in force of contraction (positive inotropy) can result either from an increase in the calcium concentration available to the contractile proteins or from an increase in the calcium sensitivity of these contractile proteins (22). All inotropic compounds now in use modulate the calcium concentration during the cardiac cycle. An increase in intracellular cAMP through an activation of the adenylate cyclase (by β-agonists, histamine or glucagon) or an inhibition of phosphodiesterase increases the release of calcium from the sarcoplasmic reticulum and the calcium influx through the cell membrane. The calcium content is also increased by the cardiac glycosides probably through an inhibition of the $(Na^+ + K^+)$-ATPase and subsequent activation of the Na/Ca-exchange system. The phosphodiesterase inhibitor, sulmazole, may additionally increase the calcium sensitivity of the contractile proteins (12). These inotropic mechanisms may be altered by cardiac diseases such as congestive heart failure, hypertrophy and mitral valve disease, for which these compounds are used therapeutically. Therefore, we have measured the concentration-response curves of inotropic agents acting through various mechanisms (β-agonists – isoprenaline, dobutamine, dopamine; histamine; phosphodiesterase inhibitors – isobutylmethylxanthine (IBMX),

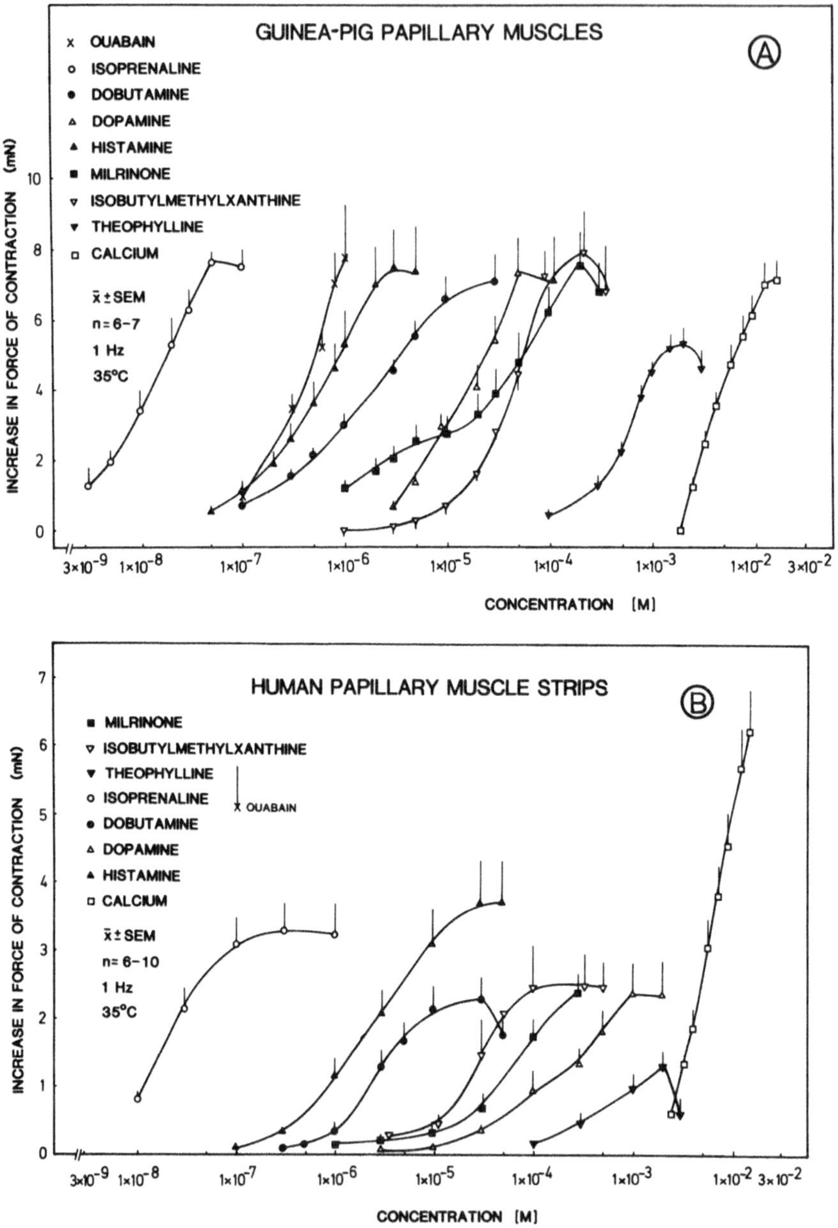

Fig. 1. Cumulative concentration-response curves for the effects on force of contraction in isolated, contracting guinea-pig papillary muscles (Fig. 1a) or human papillary muscle strips (Fig. 1b). The muscles were exposed to each drug concentration for 10–15 minutes when a stable effect had been achieved; the only exception being ouabain, where incubation times of 45 minutes or 120 minutes were used for guinea-pig or human muscles, respectively.
The pre-drug force of contraction, maximal increase in force of contraction and EC_{50} values are given in Table 1.

milrinone, theophylline; ouabain; calcium) using ventricular myocardium from healthy guinea-pigs and diseased humans. These determinations should indicate whether the potency or efficacy of these inotropes is varied in cardiac diseases.

Many studies with newer positive inotropic substances such as dopamine, dobutamine, milrinone or amrinone (1, 4, 14, 16, 17) have reported beneficial haemodynamic effects in patients optimally treated with digitalis. The improvements, notably the increased cardiac output and the decreased systemic vascular resistance suggest additional inotropic effects. In patients, purely inotropic effects can only be reliably assessed when preload, afterload and frequency are kept constant; requirements which are virtually impossible to meet. Therefore, the isolated human papillary muscle has been used as a model for inotropic effects to determine whether an almost maximal inotropic effect of ouabain or dobutamine can be further increased by the addition of other inotropic substances.

Methods

The right ventricular papillary muscles of 6–8 week old guinea-pigs were mounted as previously described (6) at 35° in Tyrode solution (75 ml) containing in mM: NaCl, 136.9; KCl, 5.4; $MgCl_2$, 1.05; $CaCl_2$, 1.8; $NaHCO_3$, 22.6; NaH_2PO_4, 0.42; glucose, 5.5; ascorbic acid, 0.28; Na_2-EDTA, 0.05; continuously aerated with 95 % O_2 + 5 % CO_2; stimulated for 5 msec at 1 Hz at 10–20 % above threshold.

Left ventricular papillary muscles were excised from 27 patients (11 male, 16 female, aged 40–69 years, mean 54.6 years) during cardiac surgery involving mitral valve replacements, as described previously (7). Papillary muscle strips (about 1 mm thick, 4–12 mm long) were mounted as described above for the guinea-pig papillary muscles.

Concentration-response curves were obtained cumulatively. All values are given as the mean ± standard error of the mean. Each compound was tested using papillary muscle strips from at least four patients.

Results

The positive inotropic effects of compounds acting either through adenylate cyclase (histamine, β-agonists such as isoprenaline, dobutamine, dopamine) or phosphodiesterase inhibition (IBMX, milrinone, theophylline) were measured for comparison with ouabain and calcium. The concentration-response curves on contracting papillary muscle are given in Figs. 1 a (healthy guinea-pig heart) and 1 b (diseased human heart). In the guinea-pig papillary muscles, the potency of these inotropic substances varied over six orders of magnitude, but the efficacy was unchanged (Table 1). This is in contrast to the results with human papil-

Table 1 Positive inotropic effects on isolated, contracting guinea-pig and human papillary muscles

	Guinea-pig papillary muscles			Human papillary muscle strips		
Compound	Initial force of contraction (mN)	EC_{50} (M)	Maximal inotropic effect (mN)	Initial force of contraction (mN)	EC_{50} (M)	Maximal inotropic effect (mN)
Ouabain	2.41 ± 0.61	3.6×10^{-7}	7.8 ± 1.4	–	–	–
Isoprenaline	2.28 ± 0.55	1.2×10^{-8}	7.6 ± 0.5	3.8 ± 0.60	2.1×10^{-8}	3.29 ± 0.42
Dobutamine	1.51 ± 0.32	1.6×10^{-6}	7.0 ± 0.9	3.71 ± 0.36	2.7×10^{-6}	2.29 ± 0.32
Dopamine	1.98 ± 0.32	1.7×10^{-5}	7.1 ± 1.0	3.8 ± 0.42	1.9×10^{-4}	2.35 ± 0.45
Histamine	1.36 ± 0.22	5.2×10^{-6}	7.5 ± 1.1	4.62 ± 0.66	2.3×10^{-6}	3.73 ± 0.60
IBMX	1.63 ± 0.37	4.2×10^{-5}	7.9 ± 1.2	3.68 ± 0.56	2.5×10^{-5}	2.45 ± 0.47
Milrinone	1.36 ± 0.18	3.0×10^{-5}	7.5 ± 1.0	3.61 ± 0.34	5.8×10^{-5}	2.38 ± 0.30
Theophylline	1.01 ± 0.20	5.8×10^{-4}	5.3 ± 0.5	4.45 ± 0.55	5.2×10^{-4}	1.29 ± 0.22
Calcium	2.20 ± 0.41	4.4×10^{-3}	7.1 ± 0.4	3.16 ± 0.40	5.8×10^{-3}	6.21 ± 0.61

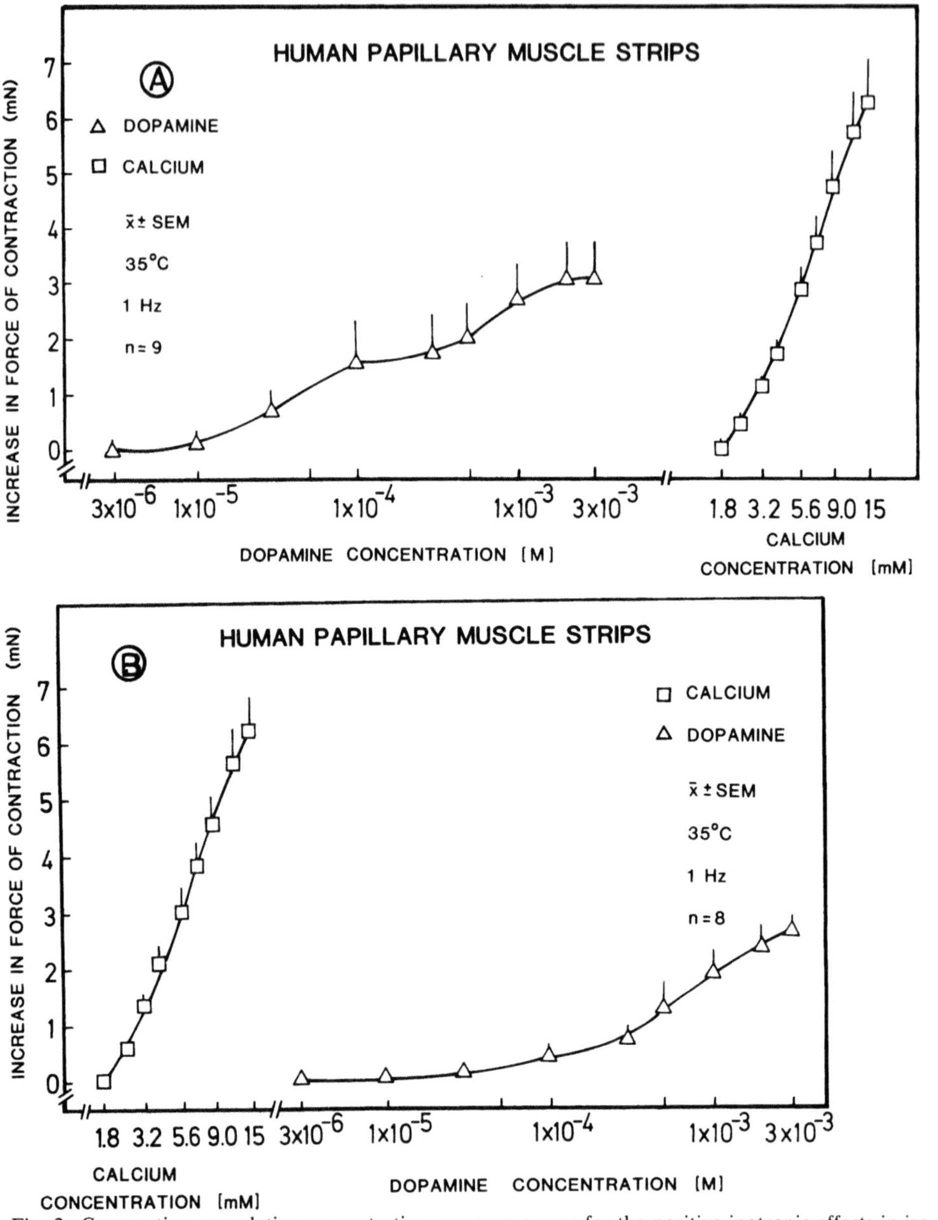

Fig. 2. Consecutive, cumulative concentration-response curves for the positive inotropic effects in isolated, contracting human papillary muscle strips of dopamine and calcium.
The pre-drug force of contraction values were:
Fig. 2a – dopamine, 3.09 ± 0.41 mN; calcium, 3.01 ± 0.29 mN;
Fig. 2b – calcium, 3.16 ± 0.40 mN; dopamine, 1.08 ± 0.17 mN.
The maximal increases in force of contraction were:
Fig. 2a – dopamine, 3.16 ± 0.67 mN; calcium, 6.23 ± 0.87 mN;
Fig. 2b – calcium, 6.20 ± 0.61 mN; dopamine, 2.73 ± 0.29 mN;

lary muscle strips where both potency and efficacy varied markedly. All of the tested compounds which increase force of contraction through an increase in cAMP (β-agonists, histamine, phosphodiesterase inhibitors) gave a reduced inotropic effect in human papillary muscle strips compared to ouabain (1×10^{-7} M) or calcium. Further, the potency of the β-agonists was reduced in the human ventricular myocardium.

It is possible that interindividual differences could account for the reduced inotropic effect measured in some human papillary muscle strips. This has been tested by measuring consecutive, cumulative concentration-response curves of dopamine and calcium (Fig. 2). The maximal inotropic effect of calcium was about twice that of dopamine when measured in the same muscle strips, independent of the order of addition of these inotropic compounds.

The concept that the maximal inotropic effect of ouabain could be increased by other inotropic substances with differing mechanisms of action was tested in guinea-pig left atria and right ventricular papillary muscles as well as in human papillary muscle strips. In guinea-pig left atria, ouabain, dobutamine, dopamine and calcium gave similar maximal inotropic effects which were greater than those produced by theophylline and histamine (Fig. 3a). The concentrations of histamine, dobutamine, dopamine and theophylline giving 25% of the maximal effect (EC_{50}) were then added to additional left atria after a stable, almost maximally effective but non-toxic positive inotropic effect of ouabain (8×10^{-7} M) had been achieved (Fig. 3b). Dobutamine (5×10^{-7} M) was the only compound to give a significant but transient positive inotropic effect after ouabain; theophylline (2×10^{-4} M) gave significant decreases. Higher concentrations (EC_{50}, EC_{90}) gave marked toxicity in most left atria with significant, transient inotropic effects seen only after the addition of dobutamine (6). In guinea-pig left atria, a second inotropic substance does not give a sustained inotropic effect additive to that of an almost maximally effective ouabain concentration.

A different experimental procedure was used for the studies with guinea-pig and human papillary muscles. Consecutive concentration-response curves were measured with no pretreatment and after a stable inotropic effect of ouabain (human, 1×10^{-7} M; guinea-pig, 8×10^{-7} M) had been achieved. These ouabain concentrations were chosen to give almost maximal inotropy with minimal toxicity. In guinea-pig papillary muscles (Fig. 4a), both dopamine and theophylline increased the positive inotropic effects of ouabain but the maximal inotropic effect after combined addition was not greater than with dopamine or theophylline alone. Dopamine and theophylline gave significantly smaller increases in force of contraction in human papillary muscle strips than ouabain (Fig. 4b). Addition of either dopamine or theophylline after the ouabain effect was stable produced increased toxicity without transient increases in force of contraction. Similar results were found after addition of isoprenaline, dobutamine, histamine, IBMX and milrinone to ouabain (1×10^{-7} M). Toxicity in the human papillary muscle strips was shown as severe decreases in force of contraction together with after-contractions and the development of contracture; in guinea-pig muscles, ectopics were usually observed.

Since the cAMP-elevating compounds gave a reduced maximal inotropic effect in human papillary muscle strips, ouabain may further increase the maximal dobutamine-induced increase in force of contraction. Fig. 5 shows that this did occur, in contrast to the decreases seen after addition of dobutamine to the maximal ouabain response.

Discussion

Since development of tension in muscle is regulated by calcium, two mechanisms are possible to produce an increased force of contraction – either an increase in the calcium con-

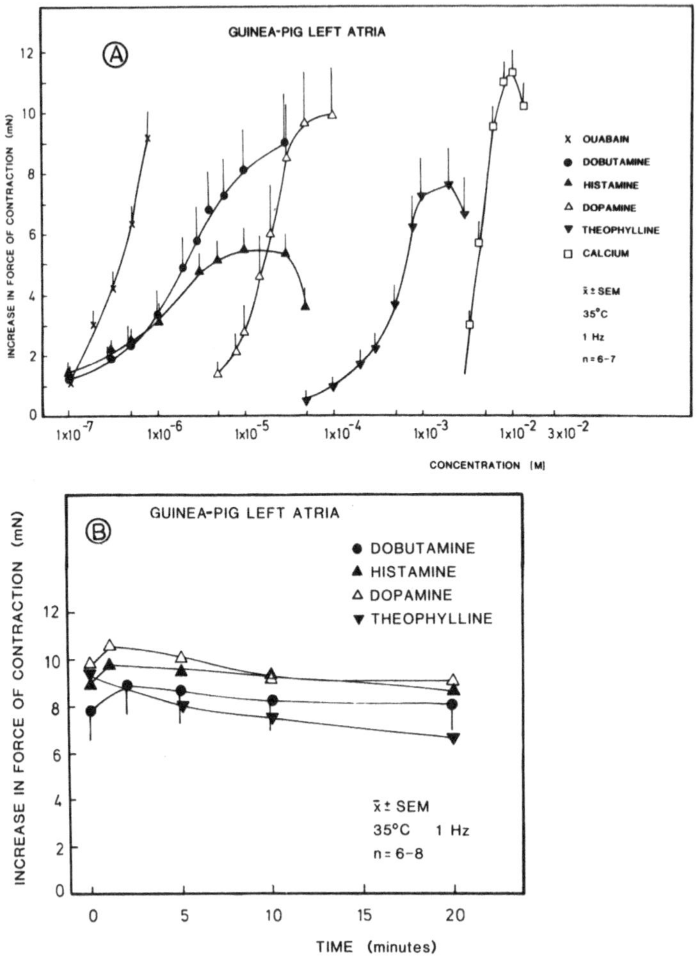

Fig. 3. Cumulative concentration-response curves for positive inotropic effects on isolated, contracting guinea-pig left atria (Fig 3 a) and the force of contraction in ouabain-treated guinea-pig left atria after addition of a second positive inotropic substance (Fig. 3 b).
The left atria were mounted in Krebs-Henseleit solution containing 2.5 mM $CaCl_2$ (6).
The initial force of contraction (in mN ± SEM) with the maximal increase in force of contraction in brackets, was:
ouabain, 4.7 ± 0.7 (9.2 ± 0.6);
dobutamine, 4.9 ± 1.1 (9.0 ± 1.5);
histamine, 5.4 ± 1.3 (5.5 ± 0.7);
dopamine, 6.6 ± 1.1 (9.9 ± 1.7);
theophylline, 4.7 ± 0.8 (7.6 ± 1.1); and
calcium, 4.1 ± 0.7 (11.3 ± 0.8).
In Fig. 3 b, the following concentrations, determined from Fig. 3 a to be the EC_{25} of each compound, were added at a stable inotropic effect of ouabain (8 x 10^{-7} M) in guinea-pig left atria; dobutamine, 5 x 10^{-7} M; histamine, 1 x 10^{-7} M; dopamine, 8 x 10^{-6} M; and theophylline, 2 x 10^{-4} M. The force of contraction was measured at the peak effect and 5, 10 and 20 minutes after addition of the EC_{25}. The standard errors of the measurements are given only for dobutamine.

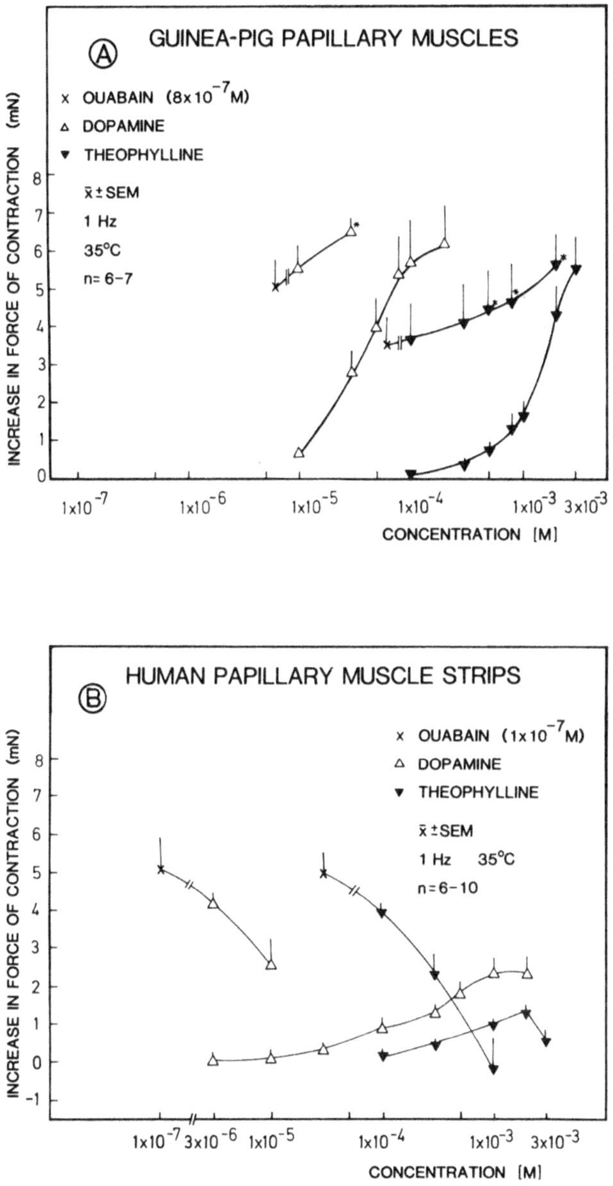

Fig. 4. Force of contraction in guinea-pig papillary muscles (Fig. 4a) and human papillary muscle strips (Fig. 4b). Guinea-pig papillary muscles were incubated in Krebs-Henseleit solution containing 2.5 mM $CaCl_2$ while human papillary muscle strips were incubated in Tyrode solution with 1.8 mM $CaCl_2$.
The muscles were treated in the following order:

1. cumulative concentration-response curve of dopamine or theophylline, then washout;
2. ouabain (guinea-pig, 8×10^{-7} M; human, 10^{-7} M); when stable,
3. repeat concentration-response curve for dopamine or theophylline.

Significant increases in force of contraction (paired t-test) are denoted by (*).

centration or an increased calcium sensitivity of the contractile proteins. All the inotropic compounds now in therapeutic use modulate changes in calcium concentration although sulmazole may additionally increase calcium sensitivity (12).

Various independent mechanisms can lead to this calcium increase (22, see also pp Scholz paper of these volume) such as increases in the cAMP concentration through adenylate cyclase activation or phosphodiesterase inhibition, inhibition of $(Na^+ + K^+)$-ATPase leading to activation of Na/Ca-exchange or receptor- or potential-operated calcium channels. These mechanisms operate in both guinea-pig and human ventricular myocardium as shown by the positive inotropic effects of the β-agonists, histamine, phosphodiesterase inhibitors, ouabain and calcium in the present study. The maximal concentrations used are certainly rather high but the isolated papillary muscle is a model to measure inotropic effects only without influence from other cardiovascular effects such as vasodilatation or from CNS effects which may affect the maximal possible concentrations in vivo. The present study clearly shows that, in human but not in guinea-pig papillary muscles, the cAMP-elevating compounds produce significantly lower maximal inotropic effects than ouabain or calcium. Further, the β-agonists have a reduced potency in diseased human ventricular myocardium. We consider these changes to disease-related rather than species differences since similar changes have been measured in animal models of cardiovascular disease (3, 15, 19, 21, 23). This alternative explanation of species differences can only be excluded by comparable measurements using normal human myocardium which is not easily available. In the present study, papillary muscles were taken from adult humans with mitral valve stenosis usually with concomitant regurgitation; most patients had a history of chronic rheumatic heart disease. All patients were severely symptomatic before surgery (NYHA class III-IV) with ejection fractions of less than 50 %. Histological examinations of the excised papillary muscles showed interstitial fibrosis. Disease-induced reductions in positive inotropy have been well documented. Wilmshurst et al. (27) showed that the phosphodiesterase inhibitor, amrinone, gave a reduced maximal inotropic effect compared to ouabain (1×10^{-7} M) in human ventricular myocardium; the extent of the reduction was more pronounced with increasing severity of heart failure. A close correlation is not to be expected since the clinical degree of congestive heart failure does not necessarily reflect the status of each myocyte. A comparison of the inotropic effects of isoprenaline in papillary muscle strips from normal and diseased humans (5, 11) showed a disease-related reduction in both inotropic response and potency. Histamine gave only a relatively small inotropic effect which was not different in diseased human myocardium (11). Animal studies have shown a reduced inotropic effect of isoprenaline in volume overload heart failure in dogs (19) as well as in the uninvolved ventricle after acute myocardial infarction in guinea-pigs (3). Hypertrophy of the rat heart decreased the inotropic response to isoprenaline (15, 21, 23) and phenylephrine (10). Neither the number of receptors nor the inotropic effects of H_2-agonists were impaired after acute myocardial infarction in guinea-pigs (2). A reduction of 26 % in the number of digitalis receptors in hypertrophied pig myocardium has been reported (13) but no inotropic measurements were given. In contrast, Prasad et al. (20) have shown increases in $(Na^+ + K^+)$-ATPase levels in dog hearts after 6 months mitral regurgitation.

High levels of circulating catecholamines reduced the β-receptor density in human hearts (5) while ischaemia and hypertrophy (9, 15, 25) reduced the adenylate cyclase activity in rat and dog hearts. These reductions adequately explain the reduced inotropic effects in diseased human ventricular myocardium of compounds acting through the adenylate cyclase system – β-agonists and histamine. But the inotropic effects of compounds acting through phosphodiesterase inhibition were also reduced in our study with human myocardium. Since the effects of phosphodiesterase inhibitors can only be indirectly influenced by adenylate cyclase reduction, we would suggest that a further biochemical defect either in the basal

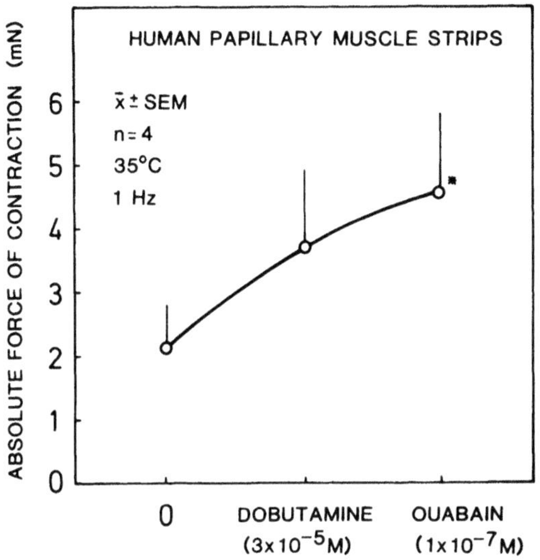

Fig. 5. Force of contraction in human papillary muscle strips.
The muscle strips were treated as follows: dobutamine (3×10^{-5} M) until stable then ouabain (1×10^{-7}). Ouabain gave a significant increase in force of contraction as shown by (*).

cAMP level or in the phosphodiesterase also occurs in congestive heart failure. This defect is reasonably specific to compounds acting through cAMP since the inotropic effects of ouabain were not diminished.

Addition of low concentrations of a second inotropic compound after a stable, almost maximally effective ouabain inotropic effect may increase the force of contraction but only to the value possible with each compound alone. Higher concentrations do not give further inotropy but only an increasing incidence of toxicity. Addition of ouabain (1×10^{-7} M) to the maximal dobutamine inotropy in diseased human papillary muscle strips is an exception: the reduced maximal effect of this β-agonist could be significantly increased by ouabain.

Is it to be expected that a maximal inotropic effect of ouabain can be further increased by compounds with differing inotropic mechanisms?

This must be considered unlikely since all the compounds used in this study finally increase the intracellular calcium without altering the calcium sensitivity of the contractile proteins. A further increase in available calcium, independent of mechanism, should produce toxicity after a maximal ouabain-induced inotropy has been achieved.

Clinical trials have produced significant haemodynamic improvements especially in cardiac output and systemic vascular resistance after combined inotropic therapy, for example with dobutamine (4), dopamine (17), amrinone (14, 16) or milrinone (1) in patients optimally treated with digitalis. Vasodilatation is produced by therapeutic levels of these β-agonists and phosphodiesterase inhibitors by increases in the cAMP content (8, 18, 24). Vasodilatators such as nitroprusside increase cardiac output and decrease systemic vascular resistance (26) without any inotropic effect. Thus, cAMP-induced vasodilatation, rather than additional positive inotropic effects, would seem to be a more likely explanation for the haemodynamic improvements in digitalis-treated patients given other inotropic compounds.

Acknowledgements

These investigations were supported by the Deutsche Forschungsgemeinschaft (Er 65/4–4).

Our special thanks are due to Prof. E. Kreuzer, Prof. B. Kemkes, Dr. C. Weinhold and their colleagues, Herzchirurgische Klinik der Universität, and Dr. K. Holper, Dr. W. Klövekorn and their colleagues, Herzchirurgische Klinik des Deutschen Herzzentrums, München, for obtaining the surgical specimens of human heart without which this study would not have been possible.

We gratefully appreciate the excellent technical assistance of Miss A. Reichl and Mr. M. Rupff.

References

1. Baim DS, McDowell AV, Cherniles J, Monrad ES, Parker JA, Edelson J, Braunwald E, Grossmann W (1983) Evaluation of a new bipyridine inotropic agent – milrinone – in patients with severe congestive heart failure. N Engl J Med 309: 748–756
2. Baumann G, Felix SB, Rieß G, Loher U, Ludwig L, Blömer H (1982) Effective stimulation of cardiac contractility and myocardial metabolism by impromidine and dimaprit – two new H_2-agonistic compounds – in the surviving, catecholamine-insensitive myocardium after coronary occlusion. J Cardiovasc Pharmacol 4: 542–554
3. Baumann G, Riess G, Erhardt WD, Felix SB, Ludwig L, Blümel G, Blömer H (1981) Impaired beta-adrenergic stimulation in the uninvolved ventricle post-acute myocardial infarction: Reversible defect due to excessive circulating catecholamine-induced decline in number and affinity of beta-receptors. Am Heart J 101: 569–581
4. Bendersky R, Chatterjee K, Parmley WW, Brundage BH, Ports TA (1981) Dobutamine in chronic heart failure: alterations in left ventricular function and coronary hemodynamics. Am J Cardiol 48: 554–558
5. Bristow MR, Ginsburg R, Minobe W, Cubicciotti RS, Sageman WS, Lurie K, Billingham ME, Harrison DC, Stinson EB (1982) Decreased catecholamine sensitivity and β-adrenergic-receptor density in failing human hearts. N Engl J Med 307: 205–211
6. Brown L, Erdmann E (1985) Concentration-response curves of positive inotropic agents before and after ouabain pretreatment. Cardiovasc Res 19: 288–298
7. Brown L, Lorenz B, Erdmann E (1985) The inotropic effects of dopamine and its precursor levodopa on isolated human ventricular myocardium. Klin Wochenschr 63: 1117–1123
8. Conrad KA, Prosnitz EH (1981) Cardiovascular effects of theophylline. Partial attenuation by beta-blockade. Eur J Clin Pharmacol 21: 109–114
9. Drummond RW, Sordahl LA (1981) Temporal changes in adenylate cyclase activity in actuely ischemic dog heart: evidence of functional subunit damage. J Mol Cell Cardiol 13: 323–330
10. Fouad FM, Shimamatsu K, Hanna MM, Khairallah PA, Tarazi RC (1985) Impaired inotropic responses to α-adrenergic stimulation in experimental left ventricular hypertrophy. Circulation 71: 1023–1028
11. Ginsburg R, Bristow MR, Billingham ME, Stinson EB, Schroeder JS, Harrison DC (1983) Study of the normal and failing isolated human heart: Decreased response of failing heart to isoproterenol. Am Heart J 106: 535–540
12. Herzig JW, Feile K, Rüegg JC (1981) Activating effects of AR-L 115 BS on the Ca^{2+} sensitive force, stiffness and unloaded shortening velocity (v_{max}) in isolated contractile structures from mammalian cardiac muscle. Arzneimittel-Forsch 31: 188–191
13. Khatter JC, Hoeschen RJ (1984) Reduced number of digitalis receptor sites in the hypertrophied pig myocardium. Basic Res Cardiol 79: 396–401
14. Klein NA, Sisking SJ, Frishman WH, Sonnenblick EH, LeJemtel TH (1981) Hemodynamic comparison of intravenous amrinone and dobutamine in patients with chronic congestive heart failure. Am J Cardiol 48: 170–175
15. Kumano K, Khairallah PA (1985) Adenylate cyclase activity during development and reversal of cardiac hypertrophy. J Mol Cell Cardiol 17: 537–548

16. LeJemtel TH, Keung E, Ribner HS, Davis R, Wexler J, Blaufox MD, Sonnenblick EH (1980) Sustained beneficial effects of oral amrinone on cardiac and renal function in patients with severe congestive heart failure. Am J Cardiol 45: 123–129
17. Leier CV, Heban PT, Huss P, Bush CA, Lewis RP (1978) Comparative systemic and regional hemodynamic effects of dopamine and dobutamine in patients with cardiomyopathic heart failure. Circulation 58: 466–475
18. Millard RW, Dube C, Grupp G, Grupp I, Alousi A, Schwartz A (1980) Direct vasodilator and positive inotropic actions of amrinone. J Mol Cell Cardiol 12: 647–652
19. Newman WH, Webb JG (1980) A differential inotropic responsiveness to isoprenaline and ouabain in dogs with heart failure. Cardiovasc Res 14: 530–536
20. Prasad K, O'Neil CL, Bharadwaj B (1984) Effects of chronic digoxin treatment on cardiac function, electrolytes, and sarcolemmal ATPase in the canine failing heart due to chronic mitral regurgitation. Am Heart J 108: 1487–1494
21. Saragoca M, Tarazi RC (1981) Impaired cardiac contractile response to isoproterenol in the spontaneously hypertensive rat. Hypertension 3: 380–385
22. Scholz H (1984) Inotropic drugs and their mechanisms of action. JACC 4: 389–397
23. Tse J, Powell JR, Baste CA, Priest RE, Kuo JF (1979) Isoproterenol-induced cardiac hypertrophy: Modifications in characteristics of β-adrenergic receptor, adenylate cyclase, and ventricular contraction. Endocrinology 105: 246–255
24. Vapaatalo H, Säynämälammi P (1980) Effects on the general hemodynamics and peripheral circulation. In: Szekeres L (ed) Handbook of experimental pharmacology, adrenergic activators and inhibitors Vol 54/I. Springer-Verlag, Berlin Heidelberg New York, pp 853–947
25. Will-Shahab L, Krause E-G, Bartel S, Schulze W, Küttner I (1985) Reversible inhibition of adenylate cyclase activity in the ischemic myocardium. J Cardiovasc Pharmacol 7 (Suppl 5): S23–S27
26. Wilmshurst PT, Thompson DS, Juul SM, Jenkins BS, Coltart DJ, Webb-Peploe MM (1984) Comparison of the effects of amrinone and sodium nitroprusside on haemodynamics, contractility, and myocardial metabolism in patients with cardiac failure due to coronary artery disease and dilated cardiomyopathy. Br Heart J 52: 38–48
27. Wilmshurst PT, Walker JM, Fry CH, Mounsey JP, Twort CHC, Williams BT, Davies MJ, Webb-Peploe MM (1984) Inotropic and vasodilator effects of amrinone on isolated human tissue. Cardiovasc Res 18: 302–309

Author's address:

Dr. Lindsay Brown, Medizinische Klinik I der Universität, Klinikum Großhadern, Marchioninistr. 15, 8000 München 70 (West Germany)

The cardiac Na^+/H^+ exchange system. Its role in inotropy

C. Frelin, P. Vigne and M. Lazdunski

Centre de Biochimie, Centre National de la Recherche Scientifique, Université de Nice, Nice (France)

Summary

Cardiac cells possess a very active Na^+/H^+ exchange system which is inhibited by amiloride and its derivatives. Under physiological conditions, the Na^+/H^+ exchange system contributes little to the regulation of the internal pH of chick cardiac cells. The antiporter then serves as an uptake system for Na^+ using the H^+ gradient created by other pH_i regulatory mechanisms. Treatment of cardiac cells with ouabain inhibits Na^+ efflux and produces an increase in intracellular Na^+ activity. Ethylisopropylamiloride was used to show that the Na^+/H^+ exchange system is the main pathway for Na^+ entry and accumulation in digitalis action. Amiloride derivatives which block Na^+ entry via the Na^+/H^+ antiporter antagonize ouabain action on cardiac cells. When the internal pH of cardiac cells is lowered, the Na^+/H^+ exchanger becomes the major pH_i regulating system. It is the essential system by which cardiac cells recover from cellular acidosis. The situation is due both to an increased activity of the exchanger at acidic pH_i and to a decreased activity of other pH_i regulatory systems. Under such conditions, the sensitivity of the cardiac cells to digitalis action is increased.

Introduction

The mechanism by which digitalis exerts its positive inotropic effect on the heart is the following. First digitalis blocks the (Na^+,K^+)-ATPase which leads to a rise in the intracellular Na^+ concentration ($[Na^+]_i$). As a result, the activity of the Na^+/Ca^{2+} exchanger is modified and more Ca^{2+} becomes available for contraction (1, 4, 7, 25, 30, 39). This paper describes the presence in cardiac cells of a Na^+/H^+ exchanger and shows that this system might be important for controlling $[Na^+]_i$ in cardiac cells and hence, the sensitivity to digitalis.

The pharmacological properties of the (Na^+,K^+)-ATPase in embryonic chick cardiac cells

Embryonic chick cardiac cells express two types of ouabain binding sites: a low affinity binding site with a dissociation constant (K_D) of 2–6 μM for the ouabain-receptor complex and a high affinity binding site with a K_D of 26–48 nM (19, 21). In this preparation, the low affinity ouabain binding sites seem to be responsible for both the cardiotonic and cardiotoxic effects of the drug (19, 21).

The Na^+/H^+ antiport: biochemical and pharmacological properties

The activity of the Na^+/H^+ antiport can be analyzed in essentially three ways (i) by measuring $^{22}Na^+$ flux, (ii) by measuring H^+ flux and (iii) by measuring pH_i variations. Detailed kinetic analysis of the Na^+/H^+ exchange system in various cell types (2, 12, 23) as well as in cardiac cells (8, 10, 11, 22, 26) have shown that the system has the following properties:

1) It is electroneutral with a 1:1 stoichiometry.

2) It is reversible. When the Na^+ gradient across the membrane is reversed, Na^+ efflux is accompanied by H^+ entry and cellular acidification, whereas under normal conditions of transmembrane Na^+ and H^+ gradients, Na^+ uptake is accompanied by H^+ efflux and cellular alkalinization.

3) Increasing $[Na^+]_o$ increases the activity of the exchanger. The K_m value for Na^+ is 20 mM. Li^+ is the only monovalent cation that can substitute for Na^+. The K_m value for Li^+ is 8 mM.

4) Increasing the external pH increases the activity of the exchanger. The half maximum activity is obtained at pH = 7.05.

5) Decreasing the internal pH increases the activity of the antiport. In contrast to the external pH dependence of the system, the internal pH dependence is very steep, its activity being all or none in less than a pH unit between 7.0 and 7.6. The pH_i for half maximum activation of the exchanger is 7.35 which is close to the pH_i maintained by cardiac cells under physiological conditions. This property has important physiological consequences.

6) Amiloride, a potent diuretic molecule, inhibits the activity of the Na^+/H^+ antiport. It acts at a Na^+ coordination site. In the presence of a low Na^+ concentration (3 mM), the dissociation constant of the amiloride – Na^+/H^+ antiport complex is 35 μM. The potency of amiloride can be increased as much as 200 fold by substitution of the 5 amino group of the molecule. Typical concentration response curves for the action of amiloride and 3 amiloride derivatives on the Na^+/H^+ antiport of chick cardiac cells are presented in Fig. 1 A. The most active analog, ethylisopropylamiloride, has a $K_{0.5}$ value of 30 nM at 3 mM Na^+. At 140 mM Na^+ this value is 150 nM. The pharmacological properties of the cardiac Na^+/H^+ exchanger are identical to those of the system in non cardiac cells (12, 35, 36). The Na^+/H^+ exchanger is also inhibited by derivatives of guanidinium that are unrelated to the amiloride series like guanochlor, guanoxan, guanisoquine and guanethidine (13). These molecules have potent antihypertensive properties (27).

Amiloride and ethylisopropylamiloride have very little effect on the cardiac Na^+/Ca^{2+} exchange system. In contrast, benzamil and dichlorobenzamil, two derivatives of amiloride with a substituted guanidino moiety are fairly potent blockers of the Na^+/Ca^{2+} exchange system (18, 22, 33) (Fig. 1 B).

Although amiloride is able to block the (Na^+,K^+)-ATPase in kidney cells (32) and in cardiac cells (20), ethylisopropylamiloride is devoid of action on the Na^+ pump at concentrations up to 0.1 mM (41). Therefore ethylisopropylamiloride can be considered to be a specific and high affinity inhibitor of the Na^+/H^+ exchanger without interference with other membrane cation transporting systems.

The antagonistic effect of digitalis compounds and of diuretics of the amiloride series

Amiloride has been shown to alter the responsiveness of the heart to toxic and therapeutic actions of digitalis. It reverses the electrocardiographic changes that are induced by digitalis and protects the heart against digitalis toxicity by delaying the onset of arrhythmias and cardiac arrest (9, 15, 17, 20, 28, 29, 38). The molecular basis for the antagonism between digitalis and amiloride on the heart has been studied using chick cardiac cells in culture (10).

The effect of ouabain on $[Na^+]_i$ is small when the cells are incubated in a medium at pH 6.0. It is increased when the incubation medium is made more alkaline. The pH dependence of ouabain action on $[Na^+]_i$ is parallel to the external pH dependence of the Na^+/H^+ exchanger and is suppressed in the presence of ethylisopropylamiloride. These two obser-

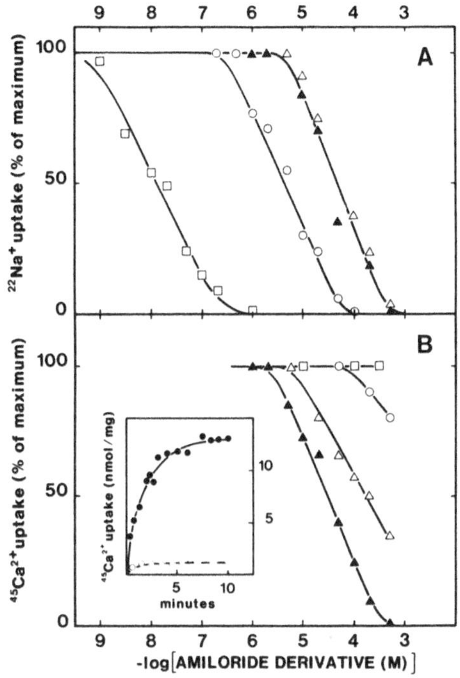

Fig. 1. The pharmacological properties of the Na^{+}/H^{+} antiport and the Na^{+}/Ca^{2+} antiport in chick cardiac cells.
Panel A: Concentration response curves for ethylisopropylamiloride (□), amiloride (○), benzamil (△) and dichlorobenzamil (▲) inhibition of the Na^{+}/H^{+} antiporter. External Na^{+}: 3 mM, external pH: 7.4. Experiments were performed in a Ca^{2+}-free medium to prevent the operation of the Na^{+}/Ca^{2+} exchange system. *Panel B:* Concentration reponse curves for ethylisopropylamiloride (□), amiloride (○), benzamil (△) and dichlorobenzamil (▲) inhibition of the initial rate of $^{45}Ca^{2+}$ uptake through the Na^{+}/Ca^{2+} exchange system. Chick cardiac cells were loaded with Na^{+} by equilibrating them in a Ca^{2+}-free, 140 mM Na^{+} Earle solution supplemented with 1 mM EGTA, 0.1 mM monensin and 0.2 mM ouabain and then shifted to a Na^{+}-free, 140 mM choline chloride medium that contained 1.8 mM Ca^{2+}, 0.2 mM ouabain and 5 μCi/ml of $^{45}Ca^{2+}$. *Inset:* Time-course of $^{45}Ca^{2+}$ uptake by chick cardiac cells that have been loaded with Na^{+} (●) or Li^{+} (○) in the presence of monensin. *Main panel:* Concentration response curves for amiloride derivatives inhibition of the initial rate of $^{45}Ca^{2+}$ uptake by Na^{+}-loaded cardiac cells. The rate of $^{45}Ca^{2+}$ uptake obtained after having loaded cells with Li^{+} was subtracted from all experimental values.
Reproduced from Lazdunski et al (1985).

vations suggest that at a normal physiological value of the pH_i, the Na^{+}/H^{+} antiport is a major uptake pathway for Na^{+} in quiescent chick cardiac cells.

The blockade of the (Na^{+},K^{+})-ATPase with ouabain promotes $^{45}Ca^{2+}$ uptake by chick cardiac cells (14, 19). Ethylisopropylamiloride, which prevents the rise in $[Na^{+}]_i$ under ouabain treatment, also prevents the increase of $^{45}Ca^{2+}$ uptake by ouabain treated cardiac cells.

In fact, in the presence of ethylisopropylamiloride, the concentration response curve for ouabain activation of the initial rate of $^{45}Ca^{2+}$ uptake is shifted to higher concentrations. The concentration response curve for the inhibition by ethylisopropylamiloride of the ouabain induced Ca^{2+} uptake is superimposable with the concentration response curve for ethylisopropylamiloride inhibition of the Na^+/H^+ antiport. Considering further that ethylisopropylamiloride does not interact with the Na^+/Ca^{2+} exchanger (10) (Fig. 1 B), it is reasonable to assume that the effect of ethylisopropylamiloride on ouabain action is primarily due to its inhibitory action on the Na^+/H^+ exchanger.

In sheep Purkinje fibers, Deitmer and Ellis (6) found that amiloride had little or no effect on the intracellular Na^+ activity changes induced by strophanthidin. This result different from the chick cardiac cells is due to the fact that the activity of the Na^+/H^+ exchange system in chick cardiac cells is about 10 times higher than in Purkinje fibers (26). It indicates that it is the relative activities of the Na^+/H^+ antiport and of the Na^+/Ca^{2+} exchanger that determines the importance of the interaction of digitalis and diuretics on the heart. Another important difference between chick cardiac cells and sheep Purkinje fibers is that ouabain produced a cell acidification in sheep Purkinje fibers (34) whereas it has no action on the pH_i of chick cardiac cells.

The coupling of cation exchange mechanisms during the recovery from an intracellular acidosis

One way of increasing the activity of the Na^+/H^+ antiport i. e. of modifying the coupling of the different cation exchange mechanisms in the membrane is to produce a cellular acidification. A cellular acidification to pH 6.65 increases the activity of the Na^+/H^+ exchanger about two fold. The increased activity of the antiporter allows pH_i recovery at the expense of the Na^+ gradient as evidenced by a transient increase in $[Na^+]_i$. This in turn activates the (Na^+,K^+)-ATPase to restore low $[Na^+]_i$. The activation of the (Na^+,K^+)-ATPase is evidenced by a 3 fold increase in the rate of ouabain sensitive $^{86}Rb^+$ uptake. The observation that ethylisopropylamiloride prevents the activation of the Na^+/H^+ antiport and also the activation of the (Na^+,K^+)-ATPase in cardiac cells that recover from an intracellular acidosis clearly indicates that the activation of the (Na^+,K^+)-ATPase is secondary to the activation of the Na^+/H^+ exchanger.

An activation of the rate of $^{45}Ca^{2+}$ uptake was not observed in chick cardiac cells that recover from an intracellular acidosis. This suggests that the activity of the (Na^+,K^+)-ATPase is sufficient to prevent an increase in $[Na^+]_i$ large enough to activate the Na^+/Ca^{2+} exchanger. However, when the reserve capacity of the (Na^+,K^+)-ATPase is reduced with ouabain, the recovery from an intracellular acidosis is followed by a 5-fold increase in the rate of $^{45}Ca^{2+}$ uptake which is through the Na^+/Ca^{2+} exchanger. Ethylisopropylamiloride, which prevents pH_i recovery and the activation of the (Na^+,K^+)-ATPase, also suppresses the activation of $^{45}Ca^{2+}$ uptake by ouabain treated cardiac cells that have been acidified.

Taken all together, these results suggest that the sensitivity of the cardiac cells to digitalis actions depends on the activity of the Na^+/H^+ exchange system. Under conditions in which the system is relatively active, e. g. when cardiac cells recover from an intracellular acidosis, cells are very sensitive to ouabain action. Under conditions in which the system is not active, e. g. after blockade with ethylisopropylamiloride, cells are less sensitive to digitalis action (10). Such a coupling of the different membrane cation transporting systems might have important consequences in the generation of arrhythmias during the recovery from an ischemic episode (22).

Protein kinase C regulates the activity of the Na^+/H^+ exchanger in chick cardiac cells

In a number of cell types, protein kinase C seems to regulate the activity of the Na^+/H^+ exchanger. This is evidenced by the finding that phorbol esters, which are potent activators of protein kinase C and diacylglycerol, the natural activator of protein kinase C both increase the activity of the Na^+/H^+ exchanger and produce a cell alkalinisation (3, 5, 16, 24, 33, 37, 40). The same regulation has been observed in chick cardiac cells. Treatment of cardiac cells with 0.1 μM phorbol myristate acetate produces an activation of the Na^+/H^+ exchanger and a 0.2 to 0.3 pH unit cell alkalinisation. The effect of phorbol esters is to shift the pH_i dependence of the Na^+/H^+ exchanger to more alkaline pH_i values, as previously reported in rat thymocytes (16) and rat skeletal myoblasts (37).

Acknowledgements

This work was supported by grants from the "Centre National de la Recherche Scientifique", the "Fondation sur les Maladies Vasculaires", the "Fondation pour la Recherche Médicale", the "Institut National de la Santé et de la Recherche Médicale" (grant no. 80.50.09).

The authors are grateful to Dr. E. J. Cragoe for supplying us with generous quantities of amiloride and amiloride derivatives and to N. Boyer, M. T. Ravier and M. Valetti for expert technical assistance.

References

1. Akera T, Brody TM (1977) The role of (Na^+,K^+)-ATPase in the inotropic action of digitalis. Pharmacol Rev 29: 187–220
2. Aronson PS (1985) Kinetic properties of the plasma membrane Na^+/H^+ exchange. Ann Rev Physiol 47: 545–560
3. Berterman JM, Cuatrecasas P (1984) Phorbol esters rapidly stimulate amiloride-sensitive Na^+/H^+ exchange in a human leukemic cell line. J Cell Biol 99: 340–343
4. Biedert S, Barry WH, Smith TW (1979) Inotropic effects and changes in sodium and calcium contents associated with inhibition of monovalent cation active transport by ouabain in cultured myocardial cells. J Gen Physiol 74: 479–494
5. Burns CP, Rozengurt E (1983) Serum, platelet derived growth factor, vasopressin and phorbol esters increase intracellular pH in Swiss 3T3 cells. Biochem Biophys Res Comm 116: 931–938
6. Deitmer JW, Ellis D (1980) Interaction between the regulation of the intracellular pH and sodium activity of sheep Purkinje fibres. J Physiol (Lond) 304: 471–488
7. Eisner DA, Lederer WJ (1982) The effects of Na^+ pump activity on the slow inward calcium current in sheep cardiac Purkinje fibres. Proc R Soc B 214: 249–262
8. Ellis D, Macleod KT (1985) Sodium dependent control of intracellular pH in Purkinje fibres of sheep heart. J Physiol (Lond) 359: 81–105
9. Floreani M, Luciani S (1984) Amiloride: relationship between cardiac effects and inhibition of Na^+/Ca^{2+} exchange. Eur J Pharmacol 105: 317–322
10. Frelin C, Vigne P, Lazdunski M (1984) The role of the Na^+/H^+ exchange system in cardiac cells in relation to the control of the internal Na^+ concentration. J Biol Chem 259: 8880–8885
11. Frelin C, Vigne P, Lazdunski M (1985) The role of the Na^+/H^+ exchange system in the regulation of the internal pH in cultured cardiac cells. Eur J Biochem 149: 1–4
12. Frelin C, Vigne P, Lazdunski M (1985) The Na^+/H^+ exchange system. Properties and role in cell function. In: Dumont JE, Hamprecht B, Nunez J (eds) Hormones and cell regulation. Elsevier Science Publishers, pp 259–268
13. Frelin C, Vigne P, Lazdunski M (1986) Interaction of guanidinium and guanidinium derivatives with the Na^+/H^+ exchange system. Eur J Biochem (in press)
14. Goshima K, Wakabayashi S (1981) Involvement of a Na^+-Ca^{2+} exchange system in genesis of ouabain-induced arrhythmias of cultured myocardial cells. J Mol Cell Cardiol 13: 489–509
15. Greeff K, Kohler E (1975) Animal experiments on the effect of triamterene and amiloride in heart and circulation and the toxicity of digoxin. Arzneimittelforschung 25: 1766
16. Grinstein S, Cohen S, Goetz JD, Rothstein A, Gelfand EW (1985) Characterization of the activation of the Na^+/H^+ exchange in lymphocytes by phorbol esters: change in cytoplasmic pH dependence of the antiport. Proc-Natl Acad Sci (USA) 82: 1429–1433

17. Jounella A, Pyorala K (1975) Effect of amiloride on digitalis-induced electrocardiographic changes. Ann Clin Res 7: 65–70
18. Kaczorowski GJ, Barros F, Dethmers JK, Trumble MJ, Cragoe EJ (1985) Inhibition of Na^+/Ca^{2+} exchange in pituitary plasma membrane vesicles by analogues of amiloride. Biochemistry 24: 1394–1403
19. Kazazoglou T, Renaud JF, Rossi B, Lazdunski M (1983) Two classes of ouabain receptors in chick ventricular cardiac cells and their relation to (Na^+,K^+)-ATPase inhibition. Intracellular Na^+ accumulation, Ca^{2+} influx and cardiotonic effect. J Biol Chem 258: 12163–12170
20. Kennedy RH, Akera T, Brody TM (1985) Suppression of positive inotropic and toxic effects of cardiac glycosides by amiloride. Eur J Pharmacol 115: 199–210
21. Lazdunski M, Kazazoglou T, Renaud JF, Rossi B (1984) Digitalis receptors in cardiac cells and their relation with positive inotropic and cardiotoxic effects. Eur Heart J 5: F281–F290
22. Lazdunski M, Frelin C, Vigne P (1985) The Na^+/H^+ exchange system in cardiac cells. Its biochemical and pharmacological properties and its role in regulating internal concentrations of Na^+ and internal pH. J Moll Cell Cardiol 17: 1029–1042
23. Mahnensmith RL, Aronson PS (1985) The plasma membrane Na^+/H^+ exchanger and its role in physiological and pathological processes. Circ Res 57: 773–788
24. Moolenaar WH, Tertoolen LGJ, Delaat SW (1984) Phorbol ester and diacylglycerol mimic growth factors in raising cytoplasmic pH. Nature 312: 371–374
25. Noble D (1980) Mechanism of action of therapeutic levels of cardiac glycosides. Cardiovasc Res 14: 495–519
26. Piwnica-Worms D, Jacob R, Horres CR, Lieberman M (1985) Na^+/H^+ exchange in cultured chick cardiac cells. pH_i regulation. J Gen Physiol 85: 43–64
27. Scriabine A (1980) Pharmacology of antihypertensive drugs. Raven Press, New York
28. Seller RH, Banach S, Namey T, Neff M, Swartz (1975) Cardiac effects of diuretic drugs. Am Heart J 89: 493–500
29. Seller RH, Greco J, Banach S, Seth R (1975) Increasing the inotropic effect and toxic dose of digitalis by the administration of antikaliuretic drugs – further evidence for a cardiac effect of diuretic agents. Am Heart J 90: 56–67
30. Sheu SS, Fozzard HA (1982) Transmembrane Na^+ and Ca^{2+} electrochemical gradients in cardiac muscle and their relationship to force development. J Gen-Physiol 80: 325–351
31. Siegl PKS, Cragoe EJ, Trumble MJ, Kaczarowski GJ (1984) Inhibition of Na^+/Ca^{2+} exchange in membrane vesicles and papillary muscle preparations from guinea-pig heart by analogs of amiloride. Proc Natl Acad Sci (USA) 81: 3238–3242
32. Soltoff SP, Mandel LJ (1983) Amiloride directly inhibits the (Na^+,K^+)-ATPase activity of rabbit proximal tubules. Science 220: 957–959
33. Vara F, Rozengurt E (1985) Stimulation of Na^+/H^+ antiport activity by epidermal growth factor and insulin occurs without activation of protein kinase C. Biochem Biophys Res Comm 130: 646–653
34. Vaughan-Jones RD, Lederer WJ, Eisner DA (1983) Ca^{2+} ions can affect intracellular pH in mammalian cardiac muscle. Nature 301: 522–525
35. Vigne P, Frelin C, Cragoe EJ, Lazdunski M (1983) Ethylisopropylamiloride: a new and highly potent derivative of amiloride for the inhibition of the Na^+/H^+ exchange system in various cells types. Biochem Biophys Res Comm 116: 86–90
36. Vigne P, Frelin C, Cragoe EJ, Lazdunski M (1984) Structure-activity relationship of amiloride and certain of its analogues in relation to the blockade of the Na^+/H^+ exchange system. Mol Pharmacol 25: 131–136
37. Vigne P, Frelin C, Lazdunski M (1985) The Na^+/H^+ antiport is activated by serum and phorbol esters in proliferating myoblasts but not in differentiated myotubes. Properties of the activation process. J Biol Chem 260: 8008–8013
38. Waldorff S, Hansen PB, Kjaergard H, Buch J, Egeblad H, Steiness E (1981) Amiloride induced changes in digoxin dynamics and kinetics. Abolition of digoxin-induced inotropism with amiloride. Clin Pharmacol Ther 30: 172–176
39. Weingart R, Kass RS, Tsien RW (1978) Is digitalis inotropy associated with enhanced slow inward calcium current? Nature London 273: 389–392

40. Whiteley B, Cassel D, Zhuang YX, Glaser L (1984) Tumor promoter phorbol 12-myristate 13-acetate inhibits mitogen stimulated Na^+/H^+exchange in human epidermoid carcinoma A 431 cells. J Cell Biol 99: 1162–1166
41. Zhuang YX, Cragoe EJ, Shaikewitz T, Glaser L, Cassel D (1984) Characterization of potent Na^+/H^+ exchange inhibitors from the amiloride series in A 431 cells. Biochemistry 23: 4481–4488

Author's address:

C. Frelin, Centre de Biochimie, Centre National de la Recherche Scientifique, Université de Nice, Parc Valrose, 06034 Nice Cedex (France)

Evidence for pharmacodynamic differences between cardiac glycosides of different polarities

P. H. Joubert

Department of Pharmacology and Therapeutics, Medunsa (South Africa)

Summary

We compared the effects of digitoxin (lipophilic) and ouabain (hydrophilic) in rats under pentobarbital anaesthaesia on inotropy (dp/dt) and atrioventricular conduction (PR-interval). Doses of the glycosides which produced a 40–50 % increase in dp/dt produced a significantly greater prolongation of the PR interval with ouabain than with digitoxin. Neither vagotomy nor atropine administration abolished this difference, but the difference could be eliminated by pre-treatment with propranolol. We concluded that the smaller effect of digitoxin on atrioventricular (AV) conduction was due to a central sympathomimetic effect which does not occur with ouabain.

A respiratory depressant effect of digitalis has been reported in humans. We therefore compared the effect of acetyldigitoxin (relatively lipophilic) to that of digoxin (relatively hydrophilic) on the hyperventilatory response to inhaled CO_2 in healthy volunteers. We found that at mean steady state levels of 12.5 ± 1.07 ng/ml (digitoxin) and 0.87 ± 0.2 ng/ml (digoxin), digitoxin appeared to stimulate ventilation as measured by minute ventilation and oxygen consumption and digoxin appeared to depress ventilation. These observations appear to be consistent with a sympathomimetic effect of the acetyldigitoxin as observed in our animal experiments. It might also be consistent with a greater vagomimetic effect with digoxin as suggested by Runge in 1977. Further research is currently being done to elucidate the mechanisms of the observed differences.

These findings suggest that when cardiac glycosides are used for supraventricular tachyarrhythmias a more hydrophilic cardiac glycoside might be indicated. If on the other hand digitalis is used for its positive inotropic effect, the use of a lipophilic cardiac glycoside might reduce the risk of atrioventricular block. It also seems that if digitalis is used in patients with respiratory impairment, a more lipophilic cardiac glycoside might be preferable.

It is well recognised that the cardiac glycosides have specific pharmacological effects. These include negative chronotropism, positive inotropism and slowing of atrioventricular conduction. The latter two properties are used clinically in the management of heart failure and supraventricular tachyarrhythmias respectively. Furthermore, it is generally accepted that these effects are common to all cardiac glycosides and that the only differences between cardiac glycosides are of a pharmacokinetic nature, resulting in differences in duration and onset of action. This notion is shown very clearly in a widely accepted text such as that by Goodman and Gilman (5) which when considering the choice of preparation discusses speed of onset, route and duration of effects, but not the desired pharmacological effect.

Investigators such as Eichna and Taube (4), Batterman, de Graff and Gutner (3) and Aravnis and Luisada (2) have compared the clinical effects of cardiac glycosides, but these investigations fall far short when seen in the light of present-day concepts of controlled clinical trials. It was however suggested that digoxin produced a greater slowing of heart rate than digitoxin (2).

Cardiac effects

About a decade ago Runge (9, 10) reported that when heart rate and systolic time intervals (QS_2c) were measured in guinea-pigs, digitoxin had a greater positive inotropic effect for the same negative chronotropism than ouabain. He suggested that hydrophilic cardiac glycosides such as ouabain were relatively more vagomimetic, whereas lipophilic cardiac glycosides such as digitoxin were more sympathomimetic. He therefore suggested that polar cardiac glycosides might be preferable in supraventricular tachyarrhythmias, whereas non-polar cardiac glycosides might be preferable when positive inotropism is desired.

Prompted by these results, we compared the effects of ouabain and digitoxin in rats with regard to dp/dt in the left ventricle and PR-interval on the electrocardiogram. We felt that dp/dt would be a more direct measurement of contractility than QS_2c and that PR-interval (an index of atrioventricular conduction) would be more appropriate than effects on heart rate. The methodology and results were reported more fully elsewhere (6).

In brief, intravenous doses of cardiac glycoside were chosen to produce a 40% increase in contractility. As can be seen in Fig. 1, we found that with similar positive inotropic responses, the PR-interval prolongation was much greater with ouabain than with digitoxin.

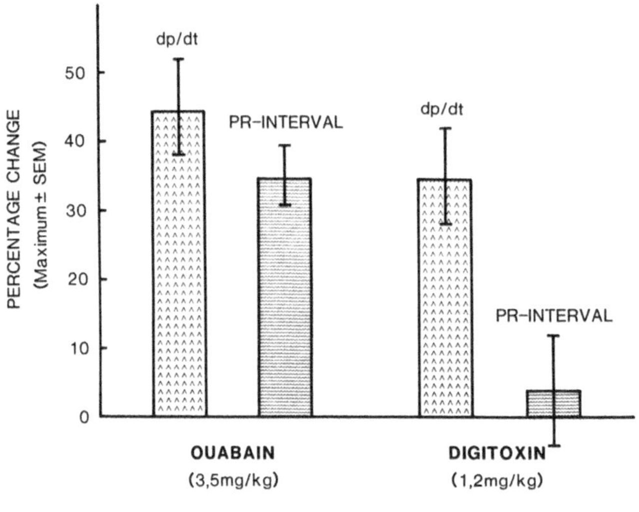

Fig. 1. Effect of intravenous ouabain and digitoxin on PR-interval and dp/dt in Wistar rats.

To try and elucidate the mechanism we firstly examined the effect of abolishing vagal influences by performing a surgical vagotomy in the neck. This produced a significant decrease in the effect of ouabain on the PR interval, but the residual effect was still markedly greater than with digitoxin (Fig. 2). When we investigated the effect of pre-treatment with propranolol, we found that the difference between ouabain and digitoxin with respect to AV-conduction was abolished (Fig. 2). We concluded that the differences that we observed fitted into the hypothesis suggested by Runge, namely that polar cardiac glycosides are relatively more vagomimetic whereas lipophilic ones are relatively more sympathomimetic.

Recently Scott Cook et al. (11) have shown that digitoxigenen in addition to its direct positive inotropic effect, affects contractility through adrenergic mechanisms, in dogs. They

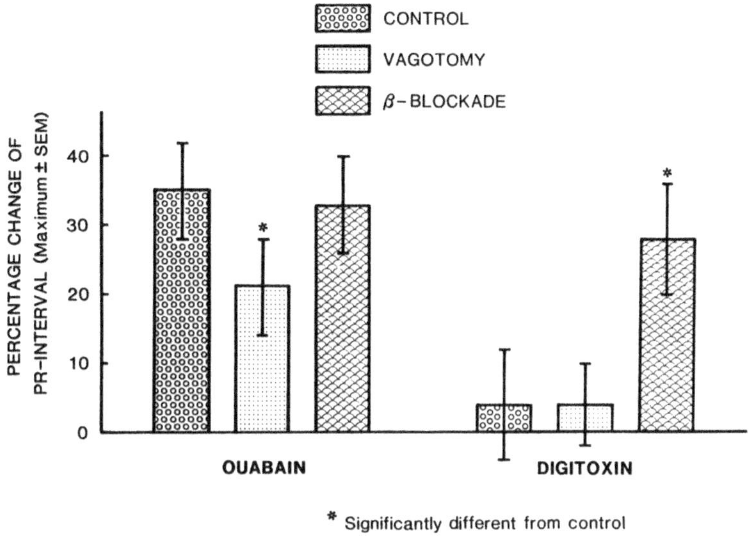

Fig. 2. Effect of vagotomy or beta-adrenergic blockade on PR-interval response to intravenous ouabain or digitoxin in Wistar rats.

concluded that alterations in the aglycone and sugar moieties can significantly alter the contribution of the autonomic nervous system to contractile response.

Evidence for pharmacodynamic differences has been found in man in studies by Alken and Belz (1). They found that with equipotent positive inotropic doses of digoxin and meproscillarin (relatively more lipophilic), the former produced a greater prolongation of the PQ-time than meproscillarin. This is in accordance with the notion that digoxin would be relatively more vagomimetic and meproscillarin relatively more sympathomimetic.

Extra-cardiac effects

Aravanis and Luisada (2) reported more nausea and vomiting with digoxin than digitoxin. Runge (9) considers this in his paper and suggests that if nausea and vomiting (vagomimetic effects) occur on digitalis therapy, the patient might do better on a more lipophilic cardiac glycoside, without loss of positive inotropic effect. Support for this comes from the paper by Alken and Belz (1), who found a relatively higher incidence of nausea (vagomimetic) with the more hydrophilic digoxin and a relatively higher incidence of diarrhoea (sympathomimetic) with the relatively more lipophilic meproscillarin.

We investigated the effects of two cardiac glycosides on the hyperventilatory response to inhaled CO_2. We found (7) that the relatively polar digoxin produced a decrease in the ventilatory response to inhaled 2.2% CO_2 (Fig. 3), as well as a depression in oxygen consumption, presumably because of decreased work by the respiratory muscles. The relatively more lipophilic acetyldigitoxin, which becomes deacylated to digitoxin, enhanced the hyperventilatory response to CO_2 and produced an increase in O_2 consumption, presumably due to increased work by the respiratory muscles. Since respiratory stimulation may be the result of a sympathomimetic effect and respiratory depression the results of a vagomimetic effect, our results are in accordance with the hypothesis already outlined.

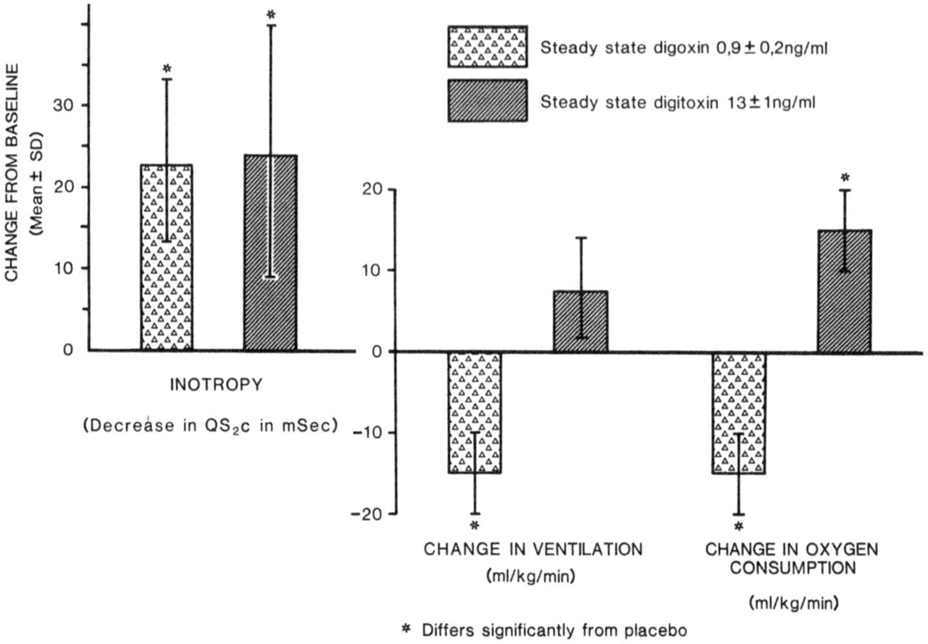

Fig. 3. Inotropic and ventilatory responses to steady state concentrations of digoxin and digitoxin.

Conclusions

From the data presented, there appears to be good evidence suggesting pharmacodynamic differences between cardiac glycosides of different polarities. Hydrophilicity or lipophilicity appear to determine the autonomic effects in terms of relative sympathomimetic or vagomimetic dominance. Autonomic influences seem to affect both cardiac and extracardiac effects.

More hydrophilic cardiac glycosides such as ouabain and to a lesser extent digoxin appear to produce a greater decrease in heart rate and greater increase in atrioventricular conduction than lipophilic ones such as meproscillarin and digitoxin. This appears to be due to a relatively greater vagomimetic effect with the polar glycosides and a relatively greater sympathomimetic effect with the non-polar glycosides. These differences also appear to influence extracardiac effects such as nausea, vomiting, diarrhoea and ventilation.

It is clear that more human research is necessary, particularly in clinical settings. Available evidence, however, suggests that hydrophilic cardiac glycosides might be preferable in the management of heart failure with sinus rhythm, particularly if there is a risk of excessive bradycardia or atrioventricular block. It is also interesting to speculate that the respiratory depressant effect of hydrophilic glycosides might be detrimental in the presence of pulmonary disease and that under these circumstances a liphophilic cardiac glycoside would be preferable.

References

1. Alken R, Belz G (1984) A comparative dose-effect study with cardiac glycosides assessing cardiac and extracardiac responses in normal subjects. J Cardiovasc Pharmacol 6: 634–640

2. Aravanis C, Luisada A (1958) Clinical comparison of six digitalis preparations by the parenteral route. Am J Cardiol 74: 706–716
3. Batterman RC et al (1950) The comparative value of gitalin and other digitalis preparations. Fed Proc 9: 256–261
4. Eicha LW, Taube H (1943) A comparison of the actions of four cardiac glycosides on a patient with congestive heart failure. Am Heart J 26: 631–644
5. Hoffman BF, Bigger JT (1980) In: Goodman and Gilman's The Pharmacological Basis of Therapeutics, 6th Edition. MacMillan, New York, pp 748–749
6. Joubert P, van der Meer L (1979) A specific cardiac glycoside for cardiac failure and another for atrial fibrillation. S Afr Med J 56: 1040–1042
7. Joubert PH et al (1985) Effects of digoxin and acetyl-digitoxin on basal and CO_2-stimulated ventilation. Eur J Clin Pharmacol 28: 155–158
8. Riggert H, Schwab M (1955) Der Einfluß von Digitoxin auf die Atmung des herzgesunden Menschen. Z ges Exp Med 125: 80–92
9. Runge T (1977) Clinical implications of differences in pharmacodynamic action of polar and non-polar cardiac glycosides. Am Heart J 93: 248–255
10. Runge T et al (1975) Pharmacodynamic distinctions between ouabain, digoxin and digitoxin. Arch Int Pharmacodyn 214: 31–45
11. Scott Cook L et al (1982) Adrenergic and cholinergic mechanisms in digitalis inotropy. J Pharm Exp Therap 223: 761–765

Author's address:

Prof. P. H. Joubert, Department of Pharmacology and Therapeutics, Medical University of Southern Africa, PO Medunsa 0204 (South Africa)

The positive inotropic action of mercury compounds on isolated heart muscle

S. Halbach and G. Schönsteiner*)

Abteilung Pharmakologie, Gesellschaft für Strahlen- und Umweltforschung München, and *) Institut für Pharmakologie der Technischen Universität München (West Germany)

Summary

The direct positive inotropic effect of Hg compounds on guinea-pig papillary muscle was investigated. The inhibition by these agents of the $Na^+ + K^+$-dependent ATP hydrolysis by sarcolemmal particles suggests the participation of the ATPase in the positive inotropic Hg effect. This effect was modified by variations of the stimulation rate and of external Na^+ and K^+ concentration in a similar way as the increase in contraction force by cardioactive glycosides. The antagonism between external K^+ and the positive inotropic action of mercurials, however, is unlikely to result from an influence of the cation on the formation of the mercaptide bond.

Introduction

The therapeutic use of mercury in potent diuretics and local antiseptics is at variance with its toxic effects in industrial and fungicidal applications. In both cases, mercury compounds exert their biological actions primarily by forming stable mercaptide bonds with protein SH-groups (14). It is known that mercurials deplete endogenous noradrenaline from adrenal medullary preparations (4) and that they are potent inhibitors of membrane ATPases (20). Both effects suggest an influence of these agents on the contractility of the isolated guinea-pig papillary muscle.

The selected agents were $HgCl_2$ as an inorganic compound; chlormerodrin, a diuretic mercurial; bromomercurihydroxypropane (BMHP), a compound used diagnostically for spleen and kidney scintigrams (7); p-chloromercuribenzoic acid (PCMB) and p-chloromercuriphenylsulfonic acid (PCMBS), a frequently employed SH-inhibitor (23), and methylmercury, an epidemiologically important pollutant and fungicide (3).

Methods

Papillary muscles with a diameter below 1 mm were excised from the right ventricle of guinea-pigs (250–350 g). They were suspended vertically in a two-chambered vessel with circulating incubation medium (19) and were prestretched with a resting force of 3.9 mN. The preparations were electrically stimulated once per s by rectangular pulses of 1 ms duration at an intensity slightly above stimulation threshold. An inductive force transducer (Q11/10p, Hottinger-Baldwin Meßtechnik, Darmstadt, F.R.G.) fed the signal of the isometric contraction force into an oscilloscope and a strip-chart recorder. The incubation medium was composed of (in mmol/l): NaCl 114.9; KCl 4.7; $CaCl_2$ 2.4; $NaHCO_3$ 24.9; KH_2PO_4 1.2; $MgSO_4$ 1.2; glucose 10.

The medium was continuously gassed and circulated by 95 % O_2 and 5 % CO_2 at 35° C. At the beginning of each experiment, the muscles were equilibrated for 1 h at a frequency of 1 Hz. Endogenous catecholamine was depleted by pretreatment of the animals with reserpine (5 mg/kg i. p.) 24 h before the experiment.

Sarcolemmal particles were prepared according to the procedure of Tada et al. (21). This method was subjected to some modifications of which the most important is the insertion of an extraction step with 0.4 mol/l LiBr. Incubations of the isolated particles with ATP and the mercurials were performed as described previously (13). Inorganic phosphate (P_i) was measured according to Fiske and Subbarow (8) and corrected for non-enzymic hydrolysis by a blank without enzyme. Protein concentration was estimated by the procedure of Lowry et al. (15). The specific activity of the $Na^+ + K^+ + Mg^{2+}$-activated ATPase is indicated in μmol P_i/mg protein/h.

PCMB and PCMBS were obtained from Sigma Chemical Co., St. Louis, U.S.A. Chlormerodrin was donated by Kali-Chemie, Hannover, F.R.G. BMHP was a generous gift from Dr. Bastian, Radiochemical laboratories of Behringwerke AG, Frankfurt, F.R.G. Methylmercuric chloride, $HgCl_2$, reserpine-HCl and all other chemicals were supplied in highest purity by Merck, Darmstadt, F.R.G.

Results

A common pattern of action emerged with all Hg compounds and consisted of a negative inotropic response followed by transient increase and a final decrease in contractility (Fig. 1). Action-potential measurements revealed that the negative inotropic effect resulted most likely from an impairment of the ionic conductivity of the sarcolemma. Addition of (±)-propranolol (1 μmol/l) to the incubation medium prior to that of mercury or pretreatment of the animals with reserpine increased the negative inotropic effect and reduced the positive inotropic response to Hg. This revealed the Hg-mediated contribution of catecholamines to the inotropic actions of the Hg compounds.

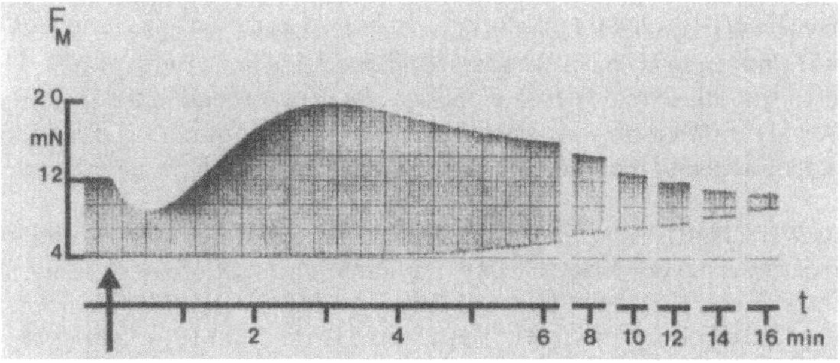

Fig. 1. The effect of $HgCl_2$ (1 mmol/l) on the contractility of the guniea-pig papillary muscle. F_M = total force = contraction force + resting force. Stimulation rate 1 Hz; no reserpine pretreatment.

The direct positive inotropic Hg effect remaining after reserpine pretreatment was investigated further. For this purpose, the influence of external cation concentration and stimulation frequency on the effect of mercury on contractility was studied. Upon lowering the concentration of extracellular sodium, the development of the positive inotropic effect was delayed and maximal contractility was diminished (Fig. 2; in these experiments the loss of osmolarity was compensated by the addition of sucrose). On the other hand, when the external potassium concentration was decreased, the time course of the positive inotropic Hg effect was accelerated (Fig. 3). Furthermore, the increase in contraction force was retarded by slowing the contraction frequency; this became particularly apparent at low Hg con-

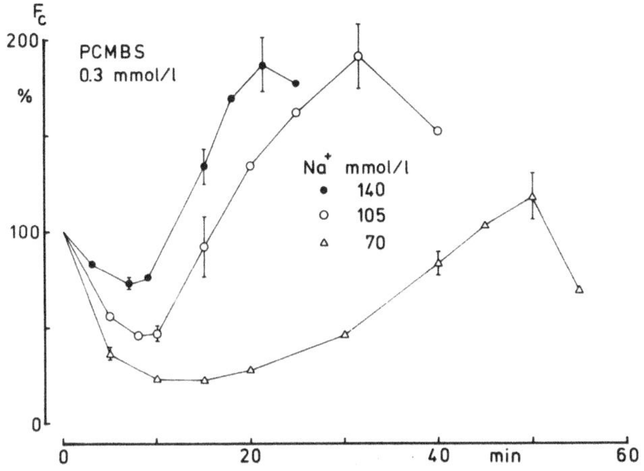

Fig. 2. The development of the positive inotropic effect of PCMBS as influenced by the external sodium concentration. The loss of osmolarity was compensated by the addition of 68.6 mmol/l of sucrose (Na^+ 105) or of 137.2 mmol/l (Na^+ 70). Ordinate: force of contraction in percent of the predrug value. Abscissa: time, min; n = 9 (Na^+ 140), otherwise n = 6 per group. Muscles from reserpine-pretreated animals, stimulation frequency 1 Hz, 5.9 mmol/l K^+.

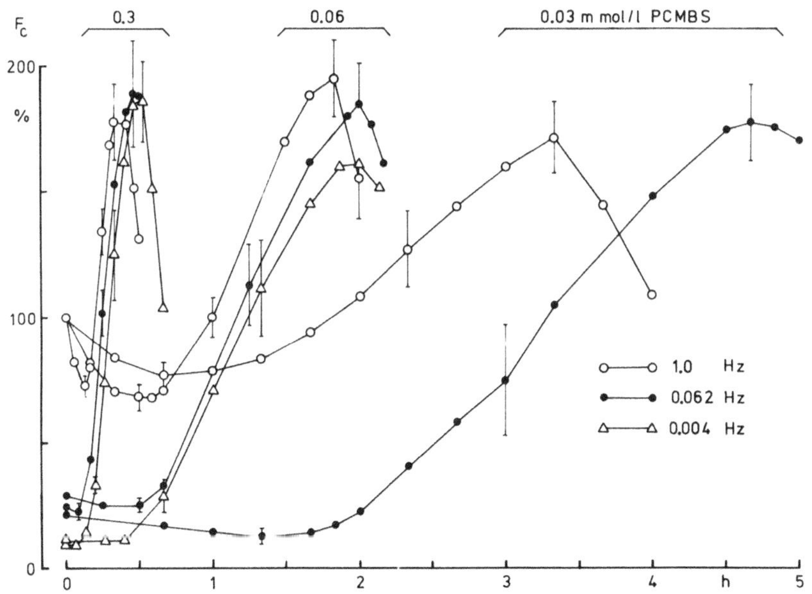

Fig. 3. The action of PCMBS on the time course of its inotropic effects dependent upon external K^+ concentration. Ordinate: normalized force of contraction. Abscissa: time in min, n = 9 per group. Preparations from reserpine-pretreated animals, stimulation rate 1 Hz.

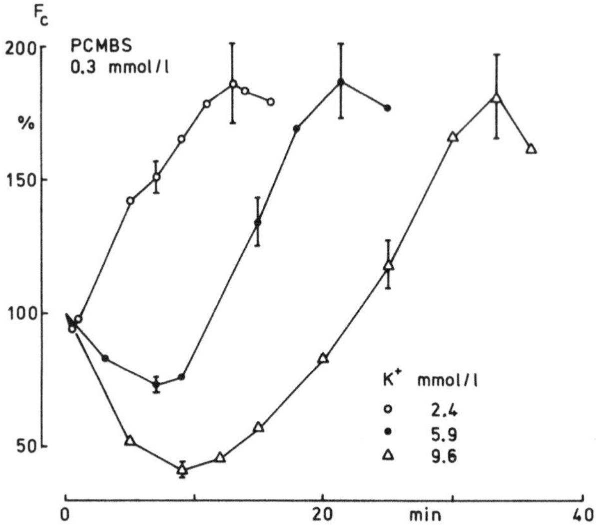

Fig. 4. The effect of PCMBS on the time course of force development in dependence on stimulation frequency. Ordinate: force of contraction in percent of the predrug value at 1 Hz. Abscissa: time in min. The values at zero time correspond to the contraction force at the specified frequency immediately prior to the addition of the mercurial. n = 6–10 per group of muscles. Muscles from reserpine-pretreated animals, 5.9 mmol/l K^+.

centrations (Fig. 4). However, the responsiveness of the positive inotropic Hg effect to these modifications varied markedly between the different compounds. For several agents, the influence of the change in cation concentration could be demonstrated only at a reduced stimulation rate. Thus, the sensitivity of the positive inotropic effect could be rated on an arbitrary scale (Table 1). It is of note that the sensitivity paralleled the hydrophilicity of the mercurials as determined from their octanol/water distribution coefficient (12).

Table 1. Susceptibility of the positive inotropic Hg effect to inotropic interventions.
The hydrophilicity (c_{HOH}/c_{oct}) is the reciprocal value of the octanol/water distribution coefficient (3).

Hg compound	Variation of K^+	Variation of Na^+	Variation of stimul. rate	Rel. sensitivity of pos. effect	c_{HOH}/c_{oct}
PCMBS	++	++	++	6	312.5
PCMB	++	++	++	6	34.5
Chlormerodrin	+	++	0	3	6.3
$HgCl_2$	++	0	0	2	1.6
BMHP	+	+	0	2	1.0
Methylmercury	+	0	0	1	0.4

++) positive inotropic Hg effect responds to variation of particular condition

+) positive inotropic Hg effect responds to cation variation only at reduced stimulation rate (0.062–0.125 Hz)

With regard to the above modifications, the positive inotropic effect of Hg compounds showed many similarities to that of the cardioactive glycosides. It was, therefore, of interest to examine the effect of mercurials on the ATP hydrolysis by isolated sarcolemmal particles. PCMBS reduced the activity of the $Na^+ + K^+ + Mg^{2+}$-ATPase to a greater extent than a maximally effective concentration of ouabain (Fig. 5); this was probably due to the inhibition of the Mg^{2+}-dependent ATPase activity by the Hg compounds.

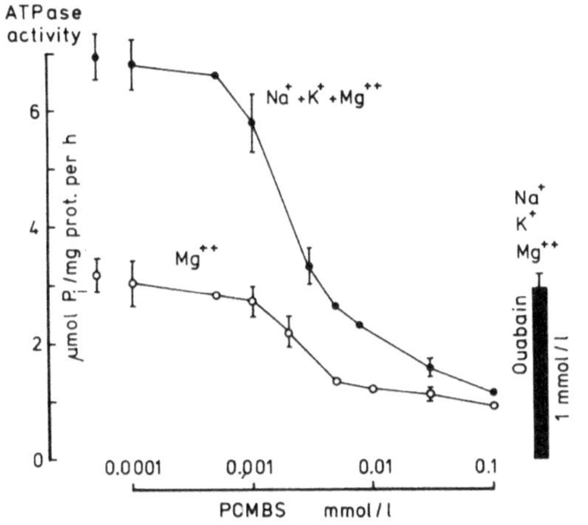

Fig. 5. Inhibition by PCMBS of the Mg^{2+}- and of the $Mg^{2+} + Na^+ + K^+$-activated ATPase of sarcolemmal particles. Ordinate: activity of the enzyme, µmol P_i/mg protein/h. Abscissa: concentration of PCMBS, mmol/l, log scale. Concentrations in the incubation medium (mmol/l): Mg^{2+} 5, Na^+ 100, K^+ 20. Control activities are indicated on the left of the curves. The column on the right represents the activity of the $Mg^{2+} + Na^+ + K^+$-ATPase in presence of 1 mmol/l ouabain. Means ± SEM of 8 ($Mg^{2+} + Na^+ + K^+$, ●) or 6 experiments (Mg^{2+}, ○).

Discussion

Based on experiments of Matsui and Schwartz (16), which showed the inhibition of a cardiac $Na^+ + K^+ + Mg^{2+}$-ATPase by PCMB, Besch et al. (2) discussed the feasibility of a positive inotropic effect of this mercurial. However, being convinced that a variety of enzymes would be affected simultaneously, they doubted the effect without performing contractility experiments. Later on, From and Probstfield (9) verified the direct positive inotropic action of the analog compound PCMBS, but did not investigate the influence on the membrane enzyme. Thereafter, the connection of both actions was attempted by Temma et al. (22) by comparing the direct positive inotropic effect of PCMB and PCMBS on atria with the inhibition of Rb^+-uptake by ventricle strips. They demonstrated that the $Na^+ + K^+$ pump was inhibited at best by 40% through positive inotropically acting concentrations of PCMB(S) and that digitoxin was by about two orders of magnitude more effective on the Rb^+ uptake than the SH-blockers. This discrepancy conveys the impression of nonalignment between ATPase inactivation and positive inotropic Hg effect. The inconsistency might result from the fact that the mercurials were added only during the preincubation of Langendorff preparations but not in the assay of Rb^+ uptake by ventricle strips excised from these hearts.

In contrast, the results reported here demonstrate that the concentrations of PCMBS required to inhibit the ATPase activity of isolated sarcolemmal particles are 1/10 to 1/100 of those necessary to produce a positive inotropic action. Assuming that the enzyme might be less accessible to PCMBS in situ, inotropic and inhibiting concentrations occupy a similar range of concentrations. The contribution to the positive inotropic effect by ATPase inhibition is further elucidated by the similar influence of the inotropic interventions on the effects of mercurials and cardioactive glycosides. Thus, the positive inotropic effectiveness of PCMBS was altered by variation of stimulation frequency and of external K^+ and Na^+ concentration in much the same way as that of the glycosides. The underlying assumption is that the positive inotropic effect of an inhibition of the sarcolemmal ATPase leads to an elevated cellular sodium load and that the Na^+ accumulated in the cell might then be exchanged for extracellular Ca^{2+} (5, 11, 17). The above changes of experimental conditions alter the background of $[Na^+]_i$ which can be raised additionally by the inhibition of the $Na^+ + K^+$ pump. Thus, a given degree of ATPase inactivation by a mercurial can result in a variable positive inotropic effect according to the variable preloading of the cells with sodium. The parallelism of the mechanisms of action of cardioactive glycosides and of Hg compounds is unlikely to extend to the interference of K^+ with the binding of the agents to their respective receptors (for glycosides: 6, 10). For the nature of mercaptide formation is much less specific than that of the binding of the glycoside to its repector which, in addition, is located on the outside of the sarcolemma whereas an inotropically relevant thiol group is possibly placed on the internal side (18). The equal contribution of the modification of glycoside binding and of the background of cellular Na^+ load to the K^+-induced reduction of the glycoside effect has recently been reported (1).

Thus, the potassium-glycoside antagonism appears to rest on a dual mode of action. In contrast, mercurials afford the possibility to separate the functional component of the K^+-versus-ATPase inhibitor antagonism as these agents are not subjected to a potassium-controlled chemical binding.

References

1. Bachmaier A, Ebner F, Reiter M (1985) Potassium changes the relationship between receptor occupancy and the inotropic effect of cardiac glycosides in guinea-pig myocardium. Br J Pharmacol 85: 755–765
2. Besch HR, Allen JC, Glick G, Schwartz A (1970) Correlation between the inotropic action of ouabain and its effects on subcellular enzyme systems from canine myocardium. J Pharmacol Exp Ther 171: 1–12
3. Clarkson TW, Amin-Zaki L, Al-Tikriti S (1976) An outbreak of methylmercury poisoning due to consumption of contaminated grain. Fed Proc 35: 2395
4. D-Iorio A (1957) The release of catecholamines from the isolated chromaffine granules of the adrenal medulla using sulphydryl inhibitors. Can J Biochem Physiol 35: 395–400
5. Ebner F, Reiter M (1977) The dependence on contraction frequency of the positive inotropic effect of dihydro-ouabain. Naunyn-Schmiedeberg's Arch Pharmacol 300: 1–9
6. Erdmann E, Schoner W (1973) Ouabain-receptor interactions in $(Na^+ + K^+)$-ATPase preparations. II. Effect of cations and nucleotides on rate constants and dissociation constants. Biochem Biophys Acta 330: 302–315
7. Fischer J, Mundschenk H, Wolf R (1965) Milzszintigraphie mit 1-Bromomercuri (^{197}Hg)-2-hydroxypropan (BMHP) Fortschr auf d Gebiet d Röntgenstrahlen 103: 349–366
8. Fiske CH, Subbarow Y (1925) The colorimetric determination of phosphorus. J Biol Chem 66: 375–400
9. From AHL, Probstfield JL (1971) p-Chloromercuribenzene sulfonic acid induced inotropism. Fed Proc 30: 2441

10. Gardner JD, Conlon Th P (1972) The effects of sodium and potassium on ouabain binding by human erythrocytes. J Gen Physiol 60: 609–629
11. Glitsch HG, Reuter H, Scholz H (1970) The effect of the internal sodium concentration on calcium fluxes in isolated guinea-pig auricles. J Physiol (London) 209: 25–43
12. Halbach S (1985) The octanol/water distribution of mercury compounds. Arch Toxicol 57: 139–141
13. Halbach S, Schönsteiner G, Ebner F, Reiter M (1981) The effects of p-chloromercuriphenylsulfonic acid (PCMBS) on force of contraction of mammalian myocardium and on ATP hydrolysis by sarcolemmal ATPase. Naunyn-Schmiedeberg's Arch Pharmacol 318: 121–129
14. Hughes WL (1950) Protein mercaptides. Cold Spring Harbour Symposia on Quantitative Biology 14: 79–84
15. Lowry OH, Rosebrough NJ, Farr AL, Randall RJ (1951) Protein measurement with the Folin phenol reagent. J Biol Chem 193: 265–275
16. Matsui H, Schwartz A (1966) Purification and properties of a highly active ouabain-sensitive $Na^+ + K^+$-dependent adenosinetriphosphatase from cardiac tissue. Biochim Biophys Acta 128: 380–390
17. Mullins LJ (1979) The generation of electric currents in cardiac fibres by Na/Ca exchange. Am J Physiol 236: C 103–110
18. Patzelt-Wenczler R, Pauls H, Erdmann E, Schoner W (1975) Evidences for a sulfhydryl group in the ATP-binding site of $(Na^+ + K^+)$-activated ATPase. Eur J Biochem 53: 301–311
19. Reiter M (1967) Die Wertbestimmung inotrop wirkender Arzneimittel am isolierten Papillarmuskel. Arzneimittel Forsch 17: 1249–1253
20. Skou JC, Hilberg C (1965) The effect of sulphydryl-blocking reagents and of urea on the $(Na^+ + K^+)$-activated enzyme system. Biochim Biophys Acta 110: 359–369
21. Tada M, Finney JO, Swartz MH, Katz AM (1972) Preparation and properties of plasma membranes from guinea-pig hearts. J Mol Cell Cardiol 4: 417–425
22. Temma K, Akera T, Ku DD, Brody TM (1978) Sodium pump inhibition by sulfhydryl inhibitors and myocardial contractility. Naunyn-Schmiedeberg's Arch Pharmacol 302: 63–71
23. Webb JL (1966) Enzyme and Metabolic Inhibitors, Vol 2. Academic Press, New York, pp 729–983

Author's address:

Dr. S. Halbach, Abteilung Pharmakologie, Gesellschaft für Strahlen- und Umweltforschung München, Ingolstädter Landstraße 1, D-8042 Neuherberg (West Germany)

Chronic pretreatment with amrinone or milrinone induces no tachyphylaxis in mammalian cardiac muscle

S. U. Sys and D. L. Brutsaert

The Department of Physiology & Medicine, University of Antwerp (Belgium)

Summary

The effect of chronic (12 days) administration of two new inotropic agents, either amrinone or milrinone, on the acute action of these drugs was investigated in papillary muscle isolated from cat heart three days after the last treatment. The orally administered dose during pretreatment was 2 mg/day/kg body weight for amrinone and 0.2 for milrinone. Blood and myocardial tissue samples taken at the time of muscle isolation did not contain any detectable amrinone or milrinone (sensitivity of the assay: 1 μg/ml).

A concentration-response curve to amrinone (from 5 up to 250μg/ml) was obtained in muscles from 8 control and 8 pretreated cats. The response to milrinone (5 and 20μg/ml) was determined for each dose in 7 control and in 7 pretreated cats. Mechanical parameters of contraction and relaxation were obtained from pre- and afterloaded isotonic, zero-load clamped and isometric contractions (12 beats/min at 37° C), recorded 10 min after administration of each dose of drug, i. e. the time of maximal response.

Both drugs induced a statistically significant concentration-dependent positive inotropic effect in all muscles. The observed differences between muscles from control and from pretreated cats were not significant. This was confirmed by analysis of variance.

Our results suggest that at least in normal cardiac muscle, chronic administration of these two inotropic drugs does not affect their positive inotropic action. The absence of tachyphylaxis contrasts with similar previous experiments which showed a diminished inotropic response to ouabain after chronic pretreatment with digoxin.

Introduction

The bipyridine derivatives amrinone and milrinone both show a marked concentration-dependent positive inotropic response in isolated ventricular myocardium from different adult mammalian species (1, 2). The inotropic response to amrinone and milrinone in isolated cat papillary muscle is characterized by (i) a concentration-dependent increase in contractility, with milrinone about five times as potent as amrinone, (ii) no effect on load dependence of relaxation, (iii) no change in timing and duration of the contraction-relaxation cycle and (iv) marked temperature dependence (14). This response is not mediated through stimulation of beta 1 –, alpha 1 –, muscarinic or histaminic (H2) receptors, nor is it modified by inhibition of the prostaglandin synthetic pathway. Unlike digitalis glycosides, these drugs do not inhibit sarcolemmal Na^+-K^+-ATPase (2, 3). Amrinone (7, 9, 13) and milrinone (1, 3) are inhibitors of cardiac adenosine 3′, 5′-monophosphate (cAMP) phosphodiesterase (PDE), more in particular of PDE III (7, 8, 11) which is highly specific for cAMP and is membrane-associated in many tissues (16).

To verify the efficacy of amrinone and milrinone for chronic clinical treatment, in the present study, the effects of amrinone and milrinone were analyzed in cat papillary muscle after chronic treatment of the cats with the drug. These data were compared with the results of a similar study (10) on the tachyphylactic effect of glycosides on contractility.

Methods

Forty-four papillary muscles from the right ventricle of adult cats were used for this study. The criteria for selection of suitable preparations and techniques for analysis were published previously (14, 15). The bathing solution contained (mM): NaCl 118; KCl 4.7; $MgSO_4 \cdot 7H_20$ 1.2; KH_2PO_4 1.1; $NaHCO_3$ 24; $CaCl_2 \cdot 2H_2O$ 2.5 and glucose 4.5. The solution was bubbled with a gas mixture of 95% O_2–5% CO_2. Homogeneous electrical stimulation (12 beats/min) was obtained by rectangular pulses of 5 ms duration about 10% above threshold through two platinum electrodes arranged longitudinally along the entire muscle. Before the actual experiment, the muscles were allowed to equilibrate for 2–3 h at 29° C. The preload on the muscle was then adjusted so that the resting muscle length corresponded to lmax, i. e. the length at which active force development was maximal and at which all experiments were performed. The basic characteristics of all muscles at 29° C, prior to the actual experiment and at initial muscle length lmax, are shown in Table 1.

Table 1. Basic characteristics of the cat papillary muscles at lmax (29° C).
Muscles were divided into groups according to the followed protocol. A1: amrinone at 37°C; A2: amrinone at 37°C after chronic pretreatment; M1: milrinone at 37° C; M2: milrinone at 37°C after chronic pretreatment. CSA = mean cross-sectional area; TT = peak total tension of the isometric twitch; RT = resting tension; MUVS = maximal unloaded velocity of shortening.

Group	n	lmax (mm)	CSA (mm^2)	TT (mN/mm^2)	RT/TT (%)	MUVS (lmax/s)
A1	8	7.2 ± 0.46	0.65 ± 0.143	89. ± 9.7	9.5 ± 1.08	1.4 ± 0.07
A2	8	6.7 ± 0.34	0.66 ± 0.128	101. ± 13.1	9.8 ± 0.94	1.5 ± 0.15
M1	14	7.4 ± 0.61	0.83 ± 0.106	90. ± 6.7	8.3 ± 0.55	1.4 ± 0.13
M2	14	6.3 ± 0.28	0.80 ± 0.109	93. ± 8.5	8.9 ± 0.52	1.5 ± 0.09

Using an electromagnetic force-length lever system (15), various parameters of the contraction (contractility) and relaxation (load dependence) phases were derived. Analysis of contractility (4) was derived from a series of three test contractions. Maximal unloaded velocity of shortening (MUVS) and its timing (tMUVS) were derived from a zero load clamped contraction (5), i. e. the first test contraction. The second test contraction, preloaded isotonic, provided values for peak shortening (PS), peak velocity of shortening (V+) and lengthening (V−) and their timing (tPS, tV+, tV−). The third test contraction was an isometric beat yielding measurements of peak total tension (TT), time to TT (tTT) and time to reach half relaxation (tHR). Load dependence of relaxation (6) was quantified by comparing force and time coordinates at the onset of the isometric relaxation phase in several afterloaded isotonic twitches with the relaxation of the isometric control twitch (15). The afterloaded isotonic contractions were separated by at least eight preloaded isotonic contractions to avoid effects of loading of preceding beats.

In the actual experiments at 37° C, either amrinone (1, 5, 25, 50, 250 and 500 μg/ml) or milrinone (5 and 20 μg/ml) was tested and the effects were compared with the effects of high $[Ca^{++}]_o$ (5 and 6 mM) in the bathing solution. The inotropic action of both drugs was tested in muscles from untreated cats and from cats after chronic oral treatment with either amrinone (2 mg/kg/day) or milrinone (0.2 mg/kg/day) for 12 days, followed by two days of washout. The actual experiment was then performed on the fifteenth day. In order to check the adequacy of a 72 hour washout period, blood samples and myocardial tissues samples were taken at the beginning of the actual experiment and assayed for amrinone and milrinone.

Mean ± SE was calculated for all groups; statistical significance of differences was computed by Student's t-test or analysis of variance.

Results

1. Effects of amrinone on isolated cat papillary muscles

1A. Acute effects of amrinone on contraction and relaxation

The acute effects of amrinone on isometric and isotonic contraction and relaxation are shown in Figs. 1 and 2. After amrinone, TT, PS and MUVS were significantly increased already at the smallest concentration, indicating a marked concentration-dependent positive inotropic effect. These parameters reach levels which at the higher concentrations are comparable to the values obtained with high $[Ca^{++}]_o$. Time to TT (tTT), time to half relaxation (tHR) of the isometric twitch and time to peak relaxation velocity (t−V) of the preloaded isotonic contraction were unaffected by addition of amrinone. These findings are in contrast to the well known, small abbreviation of the contraction-relaxation cycle duration with increasing $[Ca^{++}]_o$. No changes of load dependence of relaxation were observed, again contrasting with the increase of load dependence induced by high $[Ca^{++}]_o$.

1B. Effects of amrinone after chronic pretreatment

A separate group of 8 cats were orally treated with amrinone (2 mg/kg/day) for 12 consecutive days. Blood samples and myocardial tissue samples were taken on the third day post-treatment. Amrinone was not detected in any of the samples (sensitivity of the assay: 1 μg/ml). This indicates the adequacy of the washout period.

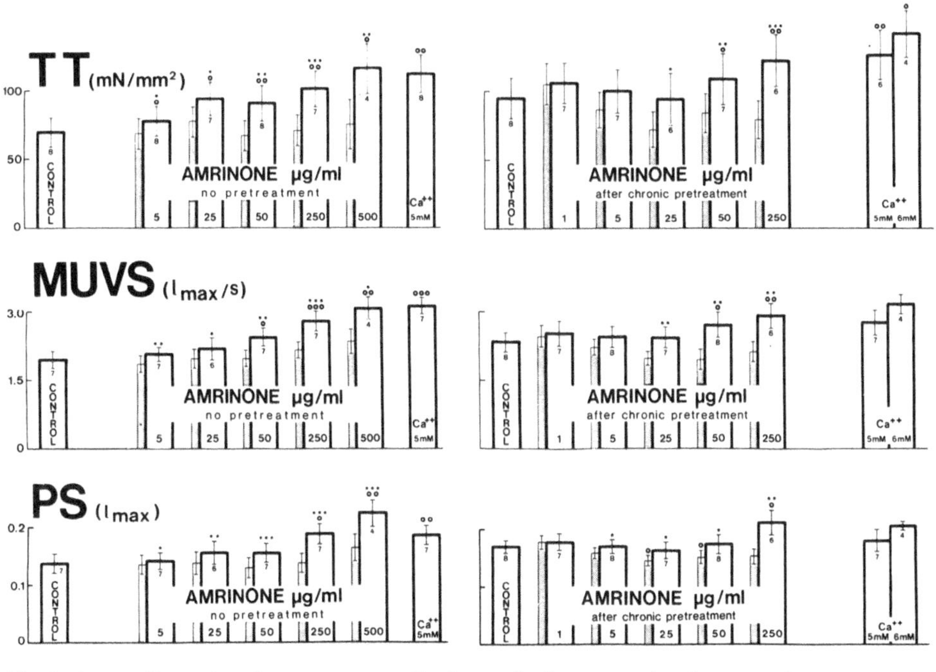

Fig. 1. Acute effects of amrinone on contractility in muscles from normal and pretreated cats.
Block diagrams were constructed for the mean absolute values of TT, directly measured maximal unloaded velocity of shortening (MUVS) and peak shortening (PS) of the isotonic contraction. The open blocks indicate the effects of amrinone and of increasing $[Ca^{++}]_o$ on muscles from control (left panels) and chronically pretreated (right panels) cats. The shaded narrow blocks to the left of the open blocks indicate the effects of the appropriate carrier lactate concentration. ∗: significance of difference between control lactate and amrinone values; ∗: significant difference with the control (no carrier, no drug) state.

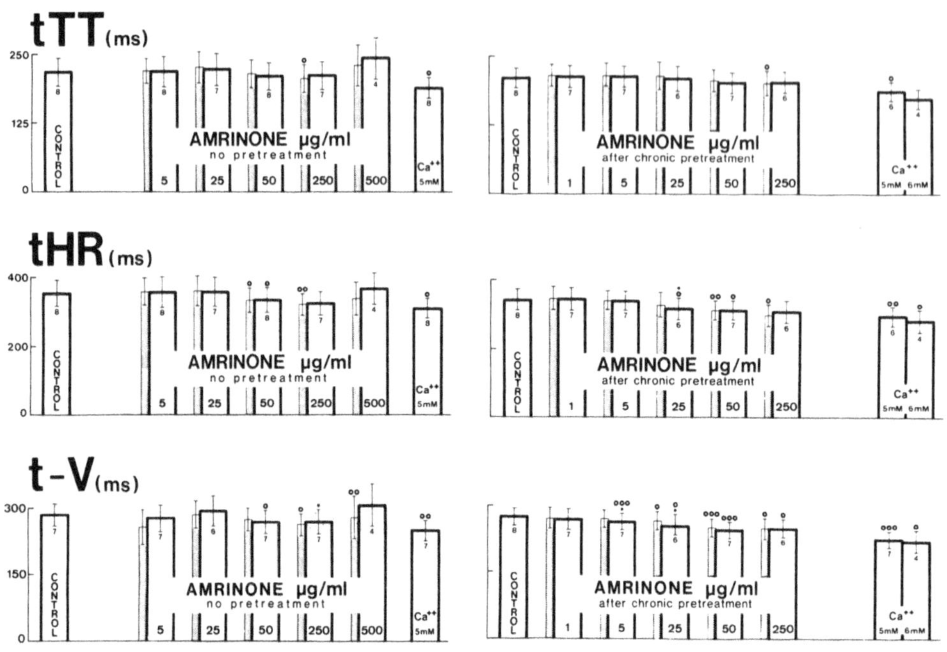

Fig. 2. Acute effects of amrinone on time indices of contraction and relaxation in muscles from normal and pretreated cats.
Block diagrams show the effects of amrinone and its carrier on the mean absolute values of time to TT (tTT), time to half relaxation (tHR) in the isometric contraction and time to peak isotonic relaxation velocity (t−V) in the preloaded isotonic contraction. Same format as in Fig. 1.

The effects of amrinone on contraction and relaxation of muscles from the pretreatment cats are shown in the right panels of Figs. 1 and 2. Pretreatment did not significantly alter the basic characteristics nor the baseline contractile performance at the end of the equilibration period (29°C) nor at 37°C at the onset of the actual experiment. After acute addition of the drug to these muscles from pretreated cats, the positive inotropic effect was not different from the effect in control muscles: when the relative increase in TT and in MUVS after the addition of amrinone, versus the control carrier solution values, were plotted as concentration response curves (Fig. 5), the observed differences were small and statistically not significant. Accordingly, amrinone-induced tachyphylaxis could not be demonstrated in our protocol.

2. *Effects of milrinone on isolated cat papillary muscle*

In order to avoid the problem of any possible milrinone-induced in vitro tachyphylaxis, as described by Alousi et al. (1), we used different muscles to study each dose of milrinone (5 and 20 µg/ml). Figs. 3 and 4 show data for milrinone obtained from two series of 7 cat papillary muscles. The acute effects of milrinone on isometric and isotonic contraction and relaxation showed significant concentration-dependent increases in TT, PS and MUVS. The inotropic level reached with 20 µg/ml milrinone was comparable to the effect of increasing $[Ca^{++}]_o$ up to 5 mM. Further increasing $[Ca^{++}]_o$ showed only minimal further increase in tension. Milrinone was about 5 times as potent as amrinone in increasing cat papillary

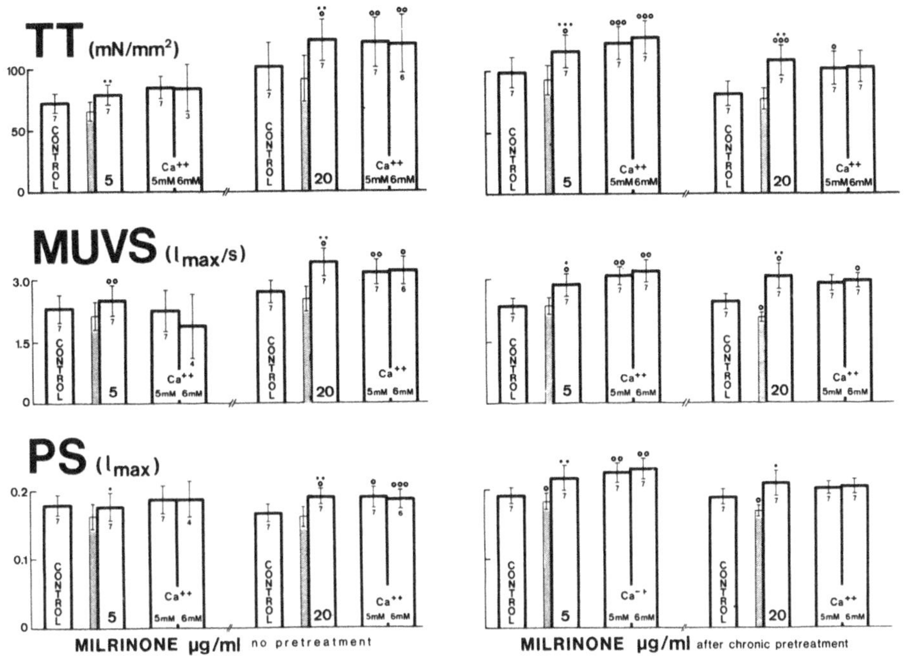

Fig. 3. Acute effects of milrinone on contractility in muscles from normal and pretreated cats. Block diagrams were constructed for the mean absolute values of the same parameters as in Fig. 1, showing the effects of milrinone (open blocks) and its carrier HCl (shaded blocks) on normal and pretreated cat papillary muscle in comparison with increasing $[Ca^{++}]_o$. * and ■ are analogous to Fig. 1.

muscles contractility in vitro, with 5 µg/ml milrinone resulting in an increased contractility comparable to 25 µg/ml amrinone and 20 µg/ml milrinone to approximately 100 µg/ml. As with amrinone, changes in tTT, tHR and t−V and in load dependence of relaxation after milrinone were negligible in comparison to the abbreviating effect on the contraction relaxation cycle and to the increased load dependence observed at higher $[Ca^{++}]_o$.

In the pretreated animals, blood and tissue samples taken on the third post-treatment day, i. e. at the onset of the in vitro experiment, contained no detectable milrinone (sensitivity of the assay: 1 µg/ml). Pretreatment with milrinone did not significantly alter the basal properties, baseline contractile performance of the papillary muscles, nor the inotropic response to the acute addition of milrinone (Fig. 5).

Conclusion

In conclusion, amrinone and milrinone did not show tachyphylaxis of the positive inotropic response in isolated ventricular myocardium from normal cats chronically treated for 12 days with these drugs. The absence of tachyphylaxis was in contrast to experiments in which a diminished positive inotropic response to ouabain was found in papillary muscles obtained from the non-failing right ventricle of cats previously subjected to a chronic (10 days treatment followed by 5 days abstinence) digoxin treatment (10).

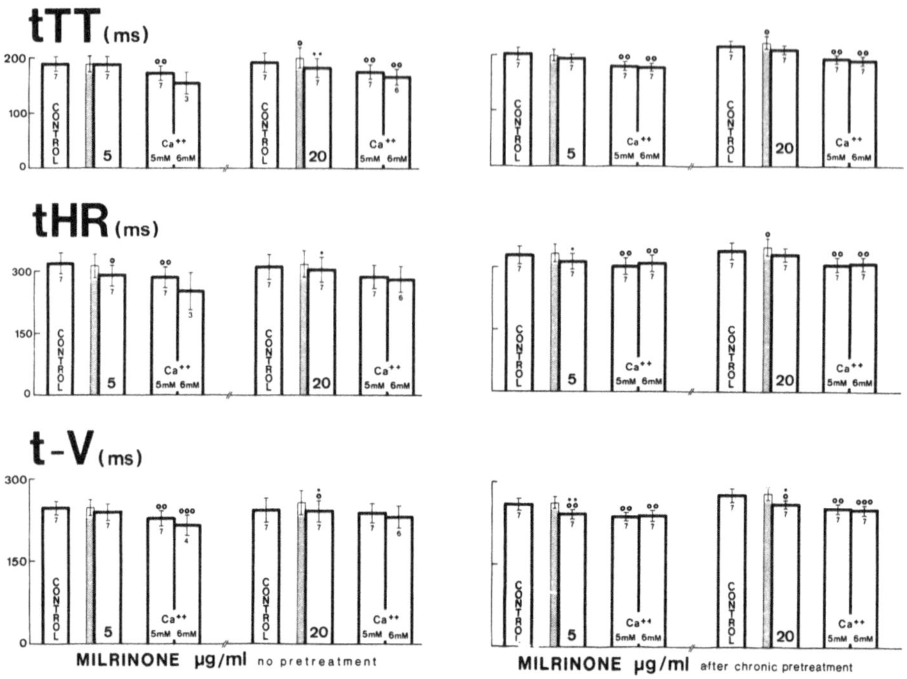

Fig. 4. Acute effects of milrinone on time indices of contraction and relaxation in muscles from normal and pretreated cats.
Block diagrams show the effects of milrinone and its carrier on the mean absolute values of time to TT (tTT), time to half relaxation (tHR) in the isometric contraction and time to peak isotonic relaxation velocity (t−V) in the preloaded isotonic contraction. Same format as in Fig. 3.

On the other hand, the absence of tachyphylaxis in our protocol is in agreement with other reports which used different protocols. Repeated treatment of isolated cat papillary muscle with the same concentration of amrinone revealed no statistically significant difference between the responses and the last addition of amrinone produced at least the same response as the first one (2). Repeated treatment of guinea-pig papillary muscle with the same dose of milrinone did not significantly reduce the inotropic response to the drug, although the tissue had to be extensively washed between treatments to reestablish its full responsiveness to milrinone (1). Moreover, in patients who received consecutive doses of amrinone at 8 hour intervals, cardiac output was increased after each dose, compared with baseline values for each administration (12).

References

1. Alousi AA, Canter JM, Montenaro MJ, Fort DJ, Ferrari RA (1983) Cardiotonic activity of milrinone, a new and potent cardiac bipyridine, on the normal and failing heart of experimental animals. J Cardiovasc Pharmacol 5 (5): 792
2. Alousi AA, Farah AE, Lesher GY, Opalka CJ Jr (1979) Cardiotonic activity of amrinone – Win 40 680 [5-amino-3,4′-bipyridine-6(1H)-one]. Circ Res 45: 666

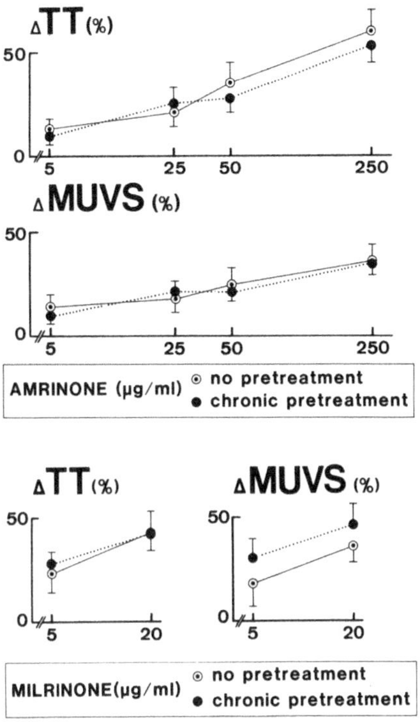

Fig. 5. Effects of pretreatment on the acute effects of amrinone and milrinone in cat papillary muscle. Dose-response curves show the influence of amrinone (upper panel) and milrinone (lower panel) on TT and MUVS of muscles from control and pretreated cats. The values are expressed as mean ± SE of the parameter values in the control carrier solution at the appropriate concentration.

3. Alousi AA, Stankus GP, Stuart JC, Walton LH (1983) Characterization of the cardiotonic effects of milrinone, a new potent cardiac bipyridine, on isolated tissues from several animal species. J Cardiovasc Pharmacol 5: 804
4. Brutsaert DL (1974) The force-velocity-length-time interrelation of cardiac muscle. In: The physiological basis of Starling's law of the heart, Ciba Foundation, Amsterdam. Elsevier, Excerpta Medica, p 155
5. Brutsaert DL, Claes VA (1974) Onset of mechanical activation of mammalian heart muscle in calcium- and strontium-containing solutions. Circ Res 35: 345
6. Brutsaert DL, Housmans PR, Goethals MA (1980) Dual control of relaxation: its role in the ventricular function in the mammalian heart. Circ Res 47: 637
7. Endoh M, Yamashita S, Taira N (1982) Positive inotropic effect of amrinone in relation to cyclic nucleotide metabolism in the canine ventricular muscle. J Pharmacol Exp Ther 221 (3): 775
8. Hayes JS, Bowling N, Boder GB, Kauffman R (1984) Molecular basis for the cardiovascular activities of amrinone and AR-L57. J Pharmacol Exp Ther 230 (1): 124
9. Honerjäger P, Schäfer-Korting M, Reiter M (1981) Involvement of cyclic AMP in the direct inotropic action of amrinone. Biochemical and functional evidence. Naunyn-Schmiedeberg's Arch Pharmacol 318: 112
10. Housmans PR, Vermeyen KM, Lewis MJ, Hanegreefs GH, Brutsaert DL (1980) Depressed contractile sensitivity of mammalian ventricular myocardium to ouabain following chronic digoxin exposure. Proc Eur Congr Cardiol: 89

11. Kariya T, Wille LJ, Dage RC (1982) Biochemical studies on the mechanism of cardiotonic activity of MDL 17,043. J Cardiovasc Pharmacol 4: 509
12. LeJemtel TH, Keung E, Ribner HS, Davis R, Wexler J, Blaufox MD, Sonnenblick EH (1980) Sustained beneficial effects of oral amrinone on cardiac and renal function in patients with severe congestive heart failure. Am J Cardiol 45: 123–129
13. Shibata S, Kondo N, Kodama I, Yamada K (1981) Electrical and mechanical effects of amrinone on isolated guinea-pig papillary muscle. Fed Proc 40: 711
14. Sys SU, Goenen MJ, Chalant ChH, Brutsaert DL (1986) Inotropic effects of amrinone and milrinone on contraction and relaxation of isolated cardiac muscle. Circulation (in press) March
15. Sys SU, Housmans PR, Van Ocken ER, Brutsaert DL (1984) Mechanisms of hypoxia-induced decrease of load dependence of relaxation in cat papillary muscle. Pflügers Arch 401: 368
16. Thompson WJ, Terasaki WL, Epstein PM, Strada SJ (1979) Assay of cyclic nucleotide phosphodiesterase and resolution of multiple molecular forms of the enzyme. Adv cyclic Nucleotide Res 10: 69

Author's address:

Prof. Dr. D. L. Brutsaert, University of Antwerp, Department of Physiology and Medicine, Groenenborgerlaan, 171, B-2020 Antwerp (Belgium)

Left ventricular function after oral milrinone in patients with congestive heart failure. A hemodynamic and angiographic study

F. Piscione*, P. W. Serruys and P. G. Hugenholtz

The Catheterization Laboratory, Thoraxcenter, Erasmus University, Rotterdam (The Netherlands)

Summary

The diastolic properties of the left ventricle (LV) are important determinants of cardiac function; alteration in diastolic function accounts for many of the clinical manifestations of congestive heart failure. Changes in myocardial relaxation and filling phases, changes in coronary blood flow, diastolic tone and ventricular shape all contribute to the abnormal diastolic compliance of the failing heart. To assess the short-term effects of a single oral dose (10 mg) of milrinone (M) on diastolic function, 14 patients (pts), mean age 59 ± 6 years, with severe congestive heart failure underwent this study. After baseline hemodynamic measurements, LV angiography with simultaneous recording of LV pressure (tip-manometry) and pressure-derived parameters was performed at paced heart rate (15 beats above spontaneous heart rate). After repeat baseline measurements, M was administered followed by LV pressure measurements at 10′ intervals to 50′ together with coronary blood flow measurement. Blood samples to determine the plasma concentration of M were withdrawn at the same intervals. Blood arterial and coronary sinus samples for catecholamine levels and oxygen saturations were withdrawn before and 50′ after M. LV angiography was repeated 60′ after M at matched atrial paced rates with simultaneous pressure recordings. During the first phase of the study, peak negative dP/dt progressively and significantly increased from 827 ± 209 to 904 ± 194 mm Hg/sec ($p < 0.01$) whereas LV systolic pressure slightly decreased. The time constant of early relaxation, Tau_1,(fit for the first 40 msec. after the occurrence of peak negative dP/dt), decreased progressively and significantly from 71 ± 17 to 58 ± 14 msec ($p < 0.01$). Minimal LV diastolic pressure and LV end-diastolic pressure (LVEDP) were significantly lowered after 50′ from 13 ± 4 to 10 ± 7 mm Hg ($p < 0.05$) and from 27 ± 7 to 20 ± 10 mm Hg ($p < 0.01$), respectively. Coronary blood flow rose progressively during the first 40′ but the increase failed to be significant at 50′. Coronary vascular resistance reciprocated these changes. Myocardial oxygen consumption remained unchanged over 50′. Noradrenaline, adrenaline and dopamine mean values also remained unchanged over 50′, thus demonstrating a lack of effect of M on the basal, elevated sympathetic tone. Mean plasma level of M rose progressively, achieving a therapeutic level after 30′. The analysis of LV volume, pressure and pressure-derived parameters during angiography showed an improved relaxation phase with an increase in peak filling rate from 290 ± 27 to 338 ± 55 ml sec. The constant of elastic chamber stiffness, measured by the simple elastic model, decreased but not significantly. Pressure-volume loops showed a downward shift in 10/14 pts, mainly because of a major reduction in LVEDP from 29 ± 10 to 20 ± 11 mm Hg ($p < 0.001$) while end-diastolic volume only slightly decreased.

Conclusion: improvement in parameters of LV diastolic relaxation and filling with changes in chamber distensibility following oral M, suggest that improved diastolic function may contribute to the beneficial effects of M on the failing heart.

* Dr. Piscione is supported by CNR-NATO, research fellowship no. 216.1095.

During the last two years several investigators have demonstrated that the administration of milrinone in patients with congestive heart failure leads to an increased inotropism together with a vasodilation, in absence or with only slight changes in heart rate and aortic pressure (1, 7, 10, 16). Recently an improvement in diastolic left ventricular performance, occurring after intravenous milrinone, has been reported (14). In the present study we evaluated the effect of a single oral dose of milrinone on central hemodynamics and on systolic and diastolic mechanics in patients with congestive heart failure.

Methods

Ten men and four women with severe chronic congestive heart failure consistent with functional Class III to IV of the New York Heart Association were studied. Patients ranged in age from 49 to 71 years (mean 59 ± 6). The etiology of heart failure was coronary artery disease in twelve patients and an idiopathic cardiomyopathy in the remaining two. No patient had valvular heart disease, hypertension, angina or other evidence of active myocardial ischemia. All patients were markedly limited in physical activity by symptoms of dyspnoea and fatigue despite optimal medical therapy with digitalis, diuretics, nitrates and/or ACE inhibitors. The nature, potential benefits and possible risks of the study were explained to all patients, who then gave informed consent. The protocol was approved by the Ethics Committee of the Erasmus University. All vasodilators were discontinued 24 hours prior to the study, diuretics 12 hours prior to the study, digoxin was witheld on the morning of the study. Patients underwent right and left heart catheterization in the fasting state. A 7F thermodilution, balloon tipped, flow directed pulmonary artery catheter was advanced to the main pulmonary artery from the right femoral vein. A pigtail 8F angiographic micromanometer was advanced to the left ventricular cavity after introduction through the right femoral artery; a tipmanometer 5F was placed in the aortic root from the left femoral artery. After placement of the catheters, baseline left ventricular angiography (RAO 30°) at atrial paced heart rate (15 beats higher than spontaneous heart rate) and simultaneous high fidelity left ventricular pressure measurements were then performed. We injected nonionic contrast medium (0.7 ml/kg metrizamide). After the first ventriculogram, a pause of 15 minutes was permitted to allow spontaneous heart rate and pressures to return to control values. Then 10 mg of milrinone was given orally. Sixty minutes after, intracardiac pressure measurements and left ventricular cineangiography were repeated. Care was taken to maintain uniform patient's position relative to X-ray equipment during angiograms which were performed with the breath in shallow inspiration.

Pressure measurement and angiograms suitable for analysis were available in each of the fourteen patients. Left ventricular pressure and derived indexes such as peak positive and peak negative dP/dt, the relation between dP/dt pressure and total pressure linearly extrapolated to pressure = 0 (Vmax) and the peak velocity of contractile element (Vce) were computed on line after date acquisition of 20 seconds (11). The left ventricular pressure decay during the isovolumic relaxation was quantified as previously described (15) using both a monoexponential and a biexponential model, where Tau 1 represents the time constant of early relaxation (during the first 40 msec after the peak negative dP/dt). Cardiac output was determined by thermodilution and expressed as cardiac index; systemic vascular resitance was calculated from standard formula, mean systolic ejection rate was calculated from the aortic pressure tracing.

Data analysis

From both cineangiograms, a complete cardiac cycle was analyzed frame by frame. The ventricular contours were detected automatically with a dedicated hard-wired system, the Contouromat (17). For each analyzed cineframe, left ventricular volume was computed according to Simpson's rule where each videoline in the right anterior oblique view is taken to represent a circular slice of the left ventricular lumen. After the end diastolic and end-systolic frames were determined, ventricular volumes, global ejection fraction and the derivative of volume relative to time (dV/dt) were calculated. End-diastole was defined with reference to the point in the pressure trace at which the derivative of left ventricular pressure first exceeded 200 mm Hg/sec. End-systole was defined, with reference to the pressure tracing, at the occurrence of the incisure of the central aortic pressure. Peak ejection rate was calculated as the lowest value of dV/dt after end-diastole, while peak filling rate was taken as the greatest dV/dt occurring early in diastole. The simultaneous recording of pressure and volume, during the angiograms, allowed to calculate the net work performed by the left ventricle ($\int$ PdV) as the pressure-volume product from a complete cardiac cycle. Furthermore in every patient a pressure-volume curve before and after milrinone was obtained.

Drug level and statistics

Plasma level of milrinone was measured using a validated high performance liquid chromatography procedure (4). All the results are expressed as mean ± standard deviation. Statistical analysis of data before and after milrinone was performed using Student's t-test for paired data.

Results

Plasma level of milrinone, measured at the end of the study, resulted above the therapeutic plasma level (150 ng/ml) (2) in 10 patients and slightly below in 1 (137 ng/ml) while in the remaining 3 it was below 50 ng/ml.

Central hemodynamics

The cycle length (613 ± 122 msec vs. 628 ± 47) as well as the aortic pressure (83 ± 12 mm Hg vs. 77 ± 19) did not change after milrinone. Cardiac index significantly increased from 2.0 ± 0.6 L/min/m^2 to 2.5 ± 0.6 (< 0.05) and concomitantly systemic vascular resistance decreased from 1600 ± 400 dynes.sec.cm^5 to 1360 ± 640 ($p < 0.05$). Stroke index increased slightly but not significantly from 30 ± 10 ml/m^2 to 34 ± 12.

Systolic mechanics (Fig. 1)

Peak left ventricular systolic pressure did not change after milrinone (113 ± 32 mm Hg vs. 107 ± 17) as well as end systolic pressure (70 ± 19 mm Hg vs. 63 ± 13). Despite a small and not significant reduction both in end-systolic and end-diastolic volume (−6% and −4% respectively) the ejection fraction significantly increased from 18 ± 8% to 22 ± 12% ($p < 0.01$). Peak positive dP/dt significantly increased from 1014 ± 365 mm Hg/sec to 1164 ± 388 ($p < 0.01$). Also Vmax and Vce significantly increased (from 27 ± 8 sec^{-1} to 36 ± 11; $p < 0.001$ and from 18 ± 5 sec^{-1} to 27 ± 10; $p < 0.001$ respectively).

Peak ejection rate increased slightly but not significantly from 262 ± 33 ml/sec to 304 ± 67. Mean systolic ejection rate showed a slight and non significant increase (from 132 ± 4 ml sec to 152 ± 5). Milrinone significantly increased the net work of left ventricle from 2222 ± 1099 gm m/m^2 beat to 2814 ± 1439 ($p < 0.01$).

Isovolumic relaxation

Peak negative dP/dt increased from 824 ± 216 mm Hg/sec to [illegible]59 ± 304 ($p < 0.001$). The time constant of left ventricular relaxation, Tau $_0$ (measured as monoexponential), decreased from 68 ± 21 msec to 58 ± 21 ($p < 0.001$). Even Tau $_1$ and Tau $_2$ significantly decreased after milrinone (by 14%; $p < 0.001$ and by 16%; $p < 0.05$ respectively) (Fig. 2).

Diastolic mechanics

Milrinone enhanced early diastolic filling in 10 patients (Fig. 3) while in 4 patients a decrease was observed. These four patients had a low plasma level of milrinone at the end

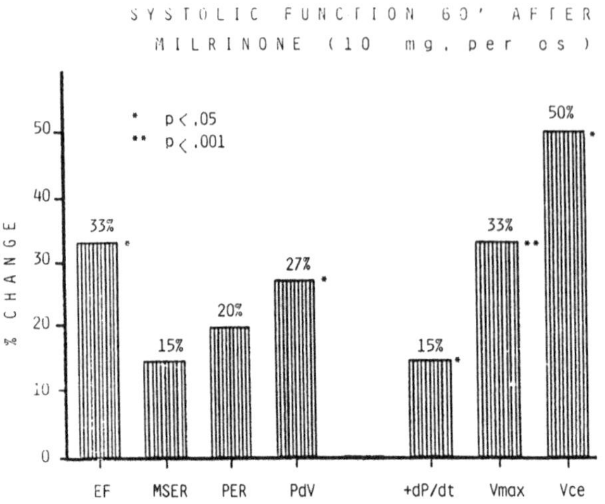

Fig. 1. Summary of percent changes in left ventricular systolic function after milrinone administration. An evident and significant improvement in the isovolumic indexes of contraction was observed (right side), whereas among the ejection phase indexes, only the ejection fraction showed a significant increase.
EF = ejection fraction; MSER = mean systolic ejection rate; PER = peak ejection rate; PdV = net work of the left ventricle.

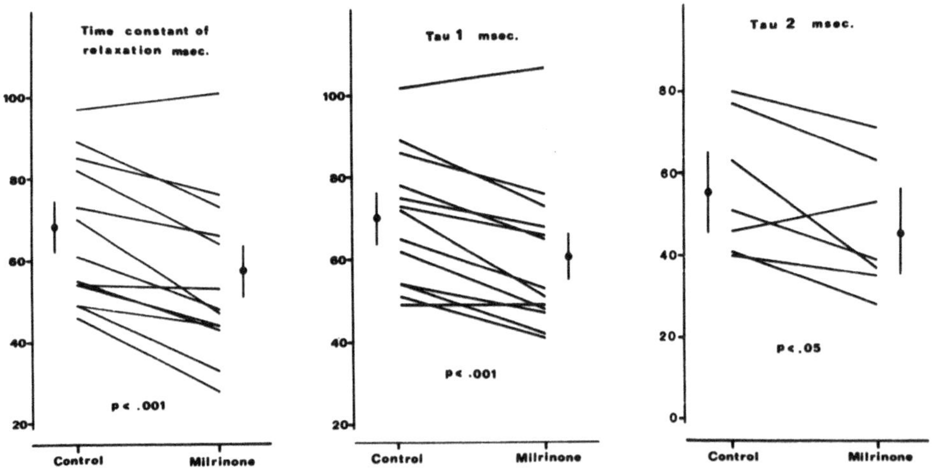

Fig. 2. Effects of milrinone on left ventricular relaxation. The time constants measured according to both the mono- and biexponential methods were severely prolonged in basal conditions. A significant decrease was observed after milrinone demonstrating that such a drug exerts a favourable effect on the relaxation rate in failing hearts.
Mean values ± SEM are shown.

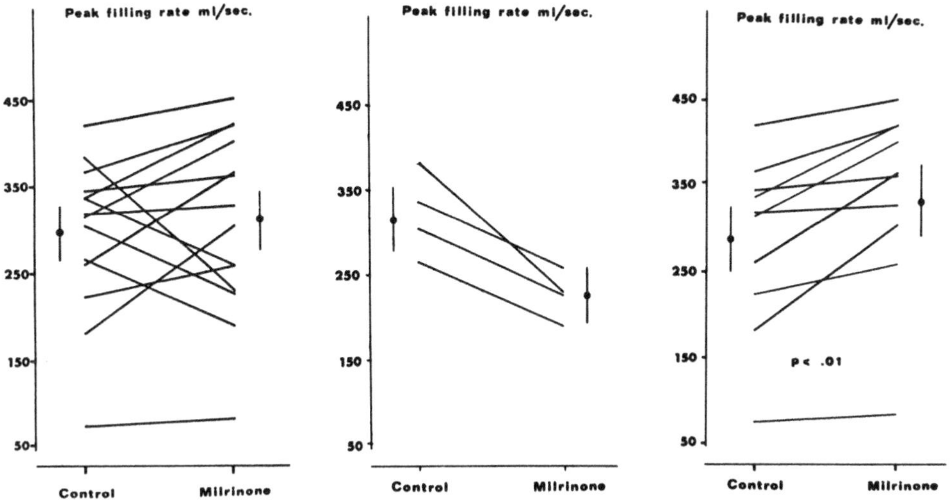

Fig. 3. Individual and group mean changes in peak filling rate after milrinone administration. The peak filling rate did not increase significantly when the whole population was considered, since in four patients a decrease was observed. All these 4 patients had a plasma level of milrinone below the therapeutic plasma level (150 ng/ml). When only the remaining 10 were considered a significant increase was observed. Mean values ± SEM are shown.

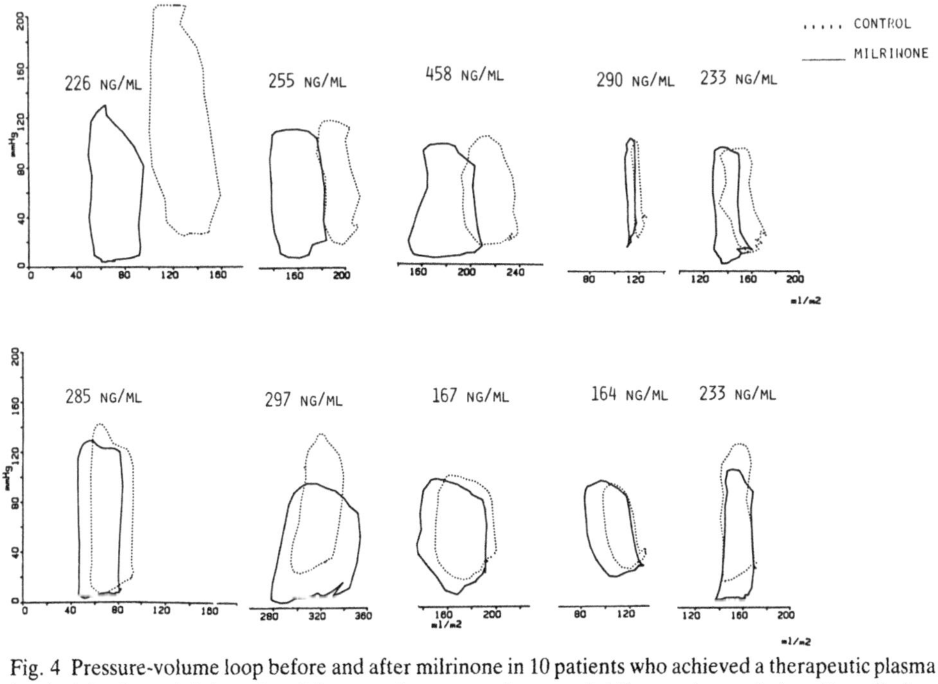

Fig. 4 Pressure-volume loop before and after milrinone in 10 patients who achieved a therapeutic plasma level of the drug. A leftward and downward or a pure downward shift was observed. A uniform finding was a lower diastolic pressure for any given diastolic volume after milrinone. Individual plasma levels of milrinone achieved at the end of the study are shown.

of the study. Significant changes were observed both in minimal and end-diastolic left ventricular pressure (from 20 ± 17 mm Hg to 10 ± 17 and from 29 ± 10 mm Hg to 20 ± 11 respectively; $p < 0.001$). Milrinone displaced leftward and in 4 patients downward the pressure-volume loop (Fig. 4) while in 4 cases a rightward shift was observed (Fig. 5). These four patients are the same in whom a decrease in peak filling rate was observed.

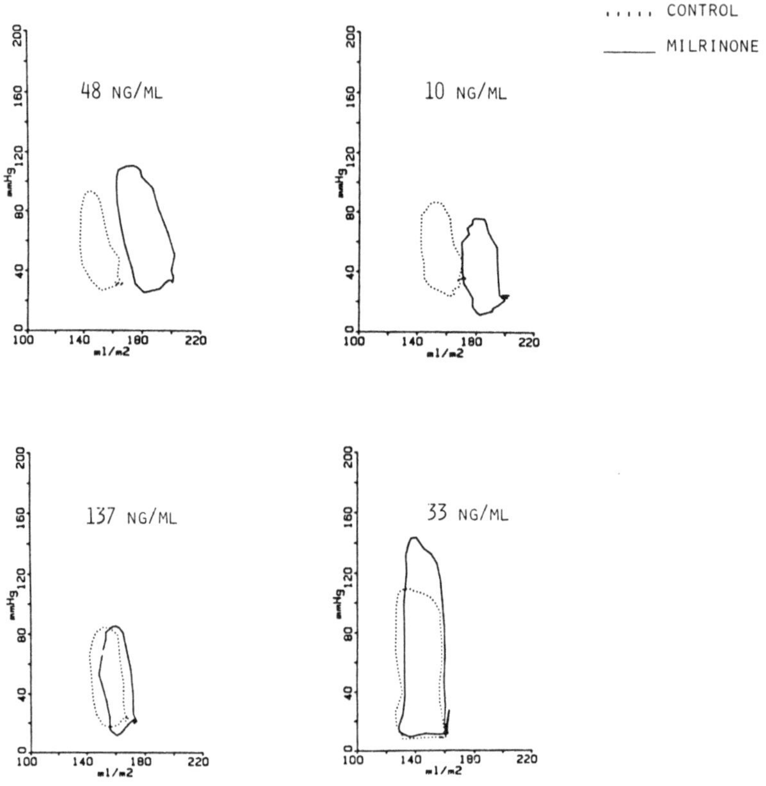

Fig. 5. The pressure-volume loops of the 4 patients with a low plasma level of milrinone showed a rightward shift with only in one case a slight downward shift. An increase in compliance with decreased distensibility was observed. Individual plasma levels of milrinone achieved at the end of the study are shown.

Discussion

The present study shows that milrinone increases myocardial contractility and improves left ventricular performance in patients with severe congestive heart failure. The augmentation in myocardial contractility was documented in our patients by an increase in the isovolumic indexes of contraction such as the peak rate of left ventricular pressure rise (+ dP/dt), the relation between dP/dt pressure and total pressure linearly extrapolated to pressure = 0 (Vmax) and the peak velocity of contractile element (Vce). We used these parameters to assess the inotropic effect of milrinone since it has been proved that they are less dependent than ejection phase indexes on change in preload and/or afterload and heart rate (5, 8, 9). The observed increase in such reliable parameters of inotropic state validate the increase observed in left ventricular ejection fraction which, if used alone, is of limited utility in the evaluation of combined inotropic-vasodilator agents such as milrinone. In fact such parameter is highly susceptible to changes in the afterload state of the left ventricle, a decrease in left ventricular afterload causing an increase in most left ventricular ejection phase indexes (9). The increase observed in the net work of left ventricle despite a fall in left ventricular end-diastolic pressure, is also suggestive of a positive inotropic action although not specific, since this combination of hemodynamic changes is also observed with pure vasodilators such as nitroprusside (12). Our protocol was not designed to assess the relative inotropic and vasodilator components of milrinone's action but a recent report (7) demonstrated that, at an equivalent afterload, milrinone increased both peak positive dP/dt and stroke work while nitroprusside did not. Thus such results strongly suggest that milrinone exerts a significant portion of its hemodynamic effects by means of a positive inotropic action. Another important effect observed after milrinone is the significant improvement in early relaxation. The peak negative dP/dt significantly increased despite a small decrease in end-systolic pressure and all the constants of left ventricular pressure decay decreased significantly. We are unable to explain the mechanism by which such improvement was determined by milrinone. It is possible that an improvement in contraction could lead to an improved relaxation although in a recent experimental study such hypothesis has been rejected (13). It is also possible that a reduced afterload could lead to an improved relaxation but we did not observe significant fall in aortic pressure. Thus it is possible to hypothesize that milrinone directly exerted a positive action on the complex phenomenon of relaxation, probably by improving the intracellular handling of calcium consistent with the inhibition of fraction III of phosphodiesterase observed in experimental studies with milrinone (6). The accelerated relaxation by milrinone may contribute to an enhancement of early diastolic filling although, in our patients, it seems that a therapeutic plasma level of milrinone is required to observe an improvement in filling. Accelerated relaxation and improved filling caused a leftward and in some cases, downward shift of the pressure volume curve, indicating improvement in left ventricular chamber distensibility. In four patients a rightward without downward shift was observed due to an increased volume in absence of significant changes in pressure. In those patients, the lack of absorption of milrinone probably led to a pure overload of the left ventricle mainly due to an increased preload consequent to the prolonged supine position in absence of any significant improvement in hemodynamics and contraction. In conclusion, oral milrinone was proved to be an active inotropic and vasodilator drug, leading to a globally improved performance of failing left ventricle, without any significant side effects. Further studies are needed to validate a prolonged and effective oral administration, which will eventually be able to substitute the use of cardiac glycosides.

References

1. Baim DS, McDowell AV, Cherniler J, Monard E, Parker JA, Edelwon J, Braunwald E, Grossmann W (1983) Evaluation of a new bipyridine inotropic agent – milrinone – in patients with severe congestive heart failure. New Engl J Med 309: 749–756
2. Borow KM, Come PC, Neumann A, Baims DS, Braunwald E, Grossmann W (1985) Physiologic assessment of the inotropic, vasodilator and afterload reducing effects of milrinone in subjects without cardiac disease. Am J Cardiol 55: 1204–1209
3. Brower RW, Meij S, Serruys PW (1983) A model of asynchronous left ventricular relaxation predicting the bi-exponential pressure decay. Cardiovasc Res 17: 482–488
4. Edelson J, Koss RF, Baker JF, Park GO (1983) High performance liquid chromatographic analysis of WIN 47.203 in plasma and urine. Intravenous pharmacokinetics in the dog. Chromatogr F276: 456–462

5. Grossmann W, Haynes F, Paraskos JA, Saltz S, Dalen JE, Dexter LE (1972) Alterations in preload and myocardial mechanics in the dog and in man. Circ Res 31: 83–94
6. Gwethmey JK, De Feo TT, Morgan JP (1984) The effects of milrinone and 2H-imidazol-2-one, 4-ethyl-1,3-didydro-5-(4-pyridimilcarbonil)-(MDL 19.205) on intracellular calcium transients in mammalian working myocardium. Fed Proc 43: 768 (abstr)
7. Jaski BE, Fifer MA, Wright RF, Braunwald E, Colucci WS (1985) Positive inotropic and vasodilator actions of milrinone in patients with severe congestive heart failure. Dose-response relationships and comparison to nitroprusside. J Clin Invest 75: 643–649
8. Mahler F, Yoran C, Ross J Jr (1974) Inotropic effect of tachycardia and post-stimulation potentiation in the conscious dog. Am J Physiol 227: 569–575
9. Mahler F, Ross J Jr, O'Rourke RA, Covell JW (1975) Effects of changes in preload, afterload and inotropic state on ejection and isovolumic phase measures of contractility in the conscious dog. Am J Cardiol 35: 626–634
10. Maskin CS, Sinoway L, Chadwick B, Sonnenblick EH, Le Jemtel TH (1983) Sustained hemodynamic and clinical effects of a new cardiotonic agent, WIN 47203, in patients with severe congestive heart failure. Circulation 67: 1065–1070
11. Meester GT, Bernard N, Zeelenberg C, Brower RW, Hugenholtz PG (1975) A computer system for real time analysis of cardiac catheterization data. Cathet Cardiovasc Diagno 1: 112–132
12. Rude RE, Grossmann W, Colucci WS, Benotti JR, Carabello BA, Wynne J, Malacoff R, Braunwald E (1981) Problems in assessment of new pharmacologic agents for the heart failure patient. Am Heart J 102: 584–590
13. Sabbah HN, Stein PD (1981) Effect of inotropic interventions on rate of change of ventricular diastolic dimensions. Am J Physiol 241: H376–380
14. Scott Monrad E, MacKay RG, Baim DS, Colucci WS, Fifer MA, Heller CV, Royal HD, Grossmann W (1984) Improvements in indexes of diastolic performance in patients with congestive heart failure treated with milrinone. Circulation 70: 1030–1037
15. Serruys PW, Wijns W, van den Brand M, Meij S, Slager CJ, Schuurbiers JCH, Hugenholtz PG, Brower RW (1984) Left ventricular performance, regional blood flow, wall motion and lactate metabolism during transluminal angioplasty. Circulation 70: 25–36
16. Simonton C, Chatterjee K, Cooly RJ, Kubo SH, Leonard D, Daly P, Rutman H (1985) Milrinone in congestive heart failure: acute and chronic hemodynamic and clinical evaluation. J Am Coll Cardiol 6: 453–459
17. Slager CJ, Reiber JHC, Schuurbiers JCH, Meester GT (1978) Contouromat, a hardwired left ventricular angio processing system. Design and application. Comput Biomed Res 11: 491–502

Author's address:

Patrick W. Serruys, Catheterization Laboratory, Thoraxcenter, P. O. Box 1738, 3000 DR Rotterdam (The Netherlands)

The importance of skeletal muscles as a distribution volume for digitalis glycosides in guinea-pigs and human subjects

K. Kjeldsen

Institute of Physiology, University of Aarhus (Denmark)

Summary

The importance of skeletal muscles as a distribution volume for digitalis glycosides was evaluated in guinea-pigs and human subjects by measuring the ^{3}H-ouabain binding capacity and the affinity of the Na, K-ATPase for ^{3}H-ouabain in vitro in skeletal and heart ventricular muscles. Furthermore, in guinea-pigs the distribution of ^{3}H-ouabain in vivo among skeletal and heart ventricular ^{3}H-ouabain receptors and plasma was investigated. The possible redistribution of ^{3}H-ouabain away from plasma and into the skeletal muscle was investigated in rats by intraperitoneal injections of ^{3}H-ouabain and β_2-adrenoceptor stimulator.

In 3–4 week old guinea pigs the total amount of ^{3}H-ouabain binding sites in skeletal and heart ventricular muscle is 42,700 and 1,200 pmol, respectively. K_D-values are 0.5×10^{-7}–0.7×10^{-7} and 1.5×10^{-7} mol/l in skeletal and heart ventricular muscles, respectively. Thus, the total binding capacity for ^{3}H-ouabain is around 36 fold larger in the skeletal muscle pool than in the entire heart ventricular muscle. The K_D-values indicate that a 10 % occupancy of ^{3}H-ouabain binding sites in the heart is associated with an occupancy of 21–30 % in the skeletal muscles. Following intraperitoneal injection of ^{3}H-ouabain it was found that a 10 % occupancy of the Na,K-pumps in the heart is associated with a 13 % occupancy in the skeletal muscles. One hour after the injection of ^{3}H-ouabain intraperitoneally the skeletal muscles contain around 5 times as much specifically bound ^{3}H-ouabain as available in the extracellular space.

In human subjects the total amount of ^{3}H-ouabain binding sites in skeletal and heart ventricular muscles is 6,100–9,000 and 185 nmol, respectively. K_D in skeletal muscles is 0.9–2.0 times that in the heart ventricular muscle. Thus, the total binding capacity for ^{3}H-ouabain is 33–49 fold larger in the skeletal muscles than in the heart ventricles. When 10 % of the ^{3}H-ouabain binding sites in the heart muscles of human subjects are occupied a fraction of 5–11 % of those in the heart can be excepted to be occupied.

The administration of ^{3}H-ouabain and a β_2-adrenoceptor stimulator intraperitoneally to rats increases the occupancy of skeletal muscle ^{3}H-ouabain binding sites by 66–82 %. This was associated with a decrease in plasma ^{3}H-activity of up to 20 %. Thus β_2-adrenoceptor stimulation can induce a redistribution of digitalis glycosides away from plasma into the skeletal muscles.

Taken together, the observations emphasize the importance of the skeletal muscles as a major pool for the distribution of cardiac glycosides. Thus, changes in skeletal muscle ^{3}H-ouabain binding capacity with age, K-depletion or thyroid status may determine the amount of digitalis glycoside available in plasma for binding to the heart during acute digitalization.

1. Introduction

The pharmacological receptor for digitalis glycosides is generally accepted to be the Na,K-ATPase of the heart cell membrane (1, 27, 38, 39). Digitalis glycoside binding is not, however, restricted to the heart, but can take place to the Na,K-ATPase throughout the body. Studies on muscle material obtained post mortem from human subjects showed a high concentration of digitalis glycosides in the heart, liver and kidney and lower concentrations in skeletal muscle (2, 8). But as the skeletal muscles can constitute as much as 40% of the body weight, considerable amounts are bound here (31). This points to the possibility that the skeletal muscle pool of Na,K-ATPase constitutes a distribution volume for digitalis glycosides of importance for the amount of the drug available in plasma for binding to the heart during digitalization.

2. ^{3}H-ouabain binding capacity in guinea-pig skeletal and heart ventricular muscle

Measurements of the ^{3}H-ouabain binding capacity in 3–4 week old guinea-pigs gave values of 303, 474 and 1,092 pmol/g wet wt. in soleus, gastrocnemius and heart ventricular muscle specimens, respectively. Assuming that skeletal and heart ventricular muscle in these animals constitute around 40 and 0.5% of the body weight, respectively, the total amount of ^{3}H-ouabain binding sites in these tissues was calculated. The ^{3}H-ouabain binding site concentration in gastrocnemius muscle was used as representative for the entire skeletal muscle pool since it is composed of a mixture of red and white fibres as is the major part of the skeletal muscles. Thus, in a 3–4 week old guinea-pig with a body weight of 225 g, the total amount of ^{3}H-ouabain binding sites in skeletal and heart ventricular muscle is 42,700 and 1,200 pmol, respectively. Hence, the total binding capacity for ^{3}H-ouabain is around 36 fold larger in the skeletal muscle pool than in the entire heart ventricular muscle in guinea-pigs. The digitalis glycoside receptor concentration determined using ^{3}H-ouabain or ^{3}H-digoxin binding was found to be in good agreement in skeletal as well as heart ventricular muscles (23).

The observation of a total ^{3}H-ouabain binding capacity of 1,092 pmol/g wet wt. in guinea-pig heart ventricular muscle was confirmed by measurements of a K-activated 3-O-methylfluorescein phosphatase (3-O-MFPase) activity corresponding to a Na,K-ATPase concentration of 1,450 pmol/g wet wt. in crude homogenates of the same tissue (34). In view of the fact that the value for 3-O-MFPase activity is based upon estimations of the molecular activity of the 3-O-MFPase in different enzyme preparations prepared using different procedures, it is in satisfactory agreement with the ^{3}H-ouabain binding measurements (23). Earlier determinations of the ^{3}H-ouabain binding capacity of papillary muscles of the guinea-pig heart have given a value of 1,900 pmol/g wet wt. (29). This discrepancy may be related to a difference between the heart ventricular and papillary muscles similar to the difference known from studies on rat skeletal muscles between skeletal muscles within one animal (20). In other studies on heart Na,K-ATPase considerably lower values were obtained (9, 10). This can be ascribed to differences in the recovery of Na,K-ATPase.

3. Relative affinities of the Na,K-ATPase for ouabain in the guinea-pig skeletal and heart ventricular muscle

Determination of the dissociation constants for ^{3}H-ouabain receptor-ligand interaction (K_D) in muscle specimens incubated in vanadate buffer gave values of 0.5×10^{-7}, 0.7×10^{-7} and 1.5×10^{-7} mol/l in soleus, gastrocnemius and heart ventricular muscle from 3–4 week old guinea-pigs, respectively (23). From these values, it could be expected that when digitalis glycosides are administered in vivo, the skeletal muscles will reach a higher level of relative occupancy than the heart. Measurements of the fraction of ^{3}H-ouabain binding sites occupied in skeletal muscle and the heart one hour after the administration of ^{3}H-ouabain intraperitoneally (0.003–0.6 nmol/g body wt.) showed that the relative occupancy of the digitalis glycoside receptors in skeletal muscle at all doses tested is somewhat higher than in the heart. Thus, at a ^{3}H-ouabain dose giving rise to an occupancy of 10% in the heart, 13%

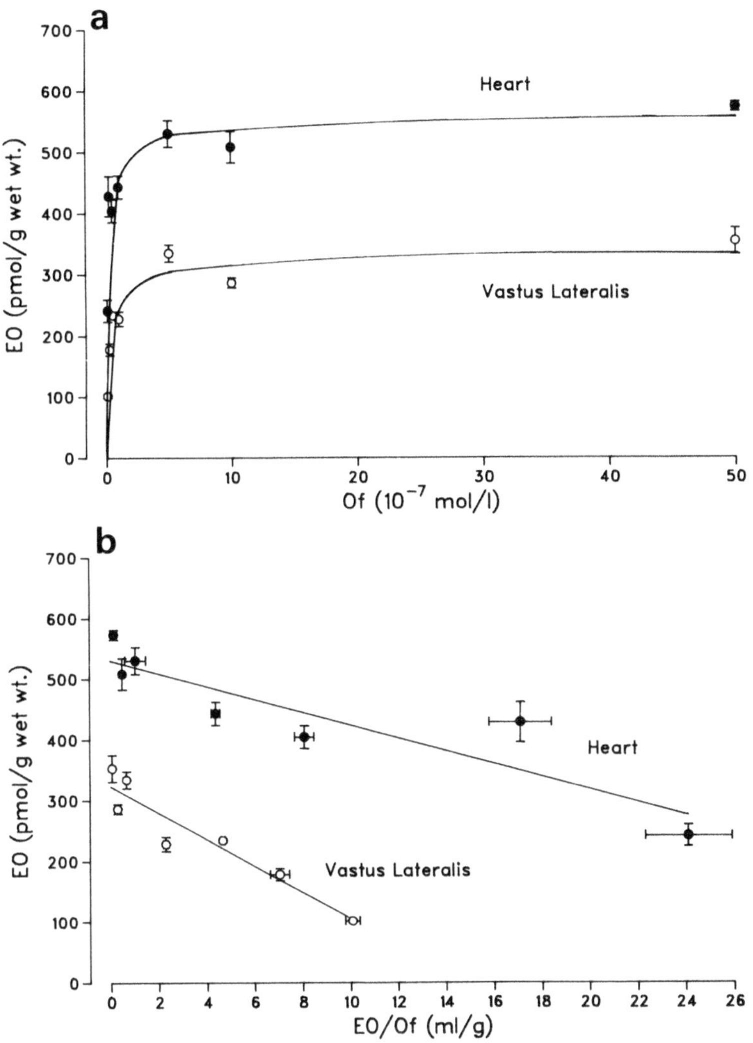

Fig. 1. Effect of the concentration of ouabain on the binding of ^{3}H-ouabain in vastus lateralis and heart ventricular muscle necropsies from a 19 year old male human subject. Cause of death was encephalitis for 1 week. Samples were obtained 8 hours post mortem. ^{3}H-ouabain binding was measured in a vanadate-tris buffer. Samples were prewashed 2 x 10 min at 0° C, incubated 6 x 60 min at 37° C in ^{3}H-ouabain and ouabain added to a final concentration of 1 x 10^{-8} to 5 x 10^{-6} mol/l and washed 4 x 30 min at 0° C. After a minor correction for unspecific uptake and retention of ^{3}H-ouabain (subtracting a relative uptake of 1–2%), values were corrected for purity of the radioisotope (multiplied by 1.05) and loss of specifically bound ^{3}H-ouabain during washout in the cold (multiplied by 1.05 or 1.02 for skeletal and heart ventricular muscles, respectively) (33, 35, 36). Panel A shows bound ^{3}H-ouabain (EO) as a function of the concentration of ouabain in the incubation medium (Of). Panel B shows EO as a function of the bound/free ^{3}H-ouabain ratio (EO/Of). The lines of this Scatchard-type plot were calculated using linear regression analysis. The calculated values for the intercepts with the ordinate – i. e. the entire ^{3}H-ouabain binding site concentration (EO_{max}) and the apparent dissociation constants for ^{3}H-ouabain ligand-receptor interaction (K_D) are given in the text. Each point represents the mean value of observations on 5 muscle samples. Bars denote S. E.

of the ^{3}H-ouabain binding sites in the soleus muscle were occupied. This corresponds to 5,550 and 120 pmol of ^{3}H-ouabain binding sites being occupied in the skeletal and heart ventricular muscle of a guinea-pig with a body weight of 225 g. From the K_D values determined in vitro, it was expected that a 10 % occupancy of ^{3}H-ouabain binding sites in the heart would be associated with an occupancy of 30 % in the soleus muscle. The deviation from the relative occupancy determined in vivo can be related to differences in blood flow and the molecular activity of the Na,K-pump. As the activity of the Na,K-pump is increased when the number of depolarisations increases, it is likely to be higher in the constantly working heart than in skeletal muscle. Based upon the observation that the binding rate of ^{3}H-ouabain to the Na,K-pump under specified conditions increases linearly with the rate of pumping (4, 17), the difference in activity can be assumed to favour binding of digitalis glycosides to the heart in vivo.

From the plasma ^{3}H–activity determined when the animals were killed by decapitation, it was calculated that 1 hour after the injection of ^{3}H-ouabain intraperitoneally the skeletal muscles contain around 5 times as much specifically bound ^{3}H-ouabain as available in the extracellular space (ECV). In contrast the heart contains only around 1/10 of the amount in ECV (23).

4. ^{3}H-ouabain binding capacity and relative affinities in human skeletal and heart ventricular muscle

Based upon the observation that the Na,K-pump concentration in human skeletal and heart muscles only decreases relatively slowly post mortem (1–2 % per hour) (35, 36), it was of relevance for the in vivo situation to study ^{3}H-ouabain binding in muscle samples obtained post mortem. To achieve optimum conditions for comparing ^{3}H-ouabain binding in samples of skeletal and heart ventricular muscles, necropsies of the two muscles were incubated within the same experiments. The evaluations were carried out on muscle samples from a 19 year old male and a 73 year old female human subject. To determine the maximum ^{3}H-ouabain binding site concentration (EO_{max}) and the apparent dissociation constants for ^{3}H-ouabain receptor-ligand interaction in vitro (K_D), samples were incubated in a vanadate-tris buffer (33, 35, 36). When the specific binding of ^{3}H-ouabain (EO) was plotted as a function of the amount of specifically bound ^{3}H-ouabain relative to the amount of free ^{3}H-ouabain in the incubation medium (EO/Of) in a Scatchard-type plot (Fig. 1b and Fig. 2b), straight lines were obtained. This indicates that human skeletal and heart ventricular muscles contain only one class of binding sites. Values for EO_{max} and K_D found in the present study were in agreement with the normal ranges earlier determined for human skeletal (35) and heart ventricular muscles (36).

In samples from the 19 year old male, EO_{max} was 323 pmol/g wet wt. and K_D 2.2×10^{-8} mol/l in the vastus lateralis muscle. In the heart, values were 530 pmol/g wet wt. and 1.1×10^{-8} mol/l (Fig. 1).

In samples from the 73 year old female, EO_{max} was 217 pmol/g wet wt. and K_D 1.3×10^{-8} mol/l in the vastus lateralis muscle. In the heart, values were 524 pmol/g wet wt. and 1.4×10^{-8} mol/l (Fig. 2).

In the skeletal muscles of the young subject EO_{max} was 1.5-times that in the old subject. As it has been shown that training in dogs is associated with an increase in the Na,K-ATPase concentration in skeletal muscles (25), it is possible that this difference is due to a higher state of muscular training in the young as compared to the old person. On the basis of a body weight of 70 kg, the total number of ^{3}H-ouabain binding sites in the skeletal and heart ventricular muscle pools were calculated, assuming that skeletal muscles and the heart of adult human subjects constitute around 40 and 0.5 % of body weight. Thus, in the young

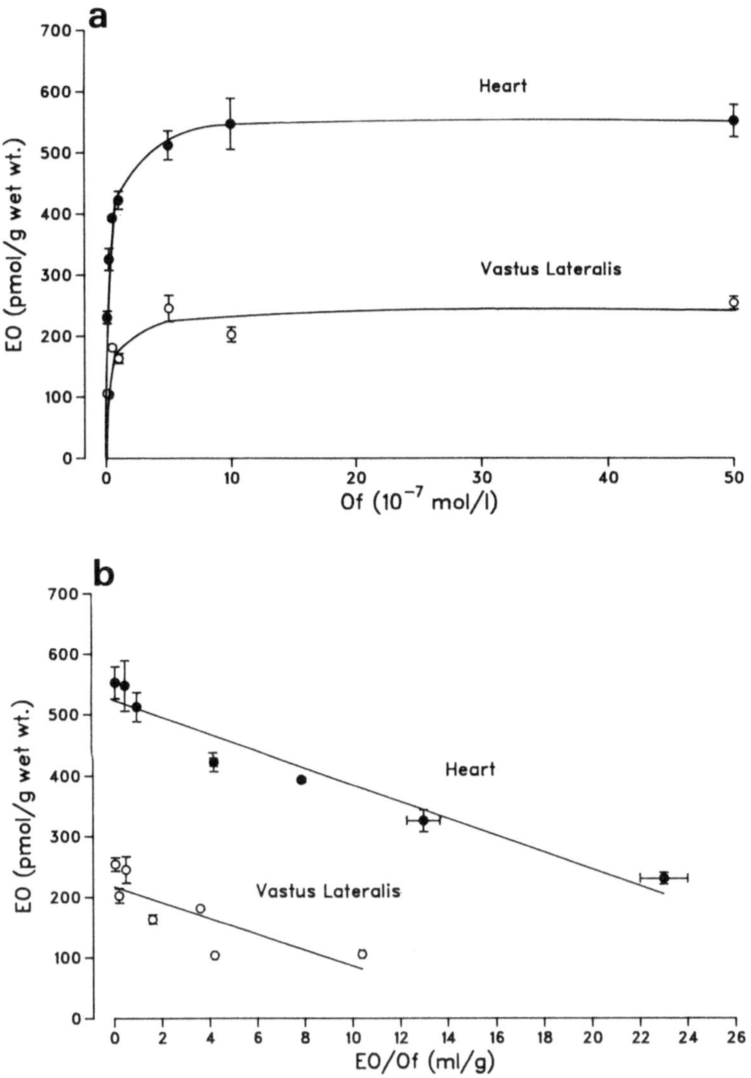

Fig. 2. Effect of the concentration of ouabain on the binding of ^{3}H-ouabain in vastus lateralis and heart ventricular muscle necropsies from a 73 year old female human subject. Cause of death was chronic obstructive lung disease for several years. Samples were obtained 10 hours post mortem. Methods as in legend to Fig. 1. Each point represents the mean value of observations on 5 muscle samples. Bars denote S. E.

subject the total amount of ^{3}H-ouabain binding sites in the skeletal and heart ventricular muscle pool is 9,000 and 185 nmol, respectively. These values are illustrated by the area of the boxes in Fig. 3. In the old subject values were 6,100 and 185 nmol, respectively. Thus, the skeletal muscle pool of digitalis receptors is 33–49 fold larger than that in the entire heart ventricular muscle in human subjects.

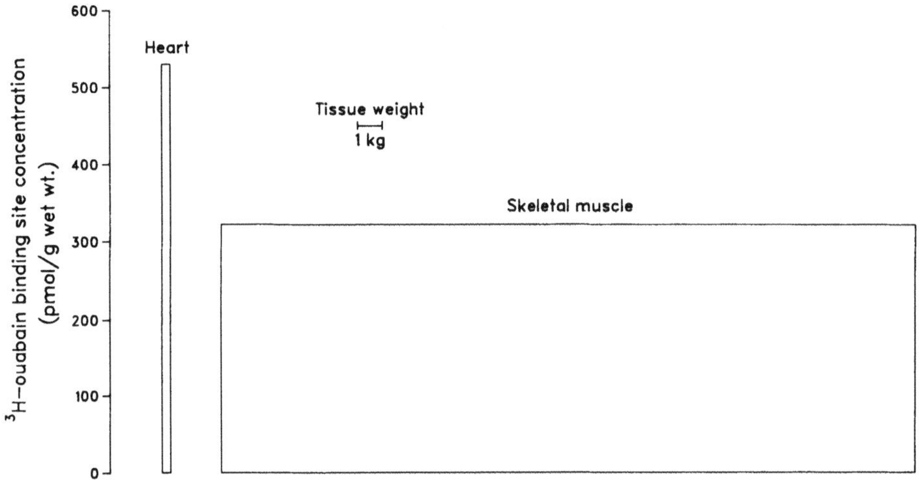

Fig. 3. Total pool of ^{3}H-ouabain binding sites in heart and skeletal muscles of a 19 year old male human subject (see legend to Fig. 1). The height of the boxes represents the concentration of ^{3}H-ouabain binding sites (EO_{max}) (see legend to Fig. 1). The lenght of the boxes represents the total mass of the tissues based upon a body weight of 70 kg and assuming that the skeletal muscles and heart ventricular tissue constitutes around 40 and 0.5 % of the body weight, respectively.

In the young subject K_D for ^{3}H-ouabain receptor-ligand interaction in skeletal muscles was twice that in the heart ventricular muscle. In the old person this factor was 0.9. Thus, when 10 % of the ^{3}H-ouabain binding sites in the heart muscles of human subjects during digitalization are occupied by digitalis glycosides, a fraction of around 5 and 11 % of the ^{3}H-ouabain binding sites in the skeletal muscles can be expected to be occupied in the young and old subject, respectively. Thus, even if it is taken into account that the heart ventricular muscle in vivo has a faster binding rate than the skeletal muscles for ^{3}H-ouabain, it is evident that the digitalis glycosides bound to the skeletal muscle receptors during digitalization can flood the plasma as well as the receptors in the heart ventricles with digitalis glycosides. Conversely, an acute increase in the rate of binding of digitalis glycosides to the muscles may decrease the concentration in plasma and lead to a net release of digitalis from the heart. Indeed, during exercise plasma digoxin was found to decrease (16, 37), probably reflecting an increased rate of active Na,K-transport in the muscles and a concomitant stimulation of the rate of digoxin binding.

5. The effect of β_2-adrenoceptor stimulator on skeletal muscle ^{3}H-ouabain binding in vivo

Because catecholamines increase the activity of the Na,K-pump (4, 30) and the rate of ^{3}H-ouabain binding was found partly to reflect the turnover rate of the Na,K-pump in skeletal muscle (4, 17), it was of interest to study if the administration of a β-adrenoceptor stimulator in vivo to digitalized animals would increase the occupancy of ^{3}H-ouabain receptors in muscles. To achieve an increase of the occupancy of digitalis glycoside receptors in the skeletal muscles without any major increase in the heart, a β_2-adrenoceptor stimulator (salbutamol) was used.

As can be seen from Fig. 4, salbutamol increased the occupancy of skeletal muscle ^{3}H-ouabain binding sites by 66–82 %. This was associated with a decrease in plasma ^{3}H-activity

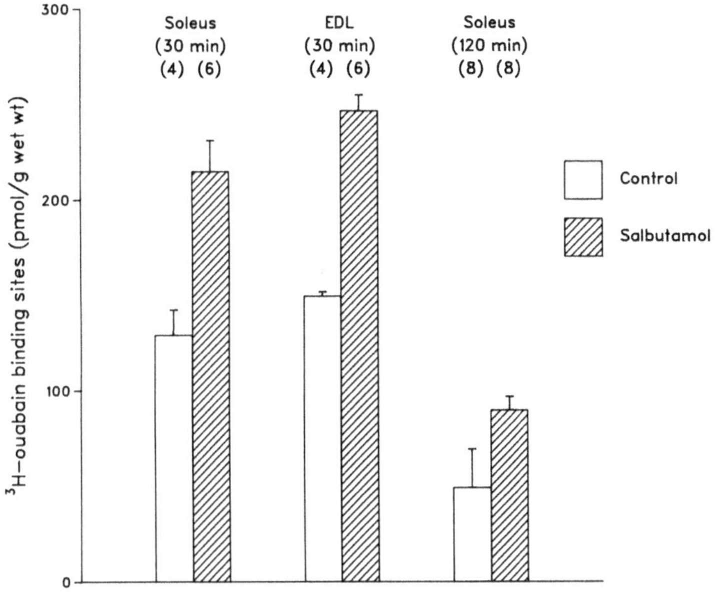

Fig. 4. Effect of β_2-adrenoceptor stimulation (salbutamol) on digitalis glycoside distribution in vivo. These evaluations were carried out on 4 week old rats given an intraperitoneal injection of ^{3}H-ouabain and ouabain of 0.75 nmol per g body weight. 30 min later they were given 2.3 μg salbutamol per g body weight intraperitoneal. 30 or 60 min later animals were killed and the ^{3}H-activity in soleus or extensor digitorum longus (e. d. l.) muscles (after washout for 4 x 30 min in the cold) and plasma determined (5). Although rats are rather insensitive to digitalis glycosides due to a low affinity of digitalis receptors in the heart for digitalis glycosides, it was reasonable to use rats in these experiments because the K_D for ^{3}H-ouabain receptor-ligand interaction in rat skeletal muscle is of the same order of magnitude as in skeletal muscles from guinea-pigs and human subjects. Each column represents the mean value on observations on 3–4 animals. Bars denote S. E.

of up to 20%. This observation indicates that β_2-adrenoceptor stimulators: 1) increase the activity of the Na,K-pump in skeletal muscles in vivo and that these drugs during digitalization 2) increase the occupancy of receptors in skeletal muscles by digitalis glycosides and 3) decrease the concentration of digitalis glycosides in plasma. Evaluation of the possible importance of β_2-adrenoceptor stimulators to obtain a temporary redistribution of digitalis glycosides away from the heart into the skeletal muscles in cases of digitalis intoxication will await studies of the occupancy of digitalis glycoside receptors in the digitalis glycoside sensitive guinea-pig heart ventricles after the administration of salbutamol to digitalized animals.

6. Importance of changes in skeletal muscle ^{3}H-ouabain binding site concentration with age, K-depletion and thyroid status for digitalis glycoside distribution

Changes in the amount of digitalis glycosides required to obtain inotropy or symptoms of intoxication occur in two different situations: a change in digitalis sensitivity occurs when the plasma concentration at which inotropy or intoxication symptoms is seen is changed. This can mainly be ascribed to changes in binding characteristics of digitalis glycosides to its receptor in the heart ventricular muscle. A change in digitalis toxicity occurs when there is a change in the dose of digitalis glycoside required to obtain an unchanged plasma concentration. Such a change can in part be ascribed to a change in distribution volume for digitalis glycosides. In some cases the two mechanisms may act antagonistically: an increase in affinity of the digitalis glycoside receptor for its ligand in the heart will increase the binding of digitalis glycosides to this tissue. If this is accompanied by an increased affinity in skeletal muscles, the binding of the drug to this tissue will increase and less digitalis glycoside will be available in plasma for binding to the heart, and vice versa for a concomitant decrease in affinity in the heart and skeletal muscles. Furthermore, an agonistic effect may be seen if the affinity of digitalis glycoside for its receptor in the heart is increased and at the same time decreased in the skeletal muscle and vice versa.

It is a well known clinical experience that digitalis toxicity changes with age and thyroid status. These changes are in part evident from differences in the amount of digitalis glycoside required per kg of body weight to achieve a given serum level (5, 6, 7, 12, 13, 40, 41). During the initial phase of any digitalization – i. e. during non-steady-state – the size of the distribution volume of the drug is important for the dose required. Furthermore, during steady state of digitalization a redistribution of the drug may take place – e. g. due to a sudden decrease or increase in the ^{3}H-ouabain binding rate of the Na,K-ATPase. Under these conditions small variations in the occupancy of ^{3}H-ouabain binding sites in skeletal muscle may determine the amount of digitalis glycoside available in plasma for specific binding to the heart.

When a bolus of ouabain was given to young and old dogs, the plasma level of ouabain was lower in the young animals (13). Clinical experience indicates that the initial dose of digoxin and digitoxin should be higher in children than in adults (6, 12). These differences have been ascribed to a larger distribution volume for digitalis glycosides in young as compared to old subjects (13, 41). In keeping herewith a more pronounced and sustained rise in plasma ^{3}H-activity after intraperitoneal injection of ^{3}H-ouabain was found in 12 week old compared to 4 week old rats receiving the same dose per kg body weight (5). From the 4th to the 12th week of life, the concentration of ^{3}H-ouabain binding sites in rat skeletal muscle decreases by 58% (19, 20), and in guinea-pigs the skeletal muscles show a decrease of around 60% from birth to maturity (23). As the reductions in ^{3}H-ouabain binding site concentration were not found to be associated with any major change in apparent K_D in rats (19, 20), the observations offer an explanation for the larger volume of distribution in young as compared to old animals. A precise evaluation should, however, take into consideration that the mass of skeletal muscles in younger animals constitutes a smaller percentage of body weight, reducing its importance as distribution volume for digitalis glycosides. Some investigators have attributed the larger volume of distribution in young subjects to changes in erythrocyte binding capacity for digitalis glycosides (14, 15, 18). It can, however, be calculated that the total amount of digitalis receptors in the erythrocyte pool of an adult human subject is around 10 nmol (11). In human skeletal muscle pool the binding capacity is 610–900 times higher. Thus, changes in erythrocyte digitalis glycoside binding seem to be of minor importance for digitalis distribution when compared to the skeletal muscle pool.

A decrease in tolerance to digitalis glycosides with hypokalemia is well documented (24, 28, 42). Together with the direct effect of hypokalemia on the action of digitalis glycosides, the decrease of up to 78% in the concentration of ^{3}H-ouabain receptors in skeletal muscle with K-depletion (21, 32) may be of importance for the amount of digitalis glycoside required to obtain a given serum level. In keeping with this, measurements of plasma ^{3}H-activity 15 min after intraperitoneal injection of ^{3}H-ouabain showed a 77% higher value in K-depleted rats as compared to controls receiving the same dose per kg body weight (5).

Following the same dose of digoxin per kg body weight, patients with hypothyroidism have higher serum levels and patients with hyperthyroidism lower levels than euthyroid subjects (7, 40). This is generally attributed to variations in the renal clearance of digoxin. Thus, in a study using digitoxin, which is mainly excreted by degradation in the liver, no major difference was observed in the mean concentration of this drug in plasma from patients with hypo- or hyperthyroidism. However, the lowest and highest digitoxin concentration was found in hyper- and hypothyroid patients, respectively (26). Thus, the large changes which occur in skeletal muscle ^{3}H-ouabain binding site concentration with thyroid status in animals (3, 21) as well as human subjects (22), probably contribute to the mechanism causing the change in tolerance to digitalis glycosides with thyroid status.

Conclusion

In adult human subjects as well as in adult guinea-pigs, the ^{3}H-ouabain binding capacity in the skeletal muscle pool is 33–49 fold larger than that in the entire heart ventricular muscle. In both species the affinity of receptors for ouabain in skeletal as well as heart ventricular muscles are of the same order of magnitude. Furthermore, β_2-adrenoceptor stimulation can cause an increase in ^{3}H-ouabain binding in skeletal muscles associated with a decrease in plasma ^{3}H-activity. Thus, skeletal muscles can be expected to constitute a distribution volume of major importance for the amount of digitalis glycoside available in plasma for binding to the heart during digitalization. Hence, changes in skeletal muscle ^{3}H-ouabain binding site concentration with age, K-depletion and hypo- or hyperthyroidism may be of importance for the change in digitalis toxicity associated with these conditions.

Acknowledgements

I wish to thank Torben Clausen and Otto Hansen for valuable discussions. Susanne Olesen and Jim Stenfatt Larsen are thanked for skilled technical assistance. The study was in part supported by the Danish Medical Research Council and the Danish Heart Foundation.

References

1. Akera T, Brody TM (1977) The role of Na$^+$,K$^+$-ATPase in the inotropic action of digitalis. Pharmacol Rev 29: 187–220
2. Andersson K-E, Bertler Å, Wettrell G (1975) Post-mortem distribution and tissue concentrations of digoxin in infants and adults. Acta Paediat Scand 64: 497–504
3. Biron R, Burger A, Chinet A, Clausen T, Dubois-Ferriere R (1979) Thyroid hormones and the energetics of active sodium-potassium transport in mammalian skeletal muscles. J Physiol 297: 47–60
4. Clausen T, Hansen O (1977) Active Na-K transport and the rate of ouabain binding. The effect of insulin and other stimuli on skeletal muscle and adipocytes. J Physiol 270: 415–430
5. Clausen T, Hansen O, Kjeldsen K, Nørgaard A (1982) Effect of age, potassium depletion and denervation on specific displaceable ^{3}H-ouabain binding in rat skeletal muscle in vivo. J Physiol 333: 367–381
6. Cree JE, Coltart DJ, Howard MR (1973) Plasma digoxin concentration in children with heart failure. Br Med J 1: 443–444
7. Doherty JE, Perkins WH (1966) Digoxin metabolism in hypo- and hyperthyroidism. Studies with tritiated digoxin in thyroid disease. Ann Intern Med 64: 489–507
8. Doherty JE, Perkins WH, Flanigan WJ (1967) The distribution and concentration of tritiated digoxin in human tissues. Ann Intern Med 66: 116–124
9. Dutta S, Goswami S, Lindower JO, Marks BH (1968) Subcellular distribution of digoxin-H^3 in isolated guinea-pig and rat hearts. J Pharmacol Exp Ther 159: 324–334

10. Erdmann E, Bolte HD, Lüderitz B (1971) The ($Na^+ + K^+$)-ATPase activity of guinea-pig heart muscle in potassium deficiency. Archs Biochem Biophys 145: 121–125
11. Erdmann E, Hasse W (1975) Quantitative aspects of ouabain binding to human erythrocyte and cardiac membranes. J Physiol 251: 671–682
12. Giardina ACV, Ehlers KH, Morrison JB, Engle MA (1975) Serum digitoxin concentrations in infants and children. Circulation 51: 713–717
13. Glantz SA, Kernoff R, Goldmann RH (1976) Age-related changes in ouabain pharmacology. Ouabain exhibits a different volume of distribution in adult and young dogs. Circ Res 39: 407–414
14. Gonsalves ML, Storstein L, Larsen A (1983) Distribution of digoxin and digitoxin between serum and erythrocytes in adults and children. In: Gillmann H, Storstein L (eds) Digitalistherapie Heute. Verlag für angewandte Wissenschaften, Munich, pp 31–34
15. Gorodischer R, Jusko WJ, Yaffe SJ (1975) Tissue and erythrocyte distribution of digoxin in infants. Clin Pharmacol Ther 19: 256–263
16. Jogestrand T, Sundqvist K (1981) Effect of physical exercise on the digoxin concentrations in skeletal muscle and serum in man. Clin Physiol 1: 99–104
17. Joiner CH, Lauf PK (1978) The correlation between ouabain binding and potassium pump inhibition in human and sheep erythrocytes. J Physiol 283: 155–175
18. Kearin M, Kelly JG, O'Malley K (1980) Digoxin "receptors" in neonates: An explanation of less sensitivity of digoxin than in adults. Clin Pharmacol Ther 28: 346–349
19. Kjeldsen K, Nørgaard A, Clausen T (1982) Age-dependent changes in the number of ^{3}H-ouabain binding sites in rat soleus muscles. Biochim Biophys Acta 686: 253–256
20. Kjeldsen K, Nørgaard A, Clausen T (1984) The age-dependent changes in the number of ^{3}H-ouabain binding sites in mammalian skeletal muscle. Pflügers Arch 402: 100–108
21. Kjeldsen K, Nørgaard A, Clausen T (1984) Effect of K-depletion on ^{3}H-ouabain binding and Na-K-contents in mammalian skeletal muscle. Acta Physiol Scand 122: 103–117
22. Kjeldsen K, Nørgaard A, Gøtzsche CO, Thomassen A, Clausen T (1984) Effect of thyroid function on number of Na-K pumps in human skeletal muscle. Lancet ii: 8–10
23. Kjeldsen K, Nørgaard A, Hansen O, Clausen T (1985) Significance of skeletal muscle digitalis receptors for ^{3}H-ouabain distribution in the guinea-pig. J Pharmacol Exp Ther 234: 720–727
24. Kleiger RE, Seta K, Vitale JJ, Lown B (1966) Effects of chronic depletion of potassium and magnesium upon the action of acetylstrophanthidin on the heart. Am J Cardiol 17: 520–527
25. Knochel JP, Blachley JD, Johnson JH, Carter NW (1985) Muscle cell electrical hyperpolarization and reduced exercise hyperkalemia in physically conditioned dogs. J Clin Invest 75: 740–745
26. Kolenda KD, Juhl-Johnson C, Kokenge F (1983) Vergleichende Untersuchungen der Serumspiegel von Digoxin und Digitoxin und der Nierenfunktion bei Patienten mit Hyper- und Hypothyreose. In: Gillmann H, Storstein L (eds) Digitalistherapie Heute. Verlag für angewandte Wissenschaften, Munich, pp 35–44
27. Langer GA (1971) The intrinsic control of myocardial contraction-ionic factors. New Engl J Med 285: 1065–1071
28. Lown B, Salzberg H, Enselberg LD, Weston RE (1951) Interrelation between potassium metabolism and digitalis toxicity in heart failure. Proc Soc Exp Biol Med 76: 797–801
29. Lüllman H, Peters T, Ravens U (1975) Studies on the kinetics of ^{3}H-ouabain uptake and exchange in the isolated papillary muscle of the guinea-pig. Br J Pharmacol 53: 99–107
30. Moore RD (1983) Effects of insulin upon ion transport. Biochim Biophys Acta 737: 1–49
31. Nyberg L (1977) Bioavailability of digoxin in man after oral administration of preparations with different dissolution rate. Acta Pharmacol Toxicol 40: (Suppl 3) 1–48
32. Nørgaard A, Kjeldsen K, Clausen T (1981) Potassium depletion decreases the number of ^{3}H-ouabain binding sites and the active Na-K transport in skeletal muscle. Nature 293: 739–741
33. Nørgaard A, Kjeldsen K, Hansen O, Clausen T (1983) A simple and rapid method for the determination of the number of ^{3}H-ouabain binding sites in biopsies of skeletal muscle. Biochem Biophys Res Comm 111: 319–325

34. Nørgaard A, Kjeldsen K, Hansen O (1985) K^+-dependent 3-O-methyl-fluorescein phosphatase activity in crude homogenate of rodent heart ventricle: Effect of K^+ depletion and changes in thyroid status. Eur J Pharmacol 113: 373–382
35. Nørgaard A, Kjeldsen K, Larsen JS, Larsen CG, Larsen FG (1985) Estimation of stability of ^{3}H-ouabain binding site concentration in rat and human skeletal muscle post mortem. Scand J Clin Lab Invest 45: 139–144
36. Nørgaard A, Kjeldsen K, Hansen O, Clausen T, Larsen CG, Larsen FG (in press) Quantification of the ^{3}H-ouabain binding site concentration in human myocardium: A post mortem study. Cardiovascular Res
37. Pedersen KE, Madsen J, Kjær K, Klitgaard NA, Hvidt S (1983) Effects of physical activity and immobilization on plasma digoxin concentration and renal digoxin clearance. Clin Pharmacol Ther 34: 303–308
38. Repke K (1964) Über den biochemischen Wirkungsmodus von Digitalis. Klin Wochenschr 41: 157–165
39. Schwartz A, Lindenmayer GE, Allen JC (1975) The sodium-potassium adenosine triphosphatase: Pharmacological, physiological and biochemical aspects. Pharmacol Rev 27: 3–134
40. Shenfield GM, Thompson J, Horn DB (1977) Plasma and urinary digoxin in thyroid dysfunction. Europ J Clin Pharmacol 12: 437–443
41. Wettrell G, Andersson KE (1977) Clinical pharmacokinetics of digoxin in infants. Clin Pharmacokinet 2: 17–31
42. Zeeman S, Hirsch S, Bellett S (1954) The effect of potassium depletion induced by desoxycorticosterone acetate on the lethal dose of lanatoside C in dogs: Relationship of plasma levels, skeletal and cardiac muscle potassium content to the lethal dose. Am J Med Sci 227: 65–73

Author's address:

Dr. Keld Kjeldsen, Institute of Physiology, University of Aarhus, DK-8000 Aarhus C (Denmark)

Relevant metabolism of cardiac glycosides

K. E. Wirth

Institute of Pharmacology, University of Düsseldorf (West Germany)

Summary

With regard to the metabolism of digitalis glycosides, recent experiments show that cleavage of the sugar residues in the liver is not simple hydrolysis but an oxidative process which is catalyzed by the cytochrome P 450 system.

The enzymatic 12 β-hydroxylation of digitoxin to digoxin is quantitatively neglisible in man.

The pattern of digitoxin and digoxin metabolites, which is found to be fairly inhomogeneous in various studies, considerably depends on the different experimental designs, for instance, on the mode of administration. The relatively low recovery of the glycosides and their metabolites in urine and (if investigated) also in feces makes the evaluation of metabolism and its magnitude more difficult. Approximately, digitoxin is metabolized to 20–50 %, digoxin to 5–50 %.

With respect to strophanthus glycosides, an insufficient recovery was also noted in most of the studies. In contrast to ouabain, cymarin and orally given k-strophanthoside were found to be highly metabolized in urine. Intravenously injected k-strophanthoside was excreted in urine mainly as unchanged drug.

Despite numerous investigations on the metabolism of cardiac glycosides in the past 30 years, there are still unsolved problems. Even some well known data, considered up to now to be reliable, have probably to be revised.

According to the original work of Okita et al. (12, 13) and Repke (14–16), metabolism of cardiac glycosides in man chiefly proceeds by the following steps:

1) cleavage of the sugar (digitoxose) residues and formation of bis- and monodigitoxosides and of aglycones (digitoxigenin and digoxigenin),
2) conjugation reactions with glucuronic and sulfuric acid of the aglycones, probably after epimerisation of the hydroxyl group at C_3, but also of the monodigitoxosides,
3) reduction of the unsaturated lactone ring and formation of dihydrodigitoxin and -digoxin.

1) Cleavage of the digitoxose residues occurs in the liver by an enzyme system which is not totally known (12–18). Recent experiments have shown that a preliminary oxidative process occurs, not just a simple hydrolysis. Schmoldt et al. were able to show in rats (20) and finally also in microsomes of human liver (21), that the sugar chain cleavage is catalyzed by the cytochrome P 450 system. As the first oxidation product, 15′-dehydrodigitoxin and -digoxin was identified by the authors (Fig. 1). It is formed by an oxidation of the axial hydroxyl group of the terminal digitoxose, after which cleavage of the sugar takes place. This also occurs with the bis and the monodigitoxoside by formation of the 9′- and the 3′-dehydro-compounds. In this study, there was no formation of the aglycone digitoxigenin and no 12 β-hydroxylation, i. e., no formation of digoxin. The former opinion that cleavage of digitoxoses is mediated by the enzyme β-glycosidase is today considered to be highly unlikely, because the enzyme has a high specifity for C_3, the hydroxyl group of which is in α (epimer) position.

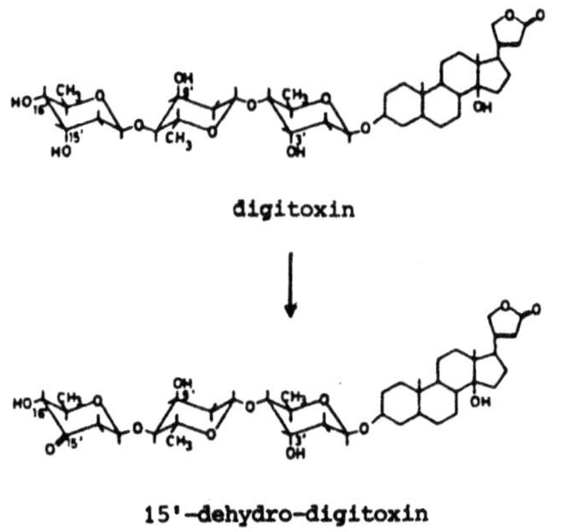

Fig. 1. Chemical formulae of digitoxin and its first oxidation product 15′-dehydro-digitoxin.

2) The reaction steps of conjugation lead to metabolites with a higher water solubility and elimination rate than the unchanged drug (16). Initially, it was assumed that conjugation with glucuronic and sulfuric acid can only take place at the C_3 of the aglycones after epimerisation of the hydroxyl group. Meanwhile, it has turned out that liver capacity is not sufficient for an adequate elimination via this pathway alone. Over the past years, conjugation reactions of the monodigitoxosides have also been fairly reliably demonstrated (5).

3) The reduction of the unsaturated lactone ring and formation of dihydro-compounds is the subject of the two following chapters.

A conversion of digitoxin to digoxin, i. e., the hydroxylation at C_{12} is not of quantitative importance in man, as has been shown several times (4, 17, 24).

There are considerable variations with respect to the magnitude of digitoxin and digoxin metabolism. For this phenomenon, one explanation seems to be that most experiments were performed exclusively on plasma and/or urine (3, 10, 24, 27). Only a few investigators also measured digitalis metabolites in feces. Besides this, in many experiments, elimination of cardiac glycosides was observed for only a few days. Thus, at least in the case of digitoxin with its long elimination period, essential information could not be taken into account. In addition, the relatively low recovery in most studies does not, strictly speaking, allow an accurate evaluation of the order of magnitude of metabolism. One example is the study of Lukas (7) in which in 5 patients after a single dose of 1.0 mg digitoxin, orally given, the elimination of glycoside was measured with a specific double isotope dilution derivative method in urine and feces for 15–21 days. The results were extrapolated to infinity. In this study, 43 % of the given dose was eliminated, 28 % by urine and 15 % by feces (Table 1). In the study of Kuhlmann (6) on 4 subjects, also after a single dose of 1.0 mg digitoxin per os, 38 % was excreted in 8 days, 23 % of which was in urine and 15 % in feces. This means that 60 % of the given dose could not be taken into consideration in these experiments.

Summarizing the various studies concerning digitoxin, it follows that in urine the total excreted amount of the glycoside consists of 50–80 % unchanged digitoxin, the remainder being essentially $CHCl_3$-insoluble metabolites, probably mainly conjugates, and the monodigitoxoside. In feces, about 40 % of the total excreted amount consists of unchanged digitoxin (24–26), in plasma the portion of unchanged digitoxin is claimed to be of about the same magnitude as in urine.

It should be mentioned that the considerable differences in the pattern of digitoxin metabolism reported by the different investigators are certainly caused by the different de-

Table 1. Urinary and fecal excretion of digitoxin, digoxin and β-methyl-digoxin according to various investigators.

	Application	n	Period of collection	(Cumulative) excretion in % of dose		Unchanged glycoside in % of excretion rate	
				urine	feces	urine	feces
Digitoxin							
Lukas (1933)	1x1.0 mg p.o.	5	0– ∞	28	15		
Vöhringer, Rietbrock (1974)	1x1.0 mg ^{3}H p.o.	6	0–192	21	13	61	42
Storstein (1977)	1x0.6 mg i.v.	5	0– 24	6 (24 h)		57	
	0.1 mg/die p.o.	9	24	35 (24 h)		87	
Kuhlmann (1981)	1x1.0mg ^{3}H i.v.	4	0–192	23	15	59	43
Digoxin							
Marcus et al. (1964)	1x1.0 mg ^{3}H i.v.	4	0–192	50	15	94^{+}	86^{+}
Lukas (1978)	0.5–0.25 mg/die p.o.	3	24			36	9
Magnussen et al. (1982)	1x0.5 mg ^{3}H p.o.	8	0– 72	42		92	
Gault et al. (1984)	maintenance or loading dose + 1x150 μCi ^{3}H	5	0– 24	18		77	
β-methyl-digoxin							
Rietbrock et al. (1975)	1x100 μCi ^{3}H i.v.	6	0–168	62	29	51 (31°)	15 (34°)*
	1x100 μCi ^{3}H p.o.	6	0–168	55	29	31 (50°)	19 (26°)*

$^{+}$) on days 1 and 2 °) digoxin *) n = 1

signs – for instance by the mode of administration – and certainly by the medical status and especially the enzymatic activity of the subjects examined.

The data on digoxin appear to be more unequivocal than those of digitoxin. Initially, it was assumed that digoxin is excreted nearly totally as unchanged drug (1, 2, 11). Recent experiments however, have yielded indications for a considerable metabolism. For instance, Lukas (8) demonstrated during a maintenance dose of daily 0.25–0.5 mg digoxin, orally given, that from the daily excreted amount in urine and feces only about 45 % is unchanged glycoside (Table 1). Magnusson et al. (9, 10) measured, for example, digoxin and its metabolites in the urine of 8 healthy subjects after a single dose of 0.5 mg digoxin + 100 μCi ^{3}H-digoxin for 72 h. During that time, an average of 42 % of radioactivity was excreted. 92 % of this consisted of unchanged glycoside, the rest was hydrolyzed compounds and dihydrodigoxin. Because there was no examination of feces and a relatively low recovery of 42 %, it is difficult to make a final conclusion about the pattern of metabolism also in this study.

From the data published so far, it can be concluded that digoxin is excreted in urine as 40–90 % unchanged drug, with the rest consisting most probably of polar, supposedly conjugated, metabolites.

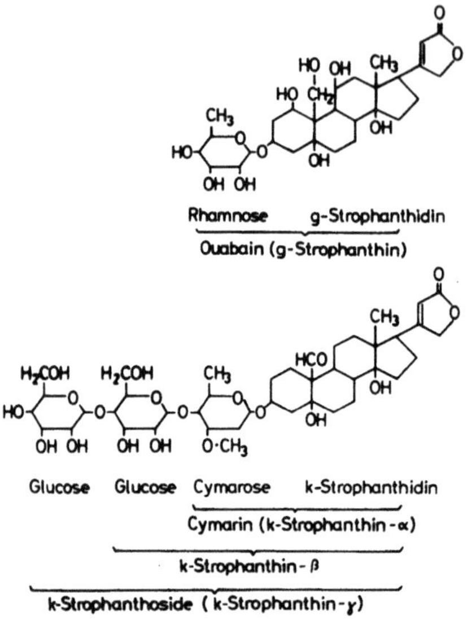

Fig. 2. Chemical formulae of ouabain and k-strophanthines.

In respect to the metabolism of strophanthus glycosides (Fig. 2), some additional data were obtained by a recently developed radioimmunoassay. Ouabain (g-strophanthin) is reported to be poorly, or not at all metabolized. In experiments performed in our institute using a combination of HPLC and RIA (19) after intravenous injection of a single dose of 0.5 mg ouabain, 80% of the urinary excreted amount could be identified as unchanged glycoside. The remaining 20% seemed to be conjugated metabolites which could not be exactly identified. Indeed, in this study, there was a total renal excretion of only 33% of the given dose (Table 2). Higher elimination rates were reported for example by Selden et al. (22), who measured during 6 days after 0.23 mg ^{3}H-ouabain, intravenously given, 40% of the dose in urine and 33% in feces.

Table 2. Urinary excretion of ouabain, cymarin and k-strophanthoside after a single intravenous and oral dose on healthy, male volunteers (19).

	Application	n	Period of collection	Cumulative urinary excretion in % of dose	Unchanged glycoside in % of excretion rate
Ouabain	1x0.5 mg i.v.	5	0–∞	33	80
(g-Strophanthin)	1x8.0 mg p.o.	6	0–∞	0.5	
Cymarin	1x0.5 mg i.v.	6	0–∞	46	1
(k-Strophanthin-α)	1x3.0 mg p.o.	6	0–∞	21	6
k-Strophanthoside	1x0.5 mg i.v.	5	0–∞	73	70
(k-Strophanthin-γ)	1x4.8 mg p.o.	5	0–∞	11	6

Cymarin (k-strophanthin-α), given intravenously (0.5 mg) and orally (3.0 mg) as single dose was excreted mainly as metabolites, especially as conjugated cymarin. Because the renal excretion consisted only of 46 % after intravenous injection and even of 21 % after oral application – extrapolated to infinity – a considerable part of the given dose could not be taken into consideration (19) (Table 2).

The metabolism of k-strophanthoside (k-strophanthin-γ) was found to depend on the route of administration: after i. v. injection of 0.5 mg, about 70 % of the detectable amount (73 %) was identified as unchanged glycoside, and about 30 % as the aglycone, k-strophanthin-β, cymarin and cymarol, mainly not conjugated. After oral administration of 4.8 mg, on the other hand, about 95 % of the total detectable amount (11 %) was conjugated metabolites, consisting of conjugates of the same substances that could be found after i. v. injection (19) (Table 2).

References

1. Doherty JE (1968) The clinical pharmacology of digitalis glycosides: A review. Am J Med Sci 255: 382–412
2. Doherty JE (1973) Digoxin: present knowledge of pharmacokinetics and pharmacodynamics. In: Storstein O (ed) Symposium on Digitalis. Gyldendal Norsk Forlag, Oslo, p 419
3. Gault MH, Longerich LL, Loo JCK, Ko PTH, Fine A, Vasder SC, Dawe MA (1984) Digoxin biotransformation. Clin Pharmacol Ther 35: 74–82
4. Herrmann I, Repke K (1963) Über die Biotransformation von Digitoxin und Digitoxigenin in der menschlichen Leber. Monatsber Dtsch Akad Wissenschaften Berlin 5: 140–143
5. Kuhlmann J, Abshagen U, Rietbrock N (1974) Pharmacokinetics and metabolism of digoxigenin-monodigitoxoside in man. Eur J Clin Pharmacol 7: 87–94
6. Kuhlmann J (1981) Metabolismus von Digitoxin. In: Digitalistherapie bei Herzinsuffizienz, S. 9. Intern Symposium in Taormina, 8.–12. 6. 1980. Urban + Schwarzenberg Verlag, Berlin Wien Baltimore
7. Lukas DS (1973) Pharmacokinetics and metabolism of digitoxin in man. In: Symposium on digitalis. Storstein O (ed). Oslo: Gyldendal Norsk Forlag, pp 84–102
8. Lukas DS (1978) Studies of the metabolism of digoxin and digitoxin using double isotope dilution derivative methods. In: Bodem G, Dengler HJ (eds) Cardiac glycosides, Part I, International Boehringer Mannheim Symposia. Springer-Verlag, Berlin Heidelberg New York, p 36
9. Magnussen JO, Bergdahl B, Bogenstoft C, Jonsson UE, Tekenbergs L (1982) Excretion of digoxin and its metabolites in urine after a single oral dose in healthy subjects. Biopharm & Drug Disp 3: 211–18
10. Magnussen JO (1983) Metabolism of digoxin after oral and intrajejunal administration. Br J Clin Pharmac 16: 741–742
11. Marcus FI, Burkhalter L, Cuccia C, Pavlovich J, Kapadia GG (1966) Administration of tritiated digoxin with and without a loading dose. A metabolic study. Circulation 24: 865–874
12. Okita GT, Talso PJ, Curry JH, Smith FD, Geiling EMK (1955) Metabolic fate of radioactive digitoxin in human subjects. J Pharmacol 115: 371–379
13. Okita GT (1964) Metabolism of radioactive cardiac glycosides. Pharmacologist 6: 45
14. Repke K (1959) Über Spaltung und Hydroxylierung von Digitoxin bei der Ratte. Naunyn Schmiedeberg's Arch Exp Path Pharmakol 237: 34–48
15. Repke K (1966) Biochemie und Klinik der Digitalis. Internist 7: 418–426
16. Repke K (1970) Stoffwechsel und Wirkung von Digitalisglykosiden. Öst Apoth Ztg 24. 515
17. Rietbrock N, Vöhringer HF, Kuhlmann J (1977) Der Metabolismus herzwirksamer Glykoside. Herz/Kreisl 9: 825–832
18. Rietbrock N, Woodcock BG (1981) Pharmacokinetics of digoxin and derivatives. In: Greeff K (Hrsg) Cardiac Glycosides. Handbuch der experimentellen Pharmakologie, Vol 56/II. Springer-Verlag, Berlin Heidelberg New York

19. Rojsathaporn K (1982) Zur Pharmakokinetik der Strophanthusglykoside beim Menschen. Inaugural-Dissertation, Düsseldorf
20. Schmoldt A, Benthe HF, Haberland G (1975) Digitoxin metabolism by rat liver microsomes. Biochem Pharmacol 24: 1639–1641
21. Schmoldt A (1984) Untersuchungen zum Metabolismus der Digitalisglykoside. Beiträge zur gerichtl Medizin 42: 95–101
22. Selden R, Margolies N, Smith TW (1974) Renal and gastrointestinal excretion of ouabain in dog and man. J Pharmacol Exp Ther 188: 615–623
23. Storstein L (1977) Studies on digitalis. VIII. Digitoxin metabolism on a maintenance regimen and after a single dose. Clin Pharmacol Ther 21: 125–140
24. Storstein L (1981) Pharmacokinetics of digitoxin. In: Cardiac Glycosides. Handbook of Experimental Pharmacology, Vol 56/II, S 3, Greeff K (ed), Springer-Verlag, Berlin Heidelberg New York
25. Vöhringer HF, Rietbrock N (1974) Metabolism and excretion of digitoxin in man. Clin Pharmacol Ther 16: 796–806
26. Vöhringer HF, Rietbrock N (1978) Pharmacokinetics and Metabolism of Digitoxin in the human. In: Bodem, G, Dengler, HJ (eds) Cardiac Glycosides, Part I. International Boehringer Mannheim Symposia. Springer-Verlag, Berlin Heidelberg New York, p 64
27. Wirth KE, Frölich JC, Hollefield JW, Falkner FC, Sweetman BS, Oates JA (1976) Metabolism of digitoxin in man and its modification by spironolactone. Eur J Clin Pharmacol 9: 345–354

Author's address:

Prof. Dr. K. E. Wirth, Pharmakologisches Institut der Universität, Moorenstraße 5, 4000 Düsseldorf (West Germany)

The dihydrogenation of digoxin

J. Lindenbaum

Departments of Medicine, Harlem Hospital Center and Columbia University, College of Physicians and Surgeons, New York (U.S.A.)

Summary

1. Digoxin is converted to cardioinactive reduced metabolites in vivo and in vitro by the human gut flora.
2. The bioavailability of digoxin preparations and metabolism by this pathway are inversely correlated.
3. Normal volunteers and patients tend to remain "in category" as metabolizers or non-metabolizers.
4. Digoxin inactivation is mediated by *Eubacterium lentum,* a normal inhabitant of the colonic flora.
5. DRP formation occurs predominantly in the cecum in rats.
6. DRP-forming *E. lentum* can be recovered in high concentration from the gut flora of metabolizers and about 40% of nonmetabolizers.
7. Infants do not metabolize digoxin by this pathway.
8. Bangladeshis and Americans differ in their tendency to reduce digoxin, regardless of their current place of residence.
9. Underdigitalization, or digitalis toxicity, may result from alterations in the intestinal flora. However, the frequency of these clinical events remains to be established.

If the double bond on the lactone ring of digoxin is reduced, a variety of cardanolides or digoxin reduction products (DRP), such as dihydrodigoxin and dihydrodigoxigenin, are produced. These reduced metabolites are markedly less cardioactive than digoxin (2, 3, 9, 21, 25). This is probably related to their reduced affinity for the Na-K-ATPase (9, 11, 20). In addition, DRP are poorly concentrated by the myocardium (13) and rapidly excreted from the body both in animals (13) and humans (4). Some patients have been described in whom markedly increased digoxin requirements were associated with the urinary excretion of substantial amounts of dihydrodigoxigenin (18) or dihydrodigoxin (22).

It was standard teaching for many years that digoxin is largely excreted unchanged in the urine and that significant metabolism is a rare event. However, earlier studies of digoxin pharmacokinetics used chromatographic methods that did not separate digoxin from dihydrodigoxin. More recent work using a variety of techniques has shown that an appreciable minority of patients on digoxin convert the drug to DRP (7, 12, 22, 23). Using separate radioimmunoassay techniques for digoxin and its reduced metabolites (4), we have found that one out of three patients or normal volunteers given digoxin excretes appreciable quantities of DRP in the urine and that one in 10 excretes massive amounts (more than 40% of the total excretion of digoxin and its metabolites) (7, 17).

Bioavailability and metabolism

The bioavailability of digoxin formulations and the extent of reductive metabolism are inversely correlated. Following single doses of tablets of slow dissolution rate, or after chronic administration in the steady state, normal subjects excrete larger amounts of reduced metabolites than after corresponding doses of tablets of higher dissolution rate (16, 17). Also, less extensive DRP formation occurs after the intravenous, compared to the oral, route of administration (16, 17). The inverse relationship between bioavailability and metabolism by this pathway has been confirmed by others (19).

When digoxin is administered as a liquid concentrate in capsule form, its bioavailability exceeds that of any tablet formulation and is close to that achieved by intravenous administration (15, 24). When single doses of currently marketed tablets and capsules were compared in 22 DRP-forming human subjects, the per cent excretion of DRP was greater after the tablets than the capsules in every instance (24). The mean percent DRP excretion after the tablets was 41.2 ± 2.7% compared to 23.5 ± 1.8% after the capsules ($p < 0.001$). A significant reduction in DRP excretion was also noted when capsules were substituted in tablets under steady state conditions in patients with heart disease (15).

Retrospective studies

After we developed a radioimmunoassay for DRP (4) which allowed for the rapid quantitation of reduced metabolites and the processing of large numbers of biological specimens, we applied the assay to over a thousand urine specimens that we had collected and stored from 131 normal volunteers who had participated in a number of bioavailability studies that we had conducted over the greater part of a decade (17). This retrospective material was useful in a number of respects. None of the subjects had taken digoxin prior to participation in the studies; 110 of them took various preparations of the cardiac glycoside on multiple occasions in crossover studies. Analysis of DRP excretion by these volunteers revealed a number of interesting patterns, in addition to the previously mentioned relationship between bioavailability and metabolism. Unlike digoxin, DRP were not excreted immediately in the urine after oral administration, but only after a delay of several hours. In contrast with the parent compound, peak excretion of DRP occurred more than 8 hours after the drug was taken (17), a similar finding to that found by Eriksson and coworkers in a single DRP-forming volunteer (10). Subjects who excreted DRP in the 24-hour period after their first ingestion of digoxin in their lifetime usually excreted DRP on every subsequent occasion that they took the drug. In contrast, the great majority of subjects who made no reduced metabolites on their first exposure failed to make them when they took digoxin over and over again in the future. In other words, with few exceptions, subjects remained "in category" as metabolizers or non-metabolizers over periods of months and years, and DRP formation was not apparently inducible by repeated drug administration. The exceptions to the rule, i. e., subjects who crossed over from one category to the other, were also provocative. Of great interest to us were six subjects who repeatedly excreted DRP in the urine after multiple digoxin doses who subsequently became nonexcretors, in two instances over two-week periods of observation. On retrospective inquiry, we ascertained that two of the volunteers had taken a course of erythromycin for an intercurrent illness immediately prior to their conversion to the status of non-metabolizers (17).

Hypotheses

The observations that bioavailability and metabolism were inversely related, the delay in appearance of the reduced derivatives, and the possible association of loss of the tendency to make DRP with therapy with an antibiotic known to be inhibitory to the human gut flora led us to formulate several hypotheses based on our retrospective findings (17). It seemed likely that DRP were formed in the gastrointestinal tract, probably lower down in the gut, which would explain the greater DRP formation when poorly bioavailable preparations were taken, the decreased amounts of metabolites formed after intravenous dosing, and their delay in appearance. If reduced derivatives were generated by enteric bacteria resident in the colon, that would also explain the conversion after antibiotic treatment. We also postulated that bacteria that reduce digoxin were only present in the gut flora of a minority of individuals, which would explain why most people do not make DRP. For any of these hypotheses to be valid, a significant fraction of an oral as well as an intravenous dose of digoxin would have to reach the colon, an idea that was consistent with the findings of some previous workers of significant biliary excretion of digoxin in humans (citied in (17)) as well as that of several groups that digoxin is directly secreted into the gastrointestinal tract from the plasma in animals (cited in (17)). Our retrospective observations and our hypotheses were also fully consistent with an earlier report by Herrmann and Repke indicating that rat cecal contents and feces, as well as human feces, could generate dihydrodigoxin from either lanatoside C or digoxin under anaerobic conditions in vitro (14).

Prospective studies

We tested these hypotheses by a series of in vitro and in vivo experiments. First we showed that the stool of DRP excretors invariably contained organisms which convert digoxin to DRP in vitro under anaerobic conditions (8, 16). We then gave five day courses of erythromycin or tetracycline to three DRP-excreting subjects under steady state conditions of digoxin administration (16). In each subject there was a dramatic fall in urinary DRP excretion within 48 hours of taking the antibiotic, as well as a corresponding increase in urinary digoxin (16). In addition, steady state serum digoxin concentrations rose in each volunteer after antibiotic therapy (16). The virtual disappearance of DRP from the urine and stool of each subject in this experiment indicated not only that DRP were generated by the gut flora, but that this was the only significant site in which the metabolites were made.

Isolation of the bacteria

We then characterized a number of isolates from stool cultures of single colonies of bacteria that converted digoxin to DRP in vitro (8). Each of these isolates was found to be a single species, *Eubacterium lentum,* a gram-positive non-sporeforming rod which is one of the most common constituents of the normal human gut flora. In addition, when we tested cultures of stock organisms representing the 150 most common inhabitants of the enteric flora in a blind fashion, only one species was found to make DRP: *Eubacterium lentum.* Although it is still possible, in view of the great variety of bacteria present in the human microflora, that other organisms will be found to reduce digoxin, at this time it appears likely that a single species is responsible for DRP formation.

Our hypothesis that DRP-forming bacteria would only be demonstrable in the gut flora of metabolizers, however, proved to be incorrect. We found that such organisms were present in stool cultures from approximately 40% of non-metabolizers, and in approximately the same concentration found in metabolizers. Furthermore, the species of DRP-forming bacteria in the stool of non-metabolizers was also *Eubacterium lentum* (8). It is likely that some difference in the in vivo metabolic milieu in the colon exists between metabolizers and non-metabolizers, which results in the in vivo expression of the tendency of the organisms to reduce digoxin only in the colon of DRP metabolizers. Digoxin reaches the colonic flora in non-metabolizers, since substantial concentrations of digoxin are found in the feces of non-metabolizers (Lindenbaum J: unpublished data).

Site of digoxin inactivation

It has not yet been shown that the colon is the site of DRP formation in humans. However, previous investigators have only very rarely isolated *Eubacterium lentum* from cultures taken from gastrointestinal sites proximal to the human cecum. Also, in studies currently in progress in our laboratory, most of an oral dose of digoxin given to rats is converted to DRP in the cecum, even though bacteria capable of reducing digoxin can be isolated higher up in the gastrointestinal tract in some animals.

Age of onset

We have recently found that DRP formation does not occur in infants given digoxin. None of 35 digitalized children below the age of 8 months excreted any detectable DRP in the urine (5). By the age of 3 years, the same proportion of children as adults (1 in 3) were DRP excretors, although few children below the age of 10 are "heavy" (i. e., more than 40%) metabolizers (5). As in adult non-metabolizers, some children below the age of 8 months carry organisms in their fecal flora which are capable of reducing digoxin. Differences in the actions of the gut flora of children during the first years of life as compared to adults have previously been found for a number of other anaerobic bacterial reactions, including the generation of methane and the dehydroxylation of bile salts.

Ethnic and geographic differences

It has long been speculated that the gut flora of individuals living in developing countries differs from that of people in Western countries. In support of this notion, we have recently found that Bangladeshi subjects (both normal volunteers and cardiac patients) have a very low prevalence of digoxin metabolism by this pathway (1). Even after immigration and residence in the United States for many years, Bangladeshis are much less likely than Americans to form reduced metabolites. In normal subjects who were employees of a hospital in Dhaka, a highly significant relationship was found between monthly salary and the tendency to reduce DRP (1). Factors related to socioeconomic status, such as differences in intake of dietary substrates for the gut flora or in gastrointestinal transit time, might conceivably account for these findings. In Americans, however, no differences in DRP excretion related to age, sex or race (black versus white) have been detected (1).

Clinical importance

Finally, the exact clinical significance of DRP formation is worthy of further study. It is unknown how often DRP production accounts for the common finding of inadequate steady state serum digoxin concentrations in clinical practice. It is also uncertain how often antibiotics may precipitate digoxin toxicity in patients. Probably this is an uncommon event, for two reasons. Such patients would be likely to be relatively underdigitalized before antibiotic therapy, which might then result in a change to adequate digitalization rather than toxicity. Secondly, not all antibiotics will interact with digoxin in this fashion; for example, the beta-lactam group of antimicrobial drugs does not appear to substantially influence DRP excretion in patients (6).

References

1. Alam AN, Saha JR, Dobkin JF, Butler VP, Lindenbaum J (1985) Americans and Bangladeshis differ in the tendency of the normal gut flora to inactivate digoxin. Clin Res 33: 599A

2. Bach EJ, Reiter M (1964) The difference in velocity between the lethal and inotropic action of dihydrodigoxin. Naunyn Schmiedeberg's Arch Exp Pathol Pharmacol 248: 437–449
3. Brown BT, Stafford A, Wright SE (1962) Chemical structure and pharmacological activity of some derivatives of digitoxigenin and digoxigenin. Br J Pharmacol 18: 311–324
4. Butler VP, Tse-Eng D, Lindenbaum J, Kalman SM, Preibisz JJ, Rund DG, Wissel PS (1982) The development and application of a radioimmunoassay for dihydrodigoxin, a digoxin metabolite. J Pharmacol Exp Ther 221: 121–131
5. Dobkin JF, Lindenbaum J (1985) Inactivation of digoxin by the gut flora and its reversal by antibiotics. In: "Pediatric Cardiology", ed Doyle et al, New York. Springer-Verlag, pp 1242–1243
6. Dobkin JF, Saha JR, Butler VP, Lindenbaum J (1982) Effects of antibiotic therapy on digoxin metabolism. Clin Res 30: 517A
7. Dobkin JF, Saha JR, Butler VP, Neu HC, Lindenbaum J (1982) Inactivation of digoxin by *Eubacterium lentum,* an anaerobe of the human gut flora. Trans Assoc Am Physicians 95: 22–29
8. Dobkin JF, Saha JR, Butler VP, Neu HC, Lindenbaum J (1983) Digoxin-inactivating bacteria: identification in human gut flora. Science 220: 325–327
9. Erdmann E, Schoner W (1974) Ouabain-receptor interactions in ($Na^{+}K^{+}$)-ATPase preparations. IV. The molecular structure of different cardioactive steroids and other substances and their affinity to the glycoside receptor. Naunyn Schmiedeberg's Arch Pharmacol 283: 335–356
10. Eriksson B-M, Tekenbergs L, Magnusson J-O, Molin L (1981) Determination of tritiated digoxin and metabolites in urine by liquid chromatography. J Chromatog 223: 401–408
11. Flasch H, Heinz N (1978) Correlation between inhibition of ($Na^{+}K^{+}$)-membrane-ATPase and positive inotropic activity of cardenolides in isolated papillary muscles of guinea-pig. Naunyn Schmiedeberg's Arch Pharmacol 304: 37–44
12. Greenwood H, Snedden W, Hayward RP, Landon L (1975) The measurement of urinary digoxin and dihydrodigoxin by radioimmunoassay and by mass spectroscopy. Clin Chim Acta 62: 213–224
13. Heinz N, Flasch H (1978) Comparison of the pharmacokinetics of digoxin and dihydrodigoxin in cats in single-dose studies. Naunyn Schmiedberg's Arch Pharmacol 303: 181–187
14. Herrmann I, Repke K (1969) Transformation of cardenolides by microorganisms of the intestine. In: Schubert K (ed) Symposium über Biochemische Aspekte der Steroidforschung. Akademie-Verlag,Berlin, pp 115–119
15. Johnson BF, Lindenbaum J, Budnitz E, Marwaha R (1986) Variability of steady-state digoxin kinetics during administration of tablets or capsules. Clin Pharmacol Ther (in press)
16. Lindenbaum J, Rund DG, Butler VP, Tse-Eng D, Saha JR (1981) Inactivation of digoxin by the gut flora: reversal by antibiotic therapy. N Engl J Med 305: 789–794
17. Lindenbaum J, Tse-Eng D, Butler VP, Rund DG (1981) Urinary excretion of reduced metabolites of digoxin. Am J Med 71: 67–74
18. Luchi RJ, Gruber JW (1968) Unusually large digitalis requirements: a study of altered digoxin metabolism. Am J Med 45: 322–328
19. Magnusson JO, Bergdall B, Bogentoft C, Johnsson UE (1982) Metabolism of digoxin and absorption site. Br J Clin Pharmac 14: 284–285
20. Marcus FI, Ryan JN, Stafford MG (1975) The reactivity of derivatives of digoxin and digitoxin as measured by the Na-K-ATPase displacement assay and by radioimmunoassay. J Lab Clin Med 85: 610–620
21. Okarma TB, Tramell P, Kalman SM (1972) The surface interaction between digoxin and cultured heart cells. J Pharmacol Exp Ther 183: 559–576
22. Peters U, Falk LC, Kalman SM (1978) Digoxin metabolism in patients. Arch Intern Med 138: 1074–1076
23. Reuning RH, Shepard TA, Morrison BE, Bockbrader HN (1985) Formation of (20R)-dihydrodigoxin in humans. Drug Metab Disposition 13: 51–57
24. Rund DG, Lindenbaum J, Dobkin JF, Butler VP, Saha JR (1983) Decreased digoxin cardioinactive-reduced metabolites after administration as an encapsulated liquid concentrate. Clin Pharmacol Ther 34: 738–743

25. Vick RL, Kahn JB, Acheson GH (1957) Effects of dihydro-ouabain, dihydrodigoxin and dihydrodigitoxin on the heart-lung preparation of the dog. J Pharmacol Exp Ther 121: 330–339

Author's address:

Prof. J. Lindenbaum, Medical Service, Harlem Hospital Center, Lenox Avenue and 135th Street, New York, N. Y. 10037 (U.S.A.)

Renal excretion and plasma levels of dihydrodigitoxin in volunteers and renally healthy and renally insufficient patients being treated with digitoxin

H. Strobach

Institut für Pharmakologie der Universität Düsseldorf (West Germany)

The question of the production of dihydrodigitoxin as a metabolite of digitoxin has up to now only been investigated by von Unruh, and Bodem et al. (1, 2, 4), using gas chromatography and mass spectroscopy: small amounts of up to 13.2% dihydrodigitoxin could be found in the plasma of 5 out of 12 renally healthy patients. In 17 renally insufficient patients, 10 of whom were undergoing chronic dialysis, the amount of dihydrodigitoxin was as much as 50–60%.

The object of our study was to estimate dihydrodigitoxin and digitoxin in volunteers as well as in patients' urine and plasma radio-immunologically. The estimations of digitoxin and dihydrodigitoxin were carried out with antisera produced according to the method of Butler (3).

Tritiated glycosides were used in the radio-immunoassay, the dextran coated charcoal method was used for the separation of bound from free glycoside. Testing the specificity of the dihydrodigitoxin antiserum, we found a weak cross-reaction with digitoxin. 3 ng digitoxin corresponded to 0.33 ng dihydrodigitoxin, 30 ng digitoxin to 0.78 ng dihydrodigitoxin and 300 ng digitoxin corresponded to 0.9 ng dihydrodigitoxin. The cross-reactivity therefore changed little by increasing the amount of digitoxin added from 30 to 300 ng. The dihydrodigitoxin antiserum contains a small amount of digitoxin antibodies.

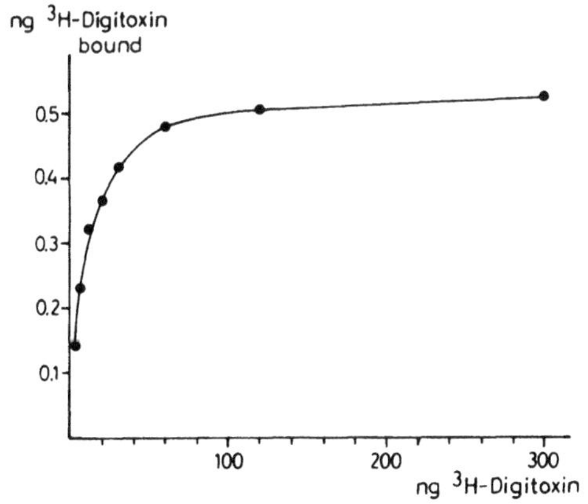

Fig. 1. Binding of ^{3}H-digitoxin to dihydrodigitoxin antibodies.

In order to find out how much digitoxin is necessary to saturate the digitoxin antibodies, increasing amounts of ^{3}H-digitoxin were added to 100 μl of the 1:5000 diluted dihydrodigitoxin antibody and the amount of digitoxin bound was calculated.

In Fig. 1, one can see that the addition of 60 ng ^{3}H-digitoxin produces a binding of 0.5 ng and that an increase in the digitoxin added leads to no further increase in binding. Therefore we added 60 ng of unlabelled digitoxin to the sample undergoing radio-inmmunological estimation for dihydrodigitoxin, so that only dihydrodigitoxin antibodies could react with dihydrodigitoxin.

The digitoxin antiserum also showed a slight cross reactivity with dihydrodigitoxin: in analogy to the previously described experiments in Fig. 2, an increasing amount of ^{3}H-dihydrodigitoxin was added to 100 μl antiserum diluted 1:500.

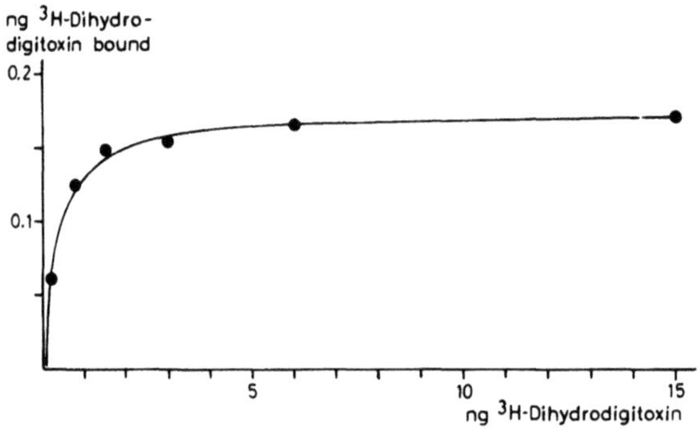

Fig. 2. Binding of ^{3}H-dihydrodigitoxin to digitoxin antibodies.

As seen in Fig. 2, the maximum binding of 0.16 ng dihydrodigitoxin is reached by the addition of 6 ng ^{3}H-dihydrodigitoxin. Therefore 6 ng of unlabelled dihydrodigitoxin was added to the radio-immunological assay of digitoxin.

The accuracy of the procedure was tested by the analysis of samples which contained 0–30 ng digitoxin or dihydrodigitoxin in varying proportions to one another. The amount of dihydrodigitoxin in percent was calculated and compared with the known value. The mean error was found to be 1.4 ± 1.1 %. To test for precision, a solution containing 15 ng digitoxin and 15 ng dihydrodigitoxin/ml was analysed 10 times. The mean content of dihydrodigitoxin was found to be 50.5 ± 0.9 %.

Pharmacokinetics of dihydrodigitoxin in healthy volunteers

To investigate as to whether the renal half-life of dihydrodigitoxin differs from that of digitoxin, 5 healthy volunteers received 0.5 mg dihydrodigitoxin in a single oral dose. Urine was collected over a period of 7 days and the cumulative excretion measured.

Fig. 3 shows the time in days as cumulative renal excretion and the amount of dihydrodigitoxin excreted as percent of the given dose. After 7 days, 32.1 ± 5.0 % of the given dose had been excreted. The mean renal half-life was 51.4 (26.4–76.8) hours. The half-life for dihydrodigitoxin is therefore shorter than that for digitoxin, which was found in an analogous experiment to be a mean of 139 hours.

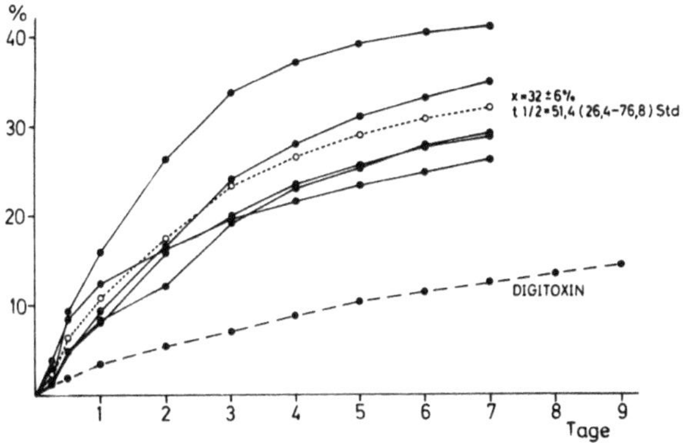

Fig. 3. Cumulative urinary excretion of dihydrodigitoxin after a single oral dose of 0.5 mg.

Excretion of dihydrodigitoxin in patients receiving digitoxin

In further investigations, we measured the excretion of dihydrodigitoxin and digitoxin in the urine of patients undergoing prolonged therapy with 0.07 or 0.1 mg digitoxin (Digimerck) daily. 62 patients were renally healthy, 43 had renal insufficiency with a serum creatinine level higher than 1.5 mg % (mean value 4.4 ± 3.6).

The dihydrodigitoxin estimated in urine as % of the total amount of digitoxin and dihydrodigitoxin in 5% intervals is shown in Fig. 4. The lenght of the columns symbolizes the number of patients in whom dihydrodigitoxin excretion lay within the values indicated. The upper half of the figure shows results obtained with renally insufficient patients and the lower half with renally healthy patients. The dihydrodigitoxin excretion was exceptionally high (50.8%) in one patient in the renally insufficient group.

Two renally healthy patients with values of 26.6 and 41.2% excreted a large amount of dihydrodigitoxin. However, the dihydrodigitoxin excretion remained under 10% in most patients. If one ignores the three extreme values, one obtains a mean excretion of 6.1 ± 4.5% in 60 renally healthy patients; significantly higher ($2p < 0.001$) than that in 42 renally insufficient patients who excrete 3.0 ± 3.3%.

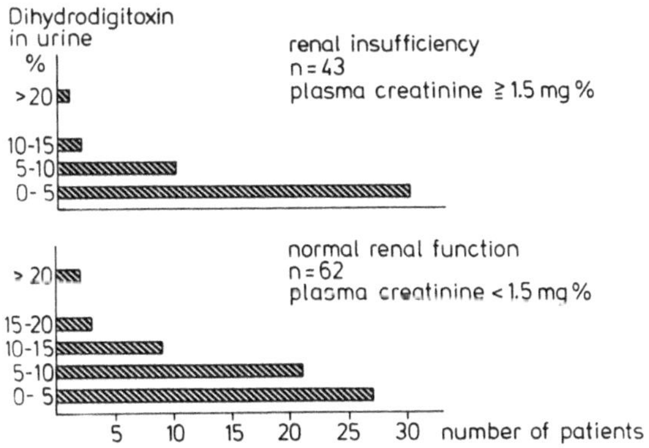

Fig. 4. Excretion of dihydrodigitoxin in patients with renal insufficiency and with normal renal function.

Table 1. Digitoxin and dihydrodigitoxin plasma levels (in ng/ml) in 11 renally healthy and 8 dialysis patients.

	Digitoxin	Dihydro-digitoxin	Dihydrodigitoxin in %
Renally healthy			
Serum creatinine:	42.2	0.9	2.1
1.1 ± 0.2 mg %	22.3	1.0	4.3
	19.9*	0.0	0.0
(n = 11)	21.7	1.3	5.7
	33.2*	0.1	0.3
	30.9*	0.5	1.6
	32.0*	1.7	5.0
	15.0*	7.5	33.3
	36.1*	1.5	4.0
	36.2*	0.4	1.1
	26.0	0.6	2.3
Dialysis patients			
Serum creatinine:	16.2	0.0	0.0
9.1 ± 3.3 mg %	11.1	0.7	5.9
	21.5	0.3	1.4
(n = 8)	33.2	1.7	4.9
	2.7	0.1	3.6
	17.9	3.9	17.9
	24.3*	0.7	2.8
	26.1	1.9	6.8

* 0.1 mg Digitoxin daily, all others 0.07 mg daily

Table 1 shows the digitoxin and dihydrodigitoxin plasma levels measured in 11 renally healthy and 8 dialysis patients.

One renally healthy patient and one dialysis patient with 33 % and 17.9 % respectively showed remarkably high dihydrodigitoxin levels. In all other patients, the amount of dihydrodigitoxin was less than 7 %. Our investigations led to the conclusion that dihydrodigitoxin is a normal metabolite of digitoxin usually produced in small amounts, which is renally excreted more rapidly than the mother substance. (Half-life 51 h compared to digitoxin 139 h).

As dihydrodigitoxin is produced to less than 10 % of the given dose of digitoxin this metabolic pathway may play no important role in the inactivation of digitoxin.

References

1. Klehr HU, Otten H, v Unruh E, Bodem G (1977) Digitoxinstoffwechsel bei niereninsuffizienten Patienten. Verh Dtsch Ges Inn Med 83: 1656
2. Bodem G, v Unruh E (1978) Dihydrodigitoxin, a metabolite of digitoxin in humans. In: Bodem G, Dengler H (eds) Cardiac Glycosides. Springer-Verlag, Berlin Heidelberg New York
3. Bodem G, v Unruh E (1979) Enhanced transformation of digitoxin to dihydrodigitoxin in humans with renal failure. J Clin Pharmacol 19: 195
4. Butler VP, Chen JP (1967) Digoxinspecific antibodies. Proc Natl Acad Sci 57: 71

Author's address:

Dipl.-Chem. H. Strobach, Pharmakologisches Institut der Universität Düsseldorf, Moorenstraße 5, 4000 Düsseldorf (West Germany)

Enzymatic basis for digitoxin metabolism and possible drug interactions in man

A. Schmoldt, L. von Meyerinck, W. Drohn and I. Blömer

Institut für Rechtsmedizin und Pharmakologisches Institut der Universität Hamburg (West Germany)

Summary

In vitro investigations on human liver tissue have shown that the digitoxin sugar chain can only be cleaved after preceding oxidation of the terminal sugar. The resulting dehydrodigitoxosyl is split off by β-elimination. The initial oxidation is catalyzed by the cytochrome P 450 system. The rate limiting step in this metabolic pathway is the first oxidation to 15'-dehydrodigitoxin. Results obtained from animal experiments suggest no inhibition by exogenous cytochrome P 450 substrates at physiological concentrations. Noncompetitive inhibition by drugs might be possible with quinidine and cimetidine and by the steroid hydroxylase inhibitor metapyrone.

In accordance with rat experiments, dt-2 and dt-1 are suitable substrates for a conjugating glucuronyltransferase.

Introduction

It is well established now that in rats the digitoxin sugar side chain is not hydrolyzed by glycosidases but split via a stepwise oxidative cleavage. This oxidation is catalyzed by the cytochrome P 450 system by which the axial OH-group of the terminal digitoxosyl is oxidized to an oxo-group. The so formed dehydrodigitoxosyl is split in a nonenzymatic β-elimination reaction which is base-catalyzed (5) (Fig. 1). Following this repeated split-off, the resulting digitoxigenin-monodigitoxoside (dt-1) is eliminated as conjugates (1, 9).

Basically, digoxin is cleaved in the same manner (7), but digoxin is a worse substrate for the enzymatic metabolic reactions, probably because of its lower lipophilicity. This may serve as an explanation why this reaction does not take place in the in vivo metabolism.

In addition to these results, experiments with rats have shown that the oxidation as well as the glucuronidation are catalyzed by specific isoenzymes. These isoenzymes are inducible with spironolactone, 16-cyano-pregnenolone and cyproterone (3, 8, 10). The UDP-glucuronyltransferase which is responsible for glucuronidation of dt-1 in the rat has recently been purified (2).

In contrast to this main metabolic pathway, the known 12β-hydroxylation of the genin moiety to the resulting digoxigenin derivatives is of minor importance.

Abbreviations		
	dt-0	= digitoxigenin
	dt-1	= digitoxigenin-monodigitoxoside
	dt-2	= digitoxigenin-bisdigitoxoside
	dt-3	= digitoxigenin-trisdigitoxoside, digitoxin
	TLC	= thin layer chromatography
	cyt	= cytochrome
	GT	= UDP-glucuronyltransferase

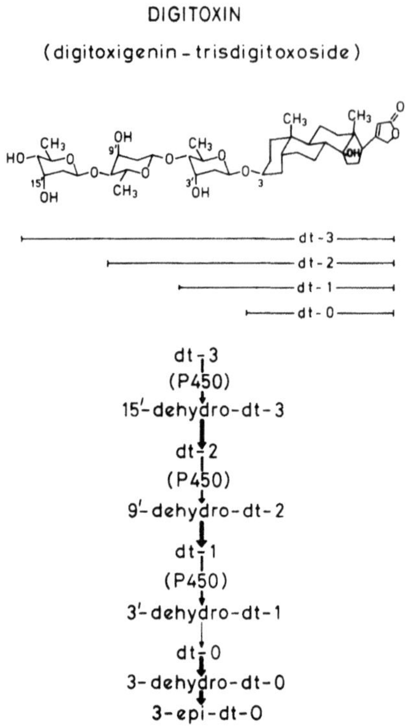

Fig. 1. Metabolic pathway of digitoxin in the rat: dt-3, dt-2, dt-1 are digitoxosides of digitoxigenin: dt-3 = digitoxin, dt-2 = digitoxigenin-bisdigitoxoside, dt-1 = digitoxigenin-monodigitoxoside, and dt-0 = digittoxigenin. P450 = cytochrome P450.
From (4).

After establishing the metabolism of digitoxin in the rat, it has been interesting to investigate the metabolic pathway of digitoxin in man. Moreover, it should be investigated whether drug interactions may occur due to inhibition of the metabolic enzymes involved and whether other enzyme inducers exist.

Methods

Preparation of microsomes, incubations of 20–22-^{3}H-digitoxigenin glycosides for cleavage or conjugation, analytical procedures for protein concentrations, cyt P450, separation and determinations of radioactive metabolites were the same as previously described (5, 8, 10).

Results

A. Oxidative cleavage of digitoxin using human liver microsomes

Incubations with digitoxin using subcellular fractions of human tissue (mitochondrial and microsomal fractions, and 100,000 x g supernatant) without NADPH have not shown cleavage products of digitoxin. This result confirms the assumption that the sugar side chain of dt-3 cannot be hydrolyzed by human glycosidases. However, after incubating digitoxin with human liver microsomes in the presence of NADPH, the TLC-separation of the radioactivity revealed radioactive fractions corresponding to dt-2 and 15′-dehydro-dt-3 (Fig. 2). The same results have been obtained using other solvent systems. In further experiments, dt-2 and dt-1 have been incubated in the presence of NADPH and human liver microsomes. Fig. 3 shows the reaction products 9′-dehydro-dt-2 and dt-1 in case of dt-2 and 3′-dehydro-dt-1 following incubation of dt-1. In good correlation with results obtained from rat liver microsomes, the cleavage product of 3′-dehydro-dt-1, which is dt-0, has not been detected. This compound seems to be resistant to β-elimination.

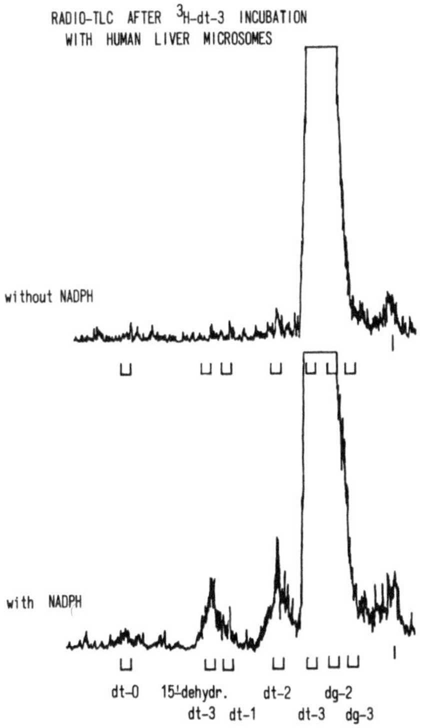

Fig. 2 .Radio thin layer chromatographic separation of radioactive metabolites after incubation of digitoxin in the presence of human liver microsomes.
Upper trace: Without addition of NADPH.
Lower trace: In the presence of NADPH and NADPH-regenerating system.
Solvent: Ethylacetate.
Running direction: Right to left.
U: unlabelled reference compounds.

The presence of CO (CO/air 2+1 v/v) caused a 51 % decrease of the dt-2 oxidation whereas under anaerobic conditions no metabolites have been detectable. From these experiments, it can be concluded that also in man the stepwise cleavage is an oxidative cyt P 450 catalyzed reaction.

The rates for dt-3, dt-2, and dt-1 have shown remarkable interindividual differences (Table 1).

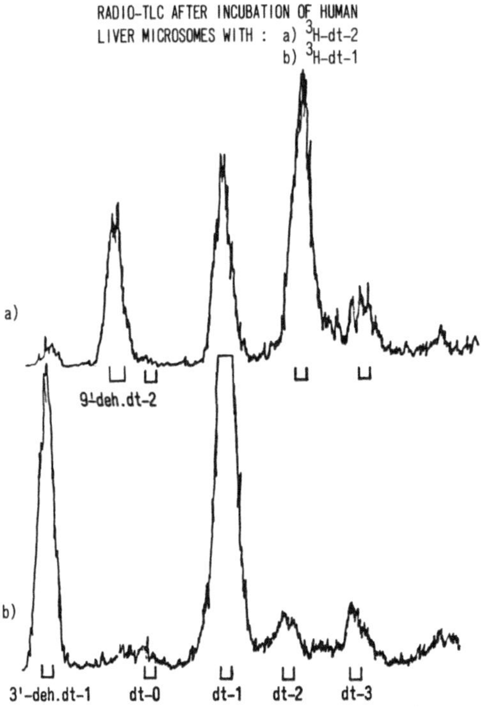

Fig. 3. Radio thin layer chromatography following incubation of ^{3}H-dt-2 (upper trace) and ^{3}H-dt-1 (lower trace). Methods and labelling as in Figs. 1 and 2.

Table 1. Oxidative cleavage of digitoxigenin of digitoxosides by human liver microsomes.

		pmoles oxidised/mg micr. prot. min				
Substrate	Patients:	A	B	C	D	E
dt-3		12	34	8,2	18	9,1
dt-2		154	422	96	270	74
dt-1		39	54	34	124	39
dt-2/dt-3		13	13	12	15	8
dt-1/dt-3		2,5	1,6	4,2	6,9	4,3

Using digitoxin, the rates obtained were between 8–32 pmol mg^{-1} min^{-1}. Intraindividually, the conversion rate for dt-2 was about 12-times higher compared to dt-3. The latter result is further evidence that the oxidative cleavage is catalyzed by the same cyt P450.

B. Drug interactions

Drug interactions are possible inhibition of the cyt P450 system with endogenous or exogenous compounds acting as competitive or noncompetitive inhibitors. Table 2 shows results with rat liver microsomes using endogenous compounds, drugs and cardioactive compounds. The best inhibitor has been found to be metapyrone. Endogenous steroids and bile acids produce non-competitive inhibition only at unphysiologically high concentra-

Table 2. Inhibition of digitoxin cleavage in rat liver microsomes.

Inhibitor	(μmol/l)	dt-3 cleavage rate in % of uninhibited controls	Ki (μmol/l)
–	–	100 ± 11	–
Aldolsterone	(50)	72 ± 11*	–
Estradiol	(75)	43 ± 7*	–
Estrone	–	–	138**
Hydrocortisone	(75)	93 ± 7*	–
Pregnandiol	(50)	78 ± 6	–
Cholate	(100)	112 ± 18	–
Chenodeoxycholate	(100)	85 ± 2*	–
Lithocholate	(100)	95 ± 5	–
Carbamazepine	(50)	84 ± 2	–
Ethylmorphine	–	–	812**
Catechin	(100)	91 ± 2*	–
Quinidine	–	–	174**
Rifampicine	(50)	104 ± 119	–
Clofibrate	(50)	104 ± 6	–
Cimetidine	–	–	558
Metyrapone	–	–	0,085**
dt-0	–	–	40***
dt-1	–	–	96***
dt-2	–	–	63***
Strophanthin	(50)	106 ± 27	–

* significant difference to control values
** noncompetitive inhibition
*** competitive inhibition

tions. From all drugs testes, only cimetidine and quinidine act as inhibitors. In addition to these experiments, the relatively new drugs tested for induction – clofibrate, diazepam, and carbamazepine – caused neither inhibition nor induction of the oxidative cleavage in rats (Table 3).

Table 3. Induction of dt-3-oxidation by rat liver microsomes.

Pretreatment with inducer	Cyt. P450 (% of controls)	dt-3 sugar oxidation (% of controls)
– (controls)	100 ± 2	100 ± 13
Clofibrate	495 ± 12	104 ± 7
Carbamazepine	130 ± 13	130 ± 48
Diazepam	119 ± 3	93 ± 28

C. Conjugation reactions forming glucuronides

Whether or not unchanged digitoxin can be glucuronidated in the human is still a matter for discussion. In our experiments, only in 4 out of 8 patients could traces of dt-3 glucuronides be detected (detection limit about 0.1–0.3 pmol min^{-1} mg^{-1}) under optimum conditions. Remarkably high glucuronidation rates have been obtained with dt-2 and dt-1 as substrates (Table 4). Conventional substrates of GTs have not been able to act as competitive inhibitors for dt-1 glucuronidation. Competitive inhibition has been found only with cardiac glycosides.

Table 4. Glucuronidation rate of digitoxigenin-digitoxosides.

	pmole Glucuronide/mg micr. prot. min							
Substrate Patients:	A	B	C	D	E	F	G	H
dt-3	0.15	0.2	n. d.	n. d.	0.4	0.4	n. d.	n. d.
dt-2	3.51	2.53	2.79	2.79	–	3.03	1.89	1.08
dt-1	11.5	17.1	–	13.1	35.9	1.39	93.4	87.6

n. d. = not detectable

Discussion

The present paper shows that digitoxin can only be metabolized in a stepwise manner under cyt P 450 catalysis. Therefore, the metabolic rate is dependent on the acitivity of this specific sugar oxidizing cyt P 450. The high interindividual differences in the activity of this enzyme give rise to the assumption that high interindividual differences in the metabolic rates occur in vivo. This might be of minor importance in general because unchanged digitoxin can be renally excreted. However, the metabolic rate becomes very important in cases of renal dysfunction. In these cases, the digitoxin elimination rate is only dependent on the metabolic rates.

Besides the potent inhibitor metapyrone, which is used simply for diagnostic purposes, high levels of quinidine, cimetidine and acute high doses of spironolactone (9) may cause drug interactions. These drugs should be investigated in clinical trials. Other cyt P 450 substrates such as steroids and bile acids should not inhibit at physiological levels in vivo.

Results obtained with rats have shown that spironolactone and cyproterone are inducers. Investigations carried out on volunteers without kidney disorders have shown spironolactone as an inducer in man (11).

Data given in Figs. 1 and 2 show the 12β-hydroxylation of the genin resulting in the digoxin derivative to be without importance for the human metabolism compared to oxidation of the sugar side chain.

In spite of insignificant inhibition of dt-1 glucuronidation in the human by other substrates, structural analogues of heart glycosides have to be tested for their inhibitory properties. These investigations may be of major importance because any inhibition of the glucuronidation of dt-1 can result in a higher concentration of the cardioactive dt-1 and accumulation of poorly glucuronidated dt-0 or 3-epi-dt-0.

References

1. Castle MC, Lage GL (1974) Toxicol Appl Pharmacol 27: 641
2. Von Meyerinck L, Schmoldt A, Green MD, Tephly TR (1985) Naunyn-Schmiedeberg's Arch Pharmacol 329: R 17
3. Schmoldt A (1978) Naunyn-Schmiedeberg's Arch Pharmacol 305: 261
4. Schmoldt A (1984) Beitr Gerichtl Med 42: 95
5. Schmoldt A, Rohloff C (1978) Naunyn-Schmiedeberg's Arch Pharmacol 305: 167
6. Schmoldt A, Rohloff C (1978) Arch Pharm 311: 833

7. Schmoldt A, Ahsendorf B (1980) Eur J Drug Metabolism 5: 225
8. Schmoldt A, Promies J (1982) Biochem Pharmacol 31: 2285
9. Schmoldt A, Benthe HF, Grote W (1974) Naunyn-Schmiedeberg's Arch Pharmacol 282: R 85
10. Schmoldt A, Buhr SL, Albersmeyer K (1980) Biochem Pharmacol 29: 405
11. Wirth KE, Fröhlich JC, Hollifield JW, Falkner FC, Sweetman BS, Oates JA (1976) Eur J Clin Pharmacol 9: 345

Acknowledgement

The authors are grateful to P.D. Dr. H. van Ackeren (Marienkrankenhaus Hamburg) for supplying them with liver specimen of hepatotomized patients.

Author's address:

Prof. Dr. A. Schmoldt, Institut für Rechtsmedizin, Universität Hamburg, Butenfeld 34, 2000 Hamburg 54 (West Germany)

Effect of quinidine on the inotropic effect of digoxin in isolated cat or guinea-pig papillary muscle

K. A. Taubert, C. Bufalino, T. Hair and C. Lafleur

Division of Biomedical Sciences, University of California at Riverside, and Departments of Physiology and Pharmacology, University of the Pacific, School of Pharmacy, Stockton, California (U.S.A.)

Summary

The effect of quinidine pretreatment on the inotropic effect and tissue content of digoxin was studied in 18 cat and 13 guinea-pig isolated, isometrically contracting papillary muscles. Muscles were divided into pretreated and non-pretreated groups. Pretreated muscles received 10^{-5} M quinidine one hour before 2 x 10^{-6} M digoxin was added. At the peak inotropic effect of digoxin, muscles in both pretreated and non-pretreated groups were assayed for digoxin content. Quinidine alone exerted a small but significant positive inotropic effect in cat papillary muscle, but did not alter guinea-pig muscle performance. Digoxin elicited a significant positive inotropic effect in all groups of muscles studied. In cat papillary muscle, however, both developed tension and the rate of tension rise (+dT/dt) were increased significantly less in the quinidine-pretreated muscles. There were no significant differences between pretreated and non-pretreated groups for the guinea-pig muscles. The non-pretreated cat papillary muscles had a muscle digoxin level of 1302 ng/gm wet weight, and this was reduced significantly to 657 ng/gm in the pretreated muscles. The non-pretreated guinea-pig muscles had a muscle digoxin level of 1599 ng/gm, and this was not significantly altered by quinidine pretreatment. These data suggest that an interaction between quinidine and digoxin exists in isolated cat papillary muscle, but this interaction is not apparent in guinea-pig muscle. Therefore, when studying the interaction between these two drugs, the choice of the animal species used may be important.

Introduction

Within the last several years a clinically significant interaction between digoxin and quinidine has been described in the literature, and reviews of these reports have been made (2, 3, 4, 6). Several animal studies have been conducted in order to further investigate this interaction (5, 7, 8, 10–13, 17–19). Some of these studies (8, 11, 13, 19) looked at the effects of quinidine and digoxin in isolated cardiac muscle and one previous study from our laboratory (8) measured muscle levels of digoxin. The current study was undertaken in order to further investigate a possible drug interaction between digoxin and quinidine on cardiac muscle itself. Both contractility and muscle digoxin levels were examined, and studies were done in two commonly used animal species, cats and guinea-pigs.

Methods

Right ventricular papillary muscles were removed from the hearts of 18 adult cats and 13 adult guinea-pigs, previously anesthetized with sodium pentobarbital (35–40 mg/kg, i. p.). The muscles were placed in a muscle bath containing a Krebs-Ringer bicarbonate buffer solution in a manner described previously

(16). Muscles were stimulated to contract isometrically 12 times per minute at a voltage 20% above threshold, and were oxygenated with 95% O_2, 5% CO_2. Muscles stabilized during the first hour, during which time Lmax was set. Experiments were then run at Lmax. A tension transducer (Statham UC2) was connected to a clip holding the top of the muscle and allowed recording of resting tension (RT), developed tension (DT), maximal rate of tension rise (+dT/dt), and decline (−dT/dt), and time to peak tension (TPT).

After the stabilization period control data were collected. At this time quinidine (10^{-5} M) was added to the bath of eight cat and six guinea-pig muscles. One hour later another set of data were collected and 2 x 10^{-6} M digoxin was added to the bath of both the pretreated and non-pretreated muscles. Experiments were terminated at the peak inotropic effect of digoxin. Muscles were then weighed, cross-sectional area was determined, and all data except TPT were normalized by dividing by cross-sectional area. Muscles were then assayed for digoxin content as we have previously described (8). Data were analyzed by a paired or grouped t-test, and results were considered significant when $p < 0.05$.

Results

Mean control parameters for the 18 cat papillary muscles were: RT, 1.0 ± 0.05 (SE) g/mm²; DT, 6.5 ± 0.5 g/mm²; +dT/dt, 29.6 ± 2.3 g/mm²/sec; −dT/dt, 18.9 ± 1.2 g/mm²/sec; and TPT, 399 ± 5.6 msec. For the 13 guinea-pig muscles, control parameters were: RT, 1.1 ± 0.06 g/mm²; DT, 0.55 ± 0.06 g/mm²; +dT/dt, 3.3 ± 0.3 g/mm²/sec; −dT/dt, 2.7 ± 0.3 g/mm²/sec; and TPT, 279 ± 15.6 msec. Within either species studied, there were no differences in control parameters between pretreated and non-pretreated groups. Resting tension declined with time throughout all experiments, and no evidence of contracture was seen.

Quinidine pretreatment alone caused small but significant increases in cat papillary muscle performance. DT increased 8% ($p < 0.05$), +dT/dt increased 8% ($p < 0.025$), and −dT/dt increased 4% ($p < 0.025$). In the guinea-pig muscles, no significant alterations were caused by quinidine.

Digoxin elicited a positive inotropic effect in all groups of muscles studied. Table 1 lists the percent change in performance parameters in the four subgroups of muscles at peak digoxin effect. This represents the percent change in performance parameters measured im-

Table 1. Percent change in performance parameters at peak digoxin effect.

Group	DT	+dT/dt	−dT/dt	TPT
Cats:				
Digoxin only	+28.6	+41.2	+8.0	−10.8
± S.E.M.	±4.28	±5.01	±1.95	±2.34
p	<0.001	<0.001	<0.005	<0.005
Digoxin + quinidine	+10.3	+19.7	+6.4	−5.0
± S.E.M.	±2.02	±3.83	±0.94	±2.58
p	<0.001	<0.005	<0.005	n. s.
Guinea-pigs:				
Digoxin only	+69.3	+75.8	+23.6	−5.0
S.E.M.	±5.32	±5.68	±5.24	±1.22
p	<0.001	<0.001	<0.005	<0.01
Digoxin + quinidine	+74.1	+85.2	+27.3	−7.1
± S.E.M.	±4.95	±6.00	±6.60	±1.25
p	<0.001	<0.001	<0.01	<0.005

mediately prior to the addition of digoxin. With one exception (TPT in the quinidine-pretreated cat muscles), all changes were significant. Fig. 1 compares the responses to digoxin between pretreated and non-pretreated cat (part A) and guinea-pig (part B) muscles. For cat muscles, DT and +dT/dt were increased less (both $p < 0.005$) when the muscles were pretreated with quinidine. In the guinea-pig muscles, however, there was no difference in the magnitude of the inotropic response to digoxin in the presence versus the absence of quinidine pretreatment.

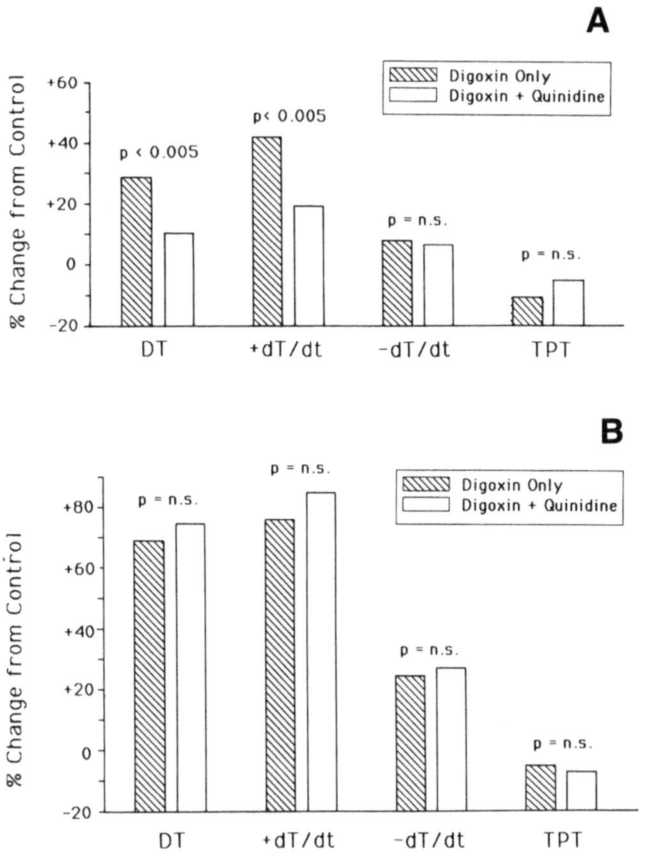

Fig. 1. Performance parameters at peak digoxin effect in quinidine-pretreated and non-pretreated cat papillary muscles (panel A) or guinea-pig muscles (panel B). Data are expressed as percent change from the control parameters obtained immediately prior to the addition of digoxin. p values are for grouped comparisons.

Table 2 lists the muscle digoxin levels (MDL) in the four subgroups. A significant decrease in MDL was seen in cat papillary muscles that had been pretreated with quinidine while there were no differences between the two subgroups of guinea-pig papillary muscles.

Table 2. Muscle digoxin levels (ng/g wet weight).

Group		Digoxin only		Digoxin + quinidine
Cats		1302 ± 186.9		657 ± 88.1
	p		< 0.01	
Guinea-pigs		1599 ± 151.5		1517 ± 113.8
	p		n. s.	

Discussion

The present study demonstrates that digoxin elicits a positive inotropic effect in isolated cardiac muscle that has been pretreated with quinidine. In cat, but not guinea-pig papillary muscle, however, the magnitude of the inotropic response was blunted in the presence of quinidine and these muscles had a reduced content of digoxin.

Quinidine alone caused a small but significant inotropic response in cat papillary muscles. These data are consistent with those reported by others. Williams and Mathew (19) demonstrated a dose-dependent effect of quinidine in isolated cat papillary muscle. 1.3×10^{-5} M quinidine caused a significant increase in DT and +dT/dt whereas twice that dose did not elicit any significant changes in these parameters. Lash and co-workers (13), using ferret papillary muscles, found that 1×10^{-5} M quinidine significantly increased muscle performance. Quinidine is capable of inhibiting Na^+,K^+-ATPase (1, 14) which could lead to a positive inotropic effect. Higher doses of quinidine also have a direct depressant effect in isolated cardiac muscle (9) and cultured heart cells (10). The current study showed no effect of quinidine on guinea-pig papillary muscles, as we have previously reported (8). Although Kim and co-workers (12) showed that 5 μM quinidine caused a negative inotropic effect in Langendorff guinea-pig hearts, Nawrath (15) reported that quinidine doses larger than 10^{-5} M were required to elicit negative inotropic effects on guinea-pig atrial preparations. Thus, the response of isolated cardiac muscle to quinidine seems to depend on the species of animal and the dose of the drug used, and perhaps on the specific muscle preparation (atrial, ventricular, or whole heart) used as well.

Whole animal studies have suggested that the observed interaction of digoxin and quinidine results in alteration of digoxin's pharmacokinetics due at least partly to changes in clearance and/or volume of distribution (5, 12, 17, 18). Goldman and colleagues (7) looked at combined administration of ouabain and quinidine in conscious dogs and found that in the presence of quinidine, ouabain's inotropic effect on the heart was reduced. They felt that this could be due to a displacement of ouabain from its cardiac sites by quinidine. Doherty et al. (5) observed that the myocardial concentration of digoxin was lowered in quinidine-pretreated dogs, although Warner and associates (17) could not confirm these findings.

The current study has demonstrated that in cat papillary muscle, tissue levels of digoxin are decreased, as is the inotropic effect of digoxin, in the presence of quinidine. No apparent interaction, however, was observed in guinea-pig muscle. Williams and Mathew (19) found a reduction of digoxin's inotropic effect in quinidine-pretreated cat and rabbit papillary muscles. On the other hand, they found that if quinidine were administrated *after* digoxin, further increases in the positive inotropic effect were seen. Lash et al. (13) also observed further inotropic effects when quinidine was added after digoxin in ferret papillary muscle. They further found that this additional inotropic effect of quinidine was abolished by atropine pretreatment. Kim, Akera, and Brody (11) showed that quinidine pretreatment did not alter the positive inotropic effect of digoxin in the Langendorff guinea-pig heart preparation. They also found that quinidine did not alter digoxin binding to Na^+,K^+-ATPase in this preparation.

It is tempting to speculate that in cat, but not in guinea-pig ventricular muscle, quinidine causes a decrease in digoxin binding at the Na^+,K^+-ATPase receptor and that this causes the reduced inotropic response. Ball and associates (1) found that quinidine was capable of decreasing purified lamb kidney Na^+,K^+-ATPase affinity for digoxin. This has also been shown for beef heart Na^+,K^+-ATPase (15). The assay used in the current study, however, measures total tissue digoxin, and not just that bound to its pharmacological receptor.

In summary, the current study demonstrates that the effects of quinidine on the inotropic response of cardiac muscle to digoxin is not the same between the two species of animals studied. It appears that quinidine pretreatment alters myocardial binding of, and myocardial response to, digoxin in cat ventricular muscle under the conditions of the current study. There is no observable interaction, however, when guinea-pig muscle is used. Thus, when interpreting the results of digoxin and quinidine in isolated cardiac muscle, it is important to consider the species of the animal used in addition to considering the dosage of the drugs used and the order in which they are administered.

Acknowledgements

This research was supported in part by N.I.H. Young Investigator Award (K. A. T.) #5R23HL22 575 and by a grant from the American Heart Association, California Affiliate.

References

1. Ball WJ, Tse-Eng D, Wallick ET, Bilezikian JP, Schwartz A, Butler VP Jr (1981) Effect of quinidine on the digoxin receptor in vitro. J Clin Invest 68: 1065–1074
2. Bigger JT, Leahey EB Jr (1982) Quinidine and digoxin – an important interaction. Drugs 24: 229–239
3. Bussey HI (1984) Update on the influence of quinidine and other agents on digitalis glycosides. Am Heart J 107: 143–146
4. Doherty JE (1982) The digoxin-quinidine interaction. Ann Rev Med 33: 163–170
5. Doherty JE, Straub KD, Murphy ML, de Soyza N, Bissett JK, Kane JJ (1980) Digoxin-quinidine interaction. Changes in canine tissue concentration from steady state with quinidine. Am J Cardiol 45: 1196–1200
6. Fichtl B, Doering W (1983) The quinidine-digoxin interaction in perspective. Clin Pharmacokinet 8: 137–154
7. Goldman S, Hager WD, Olajos M, Perrier D, Mayersohn M (1983) Effect of the ouabain-quinidine interaction on left ventricular and left atrial function in conscious dogs. Circulation 67: 1054–1058
8. Hair TL, Taubert KA (1983) Effect of quinidine on the inotropic effect of cardiac glycosides in isolated cardiac muscle. Proc West Pharmacol Soc 26: 313–316
9. Hammermeister KE, Boerth RC, Warbasse JR (1972) The comparative inotropic effects of six clinically used antiarrhythmic agents. Am Heart J 84: 643–652
10. Horowitz JD, Barry WH, Smith TW (1982) Lack of interaction between digoxin and quinidine in cultured heart cells. J Pharmacol Exp Ther 220: 488–493
11. Kim D-H, Akera T, Brody TM (1981) Effects of quinidine on the cardiac-glycoside sensitivity of guinea-pig and rat heart. J Pharmacol Exp Ther 217: 559–565
12. Kim D-H, Akera T, Brody TM (1981) Interactions between quinidine and cardiac glycosides involving mutual binding sites in the guinea-pig. J Pharmacol Exp Ther 218: 108–114
13. Lash RE, Reeves RC, Reeves DN, Hefner LL (1982) Mechanism of additive effects of digoxin and quinidine on contractility in isolated cardiac muscle. Am J Cardiol 50: 483–487
14. Lowry K, Rao SN, Pitts B Jr, Askari A (1973) Effects of quinidine on some reactions and ion transactions catalyzed on the Na^+,K^+-ATPase complex. Biochem Pharmacol 22: 1369–1377
15. Straub KD, Kane JJ, Bissett JK, Doherty JE (1978) Alteration of digitalis binding by quinidine: A mechanism of digitalis-quinidine interaction. Circulation 58: II–58 (abstract)

16. Taubert K, Templeton G, Willerson JT, Shapiro W (1976) Effects of digitalis and calcium on papillary muscles in normal and hypoxic states. Am J Physiol 231: 66–72
17. Warner NJ, Leahey EB Jr, Hougen TJ, Bigger JT, Smith TW (1983) Tissue digoxin concentrations during the quinidine-digoxin interaction. Am J Cardiol 51: 1717–1721
18. Wilkerson RD, Mockridge PB, Massing GK (1980) Effects of selected drugs on serum digoxin concentration in dogs. Am J Cardiol 45: 1201–1210
19. Williams JF, Mathew B (1981) Effect of quinidine on positive inotropic action of digoxin. Am J Cardiol 47: 1052–1055

Author's address:

Kathryn A. Taubert, M. D., American Heart Association, National Center, 7320 Greenville Avenue, Dallas, Texas 75231 (U.S.A.)

Quinidine decreases the intestinal secretion of digoxin and digitoxin in the guinea-pig

S. G. Schäfer, B. Fichtl and W. Doering[1]

Walther-Straub-Institut für Pharmakologie und Toxikologie der Ludwig-Maximilians-Universität, München, and
[1] Städtisches Krankenhaus Schwabing, München (West Germany)

Summary

The secretion of digoxin and digitoxin into in situ perfused jejunal and colonic segments of normal or quinidine treated guinea-pigs was studied. Quinidine was administered intravenously by constant rate infusion resulting in a quinidine plasma concentration of about 6 μg/ml.

After two hours digoxin or digitoxin was injected i. v. (10 μg/kg). The quinidine treatment enhanced the plasma concentration of ^{3}H-digoxin to about 140 % as compared to controls, whereas the ^{3}H-digitoxin concentration was not influenced by the quinidine infusion. Both digoxin and digitoxin were secreted against a concentration gradient into the intestinal lumen. During the experimental period of 180 minutes controls secreted 0.24 % of the administered digoxin dose per cm of jejunal and 0.13 % per cm of colonic segment. Quinidine treatment resulted in a decrease of the jejunal digoxin secretion to about 80 % of the control values. In both jejunum and colon the concentration ratio between lumen and plasma (L/P) was diminished by quinidine to 50 % as compared with controls.

The amount of ^{3}H-digitoxin secreted into the intestinal segments was decreased by quinidine from 0.19 % of the dose per cm to 0.13 % in the jejunal and from 0.17 % to 0.12 % in the colonic segments, respectively. The decrease of the L/P ratio for ^{3}H-digitoxin was more pronounced in the colon (58 %) than in the jejunum (77 % of the control values). As compared with controls the content of ^{3}H-digoxin in the jejunal as well as colonic tissue was decreased by quinidine to 60 % or 73 %, respectively. On the other hand, quinidine increased the tissue content of ^{3}H-digitoxin in jejunum (+ 56 %) and colon (+ 88 %). In conclusion, quinidine inhibits the intestinal secretion of both digoxin and digitoxin, possibly by different mechanisms.

Introduction

Coadministration of quinidine in patients on digoxin treatment results in a twofold increase of serum digoxin concentration (4). The main mechanism responsible for this clinically important drug interaction is a reduction of renal digoxin clearance (4). However, since this interaction is also observed in completely anuric patients (1, 3, 8), extrarenal clearance of digoxin must also be affected.

Cardiac glycosides can be secreted by the intestinal mucosa of several species including man (14). The quinidine induced decrease of extrarenal digoxin clearance may be due to an inhibition of intestinal secretion of digoxin (15) as demonstrated recently by animal experiments (6, 20).

At present, it is still controversial whether a clinically relevant interaction also exists between digitoxin and quinidine (4). An increase in digitoxin steady-state concentrations following administration of quinidine has been evidenced recently (12, 13), which is probably due to a decrease in extrarenal digitoxin clearance (13).

Therefore, the influence of quinidine on intestinal secretion of digoxin and digitoxin was further investigated.

Material and methods

Male guinea-pigs (body weigth 400 ± 11 g) were anesthetized with urethane (1.25 g/kg i. m.). The jugular vein and carotid artery were cannulated. The quinidine treated animals received an intravenous loading dose (12.8 mg/kg) of quinidine bisulfate tetrahydrate in saline. The controls received saline alone. In each animal, a jejunal and a colonic segment were prepared for pendular perfusion in vivo as described elsewhere (19).

Two hours after starting the infusion, the animals received either digoxin or digitoxin as an intravenous bolus dose of 10 μg/kg containing 40 μCi of the ^{3}H labelled glycosides.

1, 15, 30, 60, 90 and 180 minutes after the glycoside injection blood samples were collected and immediately centrifuged. The infusion of quinidine or saline continued throughout the experimental period.

Analytical methods

^{3}H-glycosides were determined by liquid scintillation counting. ^{3}H in intestinal tissues was measured at the end of the experiments by combustion in a Packard sample oxidizer. The amount of chloroform-insoluble metabolites of the glycosides in plasma, intestinal tissue and perfusate was determined according to (2). Quinidine concentration was measured by HPLC (7). Plasma protein binding of digitoxin was determined by equilibrium dialysis. For further experimental details see (21).

Results

Effects of quinidine on the plasma concentration of digoxin and digitoxin

In Fig. 1A, the plasma levels of ^{3}H-digoxin are shown in animals which received digoxin alone or during quinidine treatment. At 180 minutes, the quinidine infusion had resulted in an increase of the ^{3}H-digoxin concentration by about 40 % with respect to the control values. The quinidine concentration at the end of the experiments was 6.3 ± 0.4 μg/ml. In contrast, the coadministration of quinidine resulted in no enhancement of the digitoxin plasma concentration (Fig. 1B). The mean plasma concentration of quinidine in these animals was 6.4 ± 0.4 μg/ml.

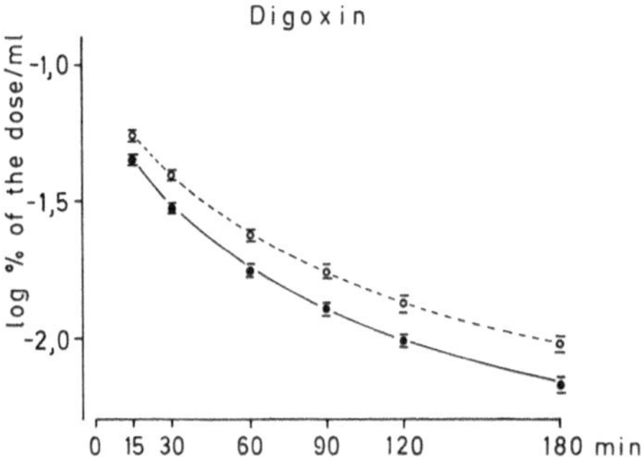

Fig. 1A. Time course of ^{3}H-digoxin concentration in plasma of controls (●) and quinidine treated (○) guinea-pigs (Mean ± SEM given as percent of dose/ml; n = 10–13).

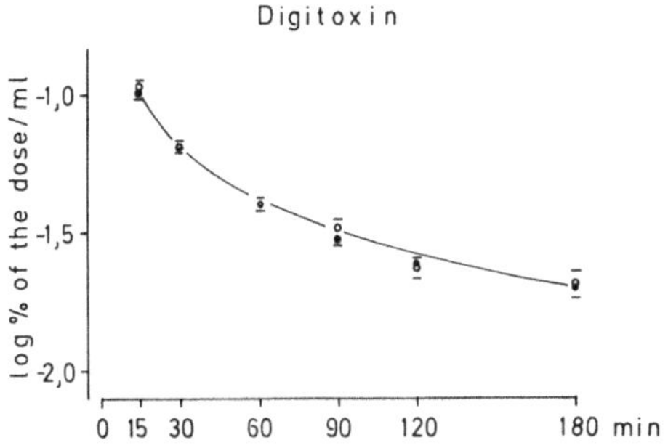

Fig. 1B. Time course of ^{3}H-digitoxin concentration in plasma of controls (●) and quinidine treated (○) guinea-pigs (Mean ± SEM given as percent of dose/ml; n = 7–10).

Total glycoside derived radioactivity was separated into chloroform-extractable (c. e. m.) and non chloroform-extractable metabolites (n. c. e. m.). Quinidine had no effect on the relative amounts of c. e. m. and n. c. e. m. in plasma.

The degree of binding of digitoxin to plasma proteins (90 ± 0.3 %) was not influenced by quinidine (91 ± 0.2 %). Binding of digoxin to guinea-pig plasma is also unaffected by quinidine (18).

Influence of quinidine on the intestinal secretion of digoxin and digitoxin

In the perfusion experiments, the movement of water across the intestinal wall was taken as an indicator of the undisturbed physiological function. Water movement was not influenced by the i. v. administration of digoxin or digitoxin nor by the simultaneous administration of one of the glycosides and quinidine. Secretion of water occurred in the jejunal segments, whereas absorption was observed in the colonic segments, which agrees with earlier findings (17). After 3 hours of perfusion, the luminal concentrations of both glycosides exceeded that in the plasma of controls and of quinidine treated animals, indicating a net secretion against a concentration gradient from plasma into intestinal lumen.

Digoxin

In controls as well as in quinidine treated animals, the amount secreted was greater in the jejunum than in the colon (Fig. 2 A). In the presence of quinidine, the jejunal secretion de-

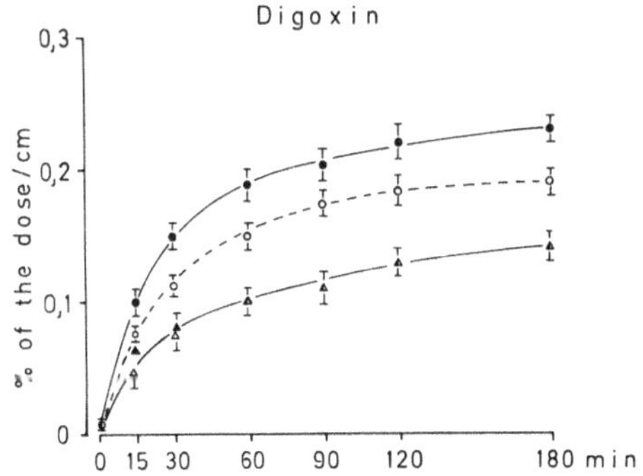

Fig. 2A. Time course of the amount of ^{3}H-digoxin secreted into the lumen of perfused intestinal segments. Jejunum: controls (●), quinidine treated (○); colon: controls (▲), quinidine treated (△). Data are given as percent of dose/cm lenght of intestinal segment (Mean ± SEM; n = 7–13).

creased to about 80 % of the control values, whereas the amount of digoxin secreted into the colon was not altered by quinidine.

Quinidine treatment did not influence the amount of either n. c. e. m. or c. e. m. of digoxin in the perfusion medium.

It should be taken into account that quinidine treatment results in a marked increase of plasma digoxin concentration. It seems reasonable, therefore, to refer the luminal digoxin concentration to the actual plasma concentration. The ratio between luminal and plasma concentration (L/P) reflects the secretory capacity of the intestine. In the jejunum as well as in the colon, L/P increased linearily with time both in controls and in the quinidine treated group (Fig. 2B). Quinidine treatment resulted in a 50 % decrease of the capacity of the intestine to maintain the concentration gradient between intestinal lumen and plasma.

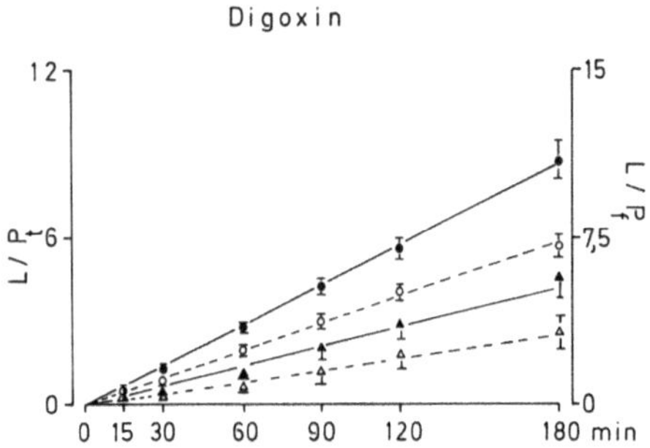

Fig. 2B. Time course of the ratio of the ^{3}H-digoxin concentrations in intestinal lumen (L) and in plasma (P). L/P_t: calculated for total plasma concentration, L/P_f: calculated for onbound concentration in plasma. Jejunum: controls (●), quinidine treated (○); colon: controls (▲), quinidine treated (△) (Mean ± SEM; n = 7–13).

Digitoxin

In controls, the digitoxin secretion into the jejunum slightly exceeds that into the colon (Fig. 3A). The quinidine treatment decreased the ^{3}H-digitoxin concentration to about 60 % and 50 % in the jejunum and colon, respectively, as compared with the controls. The amount of n. c. e. m. in the perfusion medium was increased by quinidine treatment. In both

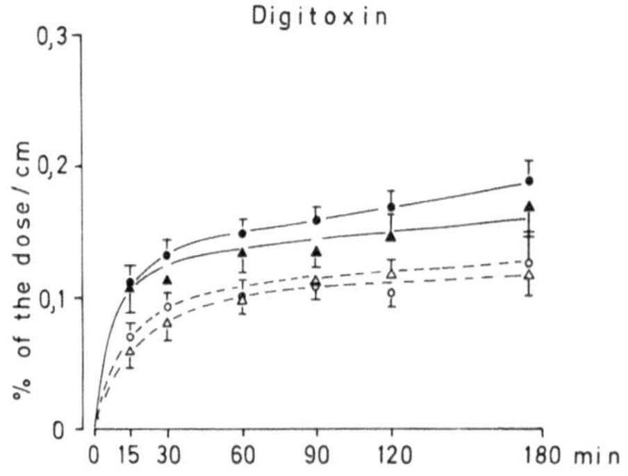

Fig. 3A. Time course of the amount of ^{3}H-digitoxin secreted into the lumen of perfused intestinal segments. Jejunum: controls (●), quinidine treated (○); Colon: controls (▲), quinidine treated (△). Data are given as percent of dose/cm length of intestinal segments (Mean ± SEM; n = 7–9).

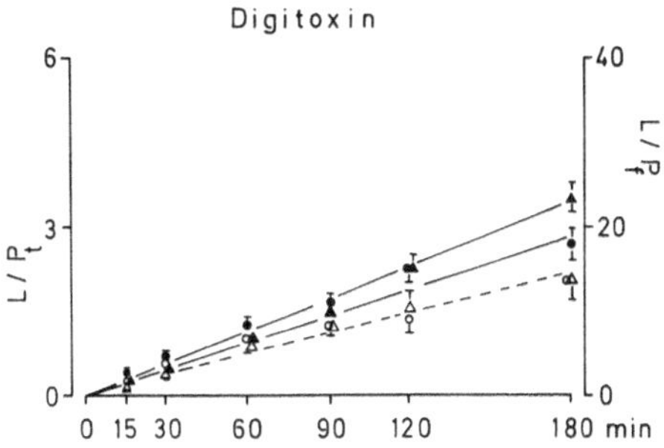

Fig. 3B. Time course of the ratio of the ^{3}H-digitoxin concentrations in intestinal lumen (L) and in plasma (P). L/P_t: calculated for total plasma concentration, L/P_f: calculated for unbound concentration in plasma. Jejunum: controls (●), quinidine treated (○); colon: controls (▲), quinidine treated (△) (Mean ± SEM; n = 7–10).

intestinal segments, L/P increased linearly during the experimental period (Fig. 3B). Quinidine coadministration decreased L/P to 77% (jejunum) and to 58% (colon).

Content of ^{3}H in the intestinal tissue

Treatment of guinea-pigs with quinidine decreased the content of ^{3}H-digoxin in the intestinal tissue to about 60% (jejunum) and 73% (colon). ^{3}H-digitoxin tissue content, however, was increased by quinidine up to 188% (jejunum) and 156% (colon) (Table 1).

Table 1. Content of ^{3}H in the tissue (% of dose/g w. w.) in controls and quinidine treated guinea-pigs 180 minutes after i. v. administration of digoxin or digitoxin (mean ± SEM; n = 6–13; *: $p < 0.01$).

	Digoxin	Dig + Qui	Digitoxin	Digit + Qui
Jejunum	0.145 ± 0.012	0.087* ± 0.009	0.095 ± 0.003	0.178* ± 0.028
Colon	0.191 ± 0.013	0.139* ± 0.015	0.112 ± 0.009	0.176* ± 0.031

Discussion

In guinea-pigs as well as in man, quinidine increases plasma digoxin (5, 10, 18, 20) and decreases volume of distribution and both renal and extrarenal clearance (10, 18).

In the present experiments, quinidine enhanced plasma digoxin but had not effect on plasma digitoxin. The latter result is at variance with the findings of Kim et al. (10). In the rat, however, this group reported (11) that quinidine increases digoxin but not digitoxin levels.

Obviously, both glycosides were secreted against a concentration gradient from plasma into the testinal lumen. An energy and temperature dependent secretion of digoxin and digitoxin by the intestinal mucosa of guinea-pigs has been demonstrated previously (9, 14).

In the colon and more pronounced in the jejunum, the L/P ratios were higher for digoxin than for digitoxin. However, when comparing the L/P ratios of both glycosides one has to take into account the extent of plasma protein binding. The unbound fraction of digitoxin (10%) is much lower than that of digoxin (60–70%) (18). If one relates the luminal concentration of digitoxin to the unbound concentration in plasma L/P ratios in the order of 20 are obtained (Fig 3 B) suggesting a higher secretory capacity of the intestinal mucosa for digitoxin.

Interestingly, quinidine had opposite effects on the concentration of the glycosides in intestinal tissue. The tissue concentration of digoxin decreased in the presence of quinidine whereas digitoxin concentration increased. At present, we can only speculate about possible reasons for the observed effect. It has been proposed (14) that the intestinal secretion of cardiac glycosides is a two step process. One transport mechanism, located at the basolateral side, is responsible for "filling up" the tissue. The glycoside is then secreted at the luminal side. It seems possible that quinidine influences these two transport mechanisms in a different way for digoxin and digitoxin.

The quantitative importance of the intestinal secretion of cardiac glycosides and its inhibition by quinidine is difficult to assess. It seems conceivable that intestinal secretion contributes considerably to the extrarenal clearance at least in the guinea-pig, because biliary elimination accounts for one fourth only of the extrarenal clearance. Whether this also applies to man remains to be determined.

Acknowledgement

This study was supported by a generous grant from E. Merck, Darmstadt.

References

1. Doering W, Fichtl B, Herrmann M, Besenfelder E (1982) Eur J Clin Pharmacol 21: 281
2. Dolphen R, Lesne M (1981) Arch Int Pharmacodyn 251: 4
3. Fenster PE, Hager WD, Powell R (1982) Am J Cardiol 49: 1026
4. Fichtl B, Doering W (1983) Clin Pharmacokinet 8: 137
5. Geiger JD, Das G, Parmar SS (1982) Pharmacology 25: 177
6. George CF, Renwick AG (1982) J Pharm Pharmacol 34: 280
7. Guentert TW, Huang J-D, Oie S (1982) J Pharm Sci 71: 812
8. Hirschberg R, Schaefer K, v Herrath D, Kewitz H (1981) Klin Wochenschr 59: 521
9. Kim D-H, Akera T, Brody TM (1981) J Pharmacol Exp Ther 217: 559
10. Kim D-H, Akera T, Brody TM (1981) J Pharmacol Exp Ther 218: 108
11. Kim D-H, Akera T, Brody TM (1981) J Pharmacol Exp Ther 218: 357
12. Kreutz G, Keller F, Gast D, Prokein E (1982) Naunyn Schmiedeberg's Arch Pharmacol (Suppl) 319: R82
13. Kuhlmann J, Dohrmann M, Marcin S (1984) Naunyn Schmiedeberg's Arch Pharmacol (Suppl) 325: R85
14. Lauterbach F (1981) In: Greeff K (ed) Handbook of Experimental Pharmacology, Vol 56/II, Cardiac Glycosides. Springer-Verlag, Berlin Heidelberg New York, p 105
15. Manninen V, Nyberg L (1981) In: Greeff K (ed) Handbook of Experimental Pharmacology, Vol 56/II, Cardic Glycosides. Springer-Verlag, Berlin Heidelberg New York, p 299
16. Misra M (1983) Naunyn Schmiedeberg's Arch Pharmacol (Suppl) 322: R15
17. Powell DW, Malwer SJ, Plotkin G (1968) Am J Physiol 215: 1226
18. Sato J, Sawada Y, Iga T, Hanano M (1983) J Pharm Sci 72: 1137
19. Schäfer SG, Forth W (1983) Biol Trace Elem Res 5: 203
20. Schäfer SG, Fichtl B, Doering W (1984) Biopharm Drug Dispos 5: 163
21. Schäfer SG, Schuhmann G, Doering W, Fichtl B Chem-Biol Interactions 55 (1985) 203

Author's address:

Dr. S. G. Schäfer, Walther-Straub-Institut für Pharmakologie und Toxikologie, Med. Fakultät d. Ludwig-Maximilians-Universität München, Nußbaumstraße 26, 8000 München 2 (West Germany)

Quinidine-induced reduction of the biliary excretion of digoxin in patients

K. Schenck-Gustafsson

Karolinska Institute, Division of Cardiology, Department of Internal Medicine, Huddinge University Hospital, Huddinge (Sweden)

Summary

In five patients with atrial fibrillation and on maintenance digoxin therapy, digoxin concentration in serum, urine and bile were determined before and during the addition of maintenance quinidine therapy. The digoxin therapy was halved when quinidine was given. The biliary clearance of digoxin was unexpectedly high before quinidine, suggesting a greater amount of extra hepatic circulation than previously has been described. When quinidine was given an about 40% (range 21–65%, $p < 0.02$) reduction of the biliary clearance of digoxin was observed. Also renal clearance was reduced as has been previously shown.

It is concluded that the mechanism behind the digoxin-quinidine interaction is not only the previously shown 50% reduction of the renal clearance of digoxin but also a quinidine induced about 40% reduction of the biliary excretion of digoxin. The influence on the intestinal secretion of digoxin is not yet determined in man.

Background

Quinidine causes an increase of plasma digoxin concentrations by about 100 per cent and with a large unpredictable interindividual variation ranging between 17 and 333 per cent (5). It has been shown that total, renal and non-renal clearance of digoxin decrease by about 50 per cent when quinidine is introduced to patients on maintenance digoxin therapy (6, 7). The main mechanism behind the interaction is the reduction of the renal clearance of digoxin due to an inhibition of the tubular secretion of digoxin shown in ten patients with atrial fibrillation and on maintenance digoxin therapy given quinidine in connection with electrocardioversion (6). The glomerular filtration rate was determined by ^{51}Cr-EDTA and endogenous creatine clearances and total renal blood flow by PAH (paraaminohypuric acid) clearance. It was found that quinidine induced a decrease of the renal clearance of digoxin by on average 55% and the clearances of ^{51}Cr-EDTA and PAH were uninfluenced, indicating that neither glomerular filtration nor total renal blood flow changed when quinidine was added. The ratio of the renal clearance of unbound digoxin to that of glomerular filtration rate was well above 1 for all 10 patients before quinidine, indicating the involvement of tubular secretion in the renal elimination of digoxin. After the administration of quinidine this ratio decreased in all patients from 1.51 ± 0.30 to 0.83 ± 0.38 (mean $\pm$ SD). Four out of ten patients had ratios well below 1, suggesting reabsorption of digoxin. From this study it is concluded that a significant portion of the renal elimination of digoxin in man results from tubular secretion and that this excretory mechanism is inhibited by quinidine (6).

That quinidine induces a 30–40% reduction of the volume of distribution of digoxin has been reported by two groups (2, 7) as well as an about 25% reduction in digoxin binding to skeletal muscle (3). Our data on reduced skeletal muscle binding of digoxin during quinidine therapy suggest that this is due to a saturation of digoxin binding site secondary to the increase in the total body load of digoxin at steady state (3). No direct displacement effect of quinidine on digoxin binding is therefore shown and this finding is supported by in vitro measurements showing that quinidine does not affect tissue binding.

The total body clearance of digoxin is the sum of renal clearance and non-renal clearance. Non-renal clearance of digoxin is made up by metabolic clearance and clearance of unknown pathways, e. g. biliary and/or intestinal secretion. Metabolic clearance is known to constitute only a minor fraction of the total clearance of digoxin. The formation of dihydrodigoxin and its degradation products is minimal and also uninfluenced by quinidine (5). Therefore it is most probable that quinidine interferes with biliary and/or intestinal excretion of digoxin. The aim of this study was to quantify the influence of quinidine on the biliary excretion of digoxin in patients by use of a special duodenal perfusion technique.

Patients and methods

Five patients with a mean age of 60 years (range 51–71) and with atrial fibrillation were scheduled for cardioversion. All were on maintenance digoxin therapy since at least one month. The patients were investigated on two occasions, the first time before the introduction of quinidine and the second time after three weeks on the combination therapy. On each investigation day blood samples were taken hourly, urine and bile were collected in one hour fractions over 6–8 hours. The biliary excretion was studied by giving the patients a double lumen catheter in duodenum using a special duodenal marker perfusion technique after a modified method developed by Grundy and Metzger (1) (Fig. 1). BSP (bromsulphtalein) was used as a marker. The patients were given a determined amount of aminoacids and glucose during the day. Cholesterol and biliary acids were analysed by a technique described by Angelin et al. (4). Digoxin concentration

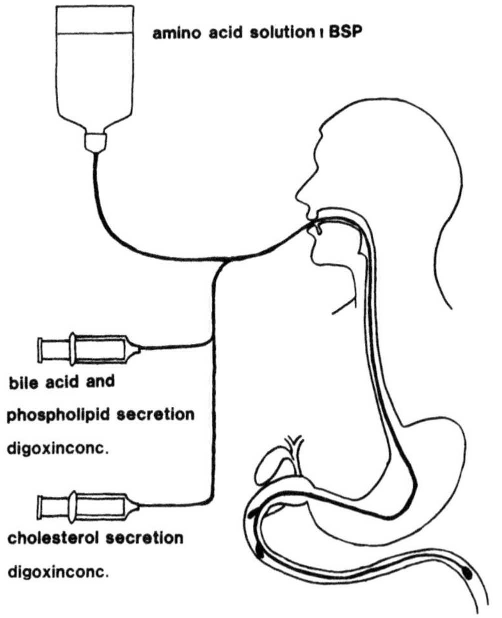

Fig. 1. Determination of digoxin concentration, bile acid, phospholipid secretion and cholesterol secretion using a special duodenal perfusion technique with BSP (bromsulphtalein) as a marker.

in plasma, urine and bile was determined by RIA-technique. Determination of digoxin concentration in bile was made after dilution. The calculation of the biliary excretion rate of digoxin was performed after:

$$\frac{\textit{Digoxin in aspirate (nmol/l)} \times \text{infusion rate of BSP } (\mu g/min)}{\text{BSP in aspirate } (\mu g/ml)}.$$

On the evening of the study day quinidine was given in a loading dose of 0.1 mg quinidine sulphate/10 kg and thereafter in a maintenance dose of 1200 mg daily. The day after the introduction of quinidine DC-shock was performed if sinus rhythm had not been obtained by quinidine. Then the digoxin dose was halved and after three weeks on the combination therapy the experiment was repeated.

Statisical methods

Values are given as mean ± SD. Student's t-test for paired observations were used as test of statistical significance. A p-value > 0.05 was considered as non-significant.

Results

The steady state serum digoxin concentrations were before quinidine 0.70 ± 0.23 nmol/l and during the combination with quinidine and with halved digoxin dose 0.88 ± 0.13 nmol/l (mean ± SD) (Table 1). Quinidine induced a reduction of about 40% of the biliary clearance of digoxin (range 21–65%, p < 0.02) (Table 1). The biliary excretion of digoxin before quinidine was unexpectedly high, suggesting a greater amount of extra hepatic circulation than previously has been described. Also renal clearance of digoxin was reduced by quinidine with a range from 9 to 54%. However, one patient was excluded for technical problems with urine collections and therefore only four patients were left, upon whom the calculation could be based.

Table 1. Effects of quinidine on the biliary clearance of digoxin

Patient		Age (yr)	Steady-state concentration of digoxin in plasma (μmol/L)		Renal clearance of digoxin (ml/min)			Biliary clearance of digoxin (ml/min)		
			D	D + Q	D	D + Q	△	D	D + Q	△
1	F	57	0.9 ± 0.1	1.0 ± 0.2	–	–	–	108 ± 11	68 ± 14	-37%
2	F	53	0.6 ± 0.1	0.7 ± 0.1	207 ± 18	160 ± 36	-23%	138 ± 42	69 ± 7	-50%
3	M	51	0.5 ± 0.1	0.8 ± 0.1	246 ± 47	113 ± 13	-54%	142 ± 12	50 ± 17	-65%
4	M	66	0.5 ± 0.1	0.9 ± 0.1	133 ± 52	110 ± 16	-17%	58 ± 12	47 ± 16	-21%
5	M	71	1.0 ± 0.2	1.0 ± 0.2	106 ± 20	97 ± 23	- 9%	105 ± 29	67 ± 17	-37%
							NS			p<0.02

The values are expressed as mean ± SD
D: treatment with digoxin alone
D + Q: treatment with digoxin + quinidine
△: difference in clearance between D and D + Q
Statistical significance determined by t-test for paired observations

Conclusions

The main mechanism behind the quinidine-digoxin interaction is the reduction of the renal clearance of digoxin, but also a reduction of the biliary excretion of digoxin seems to play a major role in the interaction. The influence on the volume of distribution is probably of minor importance.

This effect on the biliary excretion of digoxin is similar to the effect on renal tubular secretion of digoxin and suggests a similar transport system for digoxin in these organs.

References

1. Grundy SM, Metzger AL (1972) A physiological method for estimation of hepatic secretion of biliary lipids in man. Gastroenterology 62: 1200–1217
2. Hager WD, Fenster P, Mayersohn M, Perrier D, Graves P, Marcus FI, Goldman S (1979) Digoxin-quinidine interaction. Pharmacokinetic evaluation. New Engl J Med 300: 1238–1241
3. Jogestrand T, Schenck-Gustafsson K, Nordlander R, Dahlqvist R (1984) Quinidine induced changes of serum and skeletal muscle digoxin. Evidence of a saturable binding of digoxin to skeletal muscle. Eur J Clin Pharmacol 27: 571–575
4. Nilsell K, Angelin B, Leijd B, Einarsson K (1983) Comparative effect of urodeoxycholic acid and chenodeoxycholic acid on bile acid kinetics and biliary lipid secretion in humans. Gastroenterology 85: 1248–1256
5. Schenck-Gustafsson K (1982) The quinidine-digoxin interaction. Effects of quinidine on pharmacokinetics and pharmacodynamics of digoxin. Thesis, Karolinska Institute, Stockholm
6. Schenck-Gustafsson K, Juhlin-Dannfeldt A, Dahlqvist R (1982) Renal function and digoxin clearance during quinidine therapy. Clin Physiol 2: 401–408
7. Schenck-Gustafsson K, Dahlqvist R (1981) Pharmacokinetics of digoxin in patients subjected to the quinidine-digoxin interaction. Br J Clin Pharmacol 11: 181–186

Author's address:

Dr. Karin Schenck-Gustafsson, Karolinska Institute, Division of Cardiology, Department of Internal Medicine, Huddinge University Hospital, S–14186 Huddinge (Sweden)

Role of the bile acid transport system in hepatocellular ouabain uptake

E. Petzinger, K. Fischer, and H. Fasold*

Institute of Pharmacology and Toxicology, Justus-Liebig-University Gießen, Department of Veterinary Medicine, Gießen, and *Institute of Biochemistry, Johann Wolfgang Goethe-University, Frankfurt, Department of Medicine, Frankfurt (West Germany)

Summary

HeLa cells take up ^{3}H-ouabain by internalization together with Na^+/K^+-ATPase (4). Similar endocytotic uptake was suggested for rat hepatocytes which excrete ouabain into bile (10). On isolated rat hepatocytes carrier mediated uptake of ouabain has been described (5) but any identification of the transport system is still lacking. By using rauwolfia alkaloide cevadine we have proved that ouabain in contrast to HeLa cells is not taken up in rat hepatocytes by Na^+/K^+-ATPase internalization (17). We speculate additional binding proteins in rat liver cell membrane with properties for glycoside translocation and suggest the bile acid transport system to be such a candidate. Ouabain uptake is inhibited by DIDS, probenecid, furosemide and cevadine which inhibit bile acid uptake concomitantly. Bile acids are inhibitors of ouabain uptake. Cells which lack bile acid uptake, e. g. AS-30 D ascites hepatoma cells and Ehrlich ascites tumor cells do not transport ouabain. Digitoxin, digoxin and cassaine as compared to ouabain are stronger inhibitors of ^{3}H-ouabain uptake. They inhibit cholate uptake markedly.

In order to identify membrane proteins probably involved in ouabain uptake, rat hepatocytes were incubated with benzacidolysin-ouabain for photoaffinity labelling. The label was covalently attached to a 50 kDa protein which was not identified in transport deficient Ehrlich cells. With respect to recent results in the identification of the bile acid transporter from hepatocytes (14, 23) we propose ouabain as substrate for the hepatic bile acid uptake system.

Introduction

Early observations by Farah (6) indicated that ouabain is excreted in high concentrations into rat bile. Bile-to-plasma rations between 147 to 557 were found in clearance studies with intact animals. This prompted Kupferberg and Schanker (13) to postulate the active excretion of ouabain at the biliary pole of hepatocytes. Ouabain was regarded as a model compound for a transport system which is believed to enlarge the excretory capacity of the liver for neutral organic drugs (18). In addition to active canalicular excretion, active hepatocellular uptake of ouabain across the sinusoidal membrane was assumed (13). The mechanisms of both transports remained unknown, but for ouabain uptake, endocytosis of a $(Na^+ + K^+)$-ATPase/ouabain complex was proposed (10). A comparable transport mechanism was originally described for ouabain uptake into HeLa cells (3, 4). However, instead of vesicular uptake, carrier mediated transport into isolated liver cells was found (5). Recently, the participation of the $(Na^+ + K^+)$-ATPase in hepatic ouabain uptake was disproved and some evidence for a role of the bile acid transport system in hepatic ouabain uptake was presented (17). The present study is intended to clarify the importance of the bile acid transport system for hepatocellular ouabain uptake. Attempts to identify membrane proteins possibly involved in ouabain transport and distinct from $(Na^+ + K^+)$-ATPase are presented.

Methods

Rat hepatocytes were isolated according to Berry and Friend (1) after perfusion of rat livers with Ca^{++} free Krebs-Henseleit buffer, pH 7.4, 37°C, containing 0.05 % collagenase. Details of the isolation procedure have been given elsewhere (16). 2 x 10^6 hepatocytes/ml were incubated in the presence of either 55 nCi (^{3}H)-ouabain diluted with 10 μg ouabain to yield a final concentration of 15 μM or in the presence of 48 nCi (^{14}C)-cholate, final concentration 1 μM. Uptake into liver cells was measured by the silicon oil centrifugation method (12). Each experiment was performed with "n" cell preparations. Labelling of cell proteins by (^{14}C)-azidobenzoyl-lysyl-ouabain was achieved in a self-made photo flash apparatus (25). Cells which were incubated for 2 or 15 min in the presence of the photolabile label, were subjected to one to three photo flashes in a tubelike quartz cuvette. Cells were stored in liquid nitrogen and permabilized by thawing. By three washing procedures, nonbound and internalized label as well as labelled cytosolic proteins were removed. Cell ghosts were dissociated in SDS-mercaptoethanol at 95°C for 30 min and the proteins were separated in 10 % accrylamide gels (21).

Materials

(^{3}H)-ouabain, spec. activity 32 Ci/mmol and (^{14}C)-cholate, spec. activity 52 mCi/mmol were from Amersham Buchler, Braunschweig. Ouabain was from Serva, Heidelberg, ouabagenin, digitoxin, chenodeoxycholate, deoxycholate and cholate were obtained from Sigma, Munich.

The N-p-azidobenzyl-lysyl-ouabain was synthesized by cleavage of the sugar moiety of ouabain with periodate, followed by the reductive coupling of (^{3}H)-lysine with the aid of sodium borohydride. The free amino group of lysine was then reacted with p-azidobenzoic acid and DCC, and the final compound was purified over silica gel chromatography in ethanol. Initially, cold lysine and (^{14}C)-p-azido benzoic acid were used as the starting materials. The structural formula of N-p-azidobenzoyl-lysyl-ouabain with reference to Grell et al. (11) is given below.

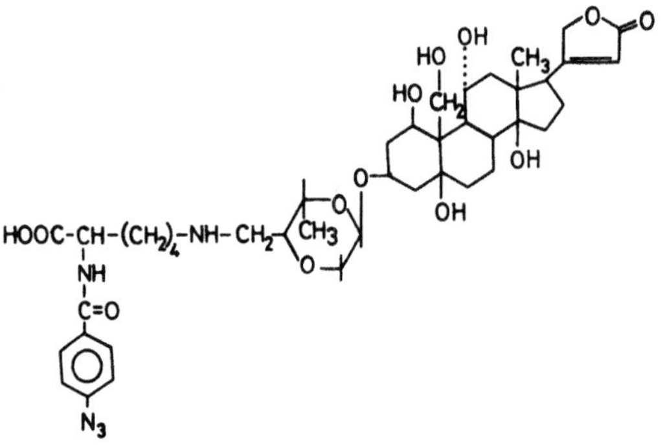

Results

Inhibition of ouabain uptake into isolated rat hepatocytes by bile acids

Schwenk et al. (20) reported competitive inhibition of ouabain uptake into liver cells by taurocholate. We tested naturally occurring bile acids of the rat for inhibition of ouabain uptake (Fig. 1). Cholate, chenodeoxycholate and deoxycholate were strong inhibitors of ouabain transport at concentrations far below their critical micellar concentration which excludes any nonspecific detergent effect on the ouabain transport proteins. In the presence of 20 μM deoxycholate, the initial rate of ouabain uptake is reduced to 18 % of the control, while the corresponding residual uptake rates in the presence of 20 μM chenodeoxycholate and 20 μM cholate are 39 % and 46 % respectively. Apparent calculated IC_{50} concentrations shown in Fig. 2 are 0.45 μM, 11 μM and 18 μM for deoxycholate, chenodeoxycholate and

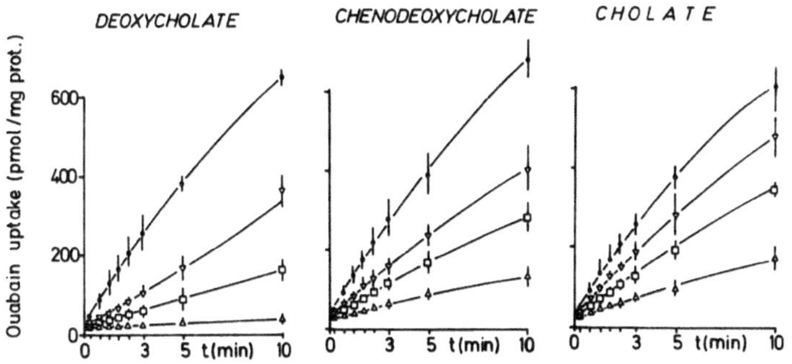

Fig. 1. Inhibition of ouabain uptake into isolated rat hepatocytes by naturally occurring portal blood bile acids. 2 x 10^6 hepatocytes/ml were pretreated for 30 sec with 10 μM (∇), 20 μM ($\square$) and 50 μM (Δ) deoxycholate, chenodeoxycholate and cholate. After adding 55 nCi (^{3}H)-ouabain/10 μg ouabain 100 μl aliquots of cell suspension were withdrawn and cells were pelleted by centrifugation at 15 000 g (Beckmann Microfuge B) through silicon oil. Radioactivity was determined after cell digestion in 3 M KOH. The amount of extracellular fluid in the cell pellet, as measured by (H)-inulin distribution, was below 1 %. The experiments were performed with four cell preparations (n = 4), $\bar{x} \pm$ SD.

cholate respectively. The affinity of cholate to the ouabain uptake system is 20 fold higher than the affinity of ouabain.

Effect of cardiac glycosides on the uptake of cholate into isolated rat hepatocytes

Ouabain, ouabagenin and digitoxin inhibited cholate uptake in isolated rat hepatocytes (Fig. 3). Digitoxin was the most effective inhibitor, while ouabagenin was the weakest. Ap-

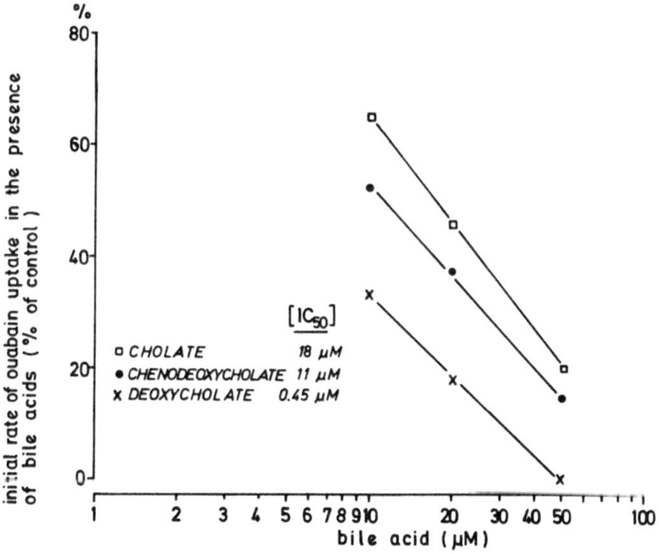

Fig. 2. Semilogarithmic plot of the inhibition of (^{3}H)-ouabain uptake by bile acids. The calculated concentrations of the bile acids for 50 % inhibition of the initial rate of ouabain uptake (IC_{50}-values) are 0.45 μM deoxycholate, 11 μM chenodeoxycholate and 18 μM cholate.

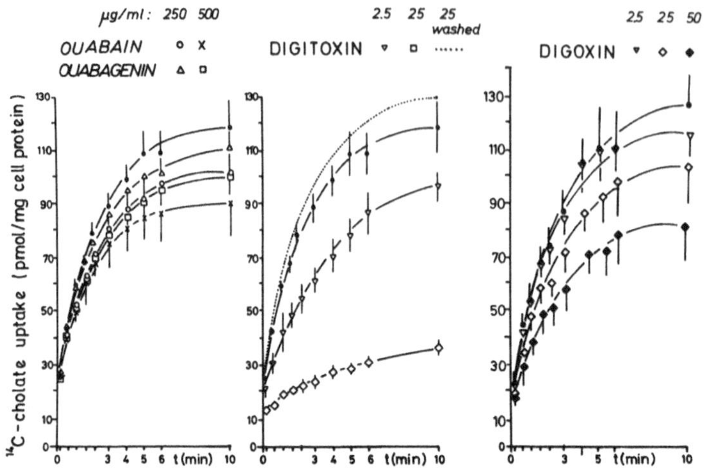

Fig. 3. Inhibition of (^{14}C)-cholate uptake into isolated rat hepatocytes by cardiac glycosides. 2 x 10^6 hepatocytes/ml were incubated for 30 sec in the presence of either 250 (○, Δ) or 500 μg (x, □) ouabain (○, x) and ouabagenin (Δ), □) or 2.5 (∇), 25 (◇) and 50 µg (◆) digitoxin or digoxin. (^{14}C)-cholate uptake (●) was determined as described in methods. Reversibility of the inhibition by digitoxin was tested by washing hepatocytes 3 times in Tyrode buffer (dotted line). $\bar{x} \pm$ SD, n = 4–5.

parent IC_{50} values were calculated to be 17 mM for ouabain, 114 mM ouabagenin, 250 μM for digoxin but 7 μM for digitoxin. The inhibition caused by digitoxin could be abolished by washing cells prior to adding (^{14}C)-cholate (Fig. 3).

Identification of proteins probably involved in (^{3}H)-ouabain uptake

By the use of the photolabile azidobenzoyl-lysyl-ouabain, the binding proteins probably involved in ouabain uptake in isolated rat hepatocytes were identified. A significant part of the transport capacity for ouabain was irreversibly inhibited after photo activation (Fig. 4). About 50% of the initial rate of ouabain uptake remained blocked after removal of non-bound label by washing.

When isolated rat liver cells were photo labelled with (^{14}C)-azidobenzoyl-lysyl-ouabain, three protein regions with covalently bound radioactivity were detected after SDS-gel-electrophoresis (Fig. 5 C). Proteins of region 1 had high molecular weights and remained on the top of the slab gel (Fig. 5 C). The proteins were also found on the top of geld rods (Fig. 6). The amount of bound radioactivity increased corresponding to the number of photo flashes (Fig. 6). Proteins of region 2 were most intensively labelled and had molecular weights of 50 kDa. A small band running behind these proteins was in the range of 57 kDa. The third region with (^{14}C)-radioactivity was between 23 and 30 kDa (Fig. 5 C). In each of the gel regions (^{14}C)-isotope incorporation was significantly reduced in the presence of an excess of nonlabeled ouabain (Fig. 5 B). Protection experiments with bile acids were hazy because only millimolar concentrations reduced significantly the incorporation of (^{14}C)-azidobenzoyl-lysyl-ouabain in these protein bands (not shown). The photo affinity label was exclusively incorporated after photo activation. No protein bound radioactivity was found in the absence of the photo flash (Fig. 5 D, E). Cross linking or degradation of liver cell proteins by photo flashes was not obvious since the protein pattern in SDS-slab-gel-electrophoresis was not altered after photo irradiation with up to three flashes (Fig. 5).

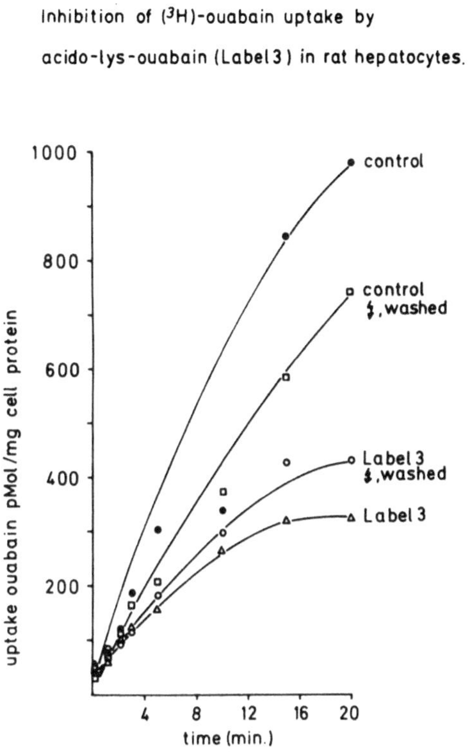

Fig. 4. Irreversibility of the inhibition of (^{3}H)-ouabain uptake by azidobenzoyl-lysyl-ouabain. 100 μg of azidobenzoyl-lysyl-ouabain (label 3) were added to 1 ml liver cell suspension 30 sec prior to addition of 55 nCi (^{3}H)-ouabain/10 μg ouabain without photo flash irradiation (Δ). Covalent binding was induced by one photo flash and nonbound label was removed by washing before (^{3}H)-ouabain/ouabain was added (○). In the absence of photo flash, inhibition by label 3 was fully reversible (not shown). Control: uptake of ouabain into liver cells which were either subjected to one (□) or no (●) photo flash (representative experiment).

Discussion

An overview of carrier mediated transport into isolated rat hepatocytes (19) indicated that properties of the sinusoidal (basolateral) cell membrane influence clearance and biliary excretion of foreign compounds. Ouabain is taken up into isolated rat liver cells by active transport without the involvement of ($Na^+ + K^+$)-ATPase (17). Recently a "multispecific" bile acid transport system in hepatocytes was shown to transport cyclic peptides e. g. phalloidin and antamanide and organic anions e. g. iodipamide (7, 8, 15). Because of their structural similarity to bile acids, cardiac glycosides would also appear to be likely substrates for this transport system. Indeed bile acids inhibit the permeation of ouabain into isolated rat liver cells at nondetergent concentrations (Fig. 1). Dihydroxy bile acids such as deoxycholate and chenodeoxycholate are stronger inhibitors than the trihydroxy bile acid cholate. Conversely, ouabain, ouabagenin, digoxin and digitoxin inhibited cholate uptake. Because cholate uptake is partially sodium dependent, one has to rule out that inhibition of the sodium pump is responsible for the observed effects. However, rat liver ($Na^+ + K^+$)-

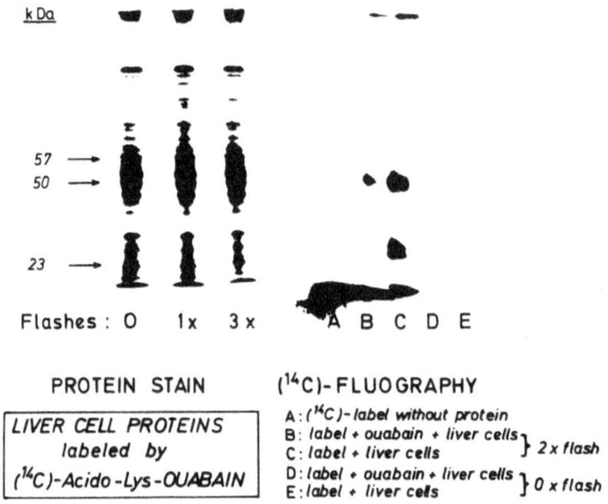

Fig. 5. Labelling of liver cell proteins by (^{14}C)-azidobenzoyl-lysyl-ouabain. 1 ml hepatocyte suspension was incubated at 37°C with 1 μCi (^{14}C)-azidobenzoyl-lysyl-ouabain either in the presence (B, D) or absence (C, E) of 1 mg ouabain. Covalent binding was induced by one photo flash (B, C). Cells were stored in liquid N_2, washed and subjected to SDS-electrophoresis in 10% acrylamid slab gels. (A): electrophoresis of a sample of (^{14}C)-azidobenzoyl-lysyl-ouabain alone.

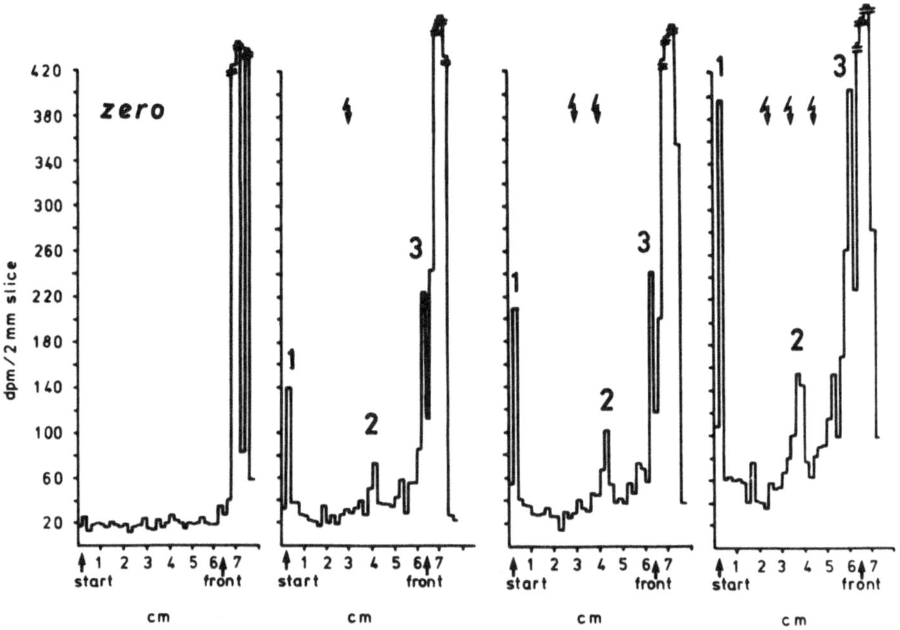

Fig. 6. Incorporation of (^{14}C)-azidobenzoyl-lysyl-ouabain into liver cell proteins by photo flash irradiation. Liver cells (2 x 10^6/ml were frozen in liquid nitrogen, thawed and washed four times in Tyrode buffer. Cell ghosts were subjected to 1, 2 and 3 photo flashes after incubation for 10 min with 1 μCi (^{14}C)-azidobenzoyl-lysyl-ouabain. Proteins were separated in 10% acrylamide rod gels and protein-bound radioactivity was measured by liquid scintillation counting of 2 mm gel slices.

Left to right: rod gels with cell samples after zero, 1, 2 and 3 photo flashes (↯, ↯↯, ↯↯↯). Protein regions 1, 2 and 3 are indicated (see text).

ATPase is remarkably insensitive to cardiac glycosides and complete inhibition is achieved at 1 mM ouabain (2). This concentration inhibited cholate uptake only moderately (Fig. 3). However, a concentration of 40 μM digitoxin blocked cholate by up to 90 % although no inhibition of the (Na^+ + K^+)-ATPase should occur at this concentration. Thus the different inhibition effects by the cardiac glycosides could merely reflect their inhibition of sodium transport, but are more probably due to different affinity for the bile acid transport system. This assumption is ascertained by the fact that inhibition by digitoxin was reversible and might therefore reflect transport competition rather than irreversible blockade of (Na^+ + K^+)-ATPase. Obviously, digitoxin exhibits a greater affinity for the bile acid transport proteins than ouabain and it was also found to have a 10000 fold higher inhibitory strength for (^{3}H)-ouabain uptake than ouabain (17).

Our attempts to identify cell proteins probably responsible for ouabain uptake were partially successful with the photolabile azidobenzoyl-lysyl-ouabain. After photolysis, the compound irreversibly inhibited ouabain uptake (Fig. 4), and by using the radioactive analog at least 3 protein regions were identified by fluography (Fig. 5). Incorporation of the label was, however, below 1 % of the added amount although intact hepatocytes trapped about 40 % of that radioactivity. Thus in gel electrophoresis most of the radioactivity migrated at the gel front (Fig. 5). Two of the protein regions overlap with those proteins which can be labelled by a taurodehydrocholate affinity label and by H_2-DIDS (22). These proteins had apparent molecular weights of 50 and 23–30 kDa. Obviously, the 95 kDa protein of the sodium pump which binds ouabain is not among those heavily labelled proteins. The 50 kDa protein was the most intensively labelled one. It appears to be a likely candidate for an integral membrane protein which constitutes the bile acid transporter (23). However, the role of the low molecular weight proteins in ouabain uptake into isolated rat hepatocytes remains unknown.

Conclusions

Cardiac glycosides and bile acids are mutual inhibitors of their hepatocellular uptake. Inhibition is neither the result of detergent effects by the bile acids nor of inhibition of the (Na^+ + K^+)-ATPase by the cardiac glycosides. It is assumed that active uptake of ouabain into isolated rat hepatocytes is mediated by proteins of the bile acid transport system. A photolabile affinity label derived from ouabain was covalently attached to cell proteins which have already been shown to be susceptible to affinity labels derived from bile acids.

Acknowledgements

The technical assistance of Mrs. S. Wilhelm is gratefully acknowledged. We thank Dr. McDowell for his kind support in preparing the manuscript. This work was supported by grant Fa 48/23–9, Deutsche Forschungsgemeinschaft, Sonderforschungsbereich 169.

References

1. Berry MS, Friend DS (1969) High-yield preparation of isolated rat liver parenchymal cells. J Cell Biol 43: 509–529
2. Boyer JL, Reno D (1975) Properties of (Na^+ + K^+)-activated ATPase in rat liver plasma membranes enriched with bile canaliculi. Biochim Biophys Acta 401: 59–72
3. Cook JS, Will PC, Proctor WR, Brake ET (1976) Turnover of ouabain binding sites and plasma membrane proteins in HeLa cells. In: Cook JS (ed) Biogenesis and Turnover of Membrane Macromolecules. Raven Press, New York, pp 15–36

4. Cook JS, Tate EH, Shaffer C (1982) Uptake of (^{3}H)-ouabain from the cell surface into the lysosomal compartment of HeLa cells. J Cellular Physiol 110: 84–92
5. Eaton DL, Klaassen CD (1978) Carrier-mediated transport of ouabain in isolated hepatocytes. J Pharmacol Exp Ther 205: 480–488
6. Farah A (1946) On the elimination of G-strophantin in the rat. J Pharmacol Exp Ther 86: 248–257
7. Frimmer M (1980) Organotropism by carrier-mediated transport. Trend Pharmacol Sci 3: 395–397
8. Frimmer M, Petzinger E, Ziegler K (1980) Protective effect of anionic cholecystographic agents against phalloidin on isolated hepatocytes by competitive inhibition of the phallotoxin uptake. Naunyn Schmiedeberg's Arch Pharmacol 313: 85–89
9. Frimmer M, Ziegler K (1985) Photoaffinity labeling of whole cells by flashed light: A simple apparatus for high-energy ultraviolet flashes. Biochim Biophys Acta (in press)
10. Graf J, Peterlik M (1976) Ouabain-mediated sodium uptake and bile formation by isolated perfused rat liver. Am J Physiol 230: 876–885
11. Grell E, Lewitzki E, Krause G, Richter WJ, Raschdorf F (1984) Mechanism of interaction between cation and ligand binding sites in Na^+,K^+-ATPase. In: Proceedings of the Third EBEC Conference, Hannover 1984. ICSU Press, Miami
12. Klingenberg M, Pfaff E (1967) Means of terminating reactions. Meth Enzymol 10: 680–684
13. Kupferberg HJ, Schanker LJ (1968) Biliary secretion of ouabain-^{3}H and its uptake by liver slices in the rat. Am J Physiol 214: 1048–1053
14. Kramer W, Bickel U, Buscher H-P, Gerok W, Kurz G (1982) Bile-salt-binding polypeptides in plasma membranes of hepatocytes revealed by photoaffinity labelling. Eur J Biochem 129: 13–24
15. Petzinger E, Joppen C, Frimmer M (1983) Common properties of hepatocellular uptake of cholate, iodipamide and antamanide, as distinct from the uptake of bromsulfophthalein. Naunyn Schmiedeberg's Arch Pharmacol 322: 174–179
16. Petzinger E, Frimmer M (1984) Driving forces in hepatocellular uptake of phalloidin and cholate. Biochim Biophys Acta 778: 539–548
17. Petzinger E, Fischer K (1985) Transport functions of the liver. Lack of correlation between hepatocellular ouabain uptake and binding to (Na^+ + K^+)-ATPase. Biochim Biophys Acta 815: 334–340
18. Schanker LS (1968) Secretion of organic compounds in the bile. In: Handbook of Physiology Vol 5, Alimentary Canal, p 2433–2449
19. Schwenk M (1980) Transport systems of isolated hepatocytes. Arch Toxicol 44: 113–126
20. Schwenk M, Wiedemann T, Remmer H (1981) Uptake, accumulation and release of ouabain by isolated rat hepatocytes. Naunyn Schmiedeberg's Arch Pharmacol 316: 340–344
21. Weber K, Osborn M (1969) The reliability of molecular weight determinations by dodecyl sulfate-polyacrylamide gel electrophoresis. J Biol Chem 244: 4406–4412
22. Ziegler K, Frimmer M, Fasold H (1984) Further characterization of membrane proteins involved in the transport of organic anions in hepatocytes. Comparison of two different affinity labels: 4,4'-diisothiocyano-1,2-diphenylethane-2,2'-disulfonic acid and brominated taurodehydrocholic acid. Biochim Biophys Acta 769: 117–129
23. Ziegler K, Frimmer M, Müller S, Fasold H (1984) 3'isothiocyanatobenzamido (^{3}H)-cholate, a new affinity label for hepatocellular membrane proteins responsible for the uptake of both bile acids and phalloidin. Biochim Biophys Acta 773: 11–22
24. Ziegler K, Frimmer M (1986) Photoaffinity labeling of whole cells by flashed light: A simple apparatus for high energy ultraviolet flashes. Biochim Biophys Acta (in press)

Author's address:

Dr. E. Petzinger, Institut für Pharmakologie und Toxikologie im Fachbereich Veterinärmedizin, Justus-Liebig-Universität Gießen, Frankfurter Straße 107, 6300 Gießen (West Germany) Reprint requests should be addressed to E. P., Max Planck Institut für Systemphysiologie, Rheinlanddamm 201, 4600 Dortmund

Distribution of digoxin, digitoxin and their cardioactive metabolites in human heart and kidney

J. Plum, T. Daldrup[1], B. Grabensee and H. Schweitzer[1]

Department of Internal Medicine, Division of Nephrology, and [1]Department of Forensic Sciences, University of Düsseldorf (West Germany)

Summary

There are still few data available about the quantitative distribution of the cardioactive metabolites of digitalis glycosides in human tissues. We therefore developed a method for the specific determination of digoxin, digitoxin, their semi-synthetic derivatives and dependent cardioactive metabolites in autopsy samples of heart and kidney. Tissue extraction and purification steps were followed by HPLC separation and RIA quantitation.

A collective of 6 patients on long-term treatment with therapeutic doses of β-acetyldigoxin had a mean myocardial digoxin content of 46.1 ± 25.0 ng/g (SD); kidney: 50.3 ± 30.3 ng/g. Digoxigenin bisdigitoxoside represented the second most important metabolite with 11.0 ± 5.1 ng/g in the heart and 13.6 ± 3.5 ng/g in the kidney. Digoxigenin monodigitoxoside and digoxigenin followed respectively with mean tissue contents below 5 ng/g.

In a collective of 7 patients on maintenance treatment with digitoxin the mean tissue levels were higher but the metabolic pattern was similar. We measured an average myocardial digitoxin content of 78.9 ± 38.4 ng/g and a renal content of 104.1 ± 44.1 ng/g. Digitoxigenin bisdigitoxoside was determined to be 15.4 ± 6.5 ng/g and 11.9 ± 1.5 ng/g respectively. The amount of digoxin formed by hydroxylation under long-term treatment with digitoxin in heart and kidney were approximately 10 ng/g.

In three cases of digoxin intoxication the myocardial digoxin content was at least 1.3 fold, the kidney level at least 2.9 fold higher than the maximum therapeutic level. The contents of digoxigenin bisdigitoxoside in the kidney also ranged high. The quotient digoxin/digoxigenin bisdigitoxoside in the myocardium may be a further parameter for the differentiation between chronic and acute digoxin ingestion (2.8:1–4.3:1 in the therapeutic group on long term treatment, 9.8:1–10.1:1 in the cases of acute intoxication).

Introduction

The content of digoxin and digitoxin in human tissues, especially in heart and kidney, has been determined by several studies in vivo (5–8) and post mortem (1, 3, 4, 8, 9, 13). But up to now few data have been available on the quantitative distribution of their cardioactive metabolites. These are formed by the splitting of the digitoxose sugar residues from the aglycone steroid (Fig. 1). The method of determination usually applied in the preceding studies was the radio-immunoassay because of its high measuring sensitivity in the ng-range. But immunologic methods have disadvantages with regard to their specificity. We found the metabolites of the digoxin and digitoxin series cross-react with a commercial RIA antibody between 64 and 160 %, referring to the parent substance. With the commonly used technique of tissue extraction, purification and immunologic measurement, the determination of tissue contents must be "cumulative" and inexact because the cross-reacting metabolites cannot be qualitatively or quantitatively differentiated. Differing results in literature

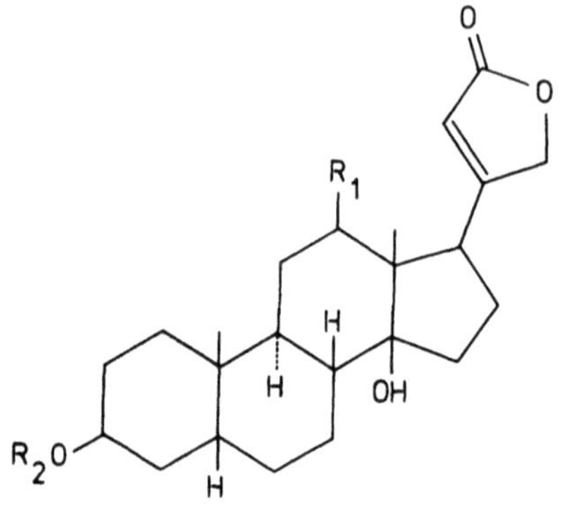

Compound	R1	R2
Digitoxigenin	H	H
Digitoxigenin monodigitoxoside	H	D
Digitoxigenin bisdigitoxoside	H	D-D
Digitoxin	H	D-D-D
Digoxigenin	OH	H
Digoxigenin monodigitoxoside	OH	D
Digoxigenin bisdigitoxoside	OH	D-D
Digoxin	OH	D-D-D
ß-Acetyldigoxin	OH	D-D-D-Ac
ß-Methyldigoxin	OH	D-D-D-Met

D = Digitoxose; Ac = acetyl; Met = methyl

Fig. 1. Structures of the digitalis glycosides and their aglycones investigated.

(1, 3, 8, 13) may thus be explained. We therefore developed a method for the specific determination of digoxin, digitoxin, their semi-synthetic derivatives and dependent cardioactive metabolites in autopsy samples of heart and kidney. Tissue extraction and purification steps were followed by high performance liquid chromatography and subsequent RIA quantitation.

Patients and methods

Autopsy samples were obtained from 9 patients on therapeutic doses of β-acetyldigoxin or β-methyldigoxin (Table 1). Seven patients were investigated under maintenance treatment with digitoxin (Table 2). Three cases of fatal digoxin intoxication were compared with the therapeutic group (Table 3). Tissue specimens were taken from the myocardium of the left ventricle and from the kidney comprising approximately equal parts of cortex and medulla. Fatty layers and pathological transformation zones were excised.

An extraction method was developed which allowed the direct chromatographic measurement of therapeutic digitalis glycoside levels in the tissue (Fig. 2). With thin-layer chromatography however, only the highly concentrated glycosides such as digoxin, digitoxin and partially their bisdigitoxosides could be detected. By the help of a HPLC gradient separation (Fig. 3) and subsequent RIA measurement, all metabolites were specifically quantitated. Elution fractions were collected and evaporated to dryness by lyophilisation. A commercial digoxin RIA (Beckman Instr., Fullerton, USA) and digitoxin RIA (Becton Dickinson, NY, USA) were used. For the assay of each cardioactive metabolite, an own standard curve was constructed. As the affinity of all substances to the corresponding digoxin and digitoxin antibody was high, quantitation could be carried out without loss in sensitivity.

Table 1 Clinical data, dosage schedule and diagnosis of patients studied after digoxin treatment.

Patient	Age	Sex	Height (m)	Weight (kg)	Time between † and autopsy (h)	Digoxin dosage before †	Diagnosis or autopsy findings
A	61	M	1.75	65	30	* 0.2 mg β-Ac-DG 60 h before † 0.4 mg	Pulmonary embolism
B	75	F	1.53	55	24	* 0.4 mg β-Ac-DG	Acute bronchopneumonia, recurrent pulmonary embolism
C	78	M	1.70	61	31	12 d 0.2 mg β-Ac-DG	Bronchial carcinoma
D	57	F	1.50	42	78	6 d 0.3 mg β-AC-DG	Cervix carcinoma, broncho-pneumonia
E	74	F	1.60	70	24	12 d 0.3 mg β-Ac-DG	Cardiac infarction
F	64	M	1.82	81	43	21 d 0.3 mg β-Ac-DG	Crush syndrome
G	78	F	1.70	68	30	* 0.2 mg β-Met-DG	Heart failure, diabetes mellitus
H	76	F	1.55	67	19	* up to 5 d before † 0.2 mg β-Met-DG 5 d before † change to DT-3	Cardiac infarction, diabetes mellitus
I	76	M	1.75	78	76	* up to 10 d before † 0.3 mg β-Ac-DG from 10 d before † no digitalisation	Bronchial carcinoma

* means long-term treatment > 30 days

Table 2. Clinical data, dosage schedule and diagnosis of patients studied after digitoxin treatment.

Patient	Age	Sex	Height (m)	Weight (kg)	Time between † and autopsy (h)	Digoxin dosage before †	Diagnosis resp. autopsy findings
K	70	F	1.55	48	28	* up to 3 d before † 0,2 mg DT-3 from 3 d before † 0,1 mg	Aortic valve stenosis and Mitral stenosis, obstructive atrium thrombus
L	75	F	1.54	54	33	11 d 0,1 mg DT-3	Hypothermia, bronchopneu-monia
M	91	F	1.45	44	34	* up to 3 d before † 0,1 mg DT-3 from 3 d before † 0,07 mg	Mammacarcinoma
N	69	F	1.52	60	33	* 0,1 mg DT-3	Abscess-forming pyelone-phritis, diabetes mellitus
O	77	F	1.53	42	35	* 0,07 mg DT-3	Hepatic abscess by perforated cholecystitis, diabetes mellitus
P	68	M	1.73	67	23	16 d 0,1 mg DT-3	Cardiac infarction, carcinoma of the prostata with metastasis
Q	81	M	1.55	60	28	17 d 0,1 mg DT-3	Acute bronchopneumonia, heart failure

* means long-term treatment > 30 days

Table 3. Clinical data and autopsy findings of persons studied after fatal digoxin intoxication.

Patient	Age	Sex	Height (m)	Weight (kg)	Time between and autopsy (h)	Digoxin dosage before	Diagnosis or autopsy findings
1	82	M	1.77	88	148	0.3 mg β-Ac-DG/d plus unknown amount by self-medication	Chronic digoxin intoxication
2	44	M	1.69	66	12	Probably 50 tablets of β-Ac-DG (à 0.3 mg)	Acute digoxin intoxication
3	35	F	1.67	107	24	Probably 300 tablets DG-3 (à 0.25 mg)	Acute digoxin intoxication

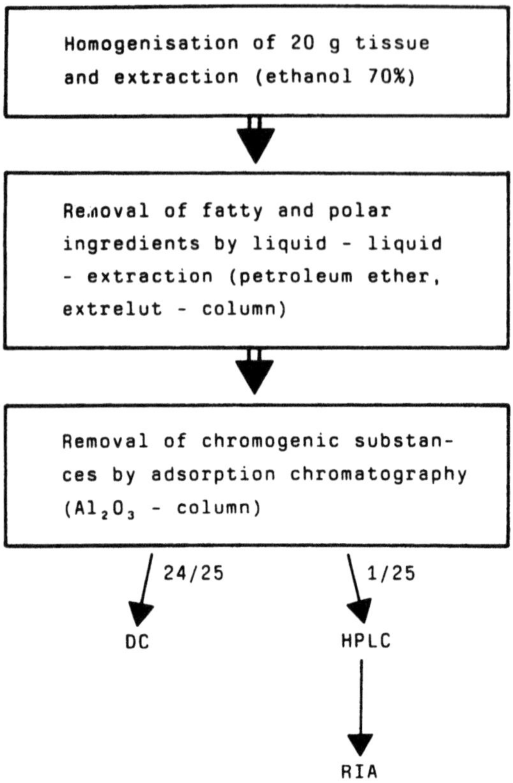

Fig. 2. Extraction and purification procedure for the analysis of digoxin, digitoxin and their cardioactive metabolites in human tissues.

Results

A collective of six patients on long-term treatment with therapeutic doses of β-acetyldigoxin (Table 1, A-F) had a mean myocardial digoxin content of 46.1 ± 25.0 ng/g (SD) and a kidney content of 50.3 ± 30.3 ng/g (Fig. 4). Digoxigenin-bisdigitoxoside represented the second most important metabolite with 11.0 ± 5.1 ng/g in the heart and 13.6 ± 3.5 ng/g in the kidney. Digoxigenin-monodigitoxoside and digoxigenin followed respectively with mean tissue contents below 5 ng/g.

In a collective of seven patients on maintenance treatment with digitoxin (Fig. 5, K–Q), the mean tissue levels were higher but the metabolic pattern was similar. We measured an average myocardial digitoxin content of 78.9 ± 38.4 ng/g and a renal content of 104.1 ± 44.1 ng/g. Digitoxigenin bisdigitoxoside was determined to be 15.4 ± 6.5 ng/g and 11.9 ± 1.5 ng/g respectively. The amount of digoxin formed by hydroxylation on long-term treatment with digitoxin in heart and kidney was approximately 10 ng/g.

Three cases of fatal digoxin intoxication were investigated (Table 3, 1–3). The digoxin content in the heart was 1.3–2.1 fold, in the kidney 2.9–8.0 fold higher than the maximum therapeutic level. Referring to the therapeutic digoxin collective (Fig. 4), the upper tolerance limit for each metabolite was calculated, under which, with a probability of 95%, at

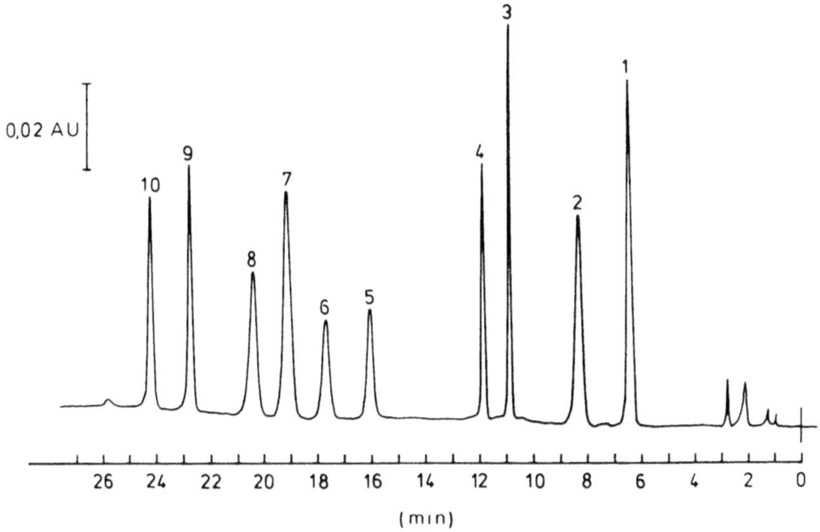

Fig. 3. High-performance chromatogram of digoxin, digitoxin, their semi-synthetic derivatives and cardioactive metabolites. Separation by a complex gradient elution.
Conditions: column, M-N Nucleosil C-18, 10 μm; mobile phase, acetonitrile-water; flow-rate, 1 ml/min; UV monitor at 220 nm; 1 μg of each compound injected in 40 μl methanol.
Peaks: 1 = digoxigenin (DG-0), 2 = digoxigenin monodigitoxoside (DG-1), 3 = digoxigenin bisdigitoxoside (DG-2), 4 = digoxin (DG-3), 5 = β-methyldigoxin (β-Met-DG), 6 = β-acetyldigoxin (β-Ac-DG), 7 = digitoxigenin (DT-0), 8 = digitoxigenin monodigitoxoside (DT-1), 9 = digitoxigenin bisdigitoxoside (DT-2), 10 = digitoxin (DT-3).

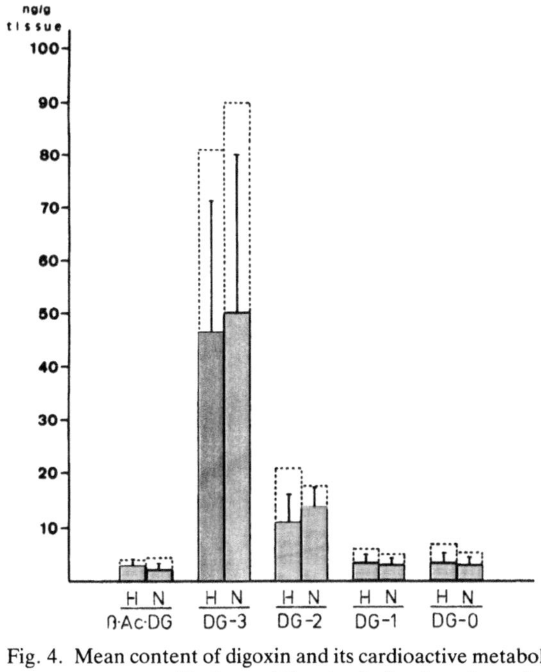

Fig. 4. Mean content of digoxin and its cardioactive metabolites in the tissues of 6 patients under maintenance treatment with β-acetyldigoxin. Indication of standard deviation and range (dashed lines). H = heart (left ventricle), N = kidney. Abbreviations of the cardenolides as in Fig. 3.

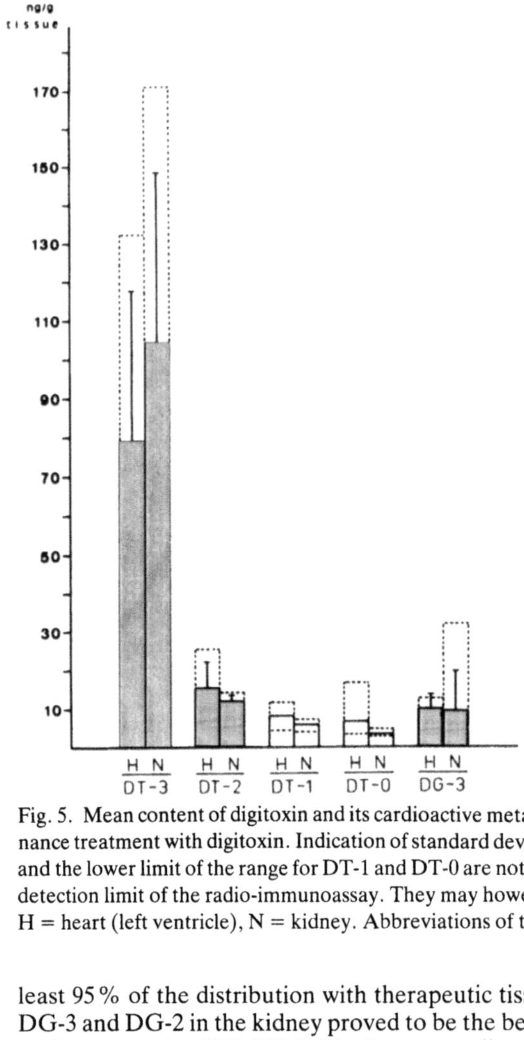

Fig. 5. Mean content of digitoxin and its cardioactive metabolites in the tissues of 7 patients under maintenance treatment with digitoxin. Indication of standard deviation and range (dashed lines). The mean value and the lower limit of the range for DT-1 and DT-0 are not exact because some tissue contents fall near the detection limit of the radio-immunoassay. They may however serve as approximation.
H = heart (left ventricle), N = kidney. Abbreviations of the cardenolides as in Fig. 3.

least 95 % of the distribution with therapeutic tissue contents fall (Fig. 6). The amount of DG-3 and DG-2 in the kidney proved to be the best parameter to indicate intoxication.

The proportion DG-3/DG-2 in the myocardium may give further information for the differentiation between chronic and acute digoxin ingestion (Fig. 7). The value is comparatively high (> 8) in the cases of acute digoxin intoxication (patients 2, 3) or after dose build-up (patient A). Under constant medication (patients C, D, E, F, G), ratios between 2.8 and 4.3 were found. In the myocardium of patients under withdrawal of the drug before death (patients H, I), we measured relatively low levels of DG-3. The elimination of the more lipophilic cardioactive metabolites DG-2, DG-1, DG-0 however was minor. Consequently, the ratios DG-3/DG-2 were low (< 2.3). Concerning the value of metabolic patterns for forensic questions, we are undertaking further studies.

Discussion

A direct comparison of the tissue contents of digoxin and digitoxin obtained in our study with preceding investigations seems problematic as we carried out a specific determination

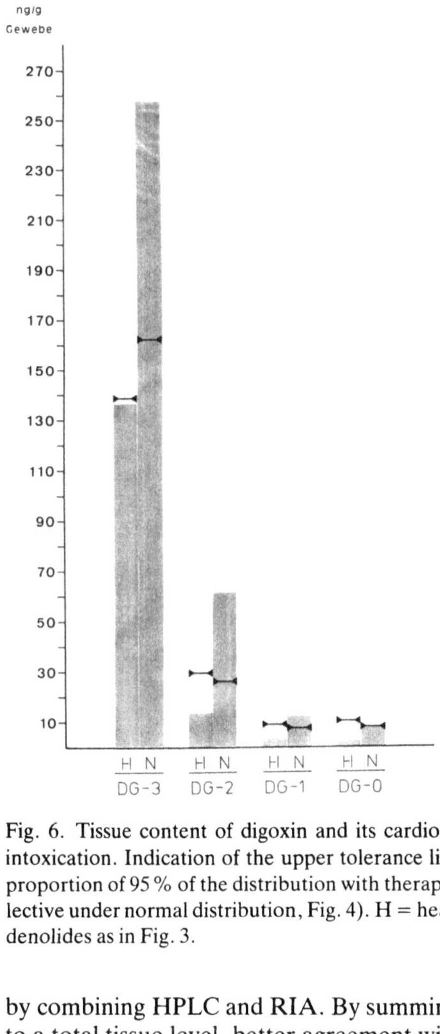

Fig. 6. Tissue content of digoxin and its cardioactive metabolites in a case (no. 3) of suicidal digoxin intoxication. Indication of the upper tolerance limit, under which, with a probability of 95%, at least a proportion of 95% of the distribution with therapeutic tissue content falls (referred to the therapeutic collective under normal distribution, Fig. 4). H = heart (left ventricle), N = kidney. Abbreviations of the cardenolides as in Fig. 3.

by combining HPLC and RIA. By summing up the contents of all metabolites determined to a total tissue level, better agreement with literature data may be obtained. With a total tissue content in the myocardium of 66.5 ± 32.9 ng/g under maintenance treatment with digoxin, our results are similar to those obtained by Weinmann et al. (13) [64.1 ± 45.1 ng/g] and Biddle et al. (3) [70.4 ± 10.0 ng/g]. But significantly higher levels were also reported (1, 8). The total content we found in the kidney of 71.9 ± 35.9 ng/g is lower than other literature data (1, 9, 13) [89.6–149 ng/g]. A higher degree of polar renal metabolites cross-reacting in the standard RIA determination may be an explanation.

The results available about the tissue content of digitoxin are still contradictory. Okita et al. (11) determined a myocardial level of unchanged digitoxin of about 20 ng/g using ^{14}C marked digitoxin. Digitoxin metabolites ranged markedly higher. Also using a tracer technique, Lukas (10) determined, in 4 cases under maintenance treatment, myocardial digitoxin contents between 99 and 205 ng/g. Aderjan et al. (2) applied the standard RIA method and found a mean tissue level of 181.7 ± 82.2 ng/g. They determined digoxin and its related compounds formed by 12 β-hydroxylation only in small amounts of 1.46 ± 1.2 ng/g.

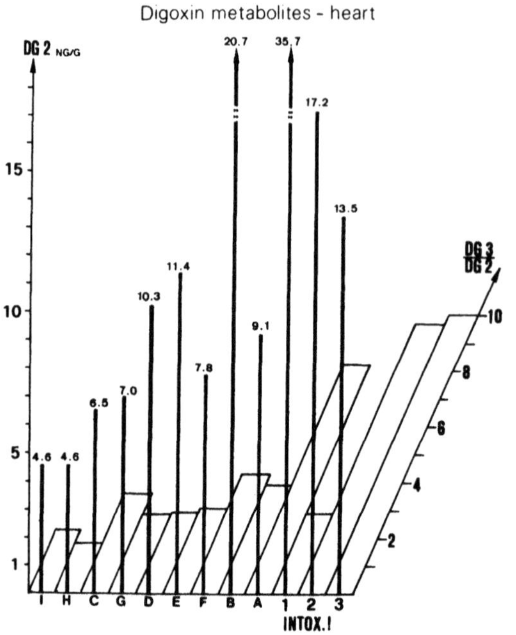

Fig. 7. Tissue content of digoxigenin bisdigitoxoside and the proportion digoxin/digoxigenin bisdigitoxoside of persons having taken digoxin in a different dosage mode.

The only study using a specific chromatographic technique (thin-layer chromatography, ^{86}Rb assay) to measure the percentage distribution of digitoxin metabolites in human tissues is by Storstein (12). In good agreement with our own results, she found unchanged digitoxin as the major compound of cardioactive metabolites (DT-3, DT-2, DT-1, DT-0, DG-3, DG-2, DG-1, DG-0). The metabolites formed by 12 β-hydroxylation (Dg-3, DG-2, DG-1, DG-0) represented 12 % of the total amount.

References

1. Aderjan R (1981) Tödliche Vergiftungen mit Herzglykosiden. Nachweis und rechtsmedizinisch toxikologische Befundbewertung. Habilitationsschrift Heidelberg
2. Aderjan R (1983) Digitoxin-Konzentration in postmortal entnommenen Blut- und Gewebeproben des Menschen nach therapeutischer Dosierung. In: Gillmann H, Storstein L (eds) Digitalistherapie heute. Verlag für angewandte Wissenschaften, München, pp 49–57
3. Biddle TL, Weintraub M, Lasagna L (1978) Relationship of serum and myocardial digoxin concentration to electrocardiographic estimation of digoxin intoxication. J Clin Pharmacol 18: 10–15
4. Brisse B, Müller KM, Bender F, Harder K (1977) Digoxinspiegel in verschiedenen Abschnitten des menschlichen Myokards. Verh Dtsch Ges Inn Med 83: 1673–1677
5. Caroll PR, Gelbart A, O'Rourke MF, Shortus J (1973) Digoxin concentrations in the serum and myocardium of digitalised patients. Aust NZJ Med 3: 400–403
6. Coltart J, Howard M, Chamberlain D (1972) Myocardial and skeletal muscle concentrations of digoxin in patients on long-term therapy. Br Med J 2: 318–319
7. Güllner HG, Stinson EB, Harrison DC, Kalman SM (1974) Correlation of serum concentrations with heart concentrations of digoxin in human subjects. Circulation 50: 653–655

8. Haasis R, Larbig D, Stunkat R, Bader H, Seboldt H (1977) Radioimmunologische Bestimmung der Glykosidkonzentration im menschlichen Gewebe. Klin Wochenschr 55: 23–30
9. Karjalainen J, Ojala K, Reissel P (1974) Tissue concentrations of digoxin in autopsy material. Acta Pharmacol Toxicol 34: 385–390
10. Lukas DS (1971) Some aspects of the distribution and disposition of digitoxin in man. Ann NY Acad Sci 179: 338–361
11. Okita GT, Talso PJ, Curry JH, Smith FD, Geiling EMK (1955) Metabolic fate of radioactive digitoxin in human subjects. J Pharmacol Exp Ther 115: 371–379
12. Storstein L (1977) Studies on digitalis. X. Digitoxin metabolites in human myocardium and relationship between myocardial and serum concentrations of digitoxin in patients on maintenance treatment. Clin Pharm Ther 21: 395–408
13. Weinmann J, Hasford J, Kuhlmann J, Bippus PH, Lichey J, Rietbrock N (1979) Digoxinkonzentrationen in Plasma und Gewebe. Med Klin 74: 613–619

Author's address:

Dr. J. Plum, M. D., Med. Einrichtungen der Universität Düsseldorf, Klinik f. d. Zentrum für innere Medizin und Neurologie, Moorenstraße 5, 4000 Düsseldorf (West Germany)

Application of HPLC-DAD instead of the guinea-pig test for the quantitative determination of cardiac glycosides

G. Tittel

LAT GmbH Dr. Tittel, Laboratorium für Analysen und Trenntechnik, Gräfelfing, Munich (West Germany)

Summary

The HPLC-DAD method can be substituted for the guinea-pig test (DAB 8) for the determination of the biological value of cardiac glycoside drugs and their preparations.

The results of the HPLC-DAD assay are correlated with those of the guinea-pig test.

Introduction

Hitherto the evaluation of cardiac glycoside drugs and their pharmaceutical preparations has been performed by the biological guinea-pig test as described in the DAB 8.

The test in the pharmacopoeia is based on the studies of Odenthal, Vogel and Görler (4): The biological assay only measures the toxicity of the preparations. The toxicity is expressed as "guinea-pig units" (MSE).

The HPLC cardiac slycosides test however allows the determination not only of the biological value but also the qualitative and quantitative distribution of the glycosides in the drug mixture.

Photodiode-Array-Detection (DAD) for HPLC was first described by Elgass and co-workers (1). Using the technique of on-line measuring and storing the UV-spectra of all peaks of the chromatographic separation, it is possible to determine selectively cardiac glycosides in complex mixtures.

The possibilities for calculation of biological values (MSE) from the results of the HPLC-DAD test are discussed in this paper.

A definitive evaluation of the HPLC-DAD assay, especially concerning a substitute for animal tests, will be published at a later date (5).

Results

Our first results with the HPLC of Urginea bufadienolides were published in 1980 (6). With the continuation of the work, some details were improved, for example the exchange of acetonitrile instead of methanol, the use of a column with 3 μm particle size and the Photodiode-Array-Detection. The methanolic extracts of Urginea drugs or preparations may be directly injected onto to the column without clean-up. Therefore, quantitative recovery rates may be expected.

The content of bufadienolides is determined by summing up the areas of all peaks which are identified as bufadienolides by their UV-spectrum; the peaks are calculated as Proscillaridin A.

All studies were performed with industrial Urginea extracts of the same origin but different date of manufacture. The HPLC assay shows that, besides Proscillaridin A, Scillarene A

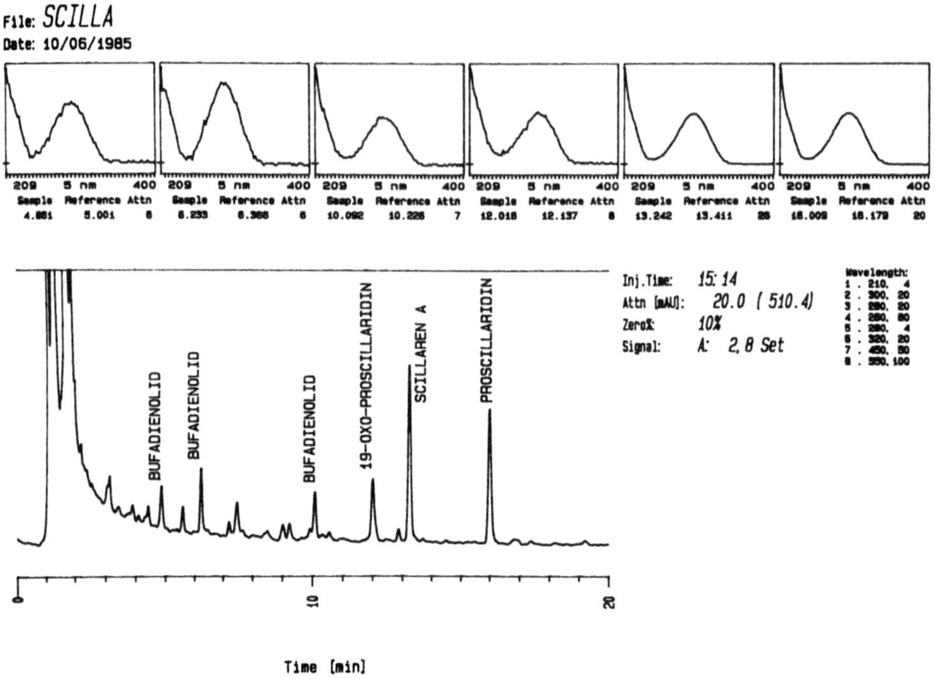

Fig. 1. HPLC-DAD assay of Scilla-extract.

and 19-Oxo-Prosicillaridin, nearly all peaks detected at 300 nm are bufadienolides (see Fig. 1).

The biological assay may be successfully replaced by a physical-chemical method where both methods have a linear correlation and are interchangeable.

In order to study the possibilities of correlations between the two methods, the following types of evaluations were chosen:

Test A:

Determination of the quantitatively most important compounds in the extract:

Proscillaridin A, Scillarene A and 19-Oxo-Proscillaridin (addition of the areas of these peaks) (see Fig. 2).

Result:

Approximately 50–80 % of the total content of bufadienolides in the examined industrial extracts were 19-Oxo-Proscillaridin, Scillarene A and Proscillaridin A. The summarized values of these three glycosides, found by HPLC assay, correspond to the value found by guinea-pig test. The correlation factor is 90 %.

Test B:

Determination of the total value of bufadienolides by means of calculating the sum of all peak areas detected at 300 nm, without registering the UV-spectra (see Fig. 3).

Results:

Summing up the areas of all peaks detected at 300 nm, the correlation factor decreases to 86 %.

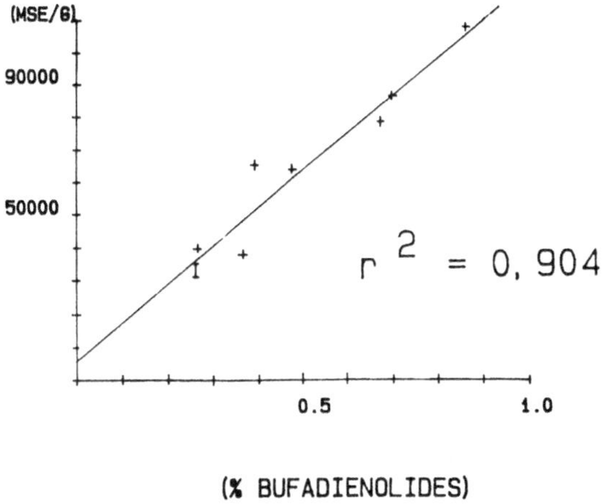

Fig. 2. Test A. Determination of bufadienolide content by addition of Proscillaridin A, Scillarene A and 19-Oxo-Proscillaridin A.

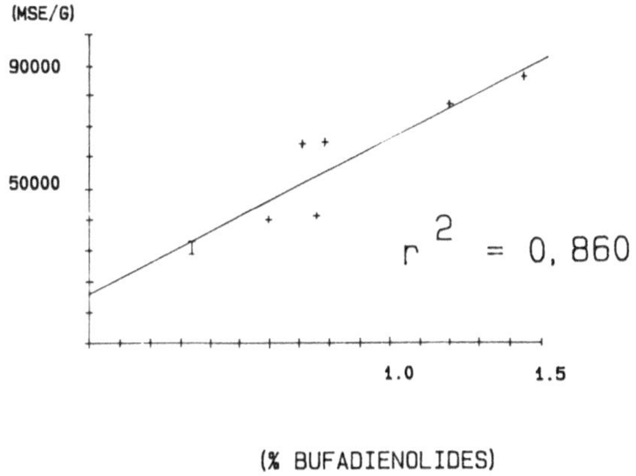

Fig. 3. Bufadienolide value determined by the addition of all peak areas detected at 300 nm.

Test C:

Determination of the total value of bufadienolides by means of summing up all peaks which had definitely been identified as bufadienolides by interpreting their UV-spectra (see Fig. 4).

Result:

Summing up all areas of peaks which had definitely been identified by HPLC-DAD, the correlation factor is 98%.

Nearly all known important Urginea bufadienolides show approximately the same biological value. Therefore, it is admissible to disregard the values of the single glycosides in these three tests.

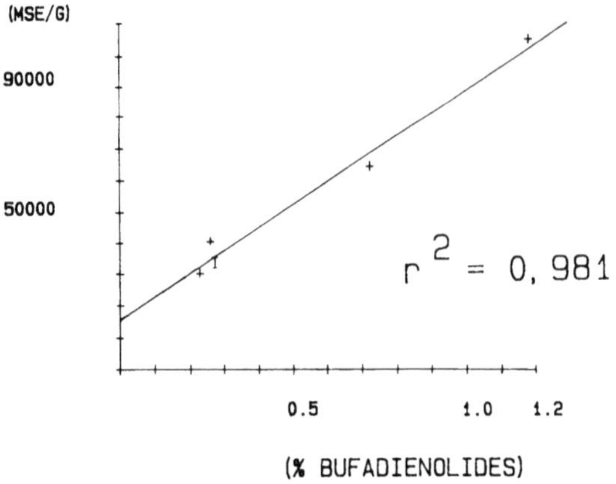

Fig. 4. Total value of bufadienolides determined by the addition of all peak areas identified by DAD.

For that reason, the diagrams for estimating the correlations may be expressed as "mg bufadienolide" as well as "MSE/g".

The data for these three tests are shown in Table 1:

Table 1. Biological value and chemical assay of industrial Scilla extracts.

Extract no.	Guinea-pig test MSE/g	HPLC assay Prosc. A %	Scillaren A %	19-Oxo-Pr. A %	Test A %	Test B %	Test C %
1	39438	0.193	0.119	0.099	0.411	0.76	–
2	77991	0.381	0.206	0.098	0.685	1.19	–
3	38404	0.188	0.099	0.077	0.364	0.59	–
4	86036	0.363	0.255	0.089	0.707	1.42	–
5	107875	0.465	0.350	0.094	0.871	1.83	1.150
6	64878	0.159	0.142	0.094	0.395	0.78	0.715
7	30995	0.091	0.059	0.070	0.220	–	0.330
8	39715	0.124	0.072	0.062	0.258	–	0.410
9	63697				0.480	0.71	–

MSE/g = unit of guinea-pig

The cardiotonic effect of May blossom (Convallaria herba) is based upon the content of cardenolides. So far 36 different glycosides have been isolated, as shown in the recent publications of Kopp and Kubelka (3).

The HPLC of Convallaria glycosides is based on own experience (7) as well as on the work of Kubelka and his team (2).

Convallatoxin is used as external standard for qualitative and quantitative determination by means of HPLC-DAD. This is possible because all Convallaria glycosides show nearly the same or similar molecular weights.

By HPLC, only the pure glycosides are determined, not the accompanying substances. That is why lower contents of cardenolides are to be found in comparison with the photometric valuation of Kedde or Baljet.

Results obtained by photometric methods may differ according to the kind of sample preparation or clean-up.

For the HPLC-DAD assay of Convallaria cardenolides (Fig. 5), the biological values of the individual cardenolides have to be considered (Table 2).

In this way, HPLC gives the absolute value of cardenolides, expressed in mg, and additional by the calculated biological value, expressed in MSE/g cardenolide in the drug or extract.

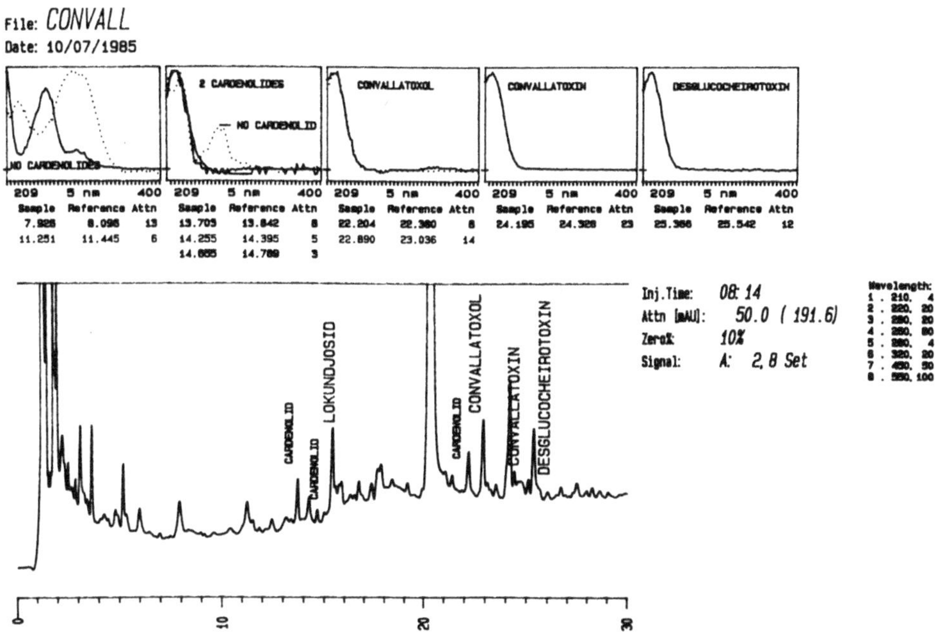

Fig. 5. HPLC-DAD assay of Convallaria cardenolides.

Table 2. Biological values of cardenolides from May blossom.

Glycoside	DL 100 (mg/kg)	MSE/mg
Lokundjoside	0.14	7142
Periplorhamnoside	0.18	5555
Convallotoxol	0.20	5000
Desglucocheirotoxol	0.22	4545
Desglucocheirotoxin	0.39	2564
Convallatoxin	0.34	2979
Convallatoxoloside	0.59	1864
Convalloside	0.59	1864

Table 3 shows the results of the investigations with 9 industrial May blossom extracts:

Table 3. Biological valuation and chemical assay of industrial May blossom extracts.

Extract no.	Animal test MSE/g	HPLC (%) glycoside	HPLC MSE/g
1	48750	0.97	32776
2	31150	0.91	32278
3	65100	1.07	42308
4	58200	1.28	46515
5	59450	0.99	38461
6	35050	0.71	28996
7	77300	1.14	46375
8	75750	1.44	60192
9	71033	1.39	57977

The biological values of the guinea-pig test listed in Table 3 are averaged values referring to the prescription of the German pharmacopoeia. Between the MSE values obtained by the animal test and the results obtained by HPLC, a regression coefficient of 0.74 is calculated. This poor factor is a result of the difficulties and faults of the biological tests, where an absolute error up to 43%, and an averaged error of 19% was observed.

On the other hand, the error of the HPLC method is about +/− 5%, including the steps of sample preparation. Correlating the MSE values measured by HPLC to those measured in the animal test, only a correlation coefficient of 0.66 may be calculated!

Correlating the MSE values measured by HPLC to the absolute glycoside content in HPLC, the correlation coefficient reaches 0.93, which means 93% linearity (Fig. 6).

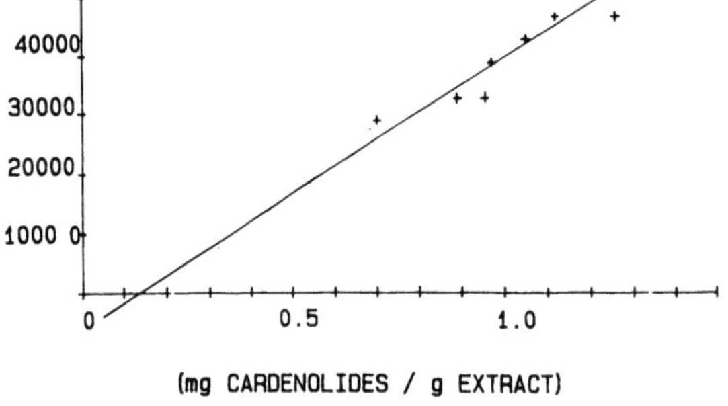

Fig. 6. Linear regression coefficient calculated by correlating MSE values measured by HPLC to the absolute glycoside content in HPLC.

These facts show that a correlation between the two methods is possible, but cannot definitively the expressed because of varying results in the animal test.

Discussion

For the first time, it is possible to standardize a complex mixture of cardiac glycosides by a physico-chemical assay, which includes simultaneous estimation of the qualitative and quantitative cardiac glycoside content and the calculation of the biological value of the preparation.

1. The HPLC-DAD method was standardized for all kinds of cardiac glycoside drugs.

In particular, preparations of Digitalis, Convallaria, Nerium, Strophanthus, and Scilla have been experimentally examined. The use of the diode array detection technique is imperative to ensure the identification and quantification of the glycoside peaks in these complexe mixtures.

2. The result of the HPLC-DAD analysis is not only an estimation of the complete cardenolide content, but also an expression of the biological value (MSE) of a plant preparation.

The MSE values estimated with HPLC correspond to those measured in the animal test (DAB 8).

To calculate the complete MSE value of a pharmaceutical, the individual MSE values of the most important glycosides (once measured in animal test) must be known. The result of the chromatographic analysis is the declaration of the milligram quantity of each cardiac glycoside in the mixture. These results are converted into the MSE values and summarized = *estimation of the biological value.*

This analytical technique implies that all glycosides of a mixture must be structurally definitely identified, and that their individual biological values are known or estimated by the guinea-pig test. The correlations between the guinea-pig test and the HPLC obtained with Scilla and Convallaria preparations prove that the HPLC may be successfully correlated to the animal test, even for other preparations containing cardiac glycosides.

3. The HPLC-DAD method may be used to standardize pharmaceuticals with plant extracts.

The HPLC results in the expression of the whole cardenolide content and in the differentiation of the biological values of all single glycosides in the mixture in one single analytical step.

This is a way to bring more efficiency and safety to phytotherapy.

4. The HPLC-DAD method substitutes the animal test with guinea-pigs.

The biological value of a preparation estimated by HPLC enables the choice of the right dose with regard to either the complete cardenolide content or the distribution of the different biological values of all the glycosides in the mixture (toxicologically important).

It is the first time that a biological determination of toxicity has been substituted by a physico-chemical technique.

References

1. Elgass H, Maute A, Martin R, George S (1982) Labor Praxis 6 (4): 2
2. Jurenitsch B, Kopp B, Bamberg-Kubelka R, Kern R, Kubelka W (1982) Chromatogr 240: 235
3. Kopp B, Kubelka W (1982) Planta med 45: 195
4. Odenthal KP, Vogel G, Görler K (1978) Pharm Ind 40: 149
5. Tittel G Pharm Ind – to be published
6. Tittel G, Wagner H (1981) Planta med 43: 252
7. Tittel G, Habermeier H, Wagner H (1982) Planta med 45: 207

Author's address:

Dr. G. Tittel, LAT GmbH Dr. Tittel, Laboratorium für Analysen und Trenntechnik, Maria-Eich-Straße 7, D–8032 Gräfelfing/München (West Germany)

Interaction between 16-acetyl-gitoxin and drugs

K.-O. Haustein, Elke Bucha, W.-D. Schlott, G. Hüller and G. Oltmanns*

Section of Clinical Pharmacology and *Department of Medicine, Medical Academy Erfurt, Erfurt (German Democratic Republic)

Summary

In 6 healthy volunteers and 11 patients suffering from congestive heart failure and rhythmic disturbances, the effects of combined pengitoxin/quinidine administration were studied. Additional investigations were carried out to examine interactions between glycoside and cholestyramine or almasilate, an aluminium-magnesium trisilicate. In the quinidine group of pengitoxin-loaded volunteers, quinidine did not affect either the systolic time intervals and ECG parameters or the basic plasma concentrations of 16-acetyl-gitoxin. In the digitalized patients, quinidine did not elevate the glycoside plasma level. Among the ECG parameters, only the mean PQ interval of 0.017 s was significantly prolonged. With the exception of one patient with affected repolarization, no signs of intoxication were observed. In 6 healthy volunteers, the mean elimination half-life and the cumulative urinary glycoside excretion of pengitoxin (16-acetyl-gitoxin) were not changed by administration of cholestyramine in daily doses of 5.0 g 3 times over 4 days. The intake of 2.5 g almasilate 30 min before and after the administration of 1.2 mg pengitoxin did not change the absorption and the elimination half-life of pengitoxin as compared with the corresponding k_a- and AUC_0^∞-values of the controls.

For treatment of congestive heart failure the cardioactive glycosides remain an important class of drugs. In the past 20 years, extensive pharmacodynamic and pharmacokinetic studies have shown distinct differences between the various glycosides, especially between digitoxin and digoxin and its derivatives. With regard to the pharmacokinetics, pengitoxin and its cardioactive metabolite 16-acetyl-gitoxin lie between digitoxin and digoxin, because pengitoxin possesses several digitoxin-like properties and an elimination half-life closely resembling that of digoxin (13, 14, 16).

The interaction of pengitoxin with several drugs such as quinidine, cholestyramine and antacids was studied in healthy volunteers and in patients in order to characterize the behaviour of pengitoxin (16-acetyl-gitoxin) in relation to digitoxin and digoxin (cf. (17)).

Patients, volunteers and methods

The interaction of pengitoxin and quinidine was investigated in 11 in-patients of the Department of Medicine of the Medical Academy Erfurt. The patients, between the ages of 39 and 70 years, took digitoxin in daily maintenance doses of 0.1 mg and quinidine in doses between 400 and 1,200 mg (Chinidin duriles®, 0.25 g capsules, Astra Sweden, or Kinilentin®, 0.2 g tablets, Medica, Finland). After a first control, digitoxin was substituted by pengitoxin (Carnacid-cor®, 0.3 mg tablets , TAD Pharmazeutisches Werk Cuxhaven) with daily maintenance doses of 0.3 mg. The plasma concentrations of potassium and creatinine were in the normal range. On the first and second days, ECGs were registered besides measurement of plasma concentrations of quinidine and digitoxin. The same parameters were evaluted on days 11 and 12 after administration of pengitoxin. The plasma concentration of quinidine was measured by GC and either glycoside was measured by a specific radio-immunoassay. During the first day of pengitoxin intake, two quinidine syncopes occured in one patient only which demanded reanimation and withdrawal of the drug, so that interruption of the study was indicated.

In an additional study in 6 healthy volunteers aged from 20 to 25 years, the quinidine effects on systolic time intervals and on the plasma level of 16-acetyl-gitoxin were studied. Before quinidine (Chinidin duriles®), was given in single oral doses of 0.25 or. 0.5 g, pengitoxin was taken in doses of 0.9 mg on two days and in a dose of 0.3 mg on 7 consecutive days. The systolic time intervals (STI) and plasma concentrations were measured before and 2 h after quinidine administration.

In the cholestyramine group, 6 healthy volunteers between the ages of 22 and 26 years participated who were loaded with pengitoxin over 3 days with doses of 0.9, 0.9 and 0.3 mg, respectively. On the 4th to 8th days instead of pengitoxin, cholestyramine was given in daily doses of 5 g (Cholestyramin powder for oral intake, VEB Chemie-Kombinat Bitterfeld, GDR) 3 times. The glycoside plasma level and the cumulative urinary glycoside excretion were measured by radio-immunoassay.

The possible interaction between pengitoxin and the aluminium-magnesium-silicate almasilate (Simagel®, powder for oral intake, VEB Arzneimittelwerk Dresden, Radebeul, GDR) was studied in 4 healthy female volunteers between the ages of 25 and 41 years. They took 2.5 g almasilate mixed with 100 ml water every 30 minutes before and after 1.2 mg pengitoxin (4 0.3 mg tablets) on an empty stomach. A light breakfast was allowed 2 h after the glycoside intake. Blood samples for measuring glycoside concentration were drawn by venipuncture over 3 days. Kinetic data such as the absorption and elimination half-life and the area under the curve were fitted by an open two-compartment model. The data were calculated by FORTRAN with a TPA-i/OS computer (Budapest, Hungary).

Results

In the quinidine group of the volunteers, there were no significant changes of the measured systolic time intervals (QS_{2c}, $LVET_c$), the ECG parameters (P wave, PQ times, QRS complex, QT time), and the glycoside plasma level 2 h after quinidine intake on 4 consecutive days (Fig. 1). The enhanced mean heart rate on day 7 was related to the heart rates of

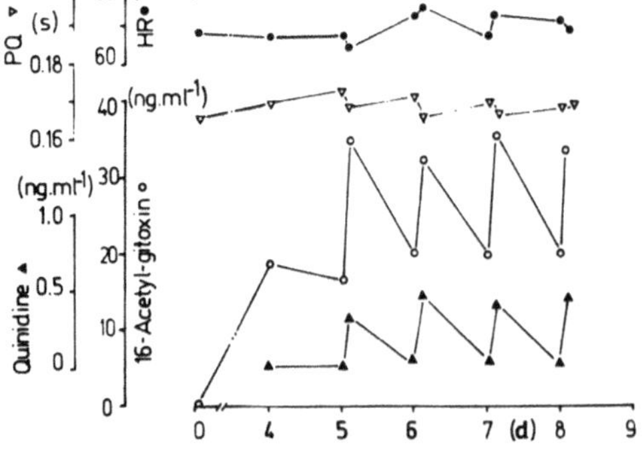

Fig. 1. The effect of daily administered quinidine (0.25 g per os) on the 16-acetyl-gitoxin (○) and quinidine plasma levels (▲) and on the time course of the heart rate (●), PQ interval (▽), $LVET_c$ (△) and QS_{2c} (▼). Mean values of 6 healthy volunteers loaded with pengitoxin (cf. Methods).

nearly 90 min^{-1} in two volunteers. On the following day, these values did not exceed the mean value of the whole group. The changes in the quinidine and 16-acetyl-gitoxin plasma levels were caused by the intake of both drugs. The mean glycoside concentration prior to the next dose did increase slightly.

Table 1. The effects of glycoside-quinidine therapy on various ECG parameters and drug plasma levels in 10 patients after substitution of digitoxin for pengitoxin. Mean values (standard deviation) of 2 controls (days 1 and 2) and of days 11 and 12 of the pengitoxin administration.

Parameter	Digitoxin day 1	Quinidine day 2	Pengitoxin day 11	Quinidine day 12	Significance
Glycoside dose (mg)	0.1	0.1	0.3	0.3	–
– Plasma level (ng.ml^{-1})	18.8 (7.6)	19.8 (9.1)	21.5 (4.3)	21.9 (4.6)	–
Quinidine dose (mg)	755 (263)	755 (263)	755 (263)	755 (263)	–
– Plasma level (μg.ml^{-1})	1.28 (0.57)	1.35 (0.71)	1.27 (0.41)	1.23 (0.37)	n. s.
P wave (s)	0.079 (0.016)	0.084 (0.014)	0.086 (0.016)	0.088 (0.015)	n. s.
PQ time (s)	0.178 (0.033)	0.182 (0.035)	0.192 (0.034)	0.199 (0.043)	$p < 0.02$
QRS complex (s)	0.071 (0.016)	0.074 (0.014)	0.078 (0.018)	0.076 (0.014)	n. s.
QT interval (s)	0.394 (0.049)	0.393 (0.044)	0.388 (0.035)	0.384 (0.040)	n. s.
Heart rate (min^{-1})	64.7 (9.0)	64.9 (6.7)	64.0 (6.4)	61.5 (8.9)	n. s.

n. s. significant; – not calculated.

In the quinidine group of patients, the 16-acetyl-gitoxin plasma level was unchanged (cf. Table 1). The mean values amounted to 21.5 and 21.9 ng per ml. Only in patient HJ did the

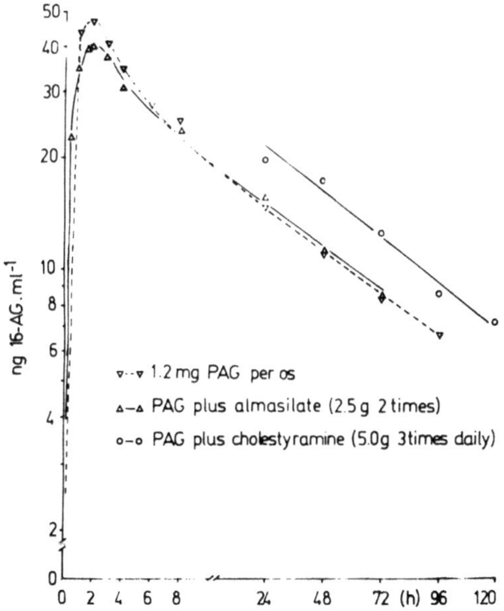

Fig. 2. Time course of the 16-acetyl-gitoxin plasma level after oral administration of 1.2 mg pengitoxin (▽) or pengitoxin plus 2.5 g almasilate 30 min before and after the glycoside (△) or cholestyramine in doses of 3 times daily 5.0 g over 4 days (○). Mean values (n = 6: ▽, o, n = 4 △). For details cf. Methods.

plasma level reach values of about 30 ng per ml accompanied by an affected repolarization. Patient LK reported nausea and anorexia a short time after the beginning of the digitoxin-quinidine therapy. After changing the glycoside, these symptoms were aggravated on the 5th day and a plasma level of 33.8 ng per ml was measured. Two days after glycoside withdrawal, the plasma level fell to 28.5 (day 7) and 23.3 (day 8) and at this time, the symptoms had disappeared. In general, the other ECG parameters showed no significant changes, with the exception of the mean PQ time which was prolonged by 17 ms (cf. the data of day 2 with those of day 12). The PQ-time was prolonged to values > 0.22 s only in one patient. In one patient it amounted to more than 0.25 s throughout the investigation.

In the cholestyramine group, the mean elimination half-life of 16-acetyl-gitoxin amounted to 63.4 h (cf. Fig. 2). Within the same period (5 days), 144.6 μg of 16-acetyl-gitoxin was excreted in the urine with a mean renal clearance of 1.28 ml/min^{-1}. The calculated half-life, the cumulated urinary glycoside excretion and the renal clearance values of the cholestyramine-treated subjects did not differ from those of untreated controls (Table 2). In the almasilate group, in 2 of the 4 subjects the glycoside appeared in plasma 15 min after oral intake. The plasma levels amounted to 17.3 and 25.4 ng per ml. In the other 2 subjects the glycoside was absorbed with a lag time of 15 min. The mean peak plasma concentration occured 1.5 h after glycoside intake (cf. Fig. 2). The peak concentration of controls slightly surpassed that of the almasilate group. The mean AUC_0^∞-value amounted to 1563.7 ng/h/ml^{-1} and was in the same order of magnitude as the corresponding values of controls (Table 2).

Table 2. AUC_0^∞-values (ng/h/ml^{-1}), half-life times ($t_{1/2\,\beta}$; h) and absorption constants (k_a; h^{-1}) after intake of pengitoxin (PAG) or PAG plus almasilate or cholestyramine.

Subject	PAG			PAG + almasilate			PAG + cholestyramine	
no.	AUC_0^∞	$t_{1/2\,\beta}$	k_a	AUC_0^∞	$t_{1/2\,\beta}$	k_a	AUC_0^∞	$t_{1/2\,h}$
1	1586.3	59.3	2.22	1494.4	71.4	2.08	2740.9	80.5
2	2065.5	66.8	2.83	2523.9	28.0	2.46	3450.5	62.4
3	1555.9	48.8	2.92	1443.4	38.3	0.58	40.61.5	53.3
4	1768.4	66.3	5.82	1283.5	46.9	0.81	2787.6	61.6
5	2494.9	83.5	1.69				2895.8	72.0
6	1934.3	72.2	0.87				3051.1	50.5
Mean	1900.9*	66.2	2.72**	1677.1*	46.2	1.48**	3164.6*	63.4
SD	351.2	11.7	1.70	574.5	18.5	0.93	508.5	11.3

*) between the mean AUC_0^∞-values are significant differences (PAG + cholestyramine vs. PAG or PAG + almasilate)

**) $p > 0.05$

Discussion

In general, no interactions were found between pengitoxin and quinidine, cholestyramine or antacids such as almasilate. With the exception of one patient with enhanced plasma level of 16-acetyl-gitoxin, no interactions were observed between pengitoxin and quinidine on the pharmacokinetic or pharmacodynamic level. The plasma levels of the volunteers increased in dependence on the administered dose, while in patients only single plasma values were supratherapeutically high. With the exception of one patient who showed disturbances of repolarisation, no patient showed signs of intoxication. Neither the systolic time intervals in the volunteers nor the ECG parameters in the patients were affected.

The data obtained with PAG corresponded to those of digitoxin rather than to those of digoxin. In the quinidine-treated patients, the glycoside plasma levels remained in an identical range independent of the administered glycoside, while in the group of volunteers high post-dose concentrations were measured without an increase in the basic level before intake of the next glycoside dose. On the basis of these studies, an interaction between pengitoxin and quinidine can be ruled out. Contradictory results have been published about the interaction with digitoxin. On the one hand, no interaction was found (21); on the other hand, an increase in the digitoxin plasma level of 31.8 per cent was observed if quinidine was taken simultaneously (22, 23). This effect can be induced by displacement of digitoxin from its tissue binding. Despite enhanced digitoxin plasma levels, no unwanted toxic digitoxin effects were observed (24). It is possible that quinidine affects the tissue binding digitoxin to a smaller extent than that of digoxin (cf. (11), (19)). In the case of the digoxin-quinidine combination, the glycoside plasma level increased up to 2 fold depending on the quinidine dose (cf. (3), (11), (14)). This dangerous interaction is due to the additional decrease in the total and renal clearance together with a two-third reduction of the apparent volume of distribution (11).

The studies in the cholestyramine-pengitoxin combination showed that the elimination half-life and the excretion properties are not affected by the resin. Thus, the reabsorption of pengitoxin from the intestinal tract seems to be without significance, because elimination was not abbreviated as in the case of digitoxin (7, 25), particularly if supratherapeutic plasma concentrations due to an intoxication were measured (cf. (2), (10)). In the case of digoxin, simultaneous administration of cholestyramine caused a dose dependent reduction in the AUC-values of digoxin (5) which can be minimized by temporal separation of the time of the administration between both drugs. Previous results of a minor enterohepatic recycling of ^{3}H-pengitoxin or ^{3}H-16-acetyl-gitoxin in patients with a cannulated bile duct underline those presented here with cholestyramine. The study does not provide information about the absorption of the glycoside to the resin which can be of importance in case of combined administration. Because of their physico-chemical properties (increase in the lipid solubility with rising partition coefficient P in octanol/water: digoxin $<$ digitoxin $<$ 16-acetyl-gitoxin $<$ pengitoxin; cf. (8)) the glycosides are bound to the resin (9, 12). Successful treatment of pengitoxin intoxication with cholestyramine has not yet been reported.

In the case of the antacids, pengitoxin behaves like digitoxin. According to the results of other authors (26), the absorption of digitoxin is not affected by silicates and other antacids (cf. also (18)). In contrast to digitoxin, the bioavailability of digoxin is reduced by 30 to 40 per cent by various antacids (cf. (1), (4), (5)).

In general, the studies showed that pengitoxin (16-acetyl-gitoxin) and digitoxin have identical properties with respect to drug interaction.

References

1. Albert KS, Elliott WJ, Abbott RD, Gilbertson TJ, Data JL (1981) Influence of kaolin-pectin suspension on steady-state plasma digoxin levels. J Clin Pharmacol 21: 449–455
2. Bazzano G, Bazzano GS (1972) Digitalis intoxication. Treatment with a new steroid binding resin. JAMA 220: 828–830
3. Bigger JT (1981) The quinidine-digoxin interaction. Int J Cardiol 1: 109–116
4. Binnion PF (1973) Absorption of different commercial preparations of digoxin in the normal human subject, and the influence of antacid, antidiarrhoeal and ion-exchange agents. In: Storstein O (ed) Symposium on Digitalis. Gyldendahl Norsk Forlag, pp 216–224

5. Brown DD, Juhl RP (1976) Decreased bioavailability of digoxin due to antacids and kaolin-pectin. New Engl J Med 295: 1034–1037
6. Brown DD, Juhl RP, Warner S (1978) Decreased bioavailability of digoxin due to hypocholesterolemic interventions. Circulation 58: 164–172
7. Carruthers SG, Dujovne CA (1980) Cholestyramine and spironolactone and their combination in digitoxin elimination. Clin Pharmacol Ther 27: 184–187
8. Cohnen E, Flasch H, Heinz N, Hempelmann FW (1978) Verteilungskoeffizienten und R_m-Werte von Cardenoliden. Arzneimittelforsch (Drug Res) 28: 2179–2182
9. Demers HG, Pabst J (1983) Digitoxin-Intoxikationen: Behandlung mit Cholestyramin. Diagnostik 16: 27–32
10. Demers HG, Pabst J, Piper Ch (1982) Digitoxin: pharmakokinetische Befunde bei der Intoxikationsbehandlung mit Cholestyramin. Dt Med Wschr 107: 1476–1480
11. Hager WD, Fenster P, Mayersohn M, Perrier D, Graves P, Marcus FI, Goldman S (1979) Digoxin-quinidine interaction. Pharmacokinetic evaluation. New Engl J Med 300: 1238–1241
12. Hall WH, Shappell SD, Doherty JE (1977) Effect of cholestyramine on digoxin absorption and excretion in man. Am J Cardiol 39: 213–216
13. Haustein KO, Alken RG, Lach HJ, Becker U, Rietbrock N (1983) On the pharmacokinetics of pengitoxin and its cardioactive derivative 16-acetyl-gitoxin. Eur J Clin Pharmacol 25: 369–373
14. Haustein KO, Oltmanns G (1981) Wechselwirkungen zwischen Digoxin und Chinidin. Z ges inn Med 35: 186–187
15. Haustein KO, Pachaly Ch, Murawski D (1978) Pharmacokinetic investigations with ^{3}H-penta-acetyl-gitoxin in volunteers and patients with respect to the occurrence of drug latentiation. Int J Clin Pharmacol 16: 285–289
16. Haustein KO, Wesser M (1982) Clinico-pharmacological and clinical studies of pengitoxin (Russ) Farmakol i Toksikol 45: 67–68
17. Kuhlmann J (1983) Grundlage der Arzneimittelinteraktionen mit Beispielen aus der Therapie mit Herzglykosiden. In: Gillmann H, Storstein L (Hrsg) Digitalistherapie heute. Verlag für angew Wissenschaften, München, S 79–97
18. Kuhlmann J (1984) Plasmaspiegel und renale Elimination von Digitoxin bei Langzeittherapie mit Aluminium-Magnesium-Hydroxid-Gel. Dt med Wschr 109: 59–61
19. Leahey EB, Reiffel JA, Giardina ECV, Bigger JT (1980) The effect of quinidine and other oral antiarrhythmic drugs on serum digoxin. Ann Int Med 92: 605–608
20. Ochs HR, Bodem G (1982) Arzneimittelinteraktionen zwischen Digitoxin und Chinidin? Dt Med Wschr 107: 673–674
21. Ochs HR, Pabst J (1981) Einfluß einer gleichzeitigen Chinidingabe auf die Kinetik von Digitoxin. In: Kochsiek K, Rietbrock N, Heusinger H (Hrsg) Digitalistherapie bei Herzinsuffizienz. Urban & Schwarzenberg, München Wien Baltimore, S 132–135
22. Peters U, Risler T, Grabensee B, Falkenstein U, Kroukou J (1980) Interaktion von Chinidin und Digitoxin beim Menschen. Dt Med Wschr 105: 438–442
23. Peters U, Risler T, Grabensee B (1980) Keine Chinidin-Interaktion mit Digitoxin? Dt Med Wschr 105: 701–702
24. Peters U, Risler T, Grabensee B, Falkenstein U (1981) Untersuchungen zur Interaktion von Chinidin und Digitoxin. In: Kochsiek K, Rietbrock N, Heusinger H (Hrsg) Digitalistherapie bei Herzinsuffizienz. Urban & Schwarzenberg, München Wien Baltimore, S 136–142
25. Pond S, Jakobs M, Marks J, Garner J, Goldschlager N, Hansen D (1981) Treatment of digitoxin overdose with oral activated charcoal. Lancet II: 1177–1178
26. Vöhringer HF, Kuhlmann J, Rietbrock N (1976) Der Einfluß von Antacida auf die Plasmakonzentration von Digoxin beim Menschen. Dt Med Wschr 101: 106–108

Author's address:

Prof. Dr. sc. med. K.-O. Haustein, Abteilung für Klinische Pharmakologie, Medizinische Akademie Erfurt, Nordhäuser Straße 74, DDR-5010 Erfurt (GDR)

Digitoxin protein binding in patients on hemodialysis

J. J. H. M. Lohman*, P. M. Hooymans*, M. L. P. Koten** and F. W. H. M. Merkus*,**

*Clinical Pharmacokinetics and Toxicology Laboratory, Hospital of Sittard, Sittard, and **Department of Biopharmaceutics, University of Amsterdam, Amsterdam (The Netherlands)

Summary

A reduction of digitoxin binding to plasma proteins during hemodialysis has been reported. Our aim was to determine whether this reduction is an in vivo effect or occurs only after blood collection as a result of heparin-induced lipolysis which increases levels of non-esterified fatty acids in vitro. Digitoxin protein binding was studied in 10 patients undergoing hemodialysis, receiving digitoxin maintenance therapy. Digitoxin free fraction was increased at the end of hemodialysis, from 2.5 ± 0.7% to 4.4 ± 1.1%, but after inhibition of in vitro lipolysis with paraoxon (2 mmol/l), a potent lipase inhibitor, there was no increase in the free fraction: 2.3 ± 0.4% before heparin and 2.4 ± 0.5% at the end of hemodialysis. Digitoxin salivary levels were also unchanged: 0.41 ± 0.08 ng/ml before heparin and 0.41 ± 0.08 ng/ml at the end of hemodialysis (n = 8). These data indicate that the binding of digitoxin in vivo is not altered during hemodialysis. The reduced binding reported elsewhere was a result of heparin-induced in vitro lipolysis.

Introduction

Because digitoxin is highly bound to plasma proteins (10), alterations in digitoxin protein binding may have important clinical implications. A marked increase in the mean digitoxin free fraction from 2.9% to 6.7% (n = 14) during hemodialysis has been reported (18). Heparin-induced release of non-esterified fatty acids (NEFAs), which cause displacement of digitoxin from the albumin binding site, was proposed as the mechanism for this interaction (19). However, Giacomini et al. suggested that the effect of heparin on plasma protein binding may be artifactual, because lipoprotein lipase activated by heparin continues to generate NEFAs after blood collection and during equilibrium dialysis (8). Our aim was to determine whether digitoxin protein binding in vivo is actually changed by heparin in patients undergoing dialysis who were receiving digitoxin maintenance therapy.

Methods

Digitoxin protein binding

Plasma protein binding of digitoxin was determined by equilibrium dialysis as described previously (11). A dialysis time of at least 9 h had been determined as the time necessary to attain equilibrium. For practical reasons, dialysis was carried out overnight (during 16 h). The digitoxin free fraction (f_u) was corrected for osmotic dilution during the equilibrium dialysis process (11). Since digitoxin binding is a non-saturable process (20), free digitoxin concentrations in the original serum sample (C_u) can be obtained from the formula:

$$C_u = f_u \cdot C_s$$

where C_s is the total concentration in the original serum sample (1, 11).

Digitoxin determination

Serum digitoxin levels were determined by radio-immunoassay (Diagnostic Products Corp.). Digitoxin concentrations in dialysis buffer and saliva were determined by extraction of 1 ml with 5 ml chloroform (Merck). Then 4 ml of the organic layer was evaporated to dryness and reconstituted with 50 μl normal human serum. Before extraction saliva was centrifuged (10 min at 1000 x g) and the clear supernatant liquid was used.

Determination of NEFAs

NEFA levels in serum were measured by a colorimetric method (NEFA C, Wako Pure Chemical Industries).

Effect of NEFA on digitoxin free fraction

Serum samples from subjects who were receiving digitoxin maintenance therapy and had not received heparin were pooled. Stearic acid was added in different amounts to produce final concentrations of 0.5 mmol/l, 1.0 mmol/l, 2.0 mmol/l, and 4.0 mmol/l. In each sample, the NEFA concentration and the digitoxin free fraction were determined in duplicate.

Effect of heparin administration on digitoxin protein binding

The influence of heparin on serum protein binding of digitoxin was studied in 10 subjects (three women and seven men) aged 51 to 76 years who received hemodialysis two or three times a week and digitoxin, 0.1 mg, five times a week. Each subject received heparin (Thromboliquine, Organon) as a bolus injection of 5000 IU and subsequently received 600 IU/h during the 3 to 4 h dialysis period. Two blood samples were drawn before heparin and again at the end of hemodialysis. One blood collection tube of each pair contained paraoxon in a final concentration of 2 mmol/l. Serum was frozen and stored at −20° C. From eight subjects, unstimulated whole saliva samples were also collected on another occasion. Saliva was centrifuged and stored at −20° C. Total digitoxin concentrations and digitoxin free fractions were determined in serum. NEFA concentrations were determined after thawing of the serum and incubation at 37° C for 120 min. Digitoxin levels were also measured in saliva.

Statistical analysis

Statistical significance of differences was determined by Student's t-test for paired observations. Data are expressed as mean ± SD.

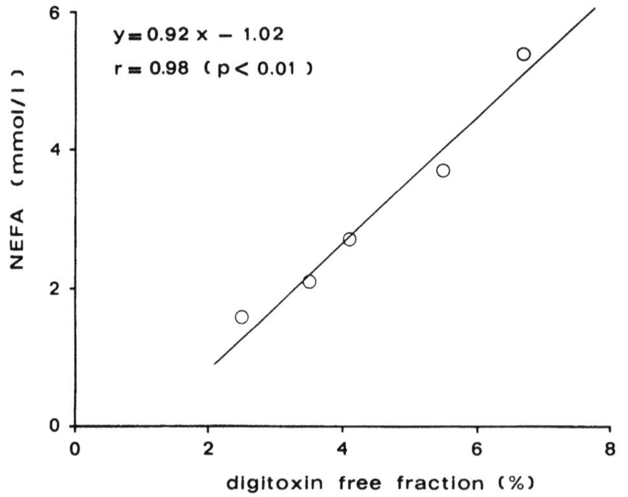

Fig. 1. Relationship between NEFA concentration and digitoxin free fraction after addition of stearic acid to pooled serum containing digitoxin.

Results

There was a strong correlation (r = 0.98) between NEFA concentration and digitoxin free fraction after stearic acid was added to pooled serum from subjects receiving digitoxin (Fig. 1). Digitoxin total concentration in pooled serum was 23 ng/ml.

Results of the study in 10 subjects undergoing hemodialysis who received digitoxin maintenance therapy are summarized in Table 1. At the end of hemodialysis, there was a small, non-significant decrease in mean total digitoxin serum concentration in the sample without paraoxon (control samples), from 19 ± 4 to 17 ± 4 ng/ml. This lipase inhibitor did not interfere with the determination of digitoxin in serum, because the same mean total digitoxin levels were found in the samples containing paraoxon as in the control samples taken at the same time. After heparin, digitoxin free fraction rose in the control samples, from 2.5 ± 0.7% to 4.4 ± 1.1%, but there was no change in digitoxin free fraction in the presence of paraoxon: 2.3 ± 0.4% before heparin, and 2.4 ± 0.5% after heparin. Determination of digitoxin protein binding was not disturbed by paraoxon, because digitoxin free fractions in the samples containing it were approximately equal to those in the control samples before heparin.

Table 1. Effects of heparin during hemodialysis on digitoxin protein binding in ten subjects receiving digitoxin maintenance therapy.[a)]

	Before heparin		After heparin	
	Control	Paraoxon	Control	Paraoxon
Total digitoxin (ng/ml)	19 ± 4	19 ± 4	17 ± 4	17 ± 3
Free fraction (%)	2.5 ± 0.7	2.3 ± 0.4	4.4 ± 1.1	2.4 ± 0.5
Free digitoxin (ng/ml)	0.46 ± 0.10	0.43 ± 0.10	0.74 ± 0.19*	0.40 ± 0.08
NEFA (mmol/l)	0.25 ± 0.11	0.19 ± 0.07	1.0 ± 0.5**	0.23 ± 0.08
Saliva level (ng/ml)	0.41 ± 0.08		0.41 ± 0.08	

a) Data are mean ± SD, n = 10; for saliva level: n = 8
* $p < 0.001$ compared to control before heparin
** $p < 0.01$ compared to control before heparin

Parallel to these observations, NEFA concentrations in the control samples rose, from 0.25 ± 0.11 mmol/l to 1.0 ± 0.5 mmol/l, and there were only minor changes in the samples containing paraoxon: 0.19 ± 0.07 mmol/l before heparin, and 0.23 ± 0.08 mmol/l after heparin. Free concentrations of digitoxin are presented graphically in Fig. 2. In control blood samples, the mean free concentration rose from 0.46 ± 0.10 ng/ml to 0.74 ± 0.19 ng/ml, but the free concentration did not change in the presence of paraoxon: 0.43 ± 0.10 ng/ml before and 0.40 ± 0.08 ng/ml after heparin. Digitoxin concentrations in saliva obtained from eight subjects were also unchanged (Fig. 3). Before heparin administration the mean level in saliva was 0.41 ± 0.08 ng/ml, and after heparin 0.41 ± 0.08 ng/ml.

Discussion

Our data demonstrate that digitoxin binding to plasma proteins is not significantly changed in vivo by heparin in subjects undergoing hemodialysis. These results are in contrast to a report which described an increased digitoxin free fraction (from 2.9 ± 0.7% to

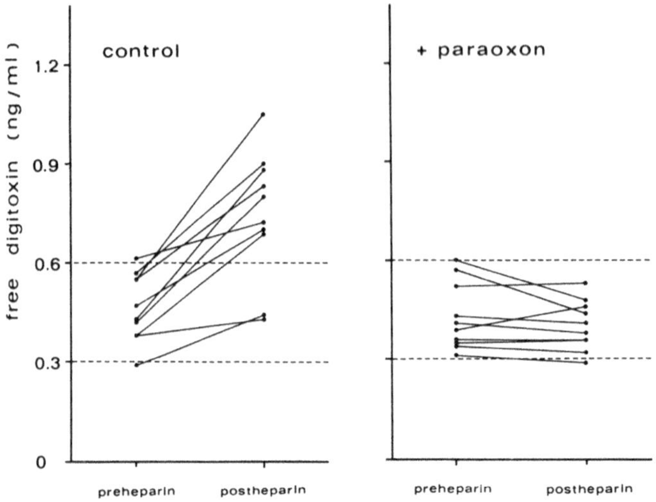

Fig. 2. Free concentrations of digitoxin in serum samples collected before heparin and at the end of hemodialysis. Values in the right panel were from blood samples containing paraoxon (2 mmol/l). Dashed lines indicate the estimated therapeutic range of free digitoxin.

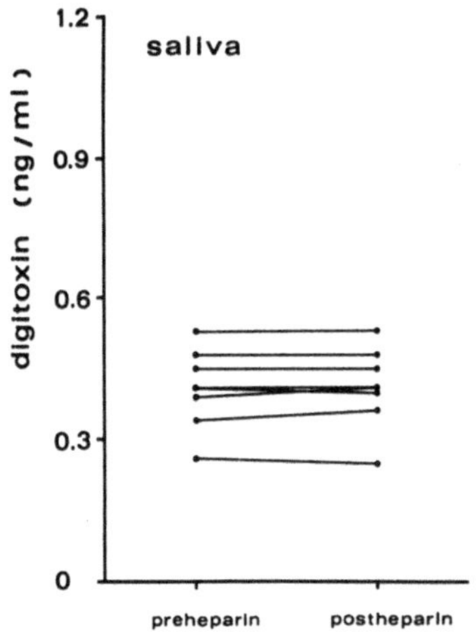

Fig. 3. Digitoxin concentrations in saliva collected before heparin and at the end of hemodialysis.

6.7 ± 3.7%, n = 14) during hemodialysis (18). In that study, no precautions were taken to prevent in vitro lipolysis. A similar increase was found in our study in the serum samples without lipoprotein lipase inhibitor (from 2.5 ± 0.7% before heparin to 4.4 ± 1.1% at the end of hemodialysis). Displacement of digitoxin from albumin binding sites by heparin-induced release of NEFAs was proposed as the mechanism for this interaction (19). The results of our experiments in which stearic acid was added to pooled serum samples from subjects on digitoxin maintenance therapy support this mechanism (Fig. 1).

After inhibition of in vitro lipolysis with paraoxon, digitoxin free fractions in the blood samples after heparin did not differ significantly from those in the samples before heparin. This indicates that the reported reduced digitoxin protein binding is caused by NEFAs produced after blood collection. In vivo, NEFAs are rapidly removed from the blood by tissue cells (Fig. 4), whereas there is no removal process in vitro (7). Apparently, the increase in NEFA concentration in vivo after heparin is small (6, 15). This is confirmed by our results: In the samples after heparin with paraoxon, NEFA concentrations were approximately equal to those in the samples before heparin.

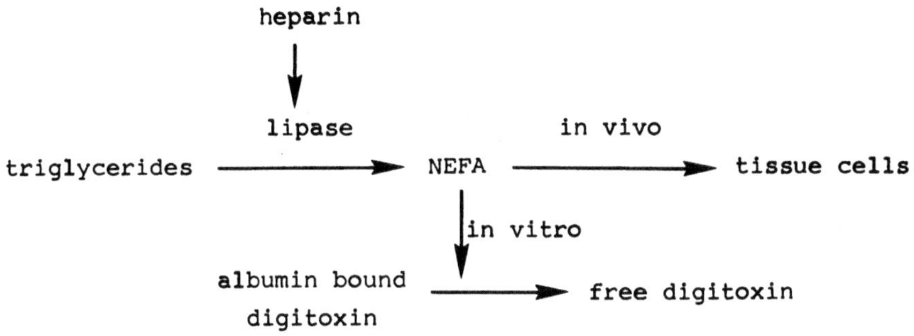

Fig. 4. Schematic representation of the effect of heparin on the binding of digitoxin to serum proteins.

The therapeutic range for total (bound plus unbound) digitoxin serum concentrations is presumably 15 to 30 ng/ml (5). The therapeutic range for free digitoxin serum levels needs to be established, but it can be estimated as 0.30 to 0.60 ng/ml, because using the same equilibrium dialysis method digitoxin free fraction in patients (n = 46) with normal renal function who did not receive heparin was 2.0 ± 0.5 %.*) After heparin, the free digitoxin concentrations apparently rose above this therapeutic range (Fig. 2, left panel), but after in vitro lipase inhibition with paraoxon free drug levels remained in the estimated therapeutic range (Fig. 2, right panel). Therefore, the recommendation that digitoxin serum levels in patients undergoing hemodialysis be adjusted between 10 and 15 ng/ml because of decreased protein binding during hemodialysis (19) is not supported.

That paraoxon is an effective inhibitor of lipoprotein lipase and lipolytic activity after heparin (14, 16) is confirmed by the the results shown in Table 1. Protamine sulfate was also tested, but inhibition of in vitro lipolytic activity was insufficient (12). Although application of paraoxon seems to be limited because drug protein binding may be disturbed (2), it was found suitable in studies of the heparin effect on protein binding of phenytoin and prazosin (7, 17). The protein binding of digitoxin is not greatly influenced, because the free fractions in the blood samples before heparin were not changed in the presence of paraoxon (Table 1). Until now, the artifactual nature of the effect of heparin on plasma protein binding of drugs in man has been established for diazepam (2), ibuprofen (13), lidocaine (2, 6), phenytoin (3, 7, 17), prazosin (7, 17), and thyroxin (21); evidence that this occurs with propranolol is less clear (2). Most of these experiments were carried out with spiked plasma samples from subjects before and after heparin. In our study, it was shown that in vivo drug protein binding of digitoxin is not significantly changed by heparin.

*) Unpublished results

It has been shown for some drugs that their concentration in saliva equals their free concentration in plasma (4). Digitoxin salivary levels are of the same order as the free serum concentrations, suggesting that digitoxin concentrations in saliva reflect free serum levels (9). In our study, no difference was found between salivary levels of digitoxin before and after heparin, supporting the statement that digitoxin protein binding in vivo is not changed during hemodialysis.

Our results indicate that the in vivo binding of digitoxin to plasma protein in patients on hemodialysis is not altered by heparin. Earlier reported reduced binding (18) was caused by continued lipolysis in vitro after blood collection and during equilibrium dialysis. A proper method to inhibit in vitro lipolysis in studies of drug protein binding in patients receiving heparin is necessary to eliminate in vitro artifacts.

References

1. Behm HL, Wagner JG (1979) Errors in interpretation of data from equilibrium dialysis protein binding experiments. Res Commun Chem Pathol Pharmacol 26: 145–160
2. Brown JE, Kitchell BB, Bjornsson TD, Shand DG (1981) The artifactual nature of heparin-induced drug protein binding alterations. Clin Pharmacol Ther 30: 636–643
3. Chou RC, Levy G (1981) Does heparin cause inhibition of drug-protein binding in vivo? Clin Pharmacol Ther 29: 236–237
4. Danhof M, Breimer DD (1978) Therapeutic drug monitoring in saliva. Clin Pharmacokinet 3: 39–57
5. Doherty JE, Kane JJ (1975) Clinical pharmacology of digitalis glycosides. Ann Rev Med 26: 159–171
6. Dubé LM, Ho-Ngoc A, Davies RF, Beanlands DS, Mousseau N, McGilveray IJ (1983) Influence of protamine on heparin-induced increases of lidocaine free fraction. Res Commun Chem Pathol Pharmacol 42: 401–415
7. Giacomini KM, Giacomini JC, Blaschke TF (1980) Absence of effect of heparin on the binding of prazosin and phenytoin to plasma proteins. Biochem Pharmacol 29: 3337–3340
8. Giacomini KM, Swezey SE, Giacomini JC, Blaschke TF (1980) Administration of heparin causes in vitro release of non-esterified fatty acids in human plasma. Life Sci 27: 771–780
9. Hooymans PM, Lohman JJHM, Overdiek JWPM, Merkus FWHM (1984) Free fraction and salivary level of digitoxin. Pharm Weekbl Sci Ed 6: 268
10. Krieglstein J (1981) Plasma protein binding of cardiac glycosides. In: Greeff K (ed) Cardiac Glycosides, Part II. Springer-Verlag, Berlin, pp 95–104
11. Lohman JJHM, Hooymans PM, Verhey MTJM, Koten MLP, Merkus FWHM (1984) Influence of volume shift in equilibrium dialysis to estimate plasma protein binding of drugs. Pharm Res 1: 187–188
12. Lohman JJHM, Hooymans PM, Koten MLP, Verhey MTJM, Merkus FWHM (1985) Effect of heparin on digitoxin protein binding. Clin Pharmacol Ther 37: 55–60
13. Pearce GA, Brown KF (1983) Heat inhibition of in vitro lipolysis and ^{14}C-ibuprofen protein binding in plasma from heparinized uraemic subjects. Life Sci 33: 1457–1466
14. Porte D, Bierman EL (1969) The effect of heparin infusion on plasma triglyceride in vivo with a method for calculating triglyceride turnover. J Lab Clin Med 73: 631–648
15. Riemersma RA, Russell DC, Oliver MF (1981) Heparin-induced lipolysis, an exaggerated risk. Lancet 2: 471
16. Rizack MA (1960) Intravascular effect of heparin on plasma non-esterified fatty acid and triglyceride during alimentary lipemia. Proc Soc Exp Biol Med 104: 111–113
17. Schulz P, Giacomini KM, Lutrell S, Turner-Tamiyasu K, Blaschke TF (1983) Effect of low doses of heparin on the plasma binding of phenytoin and prazosin in normal man. Eur J Clin Pharmacol 25: 211–214
18. Storstein L (1976) Studies on digitalis. V. The influence of impaired renal function, hemodialysis, and drug interaction on serum protein binding of digitoxin and digoxin. Clin Pharmacol Ther 20: 6–14
19. Storstein L, Janssen H (1976) Studies on digitalis. VI. The effect of heparin on serum protein binding of digitoxin and digoxin. Clin Pharmacol Ther 20: 15–23

20. Tillement JP, Zini R, Lecomte M, D'Athis P (1980) Binding of digitoxin, digoxin and gitoxin to human serum albumin. Eur J Drug Metab Pharmacokinet 5: 129–134
21. Wilkins TA, Midgley JEM, Giles AF (1982) Treatment with heparin and results for free thyroxin: An in vivo or an in vitro effect? Clin Chem 28: 2441–2443

Author's address:

Dr. J. J. H. M Lohman, Clinical Pharmacokinetics and Toxicology Laboratory, Hospital of Sittard, Walramstraat 23, 6131 BK Sittard (The Netherlands)

The effect of physical exercise on the distribution and renal excretion of digoxin

T. Jogestrand, K. Andersson, R. Dahlqvist[1] and T. Joreteg

Department of Clinical Physiology, Karolinska Hospital and [1]Department of Clinical Pharmacology, Huddinge Hospital, Stockholm (Sweden)

Summary

The effect of submaximal bicycle exercise performed during and after a single injection of digoxin, 15 μg/kg body weight, on the distribution and renal excretion of the drug was investigated in 5 healthy male volunteers. A percutaneous needle biopsy technique was used for thigh muscle sampling, and digoxin in skeletal muscle, urine and serum was analysed by radio-immunoassay. Digoxin concentration in the thigh muscle was approximately 3 times greater during exercise than at rest 2 and 10 hours after the injection of the drug. Serum digoxin concentration was significantly lower during exercise than at rest from 0–120 minutes after completion of the injection. Renal excretion of digoxin was reduced by physical exercise during the first 10 hours (approximately 20 % of the digoxin dose excreted during exercise compared to approximately 30 % at rest) without any difference in renal clearance of digoxin. The apparent volume of distribution of digoxin was approximately 25 % higher during exercise than at rest. The difference in renal excretion and apparent volume of distribution is probably caused by increased digoxin uptake in the exercising skeletal muscle. The results presented demonstrate the importance of standardization of physical activity when performing studies of the pharmacokinetics of cardiac glycosides.

Introduction

The binding of cardiac glycosides has been shown to be activity-dependent with an increased binding to muscle tissue during increased heart muscle (3, 7, 8) and skeletal muscle activity (4). Since about 50 % of the total amount of digoxin in the body is bound to skeletal muscle (10), variation in digoxin binding to this tissue is important in determining the distribution of the drug within the body. Thus, as a consequence of the varying binding of the drug to skeletal muscle, a redistribution of digoxin is obtained at steady state conditions (24 hours after the latest dose) when physical activity is changed, with an increase in skeletal muscle digoxin concentration and a decrease in serum digoxin concentration during exercise (4). This redistribution of digoxin has been shown to be related to the intensity of the exercise with an increase in muscle digoxin of approximately 10 % at a 70–90 W load and an increase of approximately 25 % at a load of 140–180 W (6). Secondary to the increased binding of digoxin to muscle, serum digoxin concentration decreased approximately 20 % at the lower and approximately 40 % at the higher load (6).

There is reason to believe that the effect of physical exercise on the distribution of digoxin within the body is more pronounced if exercise is performed during and soon after the intake of a digoxin dose instead of during steady state conditions. The purpose of the present study was therefore to investigate the effect of physical exercise, performed during and after a single injection of digoxin, on skeletal muscle uptake and renal excretion of the drug.

Subjects and procedures

Five healthy male volunteers (aged 24–32 years) with normal resting and exercise ECGs, took part in the study which was approved by the Committee of Ethics at the Karolinska Hospital. The nature and purpose of the investigation were explained to the participants and their consent was obtained.

The subjects were investigated during a period of 3 days on two different occasions, 4 weeks apart. On each occasion digoxin was injected into an antecubital vein, 15 μg/kg body weight and another cannula was inserted into an antecubital vein for blood samples. Blood samples were taken after 0, 2, 5, 10, 15, 30, 45, 60, 90 and 120 minutes and then after 4, 6, 10, 25, 34 and 49 hours for analysis of digoxin concentration. For analysis of the digoxin concentration in skeletal muscle a percutaneous biopsy was taken twice (after 2 and 10 hours) from the quadriceps femoris muscle (vastus lateralis) using the needle biopsy technique (2). Sampling of urine was performed during the first 10 hours after the injection of digoxin and then 10–25 hours and 25–49 hours after the injection for analysis of digoxin concentrations (Fig. 1).

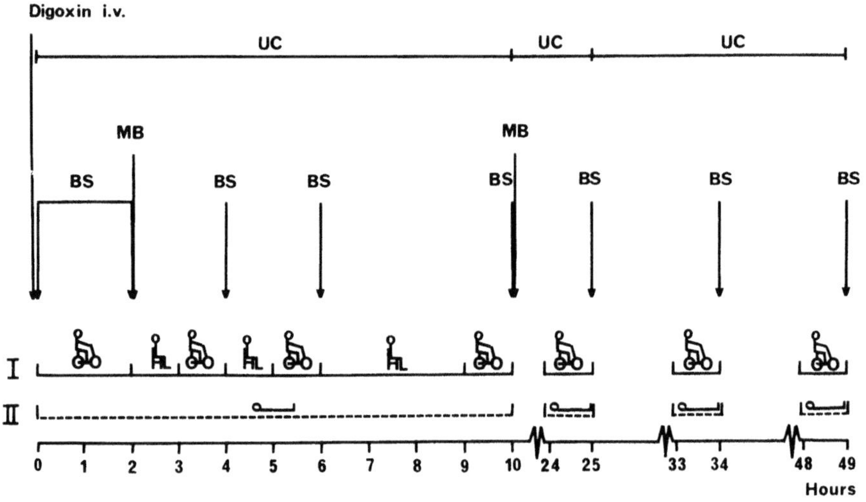

Fig. 1. Study design.
I: Investigation with the subjects exercising, and II: resting.
BS = Blood sample. MB = Muscle biopsy. UC = Urine collection.

On one of the two occasions the subjects initially exercised in the sitting position on an electrically braked bicycle ergometer (EM 380 B, Siemens-Elema, Stockholm, Sweden) during two hours at a constant work load and at a pedalling rate of 60 revolutions/min. The load was submaximal (80 or 90 W). The digoxin dose was injected during 5 minutes from the start of the exercise. Periods of exercise with the same work load and pedalling rate, with a duration of 45 minutes, were performed before each blood sample 4, 6, 10, 25, 34 and 49 hours after the injection. Muscle biopsies were taken 1–5 minutes after completion of exercise. During the first 10 hours the subjects were allowed to move around freely within the laboratory between the periods of exercise. From 10 to 49 hours they performed ordinary daily activity between the exercise periods.

On the other occasion the subjects rested supine for 10 hours after the digoxin injection and during 1.5 hours before each blood sample after 25, 34 and 49 hours. From 10 to 49 hours they performed ordinary daily activity between the resting periods. The order of the two investigations (exercise and rest) was randomized.

Methods

Skeletal muscle, urine and serum digoxin concentration

The skeletal muscle specimens were freeze-dried and carefully freed from connective tissue and fat. Two pieces from each biopsy specimen were weighed and homogenized in phosphate buffer, after which digoxin was extracted from the tissue with dichloromethane. After evaporation of the dichloromethane phase the residue was redissolved in human serum and the digoxin concentration was measured by radioimmunoassay (9) using a commerical ^{125}I-RIA-kit (New England Nuclear, Mass., USA). The method has been described in detail previously (3, 5). The urine and serum digoxin concentrations were measured with the ^{125}I-RIA-kit without any extraction procedure. Before analysis, the urine samples were diluted 1:50–1:300 with serum. The coefficients of variation for duplicate analysis for skeletal muscle, urine and serum digoxin were 11%, 5% and 6%, respectively. Skeletal muscle, urine and serum digoxin concentrations given are the mean of duplicate estimations. Digoxin concentrations are expressed as nmol/kg dry weight (d. w.) in skeletal muscle and nmol/l in urine and serum.

Pharmacokinetic calculations

Renal clearance of digoxin was obtained from the cumulated urinary excretion of the drug divided by its area under the serum concentration vs time curve for the same time interval (0–10 hours).

Non-compartmental determination (1) was used for the apparent volume of distribution. The terminal rate constant was calculated from the data 10 hours and longer after the injection.

The volume of the central compartment was assessed from parameter estimates in three-compartment open model by the AUTOAN programme.

Statistical methods

Mean values ± SD are given unless otherwise stated. Student's t-test for paired observations was used to test statistical significance.

Results

When the digoxin injection was given with the subjects resting, digoxin concentration in the thigh muscle was 44 ± 10 nmol/kg d. w. 2 hours after the injection and 37 ± 14 nmol/kg d. w. 10 hours after the injection (Fig. 2). Skeletal muscle concentrations achieved with the injection given during physical exercise were significantly higher, 137 ± 35 and 110 ± 20 nmol/kg d. w., respectively ($p < 0.05$), (Fig. 2).

The serum digoxin concentrations are given in Fig. 3. Serum concentrations were significantly lower during exercise compared to rest from 0–120 minutes after completion of the injection (except for 2 minutes after the injection when there was no significant difference in concentration). The serum digoxin concentrations were 12.4 ± 1.4, 6.6 ± 1.2 and 3.5 ± 0.6 nmol/l 30 minutes, 1 hour and 2 hours after the injection at rest, while the corresponding concentrations during physical exercise were 8.1 ± 1.5, 3.8 ± 0.8 and 1.9 ± 0.2 nmol/l ($p < 0.01$–0.001).

Renal excretion of digoxin during the first 10 hours after the injection was lower during physical exercise than at rest, 279 ± 64 and 386 ± 76 nmol, respectively ($p < 0.05$). Excretion 10–49 hours after the injection of digoxin was, however, higher when the injection was given during exercise, but the difference was not statistically significant (302 ± 36 and 255 ± 66 nmol, respectively; n. s.). There was no difference in renal clearance of digoxin

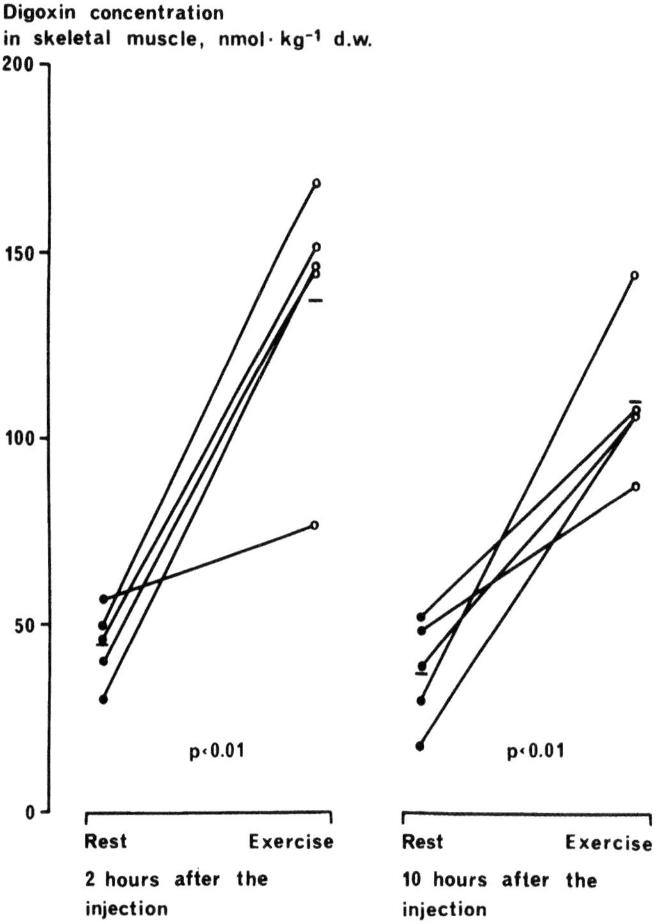

Fig. 2. Digoxin concentrations in skeletal muscle 2 and 10 hours after the injection. Arithmetic means are indicated by horizontal lines.

(0–10 hours) between the two situations,. 167 ± 24 ml/min/1.73 m^2 during exercise and 172 ± 33 ml/min/1.73 m^2 at rest (n. s.).

The apparent volume of distribution (V_{ss}) was significantly higher during exercise, 10.9 ± 3.2 l/kg, than at rest, 8.6 ± 2.3 l/kg ($p < 0.05$). A similar difference in volume of the central compartment (V_c) was seen (physical exercise: 0.31 ± 0.04 l/kg, rest: 0.24 ± 0.05 l/kg; $p < 0.01$).

Discussion

The results of the present study show that the physical activity performed during and after a loading dose of digoxin is of importance for the distribution and excretion of the drug. A moderate bicycle exercise increased the uptake of digoxin in the thigh muscle significantly, with a 3 times higher muscle digoxin concentration 2 and 10 hours after the injection of the

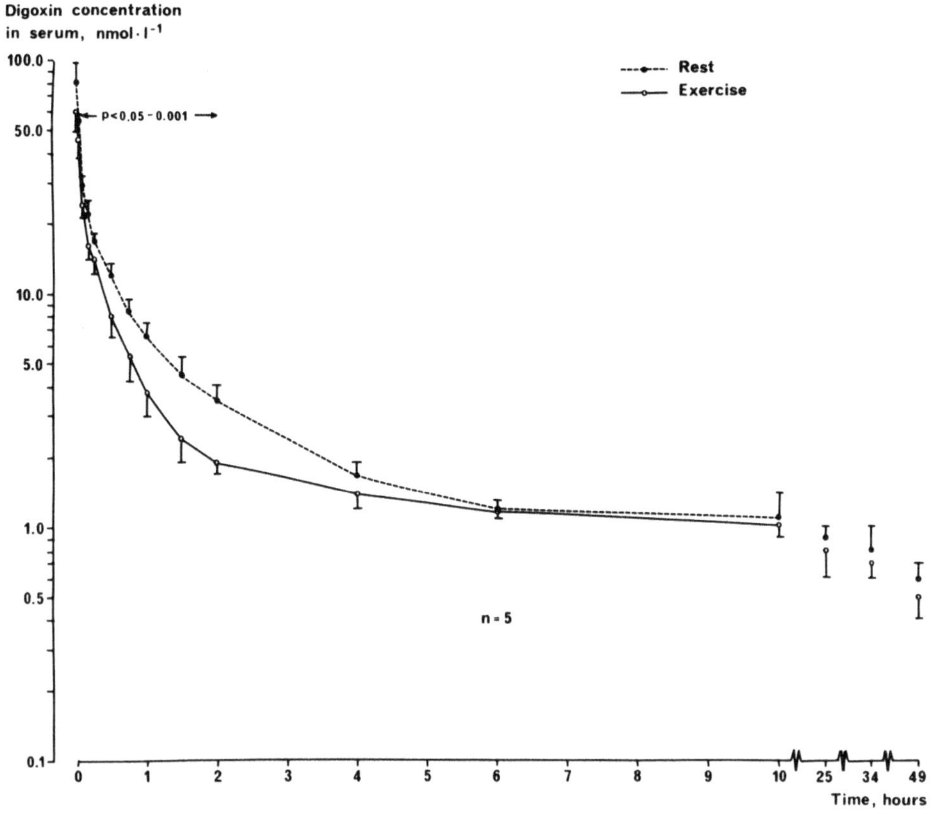

Fig. 3. Digoxin concentrations in serum (means ± SD) 0–49 hours after the injection.

drug when the injection was given during exercise instead of at rest. Furthermore, the increased uptake of digoxin in skeletal muscle during exercise is probably the main cause for the pronounced reduction of serum digoxin concentration during the first 2 hours after the injection and also for the increase in apparent volume of distribution of approximately 25%.

The renal excretion of digoxin during the first 10 hours was reduced by the physical exercise. Thus, approximately 20% of the digoxin dose was excreted during the first 10 hours during exercise compared to approximately 30% at rest. This difference in renal excretion of digoxin is obviously caused by the difference in skeletal muscle binding and volume of distribution of the drug, since renal clearance of digoxin, estimated during the first 10 hours after the injection, was not significantly different between physical exercise and rest. During the time period 10–49 hours after the injection, a somewhat greater amount of digoxin (although not statistically significant) was excreted when physical exercise was performed after the injection. During this period the difference in physical activity was small (34.5 hours of ordinary daily activity during a time period of 39 hours on both occasions) and therefore the somewhat increased excretion of digoxin is probably compensatory to the decreased excretion during the first 10 hours.

The results presented demonstrate the importance of standardization of physical activity when performing studies of the pharmacokinetics of cardiac glycosides. The clinical significance of the described effect of physical exercise on distribution and renal excretion of digoxin remains to be established.

References

1. Benet LZ, Galaezzi RL (1979) Noncompartmental determination of the steady-state volume of distribution. J Pharmacol Sci 68: 1071–1074
2. Bergström J (1962) Muscle electrolytes in man. Scand J Clin Lab Invest 14: Suppl 68
3. Jogestrand T (1980) Digoxin concentration in right atrial myocardium, skeletal muscle and serum in man – influence of atrial rhythm. Eur J Clin Pharmacol 17: 243–250
4. Jogestrand T, Sundqvist K (1981) Effect of physical exercise on the digoxin concentrations in skeletal muscle and serum in man. Clin Physiol 1: 99–104
5. Jogestrand T, Sundqvist K (1981) Skeletal muscle digoxin concentration and its relation to serum digoxin concentration and cardiac effect in healthy man. Eur J Clin Pharmacol 19: 89–95
6. Joreteg T, Jogestrand T (1983) Physical exercise and digoxin binding to skeletal muscle – relation to exercise intensity. Eur J Clin Pharmacol 25: 585–588
7. Lloyd BL, Taylor RR (1976) Influence of myocardial mechanical activity and coronary blood flow on myocardial digoxin uptake. Cardiovasc Res 10: 487–493
8. Lloyd BL, Taylor RR (1978) The effect of heart rate on myocardial ouabain uptake and on the susceptibility to ouabain cardiotoxicity in the dog. Clin Exp Pharmacol Physiol 5: 171–179
9. Smith TW, Butler VP, Haber E (1969) Determination of therapeutic and toxic serum digoxin concentrations by radioimmunoassay. N Engl J Med 281: 1212–1216
10. Steiness E (1978) Digoxin. Klinisk farmakologi. Dissertation. University of Copenhagen, Copenhagen

Author's address:

Tomas Jogestrand, M. D., Department of Clinical Physiology, Karolinska Hospital, S–10401 Stockholm (Sweden)

Ouabain binding to erythrocytes from neonates, blood cord and pregnant women

M. L. Gonçalves

Laboratório de Farmacologia, Dep. Biologia Celular, Instituto Gulbenkian de Ciência, Oeiras (Portugal)

Summary

It has been shown that in the umbilical blood cord the number of specific digoxin binding sites per red cell as well as the value obtained for the dissociation constant are both greater than two fold those observed in normal adults.

On the other hand, the number of ^{3}H-ouabain binding sites and the ouabain K_D values are not age-dependent for a population between 18 and 88 years of age.

In the present work, the receptor binding properties of ouabain to erythrocytes obtained from full-term neonates, umbilical blood cord and pregnant women were investigated. Adult nonpregnant women were taken as control.

The binding capacity for ouabain was less in neonate erythrocytes (186 ± 21 ouabain molecules/cell) than in the other groups where it ranged between 240 and 384 ouabain molecules/cell. A high affinity binding site for ouabain was found and K_D values for the binding determined by Scatchard analysis were about the same (2.26 ± 0.19–2.7 ± 0.45 nM) in all groups.

The results suggest that the decreased ouabain binding sites in neonate erythrocytes might be related to decreased (Na^+, K^+)-ATPase activity.

Introduction

The pharmacological reasons for the requirements of greater digitalis doses per kg body weight in infants and young children are not yet understood.

Pharmacokinetic studies have shown a larger apparent distribution volume of digoxin in infants (6) and of digitoxin in children (8) than in adults, but no significant changes have been observed for other pharmacokinetic parameters.

Cardiac glycosides bind specifically to (Na^+, K^+)-ATPase, an enzyme system inhibited by these compounds in various tissues including erythrocytes. In membrane erythrocytes obtained from umbilical blood cord, the number of specific digoxin binding sites per red blood cell, as well as the dissociation constant value are about two fold greater than those observed in normal adults (7). On the other hand, ^{3}H-ouabain binding studies have shown that in erythrocyte membranes, the number of ouabain binding sites and the dissociation constant values are not age-dependent for a population between 18 and 88 years of age (3).

In the present work, we have investigated the binding of ouabain to erythrocytes from full-term neonates, umbilical blood cord and pregnant women.

Material and Methods

Materials:

^{3}H-ouabain (specific activity 12 Ci/mmol) was purchased from NEN England Nuclear, Germany. All other chemicals were analytical grade and were obtained from B. D. H., England or Sigma, U.S.A. Blood was obtained from neonates (aged 5 to 10 days), umbilical cord vein, pregnant and non-pregnant (control) women.

Methods:

The binding of ^{3}H-ouabain to washed red cells was measured by the method described by Erdmann (5). In brief, the procedures were:

Washed red cells were incubated in a medium containing 1 mM Mg Cl_2, 3 mM ATP, 3 mM glucose, 4.2 nM ^{3}H-ouabain and increasing amounts (2–54 picomoles) of unlabelled ouabain in a total volume of 2 ml. Incubation time was 300 min at 37°C. The erythrocytes were then rapidly collected on Whatman GF/C glass fibre filters and rinsed with 10 ml ice-cold water. The radioactivity on the filters was determined in a scintillation counter (Beckmann LS 8100).

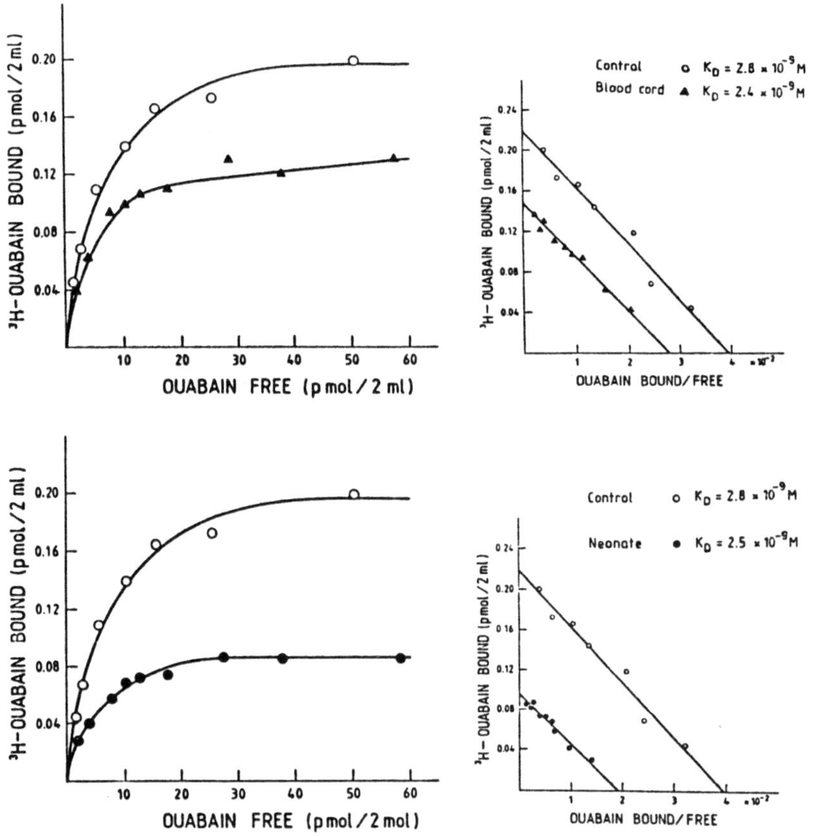

Fig. 1. Saturation curves and Scatchard plots of ^{3}H-ouabain binding to erythrocytes from neonates, blood cord and nonpregnant women (control). About 4.5 x 10^8 erythrocytes were incubated in a total volume of 2 ml incubation medium.

Results and Discussion

Equilibrium binding experiments performed with increasing ouabain concentrations in the incubation medium and saturation curves were obtained. For identical ^{3}H-ouabain concentrations, the amount of drug bound was greatly decreased in both neonate and blood cord erythrocytes (Fig. 1). Data were plotted according to Scatchard (10), to quantify the maximal ouabain binding capacity. The lines obtained from red cells of neonates and umbilical blood cord were straight lines running parallel to the control lines. This suggests that there is only one type of ouabain receptor in neonate, blood cord and adult female erythrocytes, and that the affintiy of the receptor for ouabain is similar in the three groups. The maximal extrapolated number of binding sites (B. max) the dissociation constant (K_D) and the number of ouabain molecules bound per red cell were calculated (Table 1). ^{3}H-ouabain binding in neonate and blood cord erythrocytes was about 50% less than in controls (non-pregnant women). No significant difference was found between control and pregnant women.

Table 1. Maximal extrapolated number of binding sites (B. max), dissociation constant (K_D) and number of ouabain molecules bound per red cell.

	B. max picomoles	K_D nM	Ouabain bound molecules/cell
Neonates	0.087 ± 0.0027	2.4 ± 0.08	185 ± 21
Blood cord	0.120 ± 0.012	2.4 ± 0.14	200 ± 22
Pregnant	0.185 ± 0.011	2.7 ± 0.17	384 ± 20
Non-pregnant (Control)	0.194 ± 0.009	2.5 ± 0.2	346 ± 38

The values represent the mean ± SEM (n = 5).
The maximal binding (B. max) was calculated from Scatchard plots taking the intercept on the ordinate. K_D values were calculated from the slope of the plots.

The density of ouabain receptors in erythrocytes may increase or decrease under certain conditions. In chronic hypokalaemia, Erdmann and Krawietz (4) found increases in the number of ouabain receptors in erythrocytes, without changes in affinity. In contrast, in the skeletal muscle preparations of obese mice, Lin et al. (9) observed a decreased ^{3}H-ouabain binding although the affinity remained unchanged, and in microsomal fractions from ischemic dog myocardium, decreases in (Na^+, K^+)-ATPase activity and ^{3}H-ouabain binding were observed (1). Assuming a constant relationship between ouabain binding sites and (Na^+, K^+)-ATPase activity (2) one may infer that neonate red blood cells contain a lower concentration of this enzyme than adults.

According to the present results, the decreased sensitivity of neonates to cardiac glycosides is difficult to explain in terms of interactions occurring at the receptor level.

Acknowledgements

I am greatful to Mr. Artur Costa for his technical assistance. I wish to express my gratitude to Dr. D. Fino (Maternity Alfredo da Costa, Lisboa) and Dr. L. Graça (Hospital Sta. Maria, Lisboa) for their valuable collaboration in collecting blood samples.

References

1. Beller GA, Conroy J, Smith ThW (1976) Ischemiainduced alterations in myocardial ($Na^+ + K^+$)-ATPase and cardiac glycoside binding. J Clin Invest 57: 341–350

2. Erdmann E, Schoner W (1973) Ouabain-receptor interactions in (Na^+ + K^+)-ATPase preparations from different tissues and species. Determination of kinetic constants and dissociation constants. Biochim Biophys Acta (Amst) 307: 386–398
3. Erdmann E, Hasse W (1975) Quantitative aspects of ouabain binding to human erythrocyte and cardiac membranes. J Physiol (Lond) 251: 671–682
4. Erdmann E, Krawietz W (1977) Increased number of ouabain binding sites in human erythrocyte membranes in chronic hypokalaemia. Acta Biol Med 36: 879–883
5. Erdmann E (1982) Binding studies with ^{3}H-ouabain to red cell membranes. In: Ellory JC, Young JD (eds) Red cell membranes – A methodological approach. Academic Press, London
6. Gorodischer R, Jusko WJ, Yaffe SJ (1976) Tissue and erythrocyte distribution of digoxin in infants. Clin Pharmacol Ther 19: 256–263
7. Kearin M, Kelly JG, O'Malley K (1980) Digoxin "receptors" in neonates: An explanation of less sensitivity to digoxin than in adults. Clin Pharmacol Ther 28: 346–349
8. Larsen A, Storstein L (1983) Digitoxin kinetics and renal excretion in children. Clin Pharmacol Ther 33: 717–727
9. Lin MH, Romsos DR, Akera T, Leveille GA (1978) Na^+, K^+-ATPase enzyme units in skeletal muscle from lean and obese mice. Biochem Biophys Res Commun 80: 398–404
10. Scatchard G (1949) The attractions of proteins for small molecules and ions. N Y Acad Sci 51: 660–672

Author's address:

Maria de Lurdes Gonçalves, Dr., Instituto Gulbenkian de Ciencia, Laboratorio de Farmacologia, Apartado 14–2781, Oeiras (Portugal)

Extracardiac effects of cardiac glycosides

R. A. Gillis and J. A. Quest

Department of Pharmacology, Georgetown University, Schools of Medicine and Dentistry, Washington, D. C. (U.S.A.)

1. Introduction

The extracardiac effects of cardiac glycosides which are of greatest interest to us are those that involve an interaction of these drugs with neural tissue, and result in changes in cardiovascular function. In this category, we are especially interested in interactions of cardiac glycosides with two neural sites, namely, reflexogenic areas of the cardiovascular system, and the central nervous system (i. e., CNS sites that control sympathetic outflow). In this review we will discuss evidence documenting an interaction of cardiac glycosides with reflexogenic areas and with CNS sites that influence autonomic nervous system outflow. This will be followed by a discussion of evidence indicating that neuroexcitatory effects of cardiac glycosides play an important role in the cardiovascular effects of these agents. Interaction of cardiac glycosides with reflexogenic sites can explain many of the therapeutic effects of these drugs such as their antiarrhythmic effects and effects on preload, afterload, and renal function in patients with congestive heart failure. Interaction of cardiac glycosides with the CNS, specifically CNS sites that influence sympathetic outflow, can explain in part the ventricular tachyarrhythmias that occur with toxic doses of these agents.

Most of the evidence showing that cardiac glycosides produce neural effects which result in changes in cardiovascular function has been obtained from studies in experimental animals. However, there are some data that have been obtained from humans, and experiments can be proposed which would provide important data from human studies which would help answer the question of how relevant neural effects of cardiac glycosides are to the beneficial and toxic actions of these agents. This will be discussed in the last section of this review.

2. Documentation of an interaction of cardiac glycosides with reflexogenic sites

In our previous review (20) we identified 5 reflexogenic areas that have been investigated as sites of digitalis action. These included baroreceptors in the carotid sinus and aortic arch areas, mechanoreceptors in the ventricle, mechanoreceptors in the atrium, chemoreceptors in the carotid sinus and aortic arch regions, and receptors associated with the nodose ganglia. Administration of cardiac glycosides usually by local intra-arterial injection to the specific reflexogenic site has been demonstrated to produce activation of the different receptor sites with concomitant changes in cardiovascular function (20). Of the 5 reflexogenic sites that have been studied, two appear to be of pharmacological importance with cardiac glycosides. These are the baroreceptors in the carotid sinus and aortic arch regions and mechanoreceptors in the ventricle. Both have been shown to be important in mediating decreases in heart rate, arterial pressure, and sympathic nerve activity (15, 36, 37, 41, 45, 47).

In addition, both sites are affected by concentrations of cardiac glycosides in the non-toxic therapeutic dose range (10, 28, 46, and see section 6). Effects of cardiac glycosides on baroreceptors have been described in great detail in our previous review (20). Some information of ventricular mechanoreceptors was also discussed in our review; however, new information on this topic has recently become available. In one study, Thames and colleagues (47) reported that intracoronary administration of acetylstrophanthidin produced an acute sensitization of cardiac mechanoreceptors in anesthetized dogs in which the sinoaortic baroreceptor regions were denervated. They observed that cardiac receptor-mediated reflex inhibition of renal sympathetic nerve activity, induced by expansion of the blood volume or by occlusion of the circumflex coronary artery, was significantly enhanced by intracoronary injection of the digitalis drug. In a subsequent and related study, Thames and coworkers (46) administered i. v. digoxin to sinoaortic denervated dogs for seven days, and found that the chronic digoxin treatment resulted in an augmented gain of cardiac reflex-mediated inhibition of renal sympathetic nerve activity. Of additional interest in this study was the finding that the plasma levels of digoxin associated with sensitization of the canine cardiac mechanoreceptors were within the therapeutic range of plasma levels usually observed in man.

The 3 remaining reflexogenic sites, i. e. the mechanoreceptors in the atria, chemoreceptors in the carotid sinus and aortic arch, and receptors associated with the nodose ganglia do not appear to be of pharmacological importance in the actions of cardiac glycosides. In the case of the atrial mechanoreceptors, Gilmore and Zucker (25) have clearly demonstrated that cardiac glycosides can increase the sensitivity of atrial stretch receptors in the dog. However, these investigators further noted that this reflex mechanism appears to play little, if any, physiologic role in primates.

In the case of chemoreceptors, activation of these sites would increase sympathetic activity to the heart and blood vessels. This is not the type of response observed following a non-toxic dose of cardiac glycosides which activates reflexogenic areas (36).

In the case of the nodose ganglia, although digitalis has been shown to excite these reflexes in high doses when injected directly to the ganglia (7), we found no role for these receptors in the bradycardia produced by systemically administered sub-toxic doses of digitalis (22).

3. Role of reflexogenic actions of cardiac glycosides in the cardiovascular effects of these agents

An action of cardiac glycosides on reflexogenic areas of the cardiovascular system is probably responsible for several of the beneficial effects of these drugs in patients. This is especially true for the antiarrhythmic effects of cardiac glycosides. The primary mechanism of action of digitalis drugs in treating paroxysmal supra-ventricular tachycardia, atrial flutter and atrial fibrillation is enhancement of vagal effects on the atrio-ventricular node (20). The ability of cardiac glycosides to enhance vagal tone depends entirely upon an action on the reflexes. This was shown by Pace and Gillis (34) using vagal nerve recordings. With reflexes intact, digoxin was found to increase efferent vagal nerve activity. With reflexogenic areas denervated, however, no increase in vagal nerve activity occurred.

Similarly it appears that an antiarrhythmic effect of digitalis on ventricular rhythm is related to a reflex site of action of these agents. Blumberg and colleagues (3) and Lown and coworkers (31) have reported an antiarrhythmic effect of digitalis in some patients with ventricular arrhythmias. Furthermore, Lown and coworkers (4) demonstrated in experimental studies that an antiarrhythmic effect of digitalis is mediated by an action on the arterial baroreceptors. This was demonstrated in dogs using ventricular fibrillation threshold

(VFT) measurements. With baroreceptors intact, digitalis increased the VFT by 50%, whereas with baroreceptors denervated the VFT was decreased by 42%.

Reflexogenic actions of cardiac glycosides probably play an important role in the beneficial hemodynamic effects of these agents in patients with congestive heart failure. Both arterial and venodilation occur (32), and these changes are probably due to a digitalis-induced enhancement of reflex activity leading to sympathetic withdrawal from arteries and veins. The change in arterial tone would decrease cardiac afterload and facilitate an increase in cardiac output. In addition, the change in venous tone would decrease preload and reduce pulmonary capillary wedge pressure. In support of a venodilator effect are the findings of Arnold and colleagues (2) showing that chronic digoxin therapy in patients with chronic heart failure decreases pulmonary capillary wedge pressure.

Finally, as pointed out by Thames and colleagues (47), cardiac receptors "seem to exert their greatest influence on sympathetic outflow to the kidney". When these receptors are sensitized by cardiac glycosides, withdrawal of sympathetic outflow to the kidney would result in a decrease in renin secretion and an increase in renal blood flow and diuresis. In support of cardiac glycosides bringing about these effects are the findings of Covit and colleagues (8), demonstrating that digoxin administered to patients with chronic congestive heart failure results in a reduction in plasma renin activity and plasma aldosterone levels. Reduction in renin would indirectly reduce angiotensin-mediated vasoconstriction and reduction in aldosterone would reduce aldosterone-mediated sodium retention.

4. Documentation of an interaction of cardiac glycosides with CNS sites which influence sympathetic outflow

Data from at least 5 different laboratories indicate that cardiac glycosides interact with CNS sites which influence sympathetic outflow. Abiko and colleagues (1) reported that an intravenous arrhythmogenic dose of digitalis administered to cats produced an increase in sympathetic nerve discharge recorded from preganglionic cardiac nerves. This was later confirmed by McLain (33) using recordings from postganglionic cardiac sympathetic nerves, and Gillis (15) using recordings from preganglionic sympathetic cardiac nerves of debuffered animals. Garvey (13) also reported that intravenous administration of digitalis augments preganglionic cardiac sympathetic nerve activity. Our initial finding of an effect of cardiac glycosides to increase sympathetic nerve activity appeared as a short report (15) and was confirmed in 3 later studies that appeared as full reports (18, 23, 34). In summary, investigators from 4 different laboratories have demonstrated that arrhythmogenic doses of cardiac glycosides administered by the i. v. route increase cardiac sympathetic nerve activity. In one case this was demonstrated to occur in preganglionic cardiac sympathetic nerves of an animal preparation with all reflexogenic areas denervated (15). These results suggest that cardiac glycosides act at a site in the CNS to increase sympathetic outflow.

Investigators from a fifth laboratory used a most convincing approach to demonstrate that cardiac glycosides increase central sympathetic ouflow. Garan and colleagues (12) were interested in determining whether digoxin-induced alpha-adrenergic receptor mediated coronary constriction which occurred in dogs (26) was of CNS origin. To test this, they administered an i. v. injection of digoxin into a donor dog perfusing the isolated head of a recipient animal. The only connection between the isolated head and the body of the recipient animal was the nervous system. They found that digoxin (1 mg) caused an increase in coronary resistance in the recipient animal. Most importantly, these investigators found that 1 mg of digoxin administered i. v. to the recipient animal had no effect on coronary resistance. However, this 1 mg dose of digoxin increased coronary resistance when administered i. v. to dogs with an intact cerebral circulation (12). Hence, perfusion of only the CNS of an experimental animal with digoxin results in coronary constriction whereas administration of the same dose of drug to just the periphery (with restriction from the CNS) results in no effect on the coronary vasculature. In addition, when digoxin was administered i. v. to intact dogs, an increase in cerebrospinal fluid (csf) concentration of digoxin occurred that closely followed the time action curve for the coronary constrictor effect of the drug. The peak concentration of digoxin in the csf was 2.3 ± 0.4 ng/ml fifteen minutes after a 1.0 mg i. v. dose of drug. This result clearly demonstrates that digoxin can pass the blood brain barrier. Finally, these investigators found that digoxin when administered into the lateral cerebral ventricle in one-tenth of the systemic dose (i. e., 0.1 mg) produced an increase in coronary vascular resistance. They concluded that the CNS is a major site of action of digitalis in producing neurogenic vasoconstriction in the coronary vasculature.

The specific CNS site where cardiac glycosides act to increase sympathetic outflow is unclear at this time. However, the general area of the CNS appears to be primarily in the hindbrain. Evidence for this is that increases in central sympathetic outflow occur in decerebrate animals (15), prevention of CNS mediated arrhythmogenic effects of cardiac glycosides occurs when drugs known to suppress central sympathetic ouflow are localized to the hindbrain (as compared to the forebrain) (27), and surgical transection of the brain modifies digitalis-induced arrhythmias only when the transection is at the level of the caudal medulla (43). It has been suggested by Somberg and Smith (43), that the specific CNS site of action of cardiac glycosides is the area postrema. Their evidence is based on gross transections of the cat brain rostral and caudal to the area postrema. Consistent with their suggestion are recent findings indicating that activation of neurons in the area postrema produces coronary constriction (16) and ventricular tachyarrhythmias (14). However, discrete lesions of the area postrema do not appear to modify the toxic arrhythmogenic effects of cadiac glycosides (48). Hence, it is unclear as to where in the hindbrain cardiac glycosides act to increase sympathetic outflow to the heart.

5. Role of a CNS sympathetic excitatory action of cardiac glycosides in the arrhythmogenic effect of these agents

A number of approaches have been used to test the hypothesis that the digitalis-induced increase in central sympathetic outflow is an important factor in the generation of ventricular tachyarrhythmias. One approach was to determine whether changes in cardiac sympathetic nerve discharge were correlated with the development of ventricular tachyarrhythmias. It was found that there was a clear correlation between enhanced sympathetic discharge and the initiation of arrhythmias (15, 18, 19, 23, 33, 34).

A second approach was to test the hypothesis that depression of hyperactivity in cardiac sympathetic nerves evoked by cardiac glycosides would abolish arrhythmias. It was found that administration of drugs that abolish cardiac sympathetic nerve activity such as diphenylhydantoin (19), propranolol (15, 23) and chlordiazepoxide (24) produced a concomitant abolition of the associated cardiac arrhythmias.

A third approach that has been used to determine whether cardiac glycoside-induced sympathetic neural effects contribute to the development of arrhythmias is focal stimulation of the CNS (9). A site in the posterior hypothalamus was defined from which arrhythmias could be consistently obtained. These arrhythmias were mediated by the sympathetic nervous system, since they could be prevented by pretreatment with propranolol but not with atropine, and direct recording from cardiac sympathetic nerves indicated continuous activity coincident with the arrhythmias. The stimulus strength to the posterior hypothalamus was then reduced to a degree insufficient to elicit arrhythmia but intense enough to increase sympathetic nerve activity, blood pressure, and heart rate, and ouabain was administered as a continuous infusion of 2 μg/kg/min with hypothalamic stimulation repeated at 5 minute intervals. After infusion of large doses of ouabain, arrhythmias occurred during hypothalamic stimulation and these arrhythmias were associated with significant augmentation of sympathetic nerve activity. The sympathetic neural discharge produced by hypothalamic stimulation was significantly greater than that seen during hypothalamic stimulation before administration of ouabain. Restoration of sinus rhythm occurred within 1 minute after termination of the electrical stimulation of the hypothalamus. These results indicate that ouabain and hypothalamic stimulation, each at a dose or intensity subthreshold for arrhythmias, can be arrhythmogenic when applied in combination. Hence, these observations strongly indicate that cardiac glycoside-induced increases in sympathetic nerve activity play an important role in the genesis of ventricular arrhythmias.

A fourth approach that has been used is to surgically remove portions of the CNS with which cardiac glycosides might interact to increase sympathetic outflow to the heart. This has been done by Somberg and Smith (43) by transecting the brain at several levels and determining the effect of this intervention on the dose of cardiac glycosides required to produce ventricular tachyarrhythmias. It was found that transection of the brain 2 mm caudal to the obex significantly increased the dose of digitalis to cause arrhythmias. The same was true of spinal cord transection at C-1 but not midcollicular decerebration or transection of the brain 2 mm rostral to obex (43).

A fifth approach has been to determine whether pharmacological-induced depression of central sympathetic outflow would protect against cardiac glycoside-induced ventricular tachyarrhythmias. An effective way to depress central sympathetic outflow is to administer a drug that stimulates CNS dopamine receptors (27). By administering drugs of this type as represented by apomorphrine (27), piribedil (27), and bromocriptine (29), a pronounced antiarrhythmic effect can be noted. In the case of both apomorphrine and bromocriptine, their antiarrhythmic effect was shown to be due to an action in the CNS (27, 29), and in the case of apomorphine, its antiarrhythmic effect was shown to occur in the hindbrain (27).

A sixth approach has been to utilize a digitalis preparation which does not cross the blood brain barrier. This approach has been taken by Caldwell and colleagues with a series of 4-aminosugar cardiac glycosides (5, 6, 35). Systemic administration of an aminosugar cardenolide did not produce any signs of increased cardiac sympathetic nervous system activity in dogs. Evidence for this was the lack of effect of either beta adrenergic blockade or surgical sympathectomy on the toxic dose of the aminosugar cardenolide, the character of toxicity, or the lethal dose of the digitalis preparation compared to animals with an intact sympathetic nervous system. In fact, in control dogs, the terminal event was cardiac standstill rather than ventricular fibrillation as seen in dogs intoxicated with digoxin. On the other hand, administration of the aminosugar cardenolide into the lateral brain ventricles produced signs of increased cardiac sympathetic nervous system activity.

6. Documentation of neural effects of cardiac glycosides in humans

There is evidence that administration of therapeutic doses of cardiac glycosides in man exert neural effects that result in changes in cardiovascular function. This has been demonstrated for the carotid sinus baroreceptor reflex (10). Two techniques were used to stimulate and deactivate the arterial baroreceptors. The first was to administer drugs that would either activate the baroreceptors (phenylephrine) or deactivate the baroreceptors (nitroglycerine) and note changes in heart rate. The second was the use of a variable pressure neck chamber which permitted the study of reflex changes in heart rate as well as in systemic arterial blood pressure. Responses were obtained with both techniques before and after i. v. administration of lanatoside C. It was found that the bradycardic and hypotensive effects of baroreceptor stimulation were augmented significantly after digitalis administration. However, the tachycardia and hypertensive responses produced by baroreceptor deactivation were not affected by digitalis.

Cardiac glycosides have also been shown to augment the cardiopulmonary baroreflex in humans. This was demonstrated by Imamura and colleagues (28) using lower body negative pressure to evoke the reflex. By exposing patients to lower body negative pressure there is deactivation of mechanoreceptors in the heart (and lungs) resulting in an increase in forearm vascular resistance. After i. v. administration of lanatoside C, lower body negative pressure produced greater increases in forearm vascular resistance. These data were interpreted as indicating that digitalis administration resulted in an augmentation of the tonic inhibitory influence of cardiopulmonary baroreceptors. With exposure of digitalis-treated patients with lower body negative pressure, a greater forearm vasoconstrictor effect was noted because there was more inhibitory influence of cardiopulmonary receptors to be withdrawn from forearm arterioles.

7. Summary and points to be considered for future studies

Table 1 summarizes the important neural effects of cardiac glycosides and lists the related physiological response(s) which could occur because of drug interactions with neural tissue. The consequences of the neurally-induced physiological response to changes which occur in cardiovascular function after cardiac glycosides are administered are also listed. As can be noted, a reflex site of action is of primary importance in mediating the effects of therapeutic doses of cardiac glycosides on the electrical activity of the heart. This is true in the com-

Table 1. Neural sites of action of cardiac glycosides and proposed consequences of the interaction of these agents with neural tissue

Site of neural action	Physiological response(s) occurring because of the neural interaction	Consequences of effect to cardiovascular function
1. Peripheral reflexogenic sites, especially the carotid sinus and aortic arch baroreceptors, and the cardiopulmonary receptors	1. Activation of parasympathetic and inhibition of sympathetic nervous system	1. Decrease in sinus rate 2. Antiarrhythmic effect in patients with atrial arrhythmias due to: (a) impaired conduction in the A-V node (b) decrease in automaticity of supraventricular tissue (excluding S-A mode) (c) decrease in refractory period of atrial muscle 3. Antagonism of ventricular arrhythmias 4. Decrease in arterial and venous tone (as a consequence of inhibiting the sympathetic nervous system) 5. Diuresis (as a consequence of inhibiting the sympathetic nervous system)
2. Sites in the central nervous system which control sympathetic outflow	1. Increase in central sympathetic outflow	1. Ventricular tachyarrhythmias 2. Arterial vasoconstriction (including coronary vasoconstriction)

monly approved use of these agents to treat atrial arrhythmias as well as in the less often used case of treating ventricular arrhythmias (3, 31).

Neural actions of cardiac glycosides may contribute to the effect that these agents have to decrease both arterial and venous tone in patients with congestive heart failure. Decreases in vascular tone would result from reflex-induced inhibition of sympathetic tone which is especially high in these patients. In addition, reflex-induced inhibition of sympathetic activity to the renal bed would improve renal blood flow, and decrease renin release from the juxtaglomerular cells. A decrease in renin would decrease the level of circulating angiotensin (a potent constrictor of arterioles and releaser of catecholamines and aldosterone). These indirect effects on renal blood flow and renin could cause diuresis and further reduce vascular tone.

A CNS neural site of action of cardiac glycosides (which occurs with high doses of these agents) appears to contribute to the generation of ventricular tachyarrhythmias (as discussed in section 5). In addition, a CNS site of action to increase sympathetic outflow appears to mediate part of the arterial constrictor effect of these agents. This has been documented to be the case for the coronary and skeletal muscle vascular beds (11, 12, 21, 26, 38, 42, 44). This may also be true of the mesenteric bed but this has not been adequately tested.

To obtain additional relevant information on the role of neural affects of cardiac glycosides in their overall effects on cardiovascular functions, studies need to be performed in patients with atrial arrhythmias and with congestive heart failure who also have an altered nervous system. For example, the role of neural effects of cardiac glycosides in the treatment of atrial arrhythmias in patients with a transplanted heart has been examined (30). It was found that digoxin administration in patients with atrial flutter and atrial fibrillation did not produce any change in ventricular response, indicating no effect of digoxin on the A-V node. This finding demonstrates the importance of a neural action of cardiac glycosides to an important therapeutic effect of these drugs. By using this approach in patients with transplanted hearts (or transplanted kidneys), tetraplegics (as was done with clonidine; see Reid et al. (39)), or patients with familial dysautonomia, the role of neural effects of cardiac glycosides may be better clarified.

Another approach to clarify the role of extracardiac neural effects might be to carefully examine the time-action curves of neural effects and direct cardiac effects. Our impression from studies performed in animals is that the neural effects occur almost immediately after administration (for example, see Quest and Gillis (36) and (37)), whereas direct effects require time to develop. Furthermore, we feel based on our own work (38) and work of others (for example, Garan et al. (12)), that the vasoconstrictor effect of cardiac glycosides results from activation of the sympathetic nervous system. Most importantly, Schinz and colleagues (40) have demonstrated that the time to onset of arterial constrictor effect of digitalis in humans occurs sooner than the inotropic effect. This fits with our idea of a difference in the time-action curves of neural and direct inotropic actions of these drugs. If true, one could exploit the difference in time-action curves and assess the role of neural actions of cardiac glycosides.

Acknowledgements

The authors express their thanks to Mr. George Villalon, Merck Sharp and Dohme and Ms. Sharon Stewart of Georgetown University for their assistance in the preparation of this manuscript.

References

1. Abiko Y, Mukahira K, Tanabe T (1964) Blockade with dichloroisoproterenol (DCI) of the sympathetic factor of strospeside-induced bradycardia in cats. Jap J Pharmacol 14: 212–231

2. Arnold SB, Byrd RC, Meister W, Melmon K, Cheitlin MD, Bristoro JD, Parmley WW, Chatterjee K (1980) Long term digitalis therapy improves left ventricular function in heart failure. New Eng J Med 303: 1443–1448
3. Blumberg JM, Hayes JG, Stevens M, Sullivan G, Killip T (1973) Digitalis in treatment of ventricular extrasystoles in the otherwise normal heart. Circulation 48 (Suppl IV): 8
4. Brooks WW, Verrier RL, Lown B (1979) Digitalis drugs and vulnerability to ventricular fibrillation. Eur J Pharmacol 57: 69–78
5. Caldwell RW, Nash CB (1976) Pharmacological studies of a new 4-aminosugar cardiac glycoside (ASI-222). J Pharmacol Exp Ther 197: 19–26
6. Caldwell RW, Nash CB (1977) Comparison of the therapeutic index (TI) of aminosugar cardiac glycosides with digoxin (D). Fed Proc 36: 3705
7. Chai CY, Wang HH, Hoffman BF, Wang SC (1967) Mechanisms of bradycardia induced by digitalis substances. Amer J Physiol 212: 26–34
8. Covit AB, Schaer GL, Sealey JE, Laragh JH, Cody RJ (1983) Suppression of the renin-angiotensin system by intravenous digoxin in chronic congestive heart failure. Am J Med 75: 445–447
9. Evans DE, Gillis RA (1975) Effect of ouabain and its interaction with diphenylhydantoin on cardiac arrhythmias induced by hypothalamic stimulation. J Pharmacol Exp Ther 195: 557–586
10. Ferrari A, Gregorini L, Ferrari MC, Preti L, Mancia G (1981) Digitalis and baroreceptor reflexes in man. Circulation 63: 279–285
11. Garan H, Powers ER, Powell WJ Jr (1974) Dependence of the neurogenic vascular resistance effect of digitalis on the background level of sympathetic activity in skeletal muscle. Fed Proc 33: 584
12. Garan H, Smith TW, Powell WJ Jr (1974) The central nervous system as a site of action for the coronary vasoconstrictor effect of digoxin. J Clin Invest 54: 1365–1372
13. Garvey HL (1972) Role of the autonomic nervous system in digitalis intoxication. In: Bajusz E, Rona G (eds) Recent Advances in Studies on Cardiac Structure and Metabolism, Vol 1. University Park Press, London, pp 563–573
14. Gatti PJ, Dias Souza J, DaSilva AMT, Quest JA, Gillis RA (1985) Chemical stimulation of the area postrema induces cardiorespiratory changes in the cat. Brain Res 346: 115–123
15. Gillis RA (1969) Cardiac sympathetic nerve activity: Changes induced by ouabain and propranolol. Science 166: 508–510
16. Gillis RA, Dias Souza J, Gatti PJ (1985) Mechanism of the ventricular tachyarrhythmias that occur after topical application of kainic acid to the area postrema of the cat. Neurosci Absts 11:191
17. Gillis RA, Dionne RA, Standaert FG (1972) Suppression by clonidine (St-155) of cardiac arrhythmias induced by digitalis. J Pharmacol Exp Ther 182: 218–226
18. Gillis RA, Jolson H, Thibodeaux H, Levitt B (1975) Antagonism of deslanoside-induced cardiotoxicity by combined nicotinic and muscarinic blockade of autonomic ganglia. J Pharmacol Exp Ther 195: 126–132
19. Gillis RA, McClellan JR, Sauer TS, Standaert FG (1971) Depression of cardiac sympathetic nerve activity by diphenylhydantoin. J Pharmacol Exp Ther 179: 599–610
20. Gillis RA, Quest JA (1980) The role of the nervous system in the cardiovascular effects of digitalis. Pharmacol Rev 31: 19–97
21. Gillis RA, Quest JA, Standaert FG (1969) Depression by reflexes of the pressor and cardiotoxic responses to ouabain. J Pharmacol Exp Ther 170: 294–302
22. Gillis RA, Quest JA, Thibodeaux H, Clancy MM, Evans DE (1975) Neural mechanisms involved in acetylstrophanthidin-induced bradycardia. J Pharmacol Exp Ther 193: 336–345
23. Gillis RA, Raines A, Sohn YJ, Levitt B, Standaert FG (1972) Neuroexcitatory effects of digitalis and their role in the development of cardiac arrhythmias. J Pharmacol Exp Ther 183: 154–168
24. Gillis RA, Thibodeaux H, Barr L (1974) Antiarrhythmic properties of chlordiazepoxide. Circulation 49: 272–282
25. Gilmore JP, Zucker IH (1979) Activity of atrial receptors under normal and pathological states. In: Hainsworth R, Kidd C, Linden RJ (eds) Cardiac Receptors, Cambridge Univ Press, New York, pp 139–156

26. Hamlin NP, Willerson JT, Garan H, Powell WJ Jr (1974) The neurogenic vasoconstrictor effect of digitalis on coronary vascular resistance. J Clin Invest 53: 288–296
27. Helke CJ, Gillis RA (1978) Centrally mediated protective effects of dopamine agonists on digitalis-induced ventricular arrhythmias. J Pharmacol Exp Ther 207: 263–270
28. Imamura T, Takeshita A, Ashihara T, Yamamoto K, Hoka S, Makamura M (1985) Digitalis-induced augmentation of cardiopulmonary baroreflex control of forearm vascular resistance. Circulation 71: 11–16
29. Kao A, Kriett JM, Tobler HG, Detloff BLS, Pritzker MR, Benson DW, Benditt DG (1984) Bromocriptine treatment of digitalis-induced ventricular tachyarrhythmias: Studies in a canine model. J Amer Coll Cardiol 4: 1188–1194
30. Leachman RD, Cokkinos DVP, Cabrera R, Leatherman LL, Rochelle DG (1971) Response of the transplanted, denervated human heart to cardiovascular drugs. Amer J Cardiol 27: 272–276
31. Lown B, Graboys TB, Podrid PJ, Cohen BH, Stockman MB, Gaughan CE (1977) Effect of a digitalis drug on ventricular premature beats. N Engl J Med 296: 301–306
32. Mason DT, Zelis R, Amsterdam EA (1972) Unified concept of the mechanism of action of digitalis: Influence of ventricular function and cardiac disease on hemodynamic response to fundamental contractile effect. In: Marks BH, Weissler AM (eds) Basic and Clinical Pharmacology of Digitalis. Chales C Thomas, Springfield, Ill, pp 1206–1229
33. McLain PL (1969) Effects of cardiac glycosides on spontaneous efferent activity in vagus and sympathetic nerves of cats. Int J Neuropharmacol 8: 379–387
34. Pace DG, Gillis RA (1976) Neuroexcitatory effects of digoxin in the cat. J Pharmacol Exp Ther 199: 583–600
35. Puryear SK, Nash CB, Caldwell RW (1981) Effect of cardiac β-adrenergic blockade or denervation on cardiotoxicity of digoxin and an aminosugar cardenolide. J Cardiovasc Pharmacol 3: 113–127
36. Quest JA, Gillis RA (1971): Carotid sinus reflex changes produced by digitalis. J Pharmacol Exp Ther 177: 650–661
37. Quest JA, Gillis RA (1974) Effect of digitalis on carotid sinus baroreceptor activity. Circ Res 35: 247–255
38. Quest JA, Rowles GS, Bhat HB, Gillis RA (1976) Effect of dihydroouabain on vascular tone of the perfused canine hindlimb. J Pharmacol Exp Ther 199: 255–261
39. Reid JL, Wing LMH, Mathias CJ, Frankel HL, Neill E (1977) The central hypotensive effect of clonidine. Studies in tetraplegic subjects. Clin Pharmacol Ther 21: 375–381
40. Schinz A, Klein G, Blomer H (1977) Time sequence of direct vascular and inotropic effects following intravenous administration of digixon in normal man. Int J Clin Pharmacol 15: 189–193
41. Sleight P, Lall A, Muers M (1969) Reflex cardiovascular effects of epicardial stimulation by acetylstrophanthidin in dogs. Circ Res 25: 705–711
42. Somberg JC, Kuhlman JE, Smith TW (1981) Localization of the neurally mediated coronary vasoconstrictor properties of digitalis in the cat. Circ Res 49: 226–233
43. Somberg JC, Smith TW (1979) Localization of the neurally mediated arrhythmogenic properties of digitalis. Science 204: 321–323
44. Stark JJ, Sanders CA, Powell WJ Jr (1972) Neurally mediated and direct effects of acetylstrophanthidin on canine skeletal muscle vascular resistance. Circ Res 30: 274–282
45. Thames MD (1979) Acetylstrophanthidin-induced reflex inhibition of canine renal sympathetic nerve activity mediated by cardiac receptors with vagal afferents. Circ Res 44: 8–15
46. Thames MD, Miller BD, Abboud FM (1982) Sensitization of vagal cardiopulmonary baroreflex by chronic digoxin. Am J Physiol 243 (Heart Circ Physiol 12): H 815–H 818
47. Thames MD, Vaickman LA, Abboud FM (1980) Sensitization of cardiac receptors (vagal afferents) by intracoronary acetylstrophanthidin. Am J Physiol (Heart Circ Physiol 8): H 628–H 635
48. Thron CD, Riancho JA, Borison HL (1984) Lack of protection against ouabain cardiotoxicity after chronic abtation of the area postrema in cats. Exper Neurol 85: 574–583
49. Watanabe AM, Judy WV, Cardon PV (1974) Effect of L-dopa on blood pressure and sympathetic nerve activity after decarboxylase inhibition in cats. J Pharmacol Exp Ther 188: 107–113

Author's address:

Richard A. Gillis, Georgetown University, Schools of Medicine and Dentistry, 3900 Reservoir Rd., N. W., Washington, D. C. 20007 (U.S.A.)

Effects of digitalis on renal tubular sodium reabsorption

O. M. Sejersted

Institute of Muscle Physiology, Oslo (Norway)

Introduction

Withering in his "Account of the Foxglove" categorizes digitalis as a diuretic (26). The mobilization of edema in heart failure requires of course involvement of the kidneys. However, it has been a long-standing controversy whether the diuretic effect of digitalis is secondary to improved cardiac function or whether it is due to a direct tubular effect.

In severe heart failure there is a high sympathetic tone. Stimulation of sympathetic renal nerves moves the autoregulation curve so that the renal vasodilatory capacity required for proper natriuretic response to extracellular volume expansion is greatly diminished (6, 9, 17). Hence relief of the sympathetic tone during digitalis treatment will allow the kidneys to vasodilate. The glomerular filtration rate (GFR) will remain elevated until a normal fluid balance is regained and may then return to pretreatment levels.

However, since the kidney is rich in Na,K-ATPase with high affinity for digitalis (1, 19), there is also a direct tubular effect which in fact is potentiated by the rise in GFR. The two mechanisms therefore work in concert. This direct tubular effect may not even be detectable at low GFRs.

Binding of ouabain to intact dog kidneys

Infusion of ouabain in a dose of 90 μg/kg into the renal artery of intact dog kidneys leads to inhibition of about 80 % of the renal Na,K-ATPase activity (19). Larger doses are incompatible with maintained cardiac function.

As shown in Fig. 1, the density of ouabain binding sites is less in proximal tubules (PT) than in distal tubules (DT). On average the site density in thick ascending limbs of Henle's loop (MAL) and DT is the same, but is 3–4 times higher than in the PT (24). Despite this difference in Na,K-ATPase concentration between different tubular segments, merely 21 % of the total renal content of the enzyme is located in the outer medulla (19). This is still disproportionally more than the relative volume of this part of the kidney (Fig. 2). Of the cortical 77 % of the enzymes, 61 % are located in the PT. Hence almost 50 % of the total renal amount of Na,K-ATPase is located to PT (24).

For an assessment of the functional role of the Na,K-ATPase it is important to know the amount of enzyme per unit tubular length rather than the concentration in terms of protein or wet wt. Since proximal tubular cells are so large the grain density is low (Fig. 1), but the number of grains per tubular cross section does not differ very much between the main sodium transporting segments. The number of ouabain binding sites along the dog nephron is approximately 45, 60 and 40 fmol/mm in PT, MAL and DT, respectively (24). In another study on the rabbit kidney the numbers were 11, 21 and 31 fmol/mm (14).

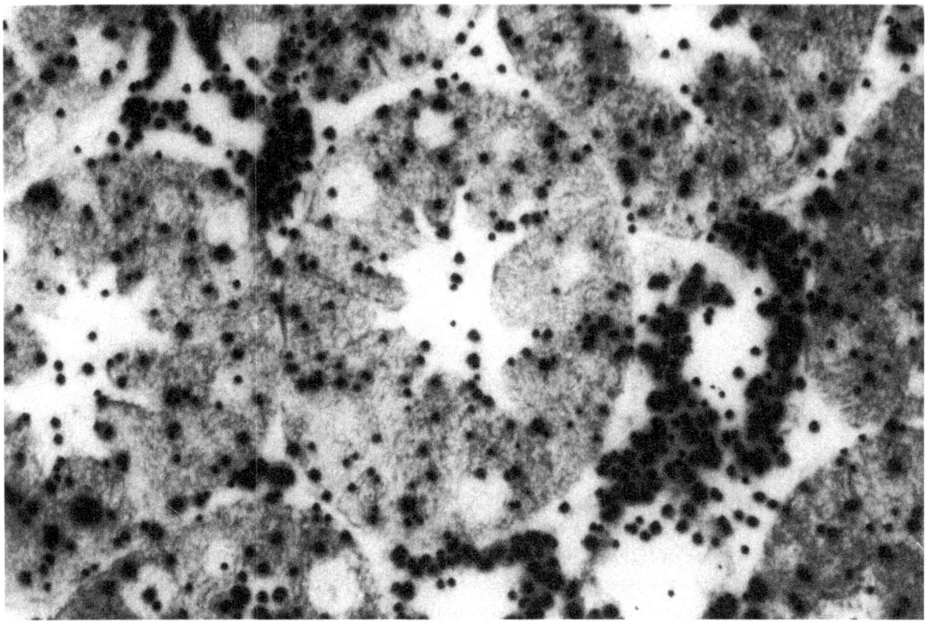

Fig. 1. Autoradiographic visualization of binding of ^{3}H-ouabain to renal tissue. ^{3}H-ouabain, 1 Ci/mmol in a dose of 120 nmol/kg (90 μg/kg), was infused as a bolus into the renal artery of intact dog kidneys in situ. To avoid vasoconstriction a continuous infusion of acetycholine 4 μg/min was given simultaneously. After the natriuretic response was well established the kidney was removed and processed as described by Sejersted et al. (24). A proximal tubule is seen in the center and a distal tubule with high grain density is seen to the right. (Reproduced with courtesy of Anne Nicolaysen.)

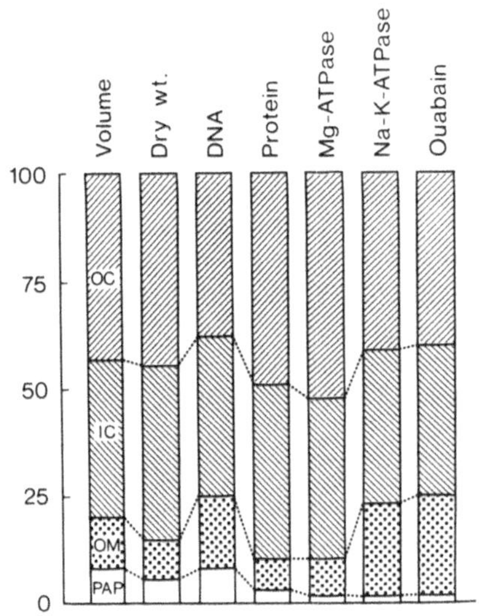

Fig. 2. Relative distribution of volume, dry wt., DNA-content, protein content, Na,K-ATPase activity and ouabain binding sites between different regions of the kidney. (Adapted from 19).

The large amount of ouabain binding sites in kidney tissue and the extensive distribution along all major nephron segments would indicate that the drug can cause extensive inhibition of tubular reabsorption. However, this is not the case. It appears that the dose response curve for inhibition of sodium reabsorption with ouabain is very different in MAL and PT. There is now reason to conclude that sodium reabsorption is not reduced in PT until more than 80 % of the Na,K-ATPase activity is extinguished. On the other hand in the MAL and DT a close and linear relationship between inhibition of Na,K-ATPase activity and reduction of sodium reabsorption can be observed at high glomerular filtration rates.

Comparison of ouabain and ethacrynic acid

One approach taken by us to identify the functional site of action for ouabain in the nephron has been comparison with the loop diuretic ethacrynic acid. In Fig. 3 is shown the response to bolus injections of these drugs. Details of the experimental procedure are given by Sejersted et al. (20).

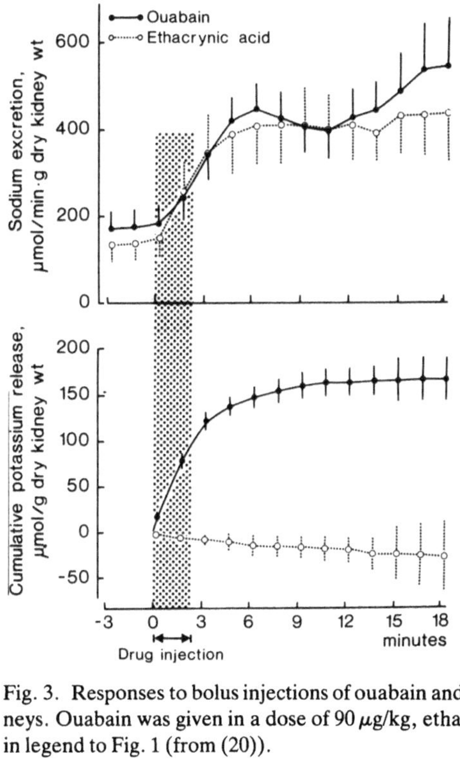

Fig. 3. Responses to bolus injections of ouabain and ethacrynic acid into the renal artery of intact dog kidneys. Ouabain was given in a dose of 90 μg/kg, ethacrynic acid in a dose of 3 mg/kg. Procedure described in legend to Fig. 1 (from (20)).

Infusion of ouabain caused a rapid rise in sodium excretion in the course of 3–4 min. The Na,K-ATPase activity was reduced by 80 % both in cortex and outer medulla. Concomitantly, the plasma potassium concentration rose transiently in the renal venous blood. Thus, the natriuretic response was parallelled by a loss of cellular potassium strictly to the venous blood. This of course, functionally localizes the Na,K-ATPase to the peritubular membrane. Thus the grains seen in Fig. 1 are localized over basolateral plasma membranes. In PT there is extensive interdigitation of neighbouring cells, in MAL and DT there is extensive invaginations of the basolateral membrane explaining why the grains are distributed over most of the tubular wall area.

The response to ethacrynic acid is also shown in Fig. 3. The time course of the natriuretic response is indistinguishable from that of ouabain. However, Na,K-ATPase activity was not reduced in these kidneys, and the renal tissue did not lose potassium as seen in the lower panel of Fig. 3.

It seems well established today that ethacrynic acid inhibits coupled Na, K and Cl transport across the luminal membrane in the MAL and DT. The net effect on the intact kidney is a pronounced and almost equimolar increase in sodium and chloride excretion (25). In addition, potassium excretion is stimulated. By the virtue of its specific action localized to one segment of the nephron – the diluting segment – ethacrynic acid is a very useful tool for the study of tubular function.

In two studies the effect of ethacrynic acid and ouabain on renal metabolic rates and sodium reabsorption were compared (10, 21). It was concluded that independent of the sequence of administration of these drugs no additive effects could be observed (Tables 1 and 2). Both drugs reduced sodium reabsorption by 30–40 %, and local metabolic rates fell by 30–40 % in the cortex and by 50–60 % in the outer medulla. The pattern of reduction in metabolic rates is compatible with an effect on the MAL and DT.

Table 1. Effects of ouabain and addition of ethacrynic acid on sodium reabsorption and heat production in dog kidneys.

	GFR ml/min	Sodium reabsorption mmol/min	Local heat production cal/cm^3 · min Cortex	Outer medulla
Control	45 ± 4	6.0 ± 0.6	0.33 ± 0.04	0.38 ± 0.06
Ouabain	39 ± 3	3.9 ± 0.4	0.19 ± 0.04	0.19 ± 0.04
Ouabain + ethacrynic acid	39 ± 1	3.8 ± 0.3	0.19 ± 0.04	0.20 ± 0.04

Adapted from Lie et al. (10). GFR and sodium reabsorption (n = 13) and local heat production (n = 5). Ouabain was given in a dose of 90 μg/kg as described in legend to Fig. 1. Ethacrynic acid was given i. v. in a dose of 3 mg/kg followed by a sustained infusion of 1.5 mg/kg · h.

Table 2. Effects of ouabain on dog kidneys already exposed to ethacrynic acid and chlorothiazide (n = 4).

	GFR ml/min	Sodium reabsorption mmol/min	Local heat production cal/cm^3 · min Cortex	Outer medulla
Control	35 ± 3	4.8 ± 0.5	0.30 ± 0.06	0.30 ± 0.03
Ethacrynic acid + chloroth.	32 ± 6	2.8 ± 0.3	0.21 ± 0.04	0.12 ± 0.02
+ ouabain	30 ± 4	2.6 ± 0.2	0.19 ± 0.02	0.09 ± 0.02

Adapted from Sejersted et al. (21). Ethacrynic acid was given i. v. and ouabain was infused into the renal artery, both drugs in doses given in Table 1. Chlorothiazide was infused i. v. in a dose of 1 mg/kg followed by 5 mg/kg · h.

Table 3. Effects of digitalis on diluting and concentrating capacity in dog kidneys.

	Control	Digitalis	Expected after digitalis
Water Diuresis (Sejersted, unpublished)			
C_{Na}	1.9	15.0	
C_{osm}	4.3	16.0	
C_{H_2O}	12.9	6.2	>15
Antidiuresis (Martinez-Maldonado et al. (12))			
C_{Na}	1.9	9.3	
C_{osm}	4.3	14.3	
T^cH_2O	2.6	3.5	>6

Clearance values (c) in ml/min. Changes in C_{H_2O} were examined after infusion of ouabain (90 μg/kg) into the renal artery of awake dogs (n = 2) and changes in T^cH_2O after infusion of digoxin (0.8 mg) into the renal artery of anesthetized dogs (n = 5).

It is well known that ethacrynic acid reduces both the concentrating and diluting capacity of the kidney (2, 4). It is less well known that digitalis has the same effect. After injection of ouabain or digoxin into the renal artery both free water clearance ($C_{H_2}O$) and negative free water clearance (T^CH_2O) fail to reach the level expected at the obtained osmolal clearance (Table 3). Again this localizes inhibition of sodium reabsorption by digitalis to the diluting segment. However, it does not preclude an effect of the drug on proximal tubular reabsorption. The lack of an additive effect to ethacrynic acid in spite of extensive Na,K-ATPase inhibition is more compelling in that regard, indicating that ouabain as well as ethacrynic acid specifically inhibits transcellular transport of NaCl in the diluting segment.

Another line of evidence linking inhibition of Na,K-ATPase strictly to transcellular transport of sodium stems from energetic considerations. When reduction in sodium reabsorption is compared to reduction in renal oxygen consumption after injection of ouabain a $\Delta Na/\Delta O_2$ ratio of 18 can be calculated (Fig. 4). Converting reduction in oxygen consump-

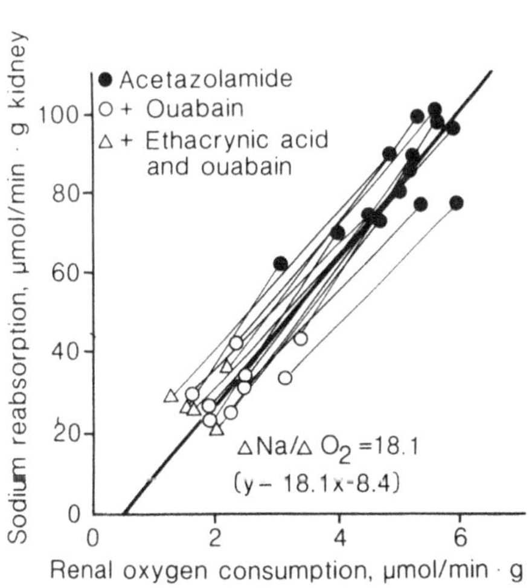

Fig. 4. Comparison of reduction in sodium reabsorption and renal oxygen consumption caused by injection of ouabain 90 μg/kg as described in legend to Fig. 1 (from (25)).

tion to reduction in ATP turnover gives a Na:ATP ratio of 3:1. This is of course equal to the normal stoichiometry of the Na,K-ATPase. Hence in intact kidneys the high energy consumption of digitalis-sensitive sodium reabsorption is compatible with an exclusive transcellular transport of sodium. This is probably the most expensive way of sodium transport across an epithelium. The overall relationship between sodium reabsorption and oxygen consumption of the intact kidney is significantly higher than 18 (8). Thus the ouabain-*insensitive* reabsorption in the nephron must be quite energy cheap. In line with this the energy cost of sodium reabsorption in PT has been estimated to have a $\Delta Na/\Delta O_2$ ratio close to 45 (13).

Digitalis does not affect sodium reabsorption in the proximal tubules of intact kidneys

How can the above energetic considerations and the lack of effect of ouabain on PT be reconciled with extensive inhibition of Na,K-ATPase activity also in this part of the nephron?

Several investigators today seem to agree that probably as much as ⅔ of the sodium reabsorbed in PT bypasses the cells and is channelled through the tight junction and intercellular clefts (3, 7, 18). Possibly, the only sodium going through the cells is what is required to accompany bicarbonate. Reabsorption of $NaHCO_3$ may proceed by the scheme outlined in Fig. 5.

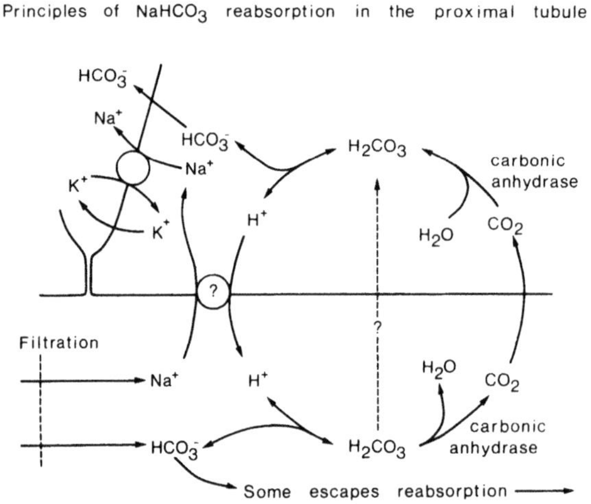

Fig. 5. Schematic diagram of main steps in the reabsorption of $NaHCO_3$ in the proximal tubules (from (22)).

Reabsorption of fluid, sodium and chloride in PT is closely linked to the reabsorption of bicarbonate. It has not been possible to dissociate bicarbonate reabsorption from reabsorption of fluid and sodium chloride except during an osmotic diuresis. So far no studies have revealed a significant effect of ouabain on bicarbonate reabsorption in intact kidneys (5, 23). Hence we may conclude that transport of $NaHCO_3$ through the cells may proceed in the face of an inhibition of 80% of the Na,K-ATPase in PT. The only way this can be accomplished is by means of a compensatory rise in intracellular sodium, stimulating remaining intact Na,K-ATPase sufficiently to maintain the reabsorption. It also implies that $NaHCO_3$ reabsorption is quite independent of the gradient for sodium across the luminal membrane. The explanation requires first that intracellular sodium activity is normally low in PT as in other cells, which has recently been shown in the Necturus kidney (11). Secondly, one would expect transport processes dependent on the sodium gradient and/or the membrane potential to suffer. This has yet to be shown.

Why is sodium reabsorption in the thick ascending limb of Henle's loop so sensitive to digitalis?

Extension of the reasoning above to the MAL and DT leads to the conclusion that a rise in intracellular sodium in these tubular cells after exposure to digitalis fails to stimulate remaining intact Na,K-pumps. The reason may be that intracellular sodium is already so high that the Na,K-ATPase is maximally stimulated.

It was recognized quite early that the natriuretic response to ouabain was most pronounced at high urine flows (21). The effect was difficult to detect in hydropenic kidneys

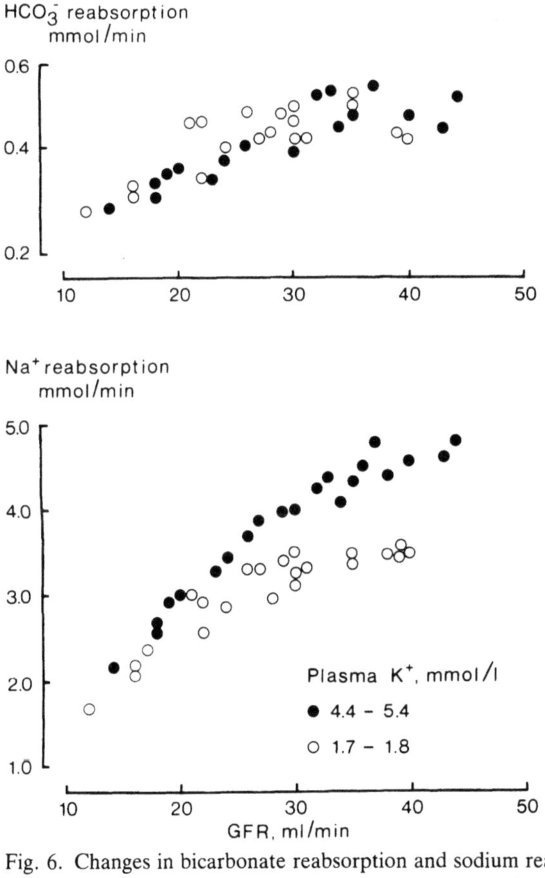

Fig. 6. Changes in bicarbonate reabsorption and sodium reabsorption during manipulation of GFR by means of an aortic clamp. Acetazolamide had been given in a dose of 100 mg/kg i. v. Hypokalemia was induced as described by Monclair (16) (from (22)).

(12). It is well known that a rise in tubular flow rate stimulates reabsorption of sodium in MAL and increases outer medullary metabolic rates (15). This effect reaches a maximum at sodium excretion rates close to 0.5 mmol/min in dog kidneys. It has therefore been proposed that intracellular sodium activity in the MAL rises along with a rising luminal sodium concentration as tubular flow is increased. Hence at low flow rates the effect of ouabain would be offset by a rise of intracellular sodium within a concentration range where the Na,K-ATPase still can be stimulated. At higher flow rates the sodium site of the enzyme approaches saturation and no compensation can occur when ouabain is given.

There are no direct data available to substantiate this hypothesis. However, we have performed experiments with hypokalemia. Fig. 6 shows the results of varying tubular flow rate by means of a clamp on the renal artery. These dogs had received acetazolamide in a dose of 100 mg/kg i. v. which inhibits a large fraction of the reabsorption in PT and renders the remaining reabsorption insensitive to changes in glomerular filtration rate (disrupts the glomerulo-tubular balance) (7). The bicarbonate reabsorption in the upper panel therefore remains constant over a wide range of GFR and does not change when plasma potassium is acutely lowered (for description of methods see (16)). The sodium reabsorption shown in the lower panel mainly takes place in MAL and DT. There is a clear flow-dependency during normokalemia, and saturation seems to occur at high tubular flow rates. After lowering of plasma potassium to 1.7–1.8 mM, reducing Na,K-ATPase activity by approximately 40 % (16), the flow dependency is diminished. Hence reabsorption of sodium at high tubular flow is reduced by 35 %, whereas no significant reduction can be observed at low tubular flows. By analogy we may propose that the effect of digitalis in the MAL and DT is most pronounced at high tubular flows. Even small doses inhibiting only a small fraction of the Na,K-ATPase activity might hence be natriuretic under these cirumstances. It is therefore likely that in the patient the rise in GFR and inhibition of renal Na,K-ATPase works in concert during digitalization promoting excretion of the excess body fluid. When GFR goes back to normal the natriuretic effect of digitalis is no longer detectable.

The inhibition of renal Na,K-ATPase activity also affects renal handling of potassium. It is clear that both reabsorption and secretion of potassium are inhibited by this drug (20, 23).

Conclusions

1) Digitalis binds to the Na,K-ATPase along the entire nephron after in vivo administration.
2) Sodium and chloride reabsorption are specifically inhibited in the thick ascending limb of Henle's loop and distal cortical tubules.
3) Bicarbonate reabsorption and hence proximal tubular reabsorption are unaffected despite localization of 50 % of the digitalis binding sites to this part of the nephron.
4) The natriuretic effect of digitalis is prominent at high urine flows, but almost completely offset when glomerular filtration rate is low.
5) Both secretion and reabsorption of potassium are also inhibited by digitalis.

References

1. Bonting SL, Simon KA, Hawkins NM (1961) Studies on sodium-potassium-activated adenosinetriphosphatase. I. Quantitative distribution in several tissues of the cat. Arch Biochem Biophys 95: 416–423
2. Early LE, Friedler RM (1964) Renal tubular effects of ethacrynic acid. J Clin Invest 43: 1495–1506
3. Frømter E, Rumrich G, Ullrich KS (1973) Phenomenological description of Na^+, Cl^- and HCO_3^- absorption from proximal tubules of the rat kidney. Pflügers Arch 343: 189–220
4. Goldberg M (1973) The renal physiology of diuretics. In: Orloff J, Berliner RW (eds) Handbook of Physiology, Section 8: Renal Physiology. American Physiological Society, Washington DC

5. Hartmann A, Holdaas H, Steen PA, Kiil F (1984) Evidence for bicarbonate-dependent lithium reabsorption in dog kidneys. Acta Physiol Scand 120: 257–264
6. Holdaas H, Langård Ø, Eide I, Kiil F (1981) Mechanism of renin release during renal nerve stimulation in dogs. Scand J Clin Lab Invest 41: 617–625
7. Kiil F (1982) Mechanism of glomerulotubular balance: The whole kidney approach. Renal Physiol 5: 209–211
8. Kjekshus J, Aukland K, Kiil F (1969) Oxygen cost of sodium reabsorption in proximal and distal parts of the nephron. Scand J Clin Lab Invest 23: 307–316
9. Langård Ø, Holdaas H, Eide I, Kiil F (1981) Conditions for humoral adrenoceptor stimulation of renin release in anaesthetized dogs. Scand J Clin Lab Invest 41: 527–534
10. Lie M, Sejersted OM, Ræder M, Kiil F (1974) Comparison of renal responses to ouabain and ethacrynic acid. Am J Physiol 226: 1221–1226
11. Lorenzen M, Lee CO, Windhager EE (1984) Cytosolic Ca^{2+} and Na^{+} activities in perfused proximal tubules of *Necturus* kidney. Am J Physiol 247: F 93–F 102
12. Martinez-Maldonado M, Allen JC, Eknoyan G, Suki W, Schwartz A (1969) Renal concentrating mechanism: Possible role for sodium-potassium activated adenosine triphosphatase. Science 165: 807–808
13. Mathisen Ø, Monclair T, Kiil F (1980) Oxygen requirement of bicarbonate-dependent sodium reabsorption in the dog kidney. Am J Physiol 238: F 175–F 180
14. el Mernissi G, Doucet A (1984) Quantitation of [^{3}H]-ouabain binding and turnover of Na,K-ATPase along the rabbit nephron. Am J Physiol 247: F 158–F 167
15. Monclair T, Mathisen Ø, Kiil F (1978) Characteristics of transcellular NaCl reabsorption in the kidney. Scand J Clin Lab Invest 38: 615–626
16. Monclair T, Sejersted OM, Kiil F (1980) Influence of plasma potassium concentration on the capacity for sodium reabsorption in the diluting segment of the kidney. Scand J Clin Lab Invest 40: 27–36
17. Ræder M, Omvik P Jr, Kiil F (1974) Effect of acute hypertension on the natriuretic response to saline loading. Am J Physiol 226: 989–995
18. Rector FC (1983) Sodium, bicarbonate and chloride absorption by the proximal tubule. Am J Physiol 244: F 461–F 471
19. Sejersted OM (1979) Estimation of ouabain specifically bound to dog renal (Na^{+} + K^{+})-ATPase in vivo. Biochim Biophys Acta 586: 33–340
20. Sejersted OM, Holdaas H, Monclair T (1978) Functional differences of ouabain and ethacrynic acid on renal potassium metabolism in dogs. Scand J Clin Lab Invest 38: 603–614
21. Sejersted OM, Lie M, Kiil F (1971) Effect of ouabain on metabolic rate in renal cortex and medulla. Am J Physiol 220: 1488–1493
22. Sejersted OM, Mathisen Ø, Monclair T, Steen PA, Kiil F (1981) Carbonic anhydrase and Na,K-ATPase in the regulation of tubular sodium transport. In: Takacs L (ed) Adv Physiol Sci Vol 11. Kidney and Body Fluids, Akadémiai Kiadó, Budapest, pp 289–298
23. Sejersted OM, Monclair T, Mathisen Ø, Hartmann A, Kiil F (1985) Dependency of renal potassium excretion on Na,K-ATPase transport rate. Acta Physiol Scand 123: 9–19
24. Sejersted OM, Nicolaysen A, Monclair T, Nicolaysen G (1985) Distribution of ouabain-binding sites along the dog nephron, Acta Physiol Scand 125: 699–710
25. Sejersted OM, Steen PA, Kiil F (1982) Energetics of sodium reabsorption sensitive to ethacrynic acid and ouabain. Am J Physiol 242: F 254–F 260
26. Withering W (1785) An Account of the Foxglove and some of its Medical Uses: with Practical Remarks on Dropsy and Other Diseases. GGJ and J Robinson, London

Author's address:

Prof. Dr. Ole Sejersted, Institute of Muscle Physiology, P. O. Box 8149 Dep, 0033 Oslo 1 (Norway)

Ouabain-induced efflux of catecholamines and metabolites from blood vessels of normotensive and hypertensive dogs*

D. Branco and W. Osswald

Laboratório de Farmacologia, Faculdade de Medicina, Porto (Portugal)

Summary

Segments of mesenteric arteries (main trunk and branches) and of saphenous veins obtained from anaesthetized dogs were incubated during 15, 30 and 60 min in oxygenated Krebs-Henseleit medium at 37 °C, either in the absence or in the presence of ouabain (1 μmol, 10 μmol or 1 mmol/liter). Endogenous noradrenaline (NA) and dihydroxyphenylglycol (DOPEG) were determined in the tissue and in the incubation medium by HPLC-EC.

In normotensive animals, ouabain (10 μmol and 1 mmol/l) increased the spontaneous efflux of NA in all vessels studied and markedly reduced the formation of DOPEG. In contrast, the lowest concentration of ouabain (1 μmol/l) had opposite effects.

The vessels of perinephritic hypertensive dogs (obtained 1 month after surgery) showed some significant differences from the vessels of normal dogs. Ouabain (1 mmol/l) had effects on the mesenteric arteries which were similar to those described above although less marked, but in the saphenous vein no such effects were observed.

It is concluded that ouabain-sensitive mechanisms play an important role in regulating the spontaneous efflux of NA and the formation of DOPEG in the adrenergic terminals of canine blood vessels; and that vessels of hypertensive animals do differ, in this aspect, from vessels of control animals.

There is ample evidence to substantiate an increase in noradrenaline outflow caused by cardenolides in incubation or perfusion experiments with adrenergically innervated tissues (review by 7). Since, as is well known, glycosides cause vasoconstriction, which is attenuated or blocked by adrenergic blocking drugs (5, 6), it is understandable that a causal connection between both phenomena has been assumed and that much attention has been paid to the mechanism(s) of action involved. Lorenz et al. (6) attributed the contraction of the dog saphenous vein caused by digitalis to displacement of noradrenaline from neuronal stores and discuss a series of possible mechanisms of action leading to an increased release of the neurotransmitter.

However, this research work has been conducted with ^{3}H-labelled noradrenaline of various types with different influences on the intraneuronal deamination, due to isotope effects (9). On the other hand, the effects of cardenolides on dopamine content and release have not yet been studied, as far as we know.

It was therefore of interest to study the spontaneous efflux of endogenous noradrenaline and its major intraneuronally formed metabolite, dihydroxyphenylglycol (DOPEG) from blood vessels as influenced by the cardenolide, ouabain. Moreover, the influence of ouabain

* Supported by a grant from Instituto Nacional de Investigação Científa (PmC1)

on the dopamine content of the blood vessels was also studied. A sensitive and specific method for the determination of catechols, as represented by the technique of high pressure liquid chromatography with electrochemical detection (HPLC-ED), is now available and in the adaptation used in this study allows for the simultaneous determination of the three catechols in tissue and fluid, thus enabling us to study, for the first time, the influence of ouabain on the efflux and content of endogenous noradrenaline, its precursor molecule dopamine and its metabolite DOPEG.

We choose to use the lateral saphenous vein and the main trunk and proximal branches of the mesenteric artery of the dog as vascular tissues, since these vessels present structural and functional differences and have been repeatedly used in our laboratory. A comparison between vessels from normotensive and hypertensive dogs was also attempted, in order to find out if there were differences in the spontaneous outflow and in its susceptibility to ouabain.

Methods

Blood vessels were obtained from mongrel dogs (12–18 kg) anesthetized with sodium pentobarbital (30 mg/kg, i.v. injected in the forelimb).

The normotensive animals had values of 122.4 ± 8.2 (n = 14) mm Hg for the diastolic pressure and 178.5 ± 15.3 (n = 14) mm Hg for the systolic pressure.

Dogs with perinephritic hypertension were used four weeks after they were made hypertensive by nephrectomy of the right kidney and wrapping of the remaining kidney with cellophane (2). The blood pressure values at this moment were of 146.4 ± 9.7 mm Hg (n = 9; diastolic pressure) and 215.8 ± 12.4 mm Hg (n = 9; systolic pressure). Strips of the lateral saphenous vein (60–80 mg in weight) and segments of the main trunk (25–40 mg) and proximal branches (70–100 mg) of the anterior mesenteric artery were incubated during 15, 30 or 60 min in 3 ml of Krebs solution (final composition in mmol/l: NaCl 118, KCl 4.7, $CaCl_2$ 2.5, $NaHPO_4$ 1.2, $MgCl_2$ 1.2, $NaHCO_3$ 25, d-glucose 11 and Na_2EDTA 0.04).

In control experiments no drug was present. In the other series of experiments ouabain was added to the Krebs solution in order to obtain a concentration of 1 mmol/l, 10 μmol/l or 1 μmol/l, respectively. After 15, 30 or 60 min of incubation period the tissue was rapidly removed, blotted, weighed and placed in 3 ml of perchloric acid 0.1 M (containing 2.5 mmol/l Na_2EDTA and 10 mmol/l Na_2SO_3). Then it was extracted during 24 hours at −20°C. Aliquots of both the acid tissue extract and the incubation fluid (also acidified with perchloric acid, final concentration 0.1 M) were used for determination of noradrenaline (NA), dopamine (DA) and dihydroxyphenylglycol (DOPEG) by high pressure liquid chromatography with electrochemical detection (HPLC-ED) after alumina adsorption, as described by Branco et al (1). Recoveries were 80.2 ± 3.0% (n = 12) for NA, 76.4 ± 3.2% (n = 12) for DA and 89.5 ± 3.5% (n = 12) for DOPEG.

Drugs used: ouabain octahydrate (Sigma) and sodium pentobarbital (Abbott).

Statistical analysis

The number (n) reported refers to the number of dogs used. The results are expressed as mean ± S.E.M. of n experiments.

Results

1. Content and spontaneous outflow in control experiments. During the 60 min incubation period, the venous and arterial tissues studied showed some important differences in their behaviour. Thus, there was no significant alteration in the noradrenaline content of the *saphenous vein* during this period and relatively low amounts of the amine appeared in the incubation fluid. DOPEG was detected in the medium in appreciable amounts, which increased with duration of incubation. The dopamine content decreased by about 50%, but only small amounts appeared in the medium.

The main trunk and proximal branches of the *mesenteric artery,* in contrast to the saphenous vein, experienced a large loss in noradrenaline content (up to almost one half) during the incubation and a much larger amount of both intact amine and its metabolite, DOPEG, appeared in the incubation medium. However, dopamine content decreased during the one hour period in the same manner as observed in the vein. The same pattern was observed to occur in the trunk and branches of the artery, although there were quantitative differences. The endogenous catecholamine content of the control vessels from

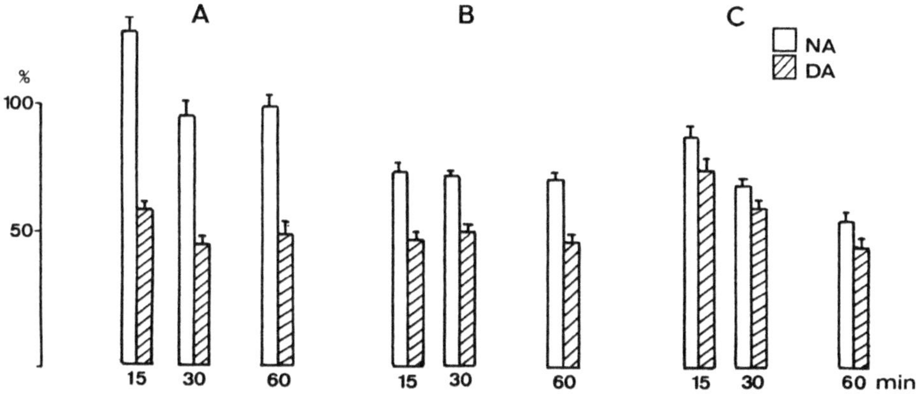

Fig. 1. Noradrenaline and dopamine content of the tissue during 15, 30 and 60 min of incubation in control experiments. Data shown expressed as percent of control values. A–saphenous vein; B–main trunk and C-proximal branches of mesenteric artery.

Table 1. Outflow of NA, DA and DOPEG (ng.g^{-1}) in control experiments (n = 6) during 15, 30 and 60 min of incubation period

		15	30	60 min
	Saphenous vein	60 ± 10	40 ± 10	30 ± 9
NA	Main trunk of mesenteric artery	184 ± 40	293 ± 50	328 ± 45
	Proximal branches of mesenteric artery	228 ± 70	376 ± 80	420 ± 100
	Saphenous vein	4 ± 1	5 ± 1	4 ± 1
DA	Main trunk of mesenteric artery	6 ± 1	4 ± 1	4 ± 1
	Proximal branches of mesenteric artery	10 ± 2	11 ± 2	7 ± 1
	Saphenous vein	16 ± 2	16 ± 3	44 ± 6
DOPEG	Main trunk of mesenteric artery	75 ± 10	95 ± 11	165 ± 20
	Proximal branches of mesenteric artery	95 ± 12	158 ± 22	182 ± 22

normotensive dogs was: saphenous vein NA 2.49 ± 0,11 μg/g (n = 16) and DA 0.050 ± 0.004 μg/g (n = 12); main trunk of mesenteric artery NA 2.12 ± 0.19 μg/g (n = 16) and DA 0.072 ± 0.009 μg/g (n = 12); proximal branches of mesenteric artery NA 4.55 ± 0.15 μg/g (n = 16) and DA 0.135 ± 0.011 μg/g (n = 12).

The data reported above are shown in Fig. 1 and Table 1.

2. *Influence of ouabain on content and spontaneous outflow.* Ouabain effects were concentration-dependent. In the highest concentration used (1 mmol/l), ouabain caused a pronounced loss of noradrenaline content of the *saphenous vein,* a total blockade of DOPEG formation and an increase in dopamine content (as opposed to the decrease observed in the control experiments referred to above). Ouabain 10 and 1 μmol/l did not alter the noradrenaline content, but reduced DOPEG and partially antagonized the decline in dopamine content, especially with the longer incubation period.

On the *arterial tissue,* ouabain 1 mmol/l somewhat intensified the spontaneous decrease of noradrenaline content and caused a marked rise in noradrenaline outflow; DOPEG was drastically reduced. Ouabain 10 μmol/l had similar, although less marked effects, but the lowest concentration (1 μmol/l) had opposite effects, preventing almost completely the spontaneous loss of noradrenaline and causing DOPEG to increase (in the branches of the artery; not in its trunk, where DOPEG was only slightly reduced). Concerning dopamine content, in the trunk ouabain in all concentrations caused it to increase, whereas in the branches of the mesenteric artery no important differences in regard to controls were found. Figure 2 and Table 2 illustrate these points.

3. *The behaviour of vessels from hypertensive dogs.* Vessels obtained from hypertensive dogs showed some differences from control vessels. Their noradrenaline content was lower (especially in the trunk of the artery) and their ability to produce DOPEG was depressed (again, the mesenteric trunk was the vascular tissue more affected). Dopamine content did not differ significantly from controls. Generally speaking, ouabain (used only in the concentration of 1 mmol/l) had effects which were similar to those shown in the control vessels – blockade of the loss in dopamine content, increase in noradrenaline outflow, reduction in DOPEG formation. However, it was evident that these effects were less pronounced and only evident when the phenomenon under study had a magnitude which approached that observed in normotensive vessels. For example, where DOPEG formation was very much depressed (mesenteric artery trunk, saphenous vein) ouabain had almost no effect, but it reduced DOPEG formation in the proximal branches of the artery (which showed a relatively well conserved capacity to form DOPEG, only slightly smaller than that of the control vessels). The endogenous noradrenaline and dopamine content of the vessels from hypertensive dogs was: saphenous vein NA 1.85 ± 0.34 μg/g (n = 3) and DA 0.042 ± 0.005 μg/g (n = 3); main trunk of mesenteric artery NA 1.49 ± 0.19 μg/g (n = 3) and DA 0.061 ± 0.009 μg/g (n = 3); and proximal branches of mesenteric artery NA 3.66 ± 0.42 μg/g (n = 3) and DA 0.13 ± 0.02 μg/g (n = 3).

The results described in this section are shown in Fig. 3 and Table 3.

4. *The noradrenaline/DOPEG ratio.* The spontaneous efflux contains more noradrenaline than DOPEG under control conditions. As shown in Table 4A, this ratio is drastically augmented by ouabain in a concentration-dependent manner (for the concentration of 10 μmol/l and 1 mmol/l; the lowest concentration –1 μmol/l – was without effect or tended to reduce this ratio). The vessels from hypertensive dogs, however, showed a marked difference (Table 4B), the ratios observed in the saphenous vein and the arterial trunk being much higher than those observed in the normotensive vessels; ouabain was devoid of significant

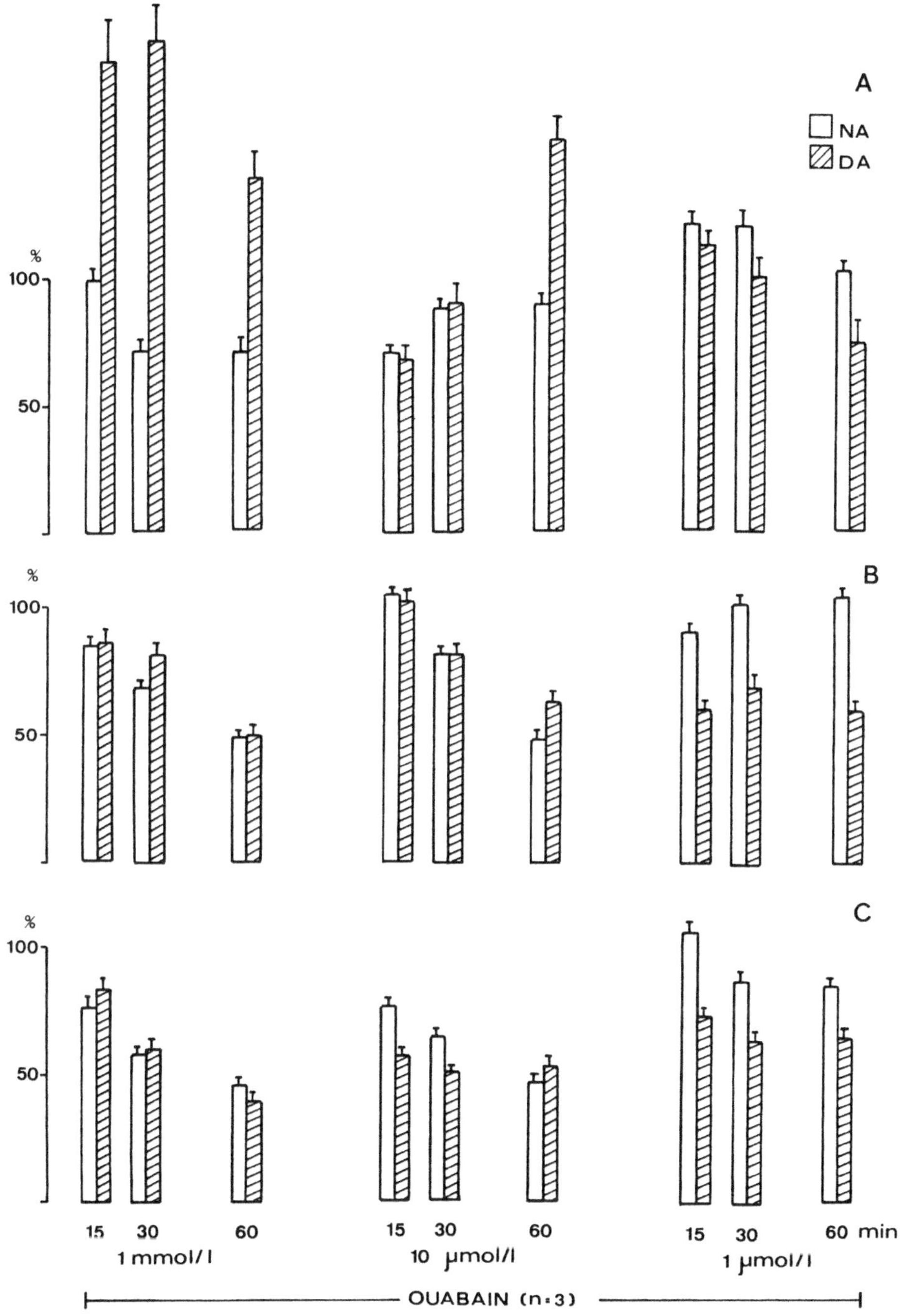

Fig. 2. Noradrenaline and dopamine content of the tissues during 15, 30 and 60 min of incubation in the presence of the ouabain. A–saphenous vein; B–main trunk and C–proximal branches of mesenteric artery.

Table 2. Outflow of NA, DA and DOPEG (ng.g^{-1}) in ouabain experiment (n = 3)

		1 mmol/l			10 μmol/l			1 μmol/l		
		15	30	60	15	30	60	15	30	60 min
NA	Saphenous vein	160 ± 18	290 ± 31	400 ± 41	28 ± 3	82 ± 9	137 ± 13	26 ± 3	33 ± 3	33 ± 4
	Main trunk of mesenteric artery	268 ± 40	494 ± 52	887 ± 110	107 ± 18	198 ± 25	269 ± 38	116 ± 21	153 ± 22	95 ± 15
	Proximal branches of mesenteric artery	950 ± 101	1370 ± 119	1620 ± 126	237 ± 28	753 ± 94	938 ± 105	178 ± 20	242 ± 29	239 ± 32
DA	Saphenous vein	8 ± 2	9 ± 2	4 ± 1	3 ± 1	3 ± 1	4 ± 1	3 ± 1	3 ± 1	3 ± 1
	Main trunk of mesenteric artery	7 ± 1	5 ± 1	4 ± 1	6 ± 1	7 ± 1	8 ± 2	4 ± 1	5 ± 1	4 ± 1
	Proximal branches of mesenteric artery	8 ± 2	13 ± 4	9 ± 2	5 ± 1	13 ± 3	19 ± 4	6 ± 1	6 ± 1	7 ± 2
DOPEG	Saphenous vein	n.d.*	n.d.	n.d.	11 ± 3	28 ± 5	9 ± 2	10 ± 2	14 ± 3	17 ± 3
	Main trunk of mesenteric artery	n.d.	n.d.	n.d.	19 ± 3	29 ± 4	12 ± 2	37 ± 5	77 ± 9	66 ± 8
	Proximal branches of mesenteric artery	36 ± 4	48 ± 5	37 ± 4	42 ± 4	114 ± 10	208 ± 21	37 ± 4	107 ± 12	273 ± 29

* n.d.: < 1 ng/g

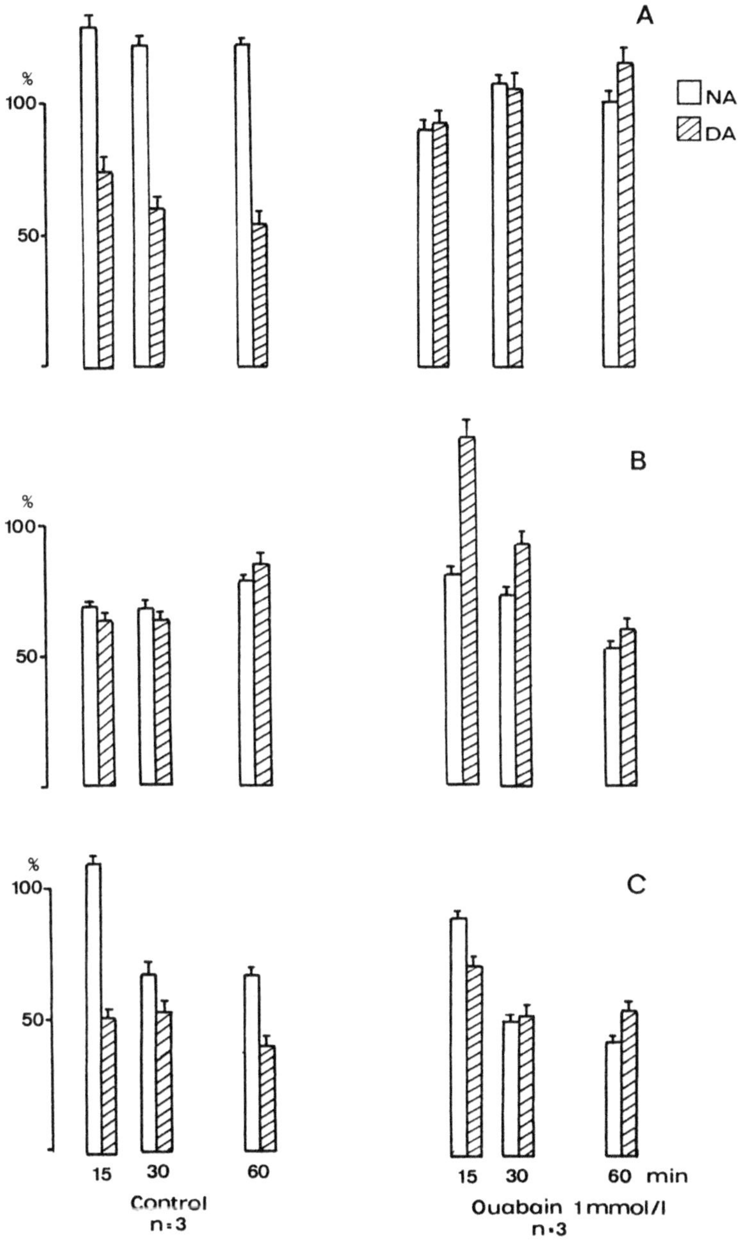

Fig. 3. Noradrenaline and dopamine content of the tissues from hypertensive dogs during 15, 30 and 60 min of incubation period in control and ouabain experiments. A–saphenous vein; B–main trunk and C–proximal branches of mesenteric artery.

Table 3. Outflow of NA, DA and DOPEG (ng.g^{-1}) from hypertensive dogs in control and ouabain experiments (n = 3)

		Control			Ouabain 1 mmol/l		
		15	30	60	15	30	60 min
NA	Saphenous vein	22 ± 6	32 ± 7	19 ± 9	26 ± 10	85 ± 20	101 ± 20
	Main trunk of mesenteric artery	83 ± 10	97 ± 11	95 ± 10	140 ± 15	220 ± 25	190 ± 20
	Proximal branches of mesenteric artery	176 ± 12	294 ± 26	305 ± 36	418 ± 170	601 ± 150	967 ± 180
DA	Saphenous vein	3 ± 1	3 ± 1	4 ± 1	3 ± 1	3 ± 1	4 ± 1
	Main trunk of mesenteric artery	5 ± 1	4 ± 1	3 ± 1	13 ± 2	8 ± 2	11 ± 2
	Proximal branches of mesenteric artery	3 ± 1	3 ± 1	3 ± 1	14 ± 2	17 ± 2	16 ± 2
DOPEG	Saphenous vein	4 ± 1	4 ± 1	5 ± 1	28 ± 5	10 ± 2	5 ± 2
	Main trunk of mesenteric artery	15 ± 2	17 ± 3	49 ± 6	23 ± 4	23 ± 5	22 ± 5
	Proximal branches of mesenteric artery	51 ± 4	72 ± 8	223 ± 25	55 ± 9	54 ± 6	55 ± 9

Table 4A. The NA/DOPEG ratio (geometric means with 95% confidence limits) from the vessels of normotensive dogs

			Saphenous vein	Main trunk of mesenteric artery	Proximal branches of mesenteric artery
Control		15	4.6 (2.8–4.6)	2.5 (1.9–3.3)	2.5 (2.0–3.1)
		30	1.9 (0.8–4.6)	3.0 (2.5–3.6)	2.8 (2.5–3.1)
		60	0.9 (0.83–0.9)	2.0 (1.5–2.8)	1.4 (1.2–1.7)
Ouabain	1 mmol/l	15	> 160	> 200	18.9 (17.5–20.4)
		30	> 290	> 350	19.4 (17.9–21.0)
		60	> 400	> 700	22.5 (19.5–26.1)
	10 μmol/l	15	2.9 (2.4–3.7)	5.6 (4.9–6.4)	8.6 (4.3–15.8)
		30	3.0 (2.4–3.7)	6.8 (5.7–8.1)	8.3 (7.8–8.8)
		60	16.6 (10.0–27.5)	22.4 (14.1–35.4)	6.3 (4.8–8.3)
	1 μmol/l	15	2.6 (2.4–2.7)	3.2 (2:3–4.5)	1.8 (1.6–1.9)
		30	2.5 (2.3–2.8)	2.0 (1.9–2.2)	1.4 (1.3–1.5)
		60	1.9 (1.7–2.2)	1.5 (1.1–1.9)	0.96 (0.90–0.98)

Table 4B. The NA/DOPEG ratio (geometric means with 95% confidence limits) from the vessels of hypertensive dogs

			Saphenous vein	Main trunk of mesenteric artery	Proximal branches of mesenteric artery
Control		15	6.3 (5.0–8.0)	5.8 (4.9–6.8)	3.8 (3.1–4.7)
		30	6.2 (3.7–10.5)	7.0 (5.7–8.8)	4.0 (3.8–4.3)
		60	4.3 (4.0–4.7)	2.3 (1.8–3.1)	1.4 (1.1–1.8)
Ouabain	1 mmol/l	15	1.47 (0.7–3.0)	6.9 (4.8–10.0)	9.3 (5.2–12.6)
		30	9.1 (6.8–12.0)	10.5 (6.9–16.0)	10.7 (9.4–12.2)
		60	25.5 (11.3–57.6)	9.9 (6.8–14.5)	17.4 (14.3–21.1)

effects in these vessels. The main mesenteric branches, however, did not show any apparent differences from their controls and ouabain increased the ratio in the usual manner.

Discussion

Noradrenaline and dopamine content, spontaneous efflux. Our results give evidence for a marked difference between the saphenous vein and the mesenteric arterial tissue: during a one hour incubation period under "physiological" conditions, the vein maintains its noradrenaline content, whereas the arterial tissue loses almost one half of its original content, as previously reported from this laboratory (3). This different behaviour does not seem to depend on a much more intense noradrenaline synthesis occurring in the vein and compensating for an eventual loss, since the decline in the precursor amine, dopamine, was very similar in arteries and vein and since the efflux of noradrenaline and its neuronally formed metabolite, DOPEG, was much larger in the case of the arteries. Thus, the difference observed appears to be related to a lesser ability of noradrenaline to permeate the neuronal membrane in the vein than in the artery, i.e., the membrane appears to be less leaky in the vein terminals than in those of the artery.

In all vessels studied the spontaneous outflow contained more noradrenaline than DOPEG, i.e. the NA/DOPEG ratio was greater than unity. This differs from what has been reported for a variety of adrenergically innervated tissues (8); however, in all these reports labelled noradrenaline has been used. The difference observed may indicate the inadequacy of exogenous, isotope-marked noradrenaline to reflect the behaviour of endogenous transmitter.

The effects of ouabain. Ouabain 1 mmol/l had very clear-cut effects, causing a decrease in noradrenaline content of the saphenous vein and an augmented loss of noradrenaline in the arterial tissue. The amounts of DOPEG formed were drastically reduced: therefore, the ratio NA/DOPEG experienced a marked increase. These results are in good agreement with those of Lorenz et al. (6) and Stute and Trendelenburg (8) and could easily be explained by a rise in intraneuronal [Na], due to blockade of neuronal membrane ATPase; Stute and Trendelenburg (8) presented evidence showing that the carrier of noradrenaline demands sodium for the transport across the membrane and that measures which result in an absolute or relative increase in $[Na]_i$, like veratridine, ouabain and a low-sodium medium result in an increased outward transport of noradrenaline and a decrease in DOPEG formation (because less noradrenaline is offered to the intraneuronal monoamine oxidase). If this is the explanation, it appears that high concentrations of ouabain are needed to inhibit the membrane ATPase, since 10 μmol/l had only discrete effects and 1 μmol/l showed opposite effects (stimulation of ATPase?).

A novel effect of ouabain consisted in the increase in dopamine content observed for the higher concentrations of ouabain used. This could be due either to a decrease in the rate of synthesis of noradrenaline or to an increase of dopamine formation by tyrosine-β-hydroxylase; our results do not allow us to speculate on the respective probabilities of these mechanisms.

The behaviour of vessels from hypertensive dogs. These vessels showed some important differences from their controls, obtained from normotensive animals. A decreased content in noradrenaline and a lower formation of DOPEG were the most apparent of these differences. For the mesenteric arteries, a former report (4) already stressed these data. We found that ouabain also differed in its action, being practically devoid of the effects shown in the saphenous vein and mesenteric trunk from normotensive animals and discussed above. The conclusion appears to be justified that these vessels from hypertensive animals appear to lose more noradrenaline and form less DOPEG than their normotensive counterparts. The results thus suggest that in the vessels of hypertensive dogs the intraneuronal sodium concentration is higher, due to an impairment of the ouabain-sensitive mechanism (probably membrane ATPase).

In *conclusion,* our data indicate that the outward-transport of noradrenaline in sympathetic nerve terminals of canine blood vessels shows marked differences between venous and arterial tissue; that ouabain facilitates this outward transport, probably by inhibition of membrane ATPase and subsequent increase in $[Na]_i$; and that vessels from hypertensive animals show an impairment of this ouabain-sensitive mechanism.

References

1. Branco D, Teixeira AA, Azevedo I, Osswald W (1984) Structural and functional alterations caused at the extraneuronal level by sympathetic denervation of blood vessels. Naunyn-Schmiedeberg's Arch Pharmacol 326: 302–313
2. Castro-Tavares J, Garrett J, Gonçalves V (1981) Cellophane perinephritic hypertension in the dog. Some biochemical, hemodynamic and pathological characteristics. Blood Vessels 18: 67–74
3. Garrett J, Branco D (1978) Different behaviour of endogenous noradrenaline in two vessels of the dog. Proc. 3rd Meeting on Adrenergic Mechanisms, Porto University, p 69
4. Garrett J, Castro-Tavares J, Branco D (1981) Uptake and metabolism of noradrenaline by blood vessels of perinephritic hypertensive dogs. Blood Vessels 18: 100–109
5. Karaki H, Urakawa N (1977) Possible role of endogenous catecholamines in the contractions induced in rabbit aorta by ouabain, sodium depletion and potassium depletion. Eur J Pharmacol 43: 65–72
6. Lorenz RR, Powis DA, Vanhoutte PM, Shepherd JT (1980) The effects of acetylstrophantidin and ouabain on the sympathetic adrenergic neuroeffector junction in canine vascular smooth muscle. Circ Res 47: 845–854
7. Powis DA (1983) Cardiac glycosides and autonomic neurotransmission. J Auton Pharmacol 3: 127–154
8. Stute N, Trendelenburg U (1984) The outward transport of axoplasmic noradrenaline induced by a rise of the sodium concentration in the adrenergic endings of the rat vas deferens. Naunyn-Schmiedeberg's Arch Pharmacol 327: 124–132
9. Trendelenburg U, Stefano FJE, Grohmann M (1983) The isotope effect of tritium in ^{3}H-noradrenaline. Naunyn-Schmiedeberg's Arch Pharmacol 323: 128–140

Author's address:

Prof. Dr. W. Osswald, Laboratório de Farmacologia, Faculdade de Medicina, 4200 Porto, Portugal

Digoxin-specific antibody treatment of digitalis toxicity: Update

T. Wenger, V. P. Butler Jr., E. Haber and T. W. Smith

Cardiovascular Division Brigham and Women's Hospital, Boston, Massachusetts (U.S.A.)

Summary

Digoxin antibody fragments (Digibind[tm], Wellcome) have now been used in 97 patients with suspected life threatening digoxin or digitoxin toxicity. In most cases therapeutic response has been prompt and complete. There have been no unexpected side effects noted in these very sick patients. Several patients developed worsened heart failure after treatment, which may have been secondary to digitalis withdrawal or may have been caused by changes in concomitant drug therapy, fluid administration, or a combination of these factors. Potassium concentrations usually fell after administration of the antibody fragments, indicating reversal of digitalis induced inactivation of (Na+K+)-ATPase. In some patients, frank hypokalemia developed after rapid reversal of digitalis effects with the antibody fragments.

Patients with renal failure did not differ clinically in their response to treatment. Infants and children have had particularly rapid reversal of toxicity. Three pediatric patients with refractory ventricular fibrillation or bradycardias requiring several hours of cardiopulmonary resuscitation were readily converted back to normal sinus rhythm within minutes of digoxin antibody fragment administration.

Patients usually responded to the antibody fragments within 30 minutes of administration, with complete response usually noted by 3 or 4 hours after administration of a neutralizing dose. Recurrence of toxicity later in the hospital course has not been a problem, except in one patient who ingested a massive amount of digoxin and received a dose of antibody fragments that was insufficient to neutralize the amount ingested.

Introduction

When antibodies to digoxin were originally developed in 1967, it was suggested they could be used to measure digoxin concentrations in body fluids and also to reverse digitalis toxicity in humans (2). Whole antibody reversal of experimentally induced digoxin toxicity was clearly demonstrated in experimental animals by 1971 (3). By the mid-1970s the technology became available to cleave antibodies enzymatically into smaller fragments that maintained their ability to recognize antigens, and advances in protein purification permitted isolation of the active Fab fragments in a form suitable to treat patients. The first clinical use of digoxin-specific antibody fragments for digitalis toxicity was described in 1976 (5). One hundred and fourteen patients have now been treated in the United States with digoxin-specific antibody fragments (Digibind[tm], Wellcome Biotechnology Limited). Previous reports of this trial involving 26 (4) and 63 patients (6) have been published.

Study design

Fab fragments had not previously been used in patients, so a cautious study design was chosen for the first clinical trial of digoxin-specific Fab to treat digitalis toxicity in man. A multicenter investigation now involving 21 medical centers geographically distributed across the United States was established by Smith, Butler and Haber (4). The entire experience with this material in man in the United States has been within the design of this multicenter trial.

The study is unblinded. Inclusion criteria require patients to have potentially life-threatening cardiac rhythm disturbances or hyperkalemia, or both, caused by digoxin or digitoxin intoxication. Initially, in order to enter the study, patients had to fail conventional therapeutic efforts to treat digitalis toxicity. By the early 1980s, the criteria were loosened slightly to permit entry of patients if they were judged highly unlikely to respond to standard measures because of massive intoxication. This facilitated earlier treatment in such cases.

The dosage of Fab was calculated to be equimolar to the amount of digoxin in the patient's body. Estimates of total digoxin load were derived from the medical history or from determinations of serum digoxin concentrations, or both. The dosage calculation was performed in one of several ways. In the case of massive acute ingestion of digoxin tablets, the presumed amount of digoxin ingested was multiplied by 0.8 to give rough estimate of the amount absorbed. This calculation is understood to be somewhat arbitrary since the absolute bioavailability of digoxin when ingested in massive quantity is unknown. In addition, many of these patients received gastric lavage, activated charcoal and other measures designed to decrease the amount of digoxin entering the circulation.

When a steady-state serum digoxin concentration was available (i.e., one obtained at least 6 to 8 hours after ingestion of digoxin) then the calculation of total body burden of digoxin was made by multiplying the serum digoxin concentration times the mean volume of distribution (5.6 l/kg) times the body weight of the individual in kg. When the serum digoxin concentration is given in ng/ml and the volume of distribution and body weight are in liters and kg respectively, the number derived can be divided by 1,000 to give a total body load of digoxin in mg.

For digitoxin, it was assumed that 100% of an ingested dose was absorbed. The calculation based on serum digitoxin concentration utilized a volume of distribution figure of 0.56 l/kg.

Once the total body load of digitalis is estimated, the equimolar dose of digoxin-specific Fab can be derived by multiplying the amount of digoxin in the body by the ratio of the molecular weight of Fab to that of digoxin. This ratio is 50,000 to 781, or approximately 64. Potency tests demonstrate that each 40 mg vial of purified Fab is capable of neutralizing approximately 0.60 mg digoxin. Thus, an alternative scheme for calculating the dose is to divide the total body load of digoxin (in mg) by 0.6 mg to determine the number of vials to administer to the patient.

Results

Patients ranged in age from 1 day to 87 years, with a mean of 50 years. Twenty-nine of the patients receiving Fab fragments had no underlying heart disease. Twelve of these were accidental ingestions and the other 17 were patients who ingested digoxin or digitoxin in suicide attempts. The remaining 85 patients had underlying heart disease. Twenty-five of these individuals ingested digitalis with suicidal intent whereas the other 60 developed life-threatening digitalis intoxication in the course of therapy for their heart disease.

Manifestations of toxicity included second or third degree atrioventricular block in 50% of the patients, hemodynamically significant ventricular tachycardia in 54%, ventricular fibrillation in 32% and hyperkalemia in 36% of the population requiring Fab treatment. Serum digoxin concentrations in this population ranged from as low as 1.2 ng/ml to > 100 ng/ml. The patient with a serum concentration of 1.2 ng/ml turned out not to have digitalis toxicity but instead had chronic stable ventricular ectopy that was misdiagnosed as digitalis toxicity in an emergency room. This patient had no response to the digoxin-specific antibody fragments. All other patients had serum concentrations > 2 ng/ml. The mean serum digoxin concentration was > 12.5 ng/ml in the entire population (this mean was calculated using a stated concentration if the value was reported as greater than that stated concentration). Over 75% of the patients had serum concentrations > 5 ng/ml.

Digoxin-specific Fab doses ranged from 4 to 1,600 mg, with an average dose of 384 mg. Patients ingesting digitalis with suicidal intent or by accident required higher Fab doses than patients with toxicity occurring during efforts to treat their underlying heart disease. Whereas patients with massive ingestions often required at least 480 mg, and sometimes 1000 mg or more, patients with digitalis toxicity resulting from therapeutic difficulties usually required 120 to 400 mg of antibody fragments to reverse the toxicity.

Of the 114 patients treated, data on renal function were available in 80 patients. Of these, only 35% had normal renal function and 65% had abnormal renal function. For those patients with abnormal renal function, the mean blood urea nitrogen was 47 ± 23 mg/dl (± SD) and the creatinine was 2.9 ± 1.7 mg/dl. At least 5 of these patients had renal failure severe enough to require dialysis.

The clinical response in the entire population of 114 patients was similar to that reported previously (Figure 1). We excluded from the analysis of response those patients who failed to meet any reasonable criteria for digitalis toxicity (n = 3), those who received a dose of

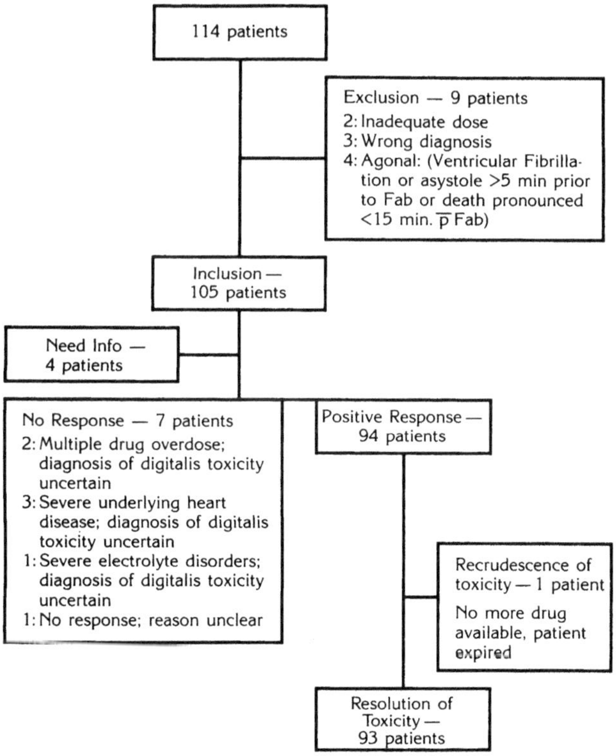

Fig. 1. Flow diagram summarizing the clinical experience of 114 patients treated with digoxin-specific Fab fragments. See text for details.

digoxin antibody fragments less than 1/10 the amount calculated to be needed (n = 2) and those who were agonal, that is to say, had ventricular fibrillation or asystole more than 5 minutes before treatment with Fab or who had cardiac arrest before Fab and were pronounced dead less than 15 minutes after administration of Fab (n = 4). This left 105 patients available for an evaluation of response. A clinical response was defined as resolution of toxic manifestations over a much more rapid time course than would be expected from withdrawal of digoxin or digitoxin and not likely to be explained by response to other treatments. The physicians caring for the patient and the investigators from each of the 21 multicenter sites in the trial were responsible for defining whether or not patients had a therapeutic response to the Fab fragments. Inadequate data were available to categorize 4 patients. Of the remaining 101 patients, 94 responded to the digoxin-specific antibody fragments and 7 did not. In all but one of these nonresponders, it seems unlikely in retrospect that the arrhythmias were actually due to digitalis.

Of the 94 patients who responded to digoxin antibody fragments, definite recurrence of severe toxicity occurred in one individual who ingested 25 mg of digoxin and received only half the dose calculated to neutralize the amount of digoxin in his body. Toxicity resolved completely in all the other patients, though transient recurrence of mild toxicity may have occurred in several cases during the first day after treatment with Fab.

Pretreatment serum potassium concentrations ranged from 2.4 to 10.3 mEq/l. In patients who had elevated serum potassium concentrations caused by digitalis toxicity, treatment with digoxin-specific antibody fragments reversed the hyperkalemia. Some of these patients received other treatment for hyperkalemia before administration of Fab.

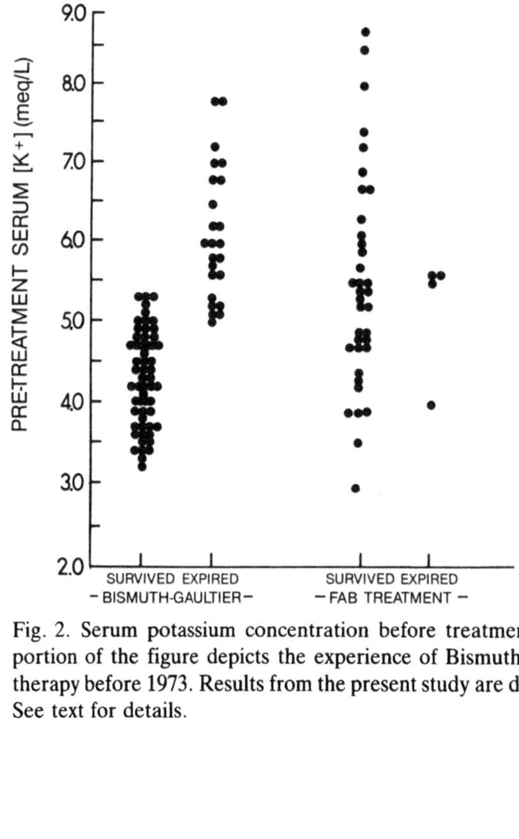

Fig. 2. Serum potassium concentration before treatment is compared with outcome. The left hand portion of the figure depicts the experience of Bismuth et al. who treated 91 patients with standard therapy before 1973. Results from the present study are depicted on the right hand portion of the figure. See text for details.

There was no relation between pretreatment hyperkalemia and outcome (Figure 2). In contrast, a clear relation between hyperkalemia and fatal outcome was demonstrated by Bismuth et al. (5) in patients who ingested large amounts of digitalis accidentally or with suicidal intent and who received standard therapy for their toxicity. They reviewed 91 episodes of acute digitalis poisoning at the Paris Poison Control Center. Patients who were receiving therapy with diuretics at the time of ingestion and those who died more than 4 days after the ingestion were excluded from their analysis. Their results are summarized and compared to the outcome of patients treated with Fab in Figure 2. To compare the results of the present study with the experience of Bismuth et al. (1), 74 patients treated with digoxin-specific antibody fragments were excluded from the analysis. Fourteen patients lacked a pretreatment serum potassium concentration determination, 52 patients were receiving concurrent diuretics, and 8 died more than 4 days after ingestion of digitalis. The comparison in Figure 2 suggests that digoxin-specific Fab substantially improved survival, and that death in patients treated with Fab may have been related to factors other than severity of digitalis intoxication.

Early in the clinical trial, digoxin antibody fragments were administered slowly intravenously over a period of 2 hours. The vast majority of patients have since been treated with infusions lasting from 15 to 30 minutes, or in some cases with bolus administration over minutes when cardiac arrest was imminent or ongoing. The time from termination of administration of the antibody fragments to an evident clinical response was usually less than half an hour. Complete reversal in most cases appeared to occur by 3 or 4 hours after completion of the infusion. This was true both for cardiac and noncardiac manifestations of toxicity. Thus, many patients had a rapid drop in serum potassium concentration, consistent with reversal of inhibition of the Na^+,K^+-pump. In some cases hypokalemia occurred, usually within 1 to 5 hours after the Fab fragments were administered. Patients with atrial fibrillation also had rapid reversal of the effects of digitalis on the atrioventricular node. In a few patients this resulted in a rapid ventricular response rate within hours after treatment with Fab fragments. Heart failure after administration of the antibody fragments has also been seen occasionally. In none of the 114 patients has there been evidence suggestive of an anaphylactic reaction or delayed serum sickness, and there have been no febrile reactions.

One patient developed slight erythema at the site of skin testing with Digibind but there was no systemic reaction to a therapeutic dose. Renal failure developed or worsened on the day of treatment with Fab in 5 patients, but in no case was the cause thought to be related to the Fab fragments. One patient had contrast dye-induced acute tubular necrosis that led to the digitalis toxicity. The other four had the onset of acute renal failure after cardiac arrests caused by digitalis toxicity preceding Fab treatment.

Twenty of the patients treated with digoxin-specific antibody fragments were infants or children. At least three of these cases were due to errors in dispensing medication. It is possible that several additional cases were caused by unrecognized or unreported errors in dispensing digitalis to infants.

Patients with renal failure required doses similar to those without renal failure and had similar responses. There was no evidence of recurrence of digitalis toxicity later in the course of hospitalization in patients with renal failure. This was so even in 5 patients who required dialysis at the time the Fab fragments were administered. We do not at present have adequate data to determine how the digoxin-Fab complex is eliminated from the body in these individuals. It appears that little is removed by dialysis.

Discussion

Continuing experience with digoxin-specific antibody fragments supports earlier findings that digitalis toxicity can be rapidly and safely reversed. The Fab fragments have had no demonstrable effects other than reversal of the effects of digitalis. Initial response after a fully neutralizing dose is usually within 30 minutes or less, and complete response is usually seen by 4 hours. A substantial experience is accumulating in patient subgroups, such as infants and children, and patients with renal failure. Concerns include exacerbation of heart failure, hypokalemia or rapid atrial fibrillation, which sometimes occur following treatment and are likely caused by reversal of digitalis effects. Following digoxin-specific antibody fragment administration, the serum digoxin concentration measurement by immunoassay will generally give spurious results. In addition, redigitalization is more complex than usual and, when necessary, should be postponed several days or more if possible, depending on renal function. To our knowledge, retreatment of patients with digoxin-specific Fab fragments has not yet occured.

Should further experience confirm the safety of digoxin-specific antibody fragments, their use may potentially be expanded to first line therapy of patients requiring definitive treatment.

References

1. Bismuth C, Gaultier M, Conso F, Efthymiou ML (1973) Hyperkalemia in acute digitalis poisoning: prognostic significance and therapeutic implications. Clin Toxicol 6: 153–162
2. Butler VP Jr, Chen JP (1967) Digoxin-specific antibodies. Proc Natl Acad Sci 57: 71–78
3. Schmidt DH, Butler VP Jr (1971) Reversal of digoxin toxicity with specific antibodies. J Clin Invest 50: 1738–1744
4. Smith TW, Butler VP Jr, Haber E et al. (1982) Treatment of life-threatening digitalis intoxication with digoxin-specific Fab antibody fragments: experience in 26 cases. N Engl J Med 307: 1357–1362
5. Smith TW, Haber E, Yeatman L, Butler VP Jr (1976) Reversal of advanced digoxin intoxication with Fab fragments of digoxin-specific antibodies. N Engl J Med 294: 797–800
6. Wenger TL, Butler VP Jr, Haber E, Smith TW (1895) Treatment of 63 severely digitalis-toxic patients with digoxin-specific antibody fragments. JACC 5: 118A–123A

Author's address:

Professor T. W. Smith, M. D. Cardiovascular Division, Brigham and Women's Hospital, 75 Francis Street, Boston, Massachusetts 02115, U.S.A.

Kinetics of the Fab fragments of digoxin antibodies and of bound digoxin in patients with severe digoxin intoxication

W. Schaumann and B. Kaufmann

Research Department Boehringer Mannheim GmbH Mannheim (West Germany)

For treatment of digitalis poisoning, high doses of the antibody are generally infused over a short period of time (2, 3). The dose of antibody is adjusted to the amount of ingested digoxin, which is either calculated from the serum concentrations or, if known, from the number of tablets swallowed. The kinetics of the Fab fragments were investigated for determining the best dosage schedule. The serum concentrations and the renal clearance of bound digoxin were assayed in 17 severely poisoned patients. In 11 of them the serum concentrations of unbound Fab fragments were also measured.

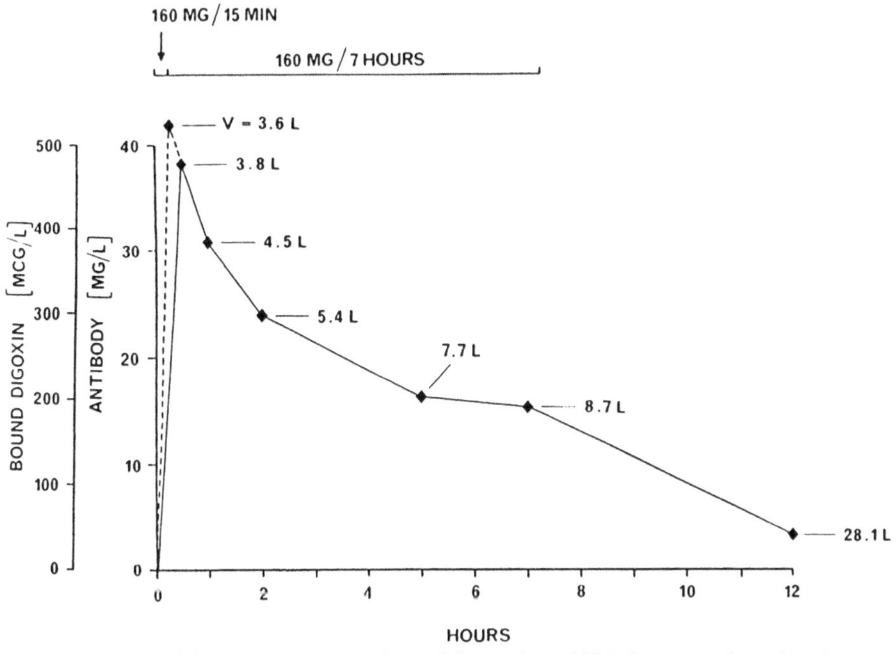

Fig. 1. Medians of the serum concentrations of free + bound Fab fragments from 6 patients treated according to the same schedule
The figure after 15 min, i.e. at the end of the loading infusion, was extrapolated from those after 0.5 and 1 h. The calculation of V is explained in the text.

In vitro, 1 molecule of digoxin stoichiometrically binds to 1 molecule antibody. Assuming that this also occurs in vivo, it has been recommended to infuse a dose of the Fab fragments which is equimolar to the amount of digoxin presumably present in the body. This assumption is not correct. In the first patient treated by Smith et al. high concentrations of free antibody were present during the first 10 hours after the infusion. The antibody is also cleared from the serum if no digoxin is present for binding. In 5 patients receiving 400–480 mg Fab fragments in 0.5–3 h, the median for the AUC under the serum concentrations of bound Fab fragments was 47.6% of the AUC of total Fab fragments. This reflects the percentage of the antibody which actually served its purpose of inactivating digoxin.

Based on these results, a different mode of administration was designed trying to keep the waste of antibody low. A loading dose of 160 mg was infused within 15 min. Another 160 mg was infused at a rate of 0.4 mg/min within the next 7 hours. Figure 1 shows the median serum concentrations from 6 patients treated in this way. In these patients the AUC of bound Fab fragments was 70.2% of the total. On average there was still a considerable excess of free antibody during the infusion. The picture looked different when the median total serum concentrations of the antibody were compared with the maximum concentrations of bound Fab fragments observed in all 17 patients (Fig. 2). In patient 5 with particularly severe poisoning a serum concentration of 30 mg/l bound Fab fragments was found after 2 hours. Although there was no free digoxin in the serum of this patient at that time, this figure was higher than the median from the 6 cases. Since treatment must be adequate for the most severe cases, 160 mg must be considered as the minimum loading dose.

After the loading dose, the desired serum concentration C is best maintained by an infusion at the constant rate I. C = I/Cl where Cl is the total body clearance. The renal

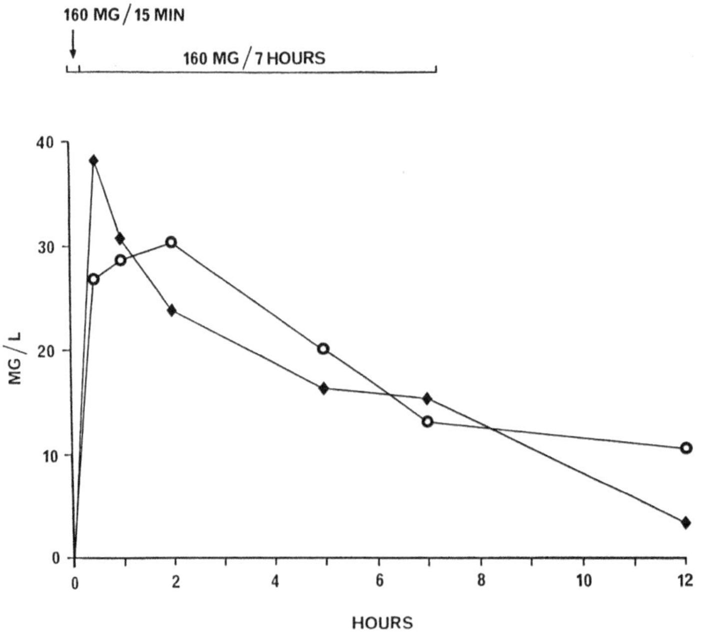

Fig. 2. Medians of the serum concentrations of total Fab fragments from Fig. 1 and maximum serum concentrations of bound Fab fragments observed in all 17 patients. ◆ total Fab fragments; ○ bound Fab fragments.

clearance of the Fab fragments was 13.6 ml/min. It amounted to 18% of the creatinine clearance. This probably reflects the percentage of antibody filtered by the glomeruli. The extrarenal clearance was 10.9 ml/min (N = 11). It is not clear at present to what extent the extrarenal clearance is due to excretion in active form or to metabolic inactivation, i.e. the protein losing its capacity to bind digoxin. Successful treatment of cases with a very low renal excretion suggests an extrarenal elimination of bound digoxin. Renal and extrarenal clearance together gave a total body clearance of 24.5 ml/min.

20 mg/l bound Fab fragments were found in patient 5 after 5 hours. With a total body clearance of 24.5 ml/min, a serum concentration of 20 mg/l would be maintained by an infusion rate of 0.5 mg/min. 3 ampoules containing together 240 mg would be sufficient to maintain this concentration over 8 hours.

Obviously the percentage of the Fab fragments binding to digoxin will decrease if the standard dose is increased from 320 to 400 mg as suggested in order to cope with the most severe cases of poisoning. Based on the results presented above 60% may be a fairly close estimate.

In some cases when the antibody was infused over a period of 30 min, free digoxin reappeared already after 3.5 h, with prolonged infusion only after 8–12 h. No free Fab fragments were detectable in the serum after that time. The maximum concentration of free digoxin in the subsequent hours was 6.8 ng/ml, which is well within the toxic range. 1 mg/l Fab fragments would be sufficient to bind this concentration. The necessary infusion rate would be 0.025 mg/min. This means that 1 additional ampoule is sufficient to prevent the reappearance of free digoxin for a whole day. Although the stability of the antibody in solution is limited, this long-term infusion is an alternative to be considered. Experience has shown, however, that 400–480 mg of the Fab fragments are sufficient to cope with an intoxication even if they are infused within 0.5 h. Apparently for the success of the treatment it is not necessary to remove all digoxin by binding it to antibodies.

The results in Fig. 1 may also serve to calculate the distribution volume of the Fab fragments. The median total body clearance in these patients was 20 ml/min = 1.2 l/h. Multiplying this figure by the partial AUC up to the time of measurement, one obtains the amounts cleared up to that time. The amount remaining in the body was calculated by deducing the amount cleared from that infused. Divided by the median serum concentration it gave the distribution volume V. Extrapolating from the medians after 0.5 and 1 h, one obtains a serum concentration of 42.2 mg/l of the total Fab fragments after 0.25 h. Assuming a linear rise during the initial infusion, the AUC up to 0.25 h multiplied by the total body clearance results in an elimination of about 10 mg. The remaining 150 mg divided by the extrapolated serum concentration of 42.2 mg/l result in a distribution volume of V = 3.6 l. This corresponds to the plasma volume showing that little antibody had diffused into the extravascular space by that time.

V increased steadily to 8.7 l at the end of the infusion. This increase may be explained by diffusion from the serum into the extracellular space. After 12 hours, however, the distribution volume had risen to 28.1 l. This agrees with the apparent distribution volume V_z of 26 l calculated from the total body clearance and the half life of 14.3 h obtained by log-linear extrapolation from 12 hours onwards. The half-life was 25.4 h and V_z was 54 l when extrapolation started after 24 hours. These figures are clearly in excess of the extracellular space. Intracellular penetration of the Fab fragments may be of practical importance when an antibody is aimed at binding antigens situated intracellularly.

The median serum concentration of 133.5 ng/ml bound digoxin after 0.5 h multiplied by a distribution volume of 3.75 l gives a serum content of 500 mcg. A mean concentration of about 10 ng/ml free digoxin before treatment multiplied by the extravascular volume of 14 l

gives an amount of 140 mcg in the extracellular space. Another 360 mcg apparently came from a superficial tissue compartment, probably the cell surface. This is also the site of action of digoxin. The concentration of free digoxin at the receptor site probably drops to zero within a few minutes of starting the infusion of the antibody. Reversal of the toxicity depends on the dissociation rate of the glycoside-receptor complex.
After the initial steep rise, the serum concentrations of bound Fab fragments remained fairly constant during the first 5 h of the infusion (Fig. 1). The decrease of the median after 7 h was mainly due to an unexplained sharp decline in case 13. Apparently the clearance of the digoxin-antibody complex from the serum was compensated by a slow re-diffusion of digoxin from a deep tissue compartment. Multiplying the renal clearance of 13.6 ml/min by a mean serum concentration of 157 ng/ml bound digoxin one obtains a re-diffusion rate of 2 mcg/min. If bound digoxin is also cleared extrarenally, this figure will be correspondingly higher.

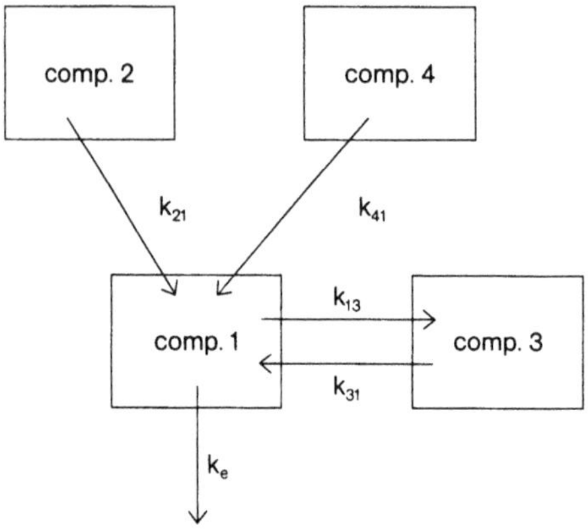

Fig. 3. Model for describing the time course of the serum concentrations of antibody-bound digoxin.

Figure 3 shows a model describing the time course of the serum concentrations of bound digoxin designed according to these findings and conclusions. The system of equations derived from the model is rather complicated (Fig. 4), but with the aid of a computer it was possible to calculate the rate constants. Table 1 summarizes the results from 11 patients in whom bound digoxin was determined for more than 24 h. Figure 5 shows the two cases with the lowest and the highest sum of squares of the deviations between experimental data and the fitted curve. In biological terms, compartment 1 is the serum. Compartment 2 consists of the extracellular space and the cell surface. The essential differences between compartments 2 and 4 is that the diffusion rate k_{41} from the intracellular compartment 4 is much lower than k_{21}. The exchange between compartments 1 and 3 represents the intracellular penetration of the Fab fragments and their re-diffusion together with bound digoxin. Although great caution is necessary in explaining biological events by mathematical models, the good agreement suggests that Figure 3 represents, in principle, the kinetics of bound digoxin and, when no more free Fab fragments are present, also those of the antibody.

Table 1. Half-lives for the exchange of digoxin binding to antibodies and of bound digoxin

Medians and range from 11 patients. Half-lives in h			
k_{21}	0.2 (0.01–2.0)	k_{31}	17.7 (4.7–55.9)
k_{41}	7.9 (3.2–301)	k_{13}	3.6 (0.3– 9.5)
k_e	3.6 (0.8–5.9)		

$$C_t = A_\beta e^{-\beta t} + A_\alpha e^{-\alpha t} + A_{21} e^{-k_{21} t} + A_{41} e^{-k_{41} t} \qquad t \leq TF$$

$$C_t = B_\beta e^{-\beta (t-TF)} + B_\alpha e^{-\alpha (t-TF)} \qquad t > TF$$

$$A_\beta = \frac{C_{02}(k_{31}-\beta)k_{21}}{(k_{21}-\beta)(\alpha-\beta)} + \frac{C_{04}(k_{31}-\beta)k_{41}}{(k_{41}-\beta)(\alpha-\beta)} + \frac{C_{i1}(k_{31}-\beta)}{\alpha-\beta}$$

$$A_\alpha = \frac{C_{02}(k_{31}-\alpha)k_{21}}{(k_{21}-\alpha)(\beta-\alpha)} + \frac{C_{04}(k_{31}-\alpha)k_{41}}{(k_{41}-\alpha)(\beta-\alpha)} + \frac{C_{i1}(k_{31}-\alpha)}{\beta-\alpha}$$

$$A_{21} = \frac{C_{02}(k_{31}-k_{21})k_{21}}{(\alpha-k_{21})(\beta-k_{21})} \qquad A_{41} = \frac{C_{04}(k_{31}-k_{41})k_{41}}{(\alpha-k_{41})(\beta-k_{41})}$$

$$B_\beta = \frac{C_{TF}(k_{31}-\beta)}{\alpha-\beta} + \frac{k_{31}C_{3F}}{\alpha-\beta} \qquad B_\alpha = C_{TF} - B_\beta$$

$$C_{TF} = A_\beta e^{-\beta TF} + A_\alpha e^{-\alpha TF} + A_{21} e^{-k_2 TF} + A_{41} e^{-k_{41} TF}$$

$$C_{3F} = C_{02}k_{21}k_{13}\left(\frac{e^{-\alpha TF}}{(k_{21}-\alpha)(\beta-\alpha)} + \frac{e^{-\beta TF}}{(k_{21}-\beta)(\alpha-\beta)} + \frac{e^{-k_{21}TF}}{(\alpha-k_{21})(\beta-k_{21})}\right)$$
$$+ C_{04}k_{41}k_{13}\left(\frac{e^{-\alpha TF}}{(k_{41}-\alpha)(\beta-\alpha)} + \frac{e^{-\beta TF}}{(k_{41}-\beta)(\alpha-\beta)} + \frac{e^{-k_{41}TF}}{(\alpha-k_{41})(\beta-k_{41})}\right)$$
$$+ C_{i1}k_{13}\left(\frac{e^{-\alpha TF}}{\beta-\alpha} + \frac{e^{-\beta TF}}{\alpha-\beta}\right)$$

$$\alpha;\ \beta = \frac{1}{2}\left(k_{13} + k_{31} + k_e \pm \sqrt{(k_{13} + k_{31} + k_e)^2 - 4 \cdot k_{31} \cdot k_e}\right)$$

Fig. 4. Equations for the calculation of the time course of the concentration of bound digoxin in serum
C_t = concentration of bound digoxin in the serum. The two partial equations for $t \leq TF$ und $t > TF$ are combined in the text as equation (1).
TF = time at which free digoxin first reappears in the serum.
C_{il} = serum concentration of digoxin before infusion of the antibody
C_{02}; C_{04} = Fictive initial concentrations in compartments 2 and 4.
k_{21}; k_{41}; k_{13}; k_{31}; k_e = first order micro rate constants according to model scheme in Fig. 1.
β, α = disposition (hybrid) rate constants.
A_β; A_α; A_{21}; A_{41} = intercepts of the pertinent exponentials (rate constants β, α, k_{21}; k_{41}) with the line $t = 0$; partial equation for $t \leq TF$.
B_β; B_α = intercepts of the pertinent exponentials (rate constants β, α) with the line $t = TF$; partial equation for $t > TF$.
C_{TF} = serum concentration at time $t = TF$.
C_{3F} = concentration in compartiment 3 at time $t = TF$.

References

1. Smith TW, Haber E, Yeatman L, Butler VP (1976) Reversal of advanced digoxin intoxication with Fab fragments of digoxin-specific antibodies. New Engl. J Med 294: 797–800

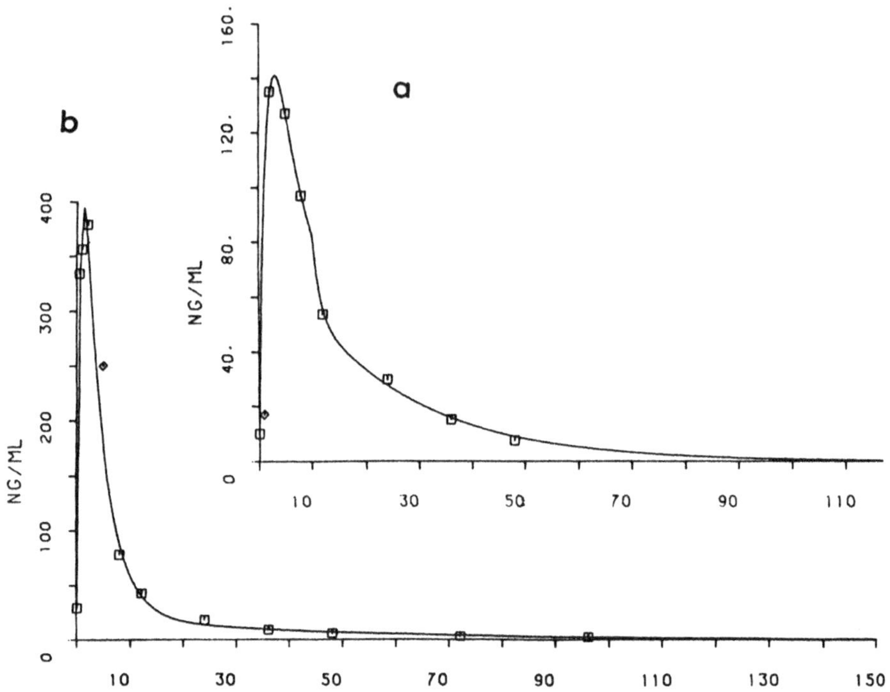

Fig. 5. Serum concentratiaons of bound digoxin calculated by the model in Fig. 3.
a = patient 2; b = patient 5.

2. Smolarz A, Abshagen U (1986) Digitalis-Antikörper-Fragmente (Fab) bei 90 schweren Glykosidvergiftungen. Klinische Erfahrungen aus einer multizentrischen Studie. Herz/Kreislauf (in press)
3. Wenger TL, Butler VP, Haber E, Smith TW (1985) Treatment of 63 severely digitalis-toxic patients with digoxin-specific antibody fragments. J Am Coll Cardiol 5: 118A–123A

Author's address:

Prof. Dr. W. Schaumann, Boehringer Mannheim, Med. Forschung, Sandhofer Straße 116, 6800 Mannheim 31, (West Germany)

Influence of tubular Na-K-ATPase activity on the development of hyperkalemia in digoxin intoxications

T. Risler, W. Jenkel*, J. Sander* and B. Grabensee*

Medizinische Klinik, University of Tübingen, and *Medizinische Klinik E University of Düsseldorf (West Germany)

Summary

Hyperkalemia is a complication of severe life threatening digoxin intoxication and is positively correlated to its fatal outcome. The purpose of this study was to investigate the influence of toxic digoxin doses on tubular Na^+-K^+-ATPase activity and renal K^+ excretion.

Six mongrel dogs received 2 ug/kg/h of digoxin as a constant infusion. We measured urine and serum K^+ as well as digoxin serum concentrations every 15 min. At the therapeutic endpoint constant ventricular tachycardia, all dogs were nephrectomized and Na^+-K^+-ATPase activity was measured in renal cortex and medulla, using the Rb uptake method. As controls we used six sham operated, non digitalized dogs, whose kidneys were removed after 60 min.

Serum K^+ concentration increased from 4.0 ± 0.3 (serum digoxin) to 4.6 meg/l ($p < 0.006$), while K^+ clearances increased 305.8 ± 258% ($p < 0.03$) in the digitalized group. Digoxin serum concentration increased to 138.6 ± 27.4 ng/ml at therapeutic endpoint. Compared to the control group, Na^+-K^+ATPase activity was significantly reduced in cortex (18.2 ± 5.4 versus 5.1 ± 3.6 nmol P_i/mg/h) ($p < 0.001$) as well as in medulla (6.0 ± 5.2 versus 0.2 ± 0.4 nmol P_i/mg per hour ($p < 0.023$).

We conclude that inhibition of renal Na^+-K^+-ATPase activity does not lead to K^+ retention. Thus, reduced renal K^+ excretion does not contribute to the development of hyperkalemia in severe digoxin interactions.

Severe, often suicidal, digitalis intoxications are a major clinical problem. Modern therapeutic approaches such as digitalis-specific Fab-fragments are available, but have a number of potential grave side-effects, which can only be tolerated in life threatening intoxications (12).

At the time of admission, the patients often have only minor cardiac symptoms. Their main complaints are nausea and vomiting. The stated digitalis dose ingested is often misleading, because of patients' incorrect statements or intense vomiting after ingestion of the drug.

Thus generally there are two objective criteria to assess the urgency of the situation: serum concentrations of the digitalis glycoside as well as potassium. While digitalis serum concentrations may be misleading, because of elevated but clinically ineffective serum concentrations during reabsorption and distribution of the drug, high potassium serum concentration is a reliable prognostic parameter (2).

Intra and extracellular potassium concentration depends on the intact sodium-potassium exchange at the cell membrane.

The enzyme responsible for a stable intracellular sodium and potassium distribution, the Na-K-ATPase, is inhibitable by cardiac glycosides. According to Bonting (3), the highest enzyme activities are located in the brain and organ tissues with a secretory function, e. g. the choroid plexus and the kidney.

The aim of our study was to define the influence of Na-K-ATPase inhibition in the kidney on renal potassium excretion. Although quantitatively muscular Na-K-ATPase inhibition should be responsible for the major increase in potassium serum concentration, the high enzyme concentration in the kidney could be suggestive of the importance of Na-K-ATPase inhibition for renal potassium excretion and the development of hyperkalemia in severe digitalis intoxication.

Methods

Twelve mongrel dogs weighing between 19 and 29 kg were anaesthetized with 30 mg/kg of pentobarbital. They were intubated and respirated to achieve normal arterial pH, P_{CO_2}, HCO_3, and P_{O_2}. A 6 lead ECG was registered continuously. Two femoral veins were cannulated to infuse drugs and to draw blood. Systemic blood pressure was monitored in one femoral artery using a Statham pressure transducer. Through a midline incision we cannulated both ureters to collect urine.

After one hour of equilibration and an infusion of 2.0 ml/kg b. w./min of normal saline, 5 dogs received 2 μg/kg/min of digoxin (Novodigal®) intravenously. Digoxin and potassium serum levels and renal potassium excretion were measured every 15 minutes until the experimental endpoint of constant ventricular tachycardia was reached. Then both kidneys were removed and Na-K-ATPase activity was measured using the rubidium uptake method.

As controls the kidneys of 5 sham operated dogs treated according to the same protocol without digoxin infusion were removed after 60 minutes.

Monovalent cation transport in renal cortex and medulla was estimated by measuring Rb^+ uptake by minor modifications of the method described by Hougen and Smith (6) as published in detail (13) for the cardiac muscle.

Digoxin serum concentrations were measured by radio-immunoassay as described by Smith et al. (11).

Potassium serum and urine concentrations were measured by flame photometry.

Renal clearances were calculated according to Balint (1) using the formula $\frac{C_u \times V_u}{C_p}$ first described by Rehberg (8) C_u = urine concentration, V_u = urine volume, C_p = serum concentration.

The significance of differences between mean values was determined by Student's t-test.

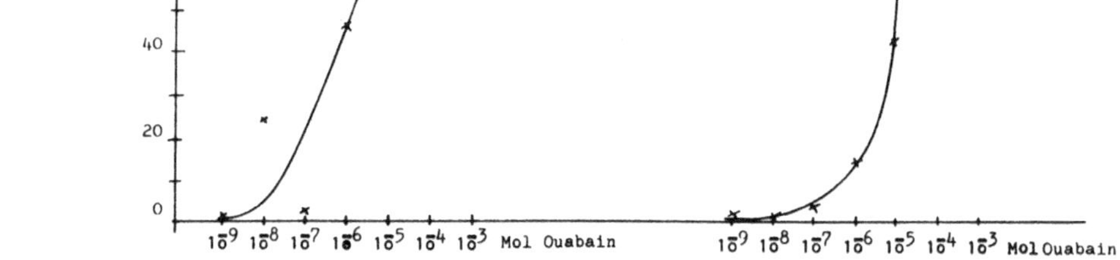

Fig. 1.

Results

Inhibition of monovalent cation transport in renal cortex and medulla by ouabain (Fig. 1). Initial experiments revealed a 50 % inhibition (I_{50}) of Rb-uptake at a ouabain concentration in the buffer between 10^{-5} M and 10^{-6} M. Maximal inhibition was reached at a ouabain concentration of 10^{-3} M.

Monovalent cation transport inhibition at the experimental endpoint of digoxin intoxication (Fig. 2). Monovalent cation transport was inhibited by 42 % in renal cortex in comparison to controls (0.158 ± 0.063 nmol Rb^+/mg/h and 0.273 ± 0.083 nmol Rb^+/mg/h [$p < 0.023$]).

In renal medulla monovalent cation transport inhibition reached 37 % in comparison to controls (0.300 ± 0.064 nmol Rb^+/mg/h and 0.188 ± 0.066 nmol/mg/h [$p < 0.014$]).

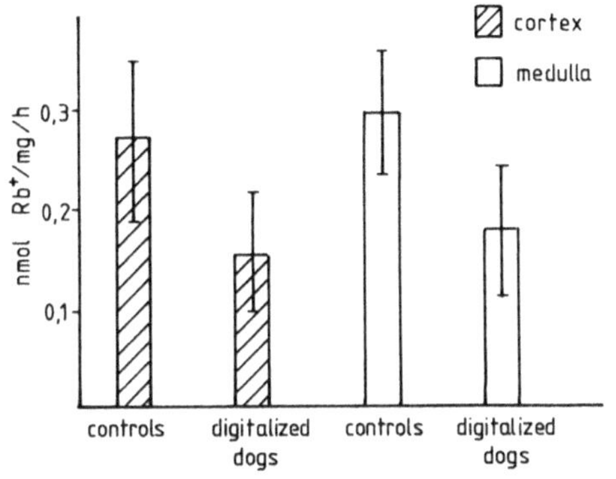

Fig. 2.

Potassium serum concentrations (Fig. 3). K^+ serum concentration increased from 4.0 ± 0.3 meq/l to 4.6 ± 0.4 meq/l ($p < 0.006$) at the experimental endpoint. There was no significant change in K^+ serum concentrations of the control group.

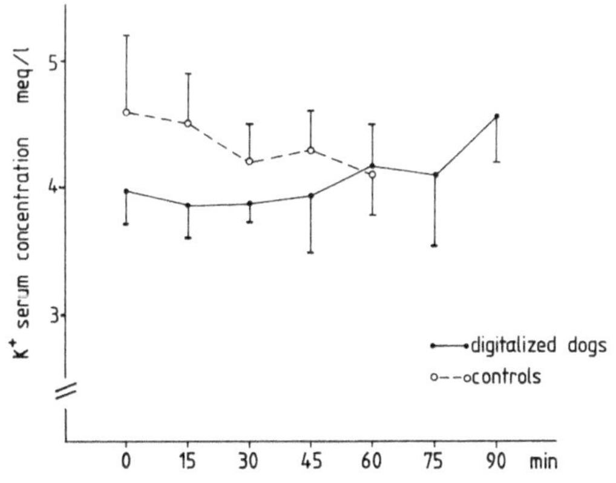

Fig. 3.

Renal potassium clearances (Fig. 4 a–d). K^+ clearances decreased slightly 30 minutes after the start of the digoxin infusion to increase significantly up to 75 minutes of the experiments. Then K^+ clearances dropped back to control values. There was no significant change in potassium clearances of the control group.

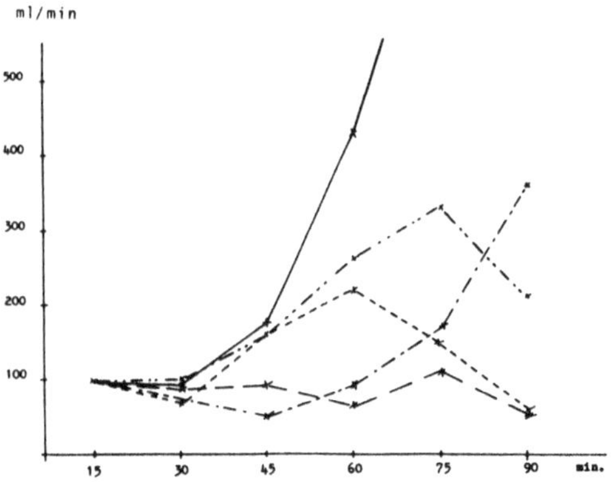

Pelative Kaliumclearance

min		15	30	45	60	75	90
dog 1		100	89	132	222	278	181
		100	111	187	295	385	236
2		100	74	188	238	146	95
		100	61	132	197	151	22
3		100	89	85	65	109	64
		100	94	103	62	117	48
4		100	82	170	435	770	1047
		100	88	176	435	770	1241
5		100	-	42	105	147	331
		100	-	59	83	185	399
	$\bar{x}$		86 ±14,6	127,4 ±53,6	213,7 ±141,2	305,8 ±258,5	366,4 ±4304,4

— n.s. —

—— n.s. —

—— p <0,032 ——

—— p < 0,035 ——

—— n.s. ——

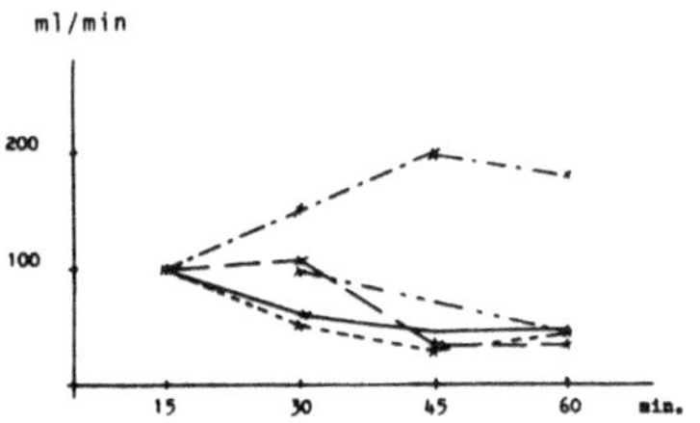

min		15	30	45	60
dog 1		-	100	-	62
		-	100	-	36
2		100	52	32	64
		100	47	26	37
3		100	49	30	33
		100	60	60	59
4		100	174	149	130
		100	118	245	222
5		100	114	33	43
		100	107	35	38
	$\bar{x}$		92,1 ±40,5 n.s.	76,3 ±79,5 n.s.	72,4 ±59,9 n.s.

Fig. 4.

Digoxin serum concentrations (Fig. 5). In five dogs, serum digoxin concentrations increased linearly up to 90 minutes of a digoxin infusion of 2 ug/kg b. w./min. At the experimental endpoint, digoxin serum concentrations had rached 138.6 ± 27.4 ng/ml.

Discussion

Hyperkalemia is a good parameter to estimate the potential severity of a digitialis intoxication. Bismuth et al. (2) could even correlate the initially measured potassium serum concentration to the prognosis of the intoxication.

There are a number of potential pathophysiologic mechanisms responsible for the development of hyperkalemia in digitalis intoxications.

Na-K-ATPase has been recognized as a receptor for cardiac glycosides, which bind to the enzyme and inhibit its physiologic action: Na^+ and K^+ transport through the cell membrane

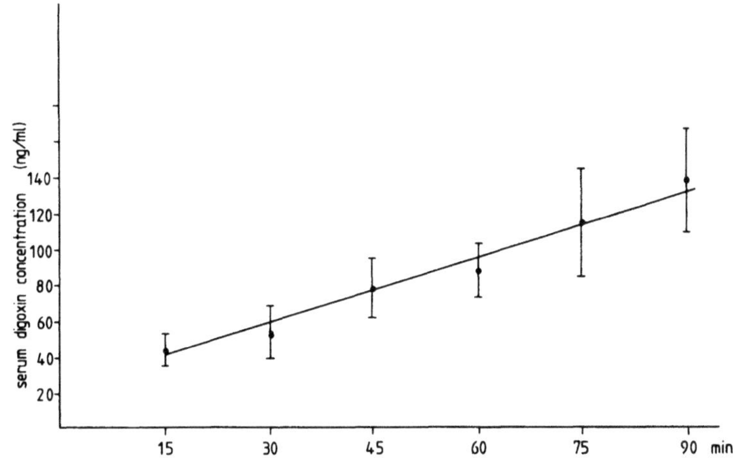

(5). The extensive distribution of Na-K-ATPase in the organism suggests a major role in digitalis induced hyperkalemia. Inhibition of the enzyme not only in cardiac but even more in skeletal muscle must induce an interstitial concentration increase of K^+, because cellular potassium re-uptake is more or less blocked at all cell membranes.

However, interstitial and consecutive intravascular increases in K^+ concentration are generally normalized by an increased renal potassium excretion. An additional K^+ load stimulates Na^+-K^+-ATPase activity in the kidney and thus increases urine potassium excretion (10). The role of Na-K-ATPase in the renal transport mechanism of K^+ is still discussed, but the importance of the enzyme for proximal tubular reabsorption as well as distal tubular secretion of potassium is accepted (4).

In our experiments, potassium clearance increased significantly during an intravenous digoxin infusion of 2 ug/kg b. w./min up to 75 minutes. Renal potassium clearances at the experimental endpoint do not give further information, because of inconsistent results probably caused by circulatory irregularities in this situation.

Although pathophysiological mechanisms of renal potassium handling cannot be deduced from intact animal experiments, there are some hypotheses to explain our experimental results.

The observed increase in renal potassium clearance cannot be explained by renal Na-K-ATPase inhibition according to Sejersted et al. (9). In comparison to these experiments, which achieved a 70 % Na-K-ATPase inhibition by injecting ouabain directly into the renal artery, we infused digoxin into the systemic circulation inducing a 40 % enzyme inhibition at the experimental endpoint. Further inhibition by systemic digoxin infusion was not compatible with survival of the animal.

This degree of Na-K-ATPase inhibition could be insufficient to produce renal potassium retention as observed by Sejersted. On the other hand, incomplete enzyme inhibition was apparently unable to blunt increased potassium clearances in the case of serum potassium concentration increase.

Another possible explanation would be a qualitative difference of minor enzyme inhibition between its impact on K^+ reabsorption (proximal tubule) or K^+ secretion (distal tubule). Renal potassium excretion could be enhanced by proximal tubular K^+ reabsorption inhibition, while at the same time Na-K-ATPase inhibition is not sufficient to hinder distal tubular potassium secretion.

Although it is impossible to resolve pathophysiologic mechanisms of renal K^+ excretion in our experiments, renal potassium retention does not contribute to the development of hyperkalemia in severe digoxin intoxications.

We conclude from our experiments:

1. A digoxin infusion of 2 ug/kg b. w./min inducing ventricular tachycardia in dogs after 90 minutes causes a serum K^+ concentration increase.
2. The given dose of digoxin induces a 42% inhibition of cortical- and a 37% inhibition of medullar Na-K-ATPase.
3. Renal K^+ clearances increase significantly up to 75 minutes of the above described digoxin infusion.
4. Renal K^+ excretion is not responsible for the development of hyperkalemia in severe digoxin intoxications.

References

1. Balint P (1969) Normale und pathologische Physiologie der Nieren. VEB Verlag Volk und Gesundheit, Berlin
2. Bismuth C, Motte G, Frejaville JP, Conso F (1973) L'hyperkaliémie dans l'intoxication digitalique massive. Valeur prognostique et implications thérapeutiques. Arch Al Coeur 12: 1537
3. Bonting SL, Simon KA, Hawkins NM (1961) Studies on "Sodium Potassium Activated Adenosine Triphosphatase." Arch Biochem Biophys 95: 416
4. Doucet A (1985) Na-K-ATPase: General considerations, role and regulation in the kidney. Advances in Nephrol 14: 87
5. Gotfraind T (1975) Cardiac glycoside receptors in the heart. Biochem Pharmacol 24: 823
6. Hougen TJ, Smith TW (1978) Inhibition of myocardial monovalent cation active transport by subtoxic doses of ouabain in the dog. Circulation Res 42: 856
7. Matsui H, Schwartz A (1968) Mechanism of cardiac glycoside inhibition of the (Na^+ + K^+) dependent ATPase from cardiac tissue. Biochem Biophys Acta 151: 655
8. Rehberg PB (1926) The rate of filtration and reabsorption in human kidney. Biochem J 20: 447
9. Sejersted OM, Monclair T, Mathisen O, Hartmann A, Kiil F (1985) Dependency of renal potassium excretion on Na,K-ATPase transport rate. Acta Physiol Scand 123: 9
10. Silva P, Hayslett JP, Epstein FH (1973) Role of Na-K-ATPase in potassium adaptation: Stimulation of enzymatic activity by potassium loading. J Clin Invest 52: 2665
11. Smith TW, Bulter VP, Haber E (1969) Determination of therapeutic and toxic serum digoxin concentrations by radio-immunoassay. New Engl J Med 281: 1212
12. Smith TW, Butler VP, Haber E, Fozzard H, Marcus FI, Bremner WF, Schulman IC, Phillips A (1982) Treatment of life-threatening digitalis intoxication with digoxin-specific Fab antibody fragments. New Engl J Med 307: 1357
13. Somberg JC, Risler T, Smith TW (1978) Neutral factors in digitalis toxicity: protective effect of C-1 spinal cord transection. Am J Physiol 235: H 531

Author's address:

Prof. Dr. Teut Risler, M. D., Medizinische Klinik der Universität, Otfried-Müller-Straße 10, 7400 Tübingen (West Germany)

24-Hour Holter ECG in digoxin intoxicated patients

W. Grille, W. Doelle, W. Johnsen and K.-D. Kolenda

II. Department of Internal Medicine, University of Kiel (West Germany)

Summary

1. About 50% of digoxin intoxicated patients reveal ventricular runs when analyzed by 24-h Holter ECG.
2. These patients can be identified as "intoxicated" from a group of patients with "toxic" SDL by conventional clinical means.

Introduction

Contrary to previous reports (2), it has been stated recently that death due to digitalis intoxication is a rather unlikely incident (3). In conformity with this, our working group observed 35 digitalis intoxicated patients, none of whom died in coincidence with intoxication (1). The aim of the present prospective study was to quantify prognostic significant arrhythmic events below lethality threshold, i. e. ventricular premature beats (VPB), by 24-h Holter ECG in a collective of patients with so-called "toxic" serum digoxin levels (SDL).

Methods

During 15 months, all digoxin treated patients of the community hospital in Kiel who fulfilled the following criteria were included.

1. SDL $\geq$ 3.0 ng/ml (radio-immunoassay),
2. no acute myocardial infarction during the preceding month,
3. no pre-existing antiarrhythmic therapy,
4. no permanent pacemaker therapy.

24-h Holter ECG was started the same day as blood samples were taken. Holter ECG was performed twice in those patients who tolerated 14 days of digoxin therapy withdrawal. Patients were divided into two groups: "intoxicated" and "non-intoxicated", independently of Holter ECG results and SDL. Diagnosis of intoxication was based only upon cardiac and extracardiac clinical symptoms and standard ECG symptoms which had been shown to be digitalis associated by a 6-day digoxin-withdrawal.

Holter ECGs were analyzed semi-automatically with visual control of all consecutive VPB.

Results

Of 26 patients who met the criteria, 16 were judged to be intoxicated, 10 were non-intoxicated.

Table 1 displays some characteristics of both groups. There was a significant difference between mean SDL.

Almost half of the intoxicated patients revealed ventricular runs (3 or more consecutive beats), whereas only one non-intoxicated patient did; $p < 0.05$, Fig. 1.

Table 1. Characteristics of patients with "toxic" SDL. (* $p \leq 0.05$; ** $\tilde{x}$ (95% confidence interval)).

	Intoxicated patients	Non-intoxicated patients
n	16	10
♂ / ♀	3/13	2/8
**Age	78.5 (73–80)	76.5 (61–83)
**Weight	61.5 (45–74)	61.5 (44–76)
**S-potassium (mval/dl)	4.4 (4.2–5.1)	4.5 (3.1–5.5)
**S-creatinine	1.45 (1.1–2.8)	1.6 (1.0–7.0)
Diuretics therapy	7	5
**Serum digoxin level (ng/ml)	4.36 (3.27–4.87)	*3.49 (3.0–4.57)

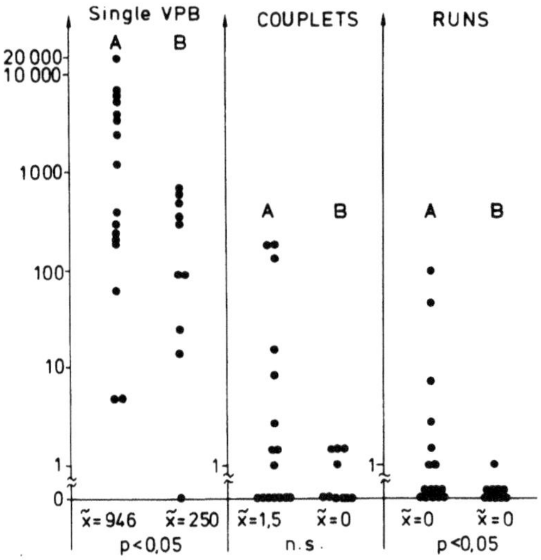

Fig. 1. Frequency of premature ventricular beats in 24-h Holter ECG: Intoxicated patients (A) vs. non-intoxicated patients (B).

Ventricular runs decreased significantly after 14 days of digoxin withdrawal (Fig. 2). Sustained ventricular tachycardias did not occur in any patient.

Intoxicated patients also revealed significantly more single VPB. However, there was no decrease in single VPB when compared with follow-up (Fig. 2). No significant differences with regard to single VPB, couplets or runs were found between non-intoxicated and intoxicated patients following digoxin withdrawal (Fig. 3). There was no significant correlation between SDL and frequency of VPB in intoxicated patients ($R_{SPEARMAN} = 0.2588$).

Discussion

Our results suggest that VPB are a common finding in patients with "toxic" SDL, when analyzed by 24-h Holter ECG. Occurrence of ventricular runs, however, is almost completely restricted to those patients judged to be intoxicated in clinical terms. Thus, clinical diagnosis of digitalis intoxication and occurence of ventricular runs analyzed by 24-h Holter

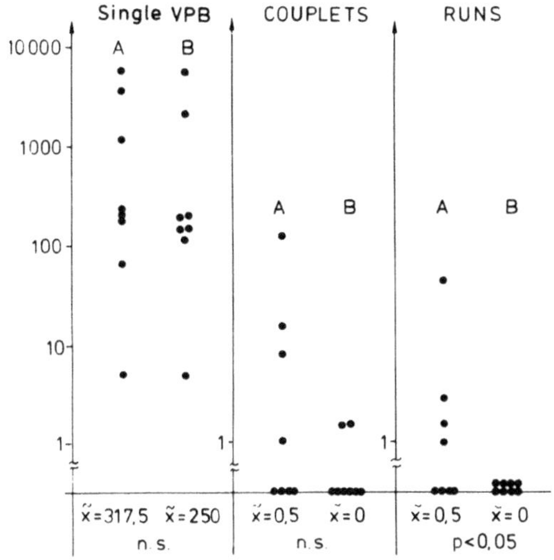

Fig. 2. Frequency of premature ventricular beats in 24-h Holter ECG: Intoxicated patients (A) vs. intoxicated patients after 14 days of digoxin-withdrawal (B), Wilcoxon-test.

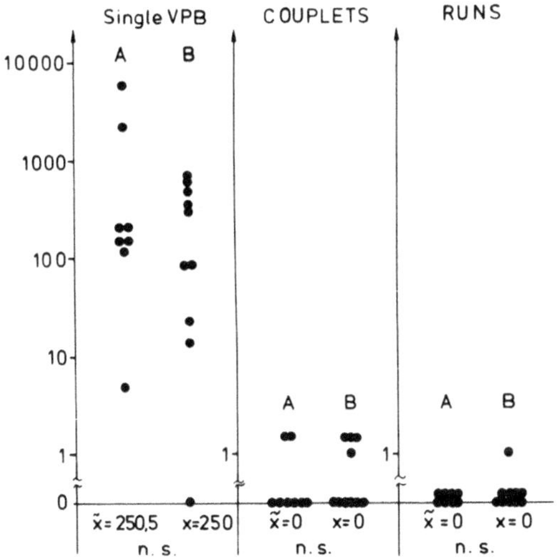

Fig. 3. Frequency of premature ventricular beats in 24-h Holter ECG: Intoxicated patients after 14 days of digoxin withdrawal (A) vs. non-intoxicated patients (B), U-test.

ECG correspond well. There was a significant difference too in single VPB between intoxicated and non-intoxicated patients, however, a lack of diminution after digoxin withdrawal in intoxicated patients. This may indicate an arrhythmogenic effect of digitalis withdrawal in some of the patients, the more so as single VPB significantly increased in some intoxicated patients following digoxin withdrawal.

References

1. Grille W, Domscheid M, Johnsen K, Kolenda K-D, Krüger J (1984) Vergleichende Untersuchungen über die Intoxikationsrate von Digoxin bzw. Digoxinderivaten und Digitoxin unter stationären und ambulanten Bedingungen. Verh Dtsch Ges inn Med 90: 878–881
2. Rodenski PL, Wassermann F (1961) Observations on digitalis intoxications. Arch Intern Med 108: 171–188
3. Smith TW, Antmann EM, Friedmann PL, Blatt CM, Marsh JD (1984) Digitalis glycosides: mechanisms and manifestations of toxicity. Progr Cardiovasc Res 26: 413–458

Author's address:

Dr. W. Grille, M. D., II. Medizinische Universitätsklinik, Metzstraße 53, 2300 Kiel (West Germany)

Clinically relevant interactions with cardiac glycosides

E. Weber

Klinische Pharmakologie, Ruprecht-Karls-Universität Heidelberg (West Germany)

It is characteristic of digitalis that
1) it can be of vital importance for patients,
2) it has a small therapeutic range.

Logically, all unintentional variations in its effects should be avoided, no matter if they are enhanced or lowered. Different factors can contribute to such variability, for instance noncompliance or overcompliance. One of the most important circumstances, however, is concomitant drug therapy which can lead to drug interactions. If we consider the clinical relevance of drug interactions, we are mostly confronted with pharmacodynamic effects in the toxic range. This does not mean that the underlying mechanisms are uniform.

Digitalis-drug interactions can lead to enhancement of pharmacodynamic effects of digitalis, intensification of its toxicity, and in variations of digitalis kinetics, resulting in an increase of digitalis serum concentrations.

The first group we have to consider is that of drugs which share one or other property with digitalis, so that administered together the clinical effect can be exaggerated. Examples are the sympathomimetics (21) and calcium (13, 40) as far as a positive inotropic effect is concerned, or the large groups of drugs which influence the electrophysiogical effects of digitalis like antiarrhythmic agents, beta-blockers, calcium antagonists, carbamazepine (24), reserpine (> 1 mg) (8, 34) and succinylcholine (49). Some of these drugs contribute to digitalis toxicity by loss of potassium, for instance potassium eliminating diuretics, corticoids, laxatives, glucose infusions (65) and amphothericine; others contribute by ectopic pacemaker activity like sympathomimetics (65) and calcium (60), or by increasing the digitalis serum concentrations. Certainly, all of these drugs can cause interactions of clinical relevance with digitalis. Since most of them are well known to physicians, this paper concentrates on the last group of drug interactions because some important and interesting new examples belong to it.

If we consider changes in digitalis serum concentration, the underlying drug interactions can be located at different sites, and several pharmacokinetic parameters can be affected. The possible sites of interactions are the gastrointestinal tract, plasma protein or tissue binding sites, and the different stages of elimination like drug metabolizing enzyme systems, biliary excretion, enterohepatic circulation, and renal excretion.

Examples for digitalis drug interactions in the gastrointestinal tract

First, we have to take note that in about 10 % of our patients digoxin is reduced to cardioinactive metabolites like dihydro digoxin and dihydro digoxigenin. The extent of this process varies a great deal from subject to subject – the variation is up to 40 % of the administered dose of digoxin. There is evidence that bacterial flora of the intestine contribute markedly to this metabolic process, so it seems logical that some antibiotic can stop it. Lindenbaum et al. (33) were the first to show that the digoxin concentration rose as much as twofold in subjects with a high conversion rate when they were treated with erythromycin or tetracyclin. Fig. 1 shows clearly that in three subjects of the 'excretor type', the urinary excretion of digoxin reduction products fell dramatically when 1 g or 2 g of erythromycin was given daily or 500 mg of tetracyclin was given four times a day.

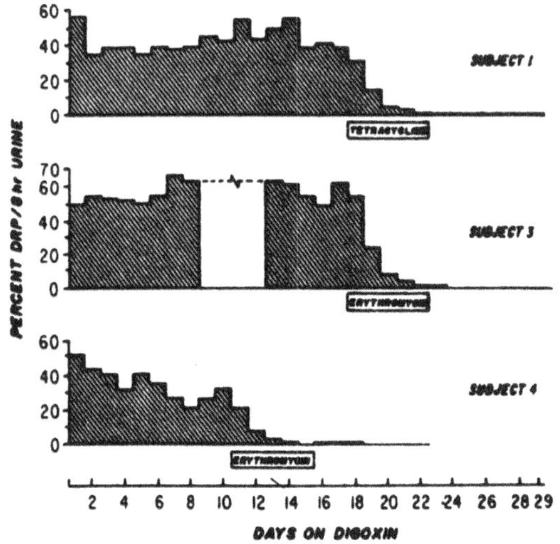

Fig. 1. Daily urinary excretion of DRPs between midnight and 8 a. m.
Three subjects who were excretors were given two 0.25 mg digoxin tablets for 22 days (Subject 4) or for 29 days (Subjects 1 and 3). (Because Subject 3 temporarily left the study, collections of his urine were interrupted on Days 9 to 12 of the pre-antibiotic period, although he continued to take digoxin.) A marked fall in DRP excretion occured in each subject during a five-day course of oral antibiotics. Urinary reduced metabolites remained at very low or undetectable levels during the week after antimicrobial treatment. DRP = Digoxin reduction products (33).

The clinical significance of the described mechanism is given only for a minority of patients, if at all. Only one case report has been published (18). However, mechanisms which can produce cardio-inactive metabolites of digoxin can be one of the factors operating with digoxin nonresponders, and can lead to hazards when these subjects are treated with antibiotics.

Digitalis elimination and drug interactions

Calcium antagonists (Table 1): Klein et al. (25, 26) were the first to report an increase of digoxin concentration in patients treated with 240 mg of verapamil a day. Other groups examined the interaction in volunteers (2, 4, 11, 46). They found an increase in digoxin concentration ranging from 30 % to 77 %, a decrease in the digoxin total clearance of about a third, and a variable decrease in the digoxin renal clearance ranging from 20 % to 68 %. As serum creatinine values and the glomerular filtration rate remained unchanged, it was concluded that verapamil diminishes the tubular secretion of digoxin. Two problems remained unsolved, namely if the effect is proportional to the dose, and if the extrarenal clearance is also impaired. Furthermore, it is not clear if the interaction is clinically relevant, nor even if it is a permanent one (46).

Table 1. The calcium antagonists-digoxin interaction. Effect on the pharmacokinetics of digoxin.
[Dig] = Digoxin serum concentration
O = unchanged

		Clearance			
	[Dig]	Total clearance	Renal clearance	Extrarenal clearance	t/2
Verapamil					
Belz et al. 1981 (2)	↑			↓	
Klein et al. 1980/82 (25, 26)				↓	
Pedersen et al. 1981 (44, 45)		↓		?	↑
Belz et al. 1982/83 (3, 4)	↑		↓	?	
Pedersen et al. 1982 (46)	↑		↓		
Doering 1983 (11)	↑				
Pedersen et al. 1983 (48)		↓			↑
Thiercelin et al. 1983 (68)	↑		O		O
Zatuchni 1984 (73)	↑				
Nifedipin					
Belz et al. 1981/83 (2, 4)	↑		↓		
Pedersen et al. 1982 (47)	O	O	↓	↑	
De Cesaris 1983 (7)	(↑)				
Kuhlmann et al. 1984 (28)	O				
Schwartz/Migliore 1984 (62)	O				
Schwartz et al. 1984 (63)	O				
Zylber-Katz 1984 (76)			↑		O
Diltiazem					
Schrager et al. 1983 (61)	O				
Oyama et al. 1984 (42)	↑		↓		
Rameis et al. 1984 (53)	↑	↓	O		↑
Yoshida et al. 1984 (72)	(↑)	↓	O		O
Kuhlmann/Marcin 1985 (29)	↑	↓	↓		
Pfeiffer 1985 (52)	O				

The first published observation concerning nifedipine came from Belz et al. (2, 4). They found a moderate increase of the digoxin concentration of 45% in nine of twelve volunteers. Others did not confirm the influence on digoxin concentration (28, 62, 68). Discrepancies also appeared with regard to the excretion of digoxin in the urine and the renal clearance under the influence of the drug (4, 28, 47, 62, 76).

Some of the reasons for the discrepancies were seen in the different doses of digoxin or nifedipine used, different time periods of urine sampling, and in the small sample sizes. Also the impact of digoxin formulations, dietary habits, and concomitant medication is unclear (64). The absorption process, however, and the glomerular filtration rate do not seem to be involved. It was claimed that the renal clearance decreases because of impairment in the tubular secretion of digoxin, and changes in the amount of reabsorption in digoxin.

Disturbing changes in pharmacodynamic and electrophysiologic parameters were not found so it can be concluded that the clinical significance of the interaction is small or lacking in the case of nifedipine (4, 28, 62).

Similar effects as depicted for verapamil have also been shown for diltiazem, but they are rather less pronounced (29, 42, 52, 53, 61, 72).

Quinidine and digoxin: Hager and his group (22) were the first to show the significant influence of quinidine on digoxin serum concentration. Subsequently a number of studies were published by different groups. What was brought to light by most of them can roughly be summarized as follows.

An increase in serum digoxin concentration by an average of 100 %, ranging from 25 % to more than 300 % and even 500 %, was found, the increase being proportional to the administered dose of quinidine. The effect starts within 24 hours and a new steady state is established in about three to six days, or in 48 hours if a loading dose of quinidine is given. More than 90 % of the patients are affected. The mechanism involves a decrease in digoxin total body clearance. More details are summarized by Bigger and Leahay (5) (Table 2). In most of the subjects the renal clearance decreases to a higher degree than the nonrenal clearance. Although there is good concordance in the results, they are poorly understood (for further references see (16)).

Table 2. The quinidine-digoxin interaction: effects of quinidine on the pharmacokinetics of digoxin (5).

Author	No. of patients	Volume of distribution	Elimination half-life	Total clearance	Renal clearance	Non-renal clearance
Hager et al. (1979) (22)	6	↓	→	⬇	⬇	→
Leahey et al. (1981) (32)	15	↓	⬆	⬇	⬇	↓
Pedersen et al. (1980) (43)	11	–	–	–	⬇	–
Schenck-Gustafson et al. (1981) (58)	5	↓	⬆	⬇	⬇	↓
Steiness et al. (1980) (66)	6	→	–	⬇	↓	⬇

In contrast to what probably holds true for the calcium antagonists, the likelihood for digoxin toxicity rises when quinidine is given concomitantly, and case reports have been published. It has been recommended that the clinical course be monitored, and the dose of digoxin halved (10). An important question is whether the inotropic effect of digoxin increases in accordance with the rise in digoxin serum concentration under the influence of quinidine. The answer is controversial. Some workers have found evidence for a parallel increase (11, 30, 59), while others did not (23, 66). However, studies have been conducted with normal subjects and the results of systolic time intervals to judge the inotropic effects have been used. The conclusion is that in any case it must be recognized that haemodynamic states may worsen during the quinidine/digoxin interaction (for further references see (16)).

Quinidine and digitoxin: An interaction between digitoxin and quinidine leading to an increase in the digoxin serum concentration has now been definitively proven (14, 19, 27, 51). Ding et al. (9) demonstrated that only patients with a particularly short half-life and consequently a large nonrenal clearance of digitoxin are at risk because the effect of quinidine depends mainly on the nonrenal clearance of digitoxin. However, digitoxin plateau concentrations from a 0.07 mg dose per day are usually such that even an increase under quinidine is still within the therapeutic range. The almost complete abolition of renal clearance is not of clinical importance because it contributes only about 15 % to the total clearance.

Other antiarrhythmic substances: Leahay et al. (31) observed an increase in digoxin concentration in one of 41 patients when these were treated concomitantly with antiarrhythmic substances like procainamide, disopyramide or mexiletine; five out of 41 patients suffered from nausea, anorexia or vomiting, while the PR intervals were unchanged. However, other work on disopyramide has not confirmed a definite increase of digoxin concentration, and so the clinical relevance of this phenomenon is not known (35, 57, 71).

In the case of amiodarone, the interaction with digoxin seems to be a more serious one (36, 37), although other authors found no increase (1, 67). Perhaps the data of Nager and Nager (39) give the explanation for the discrepancy, namely a marked variability in individual responses. Clinical relevance is a point at issue, but the interaction occurs quickly and often requires the adjustment of digoxin doses. The mechanism is unknown, but according to the data of Nademanee et al. (38) the renal as well as the nonrenal clearance of digoxin seem to be reduced.

Potassium-sparing diuretics: Some effects due to an interaction between digoxin and potassium-sparing agents can be proved; however it is not clear if these effects are of clinical relevance. Under the influence of amiloride (69), triamterene and spironolactone (70), the extrarenal clearance of digoxin decreases. The same holds for renal tubular secretion under the influence of spironolactone, whereas amiloride increases it, and triamterene probably does too. A moderate increase in plasma digoxin concentration has been found. In studying the effects on the concomitant administration of digoxin and one of these drugs on the pre-ejection period index, the same authors found that this parameter is influenced in a different manner by the three agents. Only triamterene was without effect (69, 70).

Drugs enhancing hepatic microsomal drug metabolizing enzymes

Since the publication in 1974 by Peters and Coworkers (50), it is well known that rifampicin by its property of enhancing the hepatic microsomal drug metabolizing enzymes, accelerates the metabolism of digitoxin. Under therapeutic doses of rifampicin, the digitoxin serum concentration decreases by about 50 % or more, and the clearance of digitoxin is enhanced. These findings have been confirmed by others. In 1980 Novi et al. (41) were the first to report on the corresponding phenomenon when using rifampicin and digoxin. Also Bussey et al. (6) and Gault et al. (20) described similar observations. However, all this work was done in patients dependent on chronic dialysis, a condition which can be the clue to the interpretation. Rifampicin did not induce formation of unusual metabolites nor changes in the relative amounts of the various metabolites were seen. The possible reasons discussed for lower digoxin concentrations under rifampicin were an increase of biliary secretion, nonbiliary intestinal secretion, biotransformation, and tissue binding, or a decrease in the absorption of rifampicin. Probably the institution of dialysis and the diminution in renal function (perhaps also the other drugs taken) exert a marked influence on the outcome. This seems the more likely considering a newly published study by Rameis (54) on the interaction between phenytoin and beta-acetyl-digoxin. In an open randomized cross-over trial at steady state in six volunteers, all the classical alterations were observed in those digoxin pharmacokinetic parameters which are to be expected if an induction of the drug metabolizing enzymes in the liver takes place. The half-life of digoxin decreased by about 30 %, and the total clearance increased by about 27 % mainly due to an increase in hepatic clearance. The renal clearance, the distribution volume, and the extent of absorption remained unchanged.

Nonsteroidal antirheumatics (NSAR): It has been claimed that some nonsteroidal antirheumatic substances influence digoxin concentration (55). Fortunately this has not been proven to be true in general (12, 15, 17, 56, 74, 75).

Conclusions

1. We know much more about pharmacokinetic drug interactions with cardiac glycosides in healthy volunteers than in patients. In addition we unfortunately know practically nothing about the long-term effect of drugs on plasma digoxin concentration.

2. The clinical relevance of most of the digitalis drug interactions affecting kinetic parameters of the cardiac glycosides has not been established. Probably this outcome also reflects the modulating influence of the individual patient's pathological condition on the net effect of the interaction.

References

1. Achilli A, Serra N (1981) Amiodarone increases plasma digoxin concentrations. Brit Med J 282:1630
2. Belz GG, Aust PE, Munkes R (1981) Digoxin plasma concentrations and nifedipine. Lancet I: 844–845
3. Belz GG, Matthews J, Doering W, Belz G (1982) Digoxin-antiarrhythmics: Pharmacodynamic and pharmacokinetic studies with quinidine, propafenon and verapamil. Clin Pharmacol Ther 31: 202–203
4. Belz GG, Doering W, Munkes R, Matthews J (1983) Interaction between digoxin and calcium antagonists and antiarrhythmic drugs. Clin Pharmacol Ther 33: 410–417
5. Bigger JTh Jr, Leahey EB Jr (1982) Quinidine and Digoxin. An important interaction. Drugs 24: 229–239
6. Bussey BI, Merritt GJ, Hill EG (1984) The Influence of Rifampin on Quinidine and Digoxin. Arch Intern Med 144: 1021–1023
7. De Cesaris R, Balestrazzi M, Chiarappa R, Ranieri G (1983) Increase of serum digoxin levels after combined administration of calcium antagonists and beta-methyldigoxin. G Ital Cardiol 13: 188–191
8. Dick HLH, McCawley EL, Fisher WA (1962) Reserpine-Digoxin Toxicity. Arch Int Med 109: 503–506
9. Ding R, Piper Ch, de Vries J, Weber E (1984) Nachweis der Digoxin-Chinidin-Interaktion. Verh Dtsch Ges Inn Med 90: 932–934
10. Doering W (1979) Quinidine-digoxin interaction. Pharmacokinetics, underlying mechanism and clinical implications. New Engl J Med 301: 400–404
11. Doering W (1983) Effect of coadministration of verapamil and quinidine on serum digoxin concentration. Eur J Clin Pharmacol 25: 517–521
12. Doering W, Isbary J (1983) Der Einfluß von Tiaprofensäure auf die Digoxin-Konzentration im Serum. Arzneimittelforsch. 33: 167–168
13. Erdmann E (1979) Pharmakodynamische Aspekte medikamentöser Wechselwirkungen – Rezeptorbesetzung und pharmakologische Interaktionen. Internist 20: 229–237
14. Fenster PE, Powell JR, Graves PE, Conrad KA, Hager WD, Goldman S, Marcus FI (1980) Digitoxin-Quinidine Interaction: Pharmacokinetic evaluation. Ann Intern Med 93: 698–701
15. Fenster PE, Comess KA, Dahl Hanson Ch, Finley PR (1982) Kinetics of the digoxin-aspirin combination. Clin Pharmacol Ther 32: 428–430
16. Fichtl B, Doering W (1983) The Quinidine-Digoxin Interaction in Perspective. Clin Pharmacokinetics 8: 137–154
17. Finch MB, Kelly JG, Johnston GD, McDevitt DG (1983) Evidence against a digoxin-indomethacin interaction. Br J Clin Pharmacol 16: 212 P–213 P
18. Friedman HS, Bonventre MV (1982) Erythromycin – Induced Digoxin Toxicity. Chest 82: 202
19. Garty M, Sood P, Rollins DE (1981) Digitoxin elimination reduced during quinidine therapy. Ann Intern Med 94: 35–37
20. Gault H, Longerich L, Dawe M, Fine A (1984) Digoxin-Rifampin Interaction. Clin Pharmacol Ther 35: 750–754
21. Goldstein RA, Passamani ER, Roberts R (1980) A comparison of Digoxin and dobutamine in patients with acute infarction and cardiac failure. New Engl J Med 303: 846–850
22. Hager WD, Fenster P, Mayersohn M, Perrier D, Graves P, Marcus FI, Goldman S (1979) Digoxin-Quinidine Interaction: Pharmacokinetic Evaluation. New Engl J Med 300: 1238–1241

23. Hirsh PD, Weiner HJ, North RL (1980) Further insights into digoxin-quinidine interaction: Lack of correlation between serum digoxin concentration and inotropic state of the heart. Am J Cardiol 46: 863–868
24. Killian JM, Fromm GH (1968) Carbamazepine in the treatment of neuralgia. Use and side effects. Arch Neurol 19: 129–136
25. Klein OH, Lang R, Di Segni E, Kaplinsky E (1980) Verapamil-Digoxin Interaction. New Engl J Med 303: 160
26. Klein OH, Lang R, Weiss E, Di Segni E, Libhaber C, Guerrero J, Kaplinsky E (1982) The influence of verapamil on serum digoxin concentration. Circulation 65: 998–1003
27. Kuhlmann J, Dohrmann M, Marcin S (1984) Effects of quinidine on plasma level and renal excretion of digitoxin achieving steady state conditions. Naunyn-Schmiedeberg's Arch Pharmacol 325 (Suppl): R 85 (Abstr)
28. Kuhlmann J, Marcin S, Frank KH (1984) Pharmakokinetik und kardiale Wirksamkeit von β-Acetyldigoxin und Digitoxin unter einer Kombinationstherapie mit Nifedipin. Klin Wschr 62: 451–457
29. Kuhlmann J, Marcin S (1985) Pharmakokinetic und kardiale Wirksamkeit von β-Acetyldigoxin und Digitoxin unter einer Kombinationstherapie mit Diltiazem. Klin Wschr 63: 636–642
30. Leahey EB, Reiffel JA, Heissenbuttel RH, Drusin RE, Lovejoy WP, Bigger JTh Jr (1979) Enhanced cardiac effect of digoxin during quinidine treatment. Arch Int Med 139: 519–521
31. Leahey EB Jr, Reiffel JA, Giardina E-GV, Bigger JTh Jr (1980) The effect of Quinidine and other oral antiarrhythmic drugs on serum Digoxin. Ann Intern Med 92: 605–608
32. Leahey EB Jr, Bigger JTh Jr, Butler VP Jr, Reiffel TA, O'Connell GC, Scaffidi LE, Rottman JN (1981) Quinidine-digoxin interaction. Time course and pharmacokinetics. Am J Cardiol 48: 1141–1146
33. Lindenbaum J, Rund DG, Butler VP Jr, Tse-Eng D, Ranjan Saha J (1981) Inactivation of digoxin by the gut flora: Reversal by antibiotic therapy. New Engl J Med 305: 789–794
34. Lown B, Ehrlich L, Lipschultz B, Blake J (1961) Effect of Digitalis in patients receiving reserpine. Circulation 24: 1185–1191
35. Manolas EG, Hunt D, Sloman G (1980) Effects of quinidine and disopyramide on serum digoxin concentrations. Aust NZ J Med 10: 426–429
36. Moysey JO, Jaggarao NSV, Grundy EN, Chamberlain DA (1981) Amiodarone increases plasma digoxin concentrations. Br Med J 282: 272
37. Nademanee K, Kannan R, Henrickson J, Burnam M, Kay I, Singh B (1982) Amiodarone-digoxin-interaction during treatment of resistant cardiac arrhythmias. Am J Cardiol 49: 1026
38. Nademanee K, Kannan R, Hendrickson J, Ookhtens M, Kay J, Singh BN (1984) Amiodarone-digoxin-interaction: Clinical significance, time course of development, patiental pharmacokinetic mechanisms and therapeutic implications. J Am Coll Cardiol 4: 111–116
39. Nager G, Nager F (1983) Interaktion zwischen Amiodaron und Digoxin. Schweiz Med Wschr 113: 1727–1730
40. Nola GT, Pope S, Harrison DL (1970) Assessment of the synergistic relationship between serum calcium and digitalis. Am Heart J 79: 499–507
41. Novi C, Bissoli F, Simonati V, Volpini T, Baroli A, Vignati G (1980) Rifampin and Digoxin: Possible Drug Interaction in a Dialysis Patient. JAMA 244: 2521–2522
42. Oyama Y, Fujii S, Kanda K, Akino E, Kawasaki H, Nagata M, Goto K (1984) Digoxin-Diltiazem Interaction. Am J Cardiol 53: 1480–1481
43. Pedersen KE, Hastrop J, Hvidt S (1980) The effect of quinidine on digoxin kinetics in cardiac patients. Acta Medica Scand 207: 291–295
44 Pedersen KE et al (1981) Digoxin-Verapamil Interaction. Act Med Scand 655 (Suppl): 37 (Abstr)
45. Pedersen KE, Dorph-Pedersen A, Hvidt S, Klitgaard NA, Nielsen-Kudsk F (1981) Digoxin-verapamil interaction. Clin Pharmacol Therap 30: 311–316
46. Pedersen KE, Dorph-Pedersen A, Hvidt S, Klitgaard NA, Pedersen KK (1982) The long-term effect of verapamil on plasma digoxin concentration and renal digoxin clearance in healthy subjects. Eur J Clin Pharmacol 22: 123–127

47. Pedersen KE, Dorph-Pedersen A, Hvidt S, Klitgaard NA, Kjaer K, Nielsen-Kudsk F (1982) Effect of nifedipine on digoxin kinetics in healthy subjects. Clin Pharmacol Ther 32: 562–565
48. Pedersen KE, Thayssen P, Klitgaard NA, Christiansen BD, Nielsen-Kudsk F (1983) Influence of verapamil on the inotropism and pharmacokinetics of digoxin. Eur J Clin Pharmacol 25: 155–206
49. Perez HR (1970) Cardiac arrhythmia after Succinylcholine. Anesth Analg 49: 33
50. Peters U, Hausamen T-U, Grosse-Brockhoff F (1974) Einfluß von Tuberkulostatika auf die Pharmakokinetik des Digitoxins. Dtsch Med Wschr 99: 2381–2386
51. Peters U, Risler T, Grabensee B, Falkenstein U, Kroukou J (1980) Interaktion von Chinidin und Digitoxin beim Menschen. Dtsch Med Wschr 105: 438–442
52. Pfeiffer G (1985) Gleichzeitige Gabe von Digoxin und Diltiazem bei Patienten mit Herzinsuffizienz und koronarer Herzerkrankung. Fortschr Med 103: 699–701
53. Rameis H, Magometschnigg D, Ganzinger U (1984) The diltiazem-digoxin interaction. Clin Pharmacol Ther 36: 183–189
54. Rameis H (1985) On the Interaction Between Phenytoin and Digoxin. Eur J Clin Pharmacol 29: 49–53
55. Rau R (1980) Untersuchung der Wechselwirkung von Piroxicam und Digoxin. akt rheumatol 5: 13–18
56. Rau R, Georgiopoulos G, Neumann P, Gross D (1980) Die Beeinflussung des Digoxinblutspiegels durch Antirheumatika. akt rheumatol 5: 349–358
57. Risler T, Burk M, Peters U, Grabensee B, Seipel L (1983) On the interaction between digoxin and disopyramide. Clin Pharmacol Ther 34: 176–180
58. Schenck-Gustafsson K, Dahlqvist R (1981) Pharmacokinetics of digoxin in patients subjected to the quinidine-digoxin interaction. Br J Clin Pharmacol 11: 181–186
59. Schenck-Gustafsson K, Jogestrand T, Brodin L-A, Norlander R, Dahlqvist R (1982) Cardiac effect at steady state during combined treatment with quinidine and digoxin. Br J Clin Pharmacol 14: 571 P–628 P
60. Schick D, Scheuer J (1973) Current concepts of therapy with digitalis glycosides. Part II. Am Heart J 87: 391–396
61. Schrager BR, Pina I, Frangi M, Applewhite S, Sequeira R, Chahine RA (1983) Diltiazem, Digoxin Interaction? Circulation 68 (Suppl III): 368 (Abstr)
62. Schwartz JB, Migliore PJ (1984) Effect of nifedipine on serum digoxin concentration and renal digoxin clearance. Clin Pharmacol Ther 36: 19–24
63. Schwartz JB, Raizner A, Akers S (1984) The effect of nifedipine on serum digoxin concentrations in patients. Am Heart J 107: 669–673
64. Schwartz JB (1985) Digoxin and Nifedipine. Am J Cardiol 55: 31 E–34 E
65. Sherod TR (1967) The cardiac glycosides. Hosp Practice 2: 56
66. Steiness E, Waldorff S, Hansen PB, Kjaergard H, Buch J, Egeblad H (1980) Reduction of digoxin-induced inotropism during quinidine administration. Clin Pharmacol Ther 27: 791–795
67. Theisen F, Pickardt CR, Leisner B, Theisen K, Jahrmärker H (1982) Amiodarone (A) Langzeittherapie (LT) und Nebenwirkungen (NW) bei Patienten (P) mit Lown 4 B (L4B). Z Cardiol 71: 183
68. Thiercelin JF, Hermann Ph, Warrington S, Thenot JP, Morselli PL (1983) Interaction study between Digoxin and Calcium-Antagonists: Verapamil and Diltiazem. II World Conference on Clinical Pharmacology and Therapeutics, Washington (Abstr)
69. Waldorff S, Hansen PB, Kjaergard H, Buch J, Egeblad H, Steiness E (1981) Amiloride-induced changes in digoxin dynamics and kinetics: Abolition of digoxin-induced inotropism with amiloride. Clin Pharmacol Ther 30: 172–176
70. Waldorff S, Hansen PB, Egeblad H, Berning J, Buch J, Kjaergard H, Steiness E (1983) Interactions between digoxin and potassium-sparing diuretics. Clin Pharmacol Ther 33: 418–423
71. Wellens HJJ, Gorgels AP, Braat SJ, Bär FW, Vanagt EJ, Phaf B (1980) Effect of oral disopyramide on serum digoxin levels. A prospective study. Am Heart J 100: 934–935
72. Yoshida A, Fujita M, Kurosawa A, Nioka M, Shichinohe T, Arakawa M, Fukuda R, Owada E, Ito K (1984) Effects of diltiazem on plasma level and urinary excretion of digoxin in healthy subjects. Clin Pharmacol Ther 35: 681–685

Author's address:
Prof. Dr. E. Weber, Medizinische Universitätsklinik, Abteilung Klinische Pharmakologie Heidelberg, Bergheimer Straße 58, 6900 Heidelberg 1 (West Germany)

Importance of serum glycoside concentrations

K. Kochsiek

Medizinische Universitäts Klinik, Luitpoldkrankenhaus, Würzburg (West Germany)

For drug-mediated therapy it was a great advance when dose-related response could be replaced by correlating blood level and effect. The various factors, which related to the dose could influence the blood level, the concentration at the site of effect and finally the intensity of effect, are shown in Table 1. From this synopsis it could be deduced that for digitalis glycosides some of these factors are well-known. The absorption rate of digitoxin is 100 %, for digoxin 80 % and more. It is not advisable to administer drugs with a lower absorption rate. The distribution volume is also known. For the closely protein-bound digitoxin (protein binding 92–97 %), it is about 0.5–0.7 l/kg body weight and for digoxin with a protein

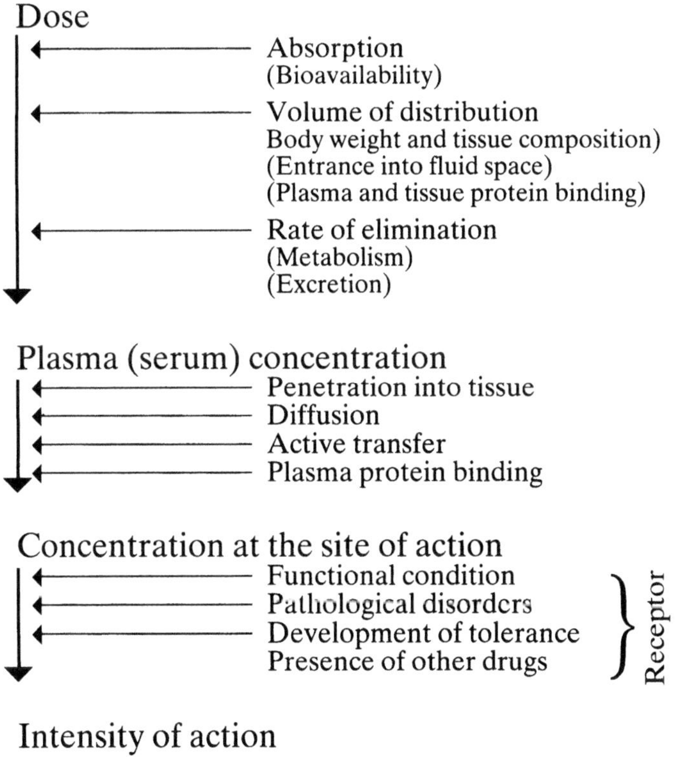

binding of 25 % about 6–10 l/kg body weight, thus for a patient of 70 kg body weight resulting in a distribution volume for digitoxin of 5 l and for digoxin of about 500–600 l respectively. Even metabolic passage and speed of elimination are well known. While digitoxin with a half-life of approximately 200 hours is mainly metabolized, digoxin is almost totally excreted via the kidney. In contrast, the significance and quantitative extent of those factors which are able to influence the concentration of glycosides at the site of action are far less known. Best studied and documented by empirical observations is the importance of pathological conditions on the intensity of effect. Here the synergism of hypokalemia and hypercalcemia and vice versa should be mentioned. The increased sensitivity for glycosides in hypoxemia is well known, especially in cor pulmonale as well as in prolonged severe heart failure, but also in hypothyroidism. In contrast, glycoside tolerance is increased in hyperthyroidism as well as in atrial fibrillation. Interactions with other drugs can influence the effect of glycosides whereas it is often unknown whether a pharmacokinetic or a pharmacodynamic interaction or both components are involved. There are, however, many factors affecting either blood level or definite effect of the drug. Additionally, as a therapeutical effect of digitalis glycosides the increase of inotropism and the influence on AV-conduction is attempted. There are also a variety of effects on other organs in the form of so-called undesired side effects, and especially these unwanted extracardiac effects are only loosely connected with the cardiac effects desired (Table 2).

Table 2. Serum glycoside concentrations with absolute and relative incidence of symptoms in 103 patients with extracardiac side-effects.

Extracardiac side-effects in 103 out of 268 patients (38 %) with glycoside intoxication:

	Number	%	Serum glycoside concentration ($\bar{x} \pm SD$) ng/ml	
Vomiting	74	72	3.9 ± 1.51	$p < 0.025$ (Vomiting vs. Nausea)
Nausea	17	16	3.1 ± 0.80	
Flickering before the eyes	5	5	3.2 ± 0.74	
Coloured vision	3	3	3.3 ± 0.72	
Psychoses	3	3	3.6 ± 1.13	
Tinnitus	1	1	2.6	

Due to these findings, it is not surprising that no threshold value for either therapeutical or toxic effect of glycosides has so determined. In consequence, some authors have denied that the blood level has any significance in the detection of digitalis intoxication (1, 12, 18). Many other authors, though, would not control glycoside administration or diagnose an intoxication without determination of blood levels (2, 14, 17, 20).

Despite very intensified research, the exact mechanism of the positively inotropic effect of cardiac glycosides is still unexplained. Today, however, it is generally accepted that the primary site of action of cardioactive steroids seems to be located in the $(Na^+ + K^+)$ ATPase system, the transport capacity of which is inhibited by very small amounts of cardiac glycosides (21). This system is therefore called digitalis receptor. Due to laboratory methods, it is still not possible to correlate the extent of inhibition of $(Na^+ + K^+)$ATPase to blood levels of glycosides. Since the further intracellular mode of action after inhibition of transmembranous $(Na^+ + K^+)$ATPase transport is not known, there is possibly little probability, even with exact knowledge of quantitative binding at the digitalis receptor, to invent

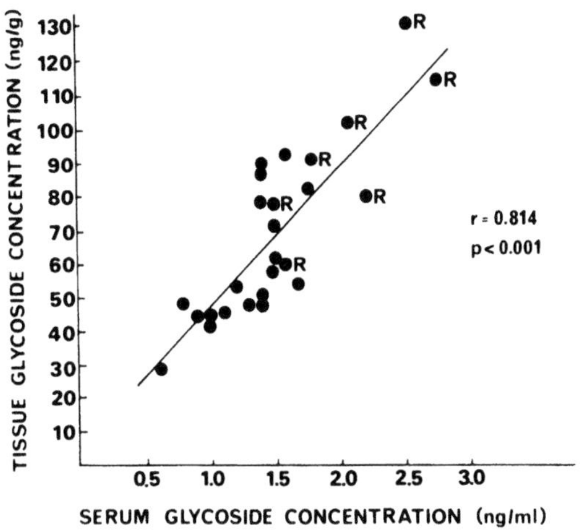

Fig. 1. Correlation between glycoside concentration in left ventricular papillary muscle and serum in 35 patients during cardiac surgery (prosthetic mitral valve replacement); $r = 0.814$; $p < 0.001$ (10). R = postoperative cardiac arrhythmias.

a close correlation to blood levels. As factors of uncertainty should be mentioned the unknown state of the contractile proteins as well as the intracellular energy available. It is all the more surprising that some authors could establish a linear relation between glycoside concentrations in the papillary muscle of the left ventricle and in the serum (Fig. 1). Nevertheless, there are great differences in the literature concerning the quotient of concentrations in the myocardium and serum levels. While Hassis et al. (10) report a mean value of 46:1, the values found by other authors vary from 24:1 to 76:1 with a range from 17:1 up to 155:1 (Table 3). It is therefore not surprising that other authors deny a correlation between concentrations in the myocardium and serum glycoside concentrations (3, 4, 6). It must be taken into account, though, that there are considerable variations in the concentrations of glycosides in different regions of the heart muscle (left and right ventricle, atria, conduction system), and that substantial changes in tissue concentrations might occur post mortem. To some extent, convincing results might be expected only by intravital examinations, for example during heart surgery (11). Findings indicating higher "therapeutic" serum glycoside concentrations in children older than one month are not controversial.

Table 3. Results of determinations of myocardial tissue to serum ratio of glycoside concentration by different authors.

Author	Mean	Range
Binnion et al. (3)	30:1	----
Caroll et al. (1973)	76:1	25:1–128:1
Coltart et al. (4)	68:1	39:1–155:1
Doherty et al. (6)	29:1	17:1– 35:1
Doherty and Perkins (1966)	43:1	35:1– 58:1
Haasis et al. (10)	46:1	----
Guellner et al. (1974)	24:1	18:1– 29:1
Karjalainen et al. (1974)	29:1	----
Selden and Neill (1975)	12:1	----

At present, however, no investigations are known of which correlate these findings to concentrations in the heart muscle.

When interpreting the results of blood level determinations, it is necessary to take into consideration the following pathogenetic mechanisms, which might be responsible for cardiac side effects of glycosides or inappropriate effects (13).

1. Too much free glycoside was supplied. The reason for this was either an overdosage or a change in glycoside elimination or glycoside metabolism. The most important example is the increased accumulation of digoxin or its derivatives in renal failure. This mechanism is recognizable by blood level determinations.
2. Glycoside binding to the receptor is either increased or decreased. This is not detectable by blood level determination but only by observing biological effects. Specific binding for example increases in hypokalemia. Acidosis intensifies the affinity to the receptor whereas alkalosis reducts it (7). Furthermore, a relation to body temperature decreases during febrile conditions (7).
3. Myocardial reaction is either augmented or diminished, e. g. in pre-existent disposition to ectopia or in hypoxemia, possibly also through the interference of drugs such as hydantoin or during stimulation or in inhibition of beta-receptors. These mechanisms are not detectable by blood level determinations either.

To make convincing interpretations of measured serum levels, further information is indispensable. The dose and intervals of administration of specific digitalis preparations must be known. Furthermore, it is very important to know the exact time of the latest dose. Many misinterpretations of serum levels are due to the fact that the distribution of the drug was not completed, which for digitalis glycosides takes 8 to 10 hours. Since constant concentrations can only be obtained in steady state conditions, pharmacokinetics and the duration of administration must also be known. The control of digitalis levels makes sense only during maintenance therapy; blood sampling should be done at the earliest 8 and at the latest 24 hours after the latest drug administration.

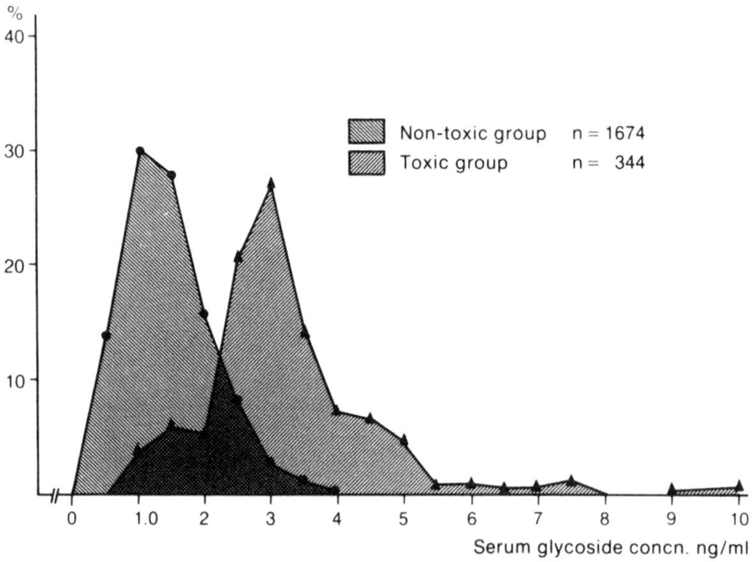

Fig. 2. Percentage distribution of serum glycoside concentrations in 1674 patients without toxic symptoms and 344 patients with symptoms of intoxication. Overlapping range: 0.6–3.8 ng/ml.

Considering the diverse influences, which are to some extent difficult to explain, it is not surprising that in the individual case the determination of serum concentrations of glycosides for control of therapy is disappointing. Nevertheless, it cannot be ignored that since the introduction of more sensitive tests for blood level determinations, digitalis therapy has not only become safer but has been based on a rationale by providing a certain security to pure empiricism by measure and figure. Numerous systematic examinations for digoxin and its derivatives and for digitoxin proved these findings (9, 17, 19, 23, 24). In our own investigations (14, 17) in 1,674 non-intoxicated and 344 intoxicated patients, a relation between blood level and extent of intoxication could be demonstrated. However variations in the individual case were considerably high. The total collective of patients is shown in Fig. 2. A large overlapping area ranging from 0.6 ng/ml up to 3.8 ng/ml can be seen. Some authors decided that serum glycoside concentrations are of no specific importance for the control of glycoside administration or the diagnosis of an intoxication (8, 12). Fig. 3 indicates that starting at a serum glycoside concentration of 2.0 ng/ml, the probability of a manifest intoxication is constantly increasing, and that in our very large collective, beginning at a serum glycoside concentration of 3.8 ng/ml all patients were intoxicated. The relatively broad overlapping area is due to a different sensitivity to digitalis, the determination of which has already been mentioned above, and which to a large extent could be fixed by specific investigations. When an intoxication is suspected, despite a not yet toxic serum glycoside concentration, one should assume toxicity, especially if some of these determinants are present. Nevertheless, in the individual case, it may be difficult to prove or exclude an intoxication if only one serum value is taken into account (Fig. 4). Additionally, it must be pointed out that patients with atrial fibrillation usually tolerate reasonably high serum levels, located in the so-called toxic range, without any signs of intoxication. Koch-Weser et al. (15) were able to show that determination of blood levels controlling glycoside administration could reduce the frequency of intoxication from 13.9% to 5.9%.

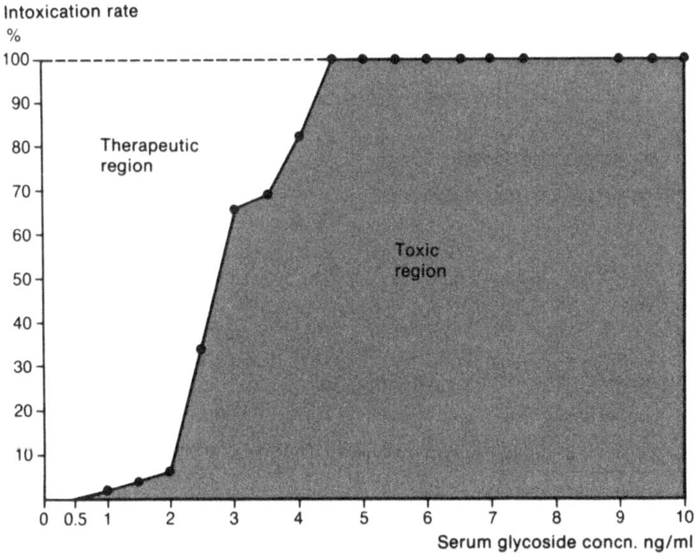

Fig. 3. Percentage intoxication rate in relation to serum glycoside concentration in 2018 patients receiving long-term treatment with digoxin, β-acetyldigoxin and β-methyldigoxin.

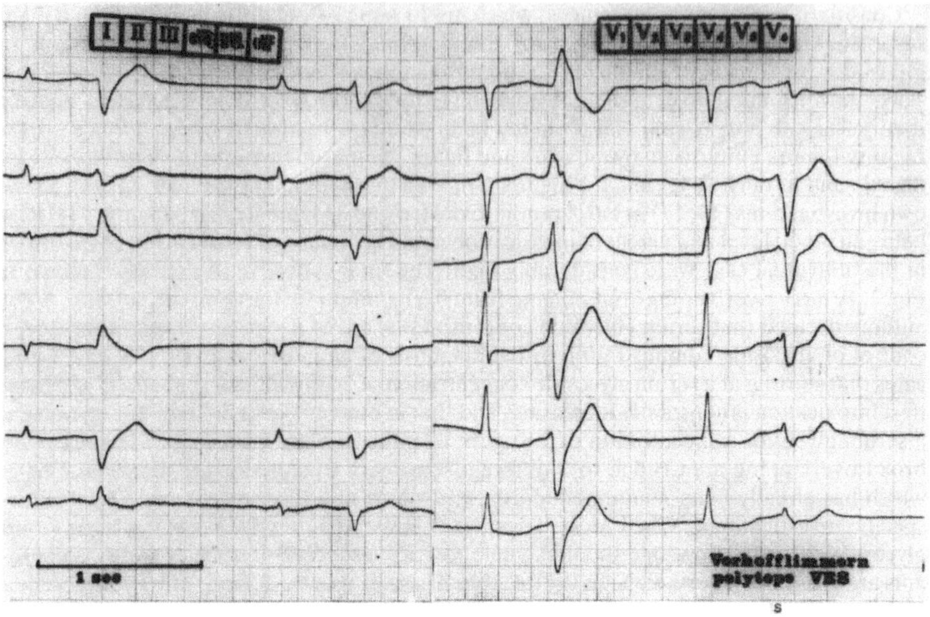

Fig. 4. ECG of a patient with digitoxin intoxication with a "normal" serum concentration of 21.3 ng/ml.

We were able to make a more detailed study of 246 patients with respect to their cardiac and 103 patients with respect to their extracardiac symptoms. Results are shown in Figs. 5–8 and Table 2. It could be demonstrated that the disturbance of AV-conduction, detectable

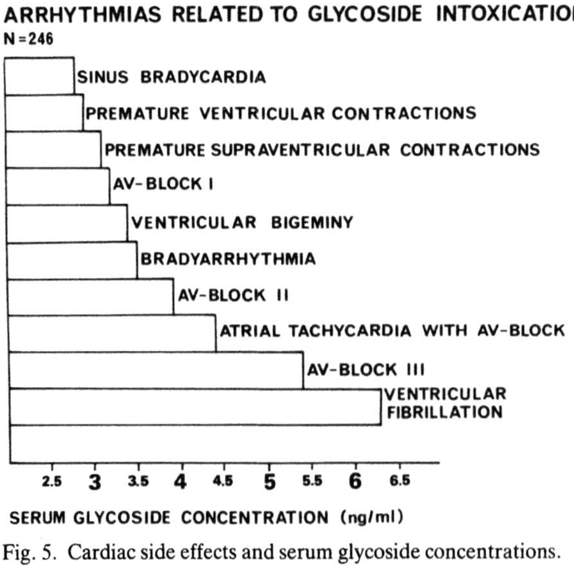

Fig. 5. Cardiac side effects and serum glycoside concentrations.

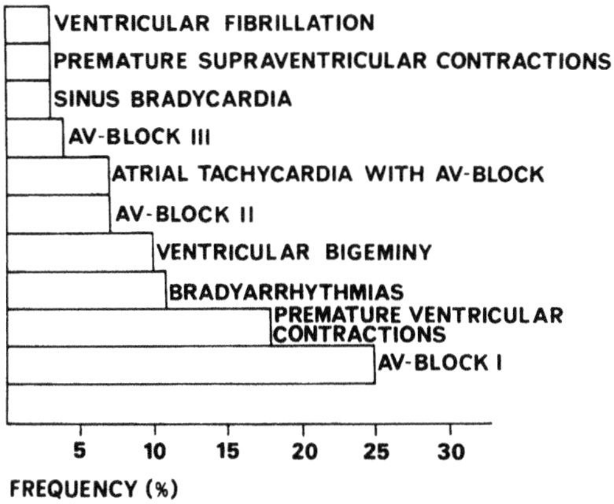

Fig. 6. Frequency of cardiac side effects.

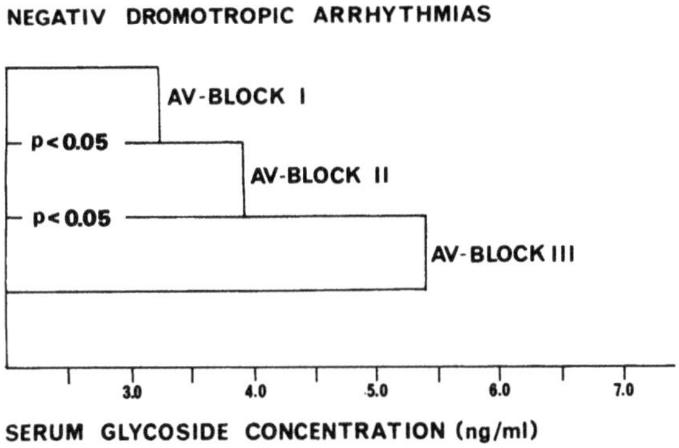

Fig. 7. Serum glycoside concentrations and negative dromotropic cardiac arrhythmias in 246 cases of glycoside intoxication confirmed by ECG.

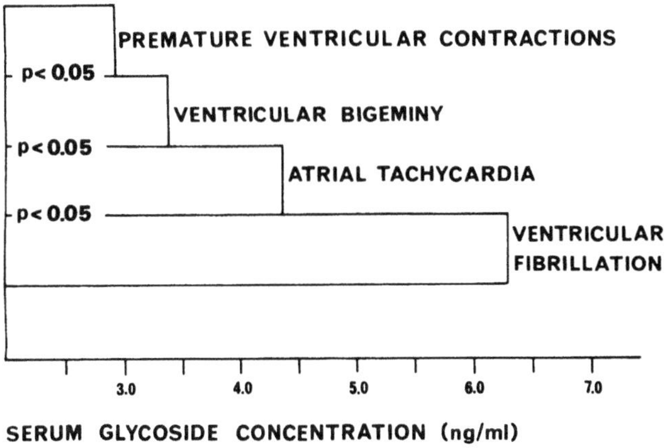

Fig. 8. Serum glycoside concentrations and positive bathmotropic cardiac arrhythmias in 246 cases of glycoside intoxication confirmed by ECG.

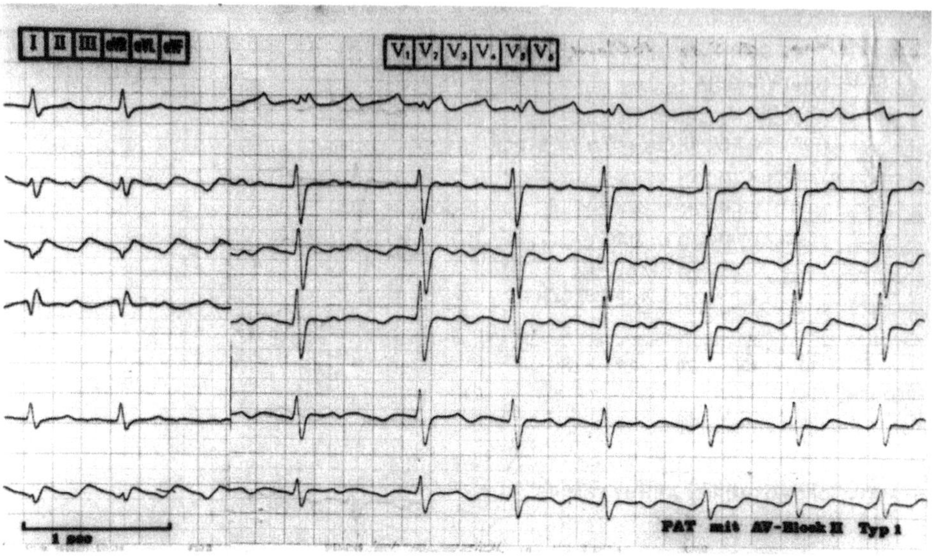

Fig. 9. ECG of a patient with paroxysmal atrial tachycardia with AV block II. Digoxin serum concentration 4.3 ng/ml.

only by ECG, is the most frequent symptom of glycoside intoxication and is much more common than ventricular extrasystoles or bigeminus. Figs. 5, 7 and 8 show impressively that the critical nature of cardiac intoxication symptoms is augmented by increasing glycoside concentrations. Mention must be made of atrial tachycardia with block (Fig. 9), which is accompanied by a relatively high mortality, and which, because of its often irregular tachycardial rhythm, remains undetected without ECG controls, thus resulting in further administration of glycosides. Figs. 7 and 8 demonstrate clearly that the hazard of disturbances of conduction or ectopic beats grows with increasing glycoside concentrations; this holds true only to some extent for patients with pacemakers in whom no conduction disturbances due to glycosides may occur. Apparently, pacemakerinduced ventricular rhythms are more rarely accompanied by glycoside-induced extrasystoles. Thus 54 pacemaker patients with "toxic" serum glycoside concentrations between 2.3 and 4.0 ng/ml behaved markedly differently with respect to their cardiac side effects than non-pacemaker patients suffering from the same "toxic" serum glycoside concentrations (Table 4). In pacemaker patients, the well-

Table 4. Glycoside-toxic effects at high serum glycoside concentration ($\geqq$ 2.3 ng/ml) during long-term therapy with digoxin and digoxin derivatives in 54 pacemaker patients in comparison to 332 patients without pacemakers (11).

Serum glycoside concentration range 2.3–4.0 ng/ml (n = 386)		
	Pacemaker patients (n = 54)	Non-pacemaker patients (n = 332)
Ventricular extrasystoles due to glycoside	7%	26%
Total arrhythmias due to glycoside	7%	63%
Extracardiac side-effects	22%	21%
No signs of glycoside intoxication	71%	33%

known warning signals indicating an increasing serum glycoside concentration are often missing, thus augmenting the danger of a sudden occurrence of fatal arrhythmias. Control of digitalis administration in these patients, therefore, is possible only by regular determination of the blood level. In Table 2 one can see the incidence of extracardiac symptoms and their specific serum glycoside concentrations. Despite high blood levels, nausea occurs relatively seldom, which may be ascribed to the fact that this symptom as an expression of digitalis intoxication has received little attention. Only if it is accompanied by vomiting will the right diagnostic conclusion be drawn.

As a consequence of the experiences of the past 15 years, the following points could be an indication to determine blood levels with diagnostic or therapeutic consequences:

1. Did the patient really swallow digitalis? This question plays an important role in unconscious patients, but it may also be necessary to estimate the patient's compliance.
2. Arrhythmias are the most common unwanted cardiac effects of digitalis glycosides. Normally, these undesired side effects are checked by ECG since there is a relatively close connection between the critical nature of rhythm disturbances and the blood level. If these typical variations of the ECG are not detectable (bundlebranch block, pacemaker), control is only possible by determination of serum glycoside concentrations.
3. Patients with impaired renal functions on digoxin administration can be safely controlled only by measuring the serum glycoside concentrations. This is similar in patients with reduced body weight because of restricted distribution volumes.
4. Drug interactions are only detectable by observing serum glycoside concentrations. At the moment, however, clinical relevance is given only to interactions between digitalis glycosides and quinidine (5).
5. Measurements of the biovailability of glycoside preparations is possible only by determination of serum glycoside concentrations. Administration of drugs with a biovailability of less than 80 % is not advisable.
6. Control of serum glycoside concentrations is indicated in each deteriorating patient whose digitalis maintenance therapy had previously been effective.

In conclusion: Nowadays therapy with digitalis glycosides without the possibility of estimating the serum glycoside concentrations is unimaginable. The physician who – irrespective of what the reasons are – orders the determination of blood levels must bear in mind that there is no fixed value for an optimal dosage. Each value needs interpretation. The knowledge of serum glycoside levels, however, facilitates the control of digitalis administration, but it can never replace the precise clinical observation of the patient, including ECG controls and other different laboratory parameters.

References

1. Arnim Th v, Krawietz W, Vogt W, Erdmann E (1980) Is the determination of serum digoxin concentration useful for the diagnosis of digitalis toxicity? Int J Clin Pharmacol Ther Toxicol 18: 261–268
2. Beller GA, Smith TW, Abelman WH, Haber E, Hood WB Jr (1971) Digitalis intoxication: A prospective clinical study with serum level correlations. New Engl J Med 284: 989
3. Binnion PF, Morgan LM, Stevenson HM (1969) Plasma and myocardial digoxin concentrations in patients on oral therapy. Brit Heart J 31: 636
4. Coltart J, Howard M, Chamberlain D (1972) Myocardial and skeletal muscle concentrations of digoxin in patients on long-term therapy. Br Med J 2: 318
5. Doering W, König E (1978) Anstieg der Digoxinkonzentration im Serum unter Chinidinmedikation. Med Klin 73: 1085
6. Doherty JE, Perkins WH, Flanigan WH (1967) The distribution and concentration of tritiated digoxin in human tissues. Ann Intern Med 66: 116

7. Erdmann E (1981) Influence of cardiac glycosides on their receptor. In: Greeff K (ed) Handbook of Experimental Pharmacology, Vol 56. Springer-Verlag Berlin Heidelberg New York, pp 337–380
8. Fogelman A, Finkelstein MS, La Monte JT, Rado E, Pearce ML (1971) Fallibillity of plasma digoxin in differentiating toxic from nontoxic patients. Lancet i: 727
9. Greeff K, Strobach H, Verspohl E (1975) Ergebnisse radioimmunologischer Bestimmungen von Digitoxin, Digoxin und g-Strophanthin am Menschen. In: Jahrmärker H (Hrsg) Digitalistherapie. Springer-Verlag, Berlin Heidelberg New York, p 52
10. Haasis R, Larbig D, Stunkat R, Bader H, Seboldt H (1977) Radioimmunologische Bestimmung der Glykosidkonzentration im menschlichen Gewebe. Klin Wschr 55: 23
11. Haasis R, Larbig D, Decot M, Futter G (1977) Zur Glykosidtherapie bei Schrittmacherpatienten. Verh Dtsch Ges Inn Med 83: 1653
12. Ingelfinger JA, Goldman P (1976) The serum digitalis concentration – does it diagnose digitalis toxicity? New Engl J Med 294: 867
13. Jahrmärker H (1975) Probleme der Digitalisintoxikation. In: Jahrmärker H (Ed) Digitalistherapie. Springer-Verlag, Berlin Heidelberg New York, p 80
14. Kochsiek K, Larbig D, Haasis R (1977) Klinik und Therapie der Digitalisintoxikation. Verh Dtsch Ges Inn Med 83: 99–115
15. Koch-Weser J (1972) Serum drug concentrations as therapeutic guides. New Engl J Med 287: 227
16. Koch-Weser J (1974) Biovailability of drugs. New Engl J Med 291: 233/503
17. Larbig D, Haasis R, Kochsiek K (1977) Die Glykosidkonzentration und ihre klinische Bedeutung. Forum cardiologicum 15, Boehringer Mannheim, 1–138
18. Lasagna L (1976) How useful are serum digitalis measurements. New Engl J Med 294: 890–899
19. Paumgartner G (1976) Die Serumkonzentration von Digitalisglykosiden als therapeutische Richtlinie. Schweiz Med Wschr 106: 909
20. Ochs HR, Greenblatt DJ, Harmatz JS, Bodem G, Dengler HJ (1981) Clinical implications of serum digoxin concentrations. Klin Wochenschr 59: 501–507
21. Smith TW, Butler VP, Haber E (1969) Determination of therapeutic and toxic serum digoxin concentrations by radio-immunoassay. New Engl J Med 281: 1212
23. Smith TW, Haber E (1970) Digoxin intoxication: The relationship of clinical presentation to serum digoxin concentration. J Clin Invest 49: 2377
24. Smith TW, Haber E (1971) The clinical value of serum digitalis glycoside concentrations in the evaluation of drug toxicity. Ann N Y Acad Sci 179: 322
25. Schatzmann HJ (1953) Herzglykoside als Hemmstoffe für den aktiven Kalium- und Natriumtransport durch die Erythrocytenmembran. Helv Physiol Acta 11: 346–354

Author's address:

Prof. Dr. K. Kochsiek, Medizinische Universitäts Klinik im Luitpoldkrankenhaus, Josef-Schneider-Straße 2, 8700 Würzburg (West Germany)

Metabolic aspects of compensatory mechanisms in cardiac failure*

W. Kübler, R. Dietz, W. Mäurer and A. Schömig

Abteilung Innere Medizin III (Kardiologie) University of Heidelberg (West Germany)

Summary

Three metabolic adaptive or compensatory mechanisms of heart failure were discussed:

1. Adaptation of energy production and energy availability in the myocardial cell. With increased myocardial oxygen demands this is achieved by a progressive displacement of the mass action ratio of the CPK reaction, so that pronounced changes in the CP/C ratio, related to myocardial oxygen consumption, are accompanied by only small changes in ATP/ADP and hence in free energy of the adeninenucleotide system.

2. Adjustment of the oxygen availability by adaptation of the hemoglobin dissociation curve due to an increase in the erythrocyte content of 2,3-DPG. This is accompanied by a swelling of erythrocytes as a consequence of an increase in the Gibbs-Donnan potential. In patients with congestive heart failure 2,3-DPG-synthesis is augmented due to respiratory alcalosis and increased concentrations of deoxygenated hemoglobin.

3. Increase in the sympathetic drive of the heart due to increased net discharge of the neurotransmitter caused by reduced neuronal reuptake of norepinephrine. The diminished myocardial norepinephrine content in heart failure is due to the diminished neuronal uptake and to insufficient de novo catecholamine synthesis in the heart. Rather than tyrosine-hydroxylase the transformation of dopamine to norepinephrine seems to be the rate limiting step for catecholamine synthesis in heart failure.

Myocardial failure due to increased workload of the heart leads to various metabolic adaptive and compensatory mechanisms:

1. Adaptation of energy production and energy availability in the myocardial cell.
2. Increase in oxygen availability by adaptation of the hemoglobin dissociation curve.
3. Increase in the sympathetic drive of the heart.

First aspect: Adaptation of energy production and energy availability in the myocardial cell

In acute heart failure the myocardial contents of high energy phosphates – such as CP and ATP – are reduced.

Figure 1 shows the relation between CP-content and myocardial oxygen consumption. If allowance is made for increased oxygen consumption, the values obtained in acute myocardial failure are within normal limits, and the same holds true for ATP (2).

* The studies have been supported by a grant from the Deutsche Forschungsgemeinschaft within the SFB 90 – kardiovaskuläres System – University of Heidelberg/Germany

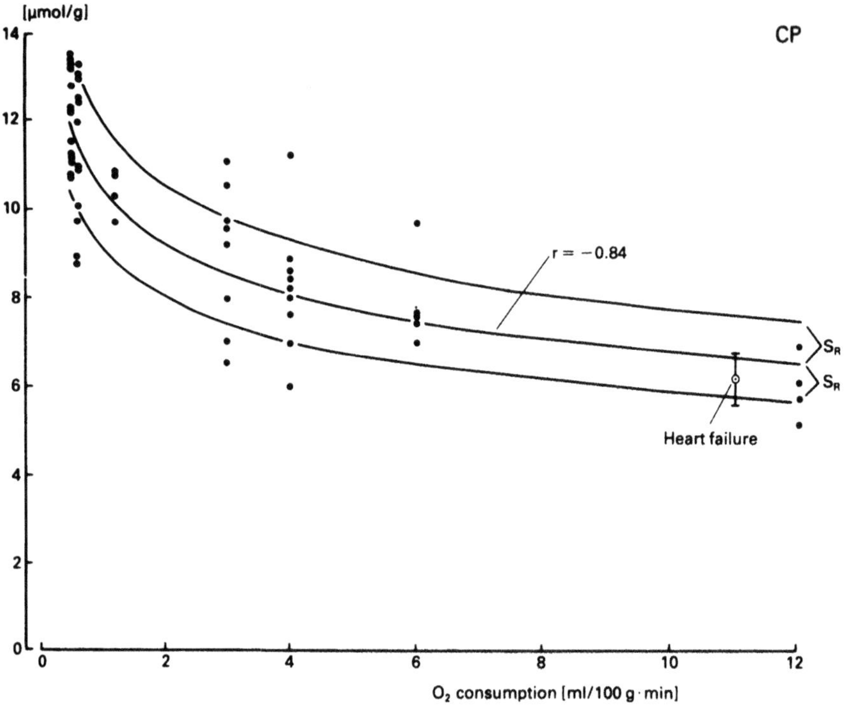

Fig. 1. The relation between myocardial CP content and myocardial oxygen consumption at normothermia. The variation of CP content and oxygen consumption was achieved by different work loads of the heart and by different forms of artificial cardiac arrest (cardioplegia). All experiments were performed in vivo using a heart-lung machine. For experimental details see (3). A hyperbolic regression line and its standard deviation are revealed.

The results suggest that the reduction in the content of the high energy phosphates – CP and ATP – in acute heart failure is an adaptive mechanism to compensate for increased myocardial oxygen consumption (2).

Stimulation of oxygen consumption by the mitochondria is regulated by reduction of the potential of high energy phosphates. This decrease, however, leads in addition to a reduction of the free energy despite increased demands. Hence, this adaptive mechanism has two controversial aspects (see Fig. 2).

Although CP is synthetized and degraded via ATP/ADP by the CPK-reaction the energy state of the myocardium is better characterized by the CP/C system, which seems to be directly related to the regulation of mitochondrial oxygen consumption in the heart for the following reasons:

– There is a specific CPK located at the mitochondrial membrane.

– The degradation of CP during ischemia is directly related to myocardial oxygen consumption, for the bulge of ATP such a relation cannot be found (Fig. 3).

– There is an almost linear relationship between C/CP quotient and myocardial oxygen consumption, explaining the close adaptation of myocardial oxygen consumption to myocardial energy demands. Even after correction for the free and unbound part of ADP, such a relation cannot be found for the ADP/ATP quotient (Fig. 4).

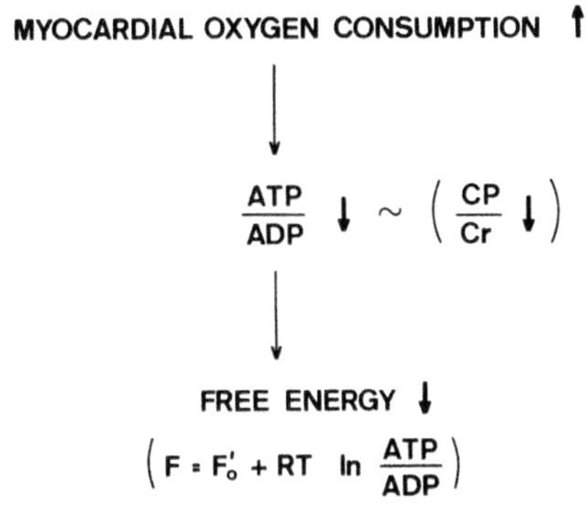

Fig. 2. See text.

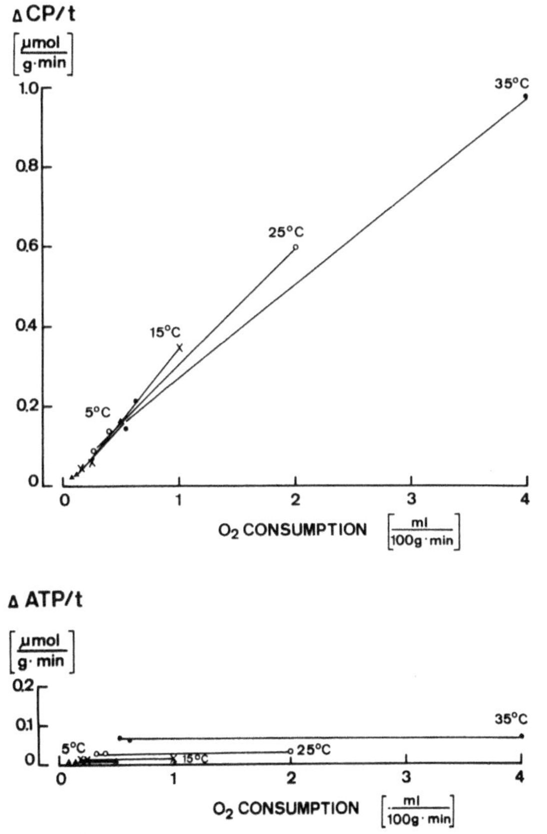

Fig. 3. (Upper part): The relation between myocardial oxygen consumption before ischemia and the velocity of CP breakdown during ischemia. Variation of myocardial oxygen consumption and the felocity of CP breakdown was achieved by various methods of cardioplegia. In addition the experiments were performed at different temperatures.
(Lower part): The relation between myocardial oxygen consumption before ischemia and the velocity of breakdown of the bulk of myocardial ATP during ischemia.

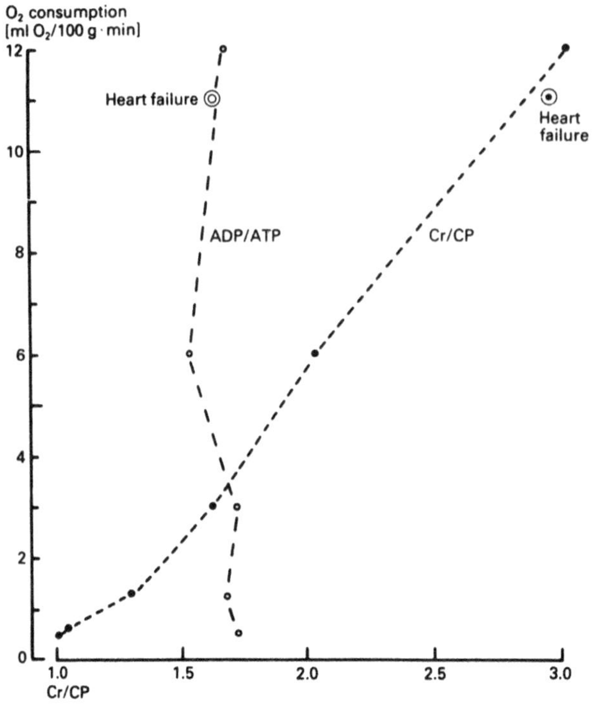

Fig. 4. The relation between myocardial oxygen consumption and the C/CP ratio as well as the ADP/ATP ratio of myocardial metabolite content are shown. For C, CP and ATP, total myocardial cell contents are given: for ADP, only the free and unbound part (taken from calculations by Achs and Garfinkel – personal conmunication) has been used for the calculations.

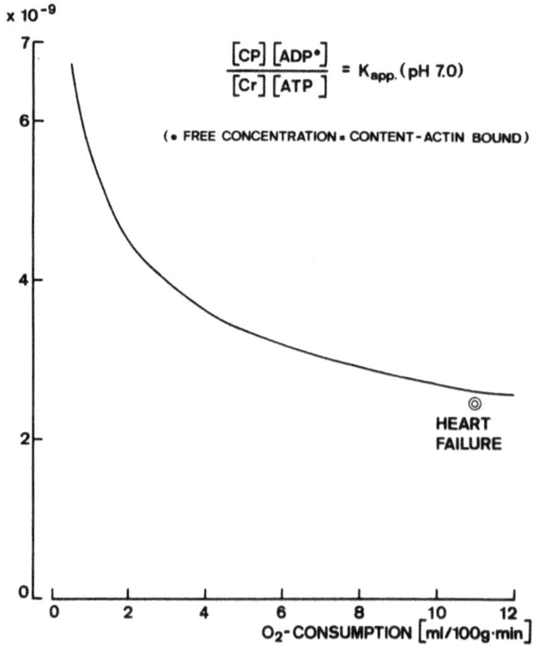

Fig. 5. The mass action ratio of the CPK reaction, calculated from the cell contents of Cr, CP, ATP and the free and unbound part of ADP. The data are given in relation to myocardial oxygen consumption.

From these data it follows that the CP/C ratio is a good index of mitochondrial respiration in the myocardium. By a more than 20 fold increase in myocardial oxygen consumption from 0.5 to 12 ml/100 g · min, it can be calculated from the cell contents of the CP/C system that the free energy of the high energy phosphate system directly related to myocardial oxygen consumption is decreased by more than 20%. For the same increase in myocardial oxygen consumption, however, the reduction in free energy for myocardial ATP/ADP amounts only to less than 4%.

This comparatively small decrease in the free energy of myocardial ATP/ADP is due to a progressive displacement of the mass action ratio of the CPK reaction – in favour of ATP synthesis – with increasing myocardial oxygen consumption. This concept does not imply compartmentalization of high energy phosphates in the myocardium, but rather changes in the kinetics of the CPK reaction with a shift in the apparent equilibrium constant (K*app) with increasing myocardial oxygen consumption (Fig. 5).

Second aspect: Increase in oxygen availability by adaptation of the hemoglobin dissociation curve

The function of the red blood cells to transport oxygen is regulated by changes in H+ concentration and by organic phosphate compounds, predominantly by 2,3-DPG, a side product of glycolysis. Increase in the 2,3-DPG content of erythrocytes leads to a rightwards shift of the hemoglobin dissociation curve so that for the same reduction in oxygen saturation a higher oxygen pressure is achieved.

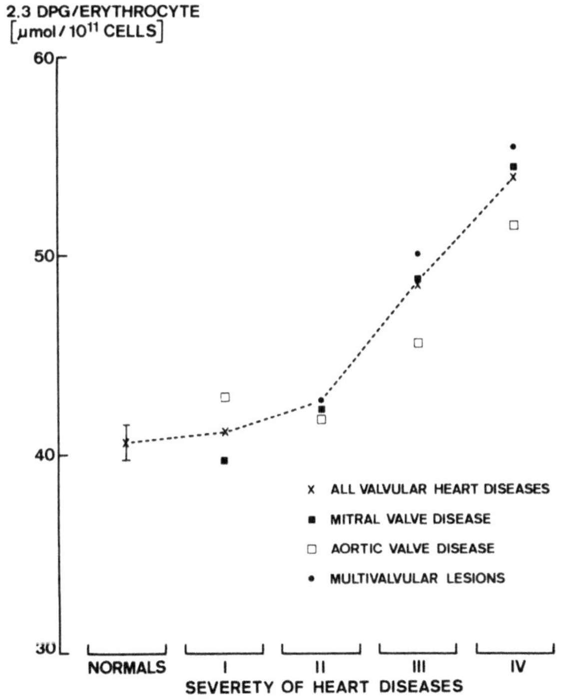

Fig. 6. The 2,3-DPG content of erythrocytes in normal persons and in patients with valvular heart diseases of Classes I–IV. The explanation of the symbols is given in the figure.

Figure 6 reveals the 2,3-DPG content of erythrocytes in 50 normal persons and in 76 patients with valvular heart diseases of NYHA Classes I–IV. Whereas the 2,3-DPG content in normals and in patients with moderate heart diseases of Classes I and II show only small, statistically not significant differences, with greater severity of heart diseases, i.e. Classes III and IV, a significant increase in the 2,3-DPG content of erythrocytes can be observed. This increase in the 2,3-DPG content is proportional to the severity and independent of the underlying disease.

The increased 2,3-DPG levels in the erythrocytes of patients with severe heart diseases are probably due to 2 mechanisms:

– By an increase in pH glycolysis and hence 2,3-DPG formation are activated. Probably due to hyperventilation in patients of Classes III and IV respiratory alcalosis exists, with a mean serum-pH in our patients of 7.46. This leads to activation of glycolysis and 2,3-DPG formation in the erythrocytes.

– 2,3-DPG is bound to deoxygenated hemoglobin, in this way the free concentration is reduced and synthesis augmented by relief of substrate inhibition of the synthetizing enzyme, 2,3-phosphoglycerate mutase. That this mechanism may be operating in patients can be assumed from the finding that there is a negative correlation between erythrocyte 2,3-DPG levels and the central venous oxygen saturation (Fig. 7).

In Fig. 8 the erythrocyte 2,3-DPG levels are again shown as a function of the severity of heart diseases. In addition, the hemoglobin content per volume erythrocytes and the mean

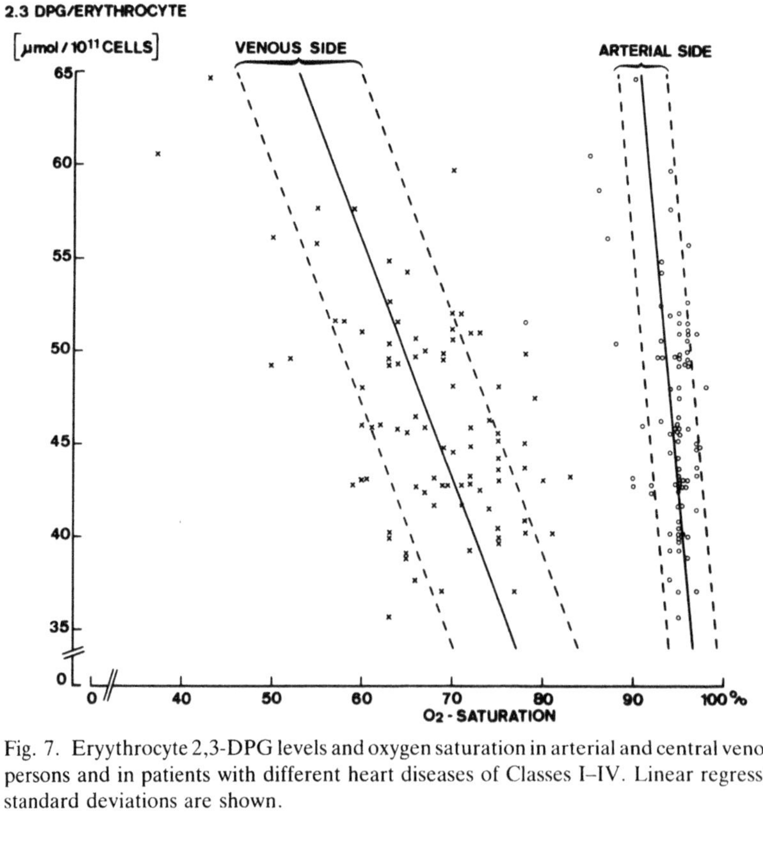

Fig. 7. Eryythrocyte 2,3-DPG levels and oxygen saturation in arterial and central venous blood in normal persons and in patients with different heart diseases of Classes I–IV. Linear regression lines and their standard deviations are shown.

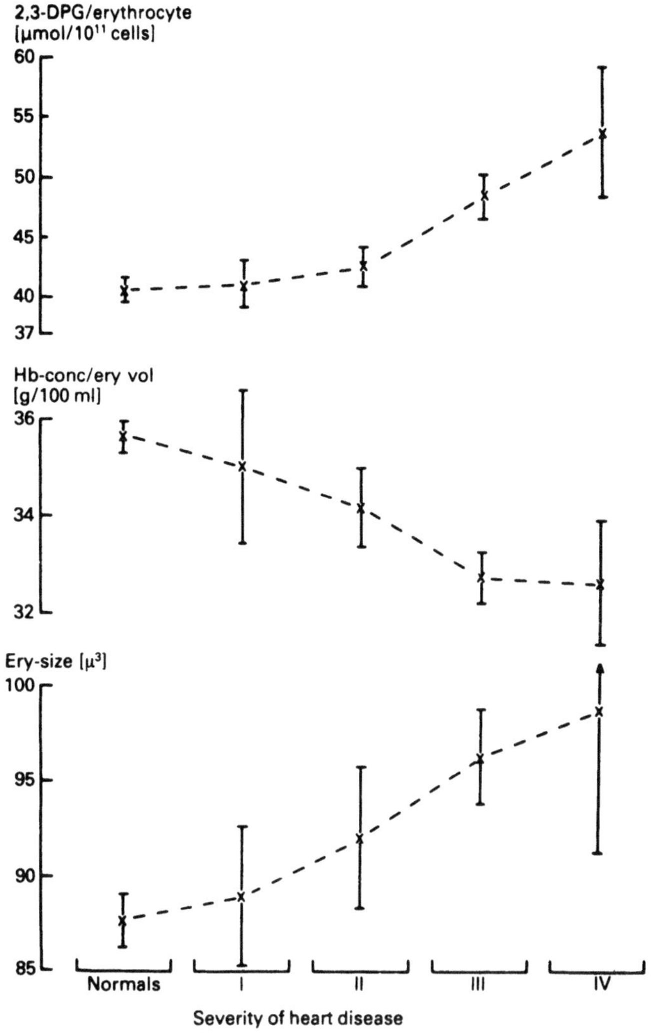

Fig. 8. (Upper part): The erythrocyte 2,3-DPG content (ordinate) in normal persons and in patients with valvular heart diseases of Classes I–IV (abscissa).
(Middle part): For the same patient groups (abscissa) the hemoglobin concentration per volume erythrocytes is given.
(Lower part): For the same patient groups (abscissa) the mean size of the single erythrocyte (ordinate) is shown.
For each value the mean and the 5% confidence limits, assuming Student's distribution, are indicated.

size of the single erythrocyte is revealed. It can be seen that with increasing severity of heart disease not only the 2,3-DPG levels but also the volume of the erythrocytes is increased, indicated by a decrease in hemoglobin concentration per volume erythrocytes and by the increase in mean size of the single erythrocyte.

From these data it follows that the size of the erythrocytes is correlated with the severity of heart diseases. This finding can be explained on the basis of the increased 2,3-DPG levels. With increasing severity of heart diseases the 2,3-DPG content and hence the intracellular concentration of impermeable anions is augmented, leading to an increase in the Gibbs-Donnan potential, and as a consequence to an increase in the equilibrating osmotic potential, which can be considered as the ultimate cause for erythrocyte swelling in patients with severe heart diseases.

Third aspect: Increase in the sympathetic drive of the heart

Since the work of Braunwald and his group (for summary see (1)) it is well established that in heart failure hyperactivity of the sympathetic system exists with increased plasma levels and increased urine excretion of norepinephrine. In severe cases also the plasma levels and urinary excretion of epinephrine, but not those of dopamine, are increased (4).

This increase in sympathetic drive has generally been attributed to depressed left ventricular function. The relationship between left ventricular ejection fraction and plasma norepinephrine, however, is not very close. High plasma levels of norepinephrine are almost always found in patients with depressed ejection fraction below 30%. In some of these patients, however, plasma norepinephrine was even found to be within normal range (Fig. 9).

If, however, depressed left ventricular function was assessed by reduced cardiac index a clearer relationship to resting levels of norepinephrine could be obtained (Fig. 9). The same holds true for epinephrine (5). These results suggest that increased plasma levels of catecholamines in heart failure may be related more to reduced peripheral perfusion than to depressed left ventricular function itself.

The increased levels of catecholamines in heart failure can be either due to an increased centrally mediated norepinephrine release, or to a reduced activity of the inactivation mechanisms for norepinephrine. Among these inactivation processes neuronal re-uptake is of the greatest importance.

Plasma catecholamine levels and inactivation processes – especially neuronal uptake – were therefore investigated in 3 different animal models of heart failure or depressed ventricular function:

1. Spontaneously hypertensive rats
2. Genetically cardiomyopathic hamsters
3. Chronically alcohol treated rats.

Spontaneously hypertensive rats at the age of 50 weeks reveal increased plasma norepinephrine concentrations compared to Wistar rats, which served as controls. At the same time neuronal norepinephrine uptake was significantly reduced in the spontaneously hypertensive rats.

Likewise in cardiomyopathic hamsters at the age of 38 weeks plasma norepinephrine levels are increased and neuronal norepinephrine uptake is reduced compared to control animals (Fig. 10).

Due to decreased neuronal uptake, enzymatic norepinephrine degradation by monoamine oxidase is also reduced in the cardiomyopathic hamster. Likewise extraneuronal catecholamine degradation due to catechol-O-methyl-transferase activity is reduced in the cardiomyopathic animals. On an quantitative basis, however, the reduction of neuronal uptake is by far of the greatest importance.

Neuronal catecholamine uptake may be further diminished by substances which are "cardiotoxic", such as alcohol in larger amounts. Acute administration of alcohol leads only

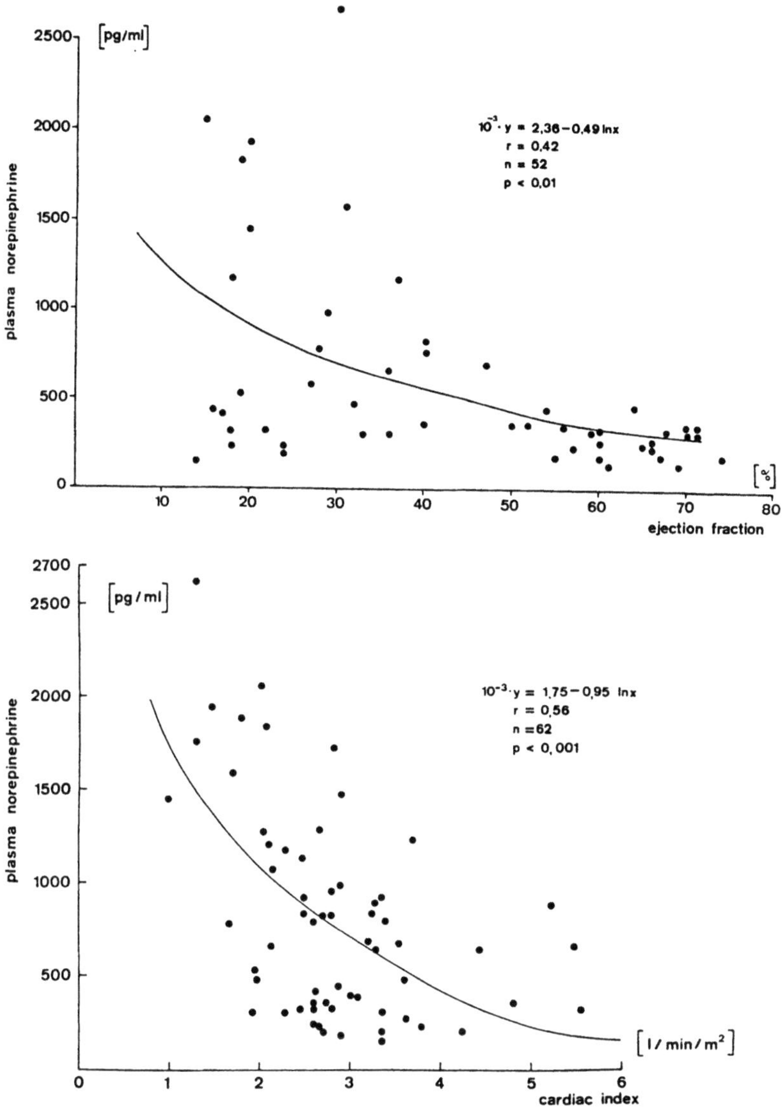

Fig. 9. (Upper panel): Relationship between plasma norepinephrine levels at rest and left ventricular ejection fraction in 52 patients with left heart disease of different origin.
(Lower panel): Relationship between plasma norepinephrine levels at rest and cardiac index in 52 patients with left heart disease of different origin.

to a small reduction in neuronal uptake, this applies to both normal Wistar rats and spontaneously hypertensive rats with and without chronic pretreatment with alcohol.

In contrast to the acute effect, chronic alcohol administration induces a far greater reduction in neuronal norepinephrine uptake in Wistar rats as well as in spontaneously hypertensive rats.

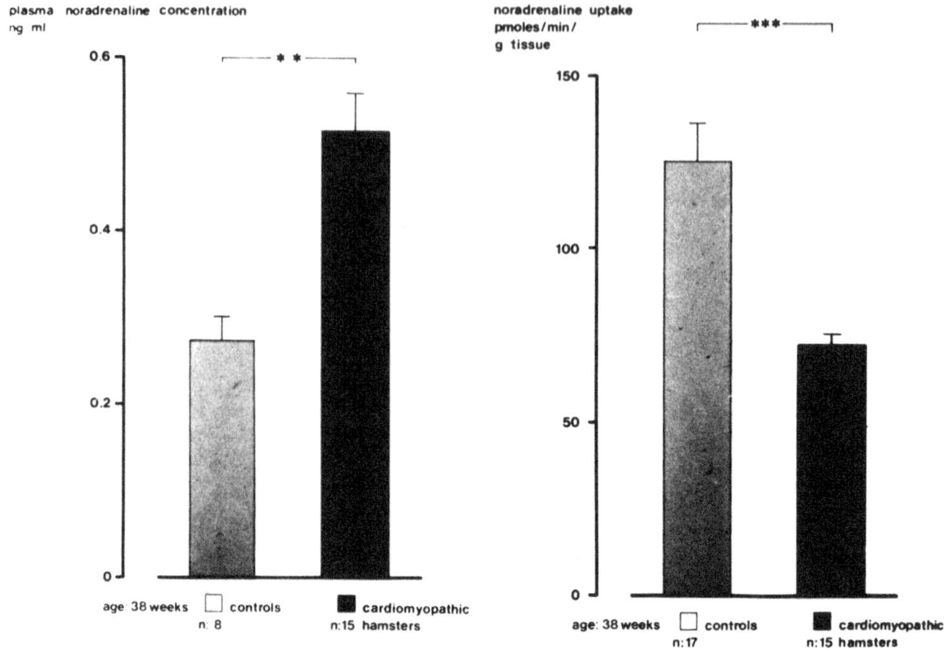

Fig. 10. (Left panel): plasma noradrenaline concentration in cardiomyopathic hamsters (n = 15) and normal controls (n = 8).
(Right panel): Neuronal uptake of noradrenaline in the hearts of cardiomyopathic hamsters (n = 15) and of normal controls (n = 17).
Mean values ± SEM;
significance levels: $^{**}p < 0.01$, $^{***}p < 0.001$.

The data obtained in these experimental models of heart failure allow the following conclusions:

1. The increased plasma levels of norepinephrine in heart failure are predominantly due to a reduced inactivation of the neurotransmitter.

2. The reduced elimination of norepinephrine by the failing heart is almost completely caused by an impairment of the neuronal uptake.

3. Factors which impair the neuronal uptake cause an increased concentration of norepinephrine at the receptor site, in this way contributing to the increased sympathetic activity.

This increased sympathetic drive of the heart is accompanied by a reduced content of the neurotransmitter in the synaptic nerve endings. Hence, in heart failure sympathetic hyperactivity is paralleled by reduced myocardial norepinephrine stores. This can be attributed to the diminished neuronal uptake as observed in heart failure, provided catecholamine de novo synthesis is not markedly increased.

As can be seen in Fig. 11, in patients with chronic aortic regurgitation in NYHA Classes III–IV, right atrial contents of norepinephrine, normetanephrine and epinephrine were significantly reduced compared to the myocardial tissue levels being found in patients with atrial septal defects. Right atrial contents of dopamine did not reveal any differences between the 2 groups.

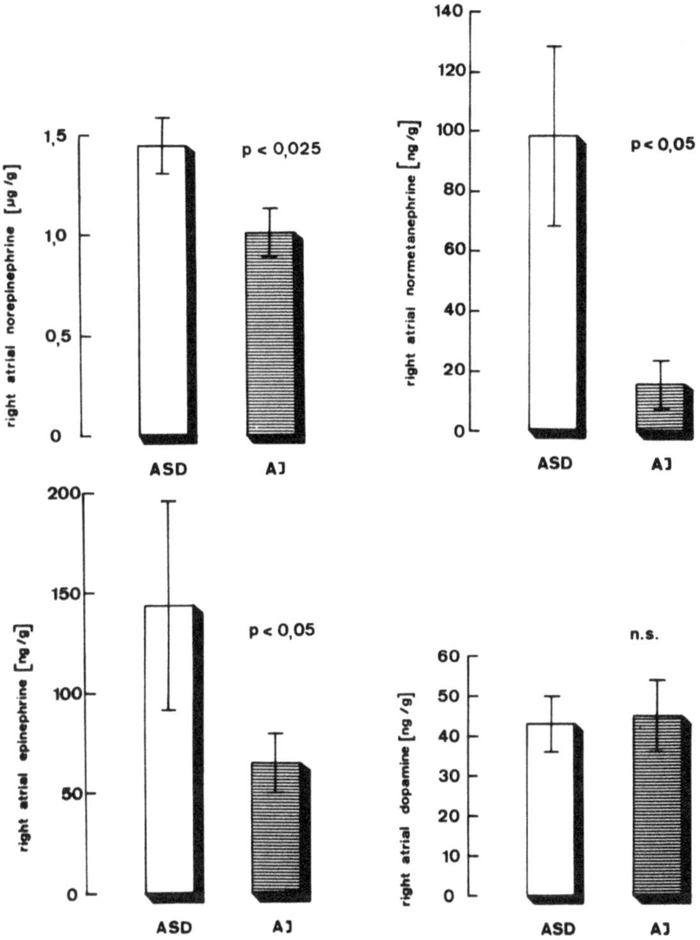

Fig. 11. Right atrial content of norepinephrine, normetanephrine, epinephrine and dopamine in patients with atrial septal defect (ASD, light columns) and in patients with chronic aortic regurgitation (AI, hatched columns). Mean values ± SEM.

The lack of change in dopamine tissue levels, in spite of the significant decrease in myocardial norenephrine content in patients with chronic aortic regurgitation, suggests that not the hydroxylation of tyrosine but rather the transformation of dopamine to norepinephrine represents a rate limiting step for myocardial norepinephrine synthesis in patients with impaired left ventricular function.

As epinephrine cannot be synthetized in the myocardium due to lack of the enzyme phenylethanol-N-methyltransferase in cardiac tissue, diminished right atrial epinephrine levels in patients with chronic aortic regurgitation can only be explained by diminished myocardial uptake, as documented before in the animal experiments.

Abbreviations:

ADP = adenosine-5'-diphosphate
ATP = adenosine-5'-triphosphate

CP = creatine phosphate
C = creatine
CPK = creatine phosphokinase
2,3-DPG = 2,3-diphosphoglycerate

References

1. Braunwald E (1974) The autonomic nervous system in heart failure. In: Braunwald E (ed) The myocardium: failure and infarction. H P Publishing Co Inc p 59
2. Kübler W, Grebe D, Orellano LE, Spieckermann PG, Bretschneider HJ (1968) Zur Bewertung des Gewebsgehaltes der energiereichen Phosphate für die Pathogenese der Herzinsuffizienz. In: Reindell H, Keul I, Doll E (eds) Herzinsuffizienz, Pathophysiologie und Klinik. Georg Thieme Verlag, Stuttgart, p 226
3. Kübler W (1969) Tierexperimentelle Untersuchungen zum Myokardstoffwechsel im Angina-pectoris-Anfall und beim Herzinfarkt. Bibliotheca cardiologica No 22 S Karger, Basel
4. Mäurer W, Yoshida Y, Kübler W (1976) Die Urinausscheidung der Katecholamine Adrenalin, Noradrenalin und Dopamin sowie der Abbauprodukte Metanephrin und Normetanephrin bei Herzkranken. Z Kardiol 65: 1124
5. Mäurer W, Tschada R, Manthey J, Ablasser A, Kübler W (1981) Catecholamines in patients with heart failure. In: Delius W, Gerlach E, Grobecker H, Kübler W (eds) Catecholamines and the heart. Springer Verlag, Berlin Heidelberg New York, p 236

Author's address:

Prof. Dr. W. Kübler, Medizinische Klinik, Abteilung Innere Medizin III (Kardiologie), Bergheimer Straße 58, 6900 Heidelberg 1, West Germany

Effects of digitalis glycosides on the systemic arterial and venous system: clinical importance in the pathophysiology of congestive heart failure

D. T. Mason, M. C. Chan and G. Lee

The Western Heart Institute, St. Mary's Hospital and Medical Center, San Francisco, California (U.S.A.)

Although the digitalis glycosides have been employed widely in clinical medicine for two centuries as the principal drug in the treatment of congestive heart failure, their pharmacologic principles have become established on a firm scientific basis only within the past three decades. The observation that the digitalis-induced stimulation of the force of contraction of the normal human heart is not translated into an increase in cardiac output (7), suggests possible extracardiac vascular effects of the glycosides (5). Indeed, studies in experimental animals have shown that the glycosides exert direct systemic vasoconstriction (10) and this finding has been extended to the peripheral arterial tree of patients during cardiopulmonary bypass in the course of heart operations (1). The purpose of this report is to delineate the clinical alterations produced by digitalis on the resistance vessels (systemic arterioles) and the capacitance beds (systemic veins) of the peripheral circulation (8), and then clarify the significance of these changes relative to the pathophysiology and treatment of congestive heart failure (9).

Normal human subjects

Systemic arteries and veins: Little information has been available concerning the extracardiac actions of the usual therapeutic doses of digitalis glycosides administered to intact human subjects. In order to characterize the effects of digitalis on a specific vascular bed, the effects of the drug on both the arteriolar resistance and venous capacitance vessels of the forearm were examined in normal subjects and in patients with congestive heart failure (8). After the vascular bed of the hand was eliminated from the measurements by inflating the sphygmomanometer cuff at the wrist to a suprasystolic pressure, venous occlusion of the forearm was produced by suddenly inflating the cuff on the upper arm to a pressure below the diastolic pressure (Fig. 1). Since blood could freely flow into the forearm but was unable to leave the venous compartment, the forearm circumference increased and was measured by a rubber strain gauge plethysmograph.

Forearm blood flow was directly proportional to the initial slope of the increase in forearm circumference (8). Forearm vascular resistance was calculated as the ratio of mean arterial pressure obtained via an indwelling arterial needle in the opposite arm to blood flow. Venous tone was determined by relating the increment in forearm venous pressure, obtained via a venous catheter, to the increment in forearm volume which occurred during venous occlusion (8). The patients were in a supine position with their forearms elevated in such a manner that the initial venous pressure was always nearly zero. Keeping the intial

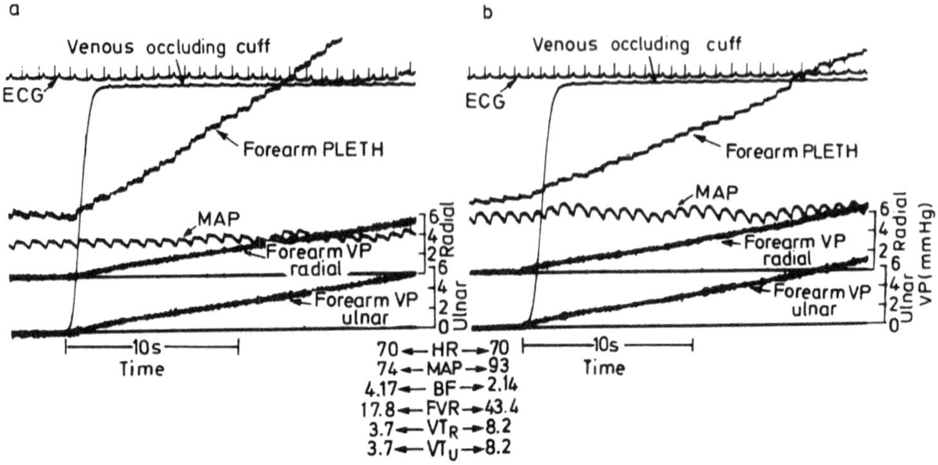

Fig. 1. Two segments of recording of venous occlusion curves in a normal subject. A: Tracing obtained during control period. B: Tracing recorded 40 min following the injection of 0.5 mg ouabain. MAP = mean arterial pressure; PLETH = forearm plethysmographic tracing; VP = venous pressure recorded simultaneously in two veins, one on the radial and the other on the ulnar side of the forearm. Data below the tracing are values of the variables (measured or calculated). HR = heart rate; BF = forearm blood flow; FVR = forearm vascular resistance (mm Hg ml^{-1}); VT_R = venous tone calculated using the vein on the radial aspect of the forearm; VT_U = venous tone calculated using the vein on the ulnar aspect of the forearm (5).

venous pressure near zero provided a constant reference point for the venous pressure-volume relationship and enabled the determinations to be made at a point where this relationship is known to be linear.

Following ouabain administration (8), forearm blood flow declined while arterial pressure rose, and thus forearm vascular resistance increased markedly (Fig. 1). Further, there was a lesser increment in forearm volume and a greater increment in venous pressure, indicating that forearm venous tone had increased. The alterations of forearm vascular dynamics caused by ouabain in a normal subject are shown in sequential fashion in Fig. 2. The drug elevated arterial pressure, reduced forearm blood flow, and increased forearm vascular resistance and venous tone (8). In these normal subjects the peak rise of resistance averaged 38% above the average value during the control period. The peak elevations of venous tone averaged 52% above the average control value. Since the cardiac output remained essentially unchanged following ouabain administration, the calculated total systemic vascular resistance was elevated in all subjects.

Direct vasoconstriction: Since the changes in the resistance and capacitance vessels produced by ouabain occurred in the presence of a stable cardiac output and an elevation of systemic arterial pressure, the drug has a direct effect on vascular smooth muscle. The constriction of the vascular beds could not have resulted from reflexes mediated through the baroreceptor reflex arc, since the elevation of mean arterial pressure that occurred would have resulted in an inhibition, rather than an augmentation, of both arterial and venous tone. That ouabain has a direct constricting effect on vascular smooth muscle was also supported by additional experiments in which forearm vascular dynamics were measured in normal subjects after prolonged treatment with the antiadrenergic drug, guanethidine (8).

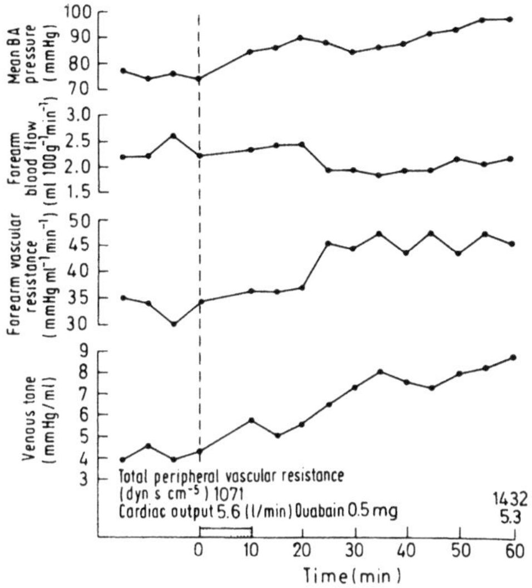

Fig. 2. Serial determinations of arterial pressure (BA), forearm blood flow, forearm vascular resistance, venous tone, cardiac output, and systemic vascular resistance before and after ouabain administration in a normal subject (5).

In these subjects, direct arteriolar constriction and venous constriction still occurred to the same degree following the administration of ouabain.

Congestive heart failure

Systemic arteries and veins: The actions of digitalis on the vascular dynamics of the forearm were also determined in patients with congestive heart failure (8). During the control period, before the administration of ouabain, patients with congestive heart failure had lower forearm blood flow, and higher calculated forearm vascular resistance and venous tone than normal subjects. The effects of ouabain in the patients with heart failure differed strikingly from those observed in the normal subjects (Fig. 3). In all patients with heart failure, the glycoside augmented forearm blood flow, while mean arterial pressure rose slightly or remained unchanged. Therefore, forearm vascular resistance declined. Venous tone decreased after ouabain administration in every patient with congestive heart failure. Again in contrast to the normal subjects, ouabain produced a marked rise in cardiac index and a decline in heart rate, central venous pressure and calculated systemic vascular resistance (Fig. 3).

Indirect vasodilation: The effects of the digitalis glycosides on the vascular dynamics of the forearm depend, therefore, on the circulatory state of the patient at the time the drug is administered. These opposite effects of ouabain on the forearm vascular resistance and venous tone in patients with and those without heart failure are of great interest. There is now considerable evidence that the peripheral vasoconstriction in patients with congestive heart failure results from increased activity of the sympathetic nervous system (2). The elevation of cardiac output induced by ouabain in patients with heart failure results in diminution of such adrenergically induced arteriolar and venous constriction in the forearm.

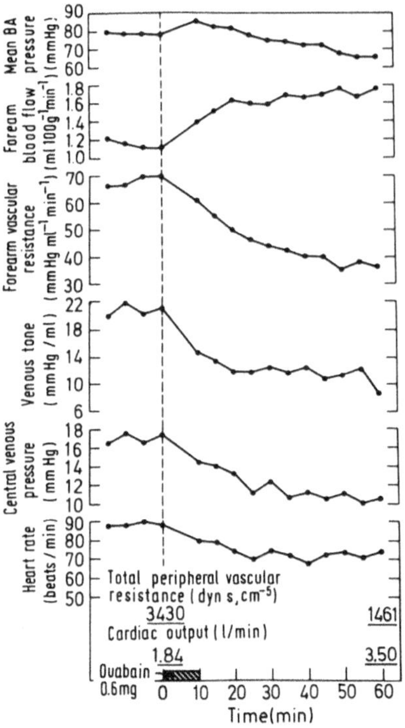

Fig. 3. Serial measurements before and after ouabain administration in a patient with rheumatic heart disease and congestive heart failure. In contrast to normal subjects, values for forearm arterial resistance and venous tone were elevated before, and are markedly reduced following drug administration, indicating arteriolar and venodilatation. Cardiac output increased, while central venous pressure fell and total systemic vascular resistance was reduced (5).

This interpretation is in agreement with the observation that the constriction of the forearm veins in patients with heart failure diminishes as the state of cardiac compensation improves (11). This conclusion is also in consort with the finding that a given level of muscular exercise produces a more intense forearm venoconstriction in patients with heart failure than in patients with compensated heart disease (12). Thus, it is postulated that, as a consequence of the improvement in cardiac dynamics induced by ouabain in patients with heart failure, the state of the resistance and capacitance vessels changes towards normal. In patients with congestive heart failure, this inhibition of sympathetic vasoconstriction (indirect vasodilation) appears to override the direct vasoconstrictive action of the drug on vascular smooth muscle, observed consistently in normal individuals.

Clinical significance

Normal versus heart failure: The effects of digitalis on circulatory dynamics in normal subjects (Fig. 4) and patients with heart failure (Fig. 5) can be summarized in the following manner (5). Mean arterial pressure tends to rise following digitalis administration in both groups of individuals; this pressor response is more pronounced in the patients with heart failure if hypotension is present. The heart rate is slowed in both groups, particularly in

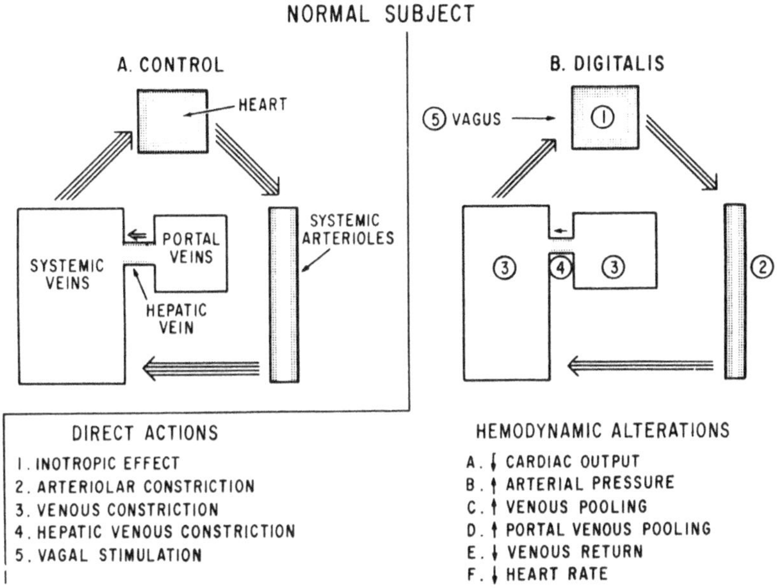

Fig. 4. Diagrammatic representation of the hemodynamic effects of digitalis in a normal subject. The cardiocirculation is shown in the control state in A and following digitalis in B. At the bottom left are shown the direct actions of the glycoside on the heart, peripheral vessels, and the vagus; at the bottom right are the hemodynamic alterations resulting from the direct actions of the agent. The circled numbers in panel B refer to direct actions of digitalis. The width of the arrows indicate the relative quantity of blood flow (6).

heart failure. The cardiac index is diminished slightly in normal subjects, while it is increased strikingly in patients with heart failure. Forearm blood flow is reduced somewhat in normal subjects, but is elevated in heart failure. Total peripheral vascular resistance increases slightly in normal subjects and falls significantly in patients with heart failure. Forearm vascular resistance increases in normal subjects but falls in patients with heart failure. Central venous pressure tends to increase in normal subjects, but declines in patients with heart failure. Finally, in normal subjects, forearm venous tone rises after digitalis administration (direct venoconstriction occurs) but in patients with heart failure forearm venous tone diminishes (indirect venodilation occurs).

Cardiac output: Following the administration of digitalis in congestive heart failure, arterial blood pressure is maintained with reduction of the augmented systemic arteriolar resistance. Cardiac output and venous return are principally improved by the drug-enhanced inotropic state in the presence of less adrenergic support of arterial and venous tone (Fig. 6). The indirect systemic arteriolar dilation also aids the elevation of the reduced cardiac output by providing less impedance to ventricular contraction (5).

Cardiac preload and energetics: The finding that digitalis causes systemic venodilation in patients with congestive heart failure provides an explanation for the clinical observation that rapid digitalization of such patients may result in a fall in elevated systemic venous pressure and in relief of pulmonary congestion before the onset of diuresis. In patients with heart failure, in spite of overall systemic venodilation and reduced central venous pressure

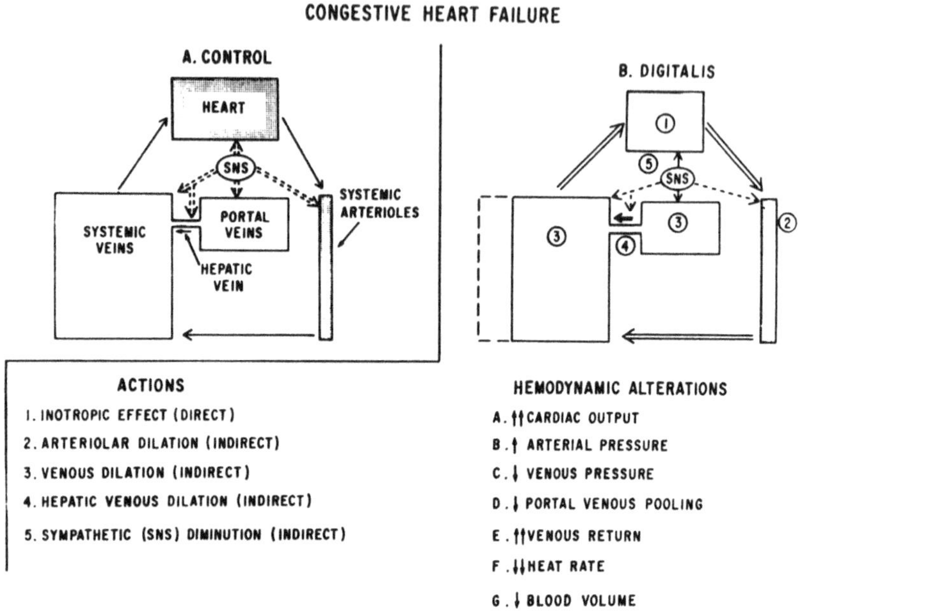

Fig. 5. Diagrammatic representation of the hemodynamic effects of digitalis in a patient with congestive heart failure. The broken arrows represent activity of the sympathetic nervous system (SNS). The circled numbers in panel B refer to the direct and indirect actions of digitalis. The broken line indicating expansion of the systemic venous reservoir in B represents an initial increase in blood volume in this compartment; later the size of this compartment is reduced following diuresis and fall in total blood volume (6).

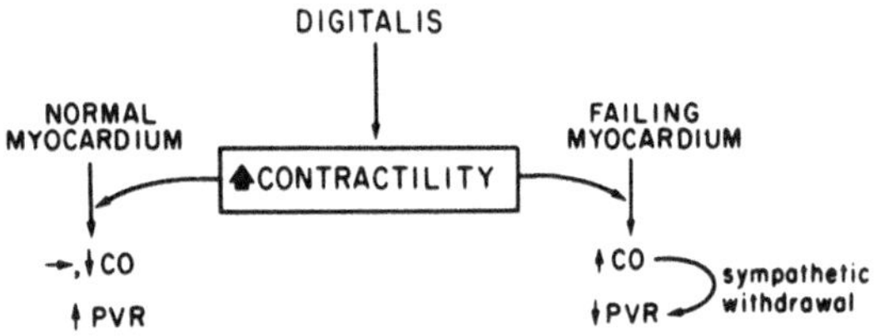

Fig. 6. Diagram showing the different translation of the fundamental digitalis positive contractile action into cardiocirculatory dynamics in the normal heart (left) and in the failing heart (right). CO = cardiac output; PVR = peripheral vascular resistance. (From Mason DT: Congestive Heart Failure. Yorke Medical Books, New York City, 1976.)

following digitalis administration, venous return is augmented and there is a shift of blood in the venous reservoirs to the systemic bed (Fig. 5). Further, in patients with ventricular dysfunction, the digitalis-induced inotropic effect provides more forceful myocardial contraction from a relatively lower filling pressure, resulting in marked enhancement of cardiac output with indirectly lowered myocardial oxygen requirements (antianginal effect) (3).

Acute pulmonary edema: The unusual development of acute pulmonary edema has been reported following rapid digitalization in patients with severe cardiac disease. This observation, which is quite rare, can be explained by the moderate vasopressor effect of intravenous digitalis taking place prior to the cardiotonic actions (1), or by the direct systemic arteriolar constriction of digitalis being unopposed by the drug's lack of improvement of ventricular dysfunction without consequent decrease of heightened adrenergic activity (6). It is also possible that pulmonary congestion might be worsened by direct drug-effected venoconstriction in the absence of the ability of the diseased heart to improve its impaired pump performance to the inotropic stimulation provided by digitalis. In addition, it is pointed out that the degree of vasoconstriction produced by the glycosides is directly related to the speed of intravenous administration (4).

Compensated heart failure: In patients with heart disease without failure, the compensatory mechanisms of ventricular dilation, myocardial hypertrophy, and sympathetic stimulation are capable of maintaining the resting cardiac output at the normal level. Although sympathetic nervous activity is increased, the degree of elevated adrenergic stimulation of the cardiovascular system is less than in overt decompensated heart failure. Thus, vasoconstriction is not severe in the resting state in heart disease without decompensation, and both the opposing direct stimulatory effect and the indirect relaxing action of digitalis on blood vessels are equally balanced. As a result, there is no alteration of vascular tone following the administration of the drug (9). Therefore, the administration of digitalis in patients with heart disease without heart failure results in no change in cardiac output, since the effects of digitalis on cardiac contractility and ventricular outflow impedance are equally balanced in terms of the drug's overall action on cardiac pump performance. In normal subjects in whom there is little sympathetic tone at rest, the cardiac output may fall slightly within the normal range after digitalis administration (Fig. 6) since the glycoside may exert a greater stimulating effect on ventricular impedance than on contractile state in this setting (5).

Conclusions

It has been shown in experimental and clinical studies that peripheral and inotropic actions play a role in the hemodynamic alterations following digitalis administration. These findings indicate that in normal subjects, digitalis constricts the arterial and venous beds of the systemic circulation by a direct stimulating action. These extracardiac vascular actions of digitalis clarify the mechanisms of essentially unaltered normal cardiac output by the agent in patients without cardiac decompensation in spite of the fact that the glycoside clearly increases myocardial contractility. The drug-induced increased impedance to ventricular ejection offsets the augmented force of myocardial contraction. In contrast to normal subjects, in patients with heart failure, the glycoside increases the reduced forearm blood flow and lowers the elevated forearm vascular resistance and venous tone. It is likely that this indirect vasodilator action, which overrides the direct stimulating effect of digitalis on the peripheral vascular beds, results from augmentation of the lowered cardiac output in heart failure. Thus, the effects of the glycoside on the peripheral vascular beds are dependent on the circulatory state of the patient at the time the drug is administered. These observations indicate that the digitalis glycosides have both cardiac and extracardiac actions and that these responses occur in normal subjects as well as in patients with congestive heart failure.

References

1. Braunwald E, Bloodwell RD, Goldberg LI, Morrow AG (1961) Studies on digitalis. IV: Observations in man on the effects of digitalis preparations on the contractility of the non-failing heart and on total vascular resistance. J Clin Invest 40: 52–60

2. Braunwald E, Chidsey CA, Pool PE, Sonnenblick EH, Ross J Jr, Mason DT, Spann JF, Covell JW (1966) Congestive heart failure: Biochemical and physiological considerations. Ann Int Med 64: 904–925
3. Covell JW, Braunwald E, Ross J Jr (1966) Studies on digitalis. XVI. Effects on myocardial oxygen consumption. J Clin Invest 45: 1535–1542
4. DeMots H, McAnulty J, Porter G, Rahimtoola S (1975) Effects of rapid and slow infusion of ouabain on systemic and coronary vascular resistance in patients without heart failure. Circulation 52 (Suppl II): 77
5. Mason DT (1974) Digitalis pharmacology and therapeutics: Recent advances. Ann Int Med 80: 520–530
6. Mason DT, Awan NA (1979) Recent advances in digitalis research. Am J Cardiol 43: 1056–1059
7. Mason DT, Braunwald E (1963) Studies on digitalis. IX: Effects of ouabain on the nonfailing human heart. J Clin Invest 42: 1105–1115
8. Mason DT, Braunwald E (1964) Studies on digitalis. X: Effect of ouabain on forearm vascular resistance and venous tone in normal subjects and in patients in heart failure. J Clin Invest 43: 532–543
9. Mason DT, Spann JF Jr, Zelis R (1969) New developments in the understanding of the actions of the digitalis glycosides. Prog. Cardiovas Dis 6: 443–478
10. Ross J Jr, Braunwald E, Waldhausen JA (1960) Studies on digitalis. II: Extracardiac effects on venous return and on the capacity of the peripheral vascular bed. J Clin Invest 39: 937–942
11. Wood JE (1962) The mechanism of the increased venous pressure with exercise in congestive heart failure. J Clin Invest 41: 2020–2032
12. Wood JE, Litter T, Wilkins RW (1956) Peripheral venoconstriction in human congestive heart failure. Circulation 13: 524–534

Author's address:

Dean T. Mason, M. D., Physician-in-Chief, Western Heart Institute, St. Mary's Hospital and Medical Center, 450 Stanyan Street, San Francisco, California 94117 (U.S.A.)

Attenuation of cardiac glycoside effects during long-term treatment*

K. Werdan, C. Reithmann, and E. Erdmann

Medizinische Klinik I der Universität München, Klinikum Großhadern, Munich (West Germany)

Summary

In cultered heart muscle cells from chicken embryos, cardiac glycoside tolerance – i. e. attenuation of the pharmacological effect during longterm treatment – develops during chronic exposure of the cells to high concentrations of ouabain, with respect to both the positive inotropic as well as the toxic, arrhythmogenic effects. Tolerance development is due to an increase in the number of sodium pump molecules per cell; however, it only comes into play in the presence of toxic, but not in the presence of positive inotropic, non-toxic cardiac glycoside concentrations. The results of clinical studies are in agreement with these in vitro data: neither the positive inotropic effect nor the therapeutic action of cardiac glycosides becomes attenuated or even lost during side-effect-free, long-term treatment of patients with congestive heart failure in sinus rhythm. Therefore, non-responsiveness to cardiac glycoside therapy cannot be attributed to a development of drug tolerance.

The beneficial effect of digitalis therapy is based on two different actions of the drug: the lowering of the ventricular rate – with subsequent prolongation of diastolic filling of the ventricles – in tachysystolic atrial fibrillation and flutter, and the positive inotropic effect in patients with heart failure in sinus rhythm. There is no doubt that patients with tachysystolic atrial fibrillation/flutter benefit from digitalis during acute as well as during long-term treatment (7). However, as to whether cardiac glycosides improve the symptoms of congestive heart disease in patients in sinus rhythm, is still a matter of controversy (15). One possible explanation for the failure of digitalis therapy in these patients could be the development of cardiac glycoside tolerance during long-term treatment, i. e. an attenuation or even loss of the positive inotropic effect of the drug during chronic application. To elucidate the problem of tolerance development, several questions have to be answered:

a) does cardiac glycoside tolerance indeed develop during long-term treatment?
b) if it develops, what is the underlying mechanism?
c) is the development of cardiac glycoside tolerance of clinical relevance?

The answers to these questions are based on the results of clinical studies as well as on experimental work with animal models and beating heart muscle cells in culture.

Development of cardiac glycoside tolerance in beating heart muscle cells in culture

Beating heart muscle cells in culture represent a suitable model to study cardiac glycoside action at the cellular level. The cells can be obtained from the hearts of 10–14 day-old chicken embryos by trypsin-collagenase treatment; after seeding in plastic culture flasks,

* Supported by Wilhelm Sander-Stiftung 78.14.3

they attach to the bottom and begin regular beating within several hours. With regular medium exchange, the cells can be kept in culture for up to several weeks. In these heart muscle cells in culture, the single steps of cardiac glycoside action can be followed in a quantitative manner: the number and affinity of cardiac glycoside receptors can be measured as well as the receptor coupling with the positive inotropic action, the toxic effects and the inhibition of active Na^+/K^+ transport (8, 17, 18): each chicken heart muscle cell possesses about 750,000 cardiac glycoside receptors (dissociation constant (K_D) for ouabain binding = 4 x 10^{-7} M); binding of ouabain to this receptor results in inhibition of active Na^+/K^+ transport (EC_{50} = 4 x 10^{-6} M, as measured by the influx of the K^+ analogue ($^{86}Rb^+$)), with subsequent rise in cell Na^+ and fall in cell K^+; receptor occupation (20–60 %) is also coupled with an increase in contraction velocity of these cells, representing a positive inotropic effect; if 60 % or more of all receptors are occupied, toxic signs – arrhythmias – arise (Fig. 1a; (18)).

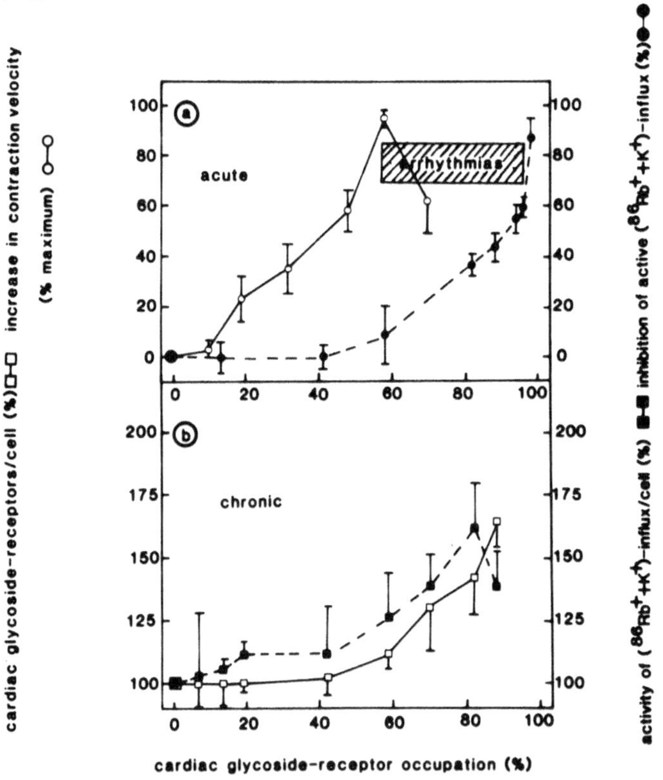

Fig. 1. Effects of acute and chronic ouabain exposure in cultured chicken heart muscle cells.

Fig. 1a. Heart muscle cells have been incubated for 2 hours at 37° C in culture medium with different ouabain concentrations. Determination of the dissociation constant for (^3H)-ouabain-receptor binding allows calculation of receptor occupancy for every ouabain concentration chosen.

Fig. 1b. Heart muscle cells have been incubated for three days at different ouabain concentrations, leading to the receptor occupancies indicated in abscissa. Thereafter, cell-bound ouabain has been removed, the number of cardiac glycoside receptors has been determined by (^3H)-ouabain binding, and sodium pump activity by ($^{86}Rb^+$ + K^+)-influx.
For further experimental details see text and (18); values are given as $\bar{x} \pm SD$, n = 3.

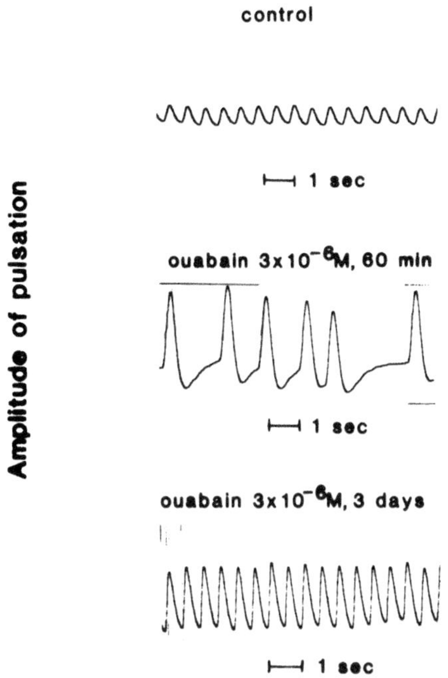

Fig. 2. Effect of acute (60 min) and chronic (3 days) ouabain exposure (3 x 10^{-6} M) on beating of cultured chicken heart muscle cells.

In these chicken heart muscle cells, chronic digitalis therapy can be simulated by cultivation of the cells in a medium supplemented with various concentrations of ouabain. At appropiate ouabain concentrations, cardiac glycoside tolerance develops with respect to both the toxic (Fig. 2), and the positive inotropic (Fig. 3) effect (9, 16, 18): acute exposure of chicken heart muscle cells to 3 x 10^{-6} M ouabain is toxic to these cells, as can be seen from the occurrence of arrhythmias (Fig. 2). However, after cultivation of the cells for three days at this toxic concentration, the arrhythmias have disappeared (Fig. 2); the threshold for toxicity has been shifted to higher ouabain concentrations, the cells have become less sensitive to ouabain. Fig. 3 presents the concentration-dependent, positive inotropic effect of ouabain – as measured by the increase in contraction velocity – in cultured chicken heart muscle cells. In control cells (●–●), the half-maximal effect is seen at 6 x 10^{-7} M ouabain. However, pretreatment of the cells for three days with 3 x 10^{-6} M ouabain (□–□) – and subsequent removal of cell-bound ouabain – clearly shifts the dose-response curve to higher ouabain concentrations, which means a decreased responsiveness of these cells to ouabain. In contrast, the catecholamine sensitivity of these cells remains unchanged (Fig. 4).

The experiments of Figs. 2–4 show without any doubt that the development of cardiac glycoside tolerance is possible. What is the mechanism underlying this tolerance development? Attenuation of positive inotropic and toxic effects during chronic ouabain exposure is paralleled by an increase in the number of cardiac glycoside receptors per chicken heart muscle cell (Fig. 1b; Table 1), while the binding properties of the receptor remain unchanged (see Table 1; (9, 16, 18)). This increase in receptor number is accomplished within 48–72 hours of ouabain exposure of the cells; it can be inhibited by cycloheximide, an inhibitor of protein synthesis; after removal of ouabain from the culture medium, the receptor

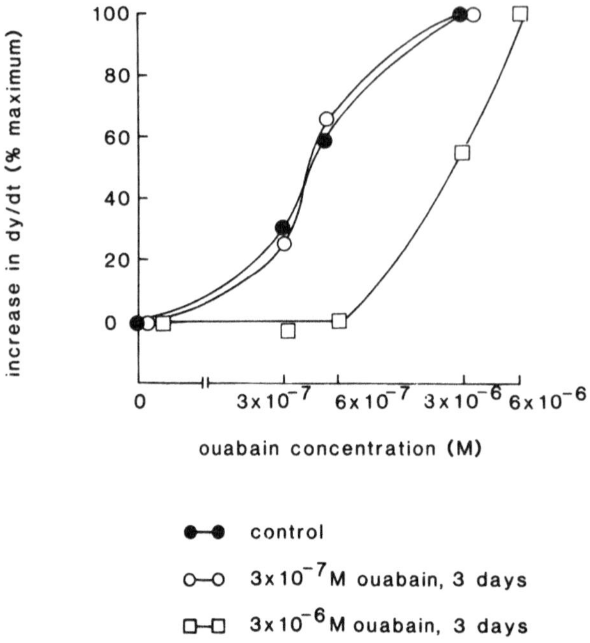

Fig. 3. Cardiac glycoside tolerance in cultured chicken heart muscle cells.
Heart muscle cells have been cultured for three days in the presence of 0, 3×10^{-7} and 3×10^{-6} M ouabain. Thereafter, cell-bound ouabain has been removed, and concentration-response curves have been carried out. For further details see text and (18).

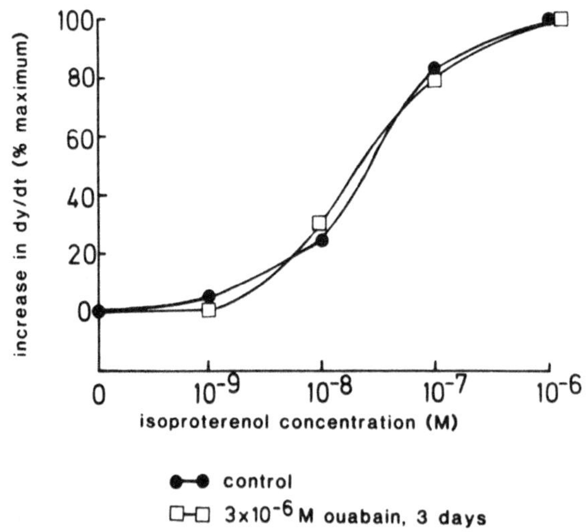

Fig. 4. Isoproterenol sensitivity of cultured chicken heart muscle cells.
Heart muscle cells have been cultured for three days in the absence (●–●) or presence (□–□) of 3×10^{-6} M ouabain. Thereafter, cell-bound ouabain has been removed, and the effect of increasing concentrations of isoproterenol on contraction velocity of the cells has been determined in a cumulative manner.

Table 1. Influence of chronic ouabain exposure of chicken heart muscle cells in culture on receptor numbers.
Heart muscle cells have been incubated for three days in absence or presence of 3×10^{-6} M ouabain. Thereafter, cell-bound ouabain has been removed, and number and affinity (dissociation constant K_D) of cardiac glycoside and beta-adrenergic receptors have been determined by concentration-dependent binding of (^{3}H)-ouabain and (^{3}H)-CGP 12 177 respectively. For further experimental details see (18).

Receptor	Control cells	Ouabain-treated cells
Cardiac glycoside receptor		
Receptors/cell	7.1×10^5	1.4×10^6
K_D (ouabain)	3.3×10^{-7} M	3.5×10^{-7} M
Beta-adrenergic receptors		
Receptors/cell	3.6×10^3	3.6×10^3
K_D (CGP 12 177)	3.6×10^{-10} M	3.0×10^{-10} M

number returns to control level within 48 hours (18). Only the number of cardiac glycoside receptors, but not that of beta-adrenergic receptors, increases during chronic ouabain exposure (Table 1). As the cardiac glycoside receptor is part of the (Na^+ + K^+)-ATPase molecule – the sodium pump – this increase in receptor number represents an increase in the number of functional sodium pump molecules per cell, as proven in the experiment of Fig. 1b. This increase in the number of functional sodium pump molecules per cell is responsible for the development of cardiac glycoside tolerance: by this adaptation process, the heart muscle cell counteracts the cardiac glycoside-induced inhibition of active Na^+/K^+ transport, thereby restoring the impaired cellular Na^+/K^+ homeostasis (18). As the pharmacological effects of digitalis are based on sodium pump inhibition, tolerance develops during this adaptation process (Fig. 5).

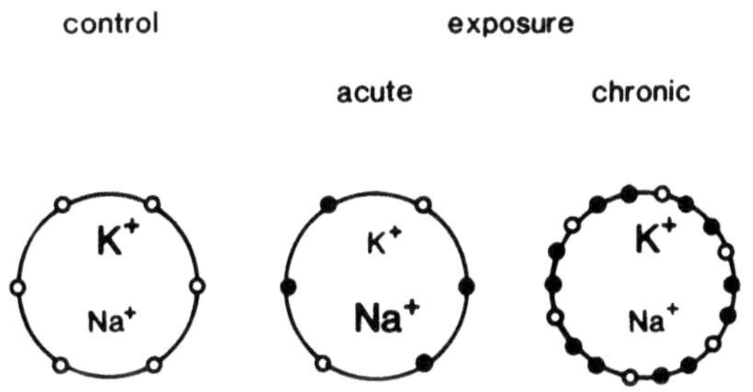

	control	acute	chronic
receptor occupancy (a)	0	2/3	2/3
(Na^++K^+)-ATPase molecules			
total (n,n')	6	6	18
inactive	0	4	12
active	6	2	6

$$n' = n \times \frac{1}{1-a}$$

Fig. 5. Cardiac glycoside tolerance in heart muscle cells.

In the experiments presented in Figs. 2–4, cardiac glycoside tolerance has been induced by cultivation of chicken heart muscle cells in toxic ouabain concentrations. However, this finding is of clinical significance only when tolerance also develops in the presence of therapeutic, non-toxic cardiac glycoside concentrations. Also this situation can be tested: according to the law of mass action, the percentage of cardiac glycoside receptors of the heart muscle cell occupied by ouabain can be calculated for every ouabain concentration in the culture medium, if the dissociation constant (K_D) for ouabain receptor-binding has been determined. A positive inotropic effect is seen when 20–60 % of all receptors are occupied by ouabain (Fig. 1a). Toxic effects – arrhythmias – are present at receptor occupations of $\geqq$ 60 % (Fig. 1a). During chronic ouabain exposure of the cells, an increase in the number of cardiac glycoside receptors is only observed when 60 % or more of all receptors are occupied by ouabain during the culture period (Fig. 1b). As this increase in the number of receptors – and therefore of sodium pump molecules – is responsible for the development of cardiac glycoside tolerance; this tolerance should only develop in the presence of toxic, but not in the presence of therapeutic, non-toxic ouabain concentrations.

This is exactly what has been found: chicken heart muscle cells have been cultured for three days in the presence of either 3 x 10^{-7} M ouabain- receptor occupancy 40 %; positive inotropic effect without toxicity; no change in the number of receptors –, or in the presence of 3 x 10^{-6} M ouabain-receptor occupancy 90 %; toxic effects; 70 % increase in the number of cardiac glycoside receptors and sodium pump molecules (Fig. 1a, b). After removal of cell-bound ouabain, cells grown at the toxic ouabain concentration have developed cardiac glycoside tolerance (Fig. 3, □–□). The dose-response curve of the cells grown at therapeutic concentration (Fig. 3, ○–○), on the other hand, is identical with that of untreated control cells (Fig. 3, ●–●). The development of cardiac glycoside tolerance, therefore, only occurs in the presence of toxic, but not in the presence of therapeutic, non-toxic ouabain concentrations.

The explanation for this surprising finding is probably the "pump reserve" of the heart muscle cell (17): during acute exposure of chicken heart muscle cells, up to 50 % of cardiac glycoside receptors can be occupied by ouabain without a high degree of inhibition of active Na^+/K^+ transport (Fig. 1a). Though sodium pump activity is slightly reduced within this therapeutic concentration range, as can be shown by the use of Na^+ sensitive, intracellularly impaled microelectrodes (12), cellular Na^+/K^+ homeostasis is not markedly impaired. As long as this compensation mechanism holds, no additional sodium pump molecules are synthetized, and cardiac glycoside sensitivity remains unchanged during chronic treatment. Only in the presence of toxic ouabain concentrations does this compensatory mechanism become exhausted (Fig. 1a). The cell now initiates a secondary adaptation process, the increase in the number of sodium pump molecules (Fig. 1b). This increase is then responsible for the attenuation of cardiac glycoside action and the development of tolerance.

Animal models of cardiac glycoside tolerance during long-term treatment

Greeff and coworkers (2, 3) were the first who paid attention to this problem. They treated guinea-pigs with digitoxin for up to 24 days and found an increase in cardiac ($Na^+ + K^+$)-ATPase activity up to 30 %. In view of the results presented in the preceding chapter, this finding would be consistent with the development of cardiac glycoside tolerance. Similar results have been obtained by Whittaker and Swaminathan (19) in pigs treated with digoxin. However, the majority of investigators (Table 2) could not find a compensatory increase either in myocardial ($Na^+ + K^+$)-ATPase activity or in ouabain binding sites in dogs and cats treated with digitalis for up to 4 months.Their findings suggest that tolerance should not develop during chronic therapy.

Table 2. Development of cardiac glycoside tolerance during long-term treatment: animal models.

Author	Animal model	Treatment	"Tolerance"*
Bluschke et al., 1976 (2)	Guinea-pig	Digitoxin, 24 days	+
Ku et al., 1977 (10)	Dog	Digoxin, 4 weeks	–
Nechay et al., 1981 (13)	Dog	Digoxin, 4 months	–
Whittaker and Swaminathan, 1983 (19)	Pig	Digoxin, 4 weeks	+
Prasad et al., 1984 (14)	Dog	Digoxin, 3 months	–
Durrett et al., 1985 (5)	Cat	Digitoxin, 30 days	–

* Experimental findings consistent with the development of cardiac glycoside tolerance.

Long-term digitalis therapy of patients with congestive heart disease in sinus rhythm

What data do we have from patients with heart failure, concerning a possible development of cardiac glycoside tolerance during long-term treatment? There are numerous studies in the literature demonstrating that at least some of the patients with heart failure in sinus rhythm do not benefit from digitalis therapy, or that clinical symptoms and hemodynamic measures do not worsen in these patients after withdrawal (for discussion see (15)). However, with respect to a possible development of digitalis tolerance during chronic treatment, most of these studies are not helpful for the following reason: the existence or absence of tolerance can only be documented when the beneficial effect of cardiac glycosides has been proven during acute treatment, and either becomes attenuated or is fully sustained during the following long-term treatment. Most of the studies mentioned do not fulfil these criteria. Those reports which do fulfil these criteria, however, strongly argue against tolerance development during chronic digitalis treatment: the improvement of congestive heart failure in patients with sinus rhythm is due to the positive inotropic action of digitalis. This positive inotropic effect – as measured by invasive as well as by noninvasive procedures – does not become attenuated during long-term treatment of up to several months. This has been shown for healthy volunteers as well as for patients with heart disease, with and without overt congestive heart failure (for references see (15)).But this sustained positive inotropic effect is only a necessary prerequisite, but not identical with a sustained clinical improvement of patients with severe heart failure in sinus rhythm. Therefore, we have to consider the clinical studies where the signs and symptoms of patients with congestive heart disease have been measured in a quantitative manner and followed during long-term digitalis treatment. Two reports are of special importance in this respect, the studies of Arnold et al. (1) and of Lee et al. (11), carried out with 9 and 25 patients in sinus rhythm respectively with congestive heart failure of different severity (NYHA Class II–III). In these studies, hemodynamic measures (e. g. cardiac output, pulmonary capillary wedge pressure) as well as clinical signs and symptoms of heart failure have been quantitated during acute, intravenous and during oral maintenance digoxin therapy up to ten weeks, as well as after withdrawal of the drug. In those patients where acute digoxin treatment demonstrated a beneficial effect, maintenance therapy was equally effective, demonstrating the absence of tolerance during long-term digitalis treatment.

On the other hand, we have to consider the adaptive changes in cation transport which occur in red cells of patients during chronic digitalis therapy (6): during initial digoxin therapy, active K^+ transport decreases, cell Na^+ rises and digoxin binding of the erythrocytes decreases, in agreement with a partial inhibition of sodium pump activity by digoxin treatment. During the later course of therapy, however, K^+ transport, cell Na^+ and digoxin binding normalize. This normalization is due to an increase in the number of sodium pump molecules of human red cells during chronic digoxin therapy. If the same adaptation also occured in the human heart, then cardiac glycoside tolerance should develop not only in the presence of toxic, but also in the presence of therapeutic cardiac glycoside concentration. However, there is increasing experimental evidence that adaptative changes in the red cell do not reflect adaptive changes of the heart, with respect to cardiac glycoside pharmacodynamics (4). Therefore, the erythrocyte is not a suitable model for studying cardiac glycoside receptor regulation and tolerance development in the heart (4).

Conclusion

Experimental work with beating chicken heart muscle cells in culture has demonstrated without any doubt that cardiac glycoside tolerance can develop during chronic exposure of the cells to ouabain. This tolerance concerns both the toxic and the positive inotropic effects of the drug (9, 16, 18). However, tolerance only develops in the presence of toxic, arrhythmogenic ouabain concentrations; in the presence of therapeutic, positive inotropic concentrations, cardiac glycoside tolerance does not develop (16, 18). Most of the animal models (Table 2) support the finding that cardiac glycoside tolerance does not develop during chronic treatment with non-toxic doses (Table 2).

Are these findings, obtained with cultured chicken heart muscle cells, also relevant for the long-term digitalis treatment of patients with congestive heart failure in sinus rhythm? At present, no experimental evidence exists which questions this relevance: cardiac glycosides exert their pharmacological effects in the same manner in both the cultured chicken heart muscle cells and the human heart. Therefore, one may conclude that cardiac glycoside tolerance will not develop during a side-effect-free, long-term treatment with cardiac glycosides. The results of the clinical studies (1, 11) are in agreement with this experimental finding. Therefore, non-responsiveness to cardiac glycoside therapy cannot be attributed to a development of drug tolerance.

Acknowledgement

The expert assistance of A. Thomschke is gratefully acknowledged.

References

1. Arnold SB, Byrd RC, Meister W, Melmon K, Cheitlin MD, Bristow JD, Parmley WW, Chatterjee K (1980) Long-term digitalis therapy improves left ventricular function in heart failure. New Engl J Med 803: 1443–1448
2. Bluschke V, Bonn R, Greeff K (1976) Increase in the ($Na^+ + K^+$)-ATPase activity in heart muscle after chronic treatment with digitoxin or potassium deficient diet. Eur J Pharmacol 37: 189–191
3. Bonn R, Greeff K (1978) The effect of chronic administration of digitoxin on the activity of the myocardial ($Na^+ + K^+$)-ATPase in guinea-pigs. Arch Int Pharmacodyn 233: 53–64
4. Brown L, Werdan K, Erdmann E The human red blood cell: a model for the ouabain receptor regulation in the heart? In "Cardiac glycosides 1785–1985"; Erdmann, Greeff, Shou (eds); Steinkopff Verlag, Darmstadt (1986) In this volume
5. Durrett LR, Mathew BP, Nechay BR, Neldon SL, Williams JF (1985) Relationship of cardiac muscle tension to Na^+,K^+-adenosine triphosphatase activity after chronic digitoxin administration in cats. J Cardiovasc Pharmacol 7: 411–416
6. Ford AR, Aronson JK, Grahame-Smith DG, Carver JG (1979) The acute changes seen in cardiac glycoside receptor sites, ^{86}rubidium uptake and intracellular sodium concentrations in the erythro-

cytes of patients during the early phases of digoxin therapy are not found during chronic therapy: pharmacological and therapeutic implications for chronic digoxin therapy. Br J Clin Pharmacol 8: 135–142
7. Hamer J (1979) The paradox of the lack of the efficacy of digitalis in congestive heart failure with sinus rhythm. Br J Clin Pharmacol 8: 109–113
8. Kim D, Barry WH, Smith TW (1984) Kinetics of ouabain binding and changes in cellular sodium content, $^{42}K^+$ transport and contractile state during ouabain exposure in cultured chick heart cells. J Pharmacol Exp Therap 231: 326–333
9. Kim D, Marsh JD, Barry WH, Smith TW (1984) Effects of growth in low potassium medium or ouabain on membrane Na,K-ATPase, cation transport, and contractility in cultured chick heart cells. Circ Res 55: 39–48
10. Ku DD, Akera T, Brody TM, Weaver LC (1977) Chronic digoxin treatment on canine myocardial Na^+,K^+-ATPase. Naunyn-Schmiedeberg's Arch Pharmacol 301: 39–47
11. Lee DC-S, Johnson RA, Bingham JB, Leahy M, Dinsmore RE, Goroll AH, Newell JB, Strauss HW, Haber E (1982) Heart failure in outpatients – a randomized trial of digoxin versus placebo. New Engl J Med 306: 699–705
12. Lee CO, Kang DH, Sokol JH, Lee KS (1980) Relation between intracellular Na ion activity and tension of sheep cardiac Purkinje fibers exposed to dihydro-ouabain. Biophys J 29: 315–330
13. Nechay BR, Jackson RE, Ziegler MG, Neldon SL, Thompson JD (1981) Effects of chronic digitalization on cardiac and renal $Na^+ + K^+$-dependent adenosine triphosphatase activity and circulating catecholamines in the dog. Circ Res 49: 655–660
14. Prasad K, O'Neil CL, Bharadwaj B (1984) Effects of chronic digoxin treatment on cardiac function, electrolytes, and sarcolemmal ATPase in the canine failing heart due to chronic mitral regurgitation. Am Heart J 108: 1487–1494
15. Smith TW (1985) Role of digitalis in long-term management of cardiac failure with sinus rhythm. Hammersmith Cardiology Workshop Series, Vol 2; ed A Maseri. Raven Press, New York, pp 151–160
16. Werdan K, Reithmann C, Krawietz W, Erdmann E (1984) Chronic exposure to toxic but not to therapeutic concentrations of ouabain increases cardiac glycoside receptors in cardiac muscle cells from chicken embryos. Biochem Pharmacol 33: 2337–2340
17. Werdan K, Wagenknecht B, Zwißler B, Brown L, Erdmann E (1984) Cardiac glycoside receptors in cultured heart cells – I: Characterization of one single class of high affinity receptors in heart muscle cells from chick embryos. Biochem Pharmacol 33: 55–70
18. Werdan K, Reithmann C, Erdmann E (1985) Cardiac glycoside tolerance in cultured chicken heart muscle cells – a dose dependent phenomenon. Klin Wochenschr 63: 1253–1264
19. Whittaker J, Swaminathan R (1983) The effect of chronic digoxin administration on the Na^+,K^+-ATPase activity in the pig. IRCS Med Sci 11: 846

Author's address:

PD Dr. K. Werdan, Medizinische Klinik I der Universität München, Klinikum Großhadern, Marchioninistraße 15, 8000 Munich 70 (West Germany)

When to start digitalis?

K. Chatterjee

University of California, San Francisco (U.S.A.)

The two major clinical applications of digitalis therapy are in the treatment of arrhythmias and heart failure. The beneficial effects of digitalis in atrial arrhythmias have remained undisputed since Mackenzie observed that digitalis slows ventricular response in the presence of atrial fibrillation (22). However, the role of digitalis therapy in the management of patients with heart failure who remain in sinus rhythm has remained controversial. A large number of studies have been performed to ascertain whether digitalis therapy is desirable in patients with acute or chronic heart failure in view of its low therapeutic/toxic ratio. It is apparent from these studies that it is difficult, if not impossible, to derive a conclusive general indication and recommendation for digitalis therapy for all patients with heart failure of different severity and diverse etiology. In any given patient, the response to the question, "When to start digitalis?" should depend on several considerations, including the type of functional abnormality causing heart failure and whether such functional abnormality can potentially be corrected by digitalis given its expected physiologic and pharmacologic effects.

The potential benefits of digitalis in heart failure are related to its positive inotropic effects (3). It is generally agreed that these positive inotropic effects result from the inhibition of the membrane Na^+-K^+ATPase activity (sodium pump) of myocardial cells which increases intracellular Na^+. An increase in the "rapidly exchangeable intracellular Ca^{++} pool" occurs due to the activation of the Na^+-Ca^{++} exchange mechanism. The increased availability of calcium to the contractile proteins appears to be the principal mechanism for the positive inotropic effects of digitalis.

That digitalis exerts a positive inotropic effect has been demonstrated in both experimental and clinical studies (3, 24, 29). In isolated papillary muscle preparations, the force/velocity curve shifts upward and to the right following the administration of digitalis. The maximum velocity of muscle shortening as well as the maximum force developed increase. The time required to develop peak tension decreases following digitalis. An increase in left ventricular dP/dt, velocity of circumferential fiber shortening and excursion of the left ventricular systolic diameter are observed in the normal conscious dog following chronic digitalis therapy. Echocardiographic studies in normal subjects have also demonstrated an increased ejection fraction and mean rate of shortening of left ventricular dimensions after chronic digitalis administration.

Despite ample evidence for increased contractility with digitalis, cardiac output may not always increase because of the counterbalancing effects of digitalis on the other determinants of cardiac output. Reduction in heart rate and vagal mediated negative inotropic effects may both curtail the expected increase in cardiac output. Furthermore, the increase in systemic vascular resistance that may occur with digitalis may decrease stroke volume by increasing the resistance to left ventricular ejection.

Digitalis enhances arterial and venous tone both in studies with isolated vascular segments and in those in intact animals. An increase in systemic vascular resistance has been observed in normal humans, as well. The increase in systemic vascular tone results from its direct effects on the smooth muscles of the peripheral vascular beds and also from the activation of the sympathetic nervous system. The peripheral vascular effects of digitalis are not uniform; nevertheless, digitalis can potentially exert adverse effects on cardiac performance. It is therefore apparent that the net effect of digitalis on left ventricular pump function and systemic hemodynamics will vary and will depend on the magnitude of positive inotropic and peripheral vascular effects in individual patients.

Digitalis in the acute phase of myocardial infarction

Left ventricular dysfunction following acute myocardial infarction is common and is frequently associated with decreased cardiac output. Left ventricular dysfunction primarily results from a decrease in contractile function of the ischemic and infarcted myocardial segments. In experimental myocardial infarction in animals, digitalis has been shown to increase the contractility of the nonischemic and borderline ischemic myocardial zones (2). However, clinical studies have failed to demonstrate any consistent and important beneficial hemodynamic effects in patients with an acute myocardial infarction. In the absence of clinical heart failure, the administration of digitalis is not associated with any significant increase in cardiac output or decrease in pulmonary capillary wedge pressure. In some patients, cardiac output may decline along with an increase in arterial pressure, which increases the left ventricular ejection impedance. In the presence of heart failure, the hemodynamic response to digitalis appears to be quite variable (Table 1) (5, 13, 15, 17, 21,

Table 1. Hemodynamic effects of digitalis in patients with heart failure complicating acute MI.

Study	Change in cardiac output (%)	Change in stroke work index	Change in left ventricular filling pressure (%)
Rahimtoola et al.	+8	+22	−35
Cohn et al.	+9	+38	−10
Hodges et al.	+6	+9	−1
Forrester et al.	−5	+6	−8
Lipp et al.	+4	+18	−4
Goldstein et al.	+9	+26	−7

27). Although left ventricular stroke work index increases in the majority of patients, no significant increase in cardiac output or stroke volume is observed. Hence, despite its ability to enhance left ventricular contractile function, digitalis cannot be expected to improve organ perfusion. The magnitude of reduction in left ventricular end-diastolic pressure or pulmonary capillary wedge pressure is also minimal in the majority of patients. In occasional patients, pulmonary capillary wedge pressure and left ventricular end-diastolic pressure may actually increase. It has also been shown in patients with acute myocardial infarction that the magnitude of digitalis-induced hemodynamic improvement is inversely related to the severity of heart failure (14). In patients without clinical evidence of left ventricular failure, and with minor hemodynamic abnormalities, ouabain produced a significant increase in cardiac output. In contrast, there was no increase in cardiac output or decrease in left ventricular filling pressure in patients with heart failure or those in cardiogenic shock. In addition, ouabain increased stroke work index only in those patients who did not have clinical heart failure. Thus, it appears that digitalis is relatively ineffective in the presence of severe left ventricular failure or cardiogenic shock complicating acute myocardial infarction. Digitalis also seems to be a relatively weaker inotropic agent than others currently available. The systemic hemodynamic effects of digitalis have been compared to those of dobutamine, a $beta_1$ adrenergic receptor agonist, in patients with acute myocardial infarction with mild to moderate heart failure. While dobutamine increased cardiac output by an average of 33 %, the increase in cardiac output with digoxin was only 9 %. Significant differences were also noted in other hemodynamic effects of dobutamine and digoxin in these patients. With dobutamine, there was a consistent decrease in pulmonary capillary wedge pressure, along with an increase in left ventricular stroke work index, indicating improved left ventricular pump function. With digoxin, however, changes in pulmonary capillary wedge pressure and stroke work index were inconsistent; and in some patients, pulmonary capillary wedge pressure even increased (15).

The effects of digitalis on coronary hemodynamics and myocardial metabolic function have not been extensively investigated. In patients with chronic coronary artery disease without heart failure, coronary blood flow and myocardial oxygen consumption usually remain unchanged despite increased contractility (10, 20). In these patients, there was a decrease in left ventricular wall tension, another determinant of myocardial oxygen demand. Decreased wall tension might have prevented the expected increase in myocardial oxygen consumption due to increased contractility. In patients with acute myocardial infarction with heart failure, however, digitalis can produce adverse effects on coronary hemodynamics. An increase in coronary blood flow and myocardial oxygen consumption following the administration of ouabain has been observed (14). The patients who showed the greatest increase in coronary blood flow and myocardial oxygen consumption also developed angina. These findings suggest that an increase in coronary blood flow was inadequate to match myocardial oxygen demand and prevent myocardial ischemia. In occasional patients, there was also myocardial lactate production, which suggests anerobic metabolism and enhanced myocardial ischemia. The influence of digitalis therapy on the extent of myocardial infarct size remains controversial. In experimental myocardial infarction, digitalis has been shown to increase the extent of ischemic injury when given in experimental animals without heart failure. Similarly, an increase in creatine phosphokinase influx in patients with acute myocardia infarction without heart failure, which is suggestive of increased infarct size, has been reported following digoxin therapy (31). On the other hand, it has also been observed that digitalis therapy may be associated with a decrease in CPK efflux in patients with an elevated pulmonary capillary wedge pressure (25). So without further investigation, the effect of digitalis on myocardial infarct size remains uncertain. It is apparent that the role of digitalis therapy in the management of pump failure complicating acute myocardial infarction is extremely limited. Digitalis therapy may not produce rapid, beneficial hemodynamic and clinical effects in patients who develop severe left ventricular failure following myocardial infarction. If inotropic therapy is indicated under such circumstances, drugs with greater potency, a more rapid onset of action and a shorter half life than digitalis should be considered. It must be emphasized that digitalis therapy is useful, however, for the treatment of supraventricular tachyarrhythmias complicating acute myocardial infarction. Atrial tachycardia, flutter and fibrillation may cause a rapid deterioration in left ventricular function in patients with myocardial infarction. Increased myocardial oxygen requirements due to tachycardia and decreased myocardial perfusion from the reduction in diastolic perfusion time may rapidly increase myocardial ischemia. Cardioversion with direct current countershock is the treatment of choice, although supraventricular tachyarrhythmias reoccur in many patients following cardioversion. In these cases, digitalis therapy is indicated to control ventricular response.

Chronic heart failure

The controversy over the beneficial effects of digitalis in patients with chronic heart failure in sinus rhythm continues despite its use for over two hundred years. Many clinical studies have suggested that digitalis therapy is associated with symptomatic benefit in patients with chronic heart failure. However, in many other studies the withdrawal of digitalis therapy was not associated with any deterioration in the clinical status of these patients (9, 16, 18, 23). It must be stressed that it is difficult to evaluate the effects of any therapeutic intervention based on subjective changes alone.

Alteration in left ventricular function (1, 4, 6, 7, 8, 11, 12, 19, 26, 28, 30, 32)

Changes in left ventricular function have been assessed both after acute and chronic digitalis therapy in patients with chronic heart failure. Measurement of systolic time intervals has demonstrated shortening of the Q-A_2 intervals as well as a decreased ratio of pre-ejection period to left ventricular ejection time. Changes in the velocity of circumferential fiber shortening have been assessed following digitalis therapy, and a significant increase in the mean velocity of circumferential fiber shortening was observed without any change in the end-diastolic volume. Increased fiber shortening occurred despite an increase in mean systolic blood pressure and a decrease in heart rate. Intravenous digitalis therapy also increases the maximum velocity of fiber shortening (V_{max}) in patients without overt heart failure.

These findings strongly suggest that the acute administration of digitalis is associated with increased contractility. Chronic digitalis therapy may also be associated with sustained increase in contractile function, although such response is not uniformly observed. An abbreviation of systolic time intervals, and increases in segmental left ventricular wall motion, the velocity of circumferential fiber shortening and ejection fraction have been observed following chronic oral digoxin therapy. In other studies, however, no consistent improvement in left ventricular function was observed. The variability of response to digitalis is likely to be related to the patient population studied and the magnitude of depression of left ventricular function prior to digitalis therapy. It is reasonable to assume that if systolic function is normal it is unlikely that digitalis therapy will produce any beneficial effects. Similarly, in patients with severely depressed left ventricular function without any contractile reserve, digitalis is unlikely to be effective.

Table 2. Effects of long-term digoxin therapy on hemodynamics at rest in nine patients with chronic heart failure.§

Parameter	Long-term digoxin (0.125–0.25 mg)	Digoxin withdrawal	Digoxin readministered (1mg)	p
Heart rate (beats/min)	77 ± 12.6	85 ± 13.3	77 ± 14.0	NS*
Mean arterial pressure (mm Hg)	92 ± 10	92 ± 12.3	98 ± 12.1	NS*
Pulmonary capillary wedge pressure (mm Hg)	21 ± 7.7	29 ± 9.5	22 ± 9.4	0.001**†
Cardiac index (liters/min/m^2)	2.4 ± 0.7	2.1 ± 0.6	2.5 ± 0.5	0.025**†
Stroke work index (g-m/m^2)	38 ± 17.7	26 ± 14.2	40 ± 14.4	0.001**†
Stroke volume index (ml/beats/m^2)	33 ± 11.6	25 ± 7.8	33 ± 6.4	0.005**†
Total systemic resistance (dyne. s. cm^{-5})	1,811 ± 781	2,041 ± 582	1,752 ± 452	NS*

§ Mean ± SD

* NS = not significant

** Probability of difference between values obtained during long-term digoxin treatment and those obtained after drug withdrawal

† Probability of difference between values obtained after digoxin withdrawal and after acute readministration of the drug to nondigitalized patients.

There is a paucity of information regarding the hemodynamic effects of long-term digitalis therapy in patients with chronic heart failure. The changes in systemic hemodynamics that occurred when patients were maintained on oral digitalis and the drug withdrawn and readministered are summarized in Table 2. In this study, the hemodynamics were determined by right heart catheterization at rest and during exercise. Some patients had idiopathic congestive cardiomyopathy while others had ischemic cardiomyopathy. Patients with severe end-stage heart failure were excluded from this study. Indications for long-term digitalis therapy were: the presence of clinical and radiologic evidence of chronic congestive heart failure with S_3 gallop, cardiomegaly, and signs and symptoms of pulmonary and systemic venous hypertension. All patients had been taking a stable dose of digoxin (0.125–0.25 mg/day) for at least two months before the first hemodynamic evaluation was performed. After withdrawal of digoxin there was a decrease in cardiac index, stroke volume index and stroke work index, along with an increase in pulmonary capillary wedge pressure. These findings suggested that withdrawal of digitalis therapy was associated with deterioration in left ventricular pump function. Following readministration of digoxin after the withdrawal, stroke volume, stroke work and cardiac indices increased and the pulmonary capillary wedge pressure fell. Thus, readministration of digitalis in these patients was associated with improved left ventricular function. Hemodynamic changes and changes in left ventricular function following withdrawal and after the readministration of digitalis during exercise are summarized in Table 3. Cardiac index, stroke volume index and stroke work index at peak exercise and at the same work load were significantly less without digoxin than while on maintenance digoxin therapy or after readministration of digoxin. Pulmonary capillary wedge pressure increased during exercise while on digoxin therapy. However, the magnitude of increase in pulmonary capillary wedge pressure during exercise was less with digoxin therapy than when digoxin therapy was withdrawn. These findings suggest that withdrawal of digoxin therapy is associated with deterioration in left ventricular function at rest and during excercise. In these patients, total systemic vascular resistance and arterial pressure did not change after withdrawal of digoxin or acute readministration of the drug. Thus, it is unlikely that improvement in left ventricular function was related to any change in left ventricular outflow resistance. As pulmonary capillary wedge pressure was lower in

Table 3. Effects of long-term digoxin therapy on hemodynamics during exercise in nine patients with chronic heart failure§.

Parameter	Long-term digoxin (0.125–0.25 mg)	Digoxin withdrawal	Digoxin readministered (1 mg)	p
Heart rate (beats/min)	111 ± 17	121 ± 16	117 ± 25	NS*
Mean arterial pressure (mm Hg)	113 ± 21	111 ± 23	116 ± 20	NS*
Pulmonary capillary wedge pressure (mm Hg)	33 ± 7	42 ± 5	35 ± 6	0.001**†
Cardiac index (liters/min/m^2)	3.5 ± 0.9	3.1 ± 1.2	3.8 ± 1.1	0.05**†
Stroke work index (g-m/m^2)	45 ± 22	33 ± 22	47 ± 29	0.005**†
Stroke volume index (ml/beats/m^2)	32 ± 10	25 ± 8	33 ± 9	0.025**†
				0.01†
Total systemic resistance (dyne. s. cm^{-5})	1,453 ± 359	1,703 ± 600	1,411 ± 459	0.05†

§ Mean ± SD
* NS = not significant
** Probability of difference between values obtained during long-term digoxin treatment and those obtained after drug withdrawal
† Probability of difference between values obtained after digoxin withdrawal and after acute readministration of the drug to nondigitalized patients.

all patients during digoxin therapy, the improvement in left ventricular function was most likely due to increase in contractility following digitalis therapy.

It needs to be stressed, however, that beneficial hemodynamic effects are not uniformly observed in all patients with chronic heart failure. Patients with moderately depressed left ventricular systolic function are more likely to benefit from digitalis therapy. Patients with normal left ventricular systolic function are unlikely to derive any benefit. This assumption is supported by the observation that patients with normal left ventricular ejection fraction or with hypertrophic cardiomyopathy do not experience any symptomatic improvement during digitalis therapy. Patients with clinical evidence of depressed left ventricular function with ventricular S_3 gallop experience symptomatic benefit. Patients with severe endstage left ventricular failure or cardiogenic shock who do not have any contractile reserve, are also unlikely to derive any benefit from digitalis therapy. Clearly, patients with depressed left ventricular systolic function with mild to moderate heart failure are the best candidates for digitalis therapy.

References

1. Arnold SB, Byrd RC, Meister W, Melmon K, Cheitlin MD, Bristow JD, Parmley WW, Chatterjee K (1980) Long-term digitalis therapy improves left ventricular function in heart failure. N Engl J Med 303: 1443–1448
2. Banka VS, Schadda KD, Bodenheimer MM, Helstadt RH (1975) Digitalis in experimental acute myocardial infarction. Am J Cardiol 35: 801
3. Braunwald E, Bloodwell RD, Goldberg LI et al (1961) Studies on digitalis. IV: Observations in man on the effects of digitalis preparations on the contractility of the non-failing heart and on total vascular resistance. J Clin Invest 40: 52–59
4. Carliner NH, Gilbert CA, Puritt AW, Goldberg LI (1974) Effects of maintenance digoxin therapy on systolic time intervals and serum digoxin concentrations. Circulation 50: 94–98
5. Cohn JN, Tristani FE, Khatri IM (1969) Cardiac and peripheral vascular effects of digitalis in clinical cardiogenic shock. Am Heart J 78: 318–330
6. Cohn K, Selzer A, Kersh ES, Karpman LS, Goldschlager N (1975) Variability of hemodynamic responses to acute digitalization in chronic cardiac failure due to cardiomyopathy and coronary artery disease. Am J Cardiol 35: 461–468
7. Crawford MH, Karliner JS, O'Rourke RA, Amon KW (1976) Favorable effects of oral maintenance digoxin therapy on left ventricular performance in normal subjects: echocardiographic study. Am J Cardiol 38: 843–847
8. Dall JLC (1978) Maintenance digoxin in elderly patients. Brit Med J 2: 705–706
9. Davidson C, Gibson D (1973) Clinical significance of positive inotropic action of digoxin in patients with left ventricular disease. Br Heart J 35: 970–976
10. DeMots H, Rahimtoola SH, Kremkau EL, Bennett W, Mahler D (1976) Effects of ouabain on myocardial oxygen supply and demand in patients with chronic coronary artery disease: a hemodynamic, volumetric, and metabolic study in patients without heart failure. J Clin Invest 58: 312–319
11. Dobbs SM, Kenyon WI, Dobbs RJ (1977) Maintenance digoxin after an episode of heart failure: placebo-controlled trial in outpatients. Brit Med J 1: 749–752
12. Fonrose HA, Ahlbaum N, Bugatch E, Cohen M, Genovese C, Kelly J (1974) The efficacy of digitalis withdrawal in an institutional aged population. J Am Geriat Soc 22: 208–211
13. Forrester J, Bezdek W, Chatterjee K et al (1972) Hemodynamic effects of digitalis in acute myocardial infarction. Ann Int Med 76: 863–864
14. Forrester JS, Chatterjee K (1974) Preservation of ischemic myocardium, In: JHK Vogel (ed) Advances in Cardiology, Basel. S Karger, pp 158–169
15. Goldstein RA, Passamani ER, Roberts R (1980) A comparison of digoxin and dobutamine in patients with acute infarction and cardiac failure. N Engl J Med 303: 846

16. Guz A, McHaffie D (1978) The use of digitalis glycosides in sinus rhythm. Clin Sci Molec Med 55: 407–421
17. Hodges M, Friesinger GC, Riggins RCK et al (1972) Effects of intravenous administered digoxin on mild left ventricular failure in acute myocardial infarction in man. Am J Cardiol 29: 749–755
18. Hull SM, Mackintosh A (1977) Discontinuation of maintenance digoxin therapy in general practice. Lancet 2: 1054–1055
19. Johnston GD, McDevitt DG (1979) Is maintenance digoxin necessary in patients with sinus rhythm? Lancet 1: 567–570
20. Kotter V, Schuren K, Schroder R (1978) Effect of digoxin on coronary blood flow and myocardial oxygen consumption in patients with chronic coronary artery disease. Am J Cardiol 42: 563
21. Lipp H, Denes P, Gametta M et al (1972) Hemodynamic responses to acute intravenous digoxin in patients with recent myocardial infarction and coronary insufficiency with and without heart failure. Chest 63: 862–867
22. Mackenzie J (1911) Digitalis. Heart 2: 273–386
23. McHaffie DJ, Purcell H, Mitchell-Heggs P, Guz A (1978) The clinical value of digoxin in patients with heart failure and sinus rhythm. Quart J Med 188: 401–419
24. Mason DT, Braunwald E (1963) Studies on digitalis. IX: Effects of ouabain on the non-failing human heart. J Clin Invest 42: 1105–1111
25. Morrison J, Pizzarello R, Reduto L, Gulotta S (1975) Effect of digitalis on predicted myocardial infarct size in man. Clin Res 23: 198 (Abstract)
26. O'Rourke RA, Henning H, Theroux P, Crawford MH, Ross J Jr (1976) Favorable effects of orally administered digoxin on left heart size and ventricular wall motion in patients with previous infarction. Am J Cardiol 37: 708–715
27. Rahimtoola SH, Sinno MZ, Chuquimia R et al (1968) Effects of ouabain on impaired left ventricular function in acute myocardial infarction. Br Heart J 30: 373–376
28. Selzer A, Malmborg RO (1962) Hemodynamic effects of digoxin in latent cardiac failure. Circulation 25: 695–702
29. Smith TW, Haber E (1973) Digitalis. N Engl J Med 259: 945–952
30. Sokolow M, Weinberg HB, Plaut JL, Katz LN (1942) Digitalis in the prevention of recurrent cardiac failure in patients with sinus rhythm. Ann Int Med 16: 427–437
31. Varomkov Y, Shell WE, Smirnov V, Jukovsky D, Chazov EI (1977) Augmentation of serum CPK activity in digitalis in patients with acute myocardial infarction. Circulation 55: 719
32. Vogel R, Frischknecht J, Steele P (1977) Short- and long-term effects of digitalis on resting and post-handgrip hemodynamics in patients with coronary artery disease. Am J Cardiol 40: 171–176

Author's address:

Dr. K. Chatterjee, University of California, Moffitt Hospital, Room 1186m, San Francisco, California 94143 (U.S.A.)

Cardiotonic effect of digitalis in sinus rhythm during exercise

L. B. Tan, R. G. Murray, A. C. Tweddel and I. Hutton

University Departments of Cardiology in Glasgow and Birmingham (United Kingtom)

Summary

Whether digitalis can exert positive inotropism in sinus rhythm is controversial. Using a newly developed analytical method which enables us to identify inotropic changes in an exercising patient, this question was studied in 10 cardiac failure patients (NYHA Classes III–IV) at rest and during exercise. Haemodynamic data were collected before treatment (Control, C), after acute intravenous ouabain (0), and after 6 weeks of maintenance treatment with digoxin (D). There was no detectable inotropic change at rest after either glycoside. During exercise the heart rates were not significantly different: 125 ± 8.6 (C), 120 ± 7.3 (0) and 122 ± 4.5 (D), but there were significant falls in LV filling pressure from 39 ± 3 mm Hg to 34 ± 3 ($p < 0.05$) after ouabain, and to 33 ± 3 ($p < 0.02$) after digoxin. The hydraulic power outputs of the hearts rose significantly from 1.54 ± 0.63 watts to 1.80 ± 0.67 watts ($p < 0.01$) and 1.96 ± 0.85 ($p < 0.01$) watts after ouabain and digoxin respectively. During exercise, 4 of the patients after ouabain, and 6 of the patients after maintained digoxin therapy, achieved power outputs which could not be accounted for by alterations in peripheral resistance, heart rate or filling pressure, i.e. positive inotropic effects were demonstrated. In conclusion, both intravenous ouabain and long-term oral digoxin can exert true positive inotropic effects during exercise in some heart failure patients with sinus rhythm.

Introduction

In this bicentenary year of William Withering's monograph on Foxglove, we have to acknowledge his monumental achievement in that after two centuries, digitalis is still the only oral inotropic agent widely used in the treatment of heart failure. Yet today, its place in clinical practice as an inotrope, as opposed to its use in the control of atrial fibrillation, is being questioned. Last year Mulrow and colleagues (4) performed a survey of the English literature on the subject and applied strict criteria to determine whether any conclusion can be drawn from the published studies. They remarked that only two double-blind placebo-controlled trials were methodoligically acceptable, and the only conclusions that can be drawn from these papers were that digoxin can be withdrawn successfully in some patients with stable congestive heart failure; and that some patients with severe heart failure with S_3 gallop may benefit from digoxin therapy. In April 1985, to mark the bicentenary anniversary of William Withering's work on digitalis, the British Cardiac Society held a Debate on the topic, and it became apparent that there have been more emotions expressed concerning its use in sinus rhythm, than reliable experimental facts.

There are still a number of issues that need to be resolved before final guidelines can be laid down regarding the use of digitalis in heart failure with sinus rhythm:

– Firstly, it cannot be over-emphasized that cardiac failure patients are a heterogeneous group and that the selection of patients for study is of paramount importance. In particular, patients with dysrhythmia, restrictive or constrictive heart failure, and obstructive structural lesions should be excluded, leaving those with essentially myocardial contractile failure and normal sinus rhythm for inclusion. In this cohort there would be some who have no contractile reserve whatsoever to respond to any kind of inotropic stimulation, and it would not be surprising to find that digoxin is ineffective in them. A means of identifying such a subgroup has to be sought, otherwise statistical analysis applied to the entire group might point to lack of benefit overall.

– Secondly, symptomatic assessments are subjective and are susceptible to a number of confounding factors. Well-designed double-blind placebo-controlled cross-over trials of sufficiently large numbers of suitable patients are required to establish a valid answer. Such a trial, in the case of a weak inotrope like digoxin, can be prohibitively difficult and costly to conduct. It also assesses the overall effects of the digitalis therapy which include the well-being of the patient as well as the direct cardiotonic effect, and the two can be disparate. The cardiotonic effect should therefore be separately ascertained by direct haemodynamic measurements. Since this approach is objective (in the sense that neither the patient nor the investigator is able subconsciously to influence the readings), there is no need to resort to blind assessments.

– Thirdly, testing the efficacy of digitalis at rest bears no relationship to its usefulness when it is most wanted to exert its effects, i.e. during exercise.

– Fourthly, for drugs, such as digitalis, which affect both the heart and the vasculature directly, there is hitherto no satisfactory analytical method to separate these two effects, as all the indices of myocardial contractility proposed so far are influenced unpredictably by afterload changes. A technique which can separate these two effects is required to identify the cardiotonic action of digitalis when studied in patients with heart failure.

It is our aim to perform a study which will detect the pressence of objective, beneficial, direct positive inotropic effects of digitalis in a subset of heart failure patients with sinus rhythm at rest and during exercise.

Methods

Patient population: Ten patients (1 female) with a mean age of 51 years (range 42–65) with cardiac failure and in sinus rhythm, in NYHA symptomatic Classification III or IV were studied. Seven of the patients had ischaemic heart disease and the remainder had dilated cardiomyopathy. None of the patients had been treated with digitalis before. All patients gave informed consent to the study.

Study protocol: On the first visit acute haemodynamic data were obtained at rest and during upright bicycle ergometric exercise prior to and 30–60 mins after 0.5 mg of intravenous ouabain. The level of exercise was identical for each patient (25–75 watts), and it was chosen to allow 3 to 5 minutes of exercise at the set level. The patients were then given oral digoxin for 6 weeks, at the end of which the haemodynamic study was repeated. The digoxin dosage was titrated so that the plasma concentration was within the therapeutic range (1.3–2.6 nmol/L). Other than diuretics, all other cardioactive medication was disallowed during the study.

Haemodynamic data were collected with the help of thermodilution Swan-Ganz catheters which gave pulmonary arterial (PAP) and wedge pressures (PAWP), and cardiac output (CO) measured in triplicate. The mean left ventricular filling pressure was accepted as either the mean pulmonary occluded arterial pressure or the pulmonary artery diastolic pressure. The heart rate was obtained from the electrocardiogram and the arterial blood pressure by sphygmomanometry. The following haemodynamic variables were calculated: Cardiac index (CI = CO/body surface area); stroke volume index (SVI = CI/HR in mls/m^2; mean arterial pressure (MAP = pulse pressure/3 + diastolic pressure in mm Hg); left ventricular stroke work index (SWI = (2/3 × arterial pulse pressure + diastolic pressure – PAWP) × SVI × 0.0136 g-m/m^2); systemic vascular resistance (SVR = MAP / CO × 80 dyn-s-cm^5).

Radionuclide left ventriculography was performed using an IGE Maxi Camera and data acquired in list mode, after in-vivo red-cell labelling with 20 mCi 99^{m}Tc Pyrolite.

Data analysis: A newly developed analytical technique which enables us to identify the true inotropic changes in an exercising patient has previously been described (5). Only the fundamental Fourier harmonic terms of flow, pressure, hydraulic power and impedance were considered. A brief description of the analytical method is given here. The mean flow and pressure during control exercise were represented by CO_C and MAP_C, and the mean cardiac power output was given by $W_C = CO_C \times MAP_C$, and the resistance facing the heart by $R_C = MAP_C/CO_C$. The variables during exercise after ouabain and digoxin were represented with subscripts O and D respectively. Referring to Fig. 1, after ouabain the cardiac power output values which could be obtained by a pure change in resistance from R_C to R_O (i.e. without any change in the pumping characteristics of the heart) would be found anywhere between W_P and W_Q. To eliminate the possibility that the increased W is due to a rise in HR, the W_O was corrected by a factor HR_C/HR_O. If W_O was found to lie above these limits, in the absence of significant increase in heart rate or LVFP, then one can infer that ouabain had contributed true positive inotropism to the exercising heart. The same approach was used for the digoxin part of the study.

Statistical analysis was by Student's paired t-test. Differences were taken as significant at $p < 0.05$. Results are expressed as means ± SEM.

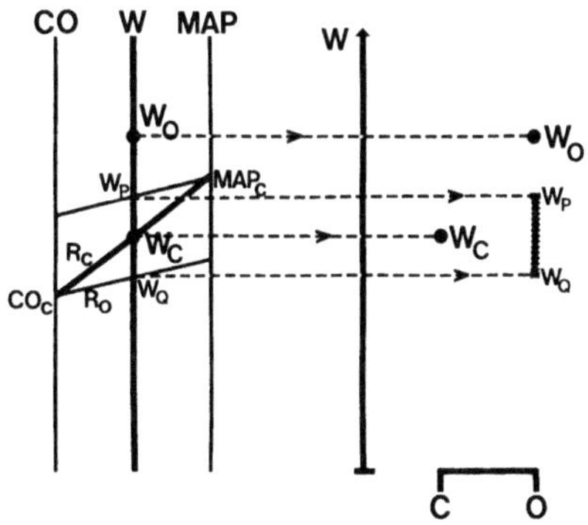

Fig. 1. Analytical technique to separate the changes in cardiac pumping performance due to alterations in vascular resistance from that due to alterations in pumping characteristics. See text for the description of the method. The format shown in the plot on the right will be the one used in the subsequent graphs.

Results

The haemodynamic data are displayed in Table 1.

Resting haemodynamics: No significant changes were noted in heart rate and LV filling pressure at rest after acute intravenous administration of ouabain or maintained oral therapy of digoxin. The systolic function of the left ventricle was not significantly altered by either the acute or chronic digitalis therapy, with the cardiac index remaining around the control values of 2.0 ± 0.2 L/min, the mean arterial pressure not significantly different from the control values of 89 ± 3.1 mm Hg, and the ejection fraction unchanged from 33 ± 2.7% at control. The calculated variables of stroke volume index, stroke work index and systemic vascular resistance were also not significantly altered.

Table 1. Haemodynamic data (mean values ± SEM)

		Control	Ouabain	Digoxin
HR	R	69 ± 3.6	67 ± 2.7	67 ± 2.9
	E	125 ± 8.6	120 ± 7.3	122 ± 4.5
MAP	R	89 ± 3.1	88 ± 3.5	91 ± 3.2
	E	109 ± 4.5	111 ± 3.3	120 ± 8.0
LVFP	R	19 ± 2.6	17 ± 1.8	18 ± 1.8
	E	39 ± 3.2	34 ± 2.6*	33 ± 3.4*
CI	R	2.0 ± 0.2	2.1 ± 0.1	2.0 ± 0.1
	E	3.3 ± 0.3	4.0 ± 0.4#	3.8 ± 0.4#
SVI	R	29 ± 3	32 ± 2	29 ± 2
	E	27 ± 2	33 ± 3#	31 ± 3*
SVR	R	2080 ± 200	1849 ± 84	2121 ± 129
	E	1460 ± 112	1291 ± 131	1416 ± 140
SWI	R	35 ± 4.2	38 ± 3.6	35 ± 3.2
	E	36 ± 4.7	47 ± 5.6#	49 ± 8.2#
LVEF	R	33 ± 2.7	32 ± 2.3	35 ± 2.3
	E	29 ± 1.7	36 ± 2.9*	36 ± 3.1*

* $p < 0.05$
# $p < 0.01$
HR = heart rate in min^{-1}; MAP = mean arterial pressure in mm Hg; LVFP = LV filling pressure in mm Hg; CI = cardiac index in L/min; SVI = stroke volume index in mls; SVR = systemic vascular resistance in $dyn\text{-}s\text{-}cm^{-5}$; SWI = stroke work index in $g\text{-}m/m^2$; LVEF = LV ejection fraction in %).

Exercise haemodynamics: Two patients were unable to sustain the minimum required exercise duration to allow haemodynamic data acquisition. The exercise data were therefore compiled from eight of the patients. At the same level of exercise for each patient, the heart rate was not significantly changed by either ouabain or digoxin. The left ventricular filling pressure, however, showed significant decrease from 39 ± 3.2 mm Hg to 34 ± 2.6 mm Hg with ouabain, and to 33 ± 3.4 mm Hg with digoxin therapy. During exercise the left ventricular systolic function improved as shown by the rise in cardiac index from 3.3 ± 0.3 L/min at control to 4.0 ± 0.4 L/min after acute ouabain and to 3.8 ± 0.4 L/min on maintenance therapy on digoxin. Since the heart rate did not change substantially, the stroke volume index also showed parallel rises. The mean arterial pressure showed a tendency to rise with the glycosides but the changes were not statistically significant taking the patient group as a whole. The left ventricular ejection fraction and stroke work index rose significantly during exercise with either cardiac glycoside.

The hydraulic power output (steady terms) of the heart rose from 1.54 (±0.23) watts during exercise at control to 1.80 (±0.24) watts with acute ouabain ($p < 0.01$), and to 1.96 (±0.30 watts) during digoxin therapy ($p < 0.01$). Taking the patients individually, and using the analytical method to separate the inotropic contributions to the rise in the power output values, from the contributions due to the changes in the peripheral resistance, it is noted in Fig. 2 that four patients demonstrated positive inotropism during exercise with acute ouabain, and six patients did so with chronic digoxin therapy.

Discussion

In this small group of patients in sinus rhythm, results were obtained which showed that digitalis did not produce significant changes in resting haemodynamics; but during exercise,

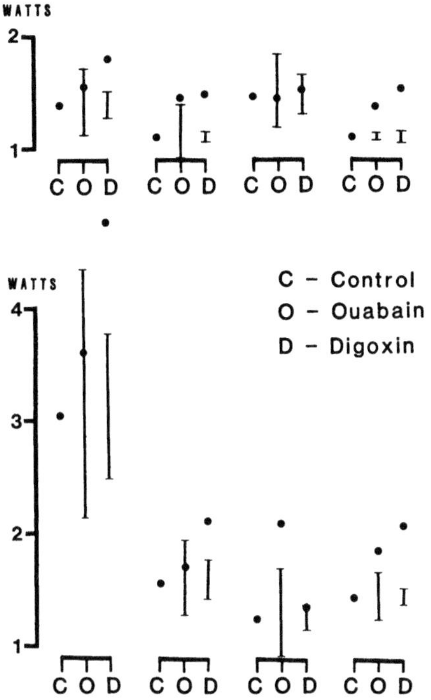

Fig. 2. the cardiac hydraulic power output values (in watts) during exercise at control (C), after IV ouabain (O), and after 6-week oral digoxin (D). The bars represent likely cardiac power output values had the changes caused by the glycosides been solely confined to changes in the peripheral vascular resistance.

direct positive inotropic effect could be detected in seven of the eight patients either with acute ouabain or chronic digoxin therapy. These results were obtained after the elimination of several possible confounding factors, particularly the subjective variability, by fixing the exercise workload for each patient constant. Variations in the heart rate and the resistance facing the heart were analytically removed from the manifestation of the positive inotropism. The preload changes could not have contributed to the increased power output as they were significantly reduced after either glycoside.

In general, patients tended to show a better response to maintained digoxin therapy than to acute ouabain. The one patient who demonstrated positive inotropism after acute ouabain administration without improved pumping performance six weeks later with maintained digoxin therapy, may be showing a progression of disease rather than the lack of drug efficacy. This demonstrates the difficulty met in evaluating long-term studies. Such a shortcoming can only be circumvented by a double-blind cross-over trial where the sample population is large.

An interesting aspect of the profile of inotropic action of digitalis in this group of patients is that there was a lack of significant increase in cardiac pumping capacity at rest, whereas the increase during exercise was quite marked in some. Such a profile is desirable for an inotrope because (a) marked increase in pumping performance at rest would be energy wasting and would reduce the pumping reserve (defined as that which can be called upon to elevate pumping performance from basal resting state to the maximally stressed state), and (b) the increased pumping capacity is available when it is actually needed.

One may speculate as to what the mechanism of such a selective inotropic action of digitalis is. The chain of events could be that digitalis inhibits the sodium pump resulting in increased calcium influx (1) and increased intracellular calcium storage pools (3). The greater release of calcium at activation from these pools may be triggered by beta-adrenergic stimulation (3), causing increased calcium availability to the myofibrils during exercise, and hence greater inotropism on exertion. Such a reaction is likely to occur in the cardiac cells because, unlike in skeletal muscle cells where the amount of calcium released is in excess of that required for maximal performance, the degree of activation in the myocardium tends to be closely regulated and submaximal (6). Also, in this experiment, the glycoside dosage may not be that which corresponds to the maximal stimulation seen in in vitro studies (2). This hypothesis does not preclude the demonstration of positive inotropism by digitalis at rest in cases where high levels of circulating catecholamines and presence of residual viable myocardium are found.

In conclusion, we have shown that digitalis can exert definite positive inotropism in a subgroup of heart failure patients, and that its action is most evident during exercise. The profile of such a digitalis action may throw some further light on the mechanism of how digitalis works as an inotrope. Whether the beneficial haemodynamic manifestation of digitalis can be translated into improved wellbeing of the patient remains to be evaluated by large scale studies. In the meantime, it is reasonable to suggest that digitalis is worth a trial in patients with heart failure in sinus rhythm.

References

1. Barry WH, Hasin Y, Smith TW (1985) Sodium pump inhibition, enhanced calcium influx via Na-Ca exchange, and positive inotropic response in cultured heart cells. Circ Res 56: 231–241
2. Brown L, Erdmann E (1985) Concentration-response curves of positive inotropic agents before and after ouabain pretreatment. Cardiovasc Res 1985 19: 288–298
3. Hunter DR, Haworth RA, Berkoff HA (1983) Modulation of cellular calcium stores in the perfused rat heart by isoproterenol and ryanodine. Circ Res 53: 703–712
4. Mulrow CD, Feussner JR, Velez R (1984) Reevaluation of digitalis efficacy. Ann Intern Med 101: 113–117
5. Tan LB, Schultz DL, Sdougos HP, Rajagopalan B, Lee G de J (1981) Separate contribution of ventricular pumping and load impedance on ventricular power output. Med Hypothesis 7: 1067–1077
6. Waser PG, Schaub MC (1981) Effects of cardiac glycosides on myofibrils. In Greeff K (ed) Cardiac Glycosides, Part I. Experimental Pharmacology. Springer Verlag, Berlin, pp 438–496

Author's address:

Dr. L. B. Tan, Department of Cardiovascular Medicine, University of Birmingham, East Birmingham Hospital, Bordesley Green East, Birmingham B9 5ST, Great Britain

Digitalis treatment in pulmonary heart disease

K. P. Schüren

Innere Abteilung, Krankenhaus Moabit, West Berlin (West Germany)

The development of pulmonary hypertension and subsequent right ventricular failure due to chronic pulmonary disease is associated with a high mortality of about 70% within 4 years (18). It would seem beneficial to improve hemodynamic performance when right ventricular function is depressed by increased afterload and disturbances in alveolar gas exchange attempting to improve clinical symptoms and long-term survival. Digitalis is frequently used in this setting, for its inotropic action on the right ventricle might show beneficial effects. With reference to this, the influence of cardiac glycosides on acute and long-term hemodynamics, pulmonary gas exchange and certain rhythm disturbances frequently associated with pulmonary heart disease have to be evaluated in detail.

1. Acute and long-term effects on right and left ventricular function

The acute hemodynamic effect of intravenously administered digoxin was studied in 10 patients without and 8 patients with congestive right heart failure secondary to severe chronic obstructive lung disease (20). The degree of airway obstruction did not differ in both groups. There were, however, marked differences in pulmonary artery pressure, pulmonary arterial wedge pressure, right ventricular end-diastolic pressure, hematocrit and total blood volume, which were significantly higher in patients with than in patients without overt right heart failure. Patients with right heart failure revealed more distinct hypoxemia and hypercapnea than patients without right ventricular dysfunction (Table 1).

Acute intravenous digitalization did not cause significant changes in right ventricular end-diastolic pressure, mean pulmonary artery pressure, pulmonary artery wedge pressure and cardiac output in patients with and without right heart failure, whereas pulmonary vascular resistance showed a slight increase in both groups (Table 2 a and b). These results confirm earlier studies of Ferrer et al. (10), Berglund et al. (5) and Jezek and Schrijen (14).

The effect of digitalis on the pulmonary vasculature arises, at least in part, from a direct local vasoconstriction of the smooth-muscle containing small pulmonary arteries, not abolished by vagotomy and sympathetic blockade (15). Any enhancement of elevated pulmonary vascular resistance, together with increased stroke volume, will increase the pressure work load of the right ventricle. Thus, an improvement in right ventricular function may be masked in some cases by a further rise in right ventricular afterload.

As the degree of abnormalities in pulmonary circulatory and right ventricular function is more pronounced under physical exercise as compared to resting conditions the hemodynamic effects of cardiac glycosides might be at least fully recognized under ergometric exercise. Figure 1, however, shows that the abnormal rise in right ventricular end-diastolic pressure in both groups could not be prevented by digoxin, although the right ventricular function curve was shifted upwards and to the left when overt right heart failure was present.

Table 1. Hemodynamic parameters and arterial blood gas tensions in patients with chronic obstructive lung disease with and without right heart failure.

		Without RHF	With RHF
HR	(min^{-1})	91	83
PAm	(mm Hg)	25	33
PCP	(mm Hg)	7	12
RVED	(mm Hg)	7	10
CI	($l/min/m^2$)	3.2	3.4
SVI	(ml/m^2)	35	41
PVR	($dyn \cdot sec \cdot cm^{-5}$)	267	276
P_aO_2	(mm Hg)	62	51
P_aCO_2	(mm Hg)	44	50
Hct	(%)	46	54
EV	(ml/kg)	27*	42**
PV	(ml/kg)	38*	43**
TBV	(ml/kg)	65*	85**

*: n = 7, **: n = 5

Abbreviations: RHF = right heart failure, HR = heart rate, PAm = pulmonary arterial mean pressure, PCP = pulmonary capillary pressure, RVED = right ventricular end-diastolic pressure, CI = cardiac index, SVI = stroke volume index, PVR = pulmonary vascular resistance, PaO_2 = arterial oxygen tension, $PaCO_2$ = arterial carbon dioxide tension, Hct = hematocrit, EV = erythrocyte volume, PV = plasma volume, TBV = total blood volume.

Table 2, a and b. Effects of intravenously administered digoxin on right ventricular and pulmonary circulatory function in ten patients without and eight patients with right heart failure due to chronic obstructive lung disease.

2a

Patients with RHF (n = 8)

		Control	Digoxin
HR	(min^{-1})	83	82
PAm	(mm Hg)	33	34
PCP	(mm Hg)	12	12
RVED	(mm Hg)	10	9
CI	($l/min/m^2$)	3.4	3.4
PVR	($dyn \cdot sec \cdot cm^{-5}$)	276	290

2b

Patients without RHF (n = 10)

		Control	Digoxin
HR	(min^{-1})	91	89
PAm	(mm Hg)	25	25
PCP	(mm Hg)	7	6
RVED	(mm Hg)	7	6
CI	($l/min/m^2$)	3.2	3.0
PVR	($dyn \cdot sec \cdot cm^{-5}$)	267	304

Abbreviations see Table 1

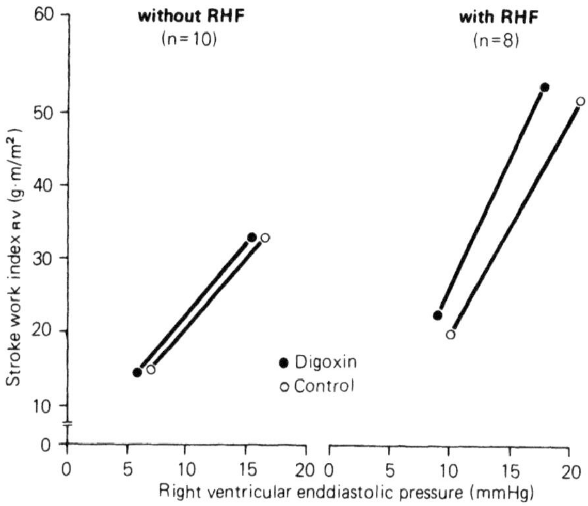

Fig. 1. Effect of intravenously administered digoxin on the right ventricular function curve at rest and during ergometric exercise in 10 patients without and 8 patients with overt right heart failure.

The presence or lack of digitalis effect did not correlate with the level of filling pressure or with cardiac output both at rest and during exercise.

Comparable small decreases in right ventricular filling pressure and increases in right ventricular stroke work at rest and during exercise could be demonstrated in patients with obstructive bronchitis and right heart failure, reported by Jezek and Schrijen (14). In a series of Berglund et al. (5) a definite fall in right ventricular enddiastolic pressure during exercise could be established only in one out of 4 patients with pulmonary heart disease after intravenous administration of lanatoside C. In 3 cases, 2 of them with severe respiratory failure, the additional administration of oxygen was at least accompanied by a significant lowering of right ventricular filling pressure, especially under exercise. This effect could be explained by a marked decrease in right ventricular afterload in only one of the 3 patients. The authors felt a more important factor, however, to be the improved oxygenation of the myocardium and thereby an improved myocardial function. A limited response of the hypoxic myocardium to acetyl-strophanthidin was demonstrated in animal experiments (9). In the present series of 18 patients with chronic obstructive lung disease with and without right heart failure as well as in the studies of Jezek and Schrijen (14) no correlation could be identified between the degree of arterial hypoxemia and hypercapnea and the digitalis-induced changes in right ventricular filling pressure (Fig. 2).

Studies on the *long-term effect* of digitalis on right and left ventricular function are rare. Mathur et al. (16) reported on the effect of digoxin on the right and left ventricular ejection fraction (EF) in 15 patients with chronic obstructive lung disease and right ventricular dysfunction. In a randomized double-blind crossover placebo-controlled trail of 16 weeks duration patients were assessed at the beginning of the study, at the eighth week (cross-over point) and at the sixteenth week, including pulmonary function studies and gated equilibrium randionuclide angiography to determine right and left ventricular ejection fractions. Patients were randomly assigned to oral digoxin 0.25 mg/day, or placebo for 8 weeks and then given the alternate therapy. No alteration to bronchodilator or diuretic treatment was made during the study. Right ventricular EF was reduced in all, left ventricular EF in 4 of the 15 patients. After 8 weeks of digoxin treatment the abnormal left ventricular EF of the 4

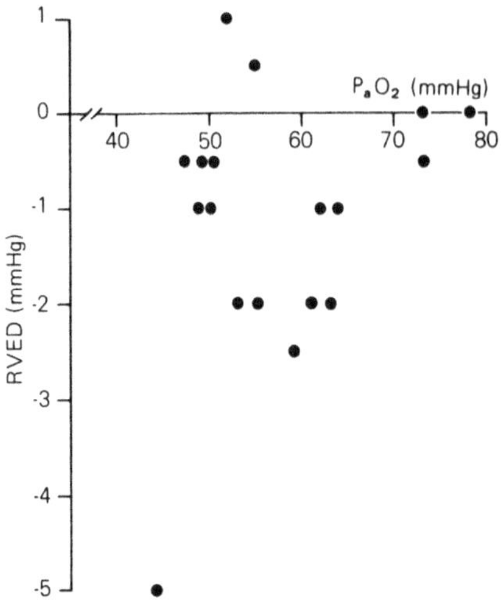

Fig. 2. Relation between arterial PO_2 and digoxin-induced changes in right ventricular end-diastolic pressure (RVED) in 18 patients with chronic obstructive lung disease.

patients had normalized. Right ventricular EF increased only in those patients who had abnormal left ventricular EF (Fig. 3). Thus, the critical factor affecting the efficacy of cardiac glycosides is the status of left ventricular function, which often is not clinically apparent. Mathur et al. (16) found that all patients had a reduced right ventricular EF despite the absence of congestive right heart failure. This observation is in keeping with the

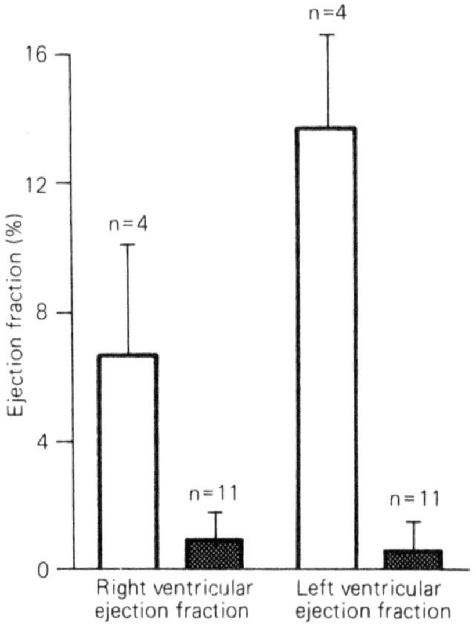

Fig. 3. Digoxin increased right ventricular ejection fraction only in the four patients with severe chronic airway obstruction whose left ventricular ejection fraction improved, but not in the eleven patients whose left ventricular ejection fraction remained unchanged (16).

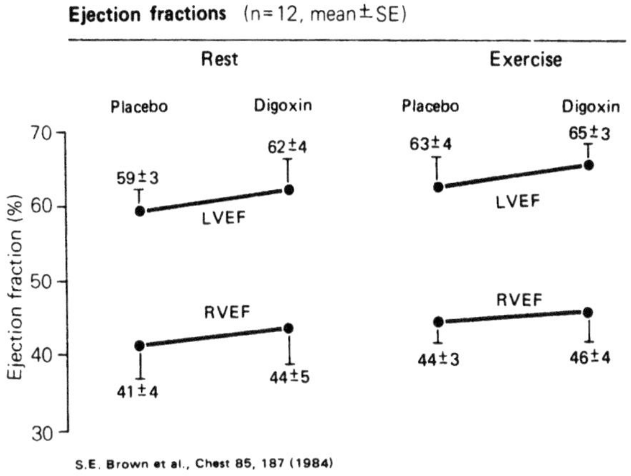

Fig. 4. Mean right and left ventricular ejection fractions at rest and during exercise before and after two weeks of therapy with oral digoxin (0.25 mg/day) or placebo in 12 patients with stable chronic airway obstruction without clinical evidence of left ventricular disease (6).

suggestion that radionuclide angiography may be sensitive to early detection of right ventricular dysfunction before the development of overt right heart failure (4). This reduction in right ventricular EF is the expected response to increased afterload as the right ventricle dilates. This does not necessarily imply an alteration in underlying myocardial response (21).

In another randomized double-blind, placebo-controlled crossover study Brown et al. (6) investigated the influence of oral digoxin (0.25 mg/day) on exercise capacity ($\dot{V}O_{2max}$) and on right ventricular performance at rest and during ergometric exercise in 12 patients with stable chronic airway obstruction and no evidence of additional left ventricular disease. Right and left ventricular EF were obtained at rest and at approximately 75% of the maximum workload by gated equilibrium radionuclide angiography. All patients had abnormal right ventricular function, manifested either by a low right ventricular EF or a subnormal response to exercise (< 5% increase). The small increase in right ventricular EF with digoxin at rest and during exercise did not achieve statistical significance (Fig. 4). With digoxin only slight increases in exercise duration (9.0 ± 1.4 min to 10.0 ± 1.5 min) and work load (42 ± 5 W to 48 ± 6 W) occurred. The left ventricular EF which was initially abnormal in 3 of the 12 patients, improved with digoxin in only one patient.

The authors conclude that digoxin does not significantly improve exercise capacity, limited by the symptom of dyspnea, nor right ventricular performance in patients with severe chronic obstructive lung disease and impaired right ventricular function.

In patients with chronic obstructive lung disease and consecutive congestive right heart failure exercise capacity is obviously limited predominantly by airway mechanics. If this is the case improved cardiac function might not result in a beneficial effect on excercise capacity (6).

2. Acute and long-term effects on pulmonary gas exchange

In our series of 18 patients with chronic obstructive lung disease with and without right heart failure digoxin did not lead to any changes in blood gas tensions. The long-term effects on alveolar gas exchange have not been analyzed, but a significant influence cannot be expected. Improvements in respiratory function that may be observed in single patients after digitalization may be primarily related to other concurrent therapeutic interventions such as the administration of β_2-sympathomimetic agents, theophylline, antibiotics and oxygen.

3. Indications for digitalis treatment in pulmonary disease

Pulmonary disease predisposes to atrial arrhythmias, presumably from both right atrial dilatation and hypoxia (12). High incidences in atrial tachycardia, atrial flutter, atrial fibrillation and junctional rhythm with rapid ventricular response could be demonstrated by several authors (7, 19). There is a legitimate indication for cardiac glycosides which are effective in controlling the atrial tachyarrhythmias by lowering increased heart rates or converting the arrhythmias into sinus rhythm. Difficulties may arise as the types of rhythm disturbances are similar in both chronic pulmonary disease and digitalis excess (Table 3) (8). In particular, atrial tachycardia with block and AV-tachycardia are relatively frequent forms of arrhythmias in cor pulmonale (7, 19) as well as in digitalis toxicity. Further difficulties may arise as patients with chronic pulmonary disease show epidemiologic evidence of enhanced sensitivity to digitalis glycosides (1, 3). The following rule often cited clinically in paroxysmal atrial tachycardia with block may be recommended: "If the patient is taking digitalis, stop it; if not, then digitalize" (8).

Table 3. Types of arrhythmias

Associated with pulmonary disease	Associated with digitalis excess
Paroxysmal atrial tachycardia with and without block	Paroxysmal atrial tachycardia with and without block
A-V junctional rhythm	A-V junctional rhythm
Atrial premature beats	Atrial premature beats
Ventricular premature beats	Ventricular premature beats
A-V dissociation	A-V dissociation
Atrial flutter	Atrial flutter (rare)
Atrial fibrillation	Atrial fibrillation (rare)

Adapted from Doherty et al. (8)

In a large series of 931 consecutive patients, who were prospectively studied by Beller et al. (3), a striking percentage (53 %) of patients who exhibited cardiac arrhythmias consistent with digitalis intoxication had acute or chronic lung disease. Morrison and Killip (17) observed an increased sensitivity in 8 patients with respiratory failure and severe hypoxemia (mean arterial PO_2 39 mm Hg) who developed rhythm disturbances induced by digitalis, although all of them had serum digoxin levels within the therapeutic range. Six patients with less profound hypoxemia (mean arterial PO_2 52 mm Hg) who showed evidence for digitalis toxicity had serum digoxin levels in a significantly higher range.

Tolerance to digitalis was specifically investigated by Baum et al. (1). In 29 patients with chronic respiratory failure 1.2 mg of acetyl-strophanthidin was injected intravenously and the electrocardiogram was monitored continuously for 30 minutes. 8 of the 29 patients exhibited evidence of increased sensitivity to digitalis, such as transient AV-dissociation, prolonged AV-conduction time and ventricular premature beats. The other 21 patients did not develop electrocardiographic abnormalities following the administration of the cardiac glycoside. The mean arterial PO_2 was significantly lower for the sensitive group than for the nonsensitive group. The authors conclude that the sensitivity to digitalis in chronic pulmonary disease is more likely to occur in the presence of severe hypoxemia. Although the total body exchangeable potassium was lower and the arterial PCO_2 was higher for the sensitive patients, these factors appeared to be poorly correlated with the presence or absence of digitalis sensitivity. Nevertheless, it is now generally accepted that hypokalemia with and without derangements in respiratory function enhances the risk of digitalis toxicity in different etiologic types of heart failure.

Green and Smith (12) stated in an excellent review on the possible mechanisms of enhanced risk of digitalis toxicity in patients with pulmonary heart disease that hypoxemia could act to lower the threshold for digitalis-induced arrhythmias through its influence on endogenous sympathetic stimulation of the heart. Hypoxemia increases the release of catecholamines from cardiac adrenergic stores (11). There is experimental evidence that catecholamine-induced enhancement of ventricular automaticity is additive to the arrhythmogenic effects of digitalis (12, 2). Particularly, a sudden deterioration in oxygenation superimposing on chronic hypoxia could provoke an acute increase in adrenergic tone (12). The severe hypoxemia observed in most patients with increased sensitivity to digitalis is suggested to be mediated by an acute deterioration of respiratory function (1, 7, 17, 19).

Although hypercapneic acidosis increases cardiac catecholamine release (11) an enhancement of sensitivity to cardiac glycosides could not be definitely established (1, 22). In acute worsening of alveolar gas exchange, however, hypercapneic acidosis may become a relevant additional factor influencing the tolerance to digitalis (13).

A further important factor predisposing to enhanced digitalis sensitivity may be the frequent exposure both to exogenuous catecholamines, especially β_2-sympathomimetic agents and theophylline preparations commonly used for the relief of bronchospasm (Table 4).

Table 4. Possible reasons for enhanced digitalis sensitivity

Hypoxia
Hypercapneic acidosis
Increase in catecholamine release
Hypokalemia
Drugs (theophylline, sympathomimetic agents, diuretics)

Conclusion

In patients with right heart failure due to chronic pulmonary disease the positive inotropic effect of digitalis on the right ventricle is of doubtful clinical value, in that it does not prevent or substantially reduce right ventricular dysfunction during exercise. The diminished effectiveness with exercise is independent of the coexistence and degree of hypoxemia or hypercapnea. Hemodynamic improvement can only be expected in patients with superimposed left ventricular failure. In those patients digitalis causes a rise in both right and left ventricular ejection fractions. The improvement in right ventricular function is consistent with the reduction of that portion of right ventricular afterload secondary to left ventricular dysfunction and supports the concept that right ventricular performance is highly dependent upon afterload which is unaffected by cardiac glycosides, unless increased afterload is in part due to left ventricular dysfunction. Even in these patients the prognostic relevance of digitalis treatment remains uncertain.

The elevated pulmonary vascular resistance in chronic cor pulmonale is at least in part reversible. As in hypertensive heart disease in which the therapy of choice includes reduction of afterload by treatment of the systemic blood pressure elevation, the main therapeutic goal of pulmonary heart disease is a reduction of right ventricular afterload. Oxygen, a potent pulmonary vasodilator, and bronchodilators, which reduce alveolar compression of the pulmonary capillaries and enhance alveolar gas exchange, lower the increased resistance to right ventricular outflow and may thereby relieve congestive right heart failure. Digitalis, however, remains the unequivocal treatment of choice, when right heart failure with or without respiratory insufficiency is accompanied by certain supraventricular rhythm disturbances with rapid ventricular response.

References

1. Baum GL, Dick MM, Blum A, Kaupe A, Carballo J (1959) Factors involved in digitalis sensitivity in chronic pulmonary insufficiency. Am Heart J 57: 460
2. Becker DJ, Nonkin PM, Bennett LD, Kimball SG, Sternberg MS, Wassermann F (1962) Effect of isoproterenol in digitalis cardiotoxicity. Am J Cardiol 10: 242
3. Beller GA, Smith TW, Abelmann WH, Haber E, Hood WB Jr (1971) Digitalis intoxication. A prospective clinical study with serum level correlations. N Engl J Med 284: 989
4. Berger HJ, Matthay RA, Loke J, Marshall RC, Gottschalk A, Zaret BL (1978) Assessment of cardiac performance with quantitative radionuclide angiography. Right ventricular ejection with reference to finding in chronic obstructive pulmonary disease. Am J Cardiol 41: 89
5. Berglund E, Widimsky J, Malmberg R (1963) Lack of effect of digitalis in patients with pulmonary disease with and without heart failure. Am J Cardiol 11: 477
6. Brown SE, Pakron FJ, Milne N, Linden GS, Stansbury DW, Fischer CE, Light RW (1984) Effects of digoxin on exercise capacity and right ventricular function during exercise in chronic airflow obstruction. Chest 85 (2): 187
7. Corazza LJ, Pastor BG (1958) Cardiac arrhythmias in chronic cor pulmonale. New Engl J Med 259: 862
8. Doherty JE, Kane JJ, Phillips JR, Adamson JS (1977) Digitalis in pulmonary disease (cor pulmonale). Drugs 13: 142
9. Donlon JV, Yu PN (1969) Cardiovascular response of hypoxic myocardium to acetyl strophanthidin. Am Heart J 78: 238
10. Ferrer MI, Harvey RM, Cathcart RT, Webster CA, Richards DW, Cournand A (1950) Some effects of digoxin upon the heart and circulation in man. Digoxin in chronic cor pulmonale. Circulation 1: 161
11. Goldman RH, Harrison DC (1970) The effects of hypoxia and hypercarbia on myocardial catecholamines. J Pharmacol Exp Ther 174: 307
12. Green LH, Smith TW (1977) The use of digitalis in patients with pulmonary disease. Ann Int Med 87: 459
13. Hudson LD, Kurt TL, Petty TL, Genton E (1973) Arrhythmias associated with acute respiratory failure in patients with chronic airway obstruction. Chest 63: 661
14. Jezek V, Schrijen F (1973) Hemodynamic effect of deslanoside at rest and during exercise in patients with chronic bronchitis. Brit Heart J 35: 2
15. Kim YS, Aviado DM (1961) Digitalis and the pulmonary circulation. Am Heart J 62: 680
16. Mathur PN, Powles P, Pugsley SO, Mc Ewan MP, Campbell EJM (1981) Effect of digoxin on right ventricular function in severe chronic airflow obstruction. Ann Int Med 95: 283
17. Morrison J, Killip T (1971) Hypoxemia and digitalis toxicity in patients with chronic lung disease (abstr.). Circulation 44 (Suppl 11): 41
18. Renzetti AD Jr., Mc Clement JH, Litt BD (1966) The Veterans Administration Cooperative Study of Pulmonary Function: III. Mortality in relation to respiratory function in chronic obstructive pulmonary disease. Am J Med 41: 115

19. Schüren KP, Calder D, Hüttemann U (1973) Rhythmusstörungen bei chronischem Cor pulmonale. Dtsch med Wschr 98: 2111
20. Schüren KP, Hüttemann U (1974) Chronisch obstruktive Lungenerkrankungen: Die haemodynamische Wirkung von Digitalis beim chronischen Cor pulmonale in Ruhe und unter Belastung. Klin Wschr 52: 736
21. Stein PD, Sabbah HN, Anbe DT, Marzilli M (1979) Performance of the failing and non failing right ventricle of patients with pulmonary hypertension. Am J Cardiol 44: 1050
22. Williams JF Jr, Boyd DL, Border JF (1968) Effect of acute hypoxia and hypercapnic acidosis on the development of acetylstrophanthidin-induced arrhythmias. J Clin Invest 47: 1885

Author's address:

Prof. Dr. med. K. P. Schüren, Innere Abteilung, Krankenhaus Moabit, Turmstraße 21, 1000 Berlin (West Germany)

Cardiac glycosides for paroxysmal atrial fibrillation?

G. Steinbeck, R. Doliwa, and P. Bach

Medical Hospital I, University of Munich (West Germany)

Summary

The antiarrhythmic efficacy of oral treatment with cardiac glycosides is evaluated in a series of 14 unselected patients with paroxysmal atrial fibrillation. Over a mean follow-up period of 20 weeks, only two patients were treated successfully (14%), derived from clinical symptoms and serial 24 h electrocardiograms. Ten patients exhibited drug failures; in the remaining two, drug efficacy could not be assessed. In contrast, success rates with common class I antiarrhythmic drugs reported in literature to convert atrial fibrillation into sinus rhythm range between 40 and 77%.

It is concluded that oral treatment with cardiac glycosides – while decreasing ventricular rate in atrial fibrillation – rarely suppresses the recurrence of paroxysmal atrial fibrillation, this efficacy probably being inferior to common class I antiarrhythmic drugs.

Introduction

Despite numerous experimental and clinical investigations, the electrophysiologic effects of cardiac glycosides on atrial fibrillation remain incompletely defined. This is mainly because of the multifactorial actions of glycosides under this circumstance, which include: the cardiac cell membrane, autonomic nervous system, baroreceptor sensitivity, coronary and peripheral vasculature, contractility and ventricular rate, central nervous system.

These various effects might contribute to the discrepancies in the data derived from animals and man.

Looking at basic electrophysiologic actions of cardiac glycosides at the atrium, both a prolongation and shortening of refractory period have been reported. Earlier experimental work indicates that a prolongation is more likely in denervated or atropinized hearts, whereas a shortening was observed under conditions which permit reflex vagal activity (10, 12). Recent studies in man using the endocardial contact injury potential technique suggest, however, that both effects can also be mediated via cardiac muscarinic receptors (7).

In the AV node, slowing of conduction and a prolongation of refractory period have been measured and attributed to an increase of vagal tone; in addition, a withdrawal of sympathetic tone, and antiadrenergic and direct effects have been thought responsible (11, 12).

This is the electrophysiologic basis of lowering ventricular rate in atrial fibrillation which without doubt is of benefit for the patient. Are cardiac glycosides also able to stop atrial fibrillation itself and protect the patient from its recurrence?

In 1978, Engel and Gonzales – using programmed stimulation – reported effects of ouabain, given intravenously in nine patients, on atrial refractory and vulnerable zone for induction of atrial fibrillation (4). These authors noted a prolongation of functional refractory period and – more importantly – a decrease of the vulnerable zone by 70%, suggesting a protective effect against atrial fibrillation.

We previously studied control patients and with sick sinus syndrome by programmed stimulation both before and during oral treatment with digoxin (Fig. 1) (14).

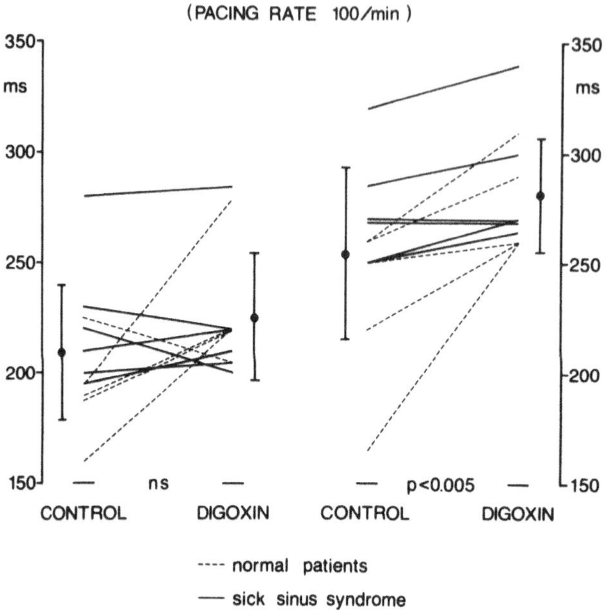

Fig. 1. Effect of oral treatment with digoxin on atrial effective refractory period (left panel) (longest S1–S2 interval not capturing the atrium during constant atrial drive with an interval S1–S1 of 600 ms) and functional refractory period (right panel) (shortest a1–a2 interval to be obtained by programmed atrial stimulation at S1–S1 of 600 ms) in five normals and six patients with sick sinus syndrome.

Also under these cirumstances, there is a tendency of the effective refractory period of the right atrium to increase; the functional refractory period is significantly increased by digoxin (Fig. 1). Apart from some early clinical observations (8) we did not become aware of any valid study assessing the role of cardiac glycosides for termination of atrial fibrillation and its further prevention.

This study evaluates the antiarrhythmic efficacy of oral treatment with cardiac glycosides for termination and further prevention of paroxysmal atrial fibrillation in a series of 14 unselected patients with paroxysmal atrial fibrillation, and results are compared with the efficacy of common antiarrhythmic drugs reported in literature.

Patients and methods

Criteria for inclusion into this study were patients with electrocardiographic documentation of paroxysmal atrial fibrillation associated with clinical symptoms such that there was an indication for antiarrhythmic therapy.

Criteria for exclusion were age above 75 years, acute myocardial infarction, hypokalemia, untreated hyperthyreoidism, and WPW syndrome. Concomitant therapy with beta-blockers, calcium antagonists or other antiarrhythmic drugs was not allowed.

Clinical data of the patients (age and sex, underlying disease, diameter of left atrium measured echocardiographically, frequency of previous arrhythmia attacks, duration of history of arrhythmias, and number of previous ineffective drug trials) are given in Table 1.

After loading, the patients received oral digoxin with a chronic daily dose of 0.375–0.5 mg.

Table 1. Clinical data of 14 patients with paroxysmal atrial fibrillation, and antiarrhythmic efficacy of oral digoxin treatment

No	Age and sex		Underlying disease	Diameter of left atrium (mm)	Frequency of atrial fibrillation	Duration of arrhyth. history (months)	Ineffective drug trials	Efficacy of digoxin	Follow-up (weeks)
1	45	♂	My	34	<1/m.	72	4	no red.	4
2	74	♂	CAD	–	>1/w.	84	4	no red.	7
3	60	♀	My, MVP	36	>1/w.	66	2	suppr.	66
4	67	♀	My	56	<1/y.	108	2	no red.	70
5	38	♂	My, CAD	48	>1/w.	84	4	no red.	5
6	51	♀	–	40	<1/w.	48	2	no red.	45
7	64	♂	RHD	26	>1/w.	24	3	no red.	1
8	65	♂	–	49	<1/m.	120	10	no red.	3
9	64	♀	–	42	<1/w.	103	3	no red.	4
10	52	♂	CAD	46	<1/y.	66	1	?	44
11	69	♀	–	48	>1/w.	120	3	no red.	10
12	58	♀	My, MVP	30	>1/w.	240	4	suppr.	11
13	48	♂	RHD	42	<1/w.	12	1	?	8
14	65	♀	MVP	43	<1/m.	30	1	no red.	3

My = history of myocarditis; CAD = coronary artery disease; MVP = mitral valve prolapse;
RHD = rheumatic heart disease;
Incidence . . . /w. = attacks per week; /m. = attacks per month; /y. = attacks per year;
Efficacy . . . suppr. = complete suppression; no red. = slight or no reduction; ? = not to be classified.

For classification of drug efficacy during long-term treatment, clinical symptoms (either in the hospital or during regular ambulatory follow-up), electrocardiogram and 24 hour electrocardiogram both before and under institution of therapy were taken together to grade efficacy as:
complete suppression,
reduction of arrhythmia attacks (so that from a clinical point of view drug therapy was a success and continued),
slight or no reduction of recurrences of atrial fibrillation (so that an additive or alternative therapy had to be chosen),
efficacy not to be classified (for example bad compliance, follow-up too short in relation to incidence of arrhythmias).

Results and discussion

The results of oral treatment with digoxin in the 14 unselected patients with paroxysmal atrial fibrillation are detailed in Table 1.

Over a follow-up period of 21.1 weeks ± 24.8 (± SD) only two patients were treated successfully (14%), as opposed to ten patients with drug failures. In the remaining two patients, drug efficacy could not be classified. As side effects during treatment we noted sinus bradycardia and sinus arrest in one patient, and nausea in another; both side effects disappeared following a decrease of drug dosage.

To compare these data, a survey is given in Table 2 (which does not claim to be complete) of the literature about the efficacy of common antiarrhythmic drugs to convert atrial fibrillation into sinus rhythm. These studies include trials with propafenone, flecainide, disopyramide, diltiazem, verapamil, quinidine, quinidine plus verapamil, and amiodarone. Of note is that while the success rate of the combination quinidine plus verapamil was reported to be rather high (6), the success rate of the calcium antagonists diltiazem or verapamil given alone is very low (13). As can be seen from Table 2, the percentage of patients in whom a sinus rhythm could be achieved, ranges between 40 and 77% with various class I (local anesthetic) and the class III antiarrhythmic drug amiodarone. These success rates are clearly higher than that of 14% obtained with cardiac glycosides in our study. Our disappointing results are in accord with the finding by others that prophylactic digitalization is of no benefit for the prevention of supraventricular tachyarrhythmias following myocardial revascularization (15).

On the other hand, a comparison of such studies as ours with those in Table 2 should be made very cautiously, since in most of the reports cited in Table 2, chronic stable atrial fibrillation was studied, instead of paroxysmal atrial fibrillation in this report. Furthermore, there may be other differences from study to study with respect to patient population, evaluation of drug efficacy, and period of follow-up.

Table 2. Antiarrhythmic efficacy in atrial fibrillation.

Authors	Drug	No. of patients	Percentage of sinus rate achieved (%)
Beck et al., 1978	Propafenone	30	46.5
Goy et al., 1985	Flecainide i. v. and orally	50	72.0
Breithardt et al., 1976	Disopyramide	23	43.5
Lerch, 1976	Disopyramide	30	40.0
Ochs et al., 1985	Diltiazem	15	6.5
	Verapamil	15	13.0
Gülker et al., 1985	Quinidine/verapamil (cardiac glycosides)	142	77.0
Cramer, 1968	Quinidine	237	62.0
Ward et al., 1980	Amiodarone	15	53.0

Despite these limitations we conclude that oral treatment with cardiac glycosides rarely suppresses the recurrence of paroxysmal atrial fibrillation, the efficacy probably being inferior to common class I antiarrhythmic drugs.

Thus, while slowing of ventricular rate in atrial fibrillation is without doubt of clinical benefit, class I antiarrhythmic agents should be added or given alternatively to cardiac glycosides if termination of the arrhythmia itself and prevention of its recurrence are the goals of therapy.

References

1. Beck OA, Lehmann H-U, Hochrein H (1978) Propafenon und Lidoflazin bei chronischem Vorhofflimmern und -flattern. Dtsch med Wschr 103: 1068
2. Breithardt G, Haerten K, Seipel L (1976) Zur antiarrhythmischen Wirksamkeit von Disopyramid bei ventrikulärer Extrasystolie und Vorhofflimmern. Z Kardiol 65: 713
3. Cramer G (1968) Early and late results of conversion of atrial fibrillation with quinidine. Acta Med Scand (Suppl) 490: 1
4. Engel TR, Gonzalez ADC (1978) Effects of digitalis on atrial vulnerability. Am J Cardiol 42: 570
5. Goy J-J, Grbic M, Hurni M, Finci L, Maendly R, Duc J, Sigwart U (1985) Conversion of supraventricular arrhythmias to sinus rhythm using flecainide. Eur Heart J 6: 518

6. Gülker H, Brisse B, Heuer H, Thale J, Bramann HU, Frenking B, Matija B, Bender F (1985) Antifibrillatorische Wirksamkeit der Kombination Chinidin/Verapamil bei Vorhofflimmern. Therapiewoche 35: 821
7. Hayward RP, Hamer J, Taggart P, Emanuel R (1983) Observations on the biphasic nature of digitalis electrophysiological actions in the human right atrium. Cardiovasc Res 17: 533
8. Jennings PB, Makous N, Vander Veer JB (1958) Reversion of atrial fibrillation to sinus rhythm with digitalis therapy. Am J Med Sci 702
9. Lerch W (1976) Bericht über eine kontrollierte Studie zur Wirksamkeit des Präparates B 712 (Disopyramid-dihydrogenphosphat) bei absoluter Arrhythmie mit Vorhofflimmern. Therapiewoche 26: 6515
10 Méndez R, Méndez C (1953) The action of cardiac glycosides on the refractory period of heart tissues. J Pharmacol Exp Ther 107: 24
11. Méndez C, Aceves J, Méndez R (1961) The anti-adrenergic action of digitalis on the refractory period of the A-V transmission system. J Pharmacol Exp Ther 131: 198
12. Moe GK, Farah AE (1970) Cardiovascular drugs: digitalis and allied cardiac glycosides. Pharmacological Basis of Therapeutics 31: 677
13. Ochs HR, Anda L, Eichelbaum M, Greenblatt DJ (1985) Diltiazem, verapamil, and quinidine in patients with chronic atrial fibrillation. J Clin Pharmacol 25: 204
14. Steinbeck G, Rosenberger W, Naumann d'Alnoncourt C, Lüderitz B (1978) Zur Glykosidwirkung beim Sinusknotensyndrom. Z Kardiol 67: 190
15. Tyras DH, Stothert JC, Kaiser GC, Barner HB, Cood JE, Willman VL (1979) Supraventricular tachyarrhythmias after myocardial revascularization: A randomized trial of prophylactic digitalization. J Thorac Cardiovasc Surg 77: 310
16. Ward DE, Camm AJ, Spurrell RAJ (1980) Clinical antiarrhythmic effects of amiodarone in patients with resistant paroxysmal tachycardias. Br Heart J 44: 91

Author's address:

Prof. Dr. G. Steinbeck, Medizinische Klinik I der Universität München, Klinikum Großhadern, Marchioninistraße 15, 8000 München 70 (West Germany)

Cardiac glycosides in anaesthesia

K. van Ackern*, L. Frey+, and H. Bardenheuer+

* Institut für Anaesthesiologie, Medizinische Universität zu Lübeck, Lübeck (West Germany)
+ Institut für Anaesthesiologie, Ludwig-Maximilians-Universität München, Munich (West Germany)

Cardioactive glycosides are used in clinical practice because of their influence on myocardial contractility and their antidysrhythmic properties. The pharmacological actions of glycosides and anaesthetics on the myocardium and their interactions during the operative period can influence the anaesthesiologic management of the patient.

Anaesthetic agents – inhalation agents as well as intravenously applied anaesthetics – can induce a dose-dependent reduction of myocardial performance. Their influence on the hemodynamic functions of the heart has been studied in isolated preparations like papillary muscles (13) and in heart lung preparations (17). Fig. 1 shows the influence of halothane on cardiac contractility when applied in the concentration range between 1 and 3 vol%. Halothane dose-dependently decreases pressure development of the heart. Other inhalative anaesthetics (enflurane, isoflurane) have also been demonstrated to exhibit similiar effects on cardiac function. However, they differ in their ability to impair myocardial contractility (30). The rank order of potency causing myocardial depression is halothane > enflurane > isoflurane.

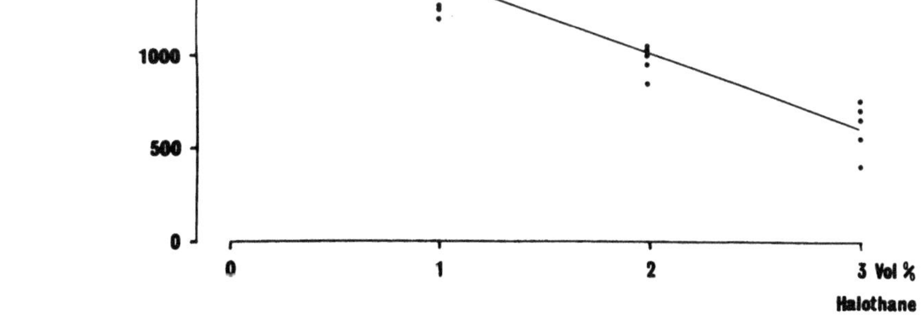

Fig. 1. Influence of halothane on tension development in the isolated perfused heart lung preparation of the dog (30).

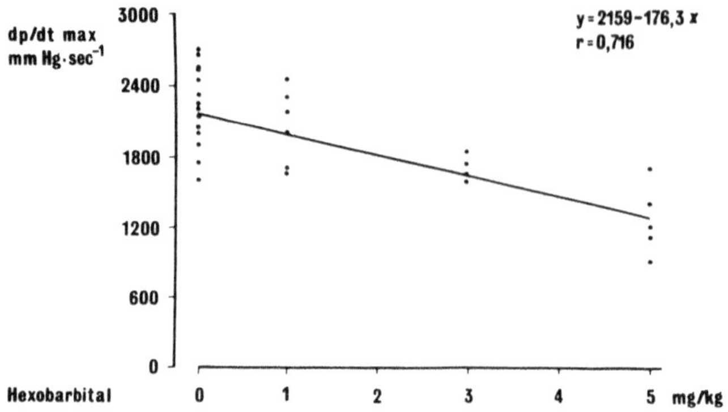

Fig. 2. Effect of hexobarbital on dP/dtmax in the isolated heart lung preparation of dogs (30).

Similar changes in myocardial contractility can be obtained with intravenously applied anaesthetics. For example, Fig. 2 shows that hexobarbital dose-dependently decreases dP/dtmax when infused in concentrations used in practical anaesthesia (29,30). As in the case of volatile substances intravenously applied anaesthetics differ in their ability to reduce heart function. Fig. 3 compares the reducing capacity of several agents on dP/dtmax (30). For example, propanidid most effectively impaired cardiac function while etomidate only slightly influenced dP/dtmax. Studies using the heart lung preparation (28) have shown that the negative inotropic effects of anaesthetic agents can be reversed by calcium or cardiac

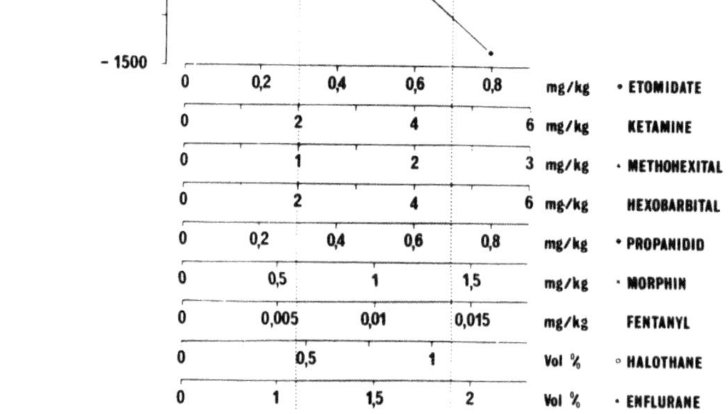

Fig. 3. Influence of different concentrations of intravenously applied and volatile anaesthetics on cardiac contractility in the heart lung preparation of dogs. The dotted lines indicate the range in which these substances are used under clinical conditions (30).

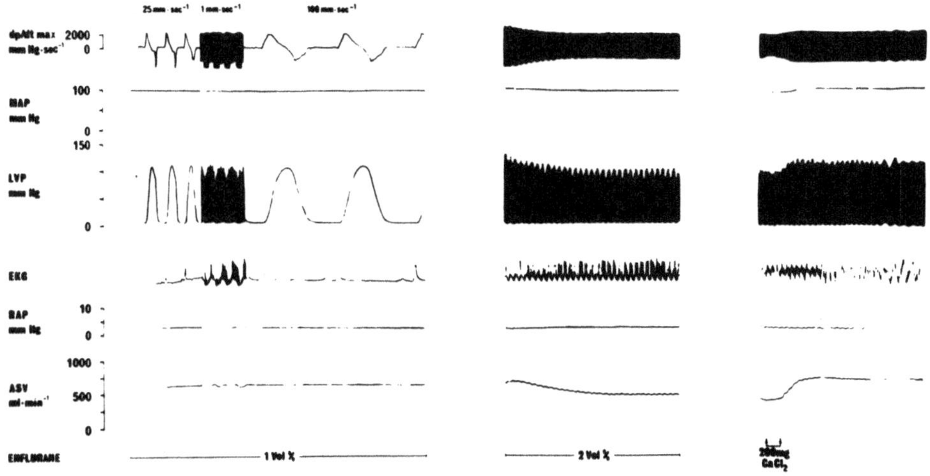

Fig. 4. Original tracing of the influence of enflurane on cardiac function and the effects of intravenously applied $CaCl_2$ on hemodynamic parameters in the isolated heart lung preparation of dogs (30). MAP: mean arterial pressure; LVP: left ventricular pressure; RAP: right atrial pressure; ASV: cardiac output.

glycosides. The original tracing (Fig. 4) demonstrates that infusion of calcium completely antagonizes the effects of 2 vol% enflurane (30). Similar improvement in cardiac function can be obtained with infusion of k-strophanthin when contractility is diminished by ketamine (Fig. 5).

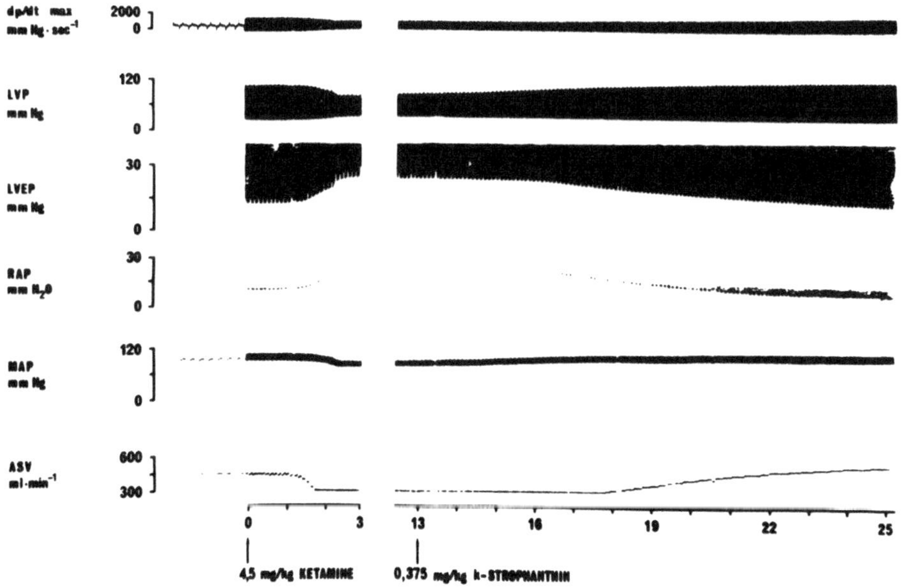

Fig. 5. Original recording illustrates the influence of ketamine on cardiac funtion and the effects of k-strophantin on hemodynamic parameters in the isolated perfused heart lung preparation of dogs (30).

The basic mechanism underlying the actions of volatile and intravenously applied anaesthetics on cardiac hemodynamics is yet not fully understood. Ca^{++} ions play an essential role in the contractile processes of cardiac muscle fibers. There is strong evidence that the negative effects of anaesthetics on cardiac contractility can be explained by interaction between those substances and excitation-contraction coupling (9, 12). The inhalational anaesthetics halothane and enflurane may modify the excitation-contraction coupling in different ways. Merin et al. (19, 20) have shown that both substances reduced the sensitivity of the actomyosinadenosine triphosphatase (ATPase) to calcium in a dose-dependent fashion. The decreased activity of the ATPase which controls the utilisation of energy by cardiac muscle could, however, be reversed when the concentration of Ca^{++} ions was enhanced. Thus, halothane and enflurane most likely act as competitive inhibitors for calcium on that enzyme system. In addition, there is further evidence that the negative inotropic action of both volatile anaesthetics is mediated by a receptor located at a component of the heavy chains of myosin.

Studies using skinned myocardial fibers have shown that halothane and enflurane dose-dependently diminished the uptake of calcium into isolated sarcoplasmic reticulum (26,27).

According to these results two mechanisms might be responsible for the effects of anaesthetic substances on cardiac contractility: First, inhibition of ATP-dependent, active transport of Ca^{++} ions into sarcoplasmic organelles; or secondly, increased calcium permeability through the sarcoplasmic membrane due to membrane leakage. From a theoretical point of view glycosides could be of advantage under clinical conditions because the negative influences of anaesthetics on myocardial contractility can be abolished in the presence of digitalis. However, studies concerning the basic mechanism of anaesthetics were performed using isolated preparations, and one has to emphasize that the isolated system greatly differs from the conditions in the intact organism. For example, the isolated heart lung preparation does not involve compensatory reflex processes which are present in the systemic circulation of the patient. In addition, the anaesthetic agents differ in their ability to depress cardiac function, and the anaesthesiologist can select the appropriate substance for anaesthesia according to the state of health of the patient. For example, etomidate which exhibited the smallest influence on cardiac contractility is valuable as an agent for the induction of anaesthesia in patients with myocardial failure. Therefore, under clinical conditions there exists no evidence that prophylactic digitalisation as any advantage (25).

Electrophysiologic effects

Cardiac glycosides directly and indirectly influence heart excitability and conductivity. The direct effects of digitalis are characterized by inhibition of the membrane-associated electrogenic pump which involves the enzyme (Na-K)-ATPase. Toxic manifestations of digitalis on the heart are due to this direct effect. The sensitivity of the heart to digitalis can be modified by several conditions that often occur in the perioperative period. Most of the antidysrhythmic effects of digitalis are caused by its indirect actions. The indirect effects on SA-node, atria, AV-conduction and AV-refractoriness are neurally mediated by increased vagal activity and a simultaneously reduced sympathetic activity.

Anaesthetic drugs can also change the electrical properties of the heart. Studies in dogs have shown that volatile anaesthetics like halothane, enflurane, and isoflurane slow the rate of the SA-node by indirect action (5). Halothane and enflurane dose-dependently prolong AV-node, His-Purkinje conduction, and ventricular activation time, while isoflurane was less depressant (2). Because anaesthetics also influence the refractory period and the latent pacemaker activity they can induce different dysrhythmias in the perioperative setting (1).

Morrow et al. investigated the tolerance of digitalis in dog experiments in the presence of

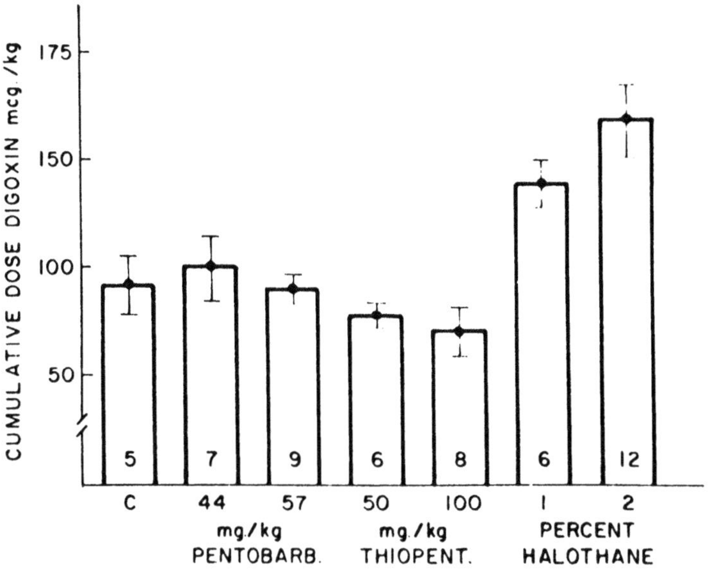

Fig. 6. Digoxin concentrations required to produce ventricular automaticity in awake dogs (C), and in anaesthetized dogs using intravenously effective and volatile anaesthetics. Numbers in the bars indicate the number of dogs in each group (22) (Mean ± SE).

pentobarbital, thiopental, and halothane (22). Under these conditions the dose of digoxin required to produce ventricular automaticity was unaltered by barbiturates but was significantly increased by halothane (Fig. 6). Ventricular automaticity was defined as the elec-

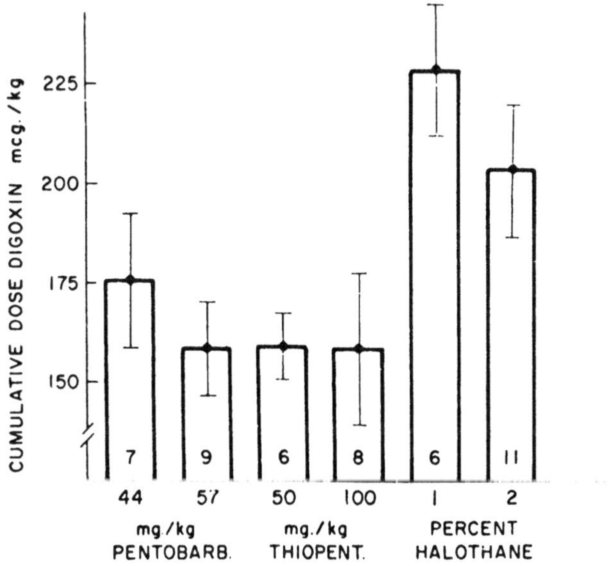

Fig. 7. Concentrations of digoxin required to produce death in dogs anaesthetized with intravenously effective and volatile anaesthetics (22) (Mean ± SE).

trocardiographic appearance of three or more consecutive premature ventricular contractions. In addition, the lethal dose of digoxin was higher in animals anaesthetized with halothane than with barbiturates (Fig. 7). Ivankovic et al. studied the ouabain tolerance in dogs anaesthetized with enflurane, isoflurane, halothane and other volatile anaesthetics (15). They found that the cumulative dose of ouabain to produce ventricular tachycardia or death was significantly higher in the presence of volatile anaesthetics than after application of pentobarbital. Ivankovic (15) additionally demonstrated that the ouabain-induced ventricular automaticity during pentobarbital anaesthesia was converted to sinus rhythm by enflurane (in 80%) and isoflurane (in 70%). Halothane, enflurane, and isolflurane which are used in clinical practice increased the tolerance to digitalis toxicity without any influence on the positive inotropic action of glycosides (21). The mechanisms by which anaesthetics cause these interactions have by no means been fully characterized. Since halothane prolongs the conductivity and the functional refractory period of the AV-node, which is not altered by vagotomy, digitalis can further lower the AV-conduction. These effects require diligence in the application of halothane in digitalized patients with AV-conduction disturbances or predisposition to heart block.

Ivankovic et al (16) investigated the tolerance to digitalis in dogs anaesthetized with the intravenous anaesthetics ketamine as well as innovar (combination of fentanyl and droperidol) (Figs. 8 and 9). Their studies demonstrate that both substances enhance the tolerance to ouabain whether they are administered before or after the appearance of digitalis toxicity. Fentanyl alone failed to convert digitalis-induced ventricular tachycardia to sinus rhythm. Droperidol and ketamine exhibit antiarrhythmic properties due to prolongation of the functional refractory period.

In the perioperative period there are several factors which modify the effects of digitalis treatment:

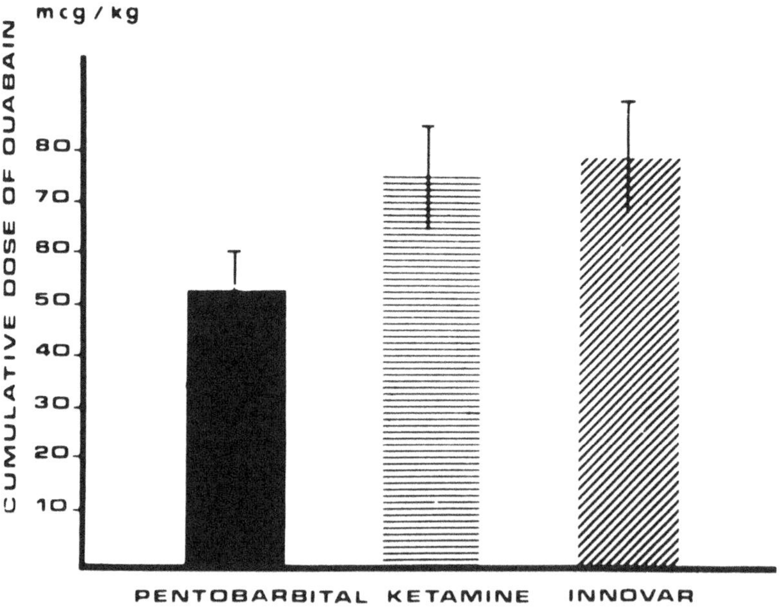

Fig. 8. Ouabain concentrations required to induce ventricular tachycardia in dogs anaesthetized using intravenously effective anaesthetics (16) (n = 10; mean ± SD).

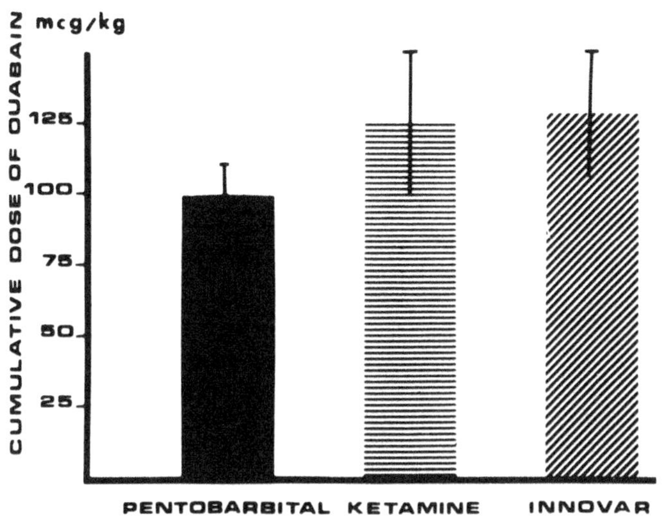

Fig. 9. Lethal ouabain concentrations required in dogs anaesthetized with different intravenously effective substances (16) (n = 10; mean ± SD).

Electrolyte disturbances

Hypokalaemia is the most important electrolyte disturbance enhancing digitalis toxicity because myocardial uptake of glycosides is increased. In contrast, hyperkalaemia lowers the affinity of digitalis to the Na-K-pump (6, 10).

Digitalis-induced dysrhythmias can be reduced by potassium because of its interaction with the receptor (11). Chronic potassium depletion seems to have less influence on digitalis toxicity than acute hypokalaemia observed during dialysis or surgical interventions. The major conditions for electrolyte disturbances during the perioperative period are as follows:

- extended abdominal surgery
- preoperative starvation
- diuretic pretreatment,
- rapid fluid replacement
- gastrointestinal disease
- tourniquet-compartment syndrome.

Under clinical conditions mostly the combination of several factors leads to electrolyte disturbance which can rapidly occur within minutes (32).

Hypercalcaemia increases the sensitivity of the heart to digitalis (24). In digitalized patients intravenous administration of Ca^{++} should be avoided. Hypercalcaemia is a real problem in patients with hyperparathyreoidism, Paget's disease of bone, metastatic breast and bronchogenic carcinoma as well as hyperthyroidism.

Respiratory changes

Chronic obstructive lung disease and acidosis enhance cardiac sensitivity to digitalis (14). Hypoxaemia as well as hypercarbia increase digitalis sensitivity and can occur during critical situations in anaesthesia due to ventilation-perfusion abnormalities or hypoventilation.

Increased sympathetic activity

Hypoxaemia and pain – mostly related to inadaequate analgesia – and surgical interventions (e. g. phaeochromocytoma) cause enhanced sympathetic activity and can counteract the antiadrenergic effects of digitalis. Catecholamines can produce serious ventricular arrhythmias in digitalized patients.

Organ failure

Organ dysfunction involved in degradation and secretion of digitalis can lead to digitalis intoxication. Most of the anaesthetic substances have enzyme inducing properties. Thus, metabolism of digitoxin in the liver can be stimulated by anaesthetics. The elimination of digoxin which is predominantly excreted by the kidney is closely related to the clearance of creatinin (8).

Non-anaesthetic drugs

The muscle relaxants succinylcholine and pancuronium can cause different dysrhythmias, including bradycardia, tachycardia, juctional rhythm and ventricular ectopy. Bartolone (3) has reported an enhanced incidence of various dysrhythmias during application of muscle relaxants in digitalized patients. Cholinesterase-blocking drugs like neostigmine can induce severe bradycardia by slowing AV conduction which is mainly due to accumulation of acetylcholine at the preganglionic neurons. Diazepam which is often used in anaesthesia can induce ventricular fibrillation under digitalis treatment (64, 23).

Indications for cardiac glycosides in anaesthesia

In summary, the interactions between the circumstances mentioned and glycosides emphasize the risks associated with digitalized patients in practical anaesthesia. Therefore the recommendation by Deutsch and Dalen (7) to digitalize all patients over the age of 60 is not generally accepted in anaesthesiology. Furthermore there is no prospective outcome study which proves the significance of prophylactic digitalization to demonstrate less perioperative complications.

The indications for cardiac glycosides in anaesthesia do not differ from those generally accepted:

- tachysystolic atrial fibrillation
- atrial flutter
- manifest myocardial failure (NYHA Classes III, and IV)

It should be emphasized that atropine frequently taken for premedication has to be avoided in digitalized patients with atrial fibrillation, because the therapeutic effects of glycosides are mainly neurally mediated and can be reversed by atropine.

A careful preoperative assessment is necessary in digitalized patients. Elective operations should be postponed when there are signs of glycoside intoxication. When cardiac failure suddenly occurs in the operative period the invasive monitoring of hemodynamic functions allows the accurate measurement of the major determinants of cardiac performance. The prompt treatment of cardiac dysfunction by influencing preload (volume, drugs), afterload (vasodilators), or myocardial contractility (dobutamine, dopamine) are better alternatives in the management of heart failure in modern anaesthesiology (18, 31).

References

1. Atlee JL, Dayer AM, Houg JC (1984) Chronic recording from the His bundle of the awake dog. Basic Res Cardiol 79: 627

2. Atlee JL (1985) Perioperative cardiac dysrhythmias: mechanisms, recognition, management. Year Book Medical Publishers, Inc, Chicago
3. Bartolone RS, Rao TLK (1983) Dysrhythmias following muscle relaxant administration in patients receiving digitalis. Anesthesiology 58: 567
4. Barret JS, Hey EB (1970) Ventricular arrhythmias associated with the use of diazepam for cardioversion. JAMA 214: 1323
5. Bonsjak ZJ, Kampine JP (1983) Effects of halothane, enflurane and isoflurane on the SA-node. Anesthesiology 58: 314
6. Chung DC (1981) Anaesthetic problems associated with the treatment of cardiovascular disease: I: Digitalis toxicity. Can Anaesth Soc J 28: 6
7. Deutsch S, Dalen JE (1967) Indications for prophylactic digitalization Anesthesiology 30: 684
8. Doherty JE, Bissett JK, Kane JJ, DeSozyza N, Murphy ML, Flanigan WJ, Dalrymple GV (1975) Titrated digoxin: Studies in renal disease in human subjects. Int J Clin Pharmacol 12: 89
9. Döring HJ (1973) Mechanismus und Therapie kardiotoxischer Narkotika-Wirkung. Intensivmed 10: 388
10. Erdmann E, Bolte HD, Lüderitz B (1971) The (Na^+-K^+)-ATPase acitvity of guinea-pig heart muscle in potassium deficiency. Arch Biochem Biophys 145: 121
11. Erdmann E, Presek P, Swozil R (1976) Über den Einfluß von Kalium auf die Bindung von Strophantin an menschlichen Herzmuskelzellmembranen. Klin Wochenschrift 54: 383
12. Fleckenstein A, Döring HJ (1967) Kammermeier: Experimental heart failure due to inhibition of utilisation of high energy phosphates, coronary circulation and energetics of the myocardium. S Karger Verlag, Basel New York, p 220
13. Goldberg AW, Ulrick WC (1967) Effects of halothane on isometric tension of isolated heart muscle. Anesthesiology 28: 838
14. Green LH, Smith TW (1977) The use of digitalis in patients with pulmonary disease. Ann Intern Med 87: 459
15. Ivankovic AD, Miletich DJ, Grossmann RK, Albrecht RF, El-Etr AA, Cairoli VJ (1976) The effect of enflurane, isoflurane, fluroxen, methoxyflurane and diethyl ether anaesthesia on ouabain tolerance in the dog. Anesth and Analg 55: 360
16. Ivankovic AD, El-Etr AA, Janeczko GF, Maronic JP (1975) The effects of ketamine and of innovar anaesthesia on digitalis tolerance in dogs. Anesth and Analog 54: 106
17. Krayer O (1931) Versuche am insuffizienten Herzen. Naunyn Schmiedebergers Arch Pharmacol 162: 1
18. Lappas DG, Powell WMJ, Willard M, Daggett WM (1977) Cardiac dysfunction in the perioperative period: Pathophysiology, diagnosis and treatment. Anesthesiology 47: 117
19. Merin RG, Kumzawa T, Honig CR (1974) Reversible interaction between halothane and Ca^{++} on cardiac actomyosin adenosine triphosphatase: Mechanisms and significance. J Pharamcol Exp Therap 190: 1
20. Merin RG, Kumuzawa T, Luka NL (1978) Enflurane depresses myocardial function, perfusion and metabolism in the dog. Anesthesiology 45: 501
21. Morrow DH, Haley JV, Logic JR (1972) Anesthesia and digitalis VII: The effect of pentobarbital, halothane and methoxyflurane on the AV-conduction and inotropic responses to ouabain. Anesth and Analg 51: 430
22. Morrow DH (1970) Anesthesia and digitalis toxicity VI: Effect of barbiturates and halothane on digoxin toxicity. Anesth and Analg 49: 305
23. Nevins MA, Mattes LM, Speritzer RC et al (1969) Ineffectivness of diazepam as an antiarrhythmic agent. J Mount Sinai Hosp NY 36: 408
24. Nola GT, Pope S, Harrison DC (1970) Assessment of the synergistic relationship between serum calcium and digitalis. Am Heart J 79: 499
25. Strauer BE (1984) Das Hochdruckherz. Springer-Verlag, Berlin
26. Su JY, Kerrik WGL (1979) Effects of halothane on caffeine-induced tension transient in functionally skinned myocardial fibers. Pflügers Arch 380: 29

27. Su JY, Kerrick WGL (1980) Effects of enflurane on functionally skinned myocardial fibers from rabbits. Anesthesiology 52: 385
28. van Ackern K, Deuster JE, Mast GJ (1972) Akute Minderung der Kontraktilität des Warmblütermyokards durch Ketamin. Z prakt Anästh 7: 309
29. van Ackern K, Brückner UB, Deuster JE, Hoyer S, Mast GJ, Mittmann U, Schmidt HD, Schmier J (1973) Negativ inotrope Wirkung von Hexobarbital, Diphenylhydantoin, Propanidid und Chlometiazol auf das vorgeschädigte Herz. Z prakt Anästh 8: 193
30. van Ackern K (1975) Beeinflussung der Kontraktilität des Warmblütermyokards durch verschiedene Narkotika. Habil-Schrift, Unversität Heidelberg
31. van Ackern K, Peter K, Franke N (1979) Combined therapy of low-output syndrom with vasodilation and positive inotropic stimulation in patients after open-heart surgery. B J Anaesth 51:4
32. Ware M (ed) (1975) Calcium, magnesium and diuretics. Br Med J 1: 170

Author's address:

Prof. Dr. K. van Ackern, Medizinische Universität zu Lübeck, Institut für Anaesthesiologie, Ratzeburger Allee 160, 2400 Lübeck (West Germany)

Effects of preoperative digitalisation on postoperative left heart performance and incidence of complications following hip surgery. First results of a prospective study

Ch. Maier and W. Knippenberg

Department of Anesthesia, Christian-Albrechts-Universität, Kiel (West Germany)

Summary

The postoperative course may be complicated by cardiac events, especially heart failure. In this study the influence of preoperative digitalisation was investigated on the occurrence of postoperative cardiac complications.

38 nondigitalised patients over 50 years undergoing hip replacement were investigated. After randomization the patients were distributed into 3 groups prospectively:

Group A: Control (without digitalis; n = 13)

Group B: Preoperative digitalisation with 1.0 mg digitoxin and maintenance dosis (MD) of 0.1 mg/die digitoxin (n = 13).

Group C: Digitalisation with 0.75 mg and MD of 0.07 mg/die digitoxin (n = 12).

Preoperative cardiac status and results of measured STI were comparable in all three groups. During the first week postoperatively the clinical course and LV-function, measured by systolic time intervals (STI), were studied according to a pre-determined schedule.

1. Severe cardiac complications (such as severe heart failure, angina pectoris, tachyarrhythmia) were significantly more frequent in the control group. In 6 patients from this group the need for digitalisation due to heart failure or tachycardia occured.
2. The STI were prolonged significantly in the nondigitalised patients 7 days postoperatively. This postoperative deterioration of LV performance was prevented by preoperative digitalisation, as demonstrated in the postoperatively unchanged STI in digitalised patients of group B and C.
3. The dose of digoxin was not significant of either the clinical course or in the alteration of STI.
4. We conclude from this study that preoperative digitalisation with a low dose of digitoxin may be useful for the reduction of postoperative cardiac risks in elderly patients.

Introduction

The postoperative course of elderly patients is primarily endangered by pulmonary and cardiac complications (1). Besides cardiac infarct, the decompensation of preoperative, usually pre-existing heart failure is to be mentioned, which is also partly responsible for a fatal course (1, 2). There are only few data about the changes of left ventricular function for a longer postoperative period. In spite of the controversial discussion of many years concerning prevention of postoperative heart failure by preoperative digitalisation, controlled prospective studies discussing "prophylactic" digitalisation are rare.

The aim of this study was to check:
- if postoperative change of cardiac function can be measurably influenced by digitalis;
- if preoperative digitalisation reduces the risk of postoperative heart failure;
- if a low dose has sufficient effect.

Methods

The examination includes 38 non-digitalised patients aged over 50 years, before and after total hip replacement. Patients with heart failure NYHA Classes III/IV were excluded. After randomisation, the patients were divided into 3 groups, preoperative:
A: control group;
B: digitoxin: high dose (saturation: 1 mg/iv; MD: 0.1 mg/die);
C: digitoxin: low dose (saturation: 0.75 mg/iv; MD: 0.07 mg/die).

Besides clinical examination and anamnesis, left ventricular function was measured by non-invasive determination of the systolic time intervals (STI), see Fig. 1. All preoperative measurements were repeated on the 1st, 3rd and 7th postoperative days. Serum levels of digitoxin (RIA) were determined on the operative and on the 7th day. The patients were examined daily, and in case of complications, medication was fixed according to clinical aspects. Evaluation of STI was made after conclusion of the study without knowledge of name and group. We used the Chi test to compile statistics frequency, and the Wilcoxon test and Friedman's rank-variance test for changes of STI. Significance was accepted for $p < 0.05$

Results

Table 1 shows the distribution of preoperative risks in the three groups. The small number of patients having no preoperative cardiac risks is conspicuous, but representative

Determination of QS2 and LVET

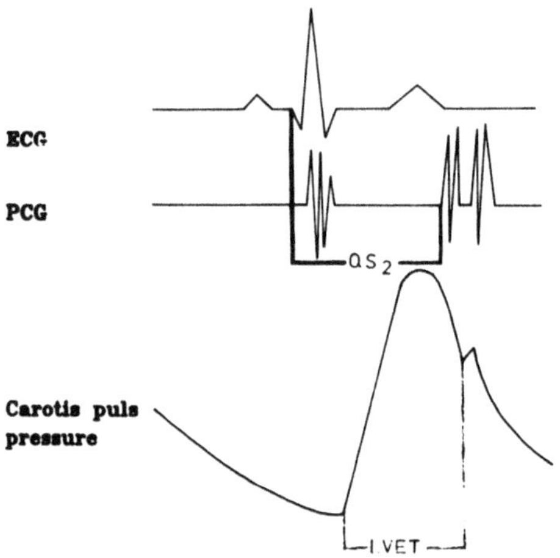

PEP = QS2 − LVET

Fig. 1. Systolic time intervals. QS 2: electromechanical systolic time; LVET: left ventricular ejection time; PEP: pre-ejection period; PEP/LVET: Weissler quotient.

for elderly patients even in elective surgery (2). There is no statistically relevant difference between the groups. The increased number of pulmonary diseases in group A as well as renal diseases in group B is noteworthy, though. All patients received general anesthesia, a combination of ethrane and fentanyl. Duration of narcosis and surgery, loss and substitution of blood, regimes of infusions as well as perioperative laboratory values were comparable in the three groups.

Table 1. Distribution of patients and preoperative risks.

Group		A	B	C
Number		13	13	12
Age ($\bar{x}$)		66	68	67
Sex	m	5	6	7
	f	8	7	5
Preoperative accompanying diseases				
None		0	1	1
Heart failure (I/II)		7	9	6
Coronary disease		4	5	4
Hypertension		8	7	6
Arrhythmia		3	4	3
Pulmonary diseases		6	3	3
Renal insufficiency		0	2	0

Clinical course

The postoperative complications shown in Tables 2 and 3 were stated on the daily visit. In only 40% of the patients did the postoperative course show no complications. Most of

Table 2. Postoperative complications.

Group	A	B	C	n
No complications	1*	6	8	15 (39%)
Mild	6	6	2	14 (37%)
Severe (#)	6*	1	2	9 (24%)
n	13	13	12	38
Type of complication				
		Note: One patient may have multiple complications		
Cardiac	9*	2	1	12 (32%)
Pulmonary	6*	1	0	7 (18%)
Pulmonary embolism	2	0	2	4 (11%)
fatal	1	0	1	2 (5%)
Renal	6	5	1	12 (32%)
Other	4	3	1	8 (21%)

* = $p < 0.05$ (Chi-Square test)
(#) = Heart failure, pulmonary embolism, etc.

Table 3. Cardiac complications.

Group	A	B	C
Patients with cardiac complications	9	2	1
Heart failure (> II)	4	0	1
Arrhythmia	2	1	0
Stenocardia	2	0	0
Edemas	6*	1	1
Dyspnea	3	0	1

these patients were digitalised; there is a significant difference to the control group A. Especially in this group severe, that is potentially life-threatening complications (heart failure Classes III/IV, stenocardia, embolism), occurred in large numbers. Two patients died of pulmonary embolism, one of them (group C) 4 weeks after operation, the other (group A) on the 10th postoperative day, after having developed severe heart failure. Six patients of group A reveiced digitoxin postoperatively after 7–14 days, 5 of whom showed clinical improvement (Fig. 3). In spite of a significant difference in the average serum levels of digitoxin (group B: 17.6; group C: 11.6 ng/ml), there was no reliable difference between group B and C. Digitalis intoxication accompanied by premature beats occurred once in

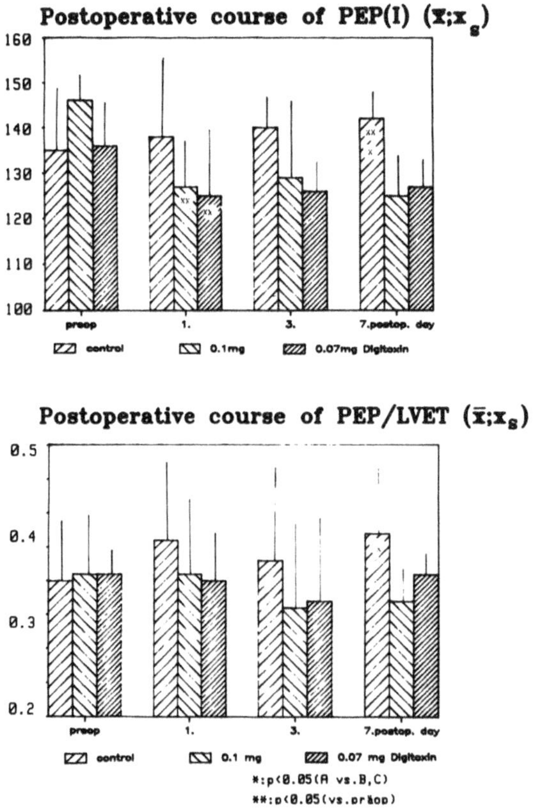

Fig. 2. Systolic time intervals: postoperative course.

group B, but did not reoccur with low dose following an interruption of digitoxin therapy. Because ECG was only recorded on the 1st, 3rd and 7th postoperative days, the incidence of arrhythmia according to different doses of digitoxin cannot be concluded from these results. In this study, the occurrence of arrhythmia was mainly related to the decrease of left heart performance, especially in the control group.

Postoperative changes of STI

As Fig. 2 shows, there is no difference in preoperative STI between the 3 groups, according to the clinical results.

The postoperative STI was mostly pathological for PEP as well as for PEP/LVET in group A. The difference in STI between groups B and C becomes significant on the 7th postoperative day (Wilcoxon test). The analysis of variance shows a significant deterioration of

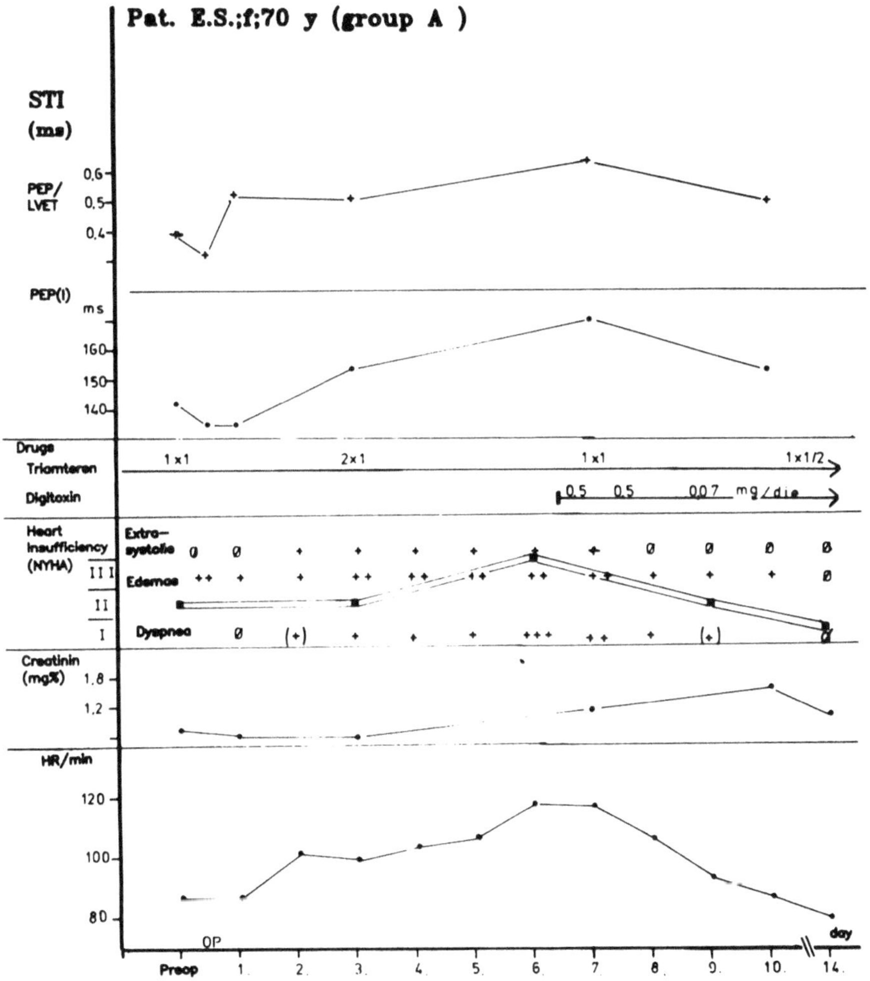

Fig. 3. Case report. Postoperative heart failure and changes of STI.

the STI on the 7th postoperative day compared to the preoperative values in group A alone. In comparison, a digitalis induced improvement on the 1st day can be noted for the patients of the digitoxin groups. This shortening of STI is not influenced by the postoperative course for the digitalised patients.

According to the clinical results, the analysis of STI also shows no essential difference between high and low dose digitalis groups.

Furthermore, a good agreement of STI with the clinical course was observed as shown in Fig. 3.

Conclusion

1. After preoperative digitalisation, severe, especially cardiac, postoperative complications occur more rarely.
2. A considerable, even increasing deterioration of left ventricular performance within the first postoperative week, can be proved by STI in non-digialised patients. Digitalisation prevents this postoperative heart failure.
3. A difference between the dose of 0.1 and 0.07 mg/day digitoxin is neither evident for clinical course nor postoperative left heart performance.
4. According to these results, preoperative digitalisation of elderly patients can be recommended before similar operations. Low dose digitalisation causes a sufficient effect, accompanied by a low risk of intoxication.

   From our measurement of serum levels, we recommend a saturation dose of 0.75 mg digitoxin in the perioperative period. The maintenance dose should be 0.07 mg or 0.1 mg/day for patients over 70 kg body weight.

References

1. Goldman L (1983) Ann Surg 198: 780
2. Hartung H-J et al (1984) Anaesthesist 33: 417

Author's address:

Dr. Ch. Maier, M. D., Department of Anesthesia, Christian-Albrechts-Universität, Schwanenweg, 2300 Kiel (West Germany)

Digoxin therapy and mortality following confirmed or suspected myocardial infarction: Experience in the MILIS study

J. E. Muller, Z. G. Turi, P. H. Stone, R. E. Rude, D. S. Raabe, A. S. Jaffe, H. K. Gold, N. Gustafson, W. K. Poole, E. Passamani, E. Braunwald, T. W. Smith, and the MILIS Group

The Cardiovascular Division, Department of Medicine, Brigham and Women's Hospital Boston, and Cooperating Institutions of the Multicenter Investigation of the Limitation of Infarct Size (MILIS) (U.S.A.)

Summary

Recent observational studies have led to controversy as to whether digoxin therapy following confirmed or suspected myocardial infaction increases mortality. We analyzed the mortality experience of 903 patients enrolled in the Multicenter Investigation of Limitation of Infarct Size (MILIS). Digoxin therapy was not randomly assigned; as in the previous observational studies, the decision to treat or not to treat with digoxin was made by the patient's personal physician on the basis of the usual clinical indications. Cumulative mortality was 28 % for the 281 digoxin-treated patients versus 11 % for the 622 patients who did not receive digoxin ($p < 0.001$; follow-up interval from 6 days to 36 months, mean = 25.1 months). However, patients treated with digoxin exhibited more baseline characteristics predictive of mortality than their counterparts. A previous myocardial infarction had occurred in 31 % of the digoxin-treated patients versus 12 % of the non-treated patients; prior angina occurred in 50 % versus 32 %, respectively, and the mean ejection fraction was 8.6 ± 1.0 % (SEM) lower for digoxin-treated patients. Adjustment for these difference was performed with three different applications of the Cox method on the basis of: 1) a composite MILIS Risk Score for prediction of mortality in the general MILIS population; 2) a special imbalance risk score composed of 19 variables for which the digoxin and non-digoxin population varied and; 3) a subset of the variables used in the CASS study. These adjustments yielded p values of 0.04, 0.34 and 0.14, respectively, for tests of difference in mortality curves, indicating an absence of significant excess mortality associated with digoxin when appropriate baseline variables were available for adjustment. Thus, findings in the MILIS Database do not support the assertion that digoxin therapy is excessively hazardous post infarction, but the existence of an undetected harmful effect can only be excluded with a randomized study. Until the results of such a study are available, we recommend 1) careful consideration if treatment of left-ventricular dysfunction is actually needed, 2) consideration of alternatives to digoxin therapy, and 3) restriction of digoxin use to the subgroup of patients with persistent CHF and evidence of cardiomegaly.

Introduction

Controversy has arisen as to whether chronic therapy with digoxin increases mortality in patients with suspected or confirmed myocardial infarction (15, 24). Reports from three large observational studies of patients post myocardial infarction (2, 16, 17) have indicated

Based on research performed by the Multicenter Investigation of the Limitation of Infarct Size (MILIS) Group pursuant to contracts N01-HV-7-2940, 7-2940, 7-2942 and 7-2979, with the National Heart, Lung and Blood Institute, National Institutes of Health, U.S. Department of Health and Human Services that mortality of digoxin-treated patients was higher than would be predicted, given their baseline characteristics. The excess mortality was reported to be concentrated in a high risk subset of patients with congestive heart failure (CHF) and complex ventricular ectopic activity (VEA).

Ryan et al. (25) and Madsen et al. (13) examined digoxin related mortality in two additional populations of patients who had sustained a myocardial infarction. Both groups of investigators concluded that the excess mortality in their populations of digoxin-treated post-infarction patients could be accounted for by baseline differences between digoxin and non-digoxin treated patients. However Madsen et al. observed that digoxin therapy was the leading predictor of four-month mortality in the subgroup of patients in whom a Holter recording was obtained and that both four-month and one-year mortality were significantly increased in the subgroup of digoxin-treated patients with CHF and *without* VEA (13). These conflicting reports, combined with differences in the types of patients studied, have led to substantial uncertainty as to the proper role of digoxin therapy in patients who have sustained a myocardial infarction.

During the past 6 years we have conducted a multicenter study of 985 patients with proven or suspected myocardial infarction (18), who were quite similar to those of Moss et al. (2, 16) and Bigger et al. (17). The primary purpose of the study, named the Multicenter Investigation of the Limitation of Infarct Size (MILIS), was to assess the efficacy of hyaluronidase and of propranolol therapy for acute myocardial infarction (18, 22). Extensive data were collected on all patients, including baseline tests of left ventricular function, detailed records of digoxin therapy and long-term mortality status.

The purpose of the present study was to determine if data from this large group of well-characterized patients could assist in resolution of the controversy regarding the safety of digoxin therapy in patients following myocardial infarction. The specific objectives were:

1) to determine whether MILIS patients who received digoxin experienced a higher mortality than those who did not;

2) to determine if any statistically significant increase in mortality that might be present could be accounted for by baseline differences between digoxin and non-digoxin treated patients;

3) to determine if MILIS patients with ejection fraction $\leq 40\%$ and ventricular couplets or runs comprise a group at high risk for a detrimental effect of digoxin therapy; and

4) to identify the origins of the seemingly conflicting results reported from previous observational studies.

Methods

I. The MILIS patient population

The structure of MILIS and a list of participating institutions are presented in Appendix A. Patients were eligible for enrolment in MILIS if they were less than 76 years of age, had at least 30 minutes of pain typical of myocardial ischemia, and demonstrated electrocardiographic (ECG) changes suggestive of acute ischemia or evolving infarction (new Q waves greater than 30 msec in width and greater than or equal to 0.2 mV in depth or ST segment elevation and/or depression greater than or equal to 0.1 mV in at least two related leads) or left bundle branch block or idioventricular rhythm. Exclusion criteria, method of informed consent, guidelines for standard care and details of the administration of hyaluronidase or propranolol have been previously reported (18).

II. Data collection

Following enrolment, but prior to randomization, baseline measurements including a 12-lead ECG and a resting radionuclide ventriculogram were obtained. Serial blood sample collection for measurement of CK and the MB-CK isoenzyme was initiated prior to randomization and continued for the length of the hospital stay, as previously reported (18). Radionuclide ventriculography for assessing ventricular function was repeated 90 minutes after initiating therapy and on the tenth day; the left ventricular ejection fraction was calculated by a semi-automated count method from the left anterior oblique view following background correction (18). A 24-hour two-channel Holter recording was obtained on the third day in patients who did not develop infarction and on the tenth day in patients who developed an infarction. Throughout the hospitalization a MILIS data coordinator recorded details of the history and physical examination, a summary of daily clinical events, vital signs and special procedures and results of routine laboratory tests. A Karnofsky Score was used to characterize the patient's functional status prior to the index infarction (18). Follow-up examinations were conducted three and six months following randomization. The vital status of all patients was ascertained at six-month intervals by a questionnaire administered by telephone. Core laboratories blinded to treatment assignment and mortality outcome analyzed total CK activity (23), MB-CK activity, ECG localization of infarction (8, 21), Holter Monitor recordings and radionuclide ventriculograms (18). Digoxin therapy was recorded at the time of randomization, at discharge from the hospitalization of the index infarction, 3 months after randomization and every 6 months following randomization until termination from the study.

Table 1. Components and coefficients of the "MILIS risk score"

Component	Coding	Coefficient	p-Value[1]
Age	Years	0.0062	0.0001
Highest systolic BP recorded (baseline MILIS physical)	1 = $\leq$ 138mm Hg 0 = $>$ 138mm Hg	0.132	0.0001
LVEF with imputation (Time 0 ECG)	Percent (46.5 substituted for missing LVEF)	−0.0052	0.0001
Education	1 = not HS grad 0 = HS grad or more	0.096	0.0006
Karnofsky score	1–10	−0.037	0.0008
Beta Blocker (last 3 weeks prior to MILIS)	1 = yes 0 = no	−0.097	0.01
ST elevation (Qualifying ECG)	1 = yes 0 = no or LBBB	−0.082	0.012
Rales, any (baseline MILIS physical)	1 = yes 0 = no	0.066	0.02
Antiarrhythmics (last 3 weeks prior to MILIS)	1 = yes 0 = no	0.135	0.03
Diabetes, ever	1 = yes 0 = no	0.077	0.04
Sex	1 = male 0 = female	−0.063	0.05
Hypertension (prior to MILIS)	1 = yes 0 = no	0.053	0.06
Heart rate	Beats per minute	0.00030	0.72
Indicator for imputation	1 = LVEF imputed 0 = LVEF not imputed	0.006	0.87

[1] Contribution to the model predicting the death or survival of a patient with particular baseline characteristics.

III. Adjustment of digoxin-associated mortality

Three separate adjustments of digoxin associated mortality were performed using the following sets of variables:

1) a composite MILIS risk score for prediction of mortality in the general MILIS population (Table 1);

2) a special Imbalance risk score composed of 19 variables for which the digoxin and non-digoxin population varied (Tables 2 and 3); and

3) a subset of the variables used by Ryan et al. (25) in the CASS study.

A. The MILIS risk score. Prior to study of the effect of digoxin on mortality post infarction, a method was developed to estimate the prognosis of MILIS patients. Baseline variables, including many previously reported by other investigators to be associated with mortality post-infarction, were entered into a step-wise linear regression model to test their predictive power in the MILIS population (3). The variables (Table 1) which emerged from this analysis were multiplied by appropriate factors derived from a least squares fit and combined to generate a "risk score" similar to a Peel (20) or Norris (19) index value for each MILIS patient. The advantage of the risk score is that it combines the predictive power of multiple variables into a single number which can then be compared to a specific variable of interest such as digoxin therapy. Comparison of digoxin therapy with multiple variables creates less stable conditions for the Cox method, since results are progressively less reliable as the number of deaths drops below 10 per variable.

B. The imbalance risk score. Baseline characteristics possibly related to mortality in the digoxin and non-digoxin population were compared in a univariate analysis (Table 2). The

Table 2. Imbalances between digoxin and non-digoxin treated patients

Baseline characteristics	Not treated with digoxin	Treated with digoxin	p-value
Propranolol acceptable	56.8% (353/622)	35.2% (99/281)	0.0001
Anterior ischemia or infarction	51.2% (318/621)	67.3% (189/281)	0.0001
Complex VEA (runs or couplets on 3- or 10 day holter	22.3% (113/506)	37.6% (88/234)	0.0001
Previous myocardial infarction	12.4% (77/622)	31.3% (88/281)	0.0001
Angina in previous 3 weeks	31.5% (196/622)	50.2% (141/281)	0.0001
History of arrhythmias	7.6% (47/622)	21.4% (60/281)	0.0001
Antiarrhythmic therapy in previous 3 weeks	1.6% (10/622)	11.4% (32/281)	0.0001
Rales at randomization	34.6% (215/622)	63.0% (177/281)	0.0001
CHF in previous 3 weeks	2.3% (14/622)	20.3% (57/281)	0.0001
ST elevation at randomization	79.7% (495/621)	67.6% (190/281)	0.0001
Cigarette smoking in previous 6 months	59.0% (367/622)	45.2% (127/281)	0.0001
Beta blockers in previous 3 weeks	14.3% (89/622)	22.5% (63/280)	0.002
Diabetes, ever	14.5% (90/622)	22.4% (63/281)	0.003
Transmural ischemia or infarction	78.4% (487/621)	70.1% (197/281)	0.007
Diuretic therapy in previous 3 weeks	23.0% (143/622)	31.0% (87/281)	0.011
	Mean (S.E.)	*Mean (S.E.)*	
Heart rate	76.9 (0.6)	85.3 (1.1.)	0.0001
LV ejection fraction	50.1 (0.5)	41.5 (0.9)	0.0001
Age	55.1 (0.4)	59.7 (0.6)	0.0001
Karnofsky score	9.14 (0.05)	8.58 (0.09)	0.0001

Table 3. Components and coefficients of the "Imbalance risk score"[1]

Components	Coding	Coefficient	p-value[2]
LVEF with imputation	Percent (46.5 substituted for missing LVEF)	−0.0029	0.003
Diuretics in previous 3 weeks	1 = yes 0 = no	0.086	0.009
Age	Years	0.0035	0.01
Beta blockers in previous 3 weeks	1 = yes 0 = no	−0.082	0.03
ST elevation at randomization	1 = yes 0 = no	−0.126	0.04
Diabetes, ever	1 = yes 0 = no	0.073	0.05
Complex VEA (runs or couplets): 3- or 10 day holter	1 = yes 0 = no	0.062	0.05
Karnofsky score	1–10	−0.021	0.06
Cigarette smoking (in last 6 months)	1 = yes 0 = no	0.046	0.10
CHF in previous 3 weeks	1 = yes 0 = no	0.083	0.14
Antiarrhythmic therapy in previous 3 weeks	1 = yes 0 = no	0.100	0.19
Rales at randomization	1 = yes 0 = no	0.035	0.23
Previous myocardial infarction	1 = yes 0 = no	0.035	0.35
Location of ischemia or infarction (time 0 or qualifying ECG)	1 = anterior 0 = inferior, non-MI, or indeterminate	−0.021	0.44
Propranolol acceptable	1 = yes 0 = no	−0.019	0.50
History of arrhythmias	1 = yes 0 = no	0.030	0.55
Heart rate	Beats per minute	−0.00034	0.67
Angina in previous 3 weeks	1 = yes 0 = no	−0.0063	0.83
Types of ischemia or infarction (time 0 or qualifying ECG)	1 = transmural 0 = other	−0.0060	0.92

1 Estimated probability of dying during study, using variables imbalanced ($p < 0.05$) for digoxin and non-digoxin patients.
2 Contribution to the model predicting the death or survival of a patient with particular baseline characteristics.

19 variables for which the populations differed were combined into an imbalance risk score which was used for the second adjustment (Table 3).

C. The Coronary Artery Surgery Study (CASS) variables

The third attempt to adjust mortality in the MILIS patients was based on elements of the set of variables used by Ryan et al. (25) to adjust digoxin-associated mortality for the CASS population which were available for MILIS patients. In addition to age, left ventricular ejection fraction, and a history of previous myocardial infarction, the following variables were used:

1) Congestive heart failure score – A single point was assigned for each of the following five characteristics; rales, pedal edema, an S_3 gallop, prescription of diuretics at discharge, and a history of CHF prior to randomization;

2) Left ventricular regional wall motion;

3) Left ventricular end-diastolic volume index.

Coronary arteriographic variables which were available to the CASS group and used in their adjustment were not available for the MILIS population.

IV. Analysis of the suspected high-risk subgroup

Patients were divided into four subgroups based on ejection fraction < or ≥ 40% at the time of randomization and the presence or absence of ventricular couplets or runs of ventricular tachycardia of 3 or more consecutive beats recorded on a 24 hour. Holter Monitor recording taken 3 days post enrolment for non-MI patients and 10 days post enrolment for MI patients. Within each subgroup, the unadjusted mortality of patients receiving digoxin was compared to that of patients not receiving digoxin. A comparison of mortality within each subgroup was also made with adjustment by the Cox method for baseline differences in the imbalance risk score between digoxin and non-digoxin treated patients.

Statistical methods

The Cox proportional hazards regression model was used to study the association between digoxin therapy and survival while taking into account the effects of baseline prognostic factors (3, 7). Unadjusted plots of the estimated cumulative mortality were produced by the Kaplan-Meier method (11). Adjusted mortality curves were based on the Kalbfleisch and Prentice discrete hazard model, the estimates of which reduce to the Kaplan-Meier when there are no covariates (10). All probability (p) values were two-tailed and a p-value of less than 0.05 was considered significant. Summary statistics for continuous data are presented as the mean ± standard error.

Results

Unadjusted mortality associated with digoxin therapy

Of the 985 patients in the MILIS study, 80 died during their initial hospitalization and 2 lacked adequate follow-up data, leaving 903 patients observed during the follow-up period. The infarction rate in this group of patients was 85.7%. Of the 903 patients, 281 were receiving digoxin at the time of hospital discharge and 622 were not. The follow-up interval ranged from 6 days to 36 months after hospital discharge (mean follow-up interval = 25.1 months). A cumulative total of 145 deaths occurred during the entire follow-up period, 78 (28% mortality) in the digoxin group and 67 (11% mortality) in the non-digoxin group ($p < 0.001$). The unadjusted increased mortality among digoxin-treated patients is shown in life-table format in Figure 1. The estimated two-year mortality for patients receiving digoxin was 27% versus 10% for those not receiving digoxin. The cross-over rate for digoxin usage from hospital discharge to 6 months was 18.3% from digoxin use to non-digoxin use and 10.9% from non-digoxin use to digoxin use.

Imbalances between digoxin and non-digoxin-treated patients

As would be expected from the customary indications for digoxin therapy, there were major differences between the group of patients which received digoxin and the group which did not (Table 2). The average age of the digoxin-treated patients was 4.6 years older than that of those who did not receive digoxin. The incidence of previous infarction was much higher for digoxin-treated patients, 31% versus 12% ($p < 0.001$), as was the incidence of angina prior to their infarction, 50% versus 32%, $p < 0.001$. The mean ejection fraction was $8.6 \pm 1.0\%$ lower in the group that received digoxin ($41.5 \pm 0.9\%$ versus $50.1 \pm 0.5\%$, $p < 0.001$).

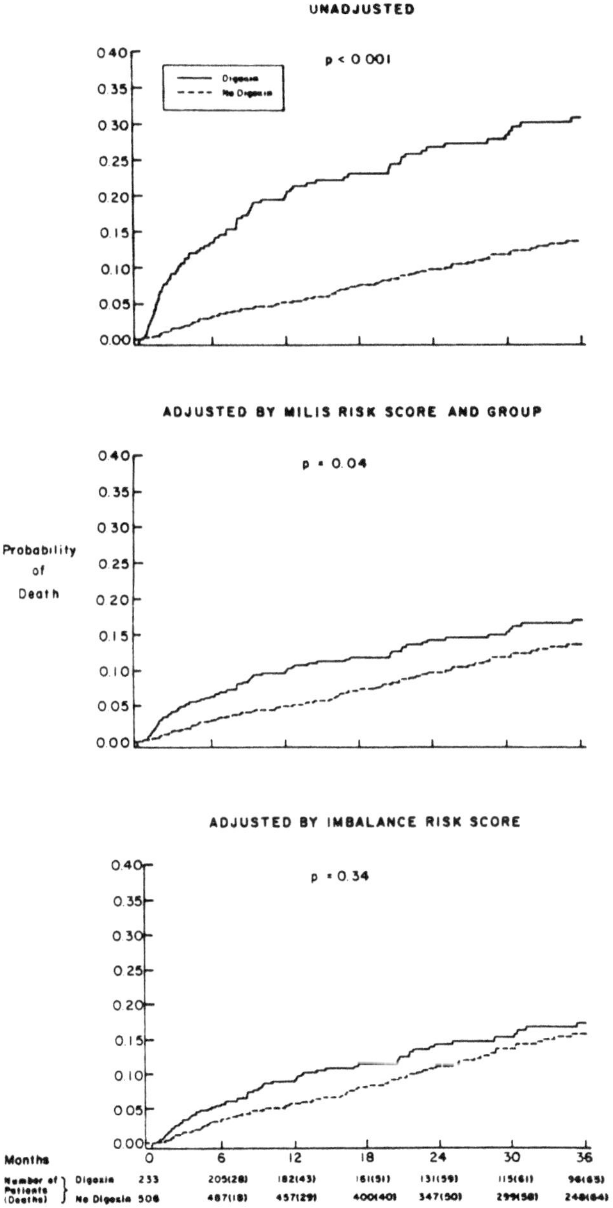

Adjustment of digoxin-associated mortality

Adjustment by the MILIS risk score and group* – Although adjustment with this composite variable narrowed the difference between mortality for digoxin and non-digoxin treated patients, the adjusted digoxin mortality curve remained significantly above that for non-digoxin patients (Figure 1, p = 0.04, Cox method).

Adjustment by the imbalance risk score – Adjustment with the use of the index of 19 imbalances shown in Tables 2 and 3 further diminished the mortality differences between digoxin and non-digoxin treated patients (Figure 1). With this method of adjustment, the curve for digoxin associated mortality remained above that for non-digoxin patients but the difference between the curves was no longer statistically significant (p = 0.34, Cox method.)

Adjustment by CASS variables – Adjustment with the subset of CASS variables produced results similar to adjustment with the imbalance risk score. Although the adjusted curve of digoxin-associated mortality remained above that of non-digoxin patients, the difference between the curves was not statistically significant (p = 0.14, Cox method.)

Results in the suspected high-risk subgroup

For investigation of the suspected high-risk subgroup, mortality was analyzed for the four-month period following infarction, the interval reported by Moss et al. (16) to show a detrimental effects of digoxin. In the subgroup of MILIS patients characterized by an ejection fraction ≤ 40%, unadjusted mortality was 17% (8 of 46) for digoxin-treated patients versus 3% (1 of 33) for patients not treated with digoxin. This difference was not statistically significant (p = 0.07), nor were the digoxin versus non-digoxin differences in mortality in the remaining three subgroups (Table 4). Furthermore, the digoxin-treated patients in all subgroups demonstrated higher imbalance risk scores than their non-digoxin treated counterparts.

Table 4. Four-month mortality results and risk scores by digoxin status in MILIS patients in the suspected high-risk subgroup

	LV ejection fraction > 40%		LV ejection fraction ≤ 40%	
	VEA absent*	VEA present	VEA absent	VEA present
Not treated with digoxin				
Proportion dying	3/269 (1%)	2/62 (3%)	1/70 (1%)	1/33 (3%)
Imbalance risk score	0.093 ± 0.007	0.201 ± 0.014	0.178 ± 0.015	0.273 ± 0.022
Treated with digoxin				
Proportion dying	3/51 (6%)	4/26 (15%)	4/59 (7%)	8/46 (17%)
Imbalance risk score	0.157 ± 0.021	0.271 ± 0.036	0.226 ± 0.016	0.390 ± 0.031
p-value for mortality for treated versus not treated (Fisher's exact test)	0.06	0.06	0.18	0.07

* VEA = ventricular couplets or runs on 24 hour Holter recording

* Adjustment by "Group A or B" accounted for mortality differences between patients considered eligible (Group A) or ineligible (Group B) for propranolol therapy at the time of enrolment

The one-year mortality for the MILIS subgroup with ejection fraction $\leq 40\%$ and ventricular couplets or runs was 9% (3 of 33) for non-digoxin-treated and 26% (12 of 46) for digoxin-treated patients ($p = 0.13$, Chi-square). Analysis with the Cox method to account for the less favorable baseline characteristics of the digoxin-treated patients yielded a p value of 0.23. Comparison of the digoxin/non-digoxin odds ratio for mortality among the four subgroups yielded no significant differences ($p = 0.78$).

Discussion

The first published report of the value of digitalis therapy carried a warning that the drug could cause death. In 1785, Withering noted that "... (digitalis) has a power over the motion of the heart, to a degree yet unobserved in any other medicine, and (that) this power may be converted to salutary ends..." However, "... (it may also cause) slow pulse, even as slow as 35 in a minute, cold sweats, convulsions, syncope, (and) death..." (30). This concern for the potential hazards of digitalis has persisted over the ensuing 200 years, and has recently been intensified regarding post-myocardial infarction patients by the reports of Moss et al. (15, 16, 17) and Bigger et al. (2).

The recent reports of increased mortality among digoxin-treated post-infarction patients have received special attention because of established difficulties with digoxin therapy. It is well appreciated that the dose of digoxin required for an adequate therapeutic effect is undesirably close to that sufficient to produce toxicity, especially in patients with advanced heart disease (27). Furthermore, Iesaka et al. reported recently that the risk of digitalis-induced ventricular tachycardia in experimental animals is greater following a myocardial infarction (9). In a group of 15 dogs with healed myocardial infarction, the mean dose of ouabain required to cause sustained ventricular tachycardia was 25% less than in a group of 10 normal dogs. The average cycle length of the tachycardia was also significantly shorter in the dogs with infarction. Morris et al. found that the dose of acetylstrophanthidin required to produce overt toxicity in pigs was one third lower after experimental infarction (14).

These studies suggesting possible increased risk associated with digoxin administration must be considered in the context of long-standing uncertainty over the benefit of digoxin therapy for certain conditions. Although there is no doubt that digoxin is a positive inotropic agent (4, 26, 28), it has been difficult to document a beneficial effect of digoxin therapy in all subsets of patients presumed to have congestive heart failure. Numerous studies have demonstrated beneficial effects of digoxin therapy in specialized tests of ventricular function (26, 29), but studies of therapeutic efficacy have yielded varying conclusions when patients with different degrees of congestive heart failure have been studied (1, 5, 6, 12). In a randomized, double-blind, placebo controlled study, Lee et al. were able to demonstrate that patients with more severe CHF and signs of cardiomegaly received a clinical benefit from digoxin therapy, while those with mild CHF and no cardiac enlargement did not (12).

Results in the MILIS patient population confirm the observation that patients receiving digoxin following confirmed or suspected myocardial infarction have a higher mortality than those not receiving digoxin (Figure 1). However, in the previous studies, as well as in the MILIS study, digoxin therapy was not randomly assigned, but was selected for patients by their individual physicians on the basis of the usual clinical indications. The digoxin-treated patients had a lower ejection fraction, a higher incidence of previous infarction, an older mean age and many other differences compared with patients who did not receive digoxin. It is certain that at least a portion of the digoxin-associated mortality results not from digoxin itself, but from the reasons that led the physicians to prescribe the drug. The goal in interpretation of the results is to determine if there is residual excess mortality in the digoxin-treated group after adjustment for such differences between digoxin and non-digoxin treated patients.

To adjust for these differences we used the Cox multiple regression technique (3). The basis of the method is the application of multivariate analysis to life-table results. Variables are added sequentially to a mathematical model that predicts survival in the population under consideration. If a given variable adds no additional predictive power to the model, it is concluded that this variable does not have an independent influence on survival. Thus digoxin or non-digoxin status is allowed to compete with other characteristics of the two populations to determine if digoxin is an independent predictor of mortality.

There are two major limitations of the Cox method: First, it is dependent on the variables submitted; it can incorrectly indicate digoxin status to be an independent predictor of mortality if important predictors that may have been associated with the physician's decision to prescribe digoxin are missing from the data set. Second, if variables are submitted for adjustment that are caused by digoxin (i.e. PVCs), digoxin will incorrectly be found to be free of a harmful effect. These inherent limitations of the method preclude definitive resolution of the issue with this method.

As noted in Figure 1, mortality in the digoxin-treated group remained significantly increased ($p = 0.04$, Cox method) after the adjustment by the MILIS risk score and group. This result is similar to that of Moss et al. (15, 16, 17), who utilized a similar group of variables for the Cox analysis. However, adjustment with the use of the imbalance risk score eliminated the statistical significance of the difference between the curves ($p = 0.34$, Cox method) although the mortality curve for digoxin remained above that for non-digoxin patients (Figure 1). Adjustment on the basis of the subset of CASS variables available in MILIS yielded a result similar to the CASS result. That is, the adjusted survival for digoxin-treated patients moved close enough to that for patients not receiving digoxin that no statistically significant difference remained ($p = 0.14$, Cox method). In other words, when a certain set of predictors is employed, such as the MILIS risk score and group assignment, digoxin intake would appear to have an independent (negative) influence on survival. However, when other sets of predictors were used, such as those found in the imbalance risk score or those used in CASS and available in MILIS, this statistically significant deleterious effect of digoxin vanished. Thus, findings in the MILIS population support the conclusion of Ryan et al. (25) and Madsen et al. (13) that digoxin-associated mortality in the general population of post-infarction patients becomes statistically insignificant when appropriate variables are submitted to the Cox analysis.

It is noteworthy that despite the absence of statistical significance, in each of the three analyses the adjusted mortality of the digoxin-treated patients in the MILIS Study remained above that of the patients who did not receive digoxin. This finding could result from a harmful effect of digoxin therapy that was too small to show statistical significance with the present sample size. Alternatively, it is possible that adjustment for baseline differences is incomplete, since numerous important variables, such as coronary anatomy, were in many cases known to the treating physician who might have used them in the decision regarding digoxin treatment, but were not available for submission to the Cox analysis.

It is important to separate the issue of increased mortality among all patients receiving digoxin post infarction from the issue of increased mortality in a subgroup of the digoxin-treated population. The original report of digoxin-associated cardiac mortality by Moss et al. (16) was based primarily on the finding that patients in a subgroup with CHF and complex VEA demonstrated a five-fold increased mortality. Four-month mortality in the population studied by Moss et al. was only 1.8% (1 of 56 patients) in the subgroup with CHF and complex VEA who were not taking digoxin compared to 38.5% (10 deaths among 26 patients) among those receiving digoxin. Adjustment for baseline characteristics yielded a predicted mortality of 7.7% for the digoxin-treated patients, if digoxin itself had no independent effect on mortality. The predicted 7.7% mortality was subtracted from the 38.5% observed mortality yielding the 30.8% absolute difference in mortality attributed to digoxin therapy in this subgroup. Results in the subgroup of MILIS patients with ejection fraction ≤ 40% and ventricular couplets or runs support the conclusion of Ryan et al. (25) that there is no statistically significant harmful effect of digoxin in the subgroup with CHF and VEA. Although the mortality of digoxin-treated patients in the subgroup was higher than that of patients not treated with digoxin, adjustment with the Cox method yielded a p value of 0.23, reflecting the preponderance of baseline characteristics predictive of a poor prognosis in the digoxin-treated patients within this subgroup.

Results in the MILIS population did not reveal excess digoxin-related mortality in either of the two subgroups for whom it was reported by Madsen et al. (13). The absence of excess digoxin-related mortality in MILIS for the subgroup with Holter monitoring is likely to result from more complete availability of Holter recordings (82% for MILIS compared with 31% for the population studied by Madsen et al. (13) and the availability in MILIS of ejection fraction data for submission to the Cox analysis. The excess digoxin-associated mortality reported by Madsen et al. (13) in the subgroup with CHF *without* VEA was not observed in the MILIS population nor was it observed in the population studied by Ryan et al. (25).

Differences in data sets complicate comparison of the conclusions of Bigger et al. (2) and Moss et al. (17) that digoxin may be harmful with those of Ryan et al. (25) and Madsen et al. (13) that digoxin is not harmful. The population studied by Ryan et al (25) was a subgroup of the patients enroled in the registry of the coronary artery surgery study (CASS). Patients were not studied as soon after infarction as were those of Moss et al. (17) but they were characterized more completely. Madsen et al. (13) note that the ratio of mortality rates at 4 months post infarction for patients receiving digoxin at discharge compared with those not receiving digoxin (odds ratio) was 3.7 in the population studied by Moss et al. (16, 17) versus only 2.6 in their own population. Madsen et al. (13) concluded that the higher strength of association in the populations studied by Moss et al. (16, 17) and Bigger et al. (2) might have led to demonstration of an independent contribution of digoxin therapy to mortality which their own study was not able to demonstrate.

The MILIS population is quite similar to the population studied by Moss et al. (17) and contains data similar to those collected by Ryan et al. (25). Furthermore, the ratio in the MILIS population of the digoxin to the non-digoxin mortality rate 4 months post infarction was 6.5, i.e., the ratio was even higher than in the population studied by Moss et al. (16, 17). Nevertheless, the results in the MILIS population do not support the assertion that a detrimental effect of digoxin therapy has been demonstrated. Mortality in patients receiving digoxin therapy was not significantly higher after adjustment for the appropriate baseline variables, nor was a significant detrimental effect observed in the subgroup with ejection fraction ≤ 40% and ventricular couplets or runs.

Although the MILIS data, supported by the data of Ryan et al. (25) and Madsen et al. (13) suggest that there is no major hazard associated with digoxin therapy, the possibility remains that the sample size was too small to detect a significant adverse effect and/or "overadjustment" occurred. These are limitations inherent in attempts to use the Cox method to determine the value of therapy in non-randomized, observational studies. A randomized trial of digoxin versus placebo following myocardial infarction is the only approach that can provide a definitive answer to the question of the safety of digoxin therapy in this setting. However, to our knowledge, such a trial has not been conducted nor is in the active planning stages.

It is therefore necessary to form a judgement about the use of digoxin post MI from the available observational studies. Given the essentially indecisive nature of the data presently available, we are in agreement with the recommendation of Bigger et al that the physician contemplating the use of digoxin post MI should first decide if therapy for left ventricular dysfunction is really necessary, and second consider if some other therapy might have a better risk-benefit ratio than digoxin. Furthermore, in the presence of uncertainty over risk, it becomes important to restrict therapy to those patients shown in a well-conducted, randomized study to be most likely to benefit from digoxin therapy-patients with chronic congestiv heart failure (as opposed to episodic), a low ejection fraction and left ventricular chamber enlargement as indicated by S_3, or a large internal diastolic diameter.

We are grateful to Mrs. Deidre Geary and to Ms. Kathleen Carney for assistance in the preparation of the manuscript.

References

1. Arnold SB, Byrd RC, Meister W, Melmon K, Cheitlin MD, Bristow JD, Parmley WM, Chatterjee K (1980) Long-term digitalis therapy improves left ventricular function in heart failure. New Engl J Med 303: 1443–1448
2. Bigger JT (1985) Effect of digitalis treatment on survival after acute myocardial infarction. Am J Cardiology 55: 623–630
3. Cox DR (1972) Regression models and life-tables. J Royal Statistical Soc B 34: 187–202
4. Crawford MH, Karliner JS, O'Rourke RA (1976) Favorable effects of oral maintenance digoxin therapy on left ventricular performance in normal subjects: Echocardiographic study. Am J Cardiol 38: 843–847
5. Dobbs SM, Kenyon WI, Dobbs RJ (1977) Maintenance digoxin after an episode of heart failure: placebo-controlled trial in outpatients. Br Med J 1: 749–752
6. Fleg JL, Gottlieb SH, Lakatta EG (1982) Is digoxin really important in treatment of compensated heart failure? A placebo-controlled cross-over study in patients with sinus rhythm. Am J Med 73: 244–250
7. Harrel FE Jr (1983) in SUGI: Supplementary Library User's Guide. SAS Institute Cary, N.C. Edition pp 267–294
8. Henry PD, Roberts R, Sobel BE (1975) Rapid separation of plasma creatine isoenzymes by batch absorption with glass beads. Clin Chem 21: 844–849
9. Iesaka Y, Aonuma K, Gosselin AJ, Pinakatt T, Stanford W, Benson J, Sampsell R, Rozanski JJ, Lister JW (1983) Susceptibility of infarcted canine hearts to digitalis-toxic ventricular tachycardia. J Am Coll Cardiol 2: 45–51
10. Kalbfleisch JD, Prentice RL The Statistical Analysis of Failure Time Data. John Wiley and Sons, New York
11. Kaplan EL, Meier P (1958) Nonparametric Estimation from Incomplete Observations. J Am Statistical Assoc 53: 457–481
12. Lee DC-S, Johnson RA, Bingham JB, Leahy M, Dinsmore RE, Goroll AH, Strauss HW, Haber E (1982) Heart failure in outpatients: A randomized trial of digoxin versus placebo. New Engl J Med 306: 699–705
13. Madsen EB, Gilpin E, Henning H, Ahnve S, LeWinter M, Mazur J, Shabetai R, Collins D, Ross J (1984) Prognostic importance of digitalis after acute myocardial infarction. J Am Coll Cardiol 3: 681–689
14. Morris JJ. Taft CV, Whalen RE, McIntosh HD (1969) Digitalis and experimental myocardial infarction. Am Heart J 77: 342–355
15. Moss AJ (1984) Caution: digitalis may be injurious to post-infarction patients. Int J Cardiol 6: 95–98
16. Moss AJ, Davis HT, Conard DL, DeCamella JJ, Odoroff CL (1981) Digitalis-associated cardiac mortality after myocardial infarction. Circulation 64: 1150–1156
17. Moss AJ, Davis H,. Conrad DL, DeCamilla J, Odoroff CL (1983) Digitalis-associated mortality in postinfarction patients. Circulation 68 (Suppl III): III–368

18. NHLBI Multicenter Investigation of the Limitation of Infarct Size (MILIS): Design and Methods of the Clinical Trial. The MILIS Study Group, Muller JE, Braunwald E, Mock MB, Mullin SM, Passamani ER, Poole WK, Scheiner E (eds) (1984) American Heart Association. Monograph (No 100)
19. Norris RM, Brandt PWT, Caughey DE, Lee AJ, Scott PJ (1969) A new coronary prognostic index. Lancet 1: 274
20. Peel AAF, Semple T, Wang I, Lancaster WM, Dall JLG (1962) A coronary prognostic index for grading the severity of infarction. Br Heart J 24: 745
21. Ritter CS, Mumm SR, Roberts R (1981) Improved radioimmunoassay for creatine kinase isoenzymes in plasma. Clin Chem 27: 1878–1887
22. Roberts R, Croft C, Gold HK, Hartwell TD, Jaffe AS, Muller JE, Mullin SM, Parker C, Passamani ER, Poole WK, Raabe DS Jr, Rude RE, Stone PH, Turi ZG, Sobel BE, Willerson JT, Braunwald E, and the MILIS Study Group (1984) Effect of propranolol on myocardial infarct size in a randomized, blinded, multicenter trial. New Engl J Med 311: 218–225
23. Rosalki SB (1967) Improved procedure for serum creatine phosphokinase determination. J Lab Clin Med 69: 696–705
24. Ryan TJ (1984) In defense of digitalis use in the coronary patient. Int J Cardiol 6: 89–94
25. Ryan TJ, Bailey KR, McCabe CH, Luk S, Fisher LD, Mock MB, Killip T (1983) The effects of digitalis on survival in high-risk patients with coronary artery disease: The coronary artery surgery study (CASS). Circulation 67: 735–742
26. Smith TW (1973) Digitalis Glycosides. New Engl J Med 288: 942–946
27. Smith TW, Antman EA, Friedman PL, Blatt CM, Marsh JD (1984) Digitalis Glycosides: Mechanisms and Manifestations of Toxicity. Progress Cardiovasc Dis Part I: 26 (5): 413–458. Part II: 26 (6): 495–540. Part III: 27 (1): 21–56
28. Sonnenblick EH, Williams JF Jr, Glick G, Mason DT, Braunwald E (1966) Studies on digitalis. XV. Effects of cardiac glycosides on myocardial force-velocity relations in the nonfailing human heart. Circulation 34: 532–539
29. Weissler AM, Schoenfeld CD (1970) Effect of digitalis on systolic time intervals in heart failure. Am J Med Sci 259: 4–20
30. Withering W An account of the foxglove. GGJ and J Robinson. London 1785: Special Edition, Gryphon, Birmingham, Alabama

Author's address:

Thomas W. Smith, M.D., Cardiovascular Division, Brigham and Women's Hospital, 75 Francis Street, Boston, Massachusetts 02115 U.S.A.

Appendix A
Multicenter investigation of the limitation of Infarct Size
Study personnel

Clinical centers

Barnes Hospital, Washington University School of Medicine, St. Louis, Missouri
Allan S. Jaffe, M.D., Principal Investigator
Robert Roberts, M.D., Principal Investigator
Edward Geltman, M.D., Co-Investigator
Dan Biello, M.D., Nuclear Medicine Coordinator
Rosanne Wettach, R.N., M.N.P. Research Nurse Coordinator
Data Coordinators: Ava Ysaguire, Susan Payne and Linda Wilson

Massachusetts General Hospital, Boston, Massachusetts
Herman K. Gold, M.D., Principal Investigator
Robert C. Leinbach, M.D., Principal Investigator

Tsunehiro Yasuda, M.D.
Research Nurse Coordinators: Wendy Werner, R.N., and Mary McHugh, R.N.
Harry Garabedian, Data Coordinator

Medical Center Hospital of Vermont, University of Vermont College of Medicine, Burlington, Vermont
Daniel S. Raabe, Jr., M.D., Principal Investigator
Walter Gundel, M.D.
Research Nurse Coordinators:
Marian Dornell, R.N., Maureen Hawley, R.N.
Patricia Beecher, R.N., Kathleen Cornell, R.N.
and Karen Helminger, R.N.
Raina Maynard, Data Coordinator

Brigham & Women's Hospital, Harvard Medical School, Boston, Massachusetts
Eugene Braunwald, M.D., Principal Investigator
Clinical Unit Directors:
Peter H. Stone, M.D.
Joseph S. Alpert, M.D. and
Robert Rude, M.D.
Research Nurse Coordinators:
Nancy E. Taplin, R.N.
Kathryn Shea, R.N.
Debbie Shiner, R.N.

Parkland Memorial Hospital, University of Texas Health Science Center at Dallas, Texas
James T. Willerson, M.D., Principal Investigator
Robert E. Rude, M.D., Clinical Unit Director
Charles Croft, M.D.
Robert Dillon, M.D.
Kervin Wheelan, M.D.
Christopher Wolfe, M.D.
Research Nurse Coordinators:
Barbara Moses, R.N., and Sandra Cochran, R.N.
Marvin Akers, R.N., Joan Reinert Corey, R.N.
Vicki Gillespie, R.N. and Barbara Fitzpatrick, R.N.
Kris Kraft, Unit Clerk

Creatine Kinase Core Laboratory

Washington University School of Medicine, St. Louis, Missouri
Burton E. Sobel, M.D., Principal Investigator
Robert Roberts, M.D., Principal Investigator
Allan Jaffe, M.D.
Cynthia Ritter, Laboratory Coordinator
Steven Mumm, Laboratory Technician

Cardiovascular Pathology Core Laboratory

Duke University Medical Center, Durham, North Carolina
Donald B. Hackel, M.D., Principal Investigator
Raymond E. Ideker, M.D., Ph.D.
Keith A. Reimer, M.D., Ph.D.
Eileen Mikat, Ph.D.

Technetium-99m Pyrophosphate Myocardial Scintigram Core Laboratory

University of Texas Health Science Center at Dallas, Dallas, Texas
James T. Willerson, M.D., Principal Investigator
Samuel E. Lewis, M.D., Laboratory Director
Robert W. Parkey, M.D., Laboratory Co-Director
Irma Dobbins, Laboratory Coordinator

Holter Recording Core Laboratory

Washington University School of Medicine, St. Louis, Missouri
Lewis J. Thomas, Jr., M.D., Principal Investigator
Robert Roberts, M.D., Co-Principal Investigator
Kenneth W. Clark, Laboratory Director
Kathleen Madden, Laboratory Coordinator
J. Phillip Miller, Biostatistician

Radionuclide Ventriculogram Core Laboratory

Massachusetts General Hospital, Boston, Massachusetts
H. William Strauss, M.D., Principal Investigator
Nathaniel M. Alpert, M.D., Co-Principal Investigator
Kenneth A. McKusick, M.D., Clinical Director, Nuclear Medicine Division
Tsunehiro Yasuda, M.D.
Karen Kelly, Laboratory Coordinator
Annali Kiers, Laboratory Coordinator and Nuclear Medicine Technician
Leander Blakeman, Laboratory Coordinator and Nuclear Medicine Technician
Merrill Griff, Laboratory Coordinator and Nuclear Medicine Technician

ECG Core Laboratory

Harvard Medical School/Brigham & Women's Hospital, Boston, Massachusetts
Eugene Braunwald, M.D., Principal Investigator
Zoltan G. Turi, M.D., Laboratory Director
John D. Rutherford, M.D., Laboratory Director
James E. Muller, M.D., Laboratory Co-Director
Peter H. Stone, M.D., Laboratory Co-Director for ETT Analysis
Laboratory Coordinators:
Gail Z. Alymer, Susan G. Albert, P.A., Jennifer Forage, and Michael Miller
Programmers/Analysts:
Neil Rhodes, Matthew Levine, Jeremy Pool, and John Rees
Computer Operators:
Jane Soukup and David Mayberry

Data Coordinating Center

Research Triangle Institute, Research Triangle Park, North Carolina
W. Kenneth Poole, Ph. D., Principal Investigator
Tyler D. Hartwell, Ph. D., Co-Principal Investigator
Corette Parker, M.S.P.H. Project Coordinator, Biostatistician
Data Coordinators: Connie Hobbs, Norma Fox, M.P.H., and Susan Warwick, R.N.
Priscilla Rigby, R.N., Data Management Coordinator
Computer Programmers: Thomas S. Farrell and Debra Fleischman
Statisticians: Nancy Gustafson, M.S., Susan K. Settergren, M.S., B. J. G. York, M.S., James H. Crowder, M.P.H., Carolyn Stuart, M.S.P.H., and Vicki Davis, M.S.
Statistical Clerk: Lee Larsen

Clinical Coordinating Center

Harvard Medical School/Brigham & Women's Hospital, Boston, Massachusetts
Eugene Braunwald, M.D., Principal Investigator
James E. Muller, M.D.
Zoltan G. Turi, M.D.
Ellenjane Scheiner, Project Coordinator
Norman R. Stein, Financial Administrator

Program Office

Cardiac Diseases Branch, Division of Heart and Vascular Diseases, National Heart, Lung, and Blood Institutes of Health, Bethesda, Maryland
Thomas Robertson, M.D., Project Officer
Eugene R. Passamani, M.D., Project Officer
Michael B. Mock, M.D., Project Officer
Suzanne M. Mullin, R.N., M.P.H., Clinical Trials Nurse Administrator
Betty Nordan, Contracting Officer
Patrick M. Sullivan, Contract Officer
Biostatisticians:
K. Gordan Lan, D.
David DeMets, Ph.D.
James Ware, Ph.D.

MILIS Policy Advisory Board

William L. Ashburn, M.D.
Eugene Braunwald, M.D.
Paul Canner, Ph.D.
Robert L. Frye, M.D. (Chairman)
Lawrence E. Hinkle, Jr., M.D.
Andrew Z. Keller, D.M.D.
Paul Meier, Ph.D.
Leroy Walters, Ph.D.
Nanette Wenger, M.D.

Improvement of ventricular function by treatment with β-methyldigoxin in patients with sinus rhythm after extended myocardial infarction: An exercise-radionuclide-ventriculography study

L. Fridrich, D. Magometschnigg[1], H. Waginger[2], A. Gaßner, G. Sommer[1] and M. Pichler
with the technical assistance of F. Eghbalian and E. Pollroß

Rehabilitation Center PVA-Ang., Grimmenstein, [1]1st Medical University Clinic Vienna, and [2]Department of Botany, University of Veterinary Medicine Vienna (Austria)

Summary

Fourteen patients with clinical signs of extended myocardial infarction were treated in the chronic state of their disease for 2 weeks with oral β-methyl-digoxin (0.3 mg/day). Supine rest and stress radionuclide ventriculography before and after therapy revealed significant changes ($p < 0.05$) for left and right ventricular ejection fractions: resting LVEF and RVEF increased from 37 ± 10 to 41 ± 11 and 49 ± 6 to 55 ± 8 % respectively; under exercise conditions, only RVEF increased significantly from 51 ± 7 to 57 ± 7 % as an effect of therapy. A placebo treated control group ($n = 10$) did not show any change. Variability in this group proved that a change in LVEF and RVEF had to be at least 5 or 8 % respectively to define an individual patient as a responder to therapy. So 11 of 14 patients responded to therapy by significant improvement of their left or right ventricular ejection fraction during rest or exercise; additionally, a significant rise of systolic blood pressure occurred at the same exercise level. Serum digoxin levels revealed 1.2 ± 0.3 mg/ml and inhibition of red blood cell ATP ase activity was 40 ± 20 %. No adverse side effects occurred. We conclude that oral therapy with β-methyldigoxin, in sufficiently high dosage, in the majority of patients with extended myocardial infarction not requiring an antianginal therapy improves left or right ventricular function and therefore might be beneficial even at a moment where no clinical signs of congestion can be detected.

Introduction

Treatment of myocardial dysfunction by cardiac glycosides has become the subject of critical discussions (22) especially with regard to patients with sinus rhythm (7). This controversy also concerns patients after myocardial infarction (17, 21).

Since patients admitted to our institution after myocardial infarction are frequently (18 %) pretreated with cardiac glycosides but often lack actual and anamnestic clinical evidence of heart insufficiency, we tried to evaluate the effect of chronic glycoside therapy in such patients by assessment of left and right ventricular function during rest and exercise by multiple gated radionuclide ventriculography (RNV).

Patients and methods

Patient selection:

Fourteen patients (12 male, 2 female) in the age range 45–72 ($\bar{x} = 56$ years), 10 with anterior and 2 with inferior myocardial infarction 4–441 (median = 5) weeks after the acute event, were selected according to the following criteria:

1. Extended myocardial infarction concluded from resting ECG (infarction signs in at least 5 leads)
2. Absence of additional ischemic signs in the exercise ECG and lack of anginal episodes indicated by the use of nitrates.
3. Absence of signs of congestion in the standard chest X-ray; no rales; no edema.
4. No anamnestic hints for episodes of cardial decompensation in the past. (One exception to this criteria was made for a patient who had a short period of decompensation following chamber fibrillation during the acute event; this patient was admitted without any cardiac therapy.)

After informed consent, all cardiac active drugs were withdrawn 1 week prior to the base-line RNV. Occurrence of symptoms during this period indicative for nitroglycerin therapy excluded the patient from the trial.

10 additional patients (all AMI) were randomly selected for placebo treatment.

Drug therapy:

Immediately after methodical sufficient base-line RNV, patients were randomised. To those receiving β-methyldigoxin (β-MD), 1 mg β-MD was injected during a period of 20 minutes; others received saline. All patients were then monitored for 4 hours to detect arrhythmias. From day 2 until day 13 a patient with 75 kg body weight principally took in 0.3 mg β-MD per day orally. The last drug intake was in the evening of day 13.

Base-line as past therapy (4h, day 14), serum-digoxin levels and red blood cell ATPase activity were controlled to guarantee therapeutic levels. Digoxin levels on day 7 were used to eventually modulate the prescribed dose.

Serum digoxin levels were determined by ELISA; inhibition of red blood cell ATPase activity by β-MD was investigated by the RB-86 method (9,25) and expressed as % of maximal possible inhibition.

Scintigraphic study:

Resting and exercise RNV were performed in the morning in the fasting patient.

The test was repeated on day 14 after therapy at the same time of day under identical conditions.

Multiple gated equilibrium radionuclide ventriculography (RNV) (11) was performed after application of 25 mCi (925 M Bqu) Tc-99-m per 75 kg body weight to the patient's red blood cells labelled in vitro. Acquisition was performed in frame mode (16 images/heart cycle) using a Siemens LFOV-ZLC-gamma camera equipped with a high sensitivity collimator. The camera was placed in a modified LAO position so that separation of heart chambers was optimal. The patient was in supine position with the legs elevated for pedalling already before resting acquisition. Acquisition modalities were standardized to achieve high reproducibility.

Acquisition time was 4 minutes during rest and at highest exercise level; exercise levels were increased in increments of 25 watts, and intermediate exercise steps lasted 3 minutes, whereby the last 2 min were also acquired, so that sufficient image statistics were obtained for ejection fraction calculation even if the patient could not pedal for an entire 4 minute period (19).

Acquisition data were stored and processed by a MDS-A^2 Computer system. For evaluation of left and right ventricular ejection fraction, clinically evaluated computer programs were used (14, 20). Data were processed for calculation of LV and RVEF by two experienced technical assistants (each had processed at least 300 rest and stress studies per year in the last 3 years) unaware of the patient's therapy. Besides scintigraphic acquisition, blood pressures and heart rates were monitored and a full 12 lead ECG was recorded in the second minute of each exercise level.

Statistical analysis:

The paired t-test (two-tailed) was applied to evaluate the significance of changes of EF values from base-line to therapy: $p < 0.05$ was considered to indicate an effect of therapy in the whole group. The unpaired t-test was used to compare base-line values of the β-MD and placebo group. The individual changes of left and right ventricular EF (RV-, LVEF) as effect of the chronic treatment are demonstrated graphically by quadrant analysis. The placebo treated control group was used to indicate the random improvement of ejection fraction for two time-different investigations caused by physiologic, pathophysiologic and methodologic variability in this subset of patients with depressed left ventricular function, by calculation of a one sided 95% confidence limit for non treatment dependent increase of ejection fraction (mean Δ EF + t_α standard deviations): for classification as responder to therapy, the increase of EF had to exceed this limiting value. All values are given as mean ± standard deviation).

Results

Control of drug therapy and side effects:

Serum digoxin levels: all base-line values were below the sensitivity of the method. 4 hours after rapid saturation and on day 14, digoxin levels averaged 1.6 ± 0.4 and 1.2 ± 0.3 ng/ml respectively. In 3 patients, intermittent control uncovered disturbed patient compliance or uncommon pharmacokinetics which could be compensated by elevation of the oral dose in 2 patients whereas 1 patient remained unsatisfactorily low (0.6 ng/ml). The average maintenance dose was 0.3 ± 0.06 mg β-MD/day for a mean body weight of 81 ± 6.5 kg.

The investigation of red blood cell ATPase activity for detection of glycoside effects on the cellular level revealed immediately, as after chronic therapy, an inhibition of 40 ± 20%. These levels might be sufficient to expect glycoside effects on the heart. Only one patient whose digoxin levels were therapeutic, showed acutely low (< 25%) and after chronic therapy no inhibition of ATPase activity. Furthermore, no treatment effect could be detected in this patient.

Telemetric ECG monitoring until four hours after intravenous therapy revealed bradycardia in 2 patients (40–50 beats/min), but no arrhythmias. One patient whose digoxin level after chronic therapy was in the lower therapeutic range showed frequent ectopic beats which disappeared after K^+– infusion. In all patients, serum electrolyte levels and serum creatinine were in the normal range.

Radionuclide ventriculography:

Resting left and right ventricular ejection fractions under therapy with β-MD were significantly higher than those under baseline conditions: 37 ± 10 to 41 ± 11% and 49 ± 6 to 55 ± 8% respectively. Under bicycle exercise, changes in LVEF failed to indicate a positive treatment effect for the whole patient group, however RVEF was significantly improved from 51 ± 7 to 57 ± 7% even under stress conditions. Collecting only individual responders according to LVEF (7 of 14 patients), a striking increase from 33 ± 12 to 39 ± 10% ($p < 0.02$) also results for stress left ventricular ejection fractions. Nevertheless, the changes in RVEF seem to be more consistent (Table 1).

Besides the response to therapy of the patient group as a whole, the evaluation of the therapeutic effect in the individual is of major clinical relevance. Therefore the individual

Table 1. Data obtained during rest and exercise radionuclide ventriculography in patients treated for 2 weeks with β-methyldigoxin or placebo (LVEF, RVEF in %; heart rate in beats per min; systolic pressure in mm Hg, achieved work load in watts).

		β-methyldigoxin n = 14			Placebo n = 10		p
		Base-line	β-MD	p	Base-line	Placebo	n.s.*
Rest	LVEF	37 ± 10	41 ± 11	< 0.02	32 ± 10	32 ± 9	n.s.
	RVEF	49 ± 6	55 ± 8	< 0.02	52 ± 4	51 ± 5	n.s.
Stress	LVEF	37 ± 11	40 ± 10	n.s.	34 ± 9	34 ± 10	n.s.
	RVEF	51 ± 7	57 ± 7	< 0.001	55 ± 5	53 ± 5	n.s.
rest	Heart rate	70 ± 10	64 ± 9	< 0.01	71 ± 9	71 ± 9	n.s.
	Systolic press	116 ± 18	118 ± 17	n.s.	114 ± 11	114 ± 11	n.s.
stress	Heart rate	114 ± 13	111 ± 13	n.s.	119 ± 13	116 ± 13	n.s.
	Systolic press	161 ± 20	172 ± 28	< 0.05	154 ± 15	147 ± 22	n.s.
Stress		77 ± 18	77 ± 18	n.s.	70 ± 28	70 ± 28	n.s.

* No significant differences could be detected in any base-line values between the treatment and placebo group.

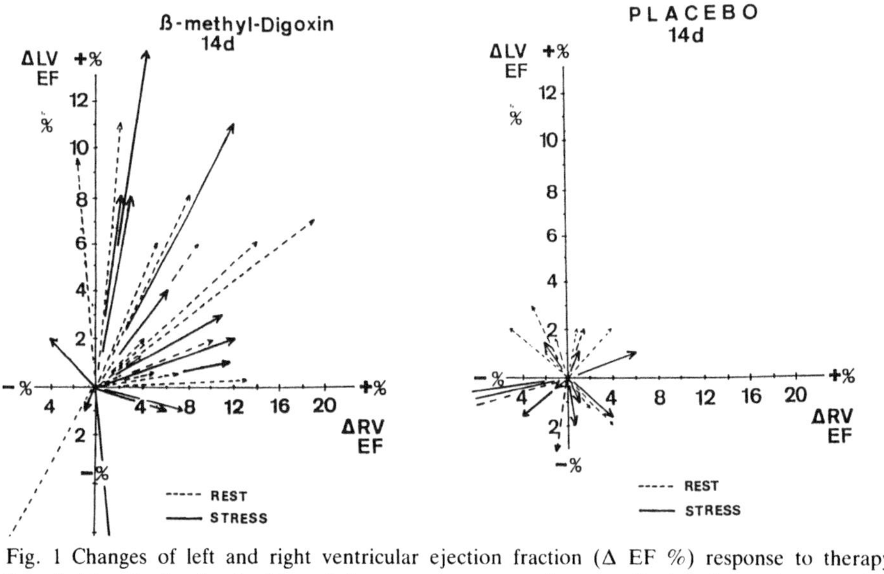

Fig. 1 Changes of left and right ventricular ejection fraction (Δ EF %) response to therapy. a: β-methyldigoxin, b: placebo (2 weeks); one arrow represents one patient during rest or exercise and demonstrates individual response of left and right ventricular ejection fraction to the applied therapy.

changes of left and right ventricular ejection fractions by chronic therapy are demonstrated in Fig. 1a by quadrant analysis.

Statistical analysis revealed that a rise of LVEF or RVEF (Δ EF %) must be at least 5 or 8 EF % respectively to identify an individual patient as responder to therapy. So we classified 7 patients as responders according to LVEF whereby in 3 patients improvement was evident during resting conditions only. This was the reason for the fact that improvement of left ventricular function during exercise conditions could not be detected for the whole group even though improvement was striking for those who responded to therapy.

For RVEF, response to therapy was evident in 7 patients. Although individual improvement was also identified mainly during resting conditions, in contrast to stress LVEF a significant effect could be proved for the whole group since a tendency to increase RVEF under stress conditions also occurred in the patients who could not be identified as individual responders.

However, altogether 11 of 14 patients revealed a significant improvement of their individual left or right ventricular ejection fractions. Five showed an increase of both (right and left ventricular) ejection fractions.

Discussion

Different authors have raised doubt as to the efficacy of chronic glycoside therapy in patients in sinus rhythm suffering from heart insufficiency and different underlying heart disease (5, 6, 10, 16, 23). On the other hand, several trials have been performed which demonstrated positive therapeutic effects of acute and chronic treatment with cardiac glycosides, but the inhomogenous patient groups under investigation suffered from chronic heart insufficiency of different origin (2, 3, 4, 12, 13, 18, 24). Nevertheless, the present study demonstrates noninvasively the positive effects of chronic therapy with β-methyldigoxin on left or right ventricular ejection fractions in a homogenous, clinically well-defined patient group in the chronic state after an extended myocardial infarction, without clinical signs of heart failure or ischemia.

The scintigraphically evaluated improvement of left and right ventricular ejection fractions is in good agreement with the results of Bussmann (3) who demonstrated by right heart catheterization the beneficial effects of β-methyldigoxin for left and right ventricular contractility in patients whose left or right ventricular end-diastolic pressures were elevated during resting or even only under exercise conditions. Especially remarkable in our results is the effect of β-methyldigoxin on right ventricular ejection fraction. An elevation of scintigraphically evaluated right ventricular ejection fraction as effect of digoxin has previously only been reported in patients with chronic obstructive pulmonary disease (15), but the authors analysed this elevation of RVEF as a subsequent effect of improved left ventricular function and therefore reduced right ventricular afterload.

However, our results showed that elevation of RVEF occurred even in patients in whom no change in LVEF was evident at all, so a direct action of β-MD on the contractility of right ventricular myocardium which was not so impaired as left ventricular has to be taken into account. Besides the positive effects demonstrated by our study it has to be remarked critically that the methodical limitations radionuclide ventriculography could result in its failure to detech an individual improvement in some patients despite a true drug effect. However, another point of view might doubt the low variability of EF changes in our placebo treated patient group. In former reproducibility studies (8), the low variability of our technique could be proved. So it has to be emphasised that a sophisticated technical procedure is necessary to detect individual therapeutic effects.

Conclusions

Patients in the chronic state after myocardial infarction showing clinical signs of impaired ventricular function, even without evidence of congestion, might benefit from therapy with β-methyldigoxin. In patients with normal renal function the oral administration of β-methyldigoxin was an uncomplicated form of therapy without severe side effects, although the applied dosage was higher than that usually prescribed. Under a maintenance dose of 0.3 mg β-MD daily for a male patient of 75 kg body weight, in the majority of patients an increase of left or right ventricular ejection fraction can be expected according to the above criteria. The control of the therapeutic effect can be performed noninvasively by scintigraphic assessment of left and right ventricular ejection fractions. The detection of a therapeutic improvement before a long lasting drug intake would be optimal. By increasing left and right ventricular ejection fraction, a decrease in end-diastolic ventricular pressures can be expected. A glycoside therapy might therefore improve the prognosis of these patients even if no clinical signs of congestion are evident.

References

1. Arnold BS, Byrd RC, Meister W et al (1980) Long-term digitalis therapy improves left ventricular function in heart failure. N Engl J Med 303: 1443
2. Balsubramanian V, Behl A, Das GS et al (1978) Effect of digoxin and duretics on high altitude left ventricular dysfunction. Circulation 57: 1180
3. Bussmann WD, Kober G, Kaltenbach M (1975) Wirkung von Digitalis auf die Kontraktilität des insuffizienten rechten und linken Ventrikels unter körperlicher Belastung. Dtsch med Wschr 100: 2265
4. Carliner NH, Gilbert CA, Pruitt AW et al (1974) Effects of maintenance digoxin-therapy on systolic time intervals and serum digoxin concentrations. Circulation 50: 94
5. Dall JLC (1970) Maintenance digoxin in elderly patients. Br Med J 2: 705
6. Dobbs SM, Kenjou WJ, Dobbs RJ (1977) Maintenance digoxin after an episode of heart failure: Placebo controlled trial in out patients. Br Med J 1: 749
7. Fleg JZ, Lakatta EG (1984) How useful is digitalis in patients with congestive heart failure and sinus rhythm? IJC 6: 295
8. Fridrich L, Pichler M, Gaßner A et al (1984) Radionuclide ventriculography: Quality control and radiation burden. Acta Med Austriaca 11 (Special issue 2): 7
9. Graham-Smith DG, Aronson JK (1978) In: Bodem G, Dengler HJ (eds) Cardiac Glycosides. Springer-Verlag, Berlin Heidelberg New York
10. Hull SM, MacIntosh A (1977) Discontinuation of maintenance digoxin therapy in general practice. Lancet 2: 1054
11. Iskandrian As, Makki AH, Depace N, Manno B, Segal BL (1983) Evaluation of left ventricular function by radionuclide angiography during exercise in normal subjects and in patients with chronic coronary heart disease. J Am Coll Cardiol 6: 1518
12. Kleiman JH, Ingels NB, Daughters G et al (1978) Left ventricular dynamics during long-term digoxin treatment in patients with stable coronary artery disease. Am J Cardiol 41: 937
13. Lee DCS, Johnson RA, Bingham BJ et al (1982) Heart failure in outpatients – a randomized trial of digoxin versus placebo. N Engl J Med 306: 699
14. Maddahi J, Berman DS, Matsuoka DT et al (1979) A new technique for assessing right ventricular ejection fraction using multigated equilibrium cardiac blood pool scintigraphy. Circulation 60: 581
15. McHaffie D, Purcell H, Mitchel-Heggs P et al (1978) The clinical value of digoxin in patients with heart failure and sinus rhythm. Quart J Med 47: 401
16. Mathur PN, Powles ACP, Pugsley O et al (1981) Effect of digoxin on right ventricular function in severe chronic airflow obstruction. Ann Int Med 95: 283
17. Moss AJ (1984) Caution: digitalis may be injurious to post infarction patients. IJC 6: 95

18. O'Rourke RA, Henning H, Theroux P et al (1976) Favourable effects of orally administered digoxin on left heart size and ventricular wall motion in patients with previous myocardial infarction. Am J Cardiol 37: 708
19. Pfisterer ME, Ricci DR, Schuler G et al (1979) Validity of left-ventricular ejection fractions masured at rest and peak-exercise by equilibrium radionuclide angiography using short acquisition times. J Nuc Med 20: 484
20. Pfisterer M, Emmenegger H (1982) Noninvasive quantification of exercise induced changes in left-ventricular function in normals and patients with one vessel coronary artery disease using radionuclide ventriculography. Eur Heart J 3: 203
21. Ryan TJ (1984) In defense of digitalis use in the coronary patient. IJC 6: 89
22. Smith TW (1982) Medical treatment of advanced congestive heart failure. In: Braunwald E, Mock MB, Watson J (eds) Heart failure. Grune & Stratton, New York London, p 261
23. Starr J, Luchi RJ (1969) Blind study on the actions of digitoxin on elderly women. Am Heart J 78: 740
24. Vogel R, Frischknecht J, Stelle P (1977) Short- and long-term effects of digitalis on resting and post hand grip hemodynamics in patients with coronary artery disease. Am J Cardiol 40: 171
25. Waginger H, Regal OS (1979) Einige neuere Methoden zum Nachweis von Digitalisglykosiden im biologischen Material. Wien Tierärztl M Schr 66: 364

Author's address:

Dr. L. Fridrich, Pensionsversicherungsanstalt d. Angestellten, Rehabilitationszentrum Hochegg, Friedrich Hillegeiststr. 2, 2840 Grimmenstein (Austria)

Effect of intravenous digoxin on left ventricular contractility at rest and during isometric exercise in intact and pharmacologically denervated heart

H. J. Partanen, M. S. Nieminen, and T. J. Pellinen

Cardiovascular Laboratory and First Department of Medicine, University Central Hospital, Helsinki (Finland)

Nine healthy volunteers (8 when denervated) were studied with echocardiography and systolic time intervals at supine rest, after 3′ isometric exercise (IE) by handgrip, during autonomic blockade (AB) by atropine 0.04 mg/kg and propranolol 0.2 mg/kg intravenously, and during IE with AB both before and after 1 mg digoxin (D) intravenously.

The results are as follows:

	No digoxin				*1 mg digoxin i. v.*			
	Rest	IE	AB	AB + IE	Rest	IE	AB	AB + IE
HR	61	73	93	94	50***	65**	91	90
mBP	85	104	95	110	87	111*	98	107
LVEDD	45.4	46.0	43.8	45.7	44.8	46.6	42.9	43.9
CO	3.2	3.8	4.1	4.0	2.8	3.5	4.1	4.0
WS	260	315	276	342	274	327	263	295
TPR	2205	2262	1949	2445	2544	2508	1981	2185
EF	56	55	50	44	61*	58	55*	53***
QS2i	522	535	561	570	500**	528	533***	538***

Abbreviations: HR = heart rate, bpm; mBP = mean blood pressure, mmHg; LVEDD = left ventricular end-diastolic diameter, mm; CO = cardiac output, l/min; WS = systolic wall stress, dyn/cm^2; TPR = total peripheral vascular resistance, dyn x sec x cm^{-5}; EF = ejection fraction, %; QS2i = electromechanical systole time index, msec. $^{*}p < 0.05$, $^{**}p < 0.01$, $^{***}p < 0.001$; values for comparisons with the respective variables without digoxin administration.

D increased contractility at rest but not during IE. D stimulated inotropic state in denervated left ventricle at rest and during IE, too. A fall in HR due to D both at rest and during IE was observed, but D induced no significant changes in HR during AB at rest or during IE. No changes occurred in loading conditions due to D.

We conclude that 1) intravenous D increases left ventricular contractility of the normal heart at rest, the loading conditions remaining unchanged; 2) D does not stimulate the inotropic state during IE; 3) the fall of HR due to intravenous D is mainly autonomically mediated; 4) the rise of contractility induced by D is direct without autonomic mediation; 5) the constancy of contractility during IE is dependent of intact autonomic nervous system. In accordance to decrease of HR due to D during IE it is suggested that with D there was sufficient rise in perfusion pressure of contracting muscles with lower activation of sympathetic tone and lesser withdrawal of vagal activity compared to those during IE without D. The increased contractility due to D and the weaker autonomic support during IE might result in no net change in inotropic state during this afterload stress.

Author's address:

H. J. Partanen, Cardiovascular Laboratory and First Department of Medicine, University Central Hospital, Helsinki (Finland)

Low versus medium dosage digoxin therapy in patients with heart failure. A randomized, double-blind, cross-over study

W. Meister, H. Schuehlen, M. Scheininger, and K. Theisen

Medizinische Klinik Innenstadt, University of Munich (West Germany)

Summary

The controversy concerning clinical efficacy, the inotropic effects of low doses in experiments, and the risk of toxicity have led to the use of lower dosages in clinical practice. Studies supporting this attitude are not available.

We compared the risk (symptoms, 24-hour Holter monitoring) and the benefit (symptoms, physical findings, radiographic cardiac volume, scintigraphic left ventricular ejection fraction) of two dosages of acetyldigoxin given cross-over for six weeks.

Fourteen patients with heart failure and sinus rhythm (pre-study ejection fraction, 30 ± 11 %, 25 to 63 years old) due to cardiomyopathy (9 patients), coronary heart disease (4 patients), and hypertensive heart disease (1 patient) completed the study. Additional therapy which was constant with two minor exceptions, included diuretics (11 patients), antiarrhythmics (7 patients), and vasodilators (6 patients).

Serum digoxin concentration was 0.8 nmol/l during "low", and 1.7 during "medium" dosage.

Marked changes in symptoms of heart failure were not reported, but three patients preferred the low, and seven the medium dosage. Clinical findings and ejection fraction did not change on average, but relevant differences were found for six individuals. Cardiac volume was 1172 ± 217 ml during the low and 1113 ± 183 ml during the medium dosage ($p = 0.10$ in the paired t-test). Holter monitoring revealed improved ventricular arrhythmias in three patients during low but in six patients during medium dosage; conduction disturbances or severe supraventricular arrhythmias did not occur.

In an overall assessment using relevant individual changes in the five areas of examination, the low dosage was superior in two, the medium dosage in eight patients ($p = 0.05$ in the sign-test), and in four patients the comparison was equivocal.

Conclusion: A medium dosage of acetyl digoxin was moderately better than a low dosage in the majority of our patients.

Increasing, but also decreasing, the dosage of digitalis glycosides may aggravate ventricular arrhythmias.

If the management of patients with heart failure remains insufficient and the serum digoxin concentration is low, increasing the dosage of cardiac glycosides should be considered as one of the alternatives of therapy changes.

Introduction

The controversy concerning the clinical efficacy of cardiac glycosides (5–8), the inotropic effects of low doses in experiments (3, 4, 9), and the risk of toxicity (14) have led to the use of lower dosages in clinical practice than previously recommended.

However, no clinical studies are available which show a superiority of a low dosage in the management of patients with cardiac failure in sinus rhythm. It is unclear whether a possible improvement of cardiac performance under a higher dosage is nullified by a dramatic increase of the risk of serious arrhythmias or other manifestations of toxicity.

Our study compared the risk-benefit ratio of two dosages of beta-acetyl digoxin, each given for six weeks in patients with heart failure. The risk was defined by symptomatology and 24-hour Holter monitoring, the benefit by symptomatology, physical findings, radiographic cardiac volume, and scintigraphic left-ventricular ejection fraction.

Methods

Patient selection
Studies were performed in outpatients of the Medizinische Klinik Innenstadt of the University of Munich who met the following criteria:
- chronic heart failure with NYHA functional Class II or III
- sinus rhythm
- left ventricular ejection fraction lower than 45%
- blood pressure lower than 160/95 mm Hg
- serum creatinine lower than 1.5 mg/dl
- presumably stable course during the study (i.e. no myocarditis or recent myocardial infarction, no consumptive disease)
- no mechanical reason for heart failure (e.g. constrictive pericarditis)

Study design
Following the selection of the patients due to baseline investigations, we obtained informed consent. Several serum digoxin concentrations under steady state yielded an estimate for the required dosage for each patient to achieve a serum digoxin concentration of 1.0 nmol/l, ("low" dosage), and of 2.0 nmol/l ("medium" dosage), respectively. Matching tablets to be administered b.i.d. were available for daily doses between 0.05 and 0.30 mg beta-acetyl digoxin for low and 0.10 and 0.60 mg for medium dosage.

Patients were then randomized to receive either the low or the medium dosage first. After six weeks we assessed the risk and benefit as defined by the study protocol (see below) as well as after a second phase of six weeks following the cross-over.

Study protocol
At the end of each of the two phases, we performed the following investigations:
- standardized symptomatology
- standardized physical examination
- laboratory data including serum digoxin concentrations
- 15-lead electrocardiogram at rest
- cardiac volume on X-ray chest film (frontal and lateral view (2))
- radionuclide ejection fraction at rest and during exercise
- ambulatory 24-hour Holter monitoring

Results

Of 19 patients who entered the study, 14 completed the protocol. Although conclusive data were lacking about the five drop-outs, we suspected that one drop-out was due to digitalis toxicity under medium dosage, another due to clinical deterioration under low dosage, the other three due to reasons independent of the study medication.

Randomization was successful, i.e. seven patients received the low dose first, seven the medium.

Patient population

Three women and eleven men completed the protocol. Their ages ranged from 25 to 63 years (median 54 years). The underlying heart disease as diagnosed by cardiac catheterization was in nine patients dilative cardiomyopathy, in four ischemic heart disease, and in one hypertensive heart disease. Eight patients were in NYHA functional Class II, six in Class III. They had been treated with digoxin for a median of 17 months. Eleven patients received diuretics, six vasodilators (mostly nitrates), and seven antiarrhythmic agents; ten patients took various other medications. All additional therapy was continued throughout the study.

Symptomatology

Seven patients preferred a medium dosage, three a low dosage. Three patients improved on medium dosage as assessed by NYHA functional class, one on low dosage, ten remained unchanged.

Specific symptoms (dyspnea, orthopnea, paroxysmal nocturnal dyspnea, nocturia, edema, angina pectoris, palpitations, fatigue, weakness, nausea) showed no consistent changes.

Clinical examination

Mean systolic and diastolic blood pressure were higher under medium dosage (129 ± 19 mm Hg to 123 ± 18 mm Hg systolic; $p = 0.046$; 84 ± 10 to 79 ± 11 mm Hg diastolic; $p = 0.048$), respiratory rate was lower (18 ± 5 to 20 ± 5/min; $p = 0.054$).

No changes were seen in mean body weight, jugular venous pressure and presence of abdominal-jugular reflux, percutory liver size, S3-gallop, pulmonary rales and peripheral edema.

Electrocardiogram

The PQ-interval was longer under medium dosage (0.18 ± 0.04 compared to 0.16 ± 0.03 sec; $p = 0.094$).

No changes could be seen for other time intervals and for the configurations of the ST-segment and the T-wave.

Laboratory findings

Mean values for serum concentrations of sodium, potassium, creatinine, urea nitrogen and hepatic enzymes as well as the blood count were unchanged. Mean serum digoxin

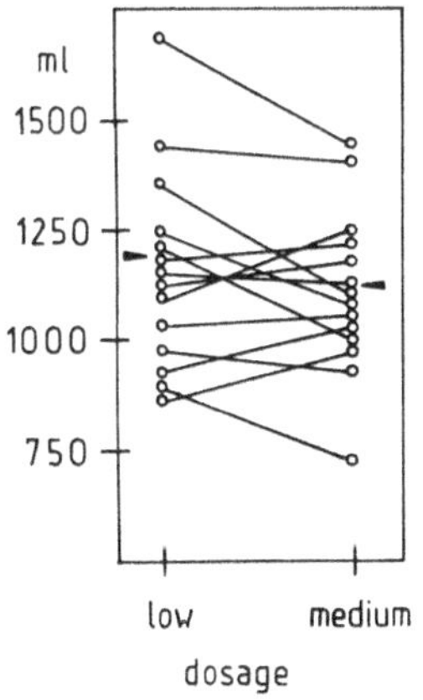

Fig. 1. Radiographic cardiac volume (mean values: ◀▶)

concentration was 0.84 ± 0.31 under low and 1.72 ± 0.72 nmol/l under medium dosage ($p < 0.0001$). No significant changes could be found for mean serum norepinephrine (636 and 517 pg/ml respectively), mean plasma renin activity (4.10 and 3.58 ngAI/ml/h) or mean urinary aldosterone excretion (17.82 and 18.22 μg/die).

Radiographic cardiac volume

Cardiac volume was enlarged in all patients but one; on average, it was significantly increased under low dosage (1172 ± 217 to 1113 ± 183ml; $p = 0.096$; Fig. 1).

Mean cardiothoracic ratio (54 ± 5 to 53 ± 5%) as well as signs of pulmonary congestion did not change.

Pleural effusion was not observed.

Scintigraphic ejection fraction

On average, left ventricular ejection fraction did not differ (28 ± 11 to 30 ± 12%; Fig. 2). Considering changes of 5% or more to be significant, we found six patients to improve under medium dosage, none under low.

All patients also underwent exercise testing. Maximal workload was higher in one patient during low and in five under medium dosage. Ejection fraction appeared to change erratically during exercise; therefore we used resting values for the overall assessment.

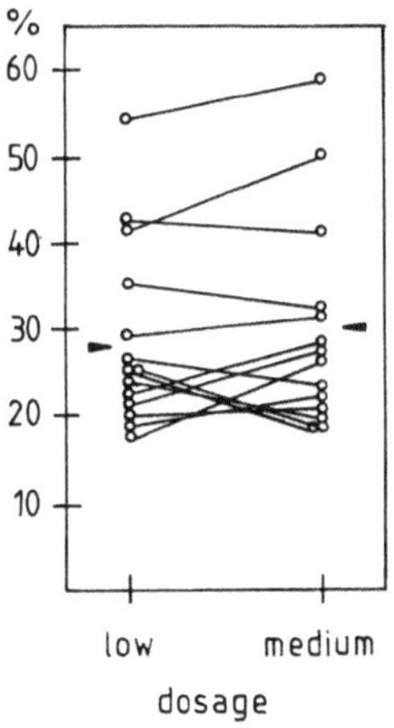

Fig. 2. Scintigraphic ejection fraction (mean values: ◀▶)

24-Hour Holter monitoring

Complex ventricular arrhythmias were present under both dosages in 12 patients despite antiarrhythmic treatment in seven of them. No consistent changes were found for median values of the number of PVCs, couplets, and VTs.

Our overall assessment was mainly based on the Lown grading system (Fig. 3). Considering in addition a change of antiarrhythmic therapy in one patient and dramatic changes in the number of PVCs under constant Lown classification in two patients, we found that three patients had improved ventricular arrhythmias under low, six under medium dosage.

Severe supraventricular arrhythmias or atrioventricular conduction disturbances were not observed.

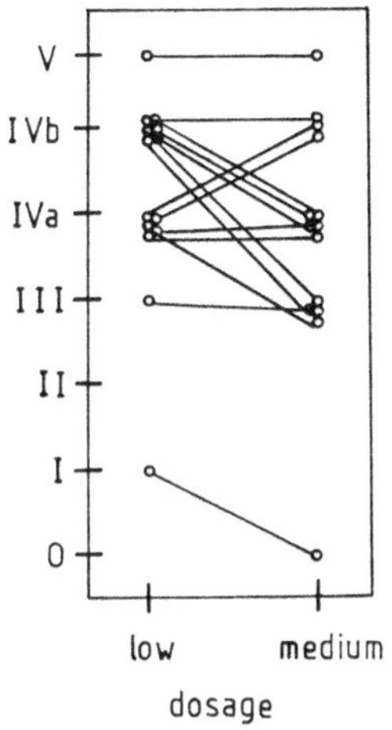

Fig. 3. Lown classification

Overall assessment for the individual patient

For a final judgement, we combined the measures for benefit (i.e. symptomatology, clinical examination, cardiac volume, left-ventricular ejection fraction) with the measures of risk (i.e. symptomatology, Holter monitoring). A final judgement was only made if detailed assessments were uniform or not contradictory.

We used the following criteria:

Symptomatology:	personal preference or NYHA functional Class
Clinical examination:	changes of respiratory rate (difference of more than 5/min), of jugular venous pressure, of percutory liver size (3cm or more) or of S3-gallop
Heart volume:	changes of more than 10%
Ejection fraction:	changes of more than 5%
Holter monitoring:	Lown grading system (for exceptions see above)

Preferences were obvious in ten of 14 patients, for eight in favour of the medium, for two in favour of the low dosage.

This result is significant by the Dixon signtest ($p = 0.05$).

Discussion

In this study, we compared two different dosages of beta-acetyl digoxin. Based on previous studies, we assumed that cardiac glycosides show a beneficial hemodynamical effect in the majority of patients with chronic congestive heart failure (1,10). Therefore, we included no placebo period which would have resulted in three cross-over phases.

Compared with the typical patient treated with cardiac glycosides, our patients were younger and suffered from more serious heart disease; furthermore, the underlying disease was more often the rather rare dilative cardiomyopathy (6). Our requirement of cardiac catheterisation probably caused this selection.

The increases of the systolic and diastolic blood pressure under medium dosage might be related to an inotropic effect as well as to a direct vasoconstriction of the cardiac glycosides (13).

The reduction of respiratory rate and cardiac volume may be considered as signs of an improvement of cardiac performance.

The Holter monitoring showed severe ventricular arrhythmias in the majority of our patients. This problem is quite common in patients with severe heart failure (12). Surprisingly, we found improved ventricular arrhythmias for three patients under a low but for six under medium dosage, most strikingly in the patients with the highest serum digoxin concentrations (2.1, 2.3, and 3.6 nmol/l). This contradicts the well-known fact that inducing arrhythmias is a typical adverse effect of digitalis, especially in toxic doses (14). On the other hand, favourable effects of therapeutic doses have also been reported (11). Our findings disprove the suggestion that a low therapeutic dose may be safer than a medium therapeutic dose. In our overall assessment, eight patients improved under a medium dosage, two showed worse performance, and four were unchanged ($p = 0.05$).

For most of our endpoints, there was no significant difference for average values. Possibly, our overall assessment amplifies the results, since in some patients differences were only marginal. On the other hand, looking at mean values may disregard individual results of patients who respond well to therapy.

Conclusions

A medium dosage of acetyl-digoxin was moderately better than a low dosage in the majority of our patients ($p = 0.05$). Therefore a general recommendation for a low dosage remains without scientific basis.

Compared with the typical patient treated with cardiac glycosides, our patients were younger and suffered from more serious heart failure; furthermore, the underlying disease was more often the rather rare dilative cardiomyopathy.

Complex ventricular arrhythmias were prevalent in most of our patients. Aggravation of these arrhythmias may occur after an increase as well as after a reduction of digoxin dosage. Therefore, alterations of digoxin therapy in both directions should be controlled by Holter monitoring in all patients at risk.

This study did not compare an increase of the dosage of digoxin with other therapeutic maneuvers such as low-salt diet, weight reduction, or changes of other medications (i.e. diuretics, antiarrhythmics).

If the management of patients with heart failure remains insufficient and the serum digoxin concentration is low, increasing the dosage of cardiac glycosides should be considered as one of the alternatives of therapy changes.

References

1. Arnold SB, Byrd RC, Meister W, Melmon K, Cheitlin MD, Bristow JD, Parmley WW, Chatterjee K (1980) Long-term digitalis therapy improves left ventricular function in heart failure. New Engl J Med 303: 1443–1448
2. Baron MG (1984) Radiologic and angiographic examination of the heart. In: Braunwald E (ed) Heart Disease. Saunders, Philadelphia, pp 146–194
3. Belz GG, Erbel R, Schumann K, Gilfrich HJ (1978) Dose-response relationships and plasma concentrations of digitalis glycosides in man. Eur J Clin Pharmacol 13: 103–111
4. Carliner NH, Gilbert CA, Pruitt AW, Goldberg LI (1974) Effects of maintenance digoxin therapy on systolic time intervals and serum digoxin concentrations. Circulation 50: 94–98
5. Erdmann E (1984) Stellenwert der Herzglykoside in der Therapie der chronischen Herzinsuffizienz. Klin Wochenschr 62: 507–511

6. Fleg JL, Lakatta EG (1984) How useful is digitalis in patients with congestive heart failure and sinus rhythmn? Int J Cardiol 6: 295–305
7. Guz A, McHaffie D (1978) The use of digitalis glycosides in sinus rhythmn. Cli Sci 55: 417–421
8. Hamer J (1979) The paradox of the lack of the efficacy of digitalis in congestive heart failure with sinus rhythmn. Br J Clin Pharmacol 8: 109–113
9. Hoeschen RJ, Cuddy TE (1975) Dose-response relation between therapeutic levels of serum digoxin and systolic time intervals. Am J Cardiol 35: 469–472
10. Lee DCS, Johnson RA, Bingham JB, Leahy M, Dinsmore RE, Goroll AH, Newell JB, Strauss HW, Haber E (1982) Heart failure in outpatients – a randomized trial of digoxin versus placebo. New Engl J Med 306: 699–705
11. Lown B, Graboys TB, Podrid PJ, Cohen BH, Stockman MB, Gaughan CE (1977) Effect of a digitalis drug on ventricular premature beats. New Engl J Med 296: 301–306
12. Maskin CS, Siskind JS, LeJemtel TH (1984) High prevalence of nonsustained ventricular tachycardia in severe congestive heart failure. Am Heart J 107: 896–901
13. Ross Jr J, Waldhausen JA, Braunwald E (1960) Studies on digitalis: direct effects on peripheral vascular resistance. J Clin Invest 39: 930
14. Smith TW, Antman EM, Friedman PL, Blatt CM, Marsh JF (1984) Digitalis glycosides: mechanisms and manifestations of toxicity. Progr Cardiovasc Dis, Part I: 26: 413–458; Part II: 26: 495–540; Part III: 27: 21–56

Author's address:

PD Dr. W. Meister, Medizinische Klinik Innenstadt der Universität München, Ziemssenstraße 1, D-8000 München 2 (West Germany)

The effects of digoxin and xamoterol in patients with moderate cardiac failure

E. Vigholt-Sørensen, B. Sterndorff, M. F. Andersen, and O. Færgeman

Medical Department I, Aarhus Amtssygehus and Specialist Clinics, Århus (Denmark)

Summary

Nineteen patients with cardiac failure (male, mean age 58 years, NYHA Class III (n = 3), Class II (n = 16) entered a double-blind randomised study comparing xamoterol (X) (200 mg b. d.) with digoxin (D) (0.1875 mg b. d.) Treatment periods were 6 weeks with run in period and washout period of 2 weeks. The mean duration of heart failure was 10.8 months (range 2–30) and the aetiology ischaemia in 17, and undiagnosed in 2. All patients were being treated with digoxin with or without diuretics. Digoxin was withdrawn and only patients who deteriorated entered the study. Four patients withdrew, 2 during the washout, 1 died on digoxin and 1 died during washout.

In 15 patients completing the study, relative to baseline (B) there was a trend for systolic time intervals to be shorter: B: 558 ± 30.1, X: 559 ± 29.6, D: 541 ± 31.8. Maximum exercise heart rate was significantly reduced by 16.7 ± 3.3 beats/min on X and unchanged on D. There was a trend for exercise duration to be increased from B at 406.7 ± 178.5 s by 168 s ± 158 s on X and 108 ± 185 on D.

In this selected group of patients who deteriorated clinically on withdrawal of digoxin, re-introduction of digoxin and careful control of digoxin plasma levels produced improvement in exercise tolerance in 13 of 15 patients. In 10 of 15 patients xamoterol also produced improvement.

Introduction

Since digitalis was introduced in 1785 by Withering (5) it has remained the most common cardiotonic drug for treatment of chronic heart failure. The effect of digitalis in patients with sinus rhythm has been controversial (1). The $beta_1$ partial adrenoceptor agnonist xamoterol (corwin) has 43 % of maximum agonist activity of that of isoprenaline (3) and has been shown to exert a positive inotropic effect acutely (4). The purpose of this study was to compare the effects of digoxin and xamoterol in patients with moderate cardiac failure.

Patients and methods

Study population

Nineteen male patients with moderate heart failure entered the study. The mean age was 58.4 (range 39–74 years). Sixteen were in New York Heart Association (NYHA) Class II and three in Class III despite treatment with digoxin and diuretics.

Study design

Table 1 shows the study design. Newly diagnosed patients with cardiac failure not receiving digoxin were included if they responded favourably to two weeks' therapy with digoxin. Other recruits were patients on regular therapy with digoxin, who entered the study if digoxin withdrawal resulted in cardial decompensa-

Table 1. Study design

Weeks	0	2	4	6	8	10	12	14	16	18
Visits:	V_1	V_2	V_3	V_4	V_5	V_6	V_7	V_8	V_9	V_{10}
Practitioner in cardiology										
Hospital visit	PC	HO	PC	PC	HO	HO	PC	PC	HO	HO
Assessments:										
Heart rate	×	×	×	×	×	×	×	×	×	
Blood pressure	×	×	×	×	×	×	×	×	×	
Clinical examination	×	×	×	×	×	×	×	×	×	
Systolic time intervals		×			×	×			×	
Echocardiography		×			×	×			×	
Exercise test					×				×	×

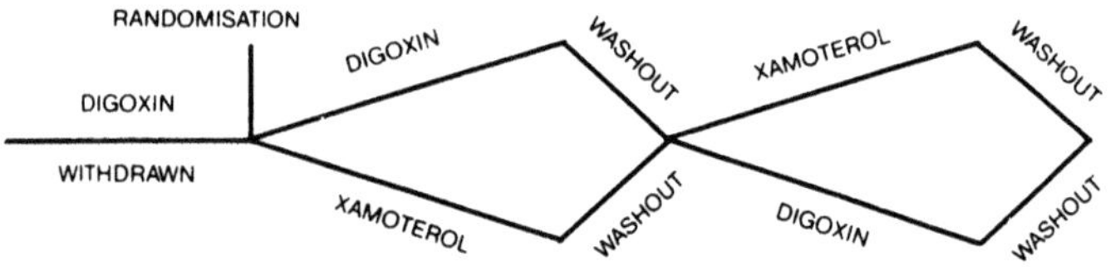

tion. The study was designed as a randomised double-blind cross-over study. The patients were randomized to either digoxin 62.5 μg x 3 or xamoterol 200 mg x 2 daily in six week treatment periods interrupted by a two week wash-out period.

Method

Clinical assessments

A clinical examination was performed at each visit. It was based on diary cards in which the patients recorded orthopnoea and gave an assessment of general well-being, dyspnoea and fatigue during the performance of usual daily activities. The physician classified the functional status according to the criteria of the NYHA.

Systolic time intervals

The systolic time intervals were measured from the simultaneous high speed (100 mm/sec) recordings of the electrocardiogram, phonocardiogram and carotid pulse. The total electromechanical systole (QS_2), left ventricular ejection time (LVET) and pre-ejection period (PEP) were determined and PEP/LVET calculated.

Echocardography

With the patients supine, an Aloka Sector Scanner was used to measure the left ventricular end-diastolic and end-systolic dimension as well as the mitral-septal distance.

Exercise test

Maximal symptom-limited exercise tests were performed with 150 Kpm load increments.

Chest radiography

Standard X-ray film taken at the start and at the end of each treatment period was evaluated.

Results (Table 2)

Nineteen patients entered and fifteen patients completed the study.

During digoxin treatment QS_2I and LVETI were significantly reduced. When compared with measurements made after the final wash-out period at visit 10, the duration of exercise

Table 2. Results of assessments

	Baseline (V2)	Digoxin (V5/9)	Baseline (V6)	Xamoterol (V5/9)	Baseline (V10)
Weight (kg)	82.9 ± 10.7 (15)	81.7 ± 11.1 (15)	84.1 ± 11.3 (15)	82.1 ± 10.9 (15)	
Exercise time (sec)		**514.3 ± 185.1 (15)		475 ± 158.3 (15)	406.7 ± 178.5 (15)
Maximal exercise heart rate (beats/min)		133.7 ± 22.3 (15)		***109.9 ± 12.3 (15)	127.7 ± 14.8 (15)
EF (%)	43.4 ± 13.4 (12)	44.2 ± 12.9 (12)	39.6 ± 9.2 (11)	39.3 ± 19 (11)	
ME (mm)	26.3 ± 8.4 (12)	25.7 ± 9.2 (12)	30.1 ± 12.3 (11)	25.5 ± 9 (11)	
QS_2I	558 ± 30.1 (15)	***541 ± 31.8 (15)	559 ± 36.3 (15)	559 ± 29.6 (15)	
LVETI	389 ± 17.5 (15)	**374 ± 21.2 (15)	393 ± 26.2 (15)	380 ± 48.3 (14)	
PEPI	169 ± 30.3 (15)	163 ± 32.6 (15)	173.5 ± 30.5 (15)	167 ± 35.2 (14)	
PEP/LVET	0.536 ± 0.13 (15)	0.532 ± 0.15 (15)	0.556 ± 0.14 (15)	0.525 ± 0.17 (14)	
NYHA	2.33 ± 0.49 (15)	2.13 ± 0.35 (15)	2.4 ± 0.51 (15)	2.26 ± 0.46 (15)	

Values are mean ± s.d. Figures in parentheses are numbers of patients.
* $p<0.05$, ** $p<0.01$, *** $p<0.001$

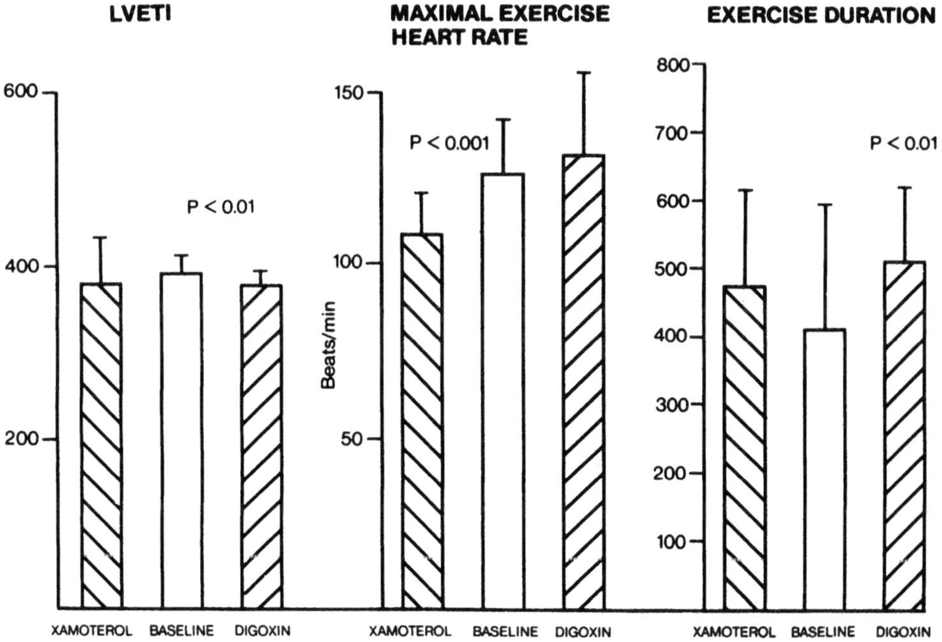

Fig. 1. Changes in LVETI, maximal exercise heart rate and exercise duration during administration of xamoterol and digoxin.

was increased by 29 and 17% during digoxin and xamoterol treatment, respectively, which was significant for digoxin ($p < 0.01$). Xamoterol caused a significant reduction in maximal exercise heart rate ($p < 0.001$) (Table 2). No adverse effects were reported by the patients.

Conclusion

In patients with chronic heart failure and sinus rhythm, studies evaluating effects of digoxin therapy have been conflicting (2). In this study, digoxin has been shown to be clinically beneficial, tending to improve symptoms and significantly increase exercise duration ($p < 0.01$).

References

1. Johnston GD, McDevitt DG (1979) Is maintenance digoxin necessary in patients with sinus rhythm? Lancet I (1): 567–570
2. Lee DC-S, Johnson RA, Bingham JB, Leahy M, Dinsmore RE, Goroll AH, Newell JB, Strauss HW, Haber E (1982) Heart failure in outpatients. A randomized trial of digoxin versus placebo. New Engl J Med 306: 699–705
3. Nuttal A, Snow HM (1982) The cardiovascular effects of ICI 118, 587: a β_1-adrenoceptor partial agonist. Br J Pharmacol 77: 381
4. Pouleur H, Rousseau MF, Mengeot P, Veriter C, Vincent MF, Brasseur LA (1982) Improvement of global and regional left ventricular function in patients with previous myocardial infarction by a new β_1-adrenoceptor partial agonist ICI 118, 587. Eur Heart J 3: 123
5. Withering W (1785) An account of the foxglove and some of its medical uses: with practical remarks on the dropsy and other diseases. GGJ & J Robinson, Birmingham

Author's address:

Dr. Else Vigholt-Sørensen, M. D., Medical Department I, Aarhus Amtssygehus, 8000 Aarhus C (Denmark)

The clinical, haemodynamic, and pharmacological consequences of withdrawal of digoxin after long-term treatment in patients with heart failure in sinus rhythm

J. K. Aronson

MRC Unit and University Department of Clinical Pharmacology, Radcliffe Infirmary, Oxford (England)

Introduction

When William Withering used digitalis in the treatment of dropsy he did so in short courses, giving digitalis, usually as powdered leaf, for about one or two days. He records several cases in which treatment of that sort was sufficient to bring about a remission of dropsy in his patients for several weeks, months, or even years. Nowhere is this better seen than in his case XLIII, that of a 70 year old woman with dropsy probably due to heart failure. Withering gave her eight separate courses of digitalis, and he describes the intervals between separate courses of digitalis, following each of which she improved and stayed well for between one and five months. The relevant extract from Withering's description is shown in Fig. 1.

It is not clear how our modern habit of continuing digitalis for life started, but it certainly was not possible before it was realised (first by Sir James Mackenzie and later by Eggleston and Gold) that after the initial loading dose of digitalis one should reduce to a smaller maintenance dose. Withering was unable to give long-term digitalis treatment in the doses he used, since he recognized that severe toxicity would occur if he did (1). The study of Rogen in 1943 (4), in which he showed that control of the ventricular rate in patients with atrial fibrillation was lost when digitalis was withdrawn, probably reinforced the long-term use of digitalis, but in recent years this use has been questioned.

Several groups of workers have withdrawn digitalis after its long-term use in patients with heart failure in sinus rhythm, and have studied the effects of its withdrawal on measures of the haemodynamic and clinical effects of digitalis (for a review see (2)). The results of such studies suggest that withdrawal of digitalis in such cases frequently does not lead to clinical deterioration, despite the apparent persistence of haemodynamic effects. It has not been clear, however, whether there is any way in which one can predict, before withdrawal, which patients will deteriorate clinically after withdrawal, and those who will not, although it has been suggested that the presence of a third heart sound discriminates the former from the latter (3).

We have therefore carried out a double-blind, cross-over, placebo-controlled comparison of digoxin and placebo after long-term digoxin therapy in order to determine:

1. What proportion of patients deteriorate following digoxin withdrawal after long-term therapy.

CASES. 1780. 43

4th. Tincture of cantharides and paregoric elixir.
28th. Infuf. Digitalis, half an ounce every morning, and one ounce every night, was now fufficient to empty her.
March 26th. Infuf. Digitalis; and when emptied, vitriol of copper twice a day.
April 1ft. A cordial mixture for occafional ufe.
Two months afterwards a purging came on, which every now and then returned, inducing great weaknefs—her appetite failed, and fhe died in in *July*.

vitriol of copper Copper sulphate. See note to p. 21.

INTERVALS.

From *July* 9th, 1780, to *December* 21ft, 171 days.
From *December* 21ft to *February* 3d, 1781, 34 days.
From *February* 3d to *March* 18th, 44 days.
From *March* 18th to *May* 24th, 66 days.
From *May* 24th to *Auguft* 11th, 79 days.
From *Auguft* 11th to *November* 16th, 98 days.
From *November* 16th to *January* 28th, 1782, 74 days.
From *January* 28th to *March* 26th, 57 days.

INTERVALS These are the intervals between successive courses of digitalis, none of which seems to have lasted for more than a few days. One could add at the end 'From March 26th to July, [at least] 98 days'. Note that some of Withering's calculations are wrong—the list should read 166, 45, 44, 68, 80, 98, 74, and 58 days. There is no evidence that Withering ever thought of giving long-term therapy, and indeed his method of dosing, if continued for more than a few days would have led quickly to toxicity—he certainly did not appreciate the concept of giving a large loading dose and following it up with small regular maintenance doses (which for digitoxin should be about one-tenth of the former).

None of the accumulations of water were at all equal to that which exifted when I firft faw her, for finding fo eafy a mode of relief, fhe became impatient under a fmall degree of preffure, and often infifted upon taking her medicine fooner than I thought it neceffary. After the 26th of *March* the degree of effufion was inconfiderable, and at the time of her death very trifling, being probably carried off by the diarrhœa.

CASE

Fig. 1. An extract from William Withering's *An account of the foxglove*... of 1785, showing his use of repeated short courses of digitalis in the patient described as Case XLIII. (Reproduced from (1)).

2. In how many of those cases digoxin can be replaced by diuretics (either freshly introduced or in increased doses).
3. Whether one could predict deterioration on the basis of any pre-withdrawal measurements, either clinical, haemodynamic, or pharmacodynamic.
4. What the clinical, haemodynamic, and pharmacological consequences of withdrawal are.

The full results of our study are being presented for publication elsewhere, and I shall therefore give here only a brief resumé of those results.

Methods

We studied 44 patients who had been taking digoxin for more than three months for heart failure in sinus rhythm. All had plasma digoxin concentrations greater than 0.8 $ng.ml^{-1}$. The patients were randomized to either digoxin or matching placebo for eight weeks, then to placebo or digoxin for a further eight weeks. Finally, all the patients were given digoxin again for eight weeks. During all these phases we made the following measurements:

1. Clinical assessment, scoring signs and symptoms according to severity.
2. Haemodynamic measurements – systolic time intervals and echocardiography.
3. Pharmacodynamic measurements – measures of erythrocytic sodium pump numbers and activity.

Results

1. Clinical deterioration during the study. During the study 28 patients (64 %) showed no evidence of deterioration. Of the other 16, 11 (25 %) deteriorated during the placebo phase and 5 (11 %) deteriorated during the digoxin phase. There was significantly more deterioration during placebo treatment periods than during digoxin treatment periods ($p < 0.04$).

2. Prediction of deterioration. We found no measurements, clinical, haemodynamic, or pharmacodynamic, which would predict deterioration. None of our patients had a third heart sound.

3. Changes in haemodynamic measurements during withdrawal and reintroduction. There was significant prolongation, in both deteriorators and non-deteriorators, of QS_2I and PEPI during withdrawal, and shortening during reintroduction. In contrast, LVETI, the ratio PEP : LVET, and the echocardiographic measurements of left ventricular function did not change during withdrawal except in the deteriorators, and only then at the time of deterioration.

4. Changes in pharmacodynamic measurements during withdrawal and reintroduction. There were decreases in the intraerythrocytic sodium concentrations in both deteriorators and non-deteriorators during withdrawal and increases during reintroduction. However, there were changes in the specific binding of [^{3}H]-digoxin to the erythrocyte membranes and in the active inward flux of ^{86}rubidium which were distinct from the changes in intraerythrocytic sodium and which differed between deteriorators and non-deteriorators.

Conclusions

We have confirmed the findings of others that withdrawal of digoxin after long-term treatment for heart failure in sinus rhythm does not lead, in the majority of cases, to clinical deterioration. In most of those in whom it does, treatment with digoxin need not be necessary, if diuretics are used appropriately.

We were unable to find any pre-withdrawal measurements, clinical, haemodynamic, or pharmacodynamic, which predicted withdrawal. That the presence of a third heart sound has been found by others to predict deterioration following withdrawal (3) may simply be because it is an index of the severity of heart failure.

The changes in haemodynamic measurements during withdrawal and reintroduction of digoxin suggest that digoxin may not be required in long-term therapy, even by patients in whom it is having pharmacological and haemodynamic effects. They also suggest that digoxin is likely to be most valuable in those in whom left ventricular function is impaired.

References

1. Aronson JK (1985) An account of the foxglove and its medical uses 1785–1985. (Incorporating a facsimile of William Withering's *An account of the foxglove*... of 1785 in a specially annotated edition) Oxford University Press, Oxford
2. Fleg JL, Lakatta EG (1984) How useful is digitalis in patients with congestive heart failure and sinus rhythm? Int J Cardiol 6: 295–305
3. Lee DC-S, Johnson RA, Bingham JB, Leahy M, Dinsmore RE, Goroll AH et al (1982) Heart failure in outpatients. A randomized trial of digoxin versus placebo. New Engl J Med 306: 699–705
4. Rogen AS (1943) Maintenance treatment with digitalis. Br Med J i: 694–695

Author's address:

Dr. J. K. Aronson, MRC Unit and University Department of Clinical Pharmacology, Radcliffe Infirmary, Woodstock Road, Oxford, OX2 6HE (England)

New positive inotropic agents: Advantages and disadvantages

K. T. Weber

Division of Cardiology, Michael Reese Hospital, University of Chicago (U.S.A.)

Heart failure secondary to ischemic heart or primary myocardial disease is a major health problem the world over. In the United States, for example, over 2 million Americans have heart failure; 400,000 new cases will be seen each year (13). The prognosis of the patient with heart failure is dismal in comparison to the normal population. This is true despite standard medical therapy with digoxin, diuretics and now more recently the vasodilator agents: 50% are dead at one year, 25–30% by year 2 (5).

Until recently, many of these patients were thought to have refractory heart failure because they remained symptomatic on digoxin and diuretic therapy. In fact it was assumed that there was little hope for the patient. This is no longer a tenable concept. We now recognize that positive inotropic drugs more potent than digoxin and with different mechanisms of action make it possible to improve ventricular pump function; the myocardium is not refractory. Many other insights into heart failure, as well as positive inotropic therapy have been gained over the past decade. The occasion of this Symposium to commemorate 200 years of digitalis glycosides is a propitious opportunity to review our current understanding of the advantages and disadvantages of the newer generation of positive inotropic agents. First, however, a word about oral digoxin therapy in the long-term management of chronic cardiac failure.

Digoxin as a positive inotropic agent

There is no question that the Na^+-K^+ATPase inhibition provided by digoxin increases myocardial contractility. Pharmacokinetic studies (6), however, have indicated the appropriate therapeutic limits for serum digoxin concentration. A serum digoxin level above 2 ng/ml is associated with the appearance of serious supraventricular and ventricular arrhythmias, as well as gastrointestinal side effects. As a result, the inotropic support provided to the failing myocardium by digoxin is limited to 1–2 ng/ml. This may represent an inadequate level of stimulation if the heart is moderately or severely depressed. Thus, greater inotropic support is necessary, much like the clinical experience with intravenous dobutamine where many patients with advanced cardiac failure require 10, 15 or even 20 mcg/kg/min of this agent; a uniform dose of 5 mcg/kg/min of dobutamine would be inadequate for many patients. This is analogous to the circumstance surrounding digoxin therapy and 1–2 ng/ml of serum digoxin.

A second generation of positive inotropic agents

A number of different biochemical mechanisms are available which increase the interaction of the intracellular contractile proteins, action and myosin, with calcium. This will lead to an enhanced contractile state of the myocardium and for the ventricle an increased ability to empty itself. The new positive inotropic agents can be categorized according to their mechanism by which they increase contractility.

Beta adrenergic receptor agonists

Intravenous catecholamines have been the mainstay in our treatment of heart failure for years. Therefore it would appear attractive to develop orally active compounds with $beta_1$ receptor agonistic properties; many of these drugs also have $beta_2$ agonistic properties.

Several of these drugs that were introduced into clinical trials included butopamine, pirbuterol, and prenalterol (15, 16). Each compound was found to increase ventricular performance (e. g., increase cardiac output and decrease filling pressure) in patients with heart failure. Each however, was subsequently withdrawn from clinical trial because they increased the frequency of ventricular arrhythmias. Catecholamine-like side effects (e. g., sweating, tremulousness) were also noted.

Adenylate cyclase activators

Glucagon is the prototype of this class of compounds. Its potency as an inotropic agent, however, is modest and it causes gastrointestinal side effects (10).

Forskolin is a recent promising development in this area (4), but as yet it has not entered clinical trials.

Phosphodiesterase inhibitors

In recent years selective type III phosphodiesterase inhibition (19), using a new generation of compounds in place of the methylxanthines, has made it possible to substantially and selectively increase myocardial contractility. Examples of these compounds include the imidazole derivatives fenoximone (MDL 17,043) and piroximone (MDL 19,205), and CI-914 (18). The bipyridine derivatives amrinone and milrinone are thought to fall into this category, but there is some question as to whether or not these compounds may also have other pharmacologic properties.

Each of these agents is undergoing clinical trials. Preliminary results using open label or non-blinded trials have provided encouraging results prompting the initiation of controlled trials. Intravenous amrinone is now approved for clinical use; its oral formulation, however, was associated with a high incidence of gastrointestinal side effects and was therefore withdrawn.

Advantages of the new generation of positive inotropic agents

The newer inotropic agents offer a number of advantages (see Table 1).

Table 1. Clinical advantages to the new generation of positive inotropic agents.

. Enhanced inotropic potency
. Improved myocardial efficiency
. Expanded pharmacologic profile
. Oral and intravenous dosing

Enhanced inotropic potency

The new inotropic agents are able to support the failing myocardium to a greater degree than digoxin. All of the new drugs, irrespective of their mechanism of action, have been found to increase cardiac output and to reduce left ventricular filling pressure in symptomatic patients with heart failure despite chronic digoxin therapy. In many cases, right atrial pressure was also found to decline with the newer drugs and thus, it is possible to unload both the right and left heart.

Improved myocardial efficiency

The work of the heart is improved, as noted above. An adverse elevation in myocardial energy expenditure or O_2 utilization has not been observed. Direct measurements of myocardial O_2 consumption ($M\dot{V}O_2$) in man have not demonstrated enhanced energy utilization accompanying the initial doses of fenoximone, piroximone, amrinone or milrinone (1, 3, 7, 9). The subacute responses to subsequent doses of piroximone also indicate that myocardial O_2 uptake to be invariant. The explanation for this response would appear to reside in the fact that even though contractility is increased, ventricular size and systolic wall stress are reduced to a greater degree, favoring a net balance of no change in energy requirement (see Table 2).

Table 2. Myocardial O_2 consumption ($M\dot{V}O_2$) of the failing, enlarged heart in response to positive inotropic agents.

Determinants of $M\dot{V}O_2$	Positive inotropic effect
1. Heart rate	→
2. Contractility	↑
3. Systolic wall stress	↓ ↓
Net balance	→

Expanded pharmacologic profile

The new generation of inotropic agents permits the clinician to utilize multiple sites of action to enhance myocardial contractility. In the failing heart, where the affinity of its beta receptors may be attenuated, the inotropic potency of the catecholamines may become limited. PDE inhibitors offer an effective alternative. In addition, they can be used in combination with the beta agonists or digoxin.

Oral and intravenous administration

With the exception of amrinone, each of the new inotropic agents is available as an oral formulation. This enhances the clinician's opportunity to provide long-term outpatient management for the patient with chronic cardiac failure.

In uncontrolled trials, long-term therapy with many of these agents has been shown to improve quality of life, including enhanced effort tolerance (18) and decreased diuretic requirements. In some cases with milder degrees of cardiac failure, the introduction of the newer inotropic agents has gradually reduced left ventricular chamber size.

Disadvantages of the new inotropic agents

Several specific adverse effects have already been identified in the foregoing discussion. The following recapitulates and summarizes other disadvantages of newer inotropic agents (see Table 3).

Table 3. Disadvantages of the new generation of positive inotropic agents.

. Arrhythmias
. Down-regulation of the beta receptor
. Dosing frequency
. Adverse effects
. Palliative therapy

Arrhythmias

The beta adrenergic agonists have been rather uniformly associated with an increased incidence of ventricular arrhythmias. As a result they have been withdrawn from clinical trials. Catecholamine-like side effects were also troublesome.

The PDE inhibitors may also have the propensity to provoke cardiac arrhythmias. The older PDE inhibitors, such as aminophylline and theophylline, are known to cause arrhythmias. To date, it is unknown whether the selective type III PDE inhibitors will also increase the frequency of supraventricular or ventricular arrhythmias. Using large oral doses of fenoximone (3–18 mg/kg), several investigators (11, 14) have suggested that there is an increased frequency of arrhythmias. This has not been our experience (17) using smaller (1–2 mg/kg) doses. Proof of such a cause and effect relationship becomes difficult to establish unless it is obvious, as it was for the beta agonists. The frequency of ventricular arrhythmias in the patient with chronic cardiac failure is quite high and variable before these drugs are given (8, 20). Nevertheless, it will be necessary to examine this issue thoroughly. Controlled clinical trials with placebo may be one means to make this assessment; electrophysiologic studies with these compounds would be another.

Down-regulation of the beta receptor

Chronic beta adrenergic receptor agonist therapy leads to a down-regulation or decreased sensitivity/affinity of the receptor for the drug itself, as well as circulating plasma norepinephrine. This would only lead to a continuous decline in myocardial contractility and an altered dose response relation to the beta agonist.

Dosing frequency

Most of the newer inotropic agents have relatively short half-lives. This is an advantage if one wishes to withdraw the pharmacologic activity, but in terms of patient compliance and quality of life it is a disadvantage. Milrinone, for example, has a short duration of action and will have to be given four times daily and in some patients even more frequently. Piroximone also needs to be given four times daily while fenoximone can be administered every 8 h. CI-914 offers the advantage of twice daily dosing. It may be desirable to consider sustained-release preparations in order to reduce dosing frequency.

Adverse effects

Gastrointestinal side effects appear common to many of the newer inotropic agents. Frequently these disturbing symptoms can be controlled by the following measures: a reduction in unit dosing, concomitant administration of the agent with meals; and the use of patient education to stagger their dosing of digoxin, diuretics, anti-arrhythmics, potassium supplements, and the new inotropic agent.

Thrombocytopenia and hepatotoxicity are also noted with several of these agents. In many cases, however, this can again be minimized by a reduction in dosing. Large doses of these agents are frequently given, based on an unrealistic goal of doubling the cardiac output of the failing heart. Dosing based on enhancing cardiac output by 30 % and/or reducing wedge pressure by a similar amount are generally satisfactory to unload the heart and to ensure the appropriate O_2 delivery to the metabolizing tissues.

Palliative therapy

Despite the optimism raised by the development of the second generation positive inotropic agents, it should be recognized that they represent palliative forms of therapy. Digoxin, diuretics and vasodilators are also palliative therapy. Until we have a better understanding of the pathophysiologic mechanisms which are responsible for the failure of the myocardium we will not be able to develop more corrective forms of therapy. In the case of coronary artery disease, prevention is also a necessary mandate for reducing the incidence of myocardial infarction and subsequent chronic cardiac failure.

Unknown issues surrounding the use of the new positive inotropic agents

Much has been learned over the past decade about heart failure and its management with the new inotropic agents. Many important questions remain however. For example, it is unclear whether these drugs have to be given continuously or if it is possible to administer these drugs intermittently, perhaps only once or several times per week (2, 12). If the syndrome of congestive heart failure arises from the salt avid kidney and an augmented activity of the adrenergic and renin-angiotensin-aldosterone systems, then ameliorating these responses by intermittent therapy (e. g., like dialysis) may be equally effective and may minimize the disadvantages of any agent.

Patient survival in response to the new drugs needs to be assessed. At the present time there is no evidence that patient survival is prolonged. The answer to this question will require a larger patient experience conducted over at least 5 years. At the same time, it needs to be determined whether or not patient survival is adversely effected by these drugs. To date, there is no evidence, in my view, that this is the case. From the opposite viewpoint, it needs to be determined whether the new drugs will favorably alter the natural history of the disease. Earlier treatments with a view toward prevention are worthwhile objectives in this regard which need to be addressed.

References

1. Amin DK, Shah PK, Hulse S, Shellock FG, Swan HJC (1984) Myocardial metabolic and hemodynamic effects of intravenous MDL 17,043, a new cardiotonic drug, in patients with chronic, severe heart failure. Am Heart J 108: 1285–1292
2. Applefeld MM, Newman KA, Grove WR, Sutton EJ, Roffman DS, Reed WP, Linberg SE (1983) Intermittent, continuous outpatient dobutamine infusion in the management of congestive heart failure. Am J Cardiol 51: 455–458
3. Bennotti J, Grossman W, Braunwald E, Carabello BA (1980) Effects of amrinone on hemodynamics and myocardial metabolism in patients with congestive heart failure from ischemic heart disease. Circulation 62: 28–34
4. Bristow MR, Ginsburg R, Strosberg A, Montgomery W, Minobe W (1984) Pharmacology and inotropic potential for forskolin in the human heart. J Clin Invest 74: 212–219
5. Cohn JN (1985) Clinical definitions and studies. Am J Cardiol 55: 11–13
6. Jelliffe RW, Buell J, Kalaba R (1972) Reduction of digitalis toxicity by computer-assisted glycoside dosage regimens. Ann Intern Med 77: 891–897
7. Martin JL, Likoff MJ, Janicki JS, Laskey WK, Hirschfeld JW, Weber KT (1984) Myocardial energetics and clinical response to the cardiotonic agent MDL 17,043 in advanced heart failure. JACC 4: 875–883

8. Maskin CS, Siskind SJ, LeJemtel TH (1984) High prevalance of nonsustained ventricular tachycardia in severe congestive heart failure. Am Heart J 107: 896–901
9. Monrad ES, Baim DS, Smith HS, Lanone AS, Braunwald E, Grossmann W (1985) Effects of milrinone on coronary hemodynamics and myocardial energetics in patients with congestive heart failure. Circulation 71: 972–979
10. Parmley WW, Glick G, Sonnenblick EH (1968) Cardiovascular effects of glucagon in man. N Engl J Med 279: 12–18
11. Rubin SA, Tabak L (1985) MDL 17,043: Short- and long-term cardiopulmonary and clinical effects in patients with heart failure. JACC 5: 1422–1427
12. Shah S, Weber KT, Maskin CS (1985) Sustained clinical and hemodynamic benefits of intermittent amrinone infusion in patients with severe chronic cardiac failure. Circulation 71
13. Smith MR (1985) Epidemology of congestive heart failure. Am J Cardiol 55: 3 A–8 A
14. Uretsky BF, Generalovich T, Verbalis JG, Valdes AM, Reddy PS (1985) MDL 17,043 therapy in severe congestive heart failure: Characterization of the early and late hemodynamic, pharmacokinetic, hormonal and clinical response. JACC 5: 1414–1421
15. Weber KT (1982) New hope for the failing heart. Am J Med 72: 665–671
16. Weber KT, Andrews V, Janicki JS, Likoff MS, Reichek N (1982) Pirbuterol, an oral beta-adrenergic receptor agonist in the treatment of chronic cardiac failure. Circulation 66: 1262–1267
17. Weber KT, Janicki JS, Jain MC (1985) Enoximone (MDL 17,043) in the treatment of stable chronic cardiac failure. Submitted
18. Weber KT, Janicki JS, Maskin CS (1986) Effects of new inotropic agents on exercise performance. Circulation, in press (March)
19. Weishaar RE, Cain MH, Bristol JA (1985) A new generation of phosphodiesterase inhibitors: multiple molecular forms of phosphodiesterase and the potential for drug selectivity. J Med Chem 28: 537–545
20. Wilson JR, Schwartz S, St John Sutton M, Ferraro N, Horowitz LN, Reichek N, Josephson ME (1983) Prognosis in severe heart failure: relation to hemodynamic measurements and ventricular ectopic activity. JACC 2: 403–410

Author's address:

Prof. Dr. K.T. Weber, Division of Cardiology, Michael Reese Hospital, University of Chicago, Chicago, IL 60616 (U.S.A.)

When should digitalis be discontinued?

J. L. Fleg

Laboratory of Cardiovascular Science, Gerontology Research Center, Baltimore, Maryland (U.S.A.)

Introduction

In a symposium commemorating the 200th anniversary of William Withering's landmark treatise entitled "An Account of the Foxglove and Some of its Medical Uses", it seems a bit improper if not sacrilegious to discuss the topic of when to stop digitalis. I am comforted, however, by Withering's sober reflections in the book's preface that "After all, in spite of opinion, prejudice or error, TIME will fix the real value upon this discovery . . ." (30). Indeed over the last few decades, a major expansion of the therapeutic armamentarium for congestive heart failure (CHF) has occurred, with the introduction of diuretics of progressively increasing efficacy and more recently the advent of vasodilator therapy. Thus I feel confident that Withering himself, were he alive today, would insist upon reappraising the relative benefits and risks of digitalis in our current setting.

Epidemiology

Digitalis continues in the 1980s to be among the most frequently prescribed medications in the United States. In the 1984 U. S. National Prescription Audit, Lanoxin®, the original digoxin formulation, ranked third in overall frequency, as it had in 1983 (29). Because digitalis is prescribed overwhelmingly for two conditions, CHF and atrial fibrillation, it may be useful to review briefly the epidemiology of these conditions. It is currently estimated that 2.3 million Americans suffer from CHF (27), although the survey on which this estimate is based is twenty years old (17). The number of hospital discharges for CHF increased from 570,000 to 1,557,000 between 1970 and 1982 (8). How much of this increase reflects longer survival with this diagnosis, better reporting, or a real increase in the incidence of CHF has not been elucidated.

The dramatic expansion of the geriatric population in America, which began early in this century and will continue well into the next, has had major ramifications for the epidemiology of cardiac glycosides. Note the exponential rise in both the prevalence and incidence of CHF with age in Fig. 1 (16, 17). The prevalence of CHF rises 200-fold between the second and eighth decades, the majority of the increase occurring from the 6th decade onward; the most common antecedents are hypertension and coronary artery disease. Atrial fibrillation (AF), the other common indication for glycoside therapy is also primarily a condition of the elderly. In the Framingham study, the cumulative 22 year incidence of AF rose from 2.6 and 2.2/1000 respectively in men and women initially aged 25–34 years to 27.9 and 29.9/1000 respectively in men and women initially 55–64 years old (15). Thus, it should be anticipated that the majority of patients for whom digitalis is likely to be prescribed (or discontinued) will be elderly.

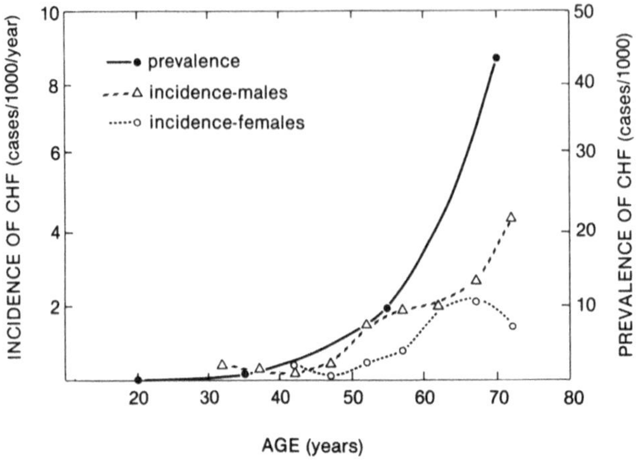

Fig. 1. Prevalence and incidence of CHF as a function of age. Prevalence of CHF was based on practising physicians' office assessments in Caledonia County, Vermont (17). Incidence of CHF was determined from 20 year follow-up of the Framingham population, using a well defined system of major and minor criteria (16). In both studies, a striking age-related increase in CHF frequency was observed.

Critical questions

Having now defined the population in which digitalis glycosides are most likely to be prescribed, we must ask the critical questions of glycoside therapy aptly enunicated by Opie: "Does the patient really need digitalis? Do the proposed benefits outweigh the possible harmful effects?" (23). These questions pertain not just to digitalis but to all therapeutic agents. I have listed below the major categories of patients in whom I believe digitalis may be withdrawn. The list begins with those situations in which agreement to discontinue digitalis should be essentially universal and progresses to those conditions in which considerable controversy exists. The case for discontinuation of digitalis in each of these five categories will be developed in the ensuing discussion.

Candidates for digitalis withdrawal

1. Initial indication for digitalis inappropriate or unknown
2. Initial indication appropriate but no longer present
3. Prior digitalis toxicity
4. Subtherapeutic serum level
5. Stable compensated CHF with sinus rhythm regardless of serum level.

1. Inappropriate initial indication

Much of difficulty in prescribing therapy for CHF lies in the diagnosis itself. Peripheral edema, rales, and sinus tachycardia are well recognized signs of CHF as are such symptoms as dyspnea and fatigue. Unfortunately, the specificity of such signs and symptoms for heart failure is poor, particularly in the population most likely to develop them, i. e. the elderly. Peripheral edema is common in old age and is frequently due to non-cardiac causes such as venous varicosities, lymphatic obstruction, hypoalbuminemia or chronic renal disease. Such edema usually requires no specific treatment or at most a mild diuretic. Factitious jugular venous distension may occur on the left side because the external jugular vein is compressed against the aortic arch. Rales may occur acutely due to a respiratory infection or atelectasis as well as chronically due to obstructive pulmonary disease. Sinus tachycardia reflects a generalized stress response whose causes are innumerable. Dyspnea and fatigue are common symptoms in the elderly which frequently are due to general deconditioning secondary to prolonged inactivity. It should be self-evident that the use of cardiac glycosides in any of these circumstances is inappropriate.

Similarly, the presence of asymptomatic cardiomegaly, manifest on chest X-ray, is often misconstrued as evidence for CHF and thence glycoside therapy. Not infrequently in such a situation echocardiography will demonstrate an enlarged left atrium or right ventricle and a normal sized left ventricular (LV) cavity. Even if the LV is enlarged, there is no evidence that digitalis, or for that matter any inotrope, will significantly reduce cardiac size or prevent the development of clinical heart failure.

Physicians in training are classically taught that the syndrome of CHF occurs only in the presence of LV systolic dysfunction. With the increasing use of echocardiography and radionuclide ventriculography, it has become apparent that a sizeable proportion of patients with classical symptoms of CHF have normal LV systolic function. Such individuals constituted 42% of patients referred for nuclear cardiac determination of LV function in a recent series (28) (Table 1). In the majority of these patients, diastolic dysfunction is present (i.e. Groups 1 and 2). Coronary artery disease and hypertension were the most common disorders, while such entities as hypertrophic or restrictive cardiopathy, which are well known to cause diastolic dysfunction, were less frequent. Note that an S_3 gallop, frequently considered a hallmark of severe systolic LV dysfunction, was present in fully one-third of these patients. A mean age approaching 70 years is common to all three subgroups and may be partially explained by the well known age-related increase in LV wall thickness and subsequent reduction in early diastolic filling rate (9).

Table 1. Characteristics of patients with CHF and normal systolic LV function.

Group	*Diastolic* Dysfunction	n	Age	EF	S_3	Etiology of CHF
I	Definite	22	68	59	36%	68% CAD, 59% HBP, 9% HCM
II	Probable	14	71	58	21%	64% CAD, 29% HBP, 21% RCM
III	None	22	66	64	41%	23% CAD, 36% LVVO, 9% Amyloid

Adapted from Soufer et al. (28).
Abbreviations: CAD = coronary artery disaese; HBP = hypertension; HCM = hypertrophic cardiomyopathy; LVVO = left ventricular volume overload; RCM = restrictive cardiomyopathy

The identification of these patients with signs and symptoms of CHF but normal systolic function has major therapeutic implications. Classical treatment with diuretics and digitalis is inappropriate in such individuals since preload is usually normal and systolic performance does not require enhancement. Attention should rather be directed to improving diastolic function. Recently calcium channel antagonists have been shown to improve LV diastolic function (3) and symptomatic status (24) in such patients. If digitalis glycosides have already been prescribed, they should clearly be withdrawn.

Is there a solid rationale for using digitalis as a primary chronic antiarrhythmic agent for conditions other than atrial tachyarrhythmias? Such therapy is not indicated in sinus tachycardia (unless due to CHF) for much the same reason that aspirin is not the primary treatment for a fever. Thus, the underlying cause of the tachycardia should be identified and specific treatment initiated. Digitalis glycosides are occasionally administered as anti-arrhythmic therapy for atrial premature beats. Although on theoretical grounds, digitalis might diminish such premature beats by a primary anti-arrhythmic effect or by improving LV function, thereby decreasing left atrial size and pressure, there are no solid data to support or refute this hypothesis. Since atrial premature beats themselves rarely cause significant symptoms or lead to major arrhythmias, therapy to suppress them is not usually necessary – with digitalis or any other agent.

Ventricular premature beats constitute an independent risk factor for mortality in patients with chronic ischemic heart disease, as well as being markers for poor LV function. Can digitalis effectively diminish ventricular ectopic beats? Although prior investigators have answered this question affirmatively (20, 25), the use of short monitoring periods and failure to assess the spontaneous variability of the ectopic beat frequency constitute major limitations in these studies. In a recent study, 24 hour ambulatory ECGs were performed on 13 patients with frequent premature ventricular beats after a placebo run-in period, again after 1 week of oral digoxin 0.25 daily and finally after 1 week of 0.375 daily digoxin (12). The frequency of ventricular ectopic beats was decreased only in the subgroup of 7 patients with normal LV ejection fraction at the higher digoxin dose. In the group with depressed LV function, no significant change occured either in the frequency or complexity of ventricular ectopic beats. Since the latter group represents the subset with a higher mortality, the failure of digitalis to exert an anti-arrhythmic effect is disappointing.

Is digitalis always needed in patients with AF? In individuals with chronic AF, the drug is employed solely to control ventricular rate. It is estimated that perhaps one-third of elderly patients with chronic AF have a ventricular response less than 90 per minute (22). Such individuals usually have atrioventricular node disease and may actually develop severe bradycardia when they are given digitalis. These patients should not be routinely maintained on cardiac glycosides.

2. Initial indication valid but no longer present

Not infrequently, digitalis therapy is initiated at a time of acute hemodynamic compromise with good reason but is then continued indefinitely. For example, a patient develops AF with a rapid ventricular response during an acute myocardial infarction or pulmonary embolus. He is treated with digoxin and rapidly reverts to normal sinus rhythm. Does such a patient require chronic therapy with digitalis? A definitive study on this issue has not been performed, but intuition would suggest that the answer is "No". After all, the most likely consequence of withdrawing maintenance digitalis in such situations would be the redevelopment of AF which could be promptly treated with digitalis again without major effect.

Analogously, a patient who develops CHF in the setting of an acute myocardial infarction might be rapidly digitalized if diuretics alone have not sufficed, although the improvement in cardiac output and reduction in LV filling pressure from glycosides in this setting are quite modest (11). It is not uncommon to see such a patient several years post infarction – still on digitalis and perhaps diuretics despite no further evidence of CHF. In this situation, it certainly seems appropriate to withdraw the glycoside and observe the patient for the unlikely reappearance of CHF. The need for maintenance digitalis could likewise be questioned whenever CHF develops during a major stress which is not likely to recur, i. e. profound blood loss, a life-threatening pneumonitis, etc.

3. Prior digitalis toxicity

Another category of patients in which the wisdom of maintenance digitalis must be questioned is the individual with prior evidence of toxicity. In this situation the numerator of the risk/benefit ratio is increased. Therefore unless the benefit to be derived from digitalis is thought to be great, the ratio will be unfavorable, justifying if not mandating discontinuation of the drug. In prior studies of hospitalized patients during the late 1960s and early 1970s, the prevalence of digitalis toxicity averaged nearly 20 percent (2, 5, 13, 26). Although toxicity is probably considerably less common than this in stable outpatients, the fact that 10 percent of 391 such patients had serum digoxin levels > 2.0 ng/ml in a general practice is indeed noteworthy (19). Because serum digoxin levels are likely to be higher in elderly patients (6) due to decreased renal function and diminished muscle mass, these individuals should not be continued on maintenance glycoside unless an unequivocal indication exists.

4. Subtherapeutic digitalis levels

Two recent studies have identified a potentially large group of patients in whom digitalis glycosides can be safely discontinued – clinically stable individuals with subtherapeutic serum levels (14, 19). In the same general practice of 391 patients cited above, fully 41 percent had serum digoxin concentration below 0.8 ng/ml (19). Of 89 such patients in sinus rhythm, none developed a recrudescence of CHF or an abnormal cardiac rhythm after drug withdrawal. These results were confirmed and extended by Johnston and McDevitt, who found in a similar population that digoxin could be successfully withdrawn in 33 of 34 outpatients with subtherapeutic serum levels but in only 8 of 15 with therapeutic levels (14). In reality, those individuals who present in well compensated status despite digoxin levels below 0.8 ng/ml have already successfully performed their own withdrawal study. Thus, the low rate of compliance which has repeatedly been documented with regard to digitalis glycosides can be used to advantage in deciding who can be safely withdrawn from these agents.

5. Compensated CHF and normal sinus rhythm

Undoubtedly the category of patients in whom the necessity for long-term digitalis therapy arouses the greatest controversy are those with well documented chronic CHF and sinus rhythm with therapeutic serum glycoside levels. Numerous studies over the past decade have looked at this important question. Arnold et al. demonstrated fairly conclusively that digoxin *could* benefit some patients with chronic CHF (1), but the relatively young age (51 years), high prevalence of idiopathic cardiomyopathy (4 of 9) and strikingly elevated wedge pressures of their 9 patients makes extrapolation to the more typical patients with CHF, i. e. older patients with coronary artery disease or hypertension and more modest LV filling pressures, somewhat hazardous. Indeed several recent studies have reached the opposite conclusion. In a double-blind crossover trial of 46 patients on maintenance digoxin, Dobbs et al. found that 65 percent showed no clinical deterioration during 6 weeks of placebo (4). Interestingly, the presence of atrial fibrillation (with ventricular rate < 120/min) had no adverse effect on the success of drug withdrawal. In another placebo controlled trial, Lee et al. demonstrated a reduction in the clinicoradiographic heart failure score in 14 of 25 outpatients in sinus rhythm (18) given digoxin for several weeks. The presence of a third heart sound and a high initial heart failure score, indicating severe CHF, were the strongest predictors of improvement on digoxin. If one redefines improvement in this study as a change in heart failure score greater than that produced by placebo, only 5 patients (whose mean EF = 16) would have been judged to improve on digoxin. Fleg et al. could find no clinical or echocardiographic deterioration in 30 outpatients in sinus rhythm after 3 months of placebo substitution for maintenance digoxin despite a 3 percent increase in echocardiographic LV diastolic dimension and a 10% decline in the velocity of circumferential fiber shortening (7). In a subset of 12 patients capable of maximal treadmill exercise, digoxin withdrawal had no effect on exercise duration or maximal heart rate. In an unblinded study in a similar population to that of Fleg et al., i. e. predominantly elderly patients with CHF due to coronary artery disease, Gheorghiade and Beller found no deterioration in clinicoradiographic score, radionuclide LV ejection fraction or treadmill exercise tolerance 1 month after cessation of maintenance digoxin (10). Studies by McHaffie et al. (21) and Aronson et al. (unpublished communication) presented during this symposium further support these observations.

Table 2. Important clinical variables in digitalis discontinuation studies.

Study	*Therapy*	*HR* (bpm)	*Wt* (kg)	S_3	*EF*	*Rate of success* (%)
Lee et al. (18)	D	81 ± 3	77 ± 3	15/25	0.30 ± 0.04	44
	P	86 ± 3⁺	78 ± 3⁺		0.29 ± 0.04	
Fleg et al. (7)	D	68.0 ± 2.5	70.8 ± 2.6	1/30	0.25 ± 0.02*	100
	P	69.8 ± 2.5	70.7 ± 2.6		0.23 ± 0.02	
Georghiade	D	75 ± 2	78.3 ± 2.7	5/27	0.41 ± 0.03	100
& Beller (10)	No D	73 ± 2	78.2 ± 2.7		0.41 ± 0.03	

⁺ $p < 0.05$ vs. placebo; * Values represent fractional shortening by M-mode echocardiography
Abbreviations: D = digoxin; EF = ejection fraction; HR = heart rate; P = placebo; S_3 = third heart sound; Wt = weight.
Rate of success denotes the percent of individuals who could be withdrawn from digitalis without clinical deterioration.

The sum total of these withdrawal studies of patients with documented CHF, sinus rhythm and therapeutic digitalis serum levels strongly suggests that the majority of such patients can be successfully maintained without cardiac glycosides, despite laboratory evidence of an inotropic effect (4, 7). Can we reconcile the more modest success rates in digitalis withdrawal in some studies (4, 18) with the striking success reported by others (7, 10, 21)? As noted in Table 2, in those studies in which body weight was unaffected by digitalis, all patients could be successfully weaned (7, 10). If weight gain was associated with glycoside withdrawal, the success rate was more modest (18). Similarly the lower ejection fraction, faster heart rate and strikingly higher prevalence of a third heart sound in the series of Lee et al. (18) would also suggest more severe LV dysfunction and a lesser degree of compensation than in those series in which digitalis withdrawal was universally successful (7, 10).

When not to discontinue digitalis

Are there, then, any categories of patients who stand to derive substantial benefit from chronic digitalis therapy? Glycosides certainly remain the agents of choice in individuals with atrial fibrillation and a rapid ventricular response (> 120/min), particularly if impaired systolic function coexists. Although beta-adrenergic blockers and calcium antagonists represent viable alternatives, their appreciably greater cost, negative inotropy and, with the latter agents, multiple daily dosage schedule represent significant drawbacks. Digitalis is likely to be useful during a subacute exacerbation of CHF, particularly if accompanied by a third heart sound. As discussed previously, many of these patients can probably be weaned from the drug once clinical compensation has been achieved. Finally, there is undoubtedly a relatively small group of patients with severe chronic CHF and sinus rhythm who derive substantial benefit from maintenance digitalis.

Conclusion

There are several diverse categories of patients in whom the adage "once on digitalis, always on digitalis" clearly is not appropriate. Clearly too there are smaller patient subsets who derive clinical benefit from cardiac glycosides. Yet despite two full centuries of use, many unanswered questions concerning these agents remain, and new questions continue to appear. Is digitalis useful in preventing exacerbations of CHF during intercurrent stresses? Does it prevent the development of *de novo* atrial fibrillation in a sizeable number of individuals? Does this agent cause an increase in mortality in post-myocardial infarction patients?

Perhaps most difficult are the questions regarding the long-term management of CHF, measured in years rather than the weeks or months covered by most studies. Should diuretics be discontinued instead of digitalis for the optimal long-term risk/benefit ratio? A most provocative question, soon to be addressed by a multicenter NIH-sponsored clinical trial, is whether the prophylactic use of inotropic drugs in asymptomatic patients with low LVEF will prevent the development of clinical CHF, thereby improving long-term prognosis (8). The final chapter on cardiac glycosides remains to be written.

Acknowledgement

I would like to acknowledge Dr. Edward G. Lakatta for his encouragement and carfeul review of the manuscript and Anne Brown for her expert secretarial assistance.

References

1. Arnold SB, Byrd RC, Meister W et al (1980) Long-term digitalis therapy improves left ventricular function in heart failure. N Engl J Med 303: 1443–1448
2. Beller GA, Smith TW, Abelman WH, Haber E, Wood WB Jr (1971) Digitalis intoxication: a prospective clinical study with serum level correlations. N Engl J Med 184: 989–997
3. Bonow RO, Bacharach SL, Green MV, Kent KM, Lipson LC, Baron BJ, Leon MB, Epstein SE (1981) Effects of verapamil on left ventricular systolic function and diastolic filling in patients with hypertrophic cardiomyopathy. Circulation 64: 787–796
4. Dobbs SM, Kenyon WI, Dobbs RJ (1977) Maintenance digoxin after an episode of heart failure: placebo-controlled trial in outpatients. Br Med J 1: 749–752
5. Evered DC, Chapman C (1971) Plasma digoxin concentrations and digoxin toxicity in hospital patients. Br Heart J 33: 540–545
6. Falch D (1973) The influence of kidney function, body size and age on plasma concentration and urinary excretion of digoxin. Acta Med Scand 194: 251–256
7. Fleg JL, Gottleib SH, Lakatta EG (1982) Is digoxin really important in treatment of compensated heart failure? A placebo-controlled crossover study in patients with sinus rhythm. Am J Med 73: 244–250
8. Furberg CD, Yusuf S, Thom TJ (1985) Potential for altering the natural history of congestive heart failure: need for large clinical trials. Am J Cardiol 55: 45 A–47 A
9. Gerstenblith G, Frederickson J, Yin FCP, Fortuin NJ, Lakatta EG, Weisfeldt ML (1977) Echocardiographic assessment of a normal adult aging population. Circulation 56: 273–278

10. Georghiade M, Beller GA (1983) Effects of discontinuing maintenance digoxin therapy in patients with ischemic heart disease and congestive heart failure in sinus rhythm. Am J Cardiol 51: 1243–1250
11. Goldstein RA, Passamani ER, Roberts R (1980) A comparison of digoxin and dobutamine in patients with acute infarction and cardiac failure. N Engl J Med 303: 846–850
12. Gradman AH, Cunningham M, Harbison MA, Berger HJ, Zaret BL (1983) Effects of oral digoxin on ventricular ectopy and its relation to left ventricular function. Am J Cardiol 51: 765–769
13. Hurwitz N, Wade OL (1969) Intensive hospital monitoring of adverse reactions to drugs. Br Med J 1: 531–536
14. Johnston GD, McDevitt DG (1979) Is maintenance digoxin necessary in patients with sinus rhythm? Lancet 1: 567–570
15. Kannel WB, Abbot RD, Savage DD, McNamara PM (1982) Epidemiologic features of chronic atrial fibrillation. N Engl J Med 306: 1018–1022
16. Kannel WB, Savage D, Castelli WP (1982) Cardiac failure in the Framingham Study: twenty-year follow-up. In: Braunwald E, Mock MB, Watson JT (eds) Congestive Heart Failure: Current Research and Clinical Applications. Grune & Stratton, New York, pp 15–30
17. Klainer LM, Gibson TD, White KL (1965) The epidemiology of cardiac failure. J Chron Dis 18: 797–814
18. Lee DC-S, Johnson RA, Bingham JB et al (1982) Heart failure in outpatients: a randomized trial of digoxin versus placebo. N Engl J Med 306: 699–705
19. Liverpool Therapeutics Group. Use of digitalis in general practice. Br J Med 2: 673–675
20. Lown B, Graboys TB, Podrid PJ, Cohen BH, Stockman MB, Gaughan CE (1977) Effect of a digitalis drug on ventricular premature beats. N Engl J Med 296: 301–306
21. McHaffie DJ, Purcell H, Mitchell-Heggs P, Guz A (1978) The clinical value of digoxin in patients with heart failure and sinus rhythm. Quart J Med 47: 401–419
22. Martin A (1977) Atrial fibrillation in the elderly (letter). Br Med J 1: 712
23. Opie LH (1980) Drugs and the heart. V. Digitalis and sympathomieetic stimulants. Lancet 1: 912–918
24. Rosing DR, Kent KM, Maron BJ, Epstein SE (1979) Verapamil therapy: a new approach to the pharmacologic treatment of hypertrophic cardiomyopathy. II. Effects of long-term administration on exercise capacity and symptomatic status. Circulation 60: 1208–1213
25. Scherf D (1960) The mechanism and treatment of extrasystoles. Prog Cardiovasc Dis 2: 370–390
26. Shapiro S, Stone D, Lewis GP, Jick H (1969) The epidemiology of digoxin: a study in three Boston hospitals. J Chron Dis 22: 361–371
27. Smith WM (1985) The epidemiology of congestive heart failure. Am J Cardiol 55: 3 A–8 A
28. Soufer R, Wohlgelernter D, Vita NA, Amuchestegui M, Sostman HD, Berger HJ, Zaret BL (1985) Intact systolic left ventricular function in clinical congestive heart failure. Am J Cardiol 55: 1032–1036
29. Top 200 drugs of 1984: 2.1 % increase in refills pushes 1984 Rx's 1.7 % ahead of 1983. Pharmacy Times 25–33, 1985
30. Withering W (1785) An Account of the Foxglove and Some of its Medical Uses. Sweeney, Birmingham

Author's address:

Jerome L. Fleg, M. D., Laboratory of Cardiovascular Science, Gerontology Research Center, 4940 Eastern Avenue, Baltimore, Maryland 21224 (U.S.A.)

Subject Index

Notizen

GPSR Compliance

The European Union's (EU) General Product Safety Regulation (GPSR) is a set of rules that requires consumer products to be safe and our obligations to ensure this.

If you have any concerns about our products, you can contact us on ProductSafety@springernature.com

In case Publisher is established outside the EU, the EU authorized representative is:

Springer Nature Customer Service Center GmbH
Europaplatz 3
69115 Heidelberg, Germany

Batch number: 10726298

Printed by Printforce, the Netherlands